Classification	Common Name(s)

Order Insectivora (429 species)

Family Solenodontidae	
Nesophontidae	(extinct)
Tenrecidae	
Chrysochloridae	
Erinaceidae	Hedgehogs
Soricidae	Shrews
Talpidae	Moles

Order Scandentia (19 species)

Family Tupaiidae	Tree shrews

Order Dermoptera (2 species)

Family Cynocephalidae	Colugos

Order Chiroptera (928 species)

Family Pteropodidae	Old World fruit bats, flying foxes
Emballonuridae	Sac-winged bats
Craseonycteridae	Kitti's hog-nosed bat
Rhinopomatidae	Mouse-tailed bats
Nycteridae	Hollow-faced bats, slit-faced bats
Megadermatidae	False vampire bats
Rhinolophidae	Horseshoe bats
Phyllostomidae	Leaf-nosed bats
Mormoopidae	Mustached bats
Noctilionidae	Bulldog bats
Mystacinidae	Short-tailed bats
Molossidae	Free-tailed bats
Myzopodidae	Sucker-footed bat
Thyropteridae	Disk-winged bats
Furipteridae	Smoky bats
Natalidae	Funnel-eared bats
Vespertilionidae	Evening bats, common bats

Order Primates (236 species)

Family Daubentoniidae	Aye-aye
Lemuridae	Lemurs
Megaladapidae	Giant lemur (extinct)
Galagonidae	Galagos
Loridae	Lorises
Cheirogaleidae	Dwarf lemurs
Indridae	Indri, sifakas, avahi
Tarsiidae	Tarsiers
Cercopithecidae	Old World monkeys

(continues on inside back cover)

MAMMALOGY

Fourth Edition

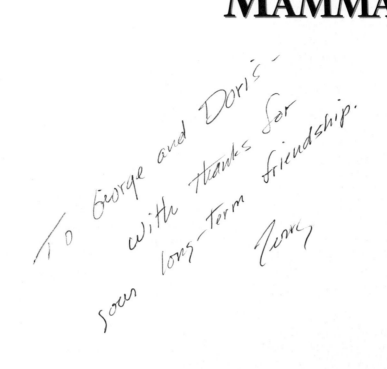

To George and Doris —
with thanks for
your long-term friendship.

Terry

MAMMALOGY

Fourth Edition

Terry A. Vaughan

Northern Arizona University
Flagstaff, Arizona

James M. Ryan

Hobart and William Smith Colleges
Geneva, New York

Nicholas J. Czaplewski

University of Oklahoma
Norman, Oklahoma

Saunders College Publishing
A division of Harcourt College Publishers

Fort Worth Philadelphia San Diego New York Orlando Austin
San Antonio Toronto Montreal London Sydney Tokyo

Publisher: Emily Barrosse
Executive Editor: Nedah Rose
Marketing Strategist: Erik Fahlgren
Developmental Editor: Lyn Knepper
Project Editor: Ted Lewis
Production Manager: Alicia Jackson
Manager of Art and Design: Carol Bleistine
Art Director and Text Designer: Kim Menning

Cover Credit: Terry A. Vaughan; back cover photo: © John Giustina/FPG International

MAMMALOGY, Fourth Edition
ISBN: 0-03-025034-X
Library of Congress Catalog Card Number: 99-61624

Requests for permission to make copies of any part of the work should be mailed to the following address:

Permissions Department
Harcourt, Inc.
6277 Sea Harbor Drive
Orlando, FL 32887-6777

Address for editorial correspondence:

Saunders College Publishing
Public Ledger Building, Suite 1250
150 S. Independence Mall West
Philadelphia, PA 19106-3412

Printed in the United States of America
9012345678 039 10 987654321

Address for domestic orders:

Saunders College Publishing
6277 Sea Harbor Drive
Orlando, FL 32887-6777
1-800-782-4479

Web Site Address:
http://www.harcourtcollege.com

Address for international orders:

International Customer Service
Harcourt, Inc.
6277 Sea Harbor Drive
Orlando, FL 32887-6777
(407) 345-3800
Fax (407) 345-4060
e-mail: hbintl@harcourtbrace.com

SAUNDERS COLLEGE PUBLISHING

soon to become

Harcourt
College Publishers

A Harcourt Higher Learning Company

Soon you will find Saunders College Publishing's distinguished innovation, leadership, and support under a different name . . . a new brand that continues our unsurpassed quality, service, and commitment to education.

We are combining the strengths of our college imprints into one worldwide brand: Harcourt Our mission is to make learning accessible to anyone, anywhere, anytime—reinforcing our commitment to lifelong learning.

We'll soon be Harcourt College Publishers. Ask for us by name.

One Company
"Where Learning
Comes to Life."

PREFACE

More than a decade has passed since the third edition of *Mammalogy*. In that time, the discipline of mammalogy has grown dramatically. The volume of literature concerning mammals has increased almost exponentially, making it impossible for us to include all of the new developments in the field; we have made every effort to include as many of the latest and most interesting discoveries as possible. Because no two instructors cover the same material in exactly the same way, we have tried to treat the biology of mammals broadly enough to make this book useful to instructors with contrasting approaches to the subject.

In the fourth edition of *Mammalogy*, most of the organization of the previous editions has been retained. Many of the chapters on the orders of mammals and the biology of mammals have been largely rewritten and supplemented with recently published material. We have not attempted to present an exhaustive review of the world literature on mammals but instead have dealt with subjects that we regard as important, interesting, and basic to an understanding of mammals.

This edition incorporates many changes reflecting the extraordinary activity in mammalian biology during the past decade. One of the most important changes is the adoption of phylogenetic systematics as the basis for determining the relationships among various mammalian groups. We have included several cladograms to illustrate the hypothesized relationships among lineages. Because phylogenetic systematics, and particularly molecular systematics, are relatively recent advances, there are many conflicting views and often many competing phylogenetic hypotheses. We point out those relationships that remain controversial and cite sources from both sides of the debate to facilitate class discussions. As a consequence of adopting a more cladistic approach, we have tried to place studies of ecology, behavior, reproduction, and physiology in an evolutionary context that better reflects the interdisciplinary nature of the field of mammalogy. For example, a new section of kin selection is used to frame the discussion of apparently altruistic behaviors of some mammals.

We have completely revised and updated our discussion of the fossil history of mammals, reorganized the chapter on echolocation with many new topics, incorporated the topics of water and temperature regulation into a single chapter focusing on the various ways that mammals solve physiological problems, completely updated the chapter on mammalian biogeography, and included many new sections in the behavior and ecology chapters. In addition, we have added many new photographs and drawings to illustrate key points made in the text, placed key terms in bold font, and added a list of web sites for student exploration. Finally, the literature citations have been brought up to date, with many citations from the period of 1990–1999. To help students access the primary literature, we have selected literature that is synthetic, accessible (in widely available journals), and current.

Another change to the fourth edition is the addition of two coauthors. James M. Ryan and Nicholas J. Czaplewski bring years of teaching and research experience with mammals to the current edition of *Mammalogy*. Jim Ryan's research interests include ecology and conservation of African mammals, functional morphology and systematics of rodents and bats, and mathematical modeling of ecological and behavioral problems. His current research focuses on establishing long-term monitoring programs for assessing the status of small mammal communities in Africa. He has conducted field research in Madagascar, West Africa, South America, and the Caribbean. Nick Czaplewski is staff curator in vertebrate paleontology at the Oklahoma Museum of Natural History. His research interests are in the evolution, paleontology, and biogeography of vertebrates, especially mammals. His current research focuses on small mammals, especially bats in North and South America, and he remains deeply interested in the conservation of biodiversity.

If this text is useful, holds the interest of students, and is respected by members of the far-flung community of mammalogists, it is primarily due to the work of the researchers on whose studies we have relied. The organization, choice of topics, and selection of illustrative material are ours, but the book should be credited largely to those whose names appear in the Bibliography.

The overall layout of this book still reflects the influence of Carl W. May, former Biology Editor of Saunders College Publishing, whose firm editorial hand guided the preparation of the first edition. Help on the earlier editions was provided by W. Clemens, E. Colbert, M. Dawson, W. Downs, J. Eisenberg, G. Goslow, Jr., E. R. Hall, E. Harwell, J. U. M. Jarvis, F. Jenkins, Jr., K. Koopman, G. Kooyman, T. Kunz, J. Lillegraven, R. MacMillen, L. Marshall, K. Norris, T. O'Shea, L. Radinsky, W. L. Robinette, N. Siepel-Hyatt, J. Stanton, H. Stanton, J. States, H. M. Van Deusen, J. Varnum, and T. Whitham. Their contributions are still evident in the present edition.

We have had a great deal of assistance on this fourth edition from many people. We would like to thank the following friends and colleagues for their especially important help: S. Altenbach, M. Adera, R. Angliss, M. Augee, R. Baxter, G. D. Bear, C. J. Bell, D. Bos, P. Boveng, R. Bowker, R. T. Bowyer, C. Brain, J. Braun, A. Brouwer, K. Catania, R. Cifelli, L. Consiglieri, M. R. Dawson, J. F. Eisenberg, J. Estes, M. Brock Fenton, T. Fleming, C. Francis, P. Freeman, M. Griffin, T. Griffiths, J. R. Henschel, I. A. Henschel, J. Hermanson, P. Hill, C. Hood, D. Hosking, T. Huels, E. W. Jameson, Jr., J. U. M. Jarvis, F. A. Jenkins, Jr., G. L. Kirkland, C. Koford, J. A. Lackey, T. A. Lawlor, W. Lawton, W. Lidicker, Jr., J. A. Lillegraven, R. Lindsay, S. L. Lindstedt, R. E. MacMillen, R. MacPhee, M. Mares, C. May, J. McDonald, S. Mizroch, T. J. O'Shea, G. Rathbun, N. Reeve, O. J. Reichman, K. Reiss, D. Rubinstein, G. Schaller, D. A. Schlitter, M. Seely, P. Stapp, R. Timm, M. Tuttle, N. Vandemey, P. Vogel, J. Waggoner, J. Wahlert, C. Wemmer, J. Wible, and K. Wilkins.

We also gratefully acknowledge the help of the staff of Saunders College Publishing. Special thanks are extended to Edith Beard Brady, Nedah Rose, Lee Marcott, Lyn Knepper, Kara Kindstrom, Kim Menning, Ted Lewis, and Christine Rickoff, each of whom contributed important advice, critical comments, and forbearance.

We offer our sincerest gratitude to our wives, Rosemary Vaughan, Donna Davenport, and Cheryl Czaplewski, who helped with the preparation of this fourth edition in many ways. Finally, we want to acknowledge the patience of our young daughters Kaylee Ryan and Jessica Czaplewski during the writing of this book.

Terry A. Vaughan

James M. Ryan

Nicholas J. Czaplewski

June, 1999

CONTENTS

1 INTRODUCTION

These are exciting times for those interested in the study of mammals. Since the first edition of *Mammal Species of the World* was published in 1982, nearly 500 new mammal species have been described as the result of recent explorations in remote parts of the world and taxonomic revisions of problematic groups. Indeed, the cumulative number of new mammal species is still on the rise. Recently discovered fossils continue to change the landscape of mammalian evolution. In addition, new tools from the field of molecular biology are helping to address questions in animal behavior and mammalian phylogeny. The study of mammalian relationships is being revolutionized by

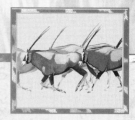

molecular data that suggest new hypotheses that remain to be tested (de Jong, 1998). Advances in radiotelemetry and the use of satellites to track far-ranging mammals increase our understanding of many species that are difficult to observe. The use of radioisotopes to measure field metabolic rates is changing the way mammalogists think about mammalian physiology. New computer models that simulate changes in populations allow mammalogists to generate testable predictions that would be impossible to formulate without such software.

Although our scientific knowledge is increasing dramatically, numerous important questions remain to be answered by the next generation of mammalogists. Most of the advances described above have resulted in new and exciting questions. For example, the discovery that hippos can communicate simultaneously above and below the water, or that kangaroo rats use foot-drumming as a form of seismic communication, has opened new areas of research on these well-studied mammals. Indeed, a combination of technological advances and innovative thinking on the part of many mammalogists has contributed to the development of entirely new areas of research. The fields of conservation biology and landscape ecology are two examples. Modern mammalogy is a dynamic and exciting field in need of curious minds.

As a discipline, mammalogy occupies the efforts of a diverse group of scientists. Vertebrate zoologists study such aspects as the structure, taxonomy, distribution, and life histories of mammals; physiologists consider mammalian hibernation and water metabolism; physicists and engineers study mammalian echolocation and locomotion; vertebrate paleontologists and geologists outline the patterns of mammalian evolution; and ecologists and psychologists consider mammalian behavior. In addition, perceptive observers without formal zoological training contribute a wealth of information.

This wide variety of scientists consider mammals worthy of study for many reasons. Practical aspects of mammalogy attract some. By studying various kinds of laboratory mammals, we gain practical knowledge about mammalian histology and about the effects of diseases and drugs. Work on domesticated breeds of mammals improves meat production, and research on game species shows how sustained yields of these animals may be achieved through appropriate management techniques. To most students and researchers, however, practical applications are not foremost. Because we human primates are mammals, we are fascinated by our relatives. Mammals are beautiful and fascinating creatures that show physiological, structural, and behavioral adaptations to an amazing array of lifestyles. Thus living mammals in their natural settings are the focal point of interest. The adaptations themselves, how they evolve, how they enable mammals efficiently to exploit demanding environmental conditions, and the interaction between mammals and their environment are all fascinating lines of inquiry. The most productive studies are the result of the intense interest of researchers in a biological relationship rather than their preoccupation with solving a practical problem. In this book, we deal primarily with the impressive literature on mammals that results from such basic research. Far from being impractical, the perspective gained from such work must guide our decisions affecting the recovery and survival of threatened species of mammals and, indeed, of entire ecosystems.

Basic research during the last half century has expanded our knowledge of mammalian biology tremendously. Echolocation (animal sonar) has been intensively studied in bats and marine mammals. The remarkable ability of some mammals to live in conditions of extreme aridity with no drinking water has been partially explained. Experiments on the circulatory and metabolic adaptations associated with temperature regulation and metabolic economy have been tested. Adaptations to deep diving in marine mammals have been examined. Hibernation and migration and the mechanisms that influence them have been studied. There have been important contributions to our knowledge of mammalian population cycles and the factors that may control them. Studies of functional morphology have increased our understanding of mammalian terrestrial, aquatic, and aerial locomotion. Probably no field has been slower to develop than that of conservation biology. This field, however, has advanced rapidly in the last decade, perhaps in response to the belated realization that time is growing short for the study of particular species in their natural environments. Studies of behavior and population genetics have contributed tremendously to our appreciation of how finely tuned mammals are to their environments. There is no doubt that mammals play vital roles in shaping regional and global ecosystems.

Mammalogists were quick to recognize that pristine ecosystems, including many of their favorite study sites, were in peril. Global **biodiversity,** the result of millions of years of evolution, is declining due to the cumulative impact of more than a quarter million new people added to the planet each day. World plant and animal species are becoming extinct at an alarming rate, primarily due to the fragmentation or outright loss of habitat. Many large or highly specialized mammals are now threatened with extinction. Since the last edition of this book was published about a decade ago, however, several new mammalian genera and species (including several primates, rodents, and ungulates) have been discovered living in remote or poorly explored parts of the world. Ironically, these discoveries were made by mammalogists attempting to describe and conserve the remaining biodiversity before it disappears.

CLASSIFICATION

In any careful study, one of the vital early steps is the organization and naming of objects. As stated by Simpson (1945) with reference to animals, "It is impossible to examine their relationships to each other and their places among the vast, incredibly complex phenomena of the universe, in short to treat them scientifically, without putting them into some sort of formal arrangement." The arrangement of organisms is the substance of **taxonomy,** but modern taxonomists, perhaps better termed **systematists,** are less interested in identifying and classifying animals than in studying their evolu-

tion. These systematists bring information from such fields as genetics, ecology, behavior, and paleontology to bear on the subjects of their research. They attempt to base their classifications on the most reliable evidence of evolutionary relationships. Excellent discussions of the importance of **systematics** to our knowledge of animal evolution are given by Simpson (1945), Mayr (1963), Wiley (1981), and McKenna and Bell (1997).

Because of difficulties that arise when a single kind of animal or plant is recognized by different common names by people in different areas, or by many common names by people in one area, scientists more than 200 years ago adopted a system of naming organisms that would be recognized by biologists throughout the world. Each known kind of organism has been given a binomial (two-part) scientific name. The first, the generic name, may be applied to a number of related kinds, but the second name refers to a specific kind, a **species.** As an example, the blacktail jackrabbit of the western United States is *Lepus californicus.* To the genus *Lepus* belong a number of similar, but distinct, long-legged species of hares, such as *L. othos* of Alaska, *L. europaeus* of Europe, and *L. capensis* of Africa. Because considerable geographic variation frequently occurs within a species, a third name is often added to designate **subspecies.** Thus, the large-eared and pale-colored subspecies of *L. californicus* that lives in the deserts of the western United States is *L. c. deserticola;* the smaller-eared and dark-colored subspecies from coastal California is *L. c. californicus.*

The species is the basic unit of classification. A once widely accepted definition of species was given by Mayr (1942): "Species are groups of actually or potentially interbreeding natural populations, which are reproductively isolated from other such groups." Each species is generally separated from all other species by a "reproductive gap," but within each species there is the possibility for gene exchange. According to Dobzhansky (1950), all members of a species "share a common gene pool."

The hierarchy of classification, based on the starting point of the species, has been developed to express degrees of **phylogenetic relationship** among species and groups of species. The taxonomic scheme includes a series of categories, each category more inclusive than the one below. Using our example of the hares, many long-legged species are included in *Lepus*. This genus and other genera containing rabbit-like mammals form the Leporidae; this family and the Ochotonidae (the pikas) share certain structural features not possessed by the other mammals and belong to the order Lagomorpha; this order and all other mammalian orders form the class Mammalia, members of which differ from all other animals in the possession of hair, mammary glands, and many other features. Mammals, birds, reptiles, amphibians, and fish all possess a bony or cartilaginous endoskeleton, and these groups (in addition to some others) form the phylum Chordata. All of the phyla of animals (Porifera, Cnidaria, Platyhelminthes, and so forth) are united in the animal kingdom. The classification of our jackrabbit can be outlined as follows:

Kingdom Animalia
    Phylum Chordata
        Class Mammalia
            Order Lagomorpha
                Family Leporidae
                    Genus *Lepus*
                        Species *Lepus californicus*
                            Subspecies *Lepus c. deserticola*

Further subdivision of this classification scheme may result from the recognition of additional intermediate categories, such as subclass, superorder, or subfamily. Most ordinal names end in -*a,* as in Carnivora; all family names end in -*idae,* and all subfamily names end in -*inae.* In this book, contractions of the names of orders, families, or subfamilies are often used as adjectives for the sake of convenience: leporid will refer to Leporidae, leporine to Leporinae, lagomorph to Lagomorpha, and so on.

Some similarities between different kinds of animals are due to **parallelism** or to **convergence,** two forms of the more general concept of **homoplasy.** Parallelism occurs when two closely related kinds of animals pursued similar modes of life and evolved similar structural adaptations. The similar specializations of the skull and dentition (elongate snouts and reduced number of teeth) that occur in a number of genera of nectar-feeding Neotropical bats are examples of parallelism. Convergence involves the development of similar adaptations to similar (or occasionally nearly identical) styles of life by species in different orders. The golden moles of Africa (p. 111) and the "marsupial moles" of Australia (p. 91) are convergently evolved. These animals belong to different mammalian infraclasses (Eutheria and Metatheria, respectively; see Table 4-1), and their lineages have been separate for more than 70 million years. Their habits are much the same, however, and structurally they resemble each other in many ways.

Chapter 4 provides an outline of the classification of mammals used in this book. It is based largely on the classification by Wilson and Reeder in 1993. We wish to stress that no universal agreement has been reached on the classification of mammals. Our knowledge of many groups of mammals is incomplete, and future study may demonstrate that some of the families listed here can be discarded because they contain animals best included in another family. Other species and families are yet to be described. The present classification, then, is not used by all mammalogists, and it is by no means immutable.

PHYLOGENY RECONSTRUCTION

Since the time of Linnaeus, taxonomists have based their systems of classification on overall similarity among organisms. In the 1960s, Willi Hennig proposed a more objective method to make the classification reflect the actual evolutionary history of the group. This system came to be known as phylogenetic systematics or **cladistics** (Wiley, 1981). In biology, therefore, the subdiscipline of systematics primarily deals with the classification and **phylogeny** of organisms. Classification is simply a way of ordering species into hierarchical groups and giving names to them so they can be recognized and discussed. In phylogenetics, researchers develop hypotheses to reconstruct the evolutionary history or relatedness among species. Classifiers of organisms, in recent decades, also attempt to incorporate phylogenetic information into classification schemes so that the named groups comprise evolutionarily related species. This is much easier said than done. There are also increasing efforts to utilize and consolidate data from a variety of sources into phylogenetic hypotheses, including molecular, paleontological, morphological, biogeographic, ecological, and behavioral data (Brooks and McLennan, 1991; Maley and Marshall, 1998; Miyamoto and Cracraft, 1991; Novacek, 1989).

Methods used to infer phylogenetic relationships are explained in Smith (1994), Maddison (1992), and Swofford (1998). Detailed descriptions of the theory and practice of classifying mammals and reconstructing their evolutionary relationships are beyond the scope of this book, but we will give a few brief notes, including some of the technical jargon. Establishing the pattern of relationships among mammals or any other group of organisms is often referred to as phylogeny reconstruction. The reconstruction of a phylogeny usually involves cladistics as a technique of analysis. The goal of cladistic analysis is to produce a hypothesis of phylogenetic relationships of a group of organisms. The hypothesis is presented as a **cladogram** (such as that in Fig. 1-1), a branching, treelike diagram in which the ends of the branches represent species or taxa, and the branching points or **nodes** indicate the point at which species separated from one another to follow their own evolutionary pathways. A cladogram may show the relationships of species of mammals within a group to one another, or it may show the relationships of a group of mammals to other living organisms. The data used to produce a cladogram of mammals are usually morphological features called **characters** that can be found on living or fossil mammals, or both.

The most important feature of cladistics is its reliance on shared derived characters (called

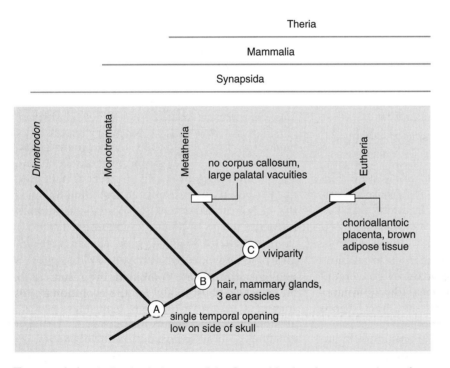

FIGURE 1-1   A simple cladogram of the Synapsida showing a nested set of monophyletic groups united by the shared derived characters (synapomorphies) listed at each node. The group Mammalia is united by the common presence of hair, mammary glands, and three middle ear ossicles. Likewise, the monotremes (Monotremata) are the sister group to the Theria. Viviparity, giving birth to live young, is a synapomorphy for the Theria. Shared primitive characters, such as the cloaca of monotremes, are symplesiomorphies. The presence of hair is a symplesiomorphy for mammals and is therefore of no use in determining eutherian relationships.

synapomorphies) to establish relationships instead of overall similarity. In building a cladogram, the characters can have various **character states** (for example, in the character "number of toes on hind feet," the possible character states might be "five toes" or "four toes" or "two toes"). The various character states must be analyzed and their **polarity** (direction of evolutionary transformation) must be determined. In this example, the trend is toward reduction from five digits, the primitive number of toes in mammals and other terrestrial vertebrates, to fewer than five digits. The best way to determine character polarity is by including in the analysis an **outgroup,** consisting of a lineage that is closely related but outside of the **ingroup** (the group being studied). The underlying assumption is that if an outgroup itself is not too **derived,** it will tend to share **ancestral (plesiomorphic)** but not derived **(apomorphic)** features with the ingroup, thereby indicating which are ancestral character states.

Primitive or ancestral features shared by the outgroup and ingroup are called **symplesiomorphies** and provide no information about phylogenetic relationships. Because all mammals have hair, for example, this shared primitive characteristic would be of no use in understanding the evolutionary relationships of rodents (or any other mammalian group). Strict application of cladistics requires that only shared derived features (synapomorphies) be used to construct a hypothesis of relationships. The presence of a small pair of peglike upper incisors immediately behind the first pair of incisors, therefore, is a synapomorphy of lagomorphs that clearly distinguishes them from rodents. Synapomorphies indicate a close relationship unless a character state arises by convergent evolution in an unrelated species. Ideally, synapomorphic character states represent **homologous** not **analogous** features, but convergent evolution of characters is very difficult to discern and frequently causes problems for systematists. Old World jerboas and New World kangaroo rats evolved elongate hindlimbs and bipedal hopping independently, a particularly good example of convergent evolution (see Fig. 18-14). Barring any convergent characters in other species or groups under study, a group that exclusively shares derived characters is said to be **monophyletic** and to form a **clade.** A monophyletic clade is a phylogenetic lineage that arose from a single ancestor and includes only the ancestor and all its descendants. The clade or branch of a cladogram nearest to a monophyletic clade is termed the **sister group** (Fig. 1-1).

A cladogram indicates only the relative timing of separation of species or lineages, but often it is desirable to know the absolute timing of divergence of lineages. For this, the cladogram must be linked with the geologic time scale. When the morphological data and resultant cladogram are combined with independent **geochronological** data from the rocks in which the fossils were buried, it is possible to estimate the time of appearance of fossil species and produce an evolutionary tree that is calibrated to the geologic time scale. Many molecular geneticists use DNA and other molecules to construct cladograms or a **molecular phylogeny** of a group of mammals. By using the so-called **molecular clock,** they can also link a time scale to the phylogeny. However, instead of measuring the time of first appearance of a recognizable member of a certain mammalian group in the fossil record, molecular systematists are able to estimate the time of divergence or lineage splitting of two species or lineages of mammals before the lineages may have become well differentiated morphologically. Often the molecular and morphological data agree, giving greater confidence in the accuracy of the phylogenetic hypothesis. The phylogenetic information, in turn, can be linked to biogeographic data and aspects of geological history, such as the past positions of the continents, islands, or tectonic plates and the timing of their collision or separation. When morphological and molecular phylogenies and various aspects of geological history are combined, the result can be an eclectic, robust model for the evolutionary history of a mammalian group (e.g., Messer et al., 1998).

Virtually all systematists agree that named taxa should share a common evolutionary origin. Naming taxa based on cladistic hypotheses is an entirely different problem from determining phylogenetic relationships, and one that is intractable and controversial. Some **cladists** prefer to name only monophyletic groups (at their nodes in a cladogram), but in practice this can result in a proliferation of names and taxa of unequivalent ranks that fall between the standard categories of Class,

Order, Family, and so on. To reconcile this problem, many systematists informally name nodes but do not provide them with a rank or category. Others name **crown groups,** which consist of the hypothetical ancestor of all the living members of a group and all its descendants, extinct and extant. For Mammalia, for example, this would include the common ancestor of Monotremata, Metatheria, and Eutheria and all of the common ancestor's descendants (Rowe, 1988; Rowe and Gauthier, 1992).

In addition to these works cited, students are referred to Wilson and Reeder (1993) and McKenna and Bell (1997). As an entry into the vast literature about the phylogeny of mammals, see Szalay et al. (1993a, b), Honeycutt and Adkins (1993), Novacek (1992a), and also "The Tree of Life" Web page that deals with mammals and their extinct relatives listed at the end of this chapter.

PLAN OF THE BOOK

The fourth edition of this text represents a significant change from the previous edition, published in 1986. Although the basic framework remains intact, we needed to incorporate more than a decade of new research. Several chapters were extensively rewritten, and others were updated with new information. In an effort to provide additional resources for students, we have included a detailed glossary and a list of Web sites that are starting points for further exploration in the field.

The first part of this book (Chapters 1 through 4) defines mammals and summarizes their origins. Chapter 2 details the structural features that characterize mammals. Chapter 3 summarizes the origin of mammals, and Chapter 4 introduces the classification of mammals. In the second part of the book (Chapters 5 through 19), the orders and families of mammals are discussed in detail. These taxonomic chapters include information on the fossil history, current distribution, morphological characteristics, and basic behavior and ecology of each family of mammals. The third part (Chapters 20 through 26) treats mammalian physiology, behavior, and ecology. From this coverage of the subject, students can gain a general understanding and an appreciation of the form and function of mammals.

The anatomical drawings, which we regard as essential to the ordinal chapters, should help students to understand the descriptions of structure and function. We have made liberal use of these drawings, most of which illustrate skulls, teeth, or feet. The profiles of skulls are usually of the right side, and most occlusal views of teeth show the right upper or the left lower tooth row. When other teeth are shown, the legend so indicates.

The bibliography has been revised extensively and includes a list of articles that are cited in the text. It provides a thorough foundation for further exploration of the literature.

WEB SITES

American Society of Mammalogists home page

> http://www.wku.edu/~asm/

Searchable electronic version of Wilson and Reeder's 1993 book *Mammal Species of the World*

> http://www.nmnh.si.edu/msw/

British Mammal Society home page

> http://www.abdn.ac.uk/mammal/

Very long list of mammal-related links

> http://www.york.biosis.org/zrdocs/zoolinfo/mam_gen.htm

A searchable index of mammalogists' email addresses

       http://nmnhgoph.si.edu/gopher-menus/MammalogistsonEmail.html

A bibliography of mammalogy books organized by topic

       http://research.amnh.org/mammalogy/mambib.html

An introduction to phylogenetic systematics and cladistics

       http://www.ucmp.berkeley.edu/clad/clad4.html

The Tree of Life: A phylogenetic tour through the animal kingdom, with information on phylogenetic methods

       http://phylogeny.arizona.edu/tree/phylogeny.html

Helpful information on taxonomy, systematics, and molecular genetics

       http://www3.ncbi.nlm.nih.gov/Taxonomy/taxpage2.html

The Smithsonian Institution site for information on mammals

       http://www.nmnh.si.edu/vert/mammals/mammals.html

Comprehensive list of mammalogy links and resources

       http://muse.bio.cornell.edu/cgi-bin/hl?mammal

2 Mammalian Characteristics

Mammals owe their spectacular success to many features. Perhaps the most important and diagnostic mammalian characteristics are those that enhance intelligence and sensory ability, promote endothermy, or increase the efficiency of reproduction or of securing and processing food. The senses of sight and smell are highly developed, and the sense of hearing has undergone greater specialization in mammals than in any other vertebrates. Efficient gathering and utilization of a wide variety of foods are aided by specializations of the dentition and the digestive system. Endothermy has allowed

mammals to remain active under a wide array of environmental conditions. Specializations of the postcranial anatomy, particularly the limbs and feet, have enabled them to make effective use of this endothermically driven physical activity. In some species, extended periods of parental care have increased the length of time during which the young can learn demanding foraging patterns and complex social behavior.

The basic structural plan of the mammalian body was inherited more than 200 million years ago from the nonmammalian synapsids (formerly known as mammal-like reptiles) of the order Therapsida (p.38). Members of this ancient order followed an evolutionary path that diverged strongly from the reptilian path from which arose the spectacular and successful Mesozoic dinosaurs, birds, and a wide diversity of other reptiles. The key to the persistence of the therapsids through the Triassic period (see Table 3-1) was perhaps their ability to move and to respond to their environment more quickly than their archosaurian contemporaries. These same abilities probably enabled the descendants of the therapsids, the mammals, to survive through the Jurassic and Cretaceous periods, when the dinosaurs dominated the terrestrial scene. Also of major importance to early mammals was their highly specialized dentition, which probably allowed them to utilize certain foods more efficiently than could reptiles.

An important morphological trend in the therapsid-mammalian line was toward skeletal simplification. In general, the therapsid skeleton evolved greater efficiency by reducing the number of parts while retaining the effective performance of a particular function. In the skull and lower jaw, which in primitive reptiles consisted of many bones, a number of bones were lost, reduced in size, or put to other uses. The limbs and limb girdles also were simplified to some extent and reduced in massiveness. The skeleton of egg-laying mammals (order Monotremata) roughly resembles or retains features similar to advanced therapsids, but the limbs of some therapsids were less laterally splayed than today's specialized monotremes.

When mammals first appeared in the Triassic period, they represented no radical structural departure from the therapsid plan but had attained a level of development (involving a dentary/squamosal jaw articulation; p. 45) that is interpreted by most vertebrate paleontologists as a key indication that the animals had crossed the non-mammalian-mammalian boundary. (And therefore, they represent an evolutionary grade, not a clade.) Many of the mammalian characters discussed in this chapter resulted from evolutionary trends clearly characteristic of therapsids. Unfortunately, the fossil record cannot directly indicate when various important features of the soft anatomy became established, and only indirect evidence can be used to judge whether advanced therapsids had such features as mammae, hair, or a four-chambered heart.

Endothermy probably began to develop in the therapsid ancestors of mammals (Bennett and Ruben, 1986; Tracy et al., 1986; Turner and Tracy, 1986). We know that today the metabolic "machinery" of mammals differs markedly from that of reptiles. Mammals have a three- to sixfold greater capacity for energy production than do reptiles and a standard metabolic rate some eight to ten times higher (Else and Hulbert, 1981). Relative to total body size, the internal organs of mammals are larger than those of reptiles. Mitochondrial membrane surface areas for the heart, kidney, and brain are far greater in mammals, as are mitochondrial enzyme activity and thyroid activity. A much greater absorptive surface area in the lungs enables a higher oxygen uptake in mammals. Similarly, an increased absorptive surface area in the digestive system enables a higher nutrient uptake in mammals than that in reptiles (Karasov and Diamond, 1994). These differences are clearly related to the mammalian capacity for high energy production (Table 2-1) and were probably part of a suite of anatomical and physiological features that allowed

TABLE 2-1   Some Contrasts Between Adult Ectothermic Reptiles (Lizards) and Endothermic Small Mammals

Ectotherms (lizards)	Endotherms (small mammals)
Lower size limit <1g	Lower size limit approx. 2g
Body shape usually elongate	Body shape more spherical
50–90% of energy from anaerobic metabolism	High level of aerobic energy production
High activity for only brief periods	High activity over long periods
Low blood pressure (30–50 mm Hg)	High blood pressure (80–200 mm Hg)
Low hematocrit	High hematocrit
Large-diameter, widely spaced capillaries	Small-diameter, closely spaced capillaries
Blood oxygen capacity 25–50% that of endotherms	
Capillary length (frogs) 155 mm/mm^3 of tissue	Capillary length (mouse) 3,500 mm/mm^3
Daily energy requirement of nontorpid lizard 3–4% that of mammal	
Can utilize highly ephemeral food source (e.g., *Heloderma* eat bird eggs; *Coleonyx* can store enough energy in 4 days to last 9 months)	Mammals need more long-term food sources
Lizards allocate 90% of energy budget to new biomass	Mammals allocate 90% of energy budget to thermoregulation
Minor consumers but often (e.g., in deserts) major producers of biomass	Major consumers but often low producers of biomass

mammals (perhaps even the earliest ones) to be nocturnal.

The characteristics proposed as being diagnostic of mammals and their closest relatives (extinct mammaliaforms, see Chapter 3) are discussed by Gauthier et al. (1988), Rowe (1988, 1996), Wible (1991), and other authors. A list of these characteristics (called "characters") of the hard and soft anatomy were repeated and discussed by McKenna and Bell (1997). Some of these are included in Table 2-2. To date, only one molecular character is included in definitions of Mammalia (Table 2-2, number 13).

SOFT ANATOMY

SKIN GLANDS

The skin of mammals contains several kinds of glands not found in other vertebrates; the most important of these are the **mammary glands** (the key feature after which Mammalia is named; Table 2-2,

number 24). These glands in females provide nourishment for the young during their postnatal period of rapid growth. Mammary glands consist of a complex system of ducts that reach the surface of the skin usually through a prominence called a nipple or teat (Fig. 2-1). During late pregnancy, the epithelium of the ducts is stimulated by endocrine secretions (estrogens and progesterone) and divides rapidly to produce secretory alveoli, which produce milk after birth. The production of milk is stimulated by secretions (prolactin and growth hormone) of the anterior lobe of the pituitary. Nursing and emptying of milk from the mammary glands result in nervous stimulation of the anterior lobe of the pituitary and continued production of prolactin and milk.

The composition of milk differs from species to species. Cow's milk contains about 85 percent water; the dry weight includes approximately 20 percent protein, 20 percent fat, and 60 percent sugars (largely lactose), as well as vitamins and salts in roughly the proportion in which they are found in blood. The milk of mammals with young that grow

TABLE 2-2 Characters Diagnosing Mammalia

1. Accessory jaw bones shifted away from the cranio-mandibular joint in adults to become associated with the cranium alone. These include the middle ear bones, the stapes (columella auris of nonmammals), incus (quadrate), and malleus (articular), as well as other bones formerly associated with the jaw.
2. Stapes very small relative to skull size
3. Atlas intercentrum and neural arches fused to form single, ring-shaped osseous structure
4. Epiphyses on the long bones and girdles
5. Heart completely divided into four chambers with a thick, compact myocardium (muscular wall)
6. Heart with atrioventricular node ("pacemaker") and Purkinje fibers
7. Single aortic trunk
8. Pulmonary artery with three semilunar valves
9. Erythrocytes lack nuclei at maturity
10. Endothermy
11. Central nervous system covered by three meninges
12. Cerebellum folded
13. Nerve filaments with three polypeptides
14. Brain with divided optic lobes
15. Strong representation of the facial nerve field in the motor cortex of the brain
16. Superficial musculature expanded onto the face and differentiated into muscle groups associated with the eye, ear, and snout
17. Muscular diaphragm encloses pleural (lung) cavities, and consequent development of diaphragmatic breathing
18. Complex lung structure with division of the lungs into lobes, bronchioles, and alveoli
19. Epiglottis
20. Skin with erector muscles and dermal papillae
21. Hair
22. Sebaceous glands
23. Sweat glands
24. Mammary glands
25. Loop of Henle in the kidney

Selected from a much longer list in McKenna and Bell (1997) and sources therein. (By this definition, Mammalia does not include the extinct near-mammals, the Mammaliaformes: Morganucodonta, Docodonta, and Haramyoidea.) The first character forms the arbitrary dividing feature separating Mammalia from other mammaliaforms and other nonmammalian synapsids.

unusually rapidly contains high levels of protein and fat. Seals, for example, produce milk with roughly 12 times as much fat and 5 times as much protein as is in cow's milk.

The period of association between the mother and her young during lactation and suckling is one that encourages close social bonds. For many mammals, especially those with complex foraging and social behavior patterns, this is an extremely important time of rapid learning by the young that prepares them for an adult life independent from their mothers. Recently, Francis et al. (1994) discovered that in some populations of the Malaysian fruit bat *Dyacopterus spadiceus,* the males also lactate.

In most mammals, the young suck milk from the projecting nipples. In monotremes, however, nipples are lacking, and the young suck milk from tufts of hair on the mammary areas (Burrell, 1927;

Ewer, 1968). Whales, dolphins, and porpoises have muscles that force milk into the mouth of the young, a seemingly necessary adaptation in underwater-feeding animals that also have no lips and are therefore unable to suckle. The number of nipples varies from 2 in many kinds of mammals to about 19 in the opossum *Marmosa* (Tate, 1933).

Other types of skin glands are also important in mammals. The watery secretion of the **sweat glands** functions primarily to promote evaporative cooling but also eliminates some waste materials. In humans and some ungulates, sweat glands are broadly distributed over the body surface, but in most mammals they are more restricted. In some insectivores, rodents, and carnivores, sweat glands occur only on the feet or on the **venter.** The glands are completely lacking in the Cetacea and in some bats and rodents. Hair follicles are supplied with **sebaceous glands** that produce an oily secretion

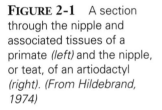

FIGURE 2-1   A section through the nipple and associated tissues of a primate *(left)* and the nipple, or teat, of an artiodactyl *(right)*. *(From Hildebrand, 1974)*

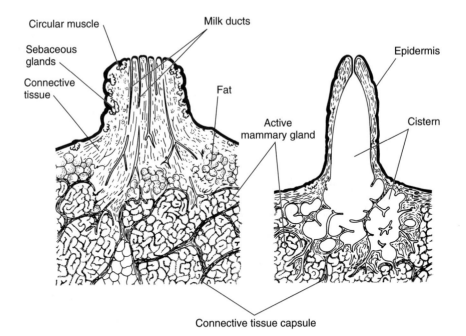

that lubricates the hair and skin (Fig. 2-2). Diverse **scent glands** and **musk glands** are found in mammals (e.g., Ewer, 1973; Ralls, 1971). These glands are variously used for attracting mates, marking territories, communication during social interactions, or protection. The smell of skunk is familiar to all but the most city-bound and has caused the temporary banishment of many a domesticated dog. A musk gland marked by a chevron-shaped patch of dark hairs occurs on the top of the tail of wolves and coyotes, as well as on the tail of many domesticated dogs. The functions of some mammalian scent glands in connection with social behavior are discussed in Chapter 23.

HAIR

The bodies of mammals are typically covered with hair, a unique mammalian feature that has no structural homolog among other vertebrates. Hair was perhaps developed by therapsids before a scaly covering was lost. In modern mammals that possess scaly tails or bony plates (such as armadillos), hairs project from beneath the scales in a regular pattern. A similar pattern of hair distribution, perhaps reflecting the ancestral condition, is also seen in mammals without scales (de Meijere, 1894).

(Note the pattern of hair projection on your own arm, for example.)

A hair consists of dead epidermal cells that are strengthened by keratin, a tough, horny tissue made of proteins. A hair grows from living cells in the hair root. Each hair consists of an outer layer of cells arranged in a scalelike pattern (the **cuticular scale**), a deeper layer of highly packed cells (the **cortex**), and, in some cases, a central core of cuboidal cells (the **medulla;** Fig. 2-3). The color of hair depends on pigment in either the medulla or the cortical cells; the cuticular scale is usually transparent.

The coat of hair, collectively termed the **pelage,** functions primarily as insulation. The dissipation of heat from the skin surface to the environment and the absorption of heat from the environment are retarded by the pelage. Seals, sea lions, and walruses, many of which live in extremely cold water, are insulated both by hair and a subcutaneous layer of blubber. Some mammals are hairless, or nearly so. These either live in warm areas or have specialized means of insulation other than hair. Essentially hairless, whales and porpoises have thick layers of blubber that provide insulation. Hair is sparse on elephants, rhinoceroses, and hippopotami. These animals live in warm areas, have

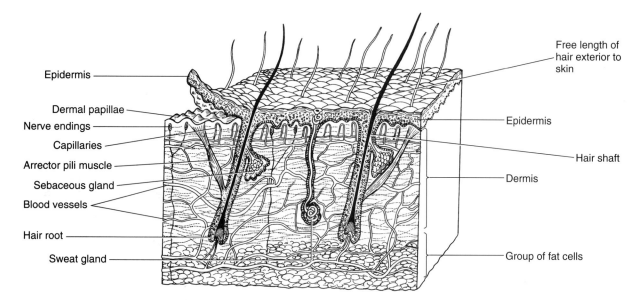

FIGURE 2-2 Generalized section of mammalian skin. The skin on different areas of the body differs in thickness as well as purpose. *(From Romer and Parsons, 1977)*

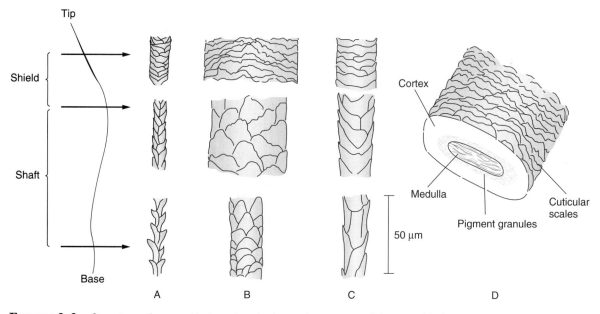

FIGURE 2-3 Structure of a guard hair and cuticular scale patterns of the guard hairs of some mammals. Shown at the left is a diagram of a single guard hair as found on the dorsal surface of a mammal, with regions labeled. Arrows indicate the positions along the hair where the highly magnified scale patterns shown in A, B, and C occur. Note how the scale pattern differs from base to tip. (A) *Pipistrellus nathusii* (pipistrelle bat); (B) *Martes martes* (pine marten); (C) *Eliomys quercinus* (garden dormouse). (D) Cross section of hair of *Meles meles* (European badger; same scale as A–C). *(Drawn after photos in Teerink, 1991)*

thick skins (Jarman, 1989) that offer some insulation, and have such favorable mass/surface ratios for heat conservation because of their large size that retention of body heat is no problem (p. 367).

Hair, being nonliving material, is subject to considerable wear and bleaching of pigments. Molts occur periodically when a juvenile mammal gains the first adult pelage, and thereafter usually once or twice a year. Old hairs are lost, and new ones replace them. These changes often occur in a regular pattern of replacement (Fig. 2-4). In many north-temperate species, the molts are in the spring and fall; the summer pelage is generally shorter and has less insulating ability than does the winter pelage. In some species that live in areas with continuous snow-cover in the winter, the summer pelage is brown and the winter coat is white. The arctic fox, several species of hares, and some weasels follow this pattern.

The color of most small terrestrial mammals closely resembles the color of the soil on which they live. In his study of concealing coloration in desert rodents of the Tularosa Basin of New Mexico, Benson (1933) found that white sands were inhabited by nearly white rodents and adjoining stretches of black lava were home to black rodents. Broadly speaking, nocturnal mammals that are active against dark substrates, such as dark forest soils, are dark-colored, whereas those foraging over light-colored soils, such as the soil of deserts, are relatively pale. In large, diurnal, open-country dwellers, such as African antelope, however, coloration may be related to temperature control. The pale bodies of the Arabian oryx (*Oryx leucoryx*) and the addax (*Addax nasomaculatus*), both desert dwellers, may be important in reflecting light and reducing the intake of solar radiation.

Countershading is a pattern common to mammals and many other vertebrates. In most lighting conditions, the back of an animal is more brightly illuminated than is the underside. If a mammal were all of a single color, the underside would appear very dark relative to the back and the form of the animal would be obvious. When the back and

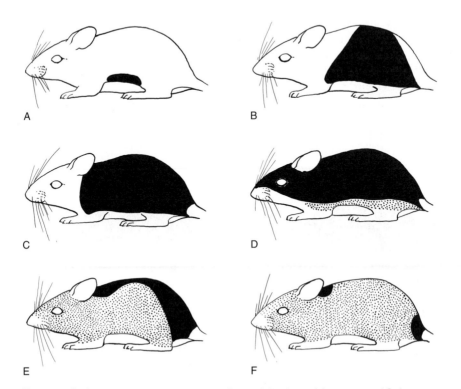

FIGURE 2-4   The pattern of postjuvenile molt in the golden mouse (*Ochrotomys nuttalli*). Black areas indicate areas in which adult hair is replacing juvenile hair. Stippled areas indicate new adult pelage. *(After Linzey and Linzey, 1967)*

sides are darkly colored and the underside and insides of the legs are white, however — an almost universal color pattern among terrestrial mammals — the well-lighted back reflects little light and the shaded white venter tends to reflect light strongly. The result is that the form of the animal becomes obscured to some extent, and the animal becomes less conspicuous.

The color patterns of mammals serve a variety of purposes. The pelages of some ungulates and some rodents are marked by white stripes that tend to obliterate the shapes of the animals when they are against broken patterns of light and shade. The eye is one of the most conspicuous and unmistakable vertebrate features; some facial markings in mammals may obviate the bold pattern of the eye by superimposing a more dominant and disruptive pattern (Fig. 2-5). If these markings only occasionally allow an animal to go unnoticed by a predator, or if they cause a predator to be indecisive in its attack for but a fraction of a second, they have adaptive value.

Even in broad daylight, the stripes of zebras cause distant herds to fade into their background, but another potentially adaptive importance has been considered by Cott (1966). The stripes are patterned so as to create an optical illusion: the animal's apparent size is increased. In the dim light in which most predators hunt, this illusion may cause a slight miscalculation of range and an occasional inaccurate leap.

What about the glaringly white rump patch of the pronghorn (see Fig. 17-19), however, or the conspicuous white-next-to-black markings of some African antelope, or the bold white eye ring of many carnivores, or the white-on-black pattern of skunks (see Fig. 12-19a)? No single explanation can be applied to these diverse patterns, but each makes an animal, or at least part of it, more obvious rather than less so. The pronghorn's markings seemingly function as warning signals to other herd members when an individual begins to run from danger; the bounding gait of some antelope shows off these markings. The eye rings of carnivores are perhaps important as accents for the eyes and face and may emphasize facial expressions used during intraspecific social interactions. The black-and-white coloration of skunks, on the other hand, makes these defensively well-endowed animals conspicuous and unmistakable to their would-be predators. This warning coloration is sometimes termed aposematic.

Numerous kinds of vertebrates — amphibians, reptiles, birds — are green, but why are there no green mammals? This frequently asked question seems to have no ready answer. Is it because of mammals' nocturnal ancestry? A few kinds of mammals actually appear greenish or olive. Tree sloths (order Xenarthra) achieve a highly camouflaged effect during the rainy season when unicellular algae grow within the flutings and cracks in individual hairs over the entire body, giving them a

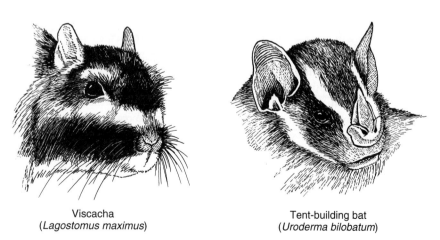

Viscacha
(*Lagostomus maximus*)

Tent-building bat
(*Uroderma bilobatum*)

FIGURE 2-5 The faces of two mammals in which facial masks reduce the conspicuousness of the eye.

green appearance (Aiello, 1985). Some mammals, particularly those in the rain forests of Australasia (e.g., New Guinean bats of the genera *Nyctimene* and *Paranyctimene,* or the Australian green ringtail possum *Pseudochirops archeri*), that sleep exposed on branches during the day (Flannery, 1995) seem to be olive or at least have a strong greenish cast to their wings or body fur. To our knowledge, these species have not been studied in order to determine whether their color might be structural (in which specific wavelengths of light, usually blues and greens, are reflected by tiny granules of melanin, as in the feathers of birds), from actual pigments, or a combination of the two.

Another unexplained pelage color phenomenon is the fact that patches of the fur in didelphid opossums reflect ultraviolet light (Pine et al., 1985). Why would ultraviolet reflectance occur in a group of nocturnal mammals that may have lost color vision (Jacobs, 1993)?

FAT AND ENERGY STORAGE

Although fat (adipose tissue) is a feature that is by no means unique to mammals, it is particularly vital in these animals. Fat serves three major functions in mammals: (1) energy storage, (2) a source of heat and water, and (3) thermal insulation. The lives of many species of mammals are punctuated by times of crisis when food is in short supply or energy demands are unusually high. For example, those mammals that hibernate must store enough energy to sustain life through periods when no food is available. Some filter-feeding whales spend their winters in plankton-poor tropical waters where they do little or no feeding. Such mammals survive by metabolizing stored fat. During times when males are competing for mates or defending territories, or when females are lactating, stored fat is often the key to survival, and those individuals with the greatest amounts of stored fat have the highest reproductive success. An example of the severity of this energy crisis in a female gray seal is given by Young (1976): during 15 days of lactation, the mother lost 45 kilograms of body weight while her young gained 27 kilograms. Desert dwellers and mammals of temperate areas often have localized fat storage (in the tail, or in the inguinal or abdominal region, for example), whereas boreal and aquatic species typically store fat subcutaneously over much of the body. This subcutaneous layer is important as insulation as well as for food storage.

CIRCULATORY SYSTEM

In keeping with their active life and their endothermic ability, mammals have a highly efficient circulatory system. A complete separation of the systemic circulation and the pulmonary circulation has been achieved. The four-chambered heart functions as a double pump: the right side of the heart receives venous blood from the body and pumps it to the lungs at low pressure (the pulmonary circulation); the left side receives oxygenated blood from the lungs and pumps it to the body at high pressure (the systemic circulation). The fascinating evolution of the mammalian heart and circulatory pattern is described in detail by Hildebrand (1974).

As might be expected because of the great size difference between the smallest and the largest mammals (the mass of a 2-gram shrew — about the same as a U.S. dime — and that of a 160,000-kilogram whale differ by a factor of 80,000,000), the heart rate is highly variable from species to species. The rate in nonhibernating mammals varies from under 20 beats per minute in seals to over 1300 in shrews (Table 2-3). Especially remarkable is the ability of some mammalian hearts to rapidly alter their rates of beat. As an extreme example, a resting big brown bat (*Eptesicus*) has a rate of about 400 beats per minute. This rate increases almost instantly to about 1000 when the bat takes flight and generally returns to the resting rate within 1 second after flight stops (Studier and Howell, 1969).

The erythrocytes (red blood cells) of mammals are biconcave disks rather than the ovoid spheres seen in other vertebrates. In all mammals except camels (Camelidae), the erythrocytes extrude their nuclei when they mature, apparently as a means of increasing oxygen-carrying capacity to support the high metabolic rate of mammals.

RESPIRATORY SYSTEM

In mammals, the lungs are large and, together with the heart, virtually fill the thoracic cavity. Air passes down the trachea, into the bronchi, and through a series of branches of diminishing size into the bronchioles, from which the alveolar

TABLE 2-3 Heart Rates of Selected Mammals

Species	Common Name	Adult Body Mass	Mean Heart Rate and/or Range (beats/min)
Sorex cinereus	Gray shrew	3–4 g	782 (588–1320)
Tamias minimus	Least chipmunk	40 g	684 (660–702)
Sciurus carolinensis	Gray squirrel	500–600 g	390
Mustela vison	Mink	0.7–1.4 kg	272–414
Erinaceus europaeus	European hedgehog	500–900 g	246 (234–264)
Phocoena phocoena	Harbor porpoise	170 kg	40–110
Ovis aries	Sheep	50 kg	70–80
Sus scrofa	Swine	100 kg	60–80
Equus caballus	Horse	380–450 kg	34–55
Elephas maximus	Asiatic elephant	2000–3000 kg	25–50
Phoca vitulina	Harbor seal	20–25 kg	18–25

Data from Altman and Dittmer, 1964:235.

ducts branch. Clustered around each alveolar duct is a series of tiny terminal chambers, the alveoli. Gas exchange between inhaled air and the bloodstream occurs in the alveoli; the thin alveolar membranes are surrounded by dense capillary beds. In humans, the lungs contain about 300 million alveoli, which provide a total respiratory surface of about 70 square meters—approximately 40 times the surface area of the body.

Air is forced into the lungs by muscular action that increases the volume of the thoracic cavity and decreases the pressure within the cavity. Some volume increase is gained by the forward and outward movement of the ribs under the control of intercostal muscles, but of greater importance is retraction and depression of the muscular diaphragm (a structure unique to mammals). When relaxed, the diaphragm is bowed forward, but when contracted its central part moves backward toward the coelomic cavity, thus increasing the volume of the thoracic cavity.

REPRODUCTIVE SYSTEM

In mammals, both ovaries are functional and the ova are fertilized in the uterine tubes. The embryo develops in the uterus within a fluid-filled amniotic sac. Nourishment for the embryo comes from the maternal bloodstream by way of the placenta. (The female reproductive cycle and the establishment of the placenta are discussed in Chapter 20.) The structure of the uterus is variable (Renfree, 1993; Tyndale-Biscoe and Renfree, 1987; Fig. 2-6).

The male copulatory organ, the penis, contains erectile tissue and is surrounded by a sheath of skin, the **prepuce.** In many species the penis contains a bone, the **os penis,** or **baculum,** which may differ markedly even between closely related species (Fig. 2-7) and therefore may be of considerable use in taxonomic studies. The tip of the penis has an extremely complicated form in some species (Fig. 2-7). The testes of mammals, instead of lying in the coelomic cavity as in other vertebrates, are typically contained in the **scrotum,** a saclike structure that lies outside the body cavity but is an extension of the coelomic cavity. The testes either descend permanently from the coelomic cavity into the scrotum when the male reaches reproductive maturity or are withdrawn into the body cavity between breeding seasons and descend when the animal again becomes fertile. In most mammals, the maturation of sperm cannot proceed normally at the usual deepbody temperature and the scrotum functions as a "cooler" for the testes and developing sperm.

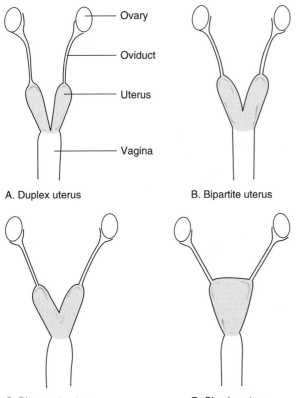

A. Duplex uterus

B. Bipartite uterus

C. Bicornuate uterus

D. Simplex uterus

FIGURE 2-6   Several types of uteri (stippled) found in eutherian mammals, showing degrees of fusion of the two "horns" of the uterus. (A) Duplex uterus occurs in the orders Lagomorpha, Rodentia, Tubulidentata, and Hyracoidea. (B) Bipartite uterus is known in the order Cetacea. (C) Bicornuate uterus is found in the order Insectivora, in some members of the orders Chiroptera and Primates, and in the orders Pholidota, Carnivora, Proboscidea, Sirenia, Perissodactyla, and Artiodactyla. (D) Simplex uterus is typical of some members of the orders Chiroptera, Primates, and Xenarthra. *(From Smith, 1960)*

BRAIN

Compared with the brains of other vertebrates, the mammalian brain is unusually large. This greater size is attributable to a tremendous increase in the size of the cerebral hemispheres. These structures were ultimately derived from a part of the brain important in lower vertebrates in receiving and relaying olfactory stimuli.

Most characteristic of the brain of higher mammals is the great development of the **neopallium,** a mantle of gray matter that first appeared as a small area in the front part of the cerebral hemispheres in some reptiles. In mammals, the neopallium has expanded over the surface of the deeper, primitive vertebrate brain. The surface area of the neopallium is vastly increased in many mammals by a complex pattern of folding (Fig. 2-8; Table 2-2, number 12). A new development in eutherian mammals is the **corpus callosum,** a large concentration of nerve fibers that passes between the two halves of the neopallium and provides additional communication between them.

The unique behavior of mammals is largely a result of the development of the neopallium, which functions as a control center that has come to dominate the original brain centers. Sensory stimuli are relayed to the neopallium, where much motor activity originates. Present actions are influenced by past experience; learning and "intelligence" are important. The size of the brain relative to total body size is not always a reliable guide to intelligence; brain size apparently need not increase in proportion to increases in body size to maintain intelligence. The degree of development of convolutions on the surface of the neopallium is perhaps a better indication of intelligence. In some groups of mammals (e.g., primates), the increased energy derived from endothermy is consumed in large part by the enlarged brain (Armstrong, 1983).

SENSE ORGANS

The sense of smell is acute in many mammals, probably resulting in part from their nocturnal ancestry. The smell sensors are primarily distributed across the mucosal surfaces of the mesethmoid and vomeronasal organ areas (Fig. 2-9). The olfactory bulbs and olfactory lobes form a great part of the brain in some insectivorans and are reasonably large in carnivorans and rodents. The sense of smell is poorly developed, and the olfactory part of the brain is strongly reduced in whales and the higher primates (which are diurnal); the olfactory system is absent in porpoises and dolphins (Kruger, 1966).

The sense of hearing is highly developed in mammals; in some 20 percent of mammalian species, hearing provides an important substitute for vision (see Chapter 22). The acuity of hearing, like olfaction, is probably related to mammals' primitively nocturnal habits. Mammals alone have

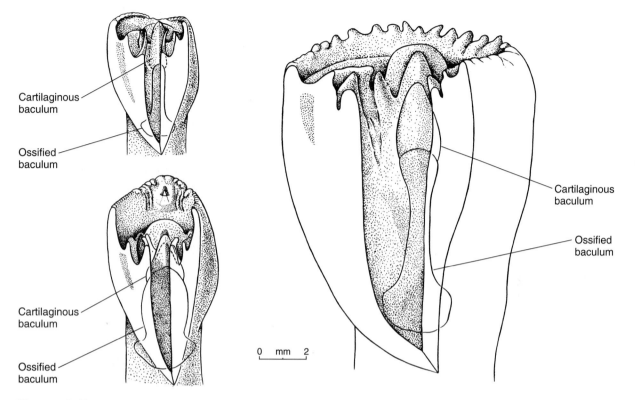

FIGURE 2-7   Ventral view of the penises of several species of New Guinean murid rodents, showing the complex structure of the organ in these mammals. Ossified and cartilaginous components of the baculum are embedded within the penis. *(From Lidicker, 1968)*

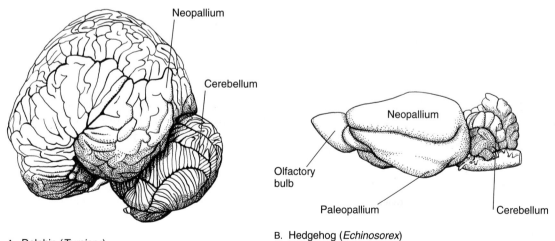

A. Dolphin (*Tursiops*)

B. Hedgehog (*Echinosorex*)

FIGURE 2-8   Left sides of the brains of (A) a dolphin and (B) a hedgehog. The neopallium is greatly enlarged and highly convoluted in the dolphin but relatively small and smooth-surfaced in the more primitive hedgehog. Anterior is to the left. (*Tursiops* after Kruger, 1966; *Hedgehog* after Romer and Parsons, 1977)

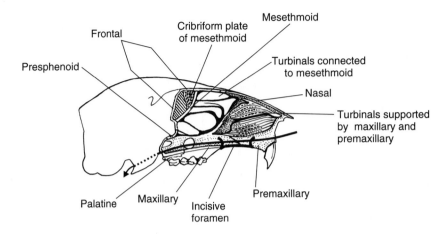

FIGURE 2-9 Cutaway view of the nasal chamber of the Abert squirrel, showing the complicated arrangement of turbinal bones. The entire right half of the nasal part of the skull is removed, exposing the left side of the nasal chamber. The arrow shows the main air path from the external to the internal nares, but some air circulates through the upper part of the chamber and over the turbinal bones. Branches of the olfactory nerve pass out of the braincase through the cribriform plate of the mesethmoid bone. The incisive foramen transmits the duct to the vomeronasal organ.

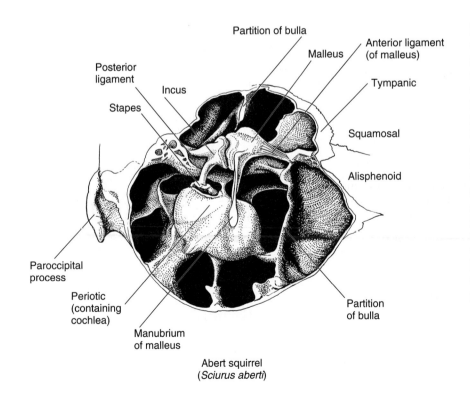

Abert squirrel
(*Sciurus aberti*)

FIGURE 2-10 Lateral view of the right middle ear chamber (anterior is to the right) of the Abert squirrel, with the auditory bulla largely removed. The complex partitioning of the air-filled bulla, the position of the ossicles of the middle ear, bones of the fluid-filled inner ear, and the ligamentous suspension of the malleus and incus are shown. In life, the manubrium of the malleus is embedded within the tympanic membrane. Much of the stirrup-shaped stapes is hidden behind the periotic in this view.

an external structure (the **pinna**) to intercept sound waves; the pinnae may be extremely large and elaborate in some mammals, particularly in bats (see Figs. 10-20 and 10-24). Pinnae are missing (presumably secondarily lost) in some insectivorans, phocid seals, and cetaceans. The **external auditory meatus,** the tube leading from the pinna to the tympanic membrane, is typically long in mammals and extremely long in cetaceans. The middle ear is an air-filled chamber that houses the three **ossicles** and is typically enclosed by a bony **bulla** (Fig. 2-10). The mammalian **cochlea** is more or less coiled in most therians; in monotremes it is angled.

The mammalian eye resembles that of most amniote vertebrates. In most nocturnal mammals, the **tapetum lucidum** is well developed. This is a reflective structure within the choroid that improves night vision by reflecting light back to the retina. (This reflection accounts for the shine when an animal's eyes are picked up by headlight beams at night.) Although in most mammals the eyes are relatively large, in some insectivorans, cetaceans, and rodents they are greatly reduced in size and function. In such species, the eyes are able to differentiate only between light and dark and may serve primarily to aid the animal in maintaining the appropriate activity or thermoregulatory cycles (Herald et al., 1969; Lund and Lund, 1965; Cooper et al., 1993). Such daily, or circadian, cycles are under the control of a self-sustaining, autonomous molecular clock in the hypothalamus of the brain (Sassone-Corsi, 1998) that is capable of being adjusted partly by the retina of the eye even in mammals with well-developed eyes (Tosini and Menaker, 1996).

Most mammals have **vibrissae.** These are the whiskers on the muzzle and the long, stiff hairs that are present on other parts of the head and the lower legs of some mammals. The vibrissae are tactile organs, and those on the face enable nocturnal species to detect obstacles near the face. The vibrissae on the muzzle generally arise from a structure termed the **mystacial pad.** They are controlled by a complex of muscles (Fig. 2-11), the superficial facial muscles, innervated by cranial nerve VII, the facial nerve. Tactile stimulation of the vibrissae is sensed by barrel structures in the somatosensory cortex of the brain.

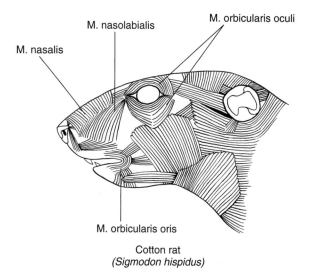

FIGURE 2-11 The superficial facial muscles of the cotton rat; these muscles almost wholly control facial expression and movement of vibrissae. *(After Rinker, 1954)*

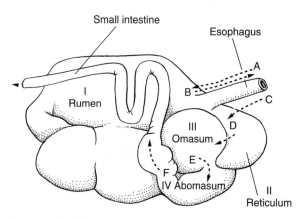

FIGURE 2-12 The four-chambered "stomach" (the first three chambers of which are esophageal in origin) of a ruminant artiodactyl. As the animal feeds, it swallows vegetation, which is then stored in the rumen. While the animal rests, it regurgitates the food from the rumen and "chews its cud" (remasticates the food). The food then goes to the reticulum, omasum, and abomasum, where digestion is aided by a diverse microbiota. *(After Storer and Usinger, 1965)*

DIGESTIVE SYSTEM

As in other vertebrates, salivary glands are present in mammals; in some ant-eating species, they are extremely large and specialized for the production of a mucilaginous material that makes the tongue sticky. The esophagus is a simple tube and the stomach is a single saclike compartment in most species, but these structures are complexly elaborated and subdivided in ruminant artiodactyls, cetaceans, and sirenians (Fig. 2-12). In herbivorous species, digestion is frequently accomplished partly by microorganisms that inhabit the stomach or the **caecum,** a blind sac that opens into the posterior end of the small intestine (see Fig. 6-28).

MUSCULAR SYSTEM

The mammalian limb and trunk musculature has been highly adaptable. Different evolutionary lines have developed muscular patterns precisely adapted to diverse modes of locomotion. Cetaceans are the fastest marine animals, certain carnivorans and ungulates are the most rapid runners, and bats as fliers are more maneuverable than most birds. Some muscular specializations favoring specific types of locomotion are described in the ordinal chapters. Intrinsic musculature of the trunk of the body controls posture and contributes to all styles of locomotion. Especially notable in mammals is the great development of dermal musculature, the superficial musculature of the skin. In many mammals, these muscles form a sheath over most of the body and allow the skin to move independently from underlying tissues. Dermal muscles have differentiated and have moved over much of the head (Fig. 2-11), where they control many essential actions. In mammals, there are no more vital voluntary muscles than those that encircle the mouth; these function during suckling and are among the first voluntary muscles to be subjected to heavy use after birth. Facial muscles move the ears, close the eyes, and control the subtle changes in expression that are so important in the social lives of many mammals.

THE SKELETON

GENERAL FEATURES

The mammalian skeleton differs from that of reptiles in several basic ways, all of which may well be related to the active style of life of mammals. The mammalian skeleton has become simplified and is more completely ossified (Figs. 2-13 and 2-14), fea-

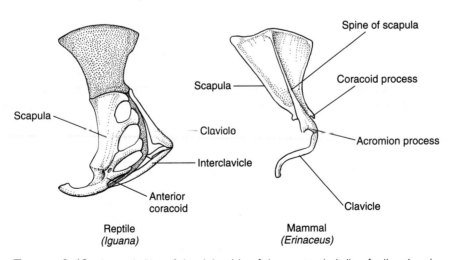

Scapula

Anterior coracoid

Claviole

Interclavicle

Reptile
(Iguana)

Spine of scapula

Coracoid process

Scapula

Acromion process

Clavicle

Mammal
(Erinaceus)

FIGURE 2-13   Lateral view of the right side of the pectoral girdle of a lizard and a hedgehog, showing the more complete ossification and reduction in number of discrete bones in the mammal. Heavily stippled areas are cartilaginous in adults.

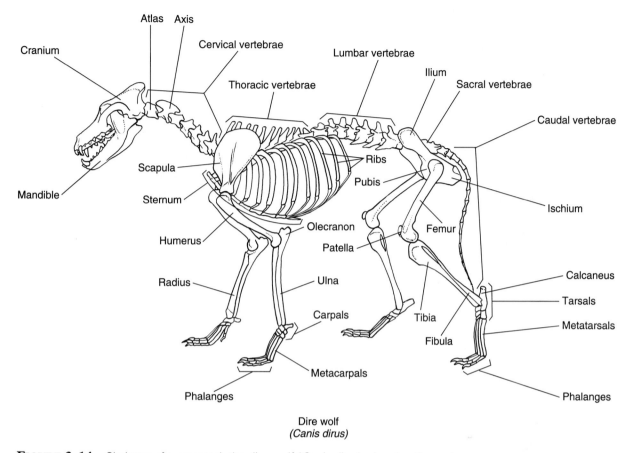

Atlas Axis
Cranium
Cervical vertebrae
Lumbar vertebrae
Ilium
Thoracic vertebrae
Sacral vertebrae
Caudal vertebrae
Ribs
Scapula
Pubis
Mandible
Sternum
Ischium
Olecranon
Femur
Humerus
Patella
Radius
Ulna
Calcaneus
Tibia
Tarsals
Carpals
Metatarsals
Fibula
Metacarpals
Phalanges
Phalanges

Dire wolf
(Canis dirus)

FIGURE 2-14 Skeleton of a mammal, the dire wolf (*Canis dirus*), showing the major elements. *(After Stock, 1949, courtesy of the Natural History Museum of Los Angeles County)*

tures perhaps associated with the need for well-braced attachments for muscles. Considerable fusion of bones has also occurred, as, for example, in the pelvic girdle. There is often great flexibility of the axial skeleton that allows the limbs greater speed and range of movement. The greater range of movement is of particular advantage to arboreal creatures, which many early mammals may have been. The simplification of the skeleton may have been advantageous in terms of metabolic economy —the less bone, the less energy invested in its development and maintenance. Further, selection may have favored a light skeleton in the interest of quick movement with relatively little expenditure of energy.

To an animal as active as a mammal, well-defined articular surfaces on limb bones and solid points of

attachment for muscles are highly advantageous during the period of skeleton growth as well as during adult life. Most mammals have evolved a pattern of bone growth very different from that of reptiles. In many reptiles, growth may continue throughout much of life. Growth in reptiles occurs at the ends of limb bones by ossification of the deep parts of a persistently growing cartilaginous cap; such a pattern limits the establishment of a clearly defined joint. In most mammals (except for some marsupials), however, skeletal growth is generally restricted to the early part of life. The articular surfaces and some points of attachment of large muscles become prominent and ossified early, while rapid growth is still under way. Growth continues at a cartilaginous zone where the end of the bone and its articular surface, the **epiphysis,** join the shaft of the bone,

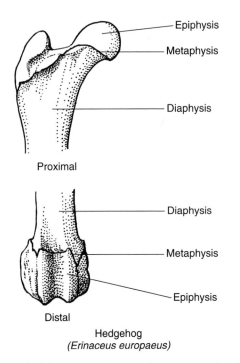

FIGURE 2-15 Anterior views of the proximal and distal ends of the right femur of a young hedgehog (*Erinaceus europaeus*), showing the epiphyses, diaphysis, and intervening cartilaginous zone or metaphysis.

the **diaphysis** (Fig. 2-15). When full growth is attained, this cartilaginous zone of growth (the **metaphysis**) becomes ossified, fusing the epiphysis and diaphysis. Because this fusion usually occurs at a certain age within a given species, the degree of closure of the "epiphyseal line" is useful in estimating the age of a mammal.

THE SKULL

Aspects of the development, structure, diversity, and function of the mammalian skull were summarized by various authors in the useful three-volume series *The Skull* (Hanken and Hall, 1993a, b, c). Unlike that of reptiles, the skull of most mammals is akinetic, that is, there are no intrinsic joints within it other than the cranio-mandibular (jaw) joint and the joints between the middle ear bones. The braincase of the mammalian skull (Fig. 2-16) is large. In addition to its primary function of protecting the brain, the braincase provides a sur-

face from which the temporal muscles originate. (In many mammals, these are the muscles that generate the force to close the jaws.) A **sagittal crest** increases the area of origin for the temporal muscles in many mammals; the **lambdoidal crest** gives origin to the temporal muscles and insertion for some cervical muscles. The **zygomatic arch** is usually present as a structure that flares outward from the skull. It protects the eyes, provides origin for the masseteric jaw muscles, and forms the surface with which the condyle of the dentary (lower jaw) bone articulates. The zygomatic arch may be reduced or lost, as in some xenarthrans, insectivorans, and cetaceans, or may be enlarged, as in those groups (such as rodents and lagomorphs) in which the masseter muscles largely supplant the temporal muscles as the major jaw muscles (see Fig. 18-2). The skull has a secondary palate (see p. 48 and Fig. 3-4), and there are usually **turbinal bones** within the nasal cavities (Fig. 2-9).

A number of **foramina** (openings) perforate the braincase and allow passage of the cranial nerves and blood vessels (Fig. 2-16). In some rodents, the infraorbital foramen, through which blood vessels and a branch of cranial nerve V (the trigeminal nerve) pass, is greatly enlarged in association with specializations of the masseter muscles (see Fig. 18-2). The incisive foramina, present in the palates of many mammals, form openings for the ducts to an olfactory organ known as the vomeronasal organ or Jacobson's organ (Fig. 2-9, 2-16B). The vomeronasal organ allows a mammal to "smell" the contents of its mouth and the airborne and/or fluid-borne pheromones of potential mates. Vomeronasal organs are widespread among vertebrates; a snake puts the tips of its forked tongue against this part of the palate after "testing" its immediate environment.

Sounds that cause vibration of the tympanic membrane are mechanically transmitted by the three ear ossicles (Fig. 2-10) through the air-filled chamber of the middle ear to a membrane at the wall of the fluid-filled inner ear. The footplate of the stapes is embedded in a membrane that fills an opening into the inner ear and, acting like a piston, transforms the movements of the ossicles to vibrations of the fluid in the cochlea. The inner ear, with the cochlea and semicircular canals, is contained by the periotic (or petrosal) bone, which is generally covered by the squamosal bone but is ex-

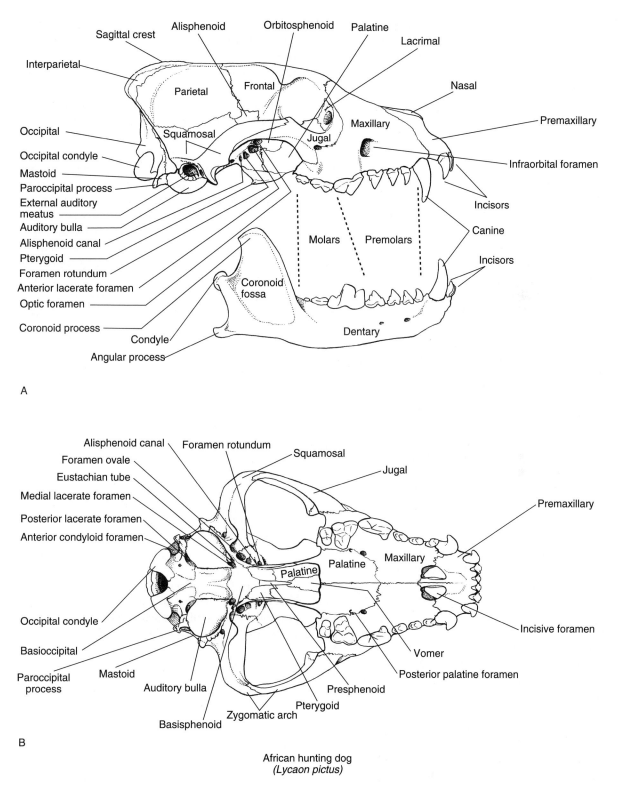

FIGURE 2-16 (A) Side and (B) palatal views of the skull of the African hunting dog (*Lycaon pictus*), showing the bones, foramina, and teeth.

posed as the mastoid bone in some mammals (Fig. 2-16). The auditory bulla is formed by the expanded tympanic bone or by the tympanic bone plus the entotympanic bone, a bone found only in mammals. The bulbous tympanic bullae are highly modified in some mammalian species in connection with specialized modes of life.

The lower jaw in Cenozoic species is formed exclusively by the **dentary** bone, which typically has a coronoid process on which the temporalis muscle inserts, a coronoid fossa, in which the masseter muscles insert, and an angular process, to which a jaw-opening muscle (the digastricus) attaches. Pterygoid muscles insert on the medial side of the dentary; these muscles originate on the pterygoid bones of the skull and are hypertrophied in mammals. The pterygoid musculature delicately controls lateral movements of the lower jaw for precise occlusion of the teeth. In some herbivores, in which the masseter muscle is enlarged at the expense of the temporalis muscle, the coronoid process is reduced or absent and the posterior part of the dentary bone becomes dorsoventrally broadened (see Fig. 18-3C).

Several skeletal elements in the throat region are highly modified homologs of the gill arches of fish. These elements, the **hyoid apparatus** and laryngeal cartilages, support the trachea, the larynx, and the base of the tongue and are often braced against the auditory bullae.

TEETH

Without doubt, one of the major keys to the success of mammals has been the possession of teeth. Fish, amphibians, reptiles, fossil birds, and mammals all have teeth, but the specialization of the dentition in mammals has exceeded anything found in the other groups. Mainly it is mammals that have dentitions capable of coping with items so difficult to prepare for digestion as dry grass and large bones. So varied are the dental specializations of mammals and so closely related are they to specific styles of feeding and to patterns of adaptations of the skull, jaws, and jaw musculature, that to know in detail the dentition of a mammal is to understand many aspects of its way of life. Much of our knowledge of the early evolution of mammals is based on studies of fossil teeth, which, because

of their extreme hardness, are often the only parts of early mammals that are preserved. The earliest known vertebrates, which were jawless, had bodies encased in bony plates. When the visceral arches anterior to those that supported the gill apparatus in primitive vertebrates became modified into jaws, teeth developed on the bony plates that bordered the mouth.

Although the teeth in various dentitions differ widely in number, structure, and function, in most mammals the dentition is **heterodont;** that is to say, it consists of teeth that vary in both structure and function. In extant mammals, teeth occur on the premaxillary, maxillary, and dentary bones (Fig. 2-16). The anteriormost teeth, the **incisors** and **canines,** are used to gather or kill food, whereas the more specialized cheek teeth, the **premolars** and **molars,** are used to grind or slice food in preparation for digestion. In many mammals, the canines

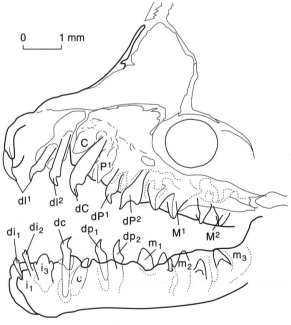

Thyroptera

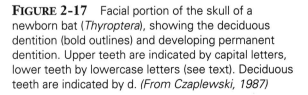

FIGURE 2-17   Facial portion of the skull of a newborn bat (*Thyroptera*), showing the deciduous dentition (bold outlines) and developing permanent dentition. Upper teeth are indicated by capital letters, lower teeth by lowercase letters (see text). Deciduous teeth are indicated by d. *(From Czaplewski, 1987)*

are used in stereotyped displays during social interactions, in addition to their essential food-procuring function. Characteristically, two sets of teeth appear in a mammal's lifetime. The **deciduous** dentition develops early and consists of incisors, canines, and premolars — but no molars. These "milk teeth" are lost and replaced by permanent teeth as the animal matures. The permanent dentition consists of a second set of incisors, canines, and premolars but also includes the molars, which normally have no deciduous counterparts but in some cases may actually represent unreplaced deciduous teeth (see p. 75). The deciduous dentition of some species bears little resemblance to the permanent dentition (Fig. 2-17).

The form, function, and origin of the cusp patterns of the cheek teeth, and especially of the molars, are of particular interest. As indicated by fossils, the molars of near-mammals and at least one lineage of early mammals underwent a transition from a primitive, cusps-in-line pattern (as in advanced cynodont synapsids, morganucodontids and triconodonts; see Chapter 3), to a tritubercular pattern with three cusps in a triangle (as in kuehneotheriids and symmetrodonts). Then a heel called a talonid was added on the lower molars (as in primitive therians, metatherians, and eutherians) (Fig. 2-18). The last type of tooth, in which the protocone of a three-cusped upper molar occludes with a well-developed basin in the talonid of a lower molar (Figs. 2-18F and 2-19), is termed **tribosphenic.** This configuration gives the tooth a crushing function during occlusion, in addition to the shearing crests already present. (The occlusal surfaces of teeth are those that contact their counterparts of the opposing jaw — the surfaces the dentist generally treats when putting in a filling.) The majority of living mammals (metatherians and eutherians) either possess tribosphenic molars or originated from groups that primitively possessed them. Indeed, tribosphenic molars probably contributed to the initial Mesozoic radiation and success of therian mammals (Butler, 1990). However, several groups of mammals that arose during the Mesozoic era did not have their ancestry in the lineage with tribosphenic teeth (see Chapter 3). This group includes the monotremes, whose extinct relatives had molars (modern monotremes lack them in adults) that are not

considered to be tribosphenic (Archer et al., 1993; Kielan-Jaworowska et al., 1987).

A stroke of luck for functional morphologists is that the American opossum (*Didelphis virginiana*) and some other living metatherians have molars that resemble those of certain of the mammals that coexisted with Cretaceous dinosaurs. Opossums are omnivorous, eating insects and other small animals as well as soft plant material; probably many Mesozoic mammals had similar diets. Careful studies of jaw action in the opossum, therefore, can probably indicate how the molars functioned in mammals more than 90 million years ago. The studies by Crompton (1971), Crompton and Jenkins (1968), and Crompton and Hiiemae (1969, 1970) provide much of the basis for the following discussion of the functional morphology of tribosphenic molars.

In the opossum, the molars serve two masticatory functions. For up to 60 percent of the time involved in chewing and throughout the initial stages of chewing, the high cusps of the upper and lower cheek teeth crush and puncture the food without coming together. After the food is pulped, it is sliced by the six matching shearing surfaces shown in Figure 2-19. This shearing is facilitated by the way in which food is trapped and steadied by the opposing molars (Fig. 2-20). Chewing occurs on but one side of the jaw at a time. During cutting strokes, the jaw action is not one of simple up-and-down movements. Instead, precise lateral adjustments of the jaw during mastication enable opposing molars to slide against each other. As shown in Figure 2-20, this movement involves a transverse as well as an upward component as the lower molars shear against the uppers. Attrition facets on the molars of Mesozoic mammals indicate that occlusion of tribosphenic teeth during chewing has always had this transverse component (Butler, 1972). Major shearing surfaces are those designated in Figure 2-19 as 1 and 2, but additional cutting occurs when the surfaces on the sides of the cusps of the quadrate posterior part (the talonid) of the lower molars shear against their counterparts on the upper molars. As a result of this complex pattern of occlusion, each time the three (in eutherians) or four (in metatherians) pairs of opposing molars of one side of the jaw come together, 18 or 24 cuts are made in the food, and thus food already pulped is rapidly

Upper teeth

Lower teeth

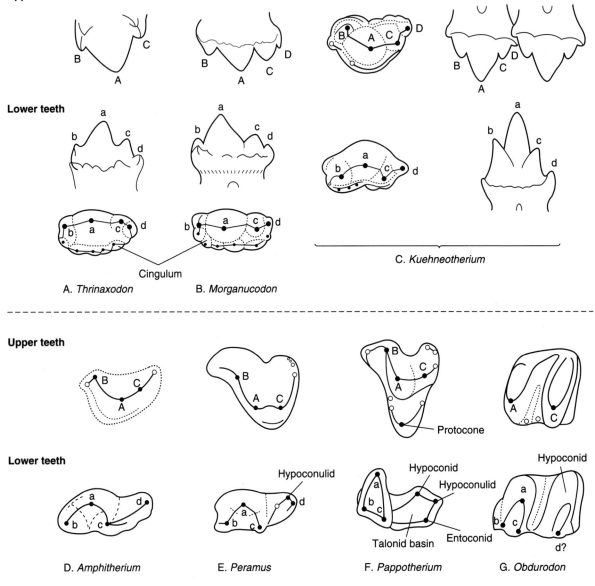

Cingulum

A. *Thrinaxodon*   B. *Morganucodon*   C. *Kuehneotherium*

Upper teeth

Protocone

Lower teeth

Hypoconulid

Hypoconid

Hypoconid

Hypoconulid

Talonid basin   Entoconid

d?

D. *Amphitherium*   E. *Peramus*   F. *Pappotherium*   G. *Obdurodon*

FIGURE 2-18   Cheek teeth of an advanced cynodont synapsid, a mammaliaform synapsid, and mammals, showing important stages in the evolution of tribosphenic molars. In each set of drawings the upper image represents right upper (maxillary) teeth; shown immediately below the upper teeth are left lower (dentary) teeth. (A) *Thrinaxodon* (a cynodont synapsid); (B) *Morganucodon* (a mammaliaform); (C) *Kuehneotherium*; (D) *Amphitherium*; (E) *Peramus* (three Mesozoic mammals); (F) *Pappotherium* (tribosphenic); (G) *Obdurodon*, a monotreme (teeth not truly tribosphenic). Cusps of lower teeth are drawn directly below the position in which they occlude with the upper teeth. Presumed homologous cusps are indicated by capital letters in upper teeth and lowercase letters in lower teeth: A=paracone; B=stylocone; C=metacone; D=metastyle?; a=protoconid; b=paraconid; c=metaconid; d=hypoconulid. Lateral views of the teeth are given in (A), (B), and (C) to show the positions of these teeth as they would be just before they occlude and shear past one another. In addition, occlusal views are shown in all cases, and in (C) through (G) the uppers are shown slightly staggered relative to the lowers, as they would be in their natural positions. In *Amphitherium* the upper teeth are unknown; this reconstruction is hypothetical. *(After Hopson, 1994)*

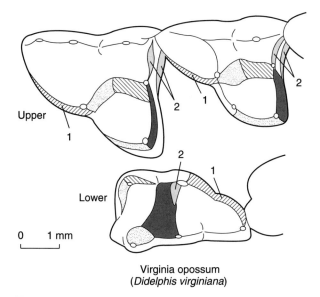

Upper

1

Lower

2

1

0   1 mm

Virginia opossum
(*Didelphis virginiana*)

FIGURE 2-19 Matching shearing planes (numbered patterns) of the occlusal surfaces of the left upper and lower molars of the Virginia opossum. Anterior is to the right. *(After Crompton and Hiiemae, 1969)*

sectioned. Natural selection has favored the evolution of efficient dentitions because time spent in masticating food means greater energy expenditure and less time available for food gathering. In some species, this also means a greater period of vulnerability to predators.

Two major evolutionary trends in tribosphenic molar structure appeared in the late Mesozoic era and became pronounced in the Cenozoic era. In carnivores, portions of some of the cheek teeth became bladelike and the vertical shearing function was elaborated. In the interest of powerful sectioning of flesh, transverse jaw action was reduced in these animals and there were a variety of associated changes in the skull, jaws, and jaw musculature. Carnivory and attendant functional changes in the teeth arose convergently in at least seven groups of flesh-eating mammals with tribosphenic teeth (Muizon and Lange-Badre, 1997). In herbivores, however, which must finely macerate plant material in preparation for digestion, the molars became quadrate; transverse or horizontal jaw action came to be of primary importance; and distinctive features of the skull, jaws, and jaw musculature favoring this action developed. The quadrate con-

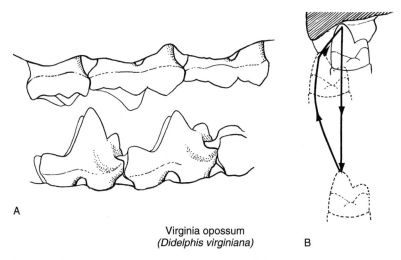

A

Virginia opossum
(*Didelphis virginiana*)   B

FIGURE 2-20 (A) Upper and lower molars of the Virginia opossum, showing the opposing cusps that steady, puncture, and crush food. View from the labial side of the right side of the animal, with anterior to the right. (B) Movement of the lower teeth as they shear against the uppers. These teeth are viewed from the rear on the left side of the animal. See text for description of action. *(B after Crompton and Hiiemae, 1969)*

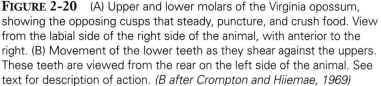

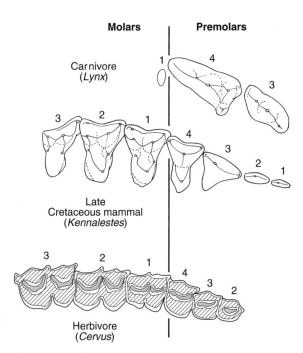

Molars | **Premolars**

Carnivore
(*Lynx*)

Late
Cretaceous mammal
(*Kennalestes*)

Herbivore
(*Cervus*)

FIGURE 2-21   Comparisons of the occlusal surfaces of the upper cheek teeth of a carnivore (*Lynx,* at top), a primitive mammal (*Kennalestes,* from Late Cretaceous, at center), and a herbivore (*Cervus,* at bottom). *(Partly after Crompton and Hiiemae, 1969)*

figuration of upper molars of mammals arose through the addition of a fourth major cusp, the hypocone, to the posterior lingual corner of the tooth. This kind of configuration is clearly an important innovation related to herbivory in therian mammals; a hypocone has evolved convergently more than 20 times among mammals during the Cenozoic era (Hunter and Jernvall, 1995). The dentitions of a modern carnivore, a modern herbivore, and a generalized tribosphenic Cretaceous mammal are compared in Figure 2-21. In the carnivore, grasping, slicing, and sometimes bone-cracking functions of the more anterior teeth (premolars and canines) are emphasized; whereas in herbivores, the grinding function of the back teeth is emphasized.

The number of teeth of each type in the dentition is designated by the **dental formula.** This is written as the number of teeth of each kind on one side of the upper jaw over the corresponding number in the lower jaw. Such a formula in carnivores may be as follows: incisors 3/3, canines 1/1, premolars 4/4, molars 2/3. Because the teeth are always listed in this order, the formula may be shortened to 3/3, 1/1, 4/4, 2/3. (The skull in Fig. 2-16 has this dental formula.) The dental formula

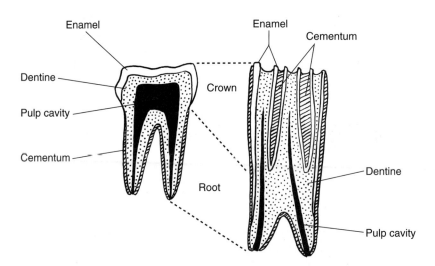

Enamel   Dentine   Pulp cavity   Cementum   Crown   Root   Enamel   Cementum   Dentine   Pulp cavity

FIGURE 2-22   Generalized sections of mammalian teeth, showing the internal structure and materials. The molar on the left is similar to that of primates and is low-crowned; the molar on the right is similar to that of a modern horse and is high-crowned.

lists the teeth of only one side; therefore, the number of teeth in the formula must be doubled to give the total number of teeth in the dentition. As an additional example, the arrangement for humans is 2/2, 1/1, 2/2, 3/3 × 2 = 32. The basic maximum number of teeth in Cenozoic eutherian mammals is 44 (3/3, 1/1, 4/4, 3/3; higher numbers are known in Cretaceous species) and in metatherians is 50 (5/4, 1/1, 3/3, 4/4). The number of adult teeth is frequently reduced, and a few eutherians lack teeth as adults. Some specialized eutherians, most notably odontocete cetaceans, have more than 44 teeth and have **homodont** dentitions (those in which all teeth are structurally alike). Individual teeth are designated by the initial for the type of tooth (**I/i** for incisor, **C/c** for canine, **P/p** for premolar, and **M/m** for molar) followed by the number for the locus (or position) of the tooth relative to others of its kind, numbered from front to back. Upper teeth are customarily represented by capital letters and lower teeth by lowercase letters. Deciduous teeth can be represented by the letter **d** in front of the tooth type. Thus, i2, dP3, and m4 can be used as shorthand ways of indicating the second lower incisor, the deciduous third upper premolar, and the fourth lower molar, respectively.

The mammalian tooth typically consists of an inner material, **dentine,** covered by a layer of **enamel** (Fig. 2-22). Dentine consists largely of hydroxyapatite $[3(Ca_3PO_4)_2\ Ca(OH)_2]$, has an organic fiber content of about 30 percent, and is harder than bone. Enamel also consists almost entirely of hydroxyapatite, which in the enamel of all living mammals except monotremes is arranged in a prismatic crystalline pattern. Enamel, the hardest mammalian (or vertebrate) tissue, is only 3 percent organic. The tooth is bound to the jaw by **cementum,** a relatively soft material that may also form part of the tooth crown. Most mammals have teeth that are **brachydont,** or short-crowned, and enamel formation ceases when the tooth erupts through the gum. Many herbivores, because their teeth are subject to rapid wear from abrasion by silica crystals in grass and by soil particles that adhere to plants, have **hypsodont,** or high-crowned, teeth (Fig. 2-22). As a further adaptation to abrasive food, in some mammals some teeth (and in some rodents and lagomorphs all teeth) grow continuously and are termed hypselodont or **ever-growing** teeth. The roots of mammalian teeth are often divided; primitively, the upper molars have three roots and the lower molars two. Incisors and canines are single-rooted in all extant mammals except colugos and elephant-shrews, and premolars may have one, two, or three roots. The dentitions of herbivores usually serve only two masticatory functions (but they may have many other functions in grooming, defense, and communication). The incisors, or the incisors and the canines, clip vegetation, and the cheek teeth grind the food. Between these teeth there is usually a space called a **diastema.** This is typical of rodents, for example.

The shape of the molar crown varies in response to the demands of different diets (Fig. 2-23). In pigs and in some rodents, carnivores, and

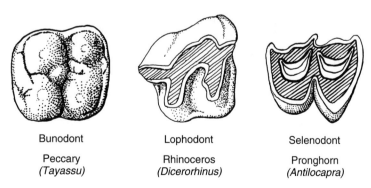

Bunodont	Lophodont	Selenodont
Peccary	Rhinoceros	Pronghorn
(*Tayassu*)	(*Dicerorhinus*)	(*Antilocapra*)

FIGURE 2-23   Three major types of right molariform teeth as defined by cusp shapes. Anterior is to the right; the outer edge of each tooth is toward the top. The cross-hatched parts are dentine.

primates, the molars are **bunodont,** which means that the cusps form separate, rounded hillocks that crush and grind food. In herbivores the molars may be **lophodont,** with cusps forming comparatively straight ridges, or **selenodont,** with cusps forming crescents; in these cases the teeth finely section and grind vegetation. In the dental batteries of many insectivores, bats, and carnivores are **sectorial** teeth. These have bladelike cutting edges that section food by slicing against the edges of their counterparts in the opposing jaw. A particularly specialized, functional pair of sectorial teeth in carnivorans are called the **carnassials;** they consist of the last upper premolar (P4) and first lower molar (m1) (see Fig. 2-16). Whether for sectioning, grinding, or slicing, the function of all of these types of teeth is based on interdental shearing and differs only in the relative vectors of movement of the teeth or other details of mechanics.

The generalized, triangular, tribosphenic upper molar (Fig. 2-24A) is marked by three major cusps, the **protocone,** the **paracone,** and the **metacone;** the apex of the triangle (the protocone) points medially. The lower molar (Fig. 2-24B) has two main anatomical components, an anterior **trigonid** and a posterior **talonid.** The trigonid is triangular; the apex of the triangle (the **protoconid**) points laterally; and the **paraconid** and **metaconid** form the medial edge. The talonid typically has three main cusps, from lateral to medial: the **hypoconid, hypoconulid,** and **entoconid.**

AXIAL SKELETON

Compared with that of quadrupedal reptiles, the mammalian vertebral column allows far greater freedom of head movement and powerful dorsoventral, rather than lateral, flexion of the spine. Most of the distinctive structural features of

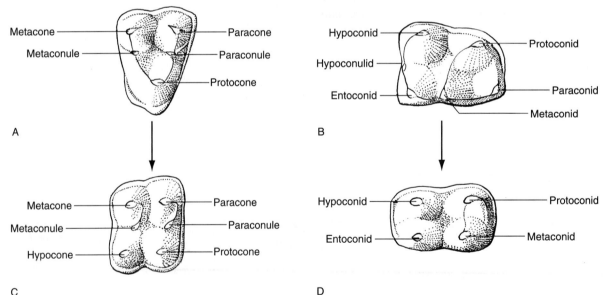

FIGURE 2-24   Basic cusp pattern of mammalian molars: (A) and (C) right upper molars; (B) and (D) left lower molars. Anterior is to the right; labial is at the top. (A) and (B) represent the generalized tribosphenic cusp pattern; this was modified in some evolutionary lines by the addition of a cusp (hypocone) in the upper tooth and the loss of a cusp (paraconid) in the lower tooth, yielding more or less quadrate teeth (C) and (D) adapted to omnivorous or herbivorous diets. *(After Romer, © 1966, Vertebrate Paleontology, University of Chicago Press)*

the mammalian vertebral column are related to these functional contrasts. The mammalian vertebral column has five well-differentiated sections: **cervical, thoracic, lumbar, sacral,** and **caudal.** There is often a sharp flexure in the vertebral column in the cervical vertebrae and another bend at the junction of the thoracic and lumbar regions, where a "switch" vertebra or diaphragmatic vertebra sometimes occurs. Only the thoracic vertebrae bear free ribs; elsewhere along the column, the ribs are fused to the vertebrae. The first two cervicals are highly modified, the sacral vertebrae are more or less fused to support the pelvic girdle, and differentiation of the vertebrae of each region is typical (Fig. 2-25; see 2-14). Usually 25 to 35 presacral vertebrae are present. All mammals, with the exception of several xenarthrans, the manatee (order Sirenia), some cetaceans, and possibly the giraffe (order Artiodactyla; Solounias, 1997) have seven cervical vertebrae.

The large, ossified, and commonly segmented sternum solidly anchors the ventral ends of the ribs, helping to form a fairly rigid rib cage. The sternum is not highly variable, but in some bats departs strongly from the typical mammalian plan.

LIMBS AND GIRDLES

In most terrestrial mammals, the main propulsive movements of the limbs are fore and aft; the toes point forward and the limb elements are usually greatly angled. In the most highly cursorial species (**cursorial** mammals are those adapted for running), the joints distal to the hip and shoulder tend to limit movement to a single plane. This allows reduction of whatever musculature does not control flexion and extension and results in lighter limbs. The mammalian pelvic girdle has a characteristic shape, with the **ilium** elongated and projecting forward and the **ischium** and **pubis** extending backward (see Fig. 2-14); these bones are solidly fused in terrestrial mammals, both among themselves and to the sacral vertebrae. In the shoulder girdle, the **coracoid** and **acromion** are usually reduced to small processes on the scapula and the reptilian interclavicle is gone (see Fig. 2-13); the clavicle is reduced or absent in some cursorial species.

In the **manus** (hand or forefoot) and **pes** (foot or hindfoot) of mammals, there is a standard pattern of bones (Fig. 2-26). However, many variations on this basic theme occur among mammals

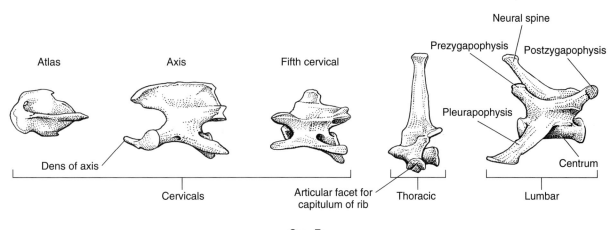

FIGURE 2-25 Vertebrae of the gray fox, showing the great structural variation in the parts of the vertebral column. The vertebrae are viewed from the left side; anterior is to the left. The atlas is the first cervical vertebra, the axis is the second; the fifth cervical vertebra is also shown.

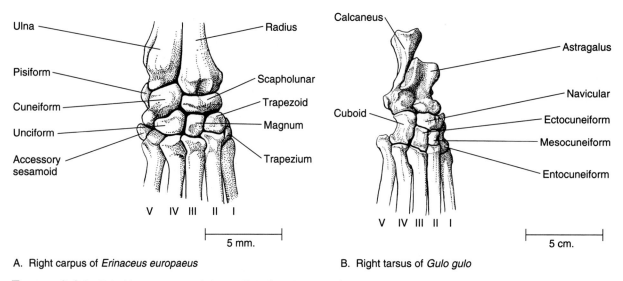

A. Right carpus of *Erinaceus europaeus*

B. Right tarsus of *Gulo gulo*

FIGURE 2-26 Primitive patterns of the podials (foot bones) of mammals as seen in anterior views: (A) right carpus of a hedgehog; (B) right tarsus of a wolverine. The centrale, a carpal element that in some mammals with primitive limbs lies proximal to the trapezoid and magnum, is missing in the hedgehog. The scapholunar in the hedgehog consists of the fused scaphoid and lunar bones. Metacarpals and metatarsals are numbered from medial to lateral.

with specialized types of locomotion, such as flight (bats), swimming (cetaceans and pinnipeds), or rapid running (ungulates, rabbits, and some carnivorans). Some of these variations are described in the chapters on orders. The primitive mammalian number of digits (five) and the basic phalangeal formula of two phalanges in the thumb (**pollex**) and first digit of the hind limb (**hallux**) and three phalanges in each of the remaining four digits (2-3-3-3-3) are retained by many mammals. Common specializations involve loss of digits, reduction in the numbers of phalanges, or, occasionally, addition of phalanges (**hyperphalangy**), as in the manus of whales and porpoises (see Fig. 13-3).

WEB SITES

The University of Michigan's Animal Diversity Web mammal pages
http://www.oit.itd.umich.edu/bio108/Chordata/Mammalia.shtml

Online key to identifying skulls of British and other European mammals
http://members.aol.com/rnorv/index.htm

3   MAMMALIAN ORIGINS

Mammals arose from a lineage of vertebrates known as synapsids that had their origin in early **amniotes** over 300 million years ago. After the first vertebrates invaded land, the group known as amphibians remained (and remain) obliged to return to water to reproduce. But a sister group to the amphibians, the Amniota, developed adapta-

tions that allowed reproduction on dry land, including the amniotic or cleidoic egg. This type of egg can be laid on dry land and is characterized by a semipermeable shell and several extraembryonic membranes. It is capable of gas exchange with the environment, contains food in the form of yolk to nourish the embryo, and stores waste products pro-

duced by the embryo until hatching. The first amniotes appeared in the Pennsylvanian period of the Paleozoic era (Table 3-1), and soon after their appearance underwent a fundamental split into two clades or phylogenetic lineages, the Sauropsida and the Synapsida (Benton, 1997; Fig. 3-1).

The sauropsid lineage eventually gave rise to turtles, lizards, snakes, dinosaurs, and birds. The synapsid lineage eventually led to mammals.

The synapsids dominated terrestrial faunas during the Permian and early Triassic periods. Many groups of early synapsids went extinct at the end

TABLE 3-1 Geologic Time Scale Since Life Became Abundant

Era	Period	Estimated Time Since Beginning of Each Epoch or Period (millions of years)	Epoch	Typical Mammals and Mammalian Ancestors
		0.001	Recent	Modern species and subspecies; extirpation of some mammals by humans
	Quaternary	1.75	Pleistocene	Appearance of modern species or their antecedents; widespread extinction of large mammals
Cenozoic				
		5.3	Pliocene	Appearance of modern genera
		23.8	Miocene	Appearance of modern subfamilies
	Tertiary	33.9	Oligocene	Appearance of modern families
		55.5	Eocene	Appearance of modern orders
		65.0	Paleocene	Adaptive radiation of metatherians and eutherians
	Cretaceous	144		Appearance of metatherians and eutherians
Mesozoic				
	Jurassic	206		Archaic mammals
	Triassic	248		Therapsids
				Appearance of mammals
	Permian	290		Appearance of therapsids, cynodonts
	Pennsylvanian	323		Appearance of synapsids
Paleozoic	Mississippian	360		
	Devonian	409		
	Silurian	438		
	Ordovician	510		
	Cambrian	570		

Mammalian events after Romer and Parsons, 1977; dates from Berggren et al., 1995; Gradstein et al., 1995

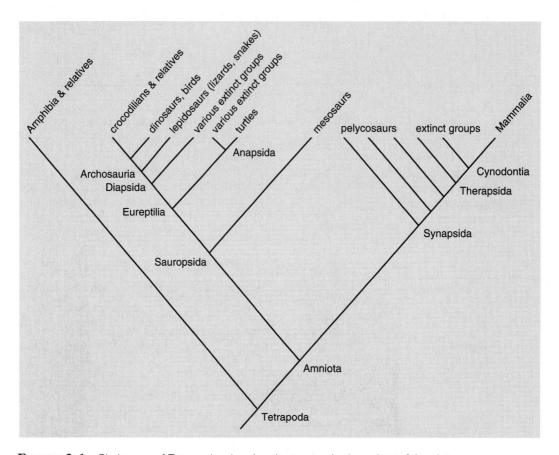

FIGURE 3-1 Phylogeny of Tetrapods, showing the two major branches of Amniotes, the Sauropsida (mesosaurs and reptiles) and the Synapsida (including mammals). More detail of the synapsid branch is shown in Figure 3-3. Among synapsids, the pelycosaurs are a paraphyletic group, artificial because the grouping includes only some of the more primitive descendants of a common ancestor and excludes the more derived descendants of that ancestor. Derived characters of the named clades (nodes) are given in the text. *(Gauthier et al., 1988; Benton, 1997; and Carroll, 1988)*

of the Permian in one of the largest extinction events of all time. However, a few lineages of more advanced synapsids (the Therapsida) survived the Permo-Triassic extinction. By the end of the Triassic, therapsids, too, dwindled in importance but not before giving rise to the cynodonts. From a cynodont ancestry, mammals arose in the late Triassic, slightly after the first appearance of dinosaurs.

Not long ago, the first two thirds of mammalian history (during the 140 million years from the late Triassic to the late Cretaceous) was regarded as the dark ages of mammalian history (Lillegraven et al., 1979). Many exciting discoveries in the last two

decades have uncovered a plethora of information about Mesozoic mammals, including their fundamental radiations, global occurrence, and paleobiology. During the vast sweep of the Mesozoic, the dinosaurs dominated the terrestrial scene, and their radiation resulted in a diverse array of herbivorous and carnivorous types, some of which were highly specialized for bipedal locomotion. Dinosaurs were the largest land animals of all time: most known species were very large, and some species reached weights of almost 100 tons (Gillette, 1994). Only a few species were smaller than 10 kilograms as adults. By comparison, Mesozoic mammals were insignificant. Most were mouse-sized, and the occasional "giant," about the size of a domestic cat, was probably not as large as the occasional species of relatively tiny dinosaur. Mesozoic mammals probably hid by day and foraged at night; the fossil record of the postcranial skeleton suggests that they adhered conservatively to a mouselike body form and to quadrupedal locomotion.

MORPHOLOGICAL FEATURES OF SYNAPSIDS

Amniote groups are characterized in part by various patterns of perforation of the temporal part of the skull (Fig. 3-2). The openings, called **temporal fenestrae,** are thought by some researchers to have developed originally to increase the freedom for expansion of the adductor muscles of the jaw; these muscles primitively attached inside the solid temporal part of the skull. From the basic split of amniotes, the Sauropsida led to animals with anap-

sid, diapsid, and other patterns of temporal fenestration.

In the Synapsida the skull is characterized by a single temporal opening low on the side of the skull, surrounded by the postorbital, squamosal, and jugal bones. A general trend in progressive synapsids was clearly toward the enlargement of the temporal opening (Fig. 3-3) and toward the movement of the jaw muscle origins from the inner surface of the temporal shield (as in pelycosaurs) to the braincase and to the zygomatic arch, the remnant of the lower part of the original temporal shield. Other derived features of Synapsida are a contact of the maxilla with the quadratojugal bone, **caniniform** maxillary teeth, and narrow neural arches on the trunk vertebrae. Many Permian synapsids form a series of outgroups to therapsids that are collectively known as pelycosaurs (a **paraphyletic** group; Fig. 3-1). Therapsids themselves seem to be monophyletic (Benton, 1997). One way of fitting the synapsids into a classification scheme is given in Table 3-2.

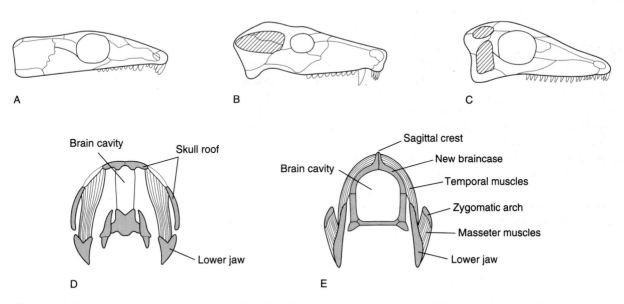

FIGURE 3-2 (A–C) Diagrammatic views of skulls of amniotes showing some of the arrangements of temporal openings. (A) Anapsid condition with no temporal opening, (B) synapsid condition with postorbital and squamosal bones meeting above a single opening, (C) diapsid condition with two temporal openings. (D and E) Cross sections of synapsid skulls showing attachments of jaw muscles. (D) Pelycosaur, with the jaw muscles originating within the remaining parts of the temporal shield ("skull roof"); the sides of the braincase are cartilaginous. (E) Mammal, with the jaw muscles originating on the new and completely ossified braincase ("new braincase," formed partly by extensions of bones that originally formed the skull roof), on the sagittal crest, and on the zygomatic arch, also a remnant of the original skull roof.

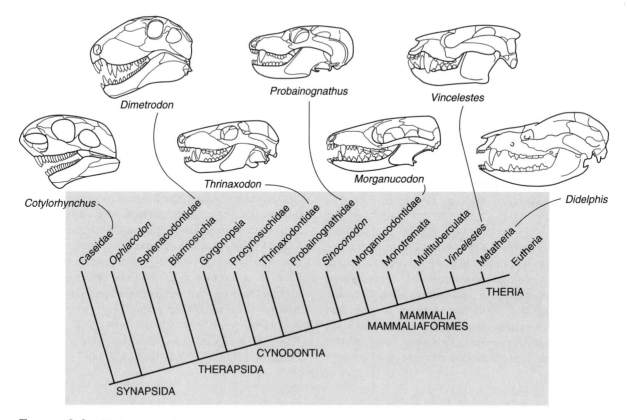

FIGURE 3-3 Phylogeny and representative skulls of synapsids. See text for definitions (synapomorphies) of named nodes. *(Phylogeny largely from Wible et al., 1995; Rowe, 1988, 1993; Benton, 1997; and others. Skull drawings from Hopson, 1994.)*

Some pelycosaurs, notably the caseids and edaphosaurids, had teeth that seem to have been designed for herbivory (e.g., in the caseid *Coty-lorhynchus;* Fig. 3-3). Most pelycosaurs, however, possessed caniniform teeth in the maxillary bone (as in the sphenacodontid *Dimetrodon;* Fig. 3-3) that were clearly designed for predation; these carnivores probably fed largely on fish and amphibians at the water's edge. Many pelycosaurs were relatively large animals compared with contemporary amphibians and sauropsids. *Cotylorhynchus* and *Dimetrodon* each reached a length of approximately 3 meters. Pelycosaurs dominated Early Permian terrestrial faunas but went extinct at the end of the Permian.

Therapsids arose from a common ancestor with the pelycosaurs, first appearing in the middle Permian. Many of the primitive forms were extinct by the end of the Permian, but others, including the early branches of cynodonts, continued into

the Jurassic. Therapsids display many morphological trends leading to the basic mammalian anatomical plan. Many of these trends are apparent in the synapomorphies of the successive nodes in Figure 3-3, and it is instructive to review these characters here. Derived features of the Therapsida include the reduction or loss of the temporal shield as the temporal opening enlarged, revealing the braincase and producing a sagittal crest and zygomatic arches. In connection with these changes to the skull, many of the origins of the jaw muscles moved to the braincase, sagittal crest, and zygomatic arches (Fig. 3-2E). There are septomaxillary bones extensively exposed on the facial part of the skull. The upper canines are enlarged, indicating a capacity for predation. There was a partial abandonment of the primitive, sprawling limb posture by modification of the limb girdles, resulting in a posture in which the limb bones are held more underneath the body. In the pelvic girdle the

acetabulum is deep. The feet are shortened. In the skull the external auditory meatus is formed within the squamosal bone. In the lower jaw the jaw joint is in line with the occiput, and the anterior coronoid bone is absent. A vertebral notochordal canal is absent in the adult.

Cynodontia includes an array of advanced, predaceous therapsids. In cynodonts, derived features include the presence of a masseteric fossa on the dentary, development of two occipital condyles (only one ventral, ball-like condyle is present in pelycosaurs), and zygomatic arches that flare laterally. At the back of the lower jaw the reflected lamina of the angular bone is reduced in size (indicating the incipience of emphasis on a hearing function of the postdentary bones; see Fig. 3-8). Teeth on the pterygoid bone are absent, the incisors are **spatulate,** and the postcanine teeth have anterior, posterior, and **lingual** accessory cusps (the last two characters result in a strongly heterodont dentition). The maxillary and palatine bones are expanded backward and toward the midline to form a partial, bony **secondary palate** (Fig. 3-4). Ribs on the lumbar vertebrae are reduced or fused to the vertebrae as pleurapophyses (Fig. 3-5). In the foot there is a distinct calcaneal heel.

Stem mammals first appear in the fossil record in the late Triassic. Many derived features, including numerous characters of the soft anatomy such as mammary glands with nipples, **viviparity** with loss of the eggshell, anal and urogenital openings separate in adults, digastricus muscle used to open the jaw, etc., unite the mammals and near-mammals into a clade (Table 2-2; McKenna and Bell, 1997). In the fossil record, only characters of the bony tissues can be used to define mammals and their nearest sister groups. Mammaliaformes possess a well-developed jaw articulation between the dentary and squamosal bones, double-rooted cheek teeth, and expansion of the brain vault in the parietal region. Elsewhere in the skull the tabular bone is absent and the occipital condyles are large and separated by a notch or groove. The medial wall of the orbit is enclosed by the orbitosphenoid and the ascending process of the palatine bone. Mesozoic Mammalia share several derived features of the dentition: the cheek teeth are divided into molars and premolars, there is precise occlusion and a consistent relationship between the upper

TABLE 3-2 A Classification of Synapsids with Emphasis on Nonmammalian Synapsids*

Kingdom Animalia
  Phylum Chordata
    Subphylum Vertebrata (Craniata)
      Superclass Tetrapoda
        Series Amniota
          Class Synapsida
            Order Pelycosauria†
              Family Eothyrididae
              Family Caseidae
              Family Varanopseidae
              Family Ophiacodontidae
              Family Edaphosauridae
              Family Sphenacodontidae
            Order Therapsida
              Suborder Biarmosuchia
              Suborder Dinocephalia
              Suborder Dicynodontia
              Suborder Gorgonopsia
              Suborder Cynodontia
                Family Procynosuchidae
                Family Galesauridae
                Family Cynognathidae
                Family Diademodontidae
                Family Chiniquodontidae
                Family Tritylodontidae
                Family Trithelodontidae
   Unnamed rank Mammaliaformes
     Class Mammalia

Modified from Benton, 1997.
*(See Table 3-3 for a Classification of Early Mammals.)
†Denotes a paraphyletic group

and lower molars, and the dentition is **diphyodont** (that is, there are two generations of teeth, juvenile and adult). In addition, the mandibular symphysis is reduced. The clade Theria is diagnosed on the basis of tribosphenic molars, the presence of a supraspinous fossa on the scapula, and, on the acetabulum, an inverted U-shaped area for articulation with the head of the femur. The cochlea is spiraled. The prootic bone lacks an anterior lamina (a specialization in the part of the skull that houses the inner ear).

The characters mentioned above and many other derived features diagnosing various synap-

FIGURE 3-4 Palatal views of synapsid skulls. (A) *Scymnognathus* (primitive therapsid); note that the internal nares open into the anterior part of the mouth. (B) *Cynognathus* (more advanced therapsid); note that the maxillaries and palatines have extended medially, forming a shelf that shunts air from the external nares to near the back of the mouth. *(After Romer, ©1966, Vertebrate Paleontology, University of Chicago Press)*

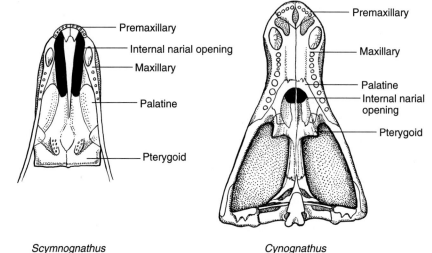

Premaxillary
Internal narial opening
Maxillary
Palatine
Pterygoid

Scymnognathus

Premaxillary
Maxillary
Palatine
Internal narial opening
Pterygoid

Cynognathus

sids (including early mammals) are taken from a flood of recent studies that have resulted from the exciting discoveries of new and informative fossils around the world. Some studies dealt with various pelycosaurs and nonmammalian therapsids (Allin and Hopson, 1992; Hopson, 1991, 1994; Hopson and Barghusen, 1986; Kemp, 1982; Laurin, 1993; Laurin and Reisz, 1995, 1996; Reisz, 1986; and Rowe, 1988). Others concentrated on the synapsids near or through the nonmammal-to-mammal transition (Crompton and Luo, 1993; Hopson and Rougier, 1993; Hu et al., 1997; Jenkins et al., 1997; Kielan-Jaworowska, 1997; Lillegraven and Hahn, 1993; Lillegraven and Krusat, 1991; Lucas and Luo, 1993; Luo and Crompton, 1994; Rich et al., 1997; Rougier et al., 1996a, b; Rowe, 1988, 1993; Wible, 1991; Wible and Hopson, 1993, 1995; and Wible et al., 1990, 1995). Almost every paper produced a cladogram slightly different from the next one, confusing most attempts at classification but constantly improving our understanding of the evolution of mammals and their ancestors. A recent classification of mammals based on putative phylogenetic relationships is provided in Table 3-3 and is followed here for the Mesozoic mammals.

Through the transition from nonmammalian to mammalian synapsids, many of the derived characters mentioned above suggest the development of unobservable physiological features in Mesozoic synapsids. For example, the expansion of the dentary bone at the expense of the other jaw elements and the establishment of a dentary-squamosal jaw joint have important implications for the development of the uniquely mammalian middle ear with three ossicles (see Figs. 3-6, 3-7, and 3-8) and for improved hearing acuity. Numerous changes in the postcranial skeleton such as reduction of the cervical and lumbar ribs; improved limb posture; alteration of the joints between the occipital

▶ **FIGURE 3-5** Reconstructed skeletons of primitive and derived synapsids, showing changes in the postcranial skeleton. (A) *Dimetrodon* (Sphenacodontidae), a pelycosaur that possessed a "sail" supported by elongated neural spines (not all pelycosaurs had sails) and a sprawled limb posture. (B) *Thrinaxodon* (Thrinaxodontidae) and (C) *Massetognathus* (Traversodontidae), cynodonts showing reduction of the lumbar ribs (due to fusion to the vertebrae) and less sprawling limb posture. (D) *Megazostrodon* (Morganucodontidae), in which the forelimbs are sprawled but the hindlimb posture indicates a mammalian stance (in which the femur swings only fore and aft, not out to the side) and the pelvis has a long, rodlike, anterodorsally oriented ilium and a large obturator foramen. (E) *Tupaia* (Tupaiidae), a modern tree shrew, showing considerable flexure in the axial skeleton (cervico-thoracic and thoraco-lumbar regions) and limbs. *(A from Romer, ©1966, Vertebrate Paleontology, University of Chicago Press; B from Jenkins, 1971; C from Jenkins 1970; D from Jenkins and Parrington, 1976; E from Jenkins, 1974)*

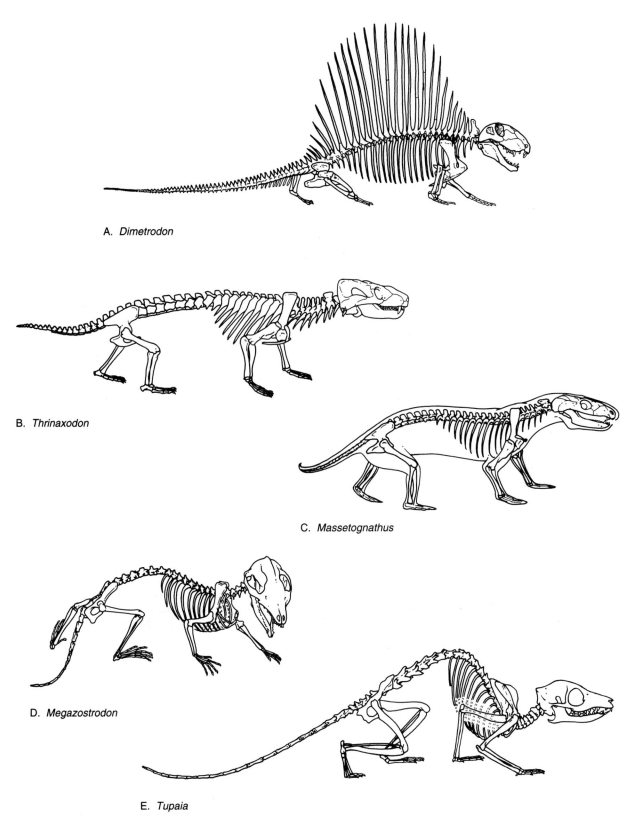

A. *Dimetrodon*

B. *Thrinaxodon*

C. *Massetognathus*

D. *Megazostrodon*

E. *Tupaia*

TABLE 3-3   Partial Classification of Mammals and Close Relatives Emphasizing Extinct Groups

Unnamed rank Mammaliaformes
>> Order Morganucodonta
>> Order Docodonta
>> Order Haramiyoidea

Class Mammalia
> Subclass Prototheria
>> Order Monotremata
> Subclass Theriiformes
>> Infraclass Allotheria
>>> Order Multituberculata
>> Infraclass Triconodonta
>> Infraclass Holotheria
>>> Superlegion Kuehneotheria
>>> Superlegion Trechnotheria
>>> Legion Symmetrodonta
>>>> Order Amphidontoidea
>>>> Order Spalacotherioidea
>>> Legion Cladotheria
>>>> Sublegion Dryolestoidea
>>>> Order Dryolestida
>>>> Order Amphitheriida
>>>> Sublegion Zatheria
>>>>> Infralegion Peramura
>>>>> Infralegion Tribosphenida
>>>>> Supercohort Aegialodontia
>>>>> Supercohort Theria
>>>>>> Order Deltatheroida
>>>>>> Order Asiadelphia
>>>>>> Cohort Marsupialia (= Metatheria)
>>>>>>> Magnorder Australidelphia
>>>>>>> Magnorder Ameridelphia
>>>>>> Cohort Placentalia
>>>>>>> Magnorder Xenarthra
>>>>>>> Magnorder Epitheria

After McKenna and Bell, 1997; Wilson and Reeder, 1993.

condyles, atlas, and axis; and simplification (by fusion of elements) of the **tarsus** and carpus probably affected many aspects of locomotion and agility. Reduction of the lumbar ribs is often used to infer the presence, first appearing in therapsids, of a muscular diaphragm. This, together with the development of a secondary palate, possibly indicates the beginnings of higher levels of metabolism in the lineage. Together with the upright posture, increased axial flexion, improved coordination of lung ventilation with locomotion, and other adaptations, these specializations eventually resulted in cursorial forms of mammals (e.g., cheetah) capable of great running speeds. Precise occlusion of the teeth and rearrangement of the jaw musculature have implications for advanced food processing (and digestive) abilities. Enlargement of the brain goes hand-in-hand with many of these adaptations. It is fair to assume that intermediate stages were present in nonmammalian synapsids. Some of these synapsid adaptations are discussed in more detail below.

IMPORTANT EVOLUTIONARY TRANSFORMATIONS

A handy and widely used landmark in nonmammalian-to-mammalian synapsid evolution is the structure of the jaw articulation. In nonmammals this joint is typically between the quadrate bone of the skull and the articular bone of the lower jaw, but in mammals the squamosal and dentary bones form this joint. The earliest mammals are regarded as mammals because of the presence of the dentary-squamosal joint, but the situation is far from simple. In cynodont therapsids, there are several stages in the transformation of the jaw joint. An intermediate stage, appearing in several cynodont families, is the development of a secondary jaw joint — in addition to the quadrate-articular joint — between the surangular bone and the squamosal bone. This secondary joint probably braced the quadrate-articular joint against backward displacement of the jaw during chewing, but it also might have functioned in sound transmission 100 million years before the appearance of the first mammals (Rowe, 1996a). In *Probainognathus* (Fig. 3-3), an articular depression (glenoid fossa) developed in the squamosal bone, and into it fit the surangular bone of the lower jaw, braced by an "articular" process of the dentary bone (Fig. 3-6).

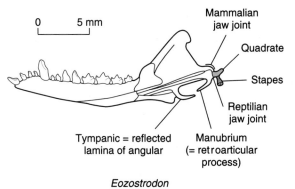

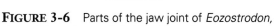
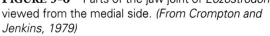

FIGURE 3-6   Parts of the jaw joint of *Eozostrodon*, viewed from the medial side. *(From Crompton and Jenkins, 1979)*

In the late Triassic, some cynodonts demonstrated the dentary-squamosal jaw joint for the first time. This involves a posterior extension of the dentary bone into the glenoid fossa of the squamosal bone, with the quadrate-articular joint also present and medial to this "mammalian" jaw joint. In this two-jointed condition, the quadrate and articular bones still formed a jaw joint but functioned together with the stapes to better transmit vibrations from the tympanic membrane to the

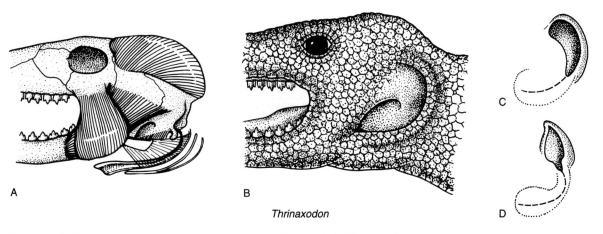

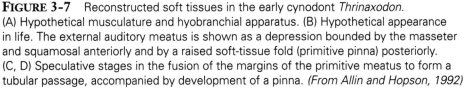

FIGURE 3-7   Reconstructed soft tissues in the early cynodont *Thrinaxodon*.
(A) Hypothetical musculature and hyobranchial apparatus. (B) Hypothetical appearance in life. The external auditory meatus is shown as a depression bounded by the masseter and squamosal anteriorly and by a raised soft-tissue fold (primitive pinna) posteriorly. (C, D) Speculative stages in the fusion of the margins of the primitive meatus to form a tubular passage, accompanied by development of a pinna. *(From Allin and Hopson, 1992)*

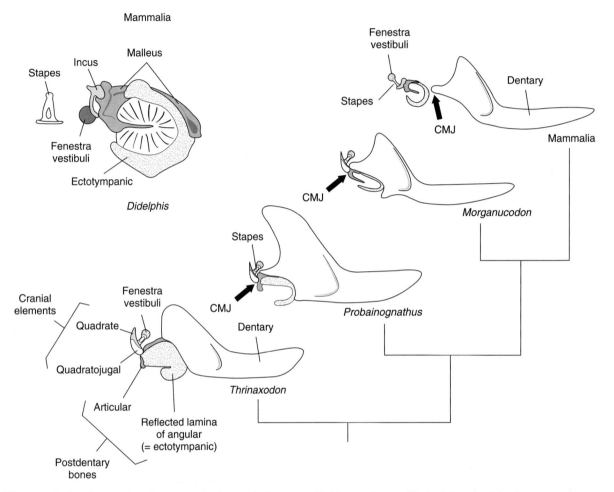

FIGURE 3-8   Selected major stages in the evolution of the mammalian jaw joint and ear region. Homologous bones are shaded similarly from one taxon to the next. View is of the lateral side. Dentary bones are shown diagrammatically, without teeth. Postdentary bones of nonmammalian synapsids became modified as middle ear bones of Mammalia (reflected lamina of the angular bone = ectotympanic; articular = malleus). Cranial bones (quadrate = incus, quadratojugal, and stapes) are shown without the rest of the cranium for simplicity. Tympanic membrane (eardrum) is shown only in the inset of *Didelphis*. The fenestra vestibuli is the opening into the inner ear in which the footplate of the stapes is normally fitted. *CMJ* = craniomandibular joint, the joint between the skull and the lower jaw; note the shift in the CMJ from an articular-quadrate joint in *Thrinaxodon* and *Probainognathus* to a condyle on the dentary in Mammalia that articulates with the squamosal of the cranium (not shown). *Morganucodon* has both these joints as jaw joints, but the dentary-squamosal joint is dominant over the reduced articular-quadrate joint. In more derived mammals, the articular-quadrate joint is maintained as a malleus-incus joint but it is removed from the jaw joint area and serves solely in hearing. Enlarged detail of ear region of *Didelphis* at upper left shows the stapes removed from the fenestra vestibuli. *(From Rowe, 1996a)*

oval window of the inner ear. Where both joints were present in one animal, the postdentary bones were much reduced in size and parts of them were loosely fitted into a groove on the medial side of the dentary (Fig. 3-6). There they apparently were capable of more delicate vibrations and more efficient transmission of sound. A part of the angular bone known as the reflected lamina supported a tympanic membrane, and the articular and quadrate bones transmitted vibrations from this membrane (via the stapes) to the inner ear (Allin, 1975; Allin and Hopson, 1992; Figs. 3-6, 3-7, and 3-8). Accordingly, still further reduction of the size of the articular and quadrate bones improved their sensitivity to vibrations and enhanced the sense of hearing.

Finally, in mammals, the articular, quadrate, and angular bones became completely detached from the lower jaw and became part of the ear apparatus (the malleus, incus, and tympanic ring, respectively). Rowe (1996a, b) argued that these phylogenetic changes in the ear region accompanied or were caused partly by a concomitant increase in brain size and the developmental onset of functionality in the jaw muscles, as revealed elegantly in **ontogenetic** stages in the postnatal development of a modern marsupial (Fig. 3-9).

The development of masseter muscles with essentially the same attachments as those of mammals was a cynodont innovation that occurred in no other therapsid line. These muscles originate on the zygomatic arch and insert on the lateral surface of the dentary bone and are powerful **adductors** of the jaws (they close the jaws). The development of these muscles resulted in several important functional refinements (Crompton and Jenkins, 1979). First, the masseter muscles formed part of a muscular sling that suspended the jaw and enhanced the precise control of transverse jaw movements. Second, these muscles increased the force of the bite. Third, the forces produced by the bite were "focused" through the point of the bite, and thus the stress on the jaw joint was reduced.

A shift in the structure and function of the dentition can be traced through the cynodont–early mammal evolutionary line. The front teeth were of primary importance in primitive therapsids and most nonmammalian cynodonts; the incisors and canines were robust and the **cheek teeth** relatively weak. The reverse was true in early mammals, in

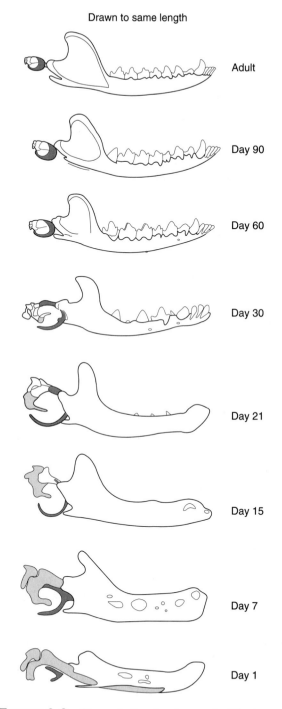

Drawn to same length

Adult

Day 90

Day 60

Day 30

Day 21

Day 15

Day 7

Day 1

FIGURE 3-9   Stages in the development of the lower jaw and ear region in a young opossum *Monodelphis* (Didelphidae), showing how ontogeny recapitulates phylogeny. Compare the ontogenetic stages with Fig. 3-8. The ectotympanic is shown in black, and gray shading indicates cartilage. Jaws are drawn to same length for ease of comparison. *(From Rowe, 1996a)*

which the cheek teeth were the more robust series. The development in cynodonts of masseter muscles and the concentration of jaw action power through the postcanine teeth attended these changes in function. Most carnivorous nonmammalian cynodonts lacked occlusion of their postcanine teeth so that powerful biting was possible only in the incisors and canines; herbivorous forms had bilateral occlusion of the postcanine teeth (Crompton, 1995). Progressively greater precision and breadth of movement of the lower jaw set the stage for the later evolution in mammals of a complex molar cusp pattern. Early mammals uniquely developed unilateral occlusion, in which chewing occurs on only one side of the mouth at a time. The postcanine teeth of some cynodonts were tricuspid and resembled those of some of the earliest known mammals. In other cynodonts, the cheek teeth were complex and double-rooted.

Several other structural features that became well developed in cynodonts are typical of both ancient and modern mammals. One such is the secondary palate, a structure formed by an inward and backward extension of the premaxillary, maxillary, and palatine bones. This bony plate lies beneath the original roof of the mouth and forms a passage that shunts air from the external nares at the front of the snout to the internal narial openings at the back of the mouth (Fig. 3-4). Such a bypass allows mammals to breathe while food is being chewed. Moreover, the presence of a secondary palate may have facilitated **suckling.** Suckling in modern mammals partly depends on the ability to form a seal between the tongue and the front portion of the soft palate. This is achieved with a muscle that originates on the pterygoid bone. The pterygoid bones of *Pachygenelus* (an advanced nonmammalian cynodont) and *Morganucodon* suggest that they had such tensor muscles, supporting the possibility that early mammals suckled their young (Crompton, 1995). The incisive foramina were large in cynodonts, a reflection perhaps of the importance of the sense of smell to these animals. The reduction in size of lumbar ribs and the retention of a thoracic rib cage in cynodonts may have been associated with the development of a muscular diaphragm and the respiratory movements typical of mammals.

The limbs and girdles of cynodonts were modified as the sprawling limb posture was partially abandoned in favor of movement in a **parasagittal** plane in the hind limbs. The ilium shifted forward, and the pubis and ischium moved backward as fore-and-aft limb movement became more important than lateral movement (see Fig. 3-5). The limbs of some cynodonts were slim and adapted to rapid running. A simplistic approach has often been taken in describing the differences between reptilian and mammalian limb postures. The reptilian posture has been characterized as sprawling, with the humerus and femur directed horizontally, whereas mammalian limbs have been described as moving directly fore and aft and being positioned nearly vertically beneath the body. Actually, this latter posture is typical only of cursorial mammals. Jenkins (1971) found that, during locomotion in a group of noncursorial species, the humerus and femur function in postures more horizontal than vertical and at oblique angles relative to the parasagittal plane. The studies of Jenkins further demonstrate that the limb postures of terrestrial mammals are extremely diverse. Certainly a trend toward a vertical limb posture can be detected in cynodonts and in early mammals, but the stereotypical picture of the vertical limb posture shared by all terrestrial mammals should be abandoned. Similarly, in certain extinct forms such as *Megazostrodon* (Fig. 3-5D), the hind limbs seem to have had a "mammalian stance" while the front limbs were probably sprawled (Jenkins and Parrington, 1976). Indeed, a sprawling posture was maintained in the forelimbs in many archaic mammals, such as morganucodontids, symmetrodonts, and multituberculates. The parasagittal posture of the forelimbs first appeared in relatively advanced (nonsymmetrodont) therians (Gambaryan and Kielan-Jaworowska, 1997; Hu et al., 1997).

Cynodonts, and therapsids in general, were active terrestrial synapsids with well-developed senses of hearing and smell, and later species were probably **endotherms** (Feder, 1981). Why, after such a long period of dominance, did the progressive cynodonts and other therapsids become extinct? A major cause may have been competition from dinosaurs (Colbert, 1982). Early therapsids were often the size of large dogs, whereas the last surviving cynodonts were squirrel-sized and the earliest mammals were no larger than mice. The therapsid–early mammal evolutionary line was apparently under intense selection for small size.

FIGURE 3-10   Reconstruction of *Eozostrodon,* a Triassic mammal of the family Morganucodontidae. The length of this animal was about 107 millimeters. *(After Crompton and Jenkins, 1968)*

As mentioned earlier in relation to Mesozoic mammals, the smaller size of later therapsids may have made available to them more retreats secure from dinosaurs.

EARLY MAMMALS

Members of the family Morganucodontidae (Figs. 3-5D, 3-10), from the late Triassic or earliest Jurassic period of Europe, represent the earliest known mammals (Mammaliaformes). The early Jurassic ones are represented in the fossil record by relatively complete skeletons and dentitions. They were small animals. Their body weight, probably 20 to 30 grams, was an order of magnitude smaller than any middle Triassic cynodont. Their cheek teeth were differentiated into premolars and molars, and the premolars were probably preceded by deciduous teeth. Chewing was on one side of the jaw at a time, and the lower jaw on the side involved in chewing followed a triangular orbit as viewed from the front (see Fig. 2-20B). During chewing, the inner surface of the upper molars sheared against the outer surface of the lower molars (Fig. 3-11). In species of like body size, the morganucodontid brain was three or four times larger than that of even the most advanced therapsids, a reflection perhaps of greater neuromuscular coordination and improved auditory and olfactory acuity.

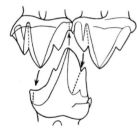

A. Late Triassic mammal   B. Opossum (*Didelphis virginiana*)

FIGURE 3-11   (A) Shearing planes of opposing molars of a primitive, late Triassic mammal; the shearing surfaces are outlined (after Crompton, 1974). (B) Occlusal view of the lower molars of an opossum *(Didelphis virginiana),* showing the tongue-in-groove fit of the anterior and posterior surfaces of adjacent teeth.

What was the lifestyle of these earliest mammals? Jenkins and Parrington (1976) regarded morganucodonts as insectivores with considerable climbing ability. The apparent ability of the hallux to move independently of the other digits indicates grasping ability, and enlargement of the foramina of the cervical vertebrae through which nerves contributing to the brachial plexus passed suggests refined neuromuscular control of the forelimbs. These mammals were likely secretive, **nocturnal** creatures that depended heavily on their well-developed senses of hearing and smell. Endothermy probably favored nocturnal activity, and the animals must have been covered with hair, but it seems unlikely that they had developed the myriad adaptations necessary for coping with the high temperatures encountered during **diurnal** activity.

These earliest mammals possessed a suite of skeletal features that marked them clearly as mammals, but what was their reproductive pattern and did they have mammary glands? Lillegraven (1979b) stated that "the development of **lactation** was probably a key feature in the origin and later success of mammals in adapting to the changing environments of the Mesozoic and Cenozoic, and was unquestionably fully functional well before the end of the Triassic." One compelling line of histologic evidence supports this view. The mammary tissue of all living mammals is essentially identical, despite the fact that the nontherian and therian evolutionary lines diverged before the end of the Triassic. In all probability, then, the histologic similarities in mammary tissue are due to inheritance by both divisions of mammals from a common late Triassic ancestor that possessed mammary glands. Further, mammary glands and deciduous dentition, which allowed the delay of the growth of the complex adult dentition in the juvenile mammal, probably occurred together. A dentition capable of masticating food can be delayed in a young mammal that is nourished by its mother's milk. During the nursing period, however, a tight social bond between mother and young is essential. Therefore, deciduous teeth, delayed adult dentition, mammary glands, lactation, maternal care, and a tight bond between mother and nursing young must have evolved in concert. When therian mammals abandoned egg-laying and began bearing living

young (viviparity) is unknown; living prototherian mammals (order Monotremata) still lay eggs, and therian mammals may have kept this pattern long after the Triassic.

Although of great interest in connection with the story of mammalian evolution, these early mammals were diminutive members of a late Triassic terrestrial fauna that was becoming increasingly dominated by dinosaurs. Yet the tiny late Triassic mammals were innovative in unspectacular ways that furthered their survival in the shadow of the dinosaurs. How very different would have been the vast sweep of post-Triassic vertebrate evolution and how altered would be the face of the earth today if the little late Triassic mammals had proven vulnerable to some contemporary reptilian predator and had relinquished the scene completely to the reptiles.

MESOZOIC MAMMALIAN RADIATIONS

Mammals clearly originated monophyletically from cynodont synapsids, and by the late Triassic and early Jurassic they had diverged into several stocks. But the relationships of these stocks are still complex and unclear because of inconsistencies in the quality and quantity of fossils of each stock. Thus, it is difficult to arrange the various Mesozoic forms and groups into a well-resolved cladogram and even more difficult to "convert" the phylogenetic relationships into a consensus classification. Benton (1997) attempted a compromise of the cladograms of several authors (Fig. 3-12), but new specimens and data will certainly continue to change this view of mammalian phylogeny. The classification followed here is from McKenna and Bell (1997) and is given in Table 3-3.

Of course, the better known taxa tend to be included in phylogenetic analyses. *Adelobasileus*, known only from a braincase from the late Triassic of North America, and the better known *Sinoconodon* (see Fig. 3-13), from the early Jurassic of Asia, are very near the nonmammal-to-mammal transition and are sometimes classified as mammals but usually are considered outside Mammalia as the nearest sister groups to Mammalia. At least some of the Morganucodontidae are very well known

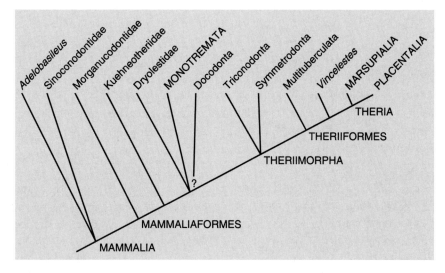

FIGURE 3-12   Hypothesis of relationships of the major groups of mammals, with emphasis on Mesozoic types. *(After Benton, 1997)*

skeletally and are usually considered to be mammals or mammaliaformes. Kuehneotheriidae (Kuehneotheria) of the early Jurassic are probably close to the ancestry of the remainder of the clade and are almost always classified as mammals, as are the Symmetrodonta and Dryolestoidea. Mesozoic mammals with unclear phylogenetic placement include the Triconodonta, Haramiyoidea (once thought to be related to the Multituberculata; Jenkins et al., 1997), Docodonta, and Multituberculata. The Monotremata are, of course, certainly mammals, but they have a poor Mesozoic fossil record and their phylogenetic relationships to other mammals are still unresolved. The South American *Vincelestes* is considered to be a sister group to the therians (Metatheria and Eutheria).

As mentioned, the first two thirds of mammalian history is documented by a spotty fossil record that leaves many geographic areas, time periods, and anatomical changes unrepresented. Nevertheless, Mesozoic mammals are now known from all continents except Antarctica. Many of their fossils have long been known from Europe, Asia, and North America, but recent discoveries in those continents and elsewhere around the world bolster our growing knowledge of early mammals and mammaliaformes, especially in the continents

that were once a part of the supercontinent Gondwana (for example, in Africa, including Madagascar: Jacobs et al., 1988; Krause et al., 1997; Sigogneau-Russell, 1989, 1991; Sigogneau-Russell and Ensom, 1997; in Australia: Archer et al., 1985; Rich et al., 1997; in South America: Bonaparte and Rougier, 1987; Krause and Bonaparte, 1993; in Asia: Khajuria and Prasad, 1998; Maschenko and Lopatin, 1998; Prasad and Manhas, 1997; Szalay and Trofimov, 1996). Eventually, their remains will probably be found in Antarctica, too, as have several kinds of late Cretaceous dinosaurs (Gasparini et al., 1996) and early Cenozoic mammals. (One of the first Amniote fossils found in Antarctica was that of a Triassic dicynodont therapsid.) Recent improvements to the fossil record of Mesozoic mammals as well as the record of feathered dinosaurs, toothed birds, and many other vertebrates provide interesting glimpses into the terrestrial biota of the Mesozoic world (Hu et al., 1997; Dashzeveg et al., 1995; Sampson et al., 1998). For example, several finds of multituberculate-like mammals known as Gondwanatheriidae in South America, Madagascar, and peninsular India (all formerly part of Gondwana) indicate the cosmopolitanism of this group, at least during the late Cretaceous (Krause et al., 1997). Gondwanatheri-

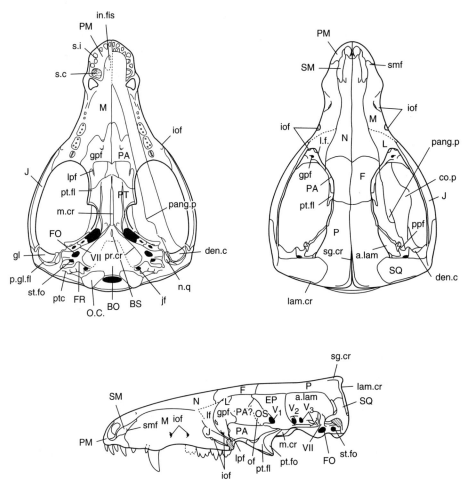

Sinoconodon

FIGURE 3-13 A reconstruction of the skull of *Sinoconodon* in ventral, dorsal, and lateral views. Ventral view includes the left half of the mandible; dorsal view includes the right half of the mandible. *Sinoconodon* is a derived synapsid considered by some authors and interpretations to be mammalian and by others to be nonmammalian.

In either case, it is near the transition, and whether it is considered a mammal depends on semantics and definition of the Mammalia. *A,* articular; *a.lam,* anterior lamina of petrosal; *BO,* basioccipital; *BS,* basisphenoid; *co.p.,* coronoid process; *den.c,* dentary condyle; *EP,* epipterygoid; *FO,* fenestra ovalis; *FR,* fenestra rotunda; *gl,* glenoid fossa for dentary condyle; *g.p.f,* greater palatine foramen; *in.fis,* incisive foramen; *iof,* infraorbital foramen; *J,* jugal; *j.f,* jugular foramen; *L,* lacrimal; *lam.cr,* lambdoid crest; *l.p.f,* lesser palatine foramen; *M,* maxilla; *N,* nasal; *n.q,* notch for quadrate (on squamosal); *o.c,* occipital condyle; *P,* parietal; *PA,* palatine; *pang.p,* pseudangular ("angular") process; *p.gl.fl,* postglenoid flange; *PM,* premaxilla; *pr.,* promontorium; *pr.cr,* promontorium crest; *pt.fl,* pterygoid flange; *Q,* quadrate; *sc,* socket for canine; *sg.cr,* sagittal crest; *si,* socket for incisor; *SM,* septomaxilla; *smf,* septomaxillary foramen; *SQ,* squamosal; *ST,* stapes; *st.fo,* stapedial muscle fossa; V_1, foramen for the ophthalmic branch of trigeminal nerve; V_2, foramen for the maxillary branch of trigeminal nerve; V_3, foramen for the mandibular branch of trigeminal nerve; *VII,* foramen for facial nerve. *(From Crompton and Luo, 1993)*

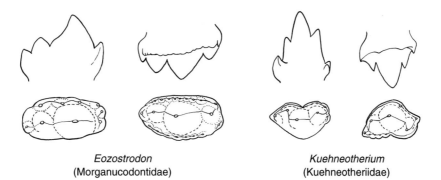

Eozostrodon
(Morganucodontidae)

Kuehneotherium
(Kuehneotheriidae)

FIGURE 3-14   Diagrams of the molars of Triassic or early Jurassic mammals. In each case, the lower molar is on the left and the upper molar is on the right. *(Modified from Crompton, 1974)*

ids are known only from a few isolated, high-crowned teeth. As a result, their relationships are unclear, but gondwanatheriids show similarities to multituberculates and might belong to the Allotheria (Krause and Bonaparte, 1993).

Mesozoic mammals are surprisingly diverse. Once they appeared, the earliest mammals clearly began to undergo a fundamental radiation that resulted in a complexity of arcane and puzzling forms. The Gondwanatheriidae just mentioned are one such group, although they appear relatively late in the era. Several other groups, enigmatic because of their poor fossil records, give us tantalizing glimpses into the hidden world of Mesozoic mammals.

Among the earliest archaic therians are the Kuehneotheria from the late Triassic or earliest Jurassic of Europe. The structure of the cheek teeth of *Kuehneotherium* indicate that it is near the ancestry of the Symmetrodonta and Dryolestoidea (Fig. 3-14). The Dryolestoidea, in turn, are tribosphenic or at least near the ancestry of the tribosphenic mammals. The Symmetrodonta themselves were a diverse group of archaic therians known from the late Triassic to the late Cretaceous. They are characterized by numerous triangular molars (up to six lower molars) with three fairly symmetrically situated cusps. Some species were the size of a small shrew and were probably insectivorous. The recent discovery of the first-known nearly complete skeleton of a symmetrodont in Asia (Hu et al., 1997) gives evidence that

symmetrodonts represent a side branch of Theria before the divergence of the Metatheria and Eutheria. The skeleton, belonging to *Zhangeotherium*, possesses an uncoiled cochlea, an interclavicle, **epipubic bones,** and a sprawled posture of its forelimbs. Isolated teeth of a symmetrodont from Morocco do not fit into previously named families and indicate a unique African lineage showing some similarities to the incipiently tribosphenic Dryolestoidea (Sigogneau-Russell, 1989).

Also stemming from among the earliest mammals were the Triconodonta, known from the late Triassic to the early Cretaceous. Triconodonts were predatory; the largest genus (*Gobiconodon*) was the size of a domesticated cat. The dentition was heterodont, with as many as 14 teeth in a dentary bone. The canines were large, and typically the molars had three primary cusps arranged in a front-to-back row. Two early Cretaceous African species of the genus *Ichthyoconodon* were speculatively considered to have been aquatic and to have fed on fish (Sigogneau-Russell, 1995); a North American middle Cretaceous triconodont *Jugulator* had an estimated body mass of 750 grams and probably was capable of preying on small vertebrates, but the true habits of these animals are yet unknown (Cifelli and Madsen, 1998).

The order Docodonta is represented by several primitive genera known from the Jurassic. Members of this group have roughly quadrate teeth, with the cusps not aligned anteroposteriorly. The braincase and postcranial skeleton seem to be on a primitive level of development. A possible docodont from Arizona (*Dinnetherium nezorum*), described by Jenkins et al. (1983), represents yet another early line of descent.

Among nontherian mammals, the order Multituberculata is especially remarkable. These were probably the first mammalian herbivores (or omnivores; Krause, 1982), and although they disappeared in the early Tertiary and left no descendants, they were highly successful. Multituberculates appear first in the late Jurassic, and their fossil record spans 100 million years. Their relationships to other mammals are still uncertain; they may have diverged very early from other mammals (Lillegraven and Hahn, 1993). These animals were widespread in both the Old World and New World and were the ecological equivalents of rodents in some ways. The strongly built lower jaw provided attachment for powerful jaw muscles; there were usually two (but sometimes three) incisors above and two below, and a diastema was present in front of the premolars (Fig. 3-15A). Typical of some advanced multituberculates were upper molars with three parallel rows of cuspules and remarkably specialized bladelike posterior lower premolars (Fig. 3-15B, C). At least one Paleocene form, *Ptilodus*, showed specializations of the postcranial skeleton that indicate that it was **arboreal.** Climbing adaptations include a hallux (first toe) that could move independently of the other toes, a highly mobile ankle as seen in mammals that can descend trees headfirst, and a possibly **prehensile** tail (Jenkins and Krause, 1983). Others were **semifossorial** or capable of jumping from a sprawled limb posture (Kielan-Jaworowska and Gambaryan, 1994; Gambaryan and Kielan-Jaworowska, 1997).

Multituberculates persistently retained several primitive features. Cervical ribs were retained in a few taxa, and throughout their history they never developed a parasagittal limb posture. Their cranial osteology indicates an unusual musculature and style of chewing that utilized a posteriorly directed power stroke of the lower jaw, opposite that of rodents and unlike any other mammals (Wall and Krause, 1992; Kielan-Jaworowska, 1997). The olfactory lobes of the brain were large, the cerebrum was smooth, the incisive foramina were large, and the cochlea of the ear was similar in size and proportion to that of extant small mammals of comparable

size, but was uncoiled. The auditory ossicles are strikingly similar in one early Tertiary genus, *Lambdopsalis,* to those of living monotremes, which suggests that the ear was ill suited for receiving high-frequency airborne sounds but well suited for low-frequency bone-conducting hearing (Meng and Wyss, 1995). By contrast, a late Cretaceous genus *Chulsanbataar* is suggested to have had high-frequency hearing but had a low sensitivity to low-decibel (quiet) sounds (Hurum, 1998). Considered together, these features suggest rather primitive mammals that could not remain long in competition with eutherians, but the fossil record indicates otherwise. For over 70 million years, multituberculates and eutherians coexisted. The decline of the

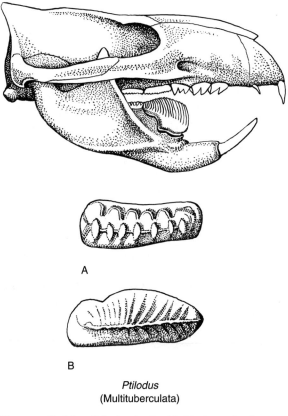

A

B

Ptilodus
(Multituberculata)

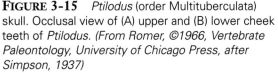

FIGURE 3-15 *Ptilodus* (order Multituberculata) skull. Occlusal view of (A) upper and (B) lower cheek teeth of *Ptilodus*. (From Romer, ©1966, *Vertebrate Paleontology, University of Chicago Press, after Simpson, 1937)*

multituberculates began in the late Paleocene and spanned 20 million years. The competition probably began with condylarths (ancestors of ungulates) in the late Cretaceous, intensified when primates became common in the Paleocene, and became overwhelming in the Eocene, when rodents became ubiquitous (Van Valen and Sloan, 1966). Multituberculates appear last in the early Oligocene fossil record of Wyoming and South Dakota.

Metatherian and eutherian mammals evolved from an ancestry within the Dryolestoidea (including the group formerly known as Pantotheria), probably among the Zatheria. The Zatheria includes the families Peramuridae and Aegialodontidae. Peramurids are known from the middle Jurassic to early Cretaceous and have lower molars with a distinct, posterior "heel" but no true talonid basin. *Aegialodon* (Aegialodontidae), of the early Cretaceous of Asia and Europe (McKenna and Bell, 1997) possesses a lower molar considered to be the earliest tribosphenic molar in the fossil record. The shape of the anterior trigonid section of the lower molar resembles the comparable part of this tooth in many eutherians and metatherians (Fig. 3-16). More important, the lower molar has a talonid with a basin into which the protocone of the upper molar fits (Fig. 3-16B), as in eutherians and metatherians.

Interestingly, pseudotribosphenic molars evolved in certain other Mesozoic mammals convergently to those of tribosphenic mammals. One such species is *Shuotherium dongi,* an Asian late Jurassic species in which the lower molars evolved a heel-like structure that is analogous to the talonid of tribosphenic mammals and that functioned in the same manner. However, in *Shuotherium* this structure occurred on the anterior instead of the posterior side of the molars (Chow and Rich, 1982).

Kokopellia represents the earliest known marsupial or marsupial-like mammal and is from the middle Cretaceous of North America (Cifelli, 1993; Cifelli and Muizon, 1997). Like many fossil vertebrates, *Kokopellia's* geological age is known by its stratigraphic association with a geochemically dated volcanic ash, in this case 98 million years old. The oldest known eutherian is *Prokennalestes* from the late Cretaceous of Asia (Kielan-Jaworowska and Dashzeveg, 1989; see Fig. 7-1). The absolute and relative ages of these specimens

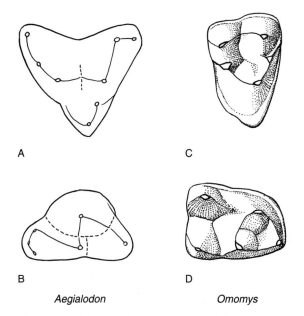

A   C

B   D

Aegialodon   Omomys

FIGURE 3-16   (A) Right upper and (B) left lower molar of *Aegialodon,* an early Cretaceous tribosphenidan (family Aegialodontidae). The upper molar is a hypothetical reconstruction. (C and D) Comparable teeth of the primitive Eocene eutherian mammal *Omomys,* a tarsier-like primate. *(C and D after Romer, ©1966, Vertebrate Paleontology, University of Chicago Press)*

suggest that marsupials and placentals had become recognizable as such by at least 100 million years ago, although their respective lineages probably split before then. As timed by the molecular clock, genetic evidence suggests that the marsupial lineage split from a therian ancestor much earlier (173 million years ago; Kumar and Hedges, 1998). This date seems exceedingly early when compared with the fossil record.

CRETACEOUS MAMMALS

A broad view of the Cretaceous — a period of great biotic change — provides a background against which the late Mesozoic evolution of mammals can be viewed. The late Jurassic was a time of considerable interchange of biotas between continents, as indicated by the occurrence in western

Europe, East Africa, and western North America of identical or closely related species of reptiles (Colbert, 1973) and by intercontinental similarities among floras (Vachrameev and Akhmet'yev, 1972). After the earliest part of the Cretaceous, however, dispersal between continents became sharply restricted. In the New World, a series of transgressions of a seaway from the Arctic Ocean to the Gulf of Mexico divided North America for much of the Cretaceous into two separate centers for the evolution of terrestrial plants and animals. In Eurasia also the dispersal of land animals was restricted; Europe was essentially an archipelago of islands during the Cretaceous, and the "Turgai Strait" Seaway separated the land faunas of Europe and Siberia. Angiosperms became established as the dominant plants in terrestrial ecosystems in the middle Cretaceous (Friis and Crepet, 1987), an event that strongly affected the evolution of land faunas. Coadaptive evolution between angiosperm flowers and insects, for example, fostered a Cretaceous insect radiation.

Most dramatic were Cretaceous changes in the fortunes of the dinosaurs. Throughout the Jurassic and early Cretaceous they were diverse and abundant and dominated the terrestrial scene. In the late Cretaceous, several herbivorous groups — the ankylosaurs, ceratopsians, and hadrosaurians — diversified in association with the increasing importance of angiosperms and decline of gymnosperms. By the close of the Cretaceous, dinosaurs were gone.

Through much of the early Cretaceous, land dwellers were barred from intercontinental movement (i.e., between the northern and southern supercontinents Laurasia and Gondwana) by oceans and seaways, but intracontinental movements would have encouraged relative homogeneity of faunas and floras. By the middle Cretaceous, Africa was isolated, but the other portions of Gondwana (South America, Antarctica, India, Australia) retained increasingly tenuous connections with one another until the late Cretaceous. Populations of mammals on different continents and subcontinents evolved in isolation under different environmental pressures. This isolation of premarsupial and preplacental stocks may well have favored their differentiation. Each group seemingly faced some comparable adaptive problems, but, as in

the case of reproduction, each group developed unique solutions to these problems.

Modern mammals (eutherians and marsupials) underwent an adaptive radiation in the Cretaceous. Similar significant radiations occurred among dinosaurs and multituberculates at this time. These radiations probably reflect the availability of fruits or seeds as an important new food source (Cifelli et al., 1997). Considerable literature (Clemens, 1970; Lillegraven, 1974; Lillegraven et al., 1979) points to the overriding importance of the early Cretaceous appearance and adaptive burst of flowering plants (angiosperms). The seeds of some angiosperms develop within an edible and nutritious fruit. Angiosperm fruits and seeds are eaten today by many mammals and were probably important to Cretaceous mammals. Mammals probably have a long history of contributing to angiosperm seed dispersal: Krassilov (1973) reported that the seed coats of some of the earliest angiosperms bore hooklets capable of tangling in the fur of foraging mammals. The Lepidoptera (moths and butterflies) appeared in the Cretaceous (MacKay, 1970), probably in response to the food offered by the flower nectar and leaves of angiosperms. The Isoptera (termites) also appeared at this time, and the Coleoptera (beetles) underwent an adaptive radiation. These insect groups are important foods for mammals today and were perhaps similarly important in the Cretaceous. The diversification of dentitions may well have enabled Cretaceous mammals not only to exploit plant foods but also to profit greatly from the expanding diversity and growing populations of insects.

For mammals, then, the Mesozoic, and especially the Cretaceous, was a time of experimentation. Natural selection, partly in the form of predation by an imposing array of reptilian carnivores and probably some avian predators and partly in the form of competition from reptiles, birds, and other mammals, affected many changes in mammalian structure and function. Behavioral, physiological, and anatomical changes that increased the efficiency of feeding, reproduction, and thermoregulation may have been critical. Various structural plans evolved and were workable for different lengths of time; some evolutionary side branches proved sterile. During the Mesozoic time of evolutionary trial and error, however, the basic mammalian structural plan was tested, retested, and perfected, and the major taxa were established. The extinction of the dinosaurs at the end of the Cretaceous, the diversity of flowering plants, and perhaps the cooling climatic trend set the stage for the dramatic adaptive burst of mammals at the start of the Cenozoic.

WEB SITES

Survey of evolution, paleontology, geologic time, and discussion of phylogenetic systematics

http://www.ucmp.berkeley.edu/history/evolution.htm

Discussion of synapsid and therapsid evolution and phylogenetic diagrams

http://phylogeny.arizona.edu/tree/eukaryotes/animals/chordata/synapsida.html

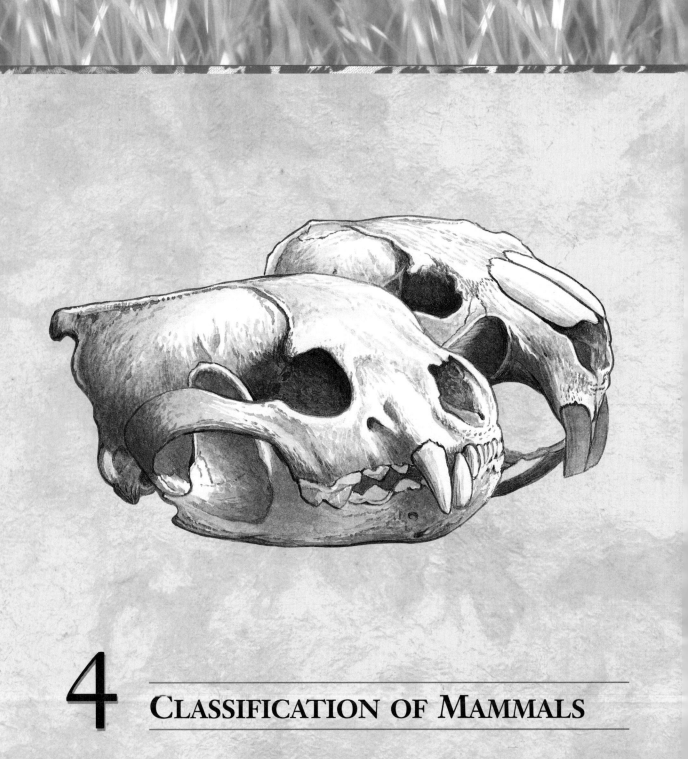

4 CLASSIFICATION OF MAMMALS

Despite their remarkable success, mammals are much less diverse than are most invertebrate groups. This is probably attributable to their far greater individual size, to the high energy requirements of endothermy, and thus to the inability of mammals to exploit great numbers of restricted ecological niches. About 1135 genera and 4630 species of living mammals are currently recognized (Wilson and Reeder, 1993 [see also searchable Internet version of their book at the web address at the end of this chapter]; Cole et al., 1994). Most species of extant mammals have

already been described, but approximately 10 to 12 new species continue to be named each year (Morell, 1996; Pine, 1994; Patterson, 1994; Raven and Wilson, 1992). When fossil mammals are considered, the numbers are more impressive. In the compendium of McKenna and Bell (1997), 5162 genera of mammals in 425 families — of which 4079 genera (79%) and 300 families (71%) are extinct — have been named. Still, the numbers of genera and species are insignificant in comparison with those for invertebrates. There are, for example, an estimated 950,000 named species of insects (perhaps 8 to 100 million undiscovered), 40,000 of protists (perhaps 100,000 to 200,000 undiscovered), and 70,000 of mollusks (perhaps 200,000 undiscovered).

Chapters 5 through 19 consider the orders and families of mammals listed in Table 4-1. In these chapters, such features as group size, present geographic distribution, time of appearance in the fossil record, structural characteristics, and brief life histories are given for each order and family. When appropriate, morphology is related to function so that the remarkable structural and functional diversity displayed by mammals can be appreciated.

We devote considerable attention to the orders and families of mammals not because we wish to put primary stress on the taxonomic aspect of mammalogy, but rather as an attempt to provide students with sufficient information on the various kinds of mammals to make the subsequent discussions of mammalian biology meaningful. Students'

interest is often dulled if they must deal with information about completely unfamiliar kinds of animals. It seems pointless to discuss water regulation in heteromyids, for example, if students have only a vague idea of what a heteromyid is. The chapters on orders, then, should serve as a background for the chapters on selected aspects of the biology of mammals.

Phylogenetic systematics is a major focus of much of the basic research conducted on mammals today. It is also an important feature of many other fields, from conservation biology to molecular genetics. Increasingly, authors attempt to construct classification schemes that reflect the presumed phylogeny of mammals and other organisms. However, reconstructing a phylogeny and producing a classification for a group of organisms are very different goals, and the results are often highly controversial. Although this chapter emphasizes classification, the reader should note that in the chapters of the book dealing with mammalian origins and with the orders of mammals, we have attempted to utilize published phylogenetic information or at least to cite pertinent references. The classification that follows is largely that of the various authors in Wilson and Reeder (1993) and partly that of McKenna and Bell (1997) but departs in several places to accommodate our preferences. No classification system yet proposed has gained universal acceptance, but Wilson and Reeder's and McKenna and Bell's are useful recent compendia.

TABLE 4-1 A Classification of Recent Mammals

Classification	Common Name(s)
Subclass Prototheria	
Order Monotremata (3 species)	
Family Tachyglossidae	Echidnas, spiny anteaters
Ornithorhynchidae	Duck-billed platypus
Subclass Theria	
Infraclass Metatheria (Marsupialia)	
Order Didelphimorphia (63 species)	
Family Didelphidae	Opossums
Order Paucituberculata (6 species)	
Family Caenolestidae	Rat opossums
Order Microbiotheria (1 species)	
Family Microbiotheriidae	Monito del monte, llaca
Order Dasyuromorphia (63 species)	
Family Thylacinidae	Thylacine (extinct)
Myrmecobiidae	Numbat
Dasyuridae	Dasyures, quolls, antechinuses, dunnarts, devil
Order Peramelemorphia (21 species)	
Family Peramelidae	Bandicoots
Peroryctidae	Bandicoots
Order Notoryctemorphia (2 species)	
Family Notoryctidae	Marsupial "mole"
Order Diprotodontia (117 species)	
Family Phascolarctidae	Koala
Vombatidae	Wombats
Phalangeridae	Cuscuses, phalangers
Potoroidae	Rat kangaroos, bettongs
Macropodidae	Kangaroos, wallabies
Burramyidae	Pygmy possums
Pseudocheiridae	Ringtailed possums
Petauridae	Gliders, striped possums
Tarsipedidae	Honey possum, noolbenger
Acrobatidae	Feathertail possum, feathertail glider
Infraclass Eutheria (Placentalia)	
Order Xenarthra (29 species)	
Family Bradypodidae	Three-toed tree sloths
Megalonychidae	Two-toed tree sloths
Dasypodidae	Armadillos
Myrmecophagidae	Anteaters
Order Insectivora (429 species)	
Family Solenodontidae	Solenodons, alamiquis
Nesophontidae	West Indian shrews (extinct)
Tenrecidae	Tenrecs
Chrysochloridae	Golden moles
Erinaceidae	Hedgehogs
Soricidae	Shrews
Talpidae	Moles
Order Scandentia (19 species)	
Family Tupaiidae	Tree shrews

TABLE 4-1 *(continued)*

Classification	Common Name(s)
Order Dermoptera (2 species)	
Family Cynocephalidae	Colugos
Order Chiroptera (928 species)	
Family Pteropodidae	Old World fruit bats, flying foxes
Emballonuridae	Sac-winged bats
Craseonycteridae	Kitti's hog-nosed bat
Rhinopomatidae	Mouse-tailed bats
Nycteridae	Hollow-faced bats, slit-faced bats
Megadermatidae	False vampire bats
Rhinolophidae	Horseshoe bats
Phyllostomidae	Leaf-nosed bats
Mormoopidae	Mustached bats
Noctilionidae	Bulldog bats
Mystacinidae	Short-tailed bats
Molossidae	Free-tailed bats
Myzopodidae	Sucker-footed bat
Thyropteridae	Disk-winged bats
Furipteridae	Smoky bats
Natalidae	Funnel-eared bats
Vespertilionidae	Evening bats, common bats
Order Primates (236 species)	
Family Daubentoniidae	Aye-aye
Lemuridae	Lemurs
Megaladapidae	Giant lemur (extinct)
Galagonidae	Galagos
Loridae	Lorises
Cheirogaleidae	Dwarf lemurs
Indridae	Indri, sifakas, avahi
Tarsiidae	Tarsiers
Cercopithecidae	Old World monkeys
Hominidae	Apes, human
Hylobatidae	Gibbons
Callitrichidae	Marmosets
Cebidae	New World monkeys
Order Carnivora (271 species)	
Family Felidae	Cats
Viverridae	Civets
Herpestidae	Mongooses
Hyaenidae	Hyenas, aardwolf
Canidae	Wolves, foxes, jackals
Ursidae	Bears, giant panda
Otariidae	Eared seals, fur seals, sea lions
Phocidae	Earless seals
Odobenidae	Walrus
Mustelidae	Weasels, skunks, badgers, otters
Procyonidae	Raccoons, ringtail cats, coatis
Order Cetacea (78 species)	
Family Balaenopteridae	Rorquals
Eschrichtiidae	Gray whale

TABLE 4-1 *(continued)*

Classification	Common Name(s)
Balaenidae	Right whales
Neobalaenidae	Pygmy right whale
Physeteridae	Sperm whales
Ziphiidae	Beaked whales
Platanistidae	River dolphins
Delphinidae	Ocean dolphins
Monodontidae	Narwhal, beluga
Phocoenidae	Porpoises
Order Sirenia (5 species)	
Family Dugongidae	Dugongs, sea cows
Trichechidae	Manatees
Order Proboscidea (2 species)	
Family Elephantidae	Elephants
Order Perissodactyla (18 species)	
Family Equidae	Horses, asses, zebras
Tapiridae	Tapirs
Rhinocerotidae	Rhinoceroses
Order Hyracoidea (6 species)	
Family Procaviidae	Hyraxes
Order Tubulidentata (1 species)	
Family Orycteropodidae	Aardvark
Order Artiodactyla (220 species)	
Family Suidae	Swine
Tayassuidae	Peccaries, javelinas
Hippopotamidae	Hippopotami
Camelidae	Camels, llamas
Tragulidae	Chevrotains
Giraffidae	Giraffe, okapi
Moschidae	Musk deer
Cervidae	Deer
Antilocapridae	Pronghorns
Bovidae	Antelope, bison, cattle, goats, sheep, etc.
Order Pholidota (7 species)	
Family Manidae	Pangolins, scaly anteaters
Order Rodentia (2024 species)	
Family Aplodontidae	Mountain beaver, sewellel
Sciuridae	Squirrels
Castoridae	Beavers
Geomyidae	Pocket gophers
Heteromyidae	Kangaroo rats, pocket mice
Dipodidae	Jumping mice, jerboas
Muridae	Rats, mice
Anomaluridae	Scaly-tailed flying squirrels
Pedetidae	Springhaas, springhare
Ctenodactylidae	Gundis
Myoxidae	Dormice
Bathyergidae	Mole rats, sand rats
Hystricidae	Old World porcupines
Petromuridae	Rock rat

TABLE 4-1   *(continued)*

Classification	Common Name(s)
Thryonomyidae	Cane rats
Erethizontidae	New World porcupines
Chinchillidae	Chinchillas, vizcachas
Dinomyidae	Pacarana
Caviidae	Cuis, cavies, Guinea pigs
Hydrochoeridae	Capybara
Dasyproctidae	Pacas
Agoutidae	Agoutis
Ctenomyidae	Tuco-tucos
Octodontidae	Degus
Abrocomidae	Chinchilla rats
Echimyidae	Spiny rats
Capromyidae	Hutias
Heptaxodontidae	Hutias (extinct)
Myocastoridae	Nutria, coypu
Order Lagomorpha (80 species)	
Family Ochotonidae	Pikas
Leporidae	Rabbits
Order Macroscelidea (15 species)	
Family Macroscelididae	Elephant-shrews

WEB SITES

"Tree of Life" home page; information about phylogenetic relationships, characteristics, and animal diversity

http://phylogeny.arizona.edu/tree/phylogeny.html

General introduction to mammals and phylogeny of mammals

http://www.ucmp.berkeley.edu/mammal/mammal.html

Searchable electronic version of Wilson and Reeder's 1993 book *Mammal Species of the World*

http://www.nmnh.si.edu/msw/

5 MONOTREMATA

The order Monotremata includes the family Tachyglossidae (echidnas, or spiny anteaters), members of which are found in Australia, Tasmania, and New Guinea, and the family Ornithorhynchidae (duck-billed platypuses), restricted to eastern Australia and Tasmania. Although represented today by only three genera, each with a single species, monotremes constitute a minor segment of the Recent mammalian fauna, but they are of great interest for several reasons. Morphologically, they closely resemble no other living mammals, and they possess several features frequently considered to be more typical of reptiles than of mammals. Monotremes lay eggs and incubate them in birdlike fashion, and yet

they have hair and suckle their young. The few surviving species thus retain some primitive features of their synapsid ancestors, yet they are highly derived (advanced) in other features.

One derived feature of monotremes is the rostrum, the "bill" of the platypus or "beak" of the echidnas, which is covered with electroreceptors and can detect the weak electrical fields of small invertebrate prey. This remarkable sensory ability is almost unique among mammals, being found also in certain moles of the family Talpidae. Among other vertebrates, electroreception is known only in certain fishes and amphibians, in which a completely different kind of receptor organ has evolved. The sensory importance of the **rhinarium** (the hairless area at the tip of the snout in mammals) is reflected in the monotreme brain, which is relatively large and extraordinarily complex (Rowe and Bohringer, 1992).

Other specializations, such as the reduction of the dentition (echidnas completely lack teeth, and platypuses retain vestigial teeth only as juveniles) and highly modified skull morphology, make it difficult to determine their phylogenetic relationships with other mammals. Most authors agree that monotremes themselves form a monophyletic group, but the two families are quite divergent and may have been separate since at least the Eocene and possibly since the late Cretaceous (Westerman and Edwards, 1992; Messer et al., 1998).

The relationship of monotremes to other mammals, however, is unresolved (Archer et al., 1993). Morphological and paleontological evidence strongly suggests that monotremes diverged from an early mammalian ancestor before the divergence of marsupials and placentals. In this view, mono-tremes are prototherians, not therians. However, recent molecular evidence contradicts that view. DNA sequences of the complete mitochondrial genomes of the platypus and therian mammals strongly suggest that monotremes lie within the Theria and are most closely related to the marsupials (Penny and Hasegawa, 1997). Other molecular studies, however, indicate that the living mono-

tremes are no more closely related to marsupials than to eutherians and that they diverged from marsupials and eutherians about 163 million to 186 million years ago (Messer et al., 1998).

MORPHOLOGY

Many structural features distinguish monotremes from other mammals. Uniquely birdlike in appearance (Fig. 5-1), the monotreme skull is toothless in living forms except in young platypuses, cranial **sutures** disappear early in life, and the elongate and beaklike rostrum is covered by a leathery sheath; this sheath is horny in birds. There are **sclerotic cartilages** in the eyes, but these do not become ossified nor form a sclerotic ring as they do in many reptiles and nonmammalian synapsids. The lacrimal and frontal bones are absent, whereas these bones are present in most therian mammals. The skull of monotremes includes a bone, the septomaxilla, that occurs in few other living mammals (Wible et al., 1990). There is no auditory bulla, but the chamber of the middle ear is partially surrounded by an oval tympanic ring. In the inner ear, the cochlea is curved but not coiled as in other mammals.

Monotreme appendages represent excellent examples of "mosaic evolution" (different parts evolving at different rates) (Crompton and Jenkins, 1973). The shoulder girdle retains a bone pattern typical of therapsids, the forelimb has a rather sprawled posture resulting in part from **fossorial** (digging) specializations, and the pelvis and posture of the hindlimbs are essentially therian. Medially directed spurs occur on the inside of the ankles of adult males.

The monotreme pectoral girdle contains an interclavicle, clavicles, precoracoids, coracoids, and scapula and provides a far more rigid connection between the shoulders and the sternum than does the girdle characteristic of therian mammals (Fig. 5-2). Large epipubic bones extend forward from the pubes in both sexes. Cervical ribs are present,

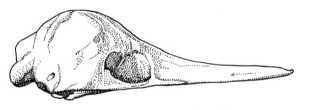

Spiny anteater
(*Tachyglossus aculeatus*)

FIGURE 5-1   Skull of the spiny anteater. Length of skull 111 millimeters.

and the thoracic ribs lack tubercles, processes that occur on the ribs of most other mammals and are braced against the transverse processes of the vertebrae.

As put by Howell (1944), no monotreme "by any strength of the imagination might be considered cursorial [strongly adapted for running]." Monotremes have retained a limb posture that is similar in some ways to that of reptiles, and in monotremes this posture is associated with limited running ability. In the Australian echidna (*Tachyglossus aculeatus*), the humerus remains roughly horizontal to the substrate during walking (Jenkins, 1970). Rotational movement of the humerus, rather than fore-and-aft movement as in most mammals, is largely responsible for propulsion. Because in reptiles the limb posture is splayed, the fore and hind feet touch the ground well to the side of the shoulder and hip joints, respectively. The limb posture in the echidna partially departs from this pattern because the forearm angles medially and the manus (forepaw) is roughly ventral to the shoulder joint; in the hindlimb the foot is

roughly ventral to the knee. The posture of the hindlimb of the echidna resembles that of many generalized therian mammals. When the echidna is in motion, its body is elevated well above the ground in nonreptilian fashion. Despite the advances in limb posture in the echidna, locomotion is slow and appears labored and awkward. The ability of monotremes to burrow and the musculoskeletal adaptations associated with burrowing (and in the platypus, swimming) confound the distinction between primitive, reptile-like features and specialized adaptations, making phylogenetic interpretation more difficult (Jenkins, 1990).

REPRODUCTION

The monotreme reproductive system and pattern are unique among mammals. As in other systems of the body, the reproductive system is a mix, including primitive features shared with amniotes and unique specializations (Renfree, 1993). Monotremes lay eggs that are **telolecithal** (the yolk is concentrated toward the vegetal pole of the ovum) and **meroblastic** (early cleavages are restricted to a small disk at the animal pole of the ovum) like bird eggs but unlike the reproductive processes of therian mammals. Only the left ovary is functional in the platypus (Asdell, 1964), as in most birds, but both ovaries are functional in the echidna. Shell glands are present in the oviducts. Monotremes are usually considered **oviparous,** because they lay rubbery-shelled eggs that are incubated and hatched outside the mother's body. The fetus employs an "egg tooth" on the egg tooth bone (caruncle) to break out of the shell. There is a **cloaca**

FIGURE 5-2   Bones of monotremes: (A) Left humerus, (B) right femur, and (C) pectoral girdle.

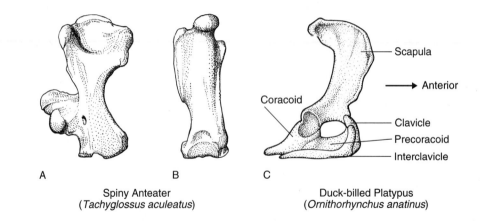

A

B

C

Scapula

Anterior

Coracoid

Clavicle

Precoracoid

Interclavicle

Spiny Anteater
(*Tachyglossus aculeatus*)

Duck-billed Platypus
(*Ornithorhynchus anatinus*)

(Fig. 5-3), and in males the penis is attached to the ventral wall of this cavity. The testes are abdominal, and seminal vesicles are absent. The female echidna temporarily develops a pouchlike structure to incubate and protect the young, but the platypus never develops one. This structure is not homologous to the pouch developed by some marsupials. The mammae lack nipples, and the young suck milk from two lobules in the pouch in the echidna (Fig. 5-4) or from the abdominal fur in the platypus.

Monotremes typically have long periods of lactation and maternal care of the young. The platypus lays usually two eggs in a leaf nest in a burrow, where incubation lasts up to 12 days. The eggs become stuck together as the mother incubates them by curling her body around them. The newly hatched young are tiny (11 millimeters in length) and nearly embryonic in appearance. The mother suckles the young and broods them (keeps them warm) for nearly 16 weeks, and they develop slowly. The first growth of hair appears seven weeks after hatching, and the eyes do not open until about nine weeks after hatching. Development of the single young echidna is similarly slow. The egg, or sometimes two or three, are laid directly into the mother's pouch. After hatching, the young resides in the pouch until about 12 weeks of age, at which time the eyes open and the juvenile leaves the pouch to live in its mother's burrow. Weaning is at about 20 weeks. Both platypuses and echidnas have low reproductive rates, apparently no more than one clutch a year.

PALEONTOLOGY

The earliest known fossil record of monotremes is from the early Cretaceous (about 110 million years ago) of Australia. *Steropodon galmani,* an ornithorhynchid, is represented only by a beautifully opalized fragment of a lower jaw with several teeth (Archer et al., 1985; 1991). During the time *Steropodon* existed, that portion of Australia was within the Antarctic Circle. An early history in cold climates may partly explain the notable thermoregulatory abilities, including hibernation, of monotremes (Archer et al., 1993).

The next oldest appearance, around 62 million years ago in early Paleocene sediments of Patagonian Argentina, is the only known occurrence of a monotreme outside Australia, indicating that

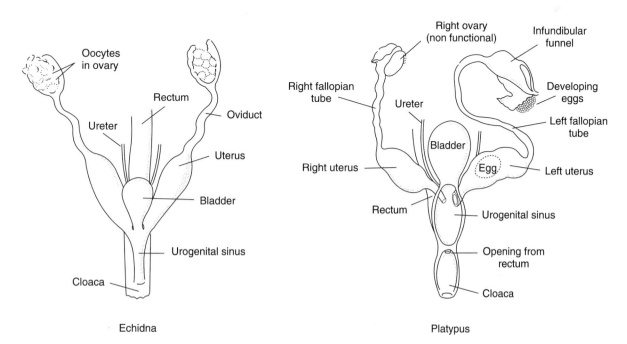

FIGURE 5-3 Anatomy of the female reproductive tract of monotremes. *(After Renfree 1993)*

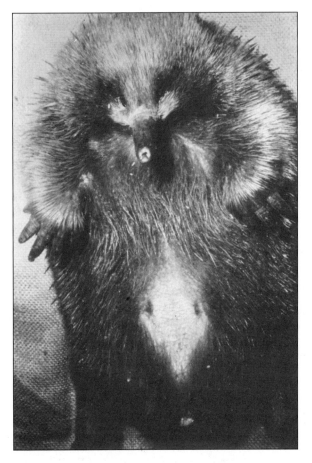

FIGURE 5-4 Ventral view of a live echidna (*Tachyglossus aculeatus*), showing the beaklike rostrum, the poorly developed pouch (typical of the nonbreeding season), and the tufts of hair at the mammary lobules. *(M. L. Augee)*

formerly this group of mammals was distributed across the southern continents (Pascual et al., 1992a, b). The Argentine species, *Monotrematum sudamericanum,* is an ornithorhynchid that had molars very similar to those of geologically younger Australian platypuses.

Two species of *Obdurodon* are known in the early Miocene of Australia (Woodburne and Tedford, 1975; Archer et al., 1992). One of these, *Obdurodon dicksoni* (Fig. 5-5) was a large platypus whose skull was a robust version of the living duck-billed platypus, *Ornithorhynchus.* Its bill included stouter septomaxillary bones and was wider than that of *Ornithorhynchus.* If it foraged in a similar manner, these skull differences and its larger size may have

enabled it to lift heavier stones underwater in searching for invertebrate prey (Archer et al., 1992, 1993). The extant genus *Ornithorhynchus* first appeared in the middle Miocene in Australia.

An insufficient fossil record and the tendency for reduction of the teeth make it difficult to determine the dental homologies between monotreme teeth and the teeth of other mammals. The dental morphology of fossil monotremes with nonvestigial teeth (*Steropodon, Monotrematum,* and *Obdurodon*) and the primitive features of the postcranial skeleton of extant species suggest that monotremes are outside the radiation of tribosphenic mammals (i.e., therians, or those with molars that bear a protocone on the uppers that fits into a talonid basin on the lowers; see Fig. 2-18F). Their molar teeth are considered "pre-tribosphenic," not close to the radiation of therians that involved the development of tribosphenic teeth. In particular, some authors have noted similarities to the Mesozoic eupantothere *Peramus* (Kielan-Jaworowska et al., 1987; Archer et al., 1993).

The fossil record of echidnas is much less complete than that of platypuses. The oldest known tachyglossid is *Zaglossus robustus* from the late Miocene of Australia. Fossils of *Zaglossus* are also known from Pleistocene deposits in New Guinea, where the genus still occurs. Certain Australian Pleistocene long-beaked echidnas are placed by some authors in a separate genus, *Megalibgwilia* (Griffiths et al., 1991). One Pleistocene echidna from southwestern Australia was much larger than any living monotreme; *Zaglossus hacketti* might have weighed 30 kilograms, twice as much as the New Guinea long-beaked echidna.

FAMILY TACHYGLOSSIDAE. Members of this group have a robust body covered with short, sturdy spines that are controlled by unusually well-developed panniculus carnosus muscles (sheet of muscles beneath the skin; Fig. 5-6). *Zaglossus bruijni,* the New Guinea long-beaked echidna, weighs from 5 to 16 kilograms, and the Australian short-beaked echidna (*Tachyglossus*) ranges from about 2.5 to 6 kilograms. The rostrum is slender and beaklike and, at least in the short-beaked echidna (*Tachyglossus*), bears electroreceptors (Gregory et al., 1989; Augee and Gooden, 1992). The dentary bones are slender and delicate, and the long,

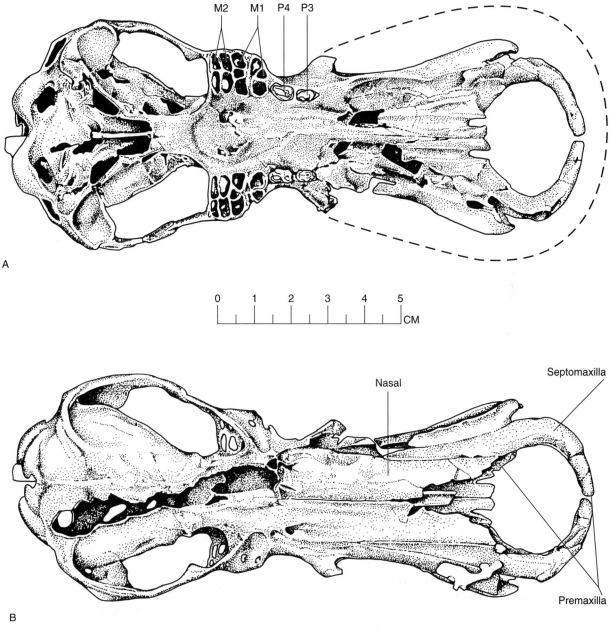

M2   M1   P4   P3

A

0   1   2   3   4   5
⌊ ⌊ ⌊ ⌊ ⌊ ⌊ CM

Nasal

Septomaxilla

Premaxilla

B

Miocene Platypus
(*Obdurodon dicksoni*)

FIGURE 5-5   Skull of the Miocene platypus *Obdurodon dicksoni*. Length of skull
137 millimeters (the length of the skull of *Ornithorhynchus* is about 108 millimeters).
(A) Palatal view. (B) Dorsal view. *M1, M2*, sockets for roots of molars; *P3*, third premolar;
P4, fourth premolar. Dashed line in A indicates restored outline of bill. *(After Archer et al., 1993)*

protrusible tongue is covered with viscous mucus secreted mostly by the enlarged submaxillary salivary glands. Food is ground between spines at the base of the tongue and adjacent transverse spiny ridges on the palate. The limbs are powerfully built and are adapted for digging. The humerus is highly modified by broad extensions of the medial and lateral epicondyles that provide unusually large surfaces for the origins of some of the powerful muscles of the forearm (Fig. 5-2). In *Zaglossus,* the number of claws is variable regionally; some animals have only three claws front and rear whereas others have a full complement of five (Flannery, 1995). In *Tachyglossus,* all digits have stout claws. The ankles of all males and some female echidnas bear medially directed spurs, the function of which is not known.

Echidnas have highly specialized modes of life. They are powerful diggers and can rapidly escape predators by burrowing. The food of *Tachyglossus* consists largely of termites and ants, whereas *Zaglossus* eats earthworms and soil arthropods. They forage by turning over stones and digging into termite and ant nests, then capturing the prey with the sticky tongue or, in *Zaglossus,* by impaling prey on a barbed tongue (Flannery, 1995).

The short-beaked echidna is a true hibernator. These animals gain weight during the warmer seasons and enter **hibernation** for three to four and a half months in winter in the mountains of southeastern Australia (Grigg et al., 1989, 1992). During periods of **torpor,** the body temperature drops as low as 3.7° C, close to the temperature in the hibernaculum, or winter residence. In laboratory conditions, hibernating echidnas showed various patterns of breathing; very slow and regular (one breath every 3 to 4 minutes), or periodic with either constant or varying tidal volume (the volume of air moved into and out of the lungs). Some of these captive hibernating animals took no breaths for periods from 20 minutes to 2 hours (Nicol et al., 1992).

FAMILY ORNITHORHYNCHIDAE. The duck-billed platypus is smaller than the echidnas, weighing from 0.5 to 2.0 kilograms. Some structural features of the platypus are associated with its semiaquatic mode of life (Fig. 5-7). The pelage is dense and velvety, and the underfur is woolly, rather like that of sea otter (*Enhydra*) or mole (Talpidae) fur. Similarly, these three kinds of mammals have fur that grows straight out of the skin (at right angles to the surface, not laid backward), so that forward and backward movement in water or in a burrow is not impeded. The external auditory meatus is tubular, as in the beaver (*Castor*). The eye and ear openings, which lack pinnae, lie in a furrow that is closed by folds of skin when the animal is submerged. The feet are webbed, but the digits retain

A

B

FIGURE 5-6   Two species of monotremes: (A) Australian spiny anteater (*Tachyglossus aculeatus*), (B) New Guinea spiny anteater (*Zaglossus bruijni*). *(Tachyglossus by M. L. Augee; Zaglossus by H. M. Van Deusen)*

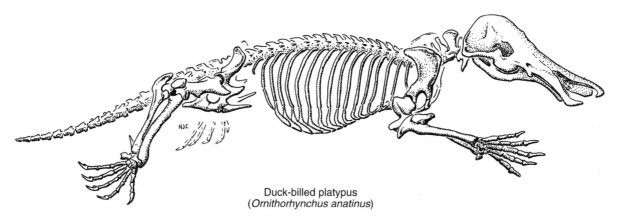

Duck-billed platypus
(*Ornithorhynchus anatinus*)

FIGURE 5-7   Skeleton of a duck-billed platypus, *Ornithorhynchus anatinus.*

claws that are used for burrowing. The web of the forefoot extends beyond the tips of the claws and is folded back against the palm when the animal is digging or when it is on land. The ankles of the male platypus have grooved and medially directed spurs that are connected to venom glands.

Although the young have teeth, the gums of the adults are toothless and covered by persistently growing, horny plates. Anteriorly, the occlusal surfaces of the plates form ridges that are used to chop food; posteriorly, the plates are flattened crushing surfaces. Some additional **mastication** is accomplished by the flattened tongue, which acts against the palate.

The elongate rostrum bears a flattened, leathery bill that contains densely packed arrays of specialized receptor organs and nerves associated with electroreception (Scheich et al., 1986). The electroreceptors are sensitive to weak electrical currents and are used by the nocturnally active platypus to locate crustacean prey underwater or in turbid water. The sensory importance of the bill is reflected in the cortex of the brain, in which receptive fields for the tactile-electrosensory neurons of the bill are enormously represented (Rowe and Bohringer, 1992).

The platypus inhabits a variety of waters, including mountain streams and slow-moving and turbid rivers, lakes, and ponds; it is primarily a bottom feeder. The platypus eats plants, aquatic crustaceans, insect larvae, and a wide variety of other animal material during dives that last for roughly 1 minute. The animals can eat up to half their body weight of food per day. The platypus takes refuge in burrows dug into banks adjacent to water. Seasonal torpor occurs in some parts of its range. Platypuses were formerly hunted for their plush fur; this hunting helped decimate populations.

WEB SITES

General information and photos of echidnas

> http://www.nexus.edu.au/schools/kingscot/pelican/monohome.htm

Information about echidnas and links to many other sites

> http://yoyo.cc.monash.edu.au/~tzvi/Echid_1.html

About monotreme biology and evolution

> http://www.ucmp.berkeley.edu/mammal/monotreme.html

6 METATHERIA

Metatherians and eutherians represent two evolutionary lines that have been separate since the middle Cretaceous, ca. 98 to 100 million years ago (Clemens, 1968; Lillegraven, 1974; Cifelli, 1993b; Springer et al., 1994). But the divergence and peak of the diversification of the extant metatherian orders probably did not occur until the latest Cretaceous or early Tertiary period, between 65 and 52 million years ago, several million years later than the peak of the diversification of extant eutherian orders (Springer et al., 1994, 1996; Springer, 1997). As a result of their long, independent history, metatherians differ structurally from eutherians in many ways.

Today, only two important strongholds for metatherians remain: the Australian region (Australia, Tasmania, New Guinea, and nearby islands) and the Neotropics (southern Mexico, Central America, and South America). Where they have been isolated from placentals for long periods, metatherians have undergone remarkable radiation. Most metatherians have functional counterparts among eutherians. Metatherians are commonly called "marsupials," for the **marsupium** or pouch in which young are carried, in order to distinguish them from eutherians, or "placentals." However, several kinds of marsupials do not develop a pouch, and a few kinds (bandicoots) have a type of placenta similar to that of eutherians.

Recent summary hypotheses of phylogenetic relationships among living marsupials have been given by Aplin and Archer (1987), Luckett (1994), Marshall et al. (1990), Retief et al. (1995), Springer et al. (1994; 1996), and Szalay (1993, 1994). One of these is shown here (Fig. 6-1). Relationships among early (Cretaceous) marsupials and other early fossil therians were proposed by Archer (1984) and Cifelli (1993a, b).

MORPHOLOGY

The metatherian skull frequently has a small, narrow braincase housing small cerebral hemispheres with simple convolutions. Ossified auditory bullae, when present, are usually formed largely by the alisphenoid bone rather than by the tympanic (both ectotympanic and entotympanic), petrosal, and/or basisphenoid bones, as occurs in most placentals. The metatherian palate characteristically has large vacuities (Fig. 6-2), and the angular process of the dentary bone is inflected medially (except in *Tarsipes* and only weakly so in *Phascolarctos* and *Myrmecobius*). The dentition is unique in that there are never equal numbers of incisors above and below, except in the family Vombatidae; the cheek teeth primitively include 3/3 premolars and 4/4 molars.

Marsupials have a unique pattern of tooth eruption and replacement in which only one tooth in each jaw (the third upper and lower premolars; P3/3) actually replaces a deciduous tooth. The pattern seems to be as follows: (1) The primary (deciduous) incisors and canines begin to develop before their secondary replacement teeth, but they are vestigial and never erupt. Instead, the deciduous precursors are resorbed when the secondary ("permanent") teeth appear. (2) The first two premolars are deciduous teeth that develop late and are not replaced by secondary teeth. (3) The deciduous third premolars are **molariform** and are replaced by secondary ones that are premolariform. (Sometimes the deciduous third premolars, dP3/3, are not immediately lost when the P3/3 erupt anterior to them, and the P3/3 and dP3/3 may coexist in the jaws for varying lengths of time.) (4) The four molars are unreplaced primary teeth (Luckett, 1993; Cifelli et al., 1996; Cifelli and Muizon, 1998).

Metatherians often have highly specialized feet associated with specialized types of locomotion (Fig. 6-3). The unusual patterns of specialization of the hind feet are probably a result of an arboreal heritage and the early development of an opposable first digit and an enlarged and powerfully clutching fourth digit. As in the monotremes, multituberculates, and early therian mammals (including two Cretaceous eutherians; Novacek et al., 1997), epipubic bones extend forward from the pubic bones in both sexes (except in the recently extinct *Thylacinus,* in which the epipubic bones were vestigial, and in the extinct Borhyaenidae, in which they were absent).

REPRODUCTION

The metatherian reproductive pattern differs sharply from that of eutherians (see Chapter 20). The females of about 50 percent of the metatherians of today have a marsupium (an abdominal pouch) or abdominal folds within which there are

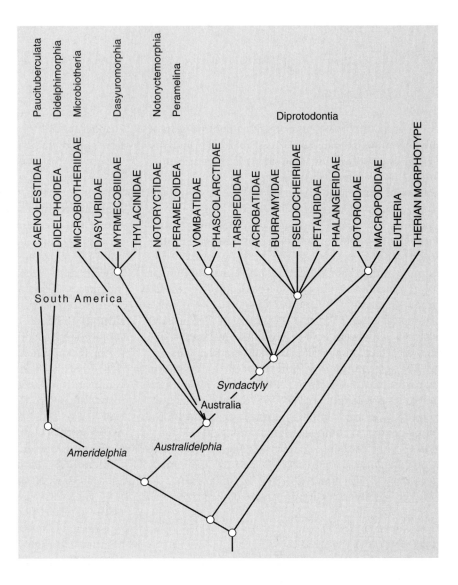

FIGURE 6-1 A phylogeny of Recent families of marsupials. *(After Luckett, 1994, and modifications by Woodburne and Case, 1996)*

nipples (Table 6-1). The number of nipples varies from 2 in the family Notoryctidae, and some members of the family Dasyuridae, to 27 in some members of the family Didelphidae. Individual variation in the number of nipples often occurs within a species. The female reproductive tract is bifid, that is, the vagina and uterus are double (Fig. 6-4). In all but the family Notoryctidae, which is adapted for digging, the testes are contained in a scrotum anterior to the penis.

The **gestation period** is characteristically short (8 to 43 days), and the young are tiny and rudimentary at birth. Newborn metatherians probably possess the minimal anatomical development allowing survival outside the uterus. Organogenesis has just begun, the separation of the ventricles of the heart is incomplete, the lungs are vascularized sacs lacking alveoli, and the kidneys lack glomeruli. Also lacking are cranial nerves II to IV and VI, eye pigments, eyelids, and cerebral commissures (nerve fiber bundles connecting the cerebral hemispheres). Despite this minimal development, however, the naked, blind, and delicate newborn is able to make its way at birth from the vulva to the marsupium. Here it attaches to a nipple and remains there for a period greatly exceeding the gestation period. The weight of the young metatherian when it leaves the pouch and that of the newborn eutherian are roughly the same in species of comparable adult size (Sharman, 1970).

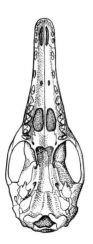

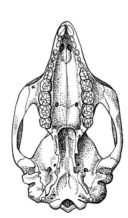

FIGURE 6-2   Ventral views of two marsupial skulls: a New Guinea bandicoot (length of skull 82 millimeters), and a ring-tail possum (length of skull 97 millimeters). *(After Tate and Archbold, 1937)*

New Guinea bandicoot
(*Peroryctes raffrayanus*, Peramelidae)

Ring-tail possum
(*Pseudochirops corinnae*, Pseudocheiridae)

Most metatherians, with the exception of the Peramelidae and Peroryctidae, have a **choriovitelline placenta** that lacks villi. These two families of bandicoots are unique among metatherians in possessing a **chorioallantoic placenta** (Hughes et al., 1990; see Chapter 20 for further discussion of metatherian reproduction).

PALEONTOLOGY

The earliest undoubted metatherian fossils are from the late Cretaceous of North America. The dinosaurs were still dominant then, and the only surviving groups of Mesozoic mammals were the Multituberculata, Triconodonta, and the Symmetrodonta. The niches previously filled by primitive Mesozoic mammals were probably being occupied in the late Cretaceous by the dominant mammalian groups of today, the metatherians and the eutherians, with the persistent multituberculates remaining important.

One of the most primitive metatherian families is the Didelphidae. Members of this family were present in North America during the late Cretaceous and have a nearly continuous fossil record there through the middle Miocene. During most of the Cenozoic, the Australian region and South America have been the two centers of metatherian diversification. In both regions, metatherian species radiated in partial or complete isolation from competition with eutherians.

Many experts believe that metatherians arose in North America and moved southward into South America in the Cretaceous and on to Australia, via Antarctica, in the late Cretaceous or earliest Tertiary (Woodburne and Case, 1996; Muizon et al., 1997). late Cretaceous fossils, possibly ancestral to the Didelphimorphia, are known from the interior of North America, and didelphoids and borhyaenoids (doglike marsupials) are known from Bolivia in the Paleocene (Muizon, 1994, 1998; Muizon et al., 1997; Woodburne and Case, 1996). The presence of metatherians in Eocene rocks in Antarctica strongly supports a southern route of their dispersal into Australia (Woodburne and Zinsmeister, 1982; Case et al., 1988; Marenssi et al., 1994; Goin and Carlini, 1995). Of course, some groups could have originated in Antarctica when it was not completely covered in ice. Metatherians reached Australia well before eutherians did and underwent a spectacular Cenozoic radiation. Fifteen living metatherian families, including 64 genera, resulted from this radiation, as did several additional extinct families and even orders, many of which were unusual and fascinating (see, for example, Archer and Clayton, 1984; Archer et al., 1991).

Anyone seeing the present Australian metatherian fauna for the first time finds the number of species and the structural extremes impressive. By comparison with the fauna of the late Pleistocene, however, the present fauna is severely depleted. Many species of large metatherians became extinct

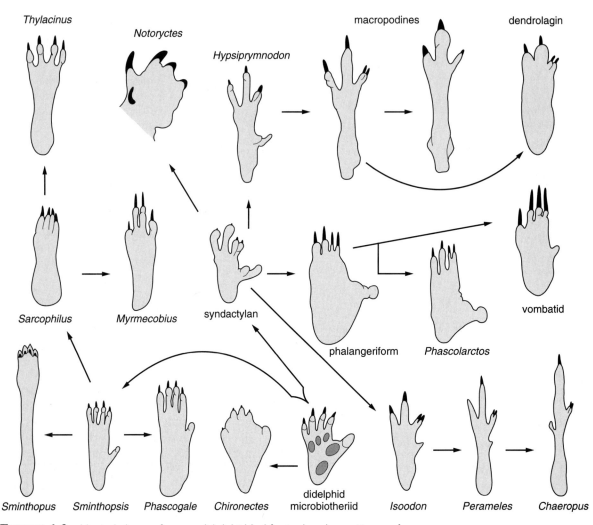

FIGURE 6-3 Ventral views of marsupial right hind feet, showing patterns of specialization associated with various styles of locomotion. At bottom center is the presumed basic arboreal type, represented by the foot of a didelphid or microbiotheriid (with plantar pads shown). The arrows indicate possible evolutionary pathways leading to greater specialization. Didelphidae: *Chironectes* (webbed aquatic foot); Dasyuridae: *Phascogale, Sminthopsis, Antechinomys, Sarcophilus;* Myrmecobiidae: *Myrmecobius;* Thylacinidae: *Thylacinus;* Notoryctidae: *Notoryctes;* Macropodidae: *Hypsiprymnodon,* macropodines, dendrolagin; Phascolarctidae: *Phascolarctos;* Vombatidae: vombatid; Peramelidae: *Isoodon, Perameles, Choeropus. (After Szalay, 1994, with permission of Cambridge University Press)*

between the late Pleistocene and historic time, and further reductions occurred in historic times. The extinct families Thylacoleonidae and Diprotodontidae are especially noteworthy. *Thylacoleo* was a predaceous Pliocene-Pleistocene genus with some species roughly the size of an African lion. The third premolars were greatly elongated shearing blades, and the strongly built front limbs had re-

tractile claws (Keast, 1972; Archer and Dawson, 1982). The Diprotodontidae were represented in the Pleistocene epoch by several genera. One (*Diprotodon*) was roughly the size of a rhinoceros and is the largest metatherian known. A number of very large kangaroos (family Macropodidae) also became extinct before historic times. Among them was a giant, *Procoptodon goliah,* an extremely short-

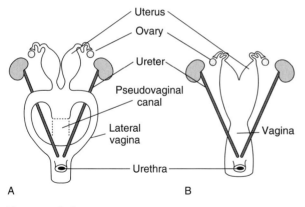

FIGURE 6-4   Diagrams of the female reproductive tracts: (A) metatherians, and (B) eutherians. *(After Sharman, 1970)*

faced macropodid (Fig. 6-5) that stood about 3 meters high.

Why did these imposing marsupials disappear? This question has been considered carefully by Choquenot and Bowman (1998), who stressed the possible influence of humans. Aborigines probably entered Australia 35,000 years ago, in the late Pleistocene, when the metatherian fauna included the large species just mentioned. Aborigines used fire extensively and perhaps used grass fires to hunt the large metatherians. After the Pleistocene, aridity and human-made fires may have tipped the balance against the large marsupials. In any case, only modern faunas are found in archeological sites spanning the last 20,000 years.

Further reductions in the number of species of metatherians began with the coming of European peoples to Australia. The combined effects of heavy grazing by livestock, clearing of land for agriculture, and the introduction of the Old World rabbit (*Oryctolagus cuniculus*), red fox (*Vulpes vulpes*), and feral cat (*Felis silvestris*) caused widespread declines in the abundance of many species and a number of extinctions (Calaby, 1971; Lever, 1985). As an example, of about 45 species of kangaroos that occupied Australia just before the entry of Europeans, three are extinct and the populations of roughly a dozen have declined drastically, although some species may have become rare before historic times.

The marsupial radiation in South America rivaled that in Australia. There are more than 130 liv-

TABLE 6-1   Marsupium Type and Number of Teats in Metatherians

Family	Marsupium	Position of Opening	Number of Teats
Didelphidae	Absent–well developed	—	4–27
Caenolestidae	Absent	—	4–5
Microbiotheriidae	Well developed	Anteroventral	4
Dasyuridae	Usually absent	Posterior	2–12
Thylacinidae	Crescent-shaped flap	Posterior	4
Myrmecobiidae	Absent	—	4
Peramelidae	Well developed	Posterior	6–10
Peroryctidae	Well developed	Posterior	—
Notoryctidae	Small	Posterior	2
Acrobatidae	Well developed	Anterior	1-4
Burramyidae	Well developed	Anterior	4–6
Macropodidae	Well developed	Anterior	4
Petauridae	Well developed	Anterior	2–4
Phalangeridae	Well developed	Anterior	2–4
Phascolarctidae	Well developed	Posterior	2
Potoroidae	Well developed	Anterior	4
Pseudocheiridae	Well developed	Anterior	2–4
Tarsipedidae	Well developed	Anterior	4
Vombatidae	Well developed	Posterior	2

From Marshall et al., 1990

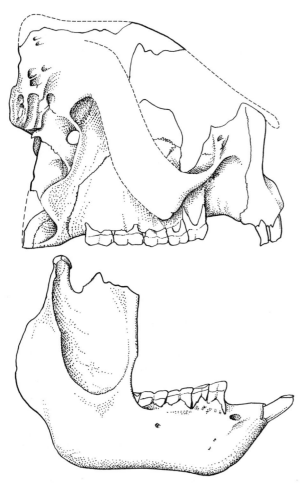

Procoptodon goliah

FIGURE 6-5 *Procoptodon goliah*, a 3-meter-long, grazing, macropodid marsupial from the Pleistocene fossil record of Australia. Length of skull 218 millimeters. *(After Tedford, 1967)*

ing and fossil genera of metatherians known from South America and more than 130 from Australia (McKenna and Bell, 1997). Accompanying the striking structural diversification of the marsupials in response to a wealth of available habitats was a convergence by many types toward eutherians. By the middle Tertiary, there were metatherians that structurally (and undoubtedly functionally) resembled shrews, moles, rodents, and carnivores. In the early Tertiary, several groups of eutherians were in South America. Xenarthrans and various ungulate groups seemingly "owned" the large herbivore adaptive zone throughout much of the Tertiary, to the complete exclusion of metatherians.

All of the South American marsupials presumably evolved from a basal, insectivorous-omnivorous metatherian stock. The occurrence of five orders in the early Paleocene in Bolivia and six orders at a late Paleocene fossil locality in Brazil documents an early Tertiary radiation (Muizon and Brito, 1993). Perhaps the most spectacular Tertiary marsupials were the doglike family Borhyaenidae and the saber-tooth family Thylacosmilidae.

The Borhyaenidae includes a number of marsupials with dentitions that suggest styles of feeding ranging from omnivorous to carnivorous. One borhyaenid, *Stylocynus*, was roughly the size of a bear and presumably omnivorous; a number of small and medium-size species were also omnivores. In some borhyaenids, including one of the earliest known (early Paleocene) species, *Mayulestes ferox*, numerous skeletal features are indicative of semiarboreal habits (Muizon, 1998). All known borhyaenids are rather short-legged, and the terrestrial types lack marked cursorial specializations. Members of the genus *Borhyaena* had a skull rather like that of a wolf (Fig. 6-6A).

The Miocene-Pliocene family Thylacosmilidae includes two genera that had recurved, saber-like upper canines. The roots of these teeth extend nearly to the occipital part of the skull, and their tips are protected by a flange on the dentary bone (Fig. 6-6B). Although basically well adapted to a carnivorous life, both borhyaenids and thylacosmilids had canines that wore rapidly because they had only a thin layer of enamel. In compensation, the roots in some forms remained open through much of life, permitting continued growth (Patterson and Pascual, 1972). The shape of the sabers of *Thylacosmilus* and skeletal features usually associated with powerful neck musculature indicate that the sabers were used like those of eutherian saber-tooth cats (p. 204); Akersten, 1985; Goin and Pascual, 1987). The limbs of thylacosmilids are short, and most prey was probably captured by a surprise attack rather than a chase.

The complex story of the decline and eventual extinction of the Borhyaenidae is told by Marshall (1977). Many early Eocene borhyaenids resembled didelphid marsupials and were of moderate size; by the early Oligocene, however, several very large forms had evolved. The largest of all, *Proborhyaena gigantea*, which had a skull roughly 1 meter long, disappeared by the end of the early Oligocene,

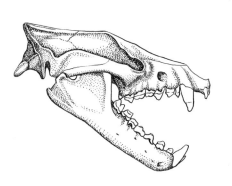

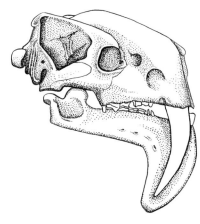

FIGURE 6-6   Skulls of members of the extinct marsupial families (A) Borhyaenidae and (B) Thylacosmilidae. (A) length of skull 230 millimeters, (B) length of skull 232 millimeters. *(After Romer, 1966 ©Vertebrate Paleontology, University of Chicago Press)*

A. *Borhyaena*   B. *Thylacosmilus*

and with its extinction a trend began toward reduced size in carnivorous borhyaenids. This evolutionary step moved counter to the trend toward large size that is common to many mammalian lineages. Further, the lack of cursorial adaptations in borhyaenids is unusual for a group of large carnivorous mammals and may have resulted from competition with large carnivorous birds.

During the early Tertiary in South America, the adaptive zones available to carnivores were probably not controlled or dominated by any one group. From early Oligocene through Pliocene times, three families of large carnivorous birds shared with borhyaenids a predatory mode of life. *Phorusrhacos,* a predaceous member of the family Phorusrhacidae (order Gruiformes), stood 1.5 to 3 meters tall and had a heavily built, hooked bill mounted on a skull about the size of that of a horse. The wings were tiny and the bird was flightless, but the hind legs were long and slim and the bird must have been a swift runner. Probably the phorusrhacid birds and the borhyaenid carnivores partitioned the resources available to carnivores: the phorusrhacids developed and maintained large size, occupied open savanna areas, and used their cursorial ability in pursuing swift prey; the borhyaenids were of more moderate size, occupied wooded country, and largely killed slow-moving prey. The late Tertiary reduction in the size of borhyaenids may well have been part of an adaptive trend toward reduction of competition with the imposing phorusrhacids (Patterson and Pascual, 1972).

Further declines in the fortunes of borhyaenids were apparently associated with the mid-Miocene arrival in South America of members of the thereto-

fore entirely North American raccoon family (Procyonidae). The fossil record indicates that replacement of the larger omnivorous borhyaenids by the omnivorous procyonids was complete by the late Pliocene. Extinction of other borhyaenids is thought by Marshall (1977) to be related to possible competition with marsupials of the family Didelphidae. The small-to-medium-size borhyaenid omnivores declined in the middle Pliocene and became extinct by the late Pliocene, whereas didelphids with similar adaptations appeared in the middle Pliocene and underwent a striking adaptive radiation in the late Pliocene.

The history of the Borhyaenidae was thus intertwined with that of the phorusrhacid birds, the immigrant South American procyonids, and the didelphid metatherians. By the late Pliocene, the borhyaenids were gone and the three latter groups prevailed.

An event of major importance to the South American mammalian fauna occurred in the late Pliocene. About 2.5 million years ago, the **land bridge** connecting North and South America was established and a flood of northern eutherians moved southward; some southern mammals moved northward (p. 547). Although the Borhyaenidae were already extinct and were thus not affected by this collision of faunas, the saber-tooth thylacosmilids were still present and were perhaps decisively affected. The occurrence of eutherian saber-tooth cats in early Pleistocene strata immediately above late Pliocene beds bearing thylacosmilids suggests the possibility of competitive replacement. The early interpretation was that competitively superior North American eutherians entered South

America over the Panamanian land bridge, leading to an eventual decline in metatherian diversity. This view has been challenged by many, including Marshall (1988) and Stehli and Webb (1985), who showed that the apparent decline in metatherians may not be directly related to the faunal interchange between the two continents. Webb (1991) further argued that ecological and geographic factors affected the different fates of the land mammals on either side of the land bridge.

The carnivory characterizing a number of the fossil types discussed above was just one of a number of feeding patterns of South American marsupials. *Necrolestes* from the middle Miocene was probably insectivorous and fossorial and may have had a mode of life similar to that of the Australian marsupial "moles" (Notoryctidae).

The family Caenolestidae, which still persists in relict populations along the Andes Cordillera, appeared first in the Oligocene of South America. Some caenolestids were convergent toward multituberculates in a series of features, including general skull form, enlargement of the anterior incisors, and structure of the **serrate,** lower pair of cheek teeth (Fig. 6-7). Both multituberculates and the multituberculate-like caenolestids (subfamily Abderitinae) probably resembled rodents in feeding habits.

The most rodent-like marsupials yet known are two South American species from the early Tertiary. These species compose the family Groeberiidae and are remarkable in having such features as enlarged incisors with enamel only on the anterior surfaces, a simplified pattern on the crowns of the cheek teeth, and a short, deep rostrum and mandible (Pascual et al., 1994).

Of special interest, the family Argyrolagidae is a supreme example of evolutionary convergence. This unique family, considered in detail by Simpson (1970), is known from the Oligocene to Pliocene of South America. Argyrolagids did not resemble closely any other group of marsupials, but they possessed a series of morphological characters that are found today in such specialized rodents as kangaroo rats (Heteromyidae) and jerboas (Dipodidae). These rodents occupy mainly sparsely vegetated desert or semiarid areas; all are **saltatorial** (use a jumping style of locomotion), and all share certain distinctive morphological features. The hindlimbs are long, and the hind feet

FIGURE 6-7   Jaw of a Miocene caenolestid, showing the highly specialized, trenchant cheek tooth. (*After Romer, 1966, ©Vertebrate Paleontology, University of Chicago Press*)

are modified by the loss of digits and, in some cases, by the fusion of metatarsal bones (see Fig. 18-16D). The long hindlimbs of argyrolagids are highly specialized along similar lines: only the third and fourth digits of the foot are retained, and the metatarsals are closely appressed and resemble to some extent the cannon bone of an artiodactyl (Fig. 6-8).

All of these animals have cheek teeth adapted to grinding and incisors suited for gnawing. In kangaroo rats and argyrolagids the cheek teeth are rootless and have a simple occlusal surface. The cheek teeth have also been considered convergent with those of elephant-shrews (Sanchez-Villagra and Kay, 1997).

Condensation of moisture on the cool nasal mucosa during exhalation is a means of reducing pulmonary water loss in kangaroo rats, and the unusual tubular extension of the nasal cavity anterior to the incisors in these animals (see Fig. 18-15) is associated with improved water conservation (p. 384). In the argyrolagids, there is an even more elongate extension of the nasal cavity. Here, too, this specialization may have facilitated the maintenance of water balance in an arid environment.

The entire form of the skull in kangaroo rats and jerboas is modified by the enormous auditory bullae, and in argyrolagids the bullae are also inflated. This enlargement of the bullae has been shown to be one of a remarkable series of specializations that allow kangaroo rats to detect faint, low-frequency sounds made by their predators. The enlarged bullae of the argyrolagids probably served a similar end.

It seems, then, that in two lineages that have been separate since at least the middle Cretaceous, but which occupy (or occupied) similar dry habitats, nearly identical suites of characteristics have evolved.

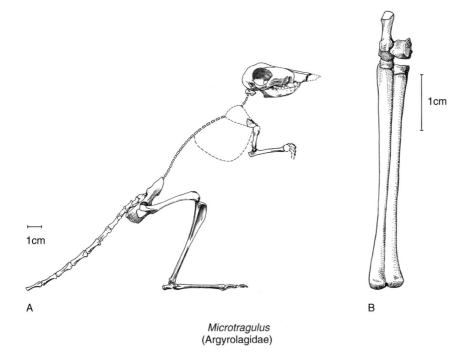

FIGURE 6-8 *Microtragulus* (Argyrolagidae), an extinct bipedal metatherian from the Mio-Pliocene of Argentina. (A) Reconstruction of the skeleton. Length of skull 55 millimeters. (B) Enlarged view of partial hind foot. Note that the appressed metatarsals of digits three and four form a structure resembling the cannon bone of artiodactyls. *(After Simpson, 1970)*

A

B

Microtragulus
(Argyrolagidae)

One cannot help but wonder why argyrolagids became extinct for "by all rules of analogy and theories of extinction, they should have survived, as did their close ecological analogs in North America, Asia, Africa, and Australia" (Simpson, 1970). At present, the mystery of the extinction of the argyrolagids remains unsolved. As the fossil record of South American mammals becomes more complete, information on the extinction of the argyrolagids and other marsupial groups will probably come to light.

METATHERIANS VERSUS EUTHERIANS: RELATIVE COMPETITIVE ABILITIES

When one views the course of mammalian evolution since metatherians and eutherians diverged from a common ancestor in middle Cretaceous times, the general impression is that the two groups are not of equal adaptive ability. Several lines of evidence can be cited.

1. Metatherians have not equaled the remarkable functional radiation of eutherians. There are no flying or marine metatherians, and some extremely productive food sources have never been tapped. Marine plankton, utilized by two orders of eutherians (Cetacea and Carnivora), and flying insects, eaten by bats (Chiroptera), have never been part of the marsupial diet.

2. Metatherians have been far more conservative in structural plan. None have modified limbs into fins or wings as eutherians have.

3. Metatherians have not been able to exploit great size. Although there were several large metatherians in the Pleistocene, the largest living metatherian (the red kangaroo) is only 1/1300 the size of the largest eutherian (the blue whale).

4. Metatherians have never evolved highly social behavior.

5. Metatherians have not developed the systematic diversity of eutherians. Only about 6 percent (272) of the total number of species of living mammals (4630) are metatherians.

Although sadly incomplete and equivocal, the fossil record does suggest a competitive edge for eutherians. In North America, where metatherians appeared before eutherians, metatherians had declined seriously by the latest Cretaceous but eutherians had radiated. In South America, similarly, where a Tertiary metatherian radiation occurred in isolation from a balanced assemblage of eutherians, the diversity of marsupials declined late in the Tertiary and in the Pleistocene, perhaps at first because of the entry into South America of only a few eutherians and then (with the emergence of the land bridge connecting North and South America) because of an invasion of eutherians from the north. Especially impressive is the total extinction of the South American metatherian carnivores (of the families Borhyaenidae and Thylacosmilidae) and their ultimate replacement by eutherian carnivores. (These examples lose force if considered in light of the hypothesis of Matthew [1915] that mammals that evolved on a large continent are competitively superior to those that originated on a smaller one.)

Although there is no general consensus on the matter, many scholars consider metatherians adaptively and competitively inferior to eutherians. Here we will briefly catalog some basic differences between the two groups to introduce the conflicting views that appear in the literature and to avoid offering a resolution for an unresolved problem.

Each of the following may be associated with adaptive-competitive differences between the two groups:

1. The metatherian mode of reproduction is probably more like that of early mammals than eutherians. Metatherians have a brief gestation period and bear almost embryonic young that have precocious forelimbs used to climb the mother's hair to reach the nipple. They undergo most of their basic development during a long period of lactation while attached to the mother's nipple and nourished by milk. Eutherians, in contrast, have long gestation periods and the young are much more developed at birth; they have a relatively short period of lactation. The metatherian need for precocious grasping forelimbs at birth may constrain their adaptability relative to placentals, precluding the development of wings, flippers, or other specializations of the forelimb.

2. The cerebral cortex develops more rapidly and attains greater volume in eutherians than in marsupials (Muller, 1969). The brain and nervous system develop best in the highly nutritive, stable, oxygen-rich environment of the uterus, possibly giving an advantage to eutherians over metatherians. Eisenberg and Wilson (1981), however, reported that overall cranial capacities are similar to eutherians for many didelphid species.

3. Behavioral plasticity is greater in eutherians: social groups with long-term dominance hierarchies and cooperative rearing of young occur only among eutherians. Territoriality, an important aspect of eutherian behavior, is uncommon in metatherians.

4. Antipredator behavior is more highly developed in eutherians: unified herd action, cooperative defense of young, complex vocal and visual communication, and sustained high-speed running are known only among eutherians.

5. The relatively low **diploid number** of chromosomes for marsupials has been mentioned (Hayman, 1977; Lillegraven, 1975) as possibly being related to their lack of evolutionary flexibility. Marsupials may be capable of far less genetic variability than placentals. The exact reasons for this are unknown but are made more apparent by a lower reproductive rate.

6. The extended gestation of eutherians produces young that are far more endothermic than metatherians and allows for better exploitation of colder climates.

7. The investment of energy by the mother is probably lower in metatherians than in eutherians (Parker, 1977), but eutherians are seemingly able to reproduce more rapidly. Unlike eutherians, metatherians show little connection between metabolic rate and rate of reproduction. This is especially evident in small metatherians with high metabolism (Lillegraven et al., 1987). These small metatherians retain a very long lactation period, thereby missing the small-size advantage that placentals often utilize by producing several litters per year. Metatherians do tend to have larger litter sizes than eutherians, but eutherians dominate by increasing their population sizes much more rapidly.

The view that the metatherian style of reproduction and metatherian biology in general represent an alternative, but not inferior, solution to survival problems has been discussed by a number of researchers (including Kirsch, 1977; Lillegraven et al., 1987; Parker, 1977; Pond, 1977; Tyndale-Biscoe and Renfree, 1987). For comparison of the adaptiveness of the metatherian and eutherian patterns of reproduction, see Lillegraven et al. (1987) and Tyndale-Biscoe and Renfree (1987).

SOUTH AMERICAN METATHERIANS

South American marsupials belong to three separate orders: Didelphimorphia, Paucituberculata, and Microbiotheria. Members of the order Didelphimorphia, opossums and mouse opossums, are the most generalized metatherians and constitute one of the oldest known groups, dating from the early part of the late Cretaceous (Springer, 1997). As mentioned, this is a basal group of the metatherian radiation (Muizon et al., 1997). The order Paucituberculata includes the small rat opossums, whereas the Monito del Monte (*Dromiciops gliroides*) is the sole living member of the order Microbiotheria.

ORDER DIDELPHIMORPHIA

FAMILY DIDELPHIDAE. The family Didelphidae includes 15 Recent genera comprising 63 species that range from southeastern Canada, with the American opossum (*Didelphis virginiana*), to southern Argentina, with the Patagonian opossum (*Lestodelphys halli*). In these New World opossums, the rostrum is long (Fig. 6-9A), the braincase is usually narrow, and the sagittal crest is prominent. The dental formula is 5/4, 1/1, 3/3, 4/4 = 50. The incisors are small and unspecialized, and the canines are large. The upper molars are basically tritubercular with sharp cusps, and the lower molars have a trigonid and a talonid (Fig. 6-10A, B).

Except for the opposable and clawless hallux in all species, a feature probably inherited from arboreal ancestral stock, and the webbed hind feet in the yapok or water opossum (*Chironectes minimus*), the feet are unspecialized, with no loss of digits or **syndactyly** (the condition in which two digits are attached by skin, as shown by a number of examples in Fig. 6-3). The foot posture is **plantigrade.** A marsupium is present in some didelphids but is represented by folds of skin protecting the nipples

in others and is absent in some. The tail is long and usually prehensile.

Although they occupy a wide range of habitats, didelphids are primarily inhabitants of tropical or subtropical areas, where they are often locally abundant. Most didelphids are partly arboreal and are omnivorous. The water opossum, however, is semiaquatic and carnivorous, and the woolly opossum (*Caluromys* spp.) is largely herbivorous. The small mouse opossum (*Marmosa*), a widespread Neotropical didelphid, is one of the most abundant small mammals in some parts of Mexico. Mouselike in general appearance, it seems to be largely insectivorous in some areas, at least during the summer (R. Smith, 1971). Poorly defined folds of skin protect the nipples of *Marmosa,* and the young simply hang on to the nipples and the mother's venter as best they can.

ORDER PAUCITUBERCULATA

FAMILY CAENOLESTIDAE. Recent members of this family (three genera and six species) bear the common name of shrew-opossum or rat opossum. The three genera have disjunct distributions from one another, each relict population occupying forested areas in a different region of the Andes Mountains of northern and western South America. *Rhyncholestes* and *Lestoros,* each with one species, occupy parts of the southern and central Andes, respectively; *Caenolestes,* with four species, occupies parts of the northern Andes (Albuja and Patterson, 1996). The earliest known cae-nolestids are found in the late Oligocene fossil record of South America. In the Oligocene and Miocene, a diverse group of caenolestids appeared, including some highly specialized types (Fig. 6-7).

Caenolestids resemble shrews because of their elongate heads and small eyes (Fig. 6-11). The skull is elongate and the brain primitive; the olfactory bulbs are large and the cerebrum lacks fissures. The dental formula is 4/3-4, 1/1, 3/3, 4/4 = 46 or 48; the first lower incisors are large and **procumbent,** and the remaining lower incisors, the canine, and the first premolar are **unicuspid.** Kirsch (1977) found that the lower incisors are used like rapiers to stab prey. The atlas bears a movable cervical rib. The feet are unspecialized, the tail is long but not prehensile, and there is no marsupium.

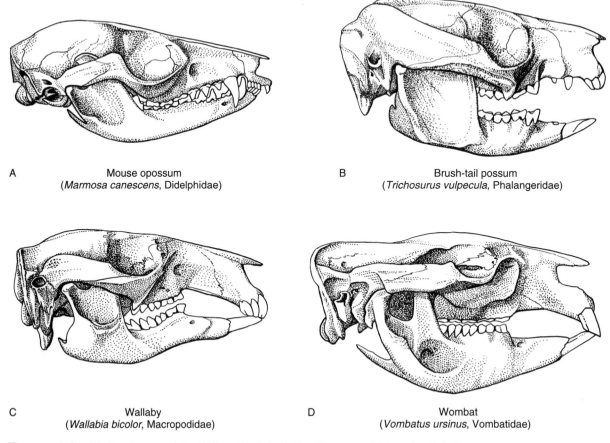

A   Mouse opossum
(*Marmosa canescens*, Didelphidae)

B   Brush-tail possum
(*Trichosurus vulpecula*, Phalangeridae)

C   Wallaby
(*Wallabia bicolor*, Macropodidae)

D   Wombat
(*Vombatus ursinus*, Vombatidae)

FIGURE 6-9   Skulls of marsupials: (A) length of skull 35 millimeters, (B) length of skull 87 millimeters, (C) length of skull 135 millimeters, (D) length of skull 180 millimeters.

ORDER MICROBIOTHERIA

FAMILY   MICROBIOTHERIIDAE. One species (*Dromiciops australis*) that inhabits cool, southern Andean forests is the single living member of the family Microbiotheriidae (Gardner, 1993). Although the species occurs in South America, the order to which it belongs is considered by many authors to be a relict of an Australian or Antarctic radiation of metatherians known as the Australidelphia (Luckett, 1994; Spotorno et al., 1997; Springer et al., 1994, 1996; Szalay, 1982; Woodburne and Case, 1996). Fossil microbiotheriids are known from the early Paleocene in South America and the middle Eocene in Antarctica (Carlini et al., 1990; Woodburne and Case, 1996).

The tiny *Dromiciops* is restricted to regions in south-central Chile. The dental formula is the same as that of the Didelphidae. The tympanic bullae are greatly enlarged and include an entotympanic bone, a component otherwise unknown among metatherians (Hershkovitz, 1992). The cloaca opens on the ventral side of the base of the tail, as in monotremes but unlike any other metatherian. The "monito del monte," as it is called, is arboreal and nocturnal, preferring habitats that include dense stands of Chilean bamboo (*Chusquea* sp.). The species feeds primarily on insects but will eat vegetation. In times of food scarcity or cold temperatures, it may store fat in the tail and enter a period of hibernation (Rageot, 1978).

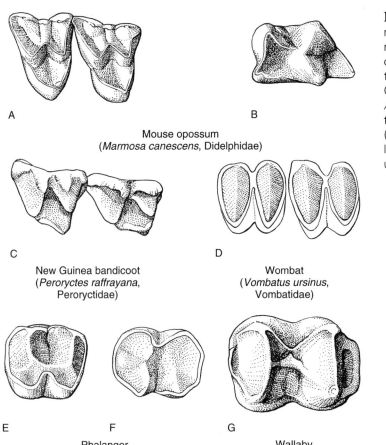

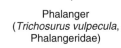

A

B

Mouse opossum
(*Marmosa canescens*, Didelphidae)

C

D

New Guinea bandicoot
(*Peroryctes raffrayana*,
Peroryctidae)

Wombat
(*Vombatus ursinus*,
Vombatidae)

E   F

G

Phalanger
(*Trichosurus vulpecula*,
Phalangeridae)

Wallaby
(*Wallabia bicolor*,
Macropodidae)

FIGURE 6-10   Occlusal views of metatherian molars: (A) Second and third right upper and (B) third lower left molar of a mouse opossum. (C) Second and third upper right molars of the New Guinea bandicoot (*After Tate and Archbold, 1937*). (D) Second and third right upper molars of a wombat. (E) Second upper right and (F) third lower left molar of a phalanger. (G) Second upper right molar of a wallaby.

FIGURE 6-11   A caenolestid marsupial *(Lestoros inca).* This individual came from Peru, at an elevation of 3530 meters. *(J. Kirsch)*

AUSTRALIAN METATHERIANS

The second great radiation of metatherians occurred primarily in the Australasian region. Often placed in the group Australidelphia to reflect their independent radiation, this assemblage includes five orders: Dasyuromorphia, Peramelemorphia, Notoryctemorphia, Diprotodontia, and Microbiotheria. The dasyuromorphs, often called carnivorous marsupials, include a large group of unspecialized small carnivores (e.g., quolls, antechinuses, dunnarts, Tasmanian devil, etc.) in the family Dasyuridae, along with the recently extinct Tasmanian thylacine (family Thylacinidae) and the specialized termite-eating numbat (family Myrmecobiidae). Bandicoots and bilbies, divided into two families, comprise the Peramelemorphia. The relationship of the marsupial mole to other metatherians is unclear, and for this reason it is placed in its own order, Notoryctemorphia. Finally, the largest order, Diprotodontia, includes ten families and 117 species of easily recognizable metatherians: kangaroos, wallabies, wombats, a variety of possums and gliders, and the koala.

ORDER DASYUROMORPHIA

FAMILY DASYURIDAE. Dasyurids (Fig. 6-12) are more progressive than didelphids, both dentally and with regard to limb structure. Although the earliest known dasyurid is from the Australian early or middle Miocene, the family must have arisen at a far earlier time. Recent members of this family include 15 genera and 61 species, and the geographic range includes Australia, New Guinea, Tasmania, the Aru Islands, and Normanby Island (Flannery, 1995; Strahan, 1995; Van Deusen and Jones, 1967).

Many of the major characters of dasyurids are shared by other metatherians, but several features are diagnostic of the former. The dental formula is 4/3, 1/1, 2-3/2-3, 4/4 = 42-46. The incisors are usually small and either pointed or bladelike, the canines are large and have a sharp edge, and the upper molars have three sharp cusps adapted to an insectivorous and carnivorous diet. The skulls of some dasyurids resemble rather closely those of didelphids (Fig. 6-13). The forefoot has five digits, and the hind foot has four or five digits. The hallux is clawless and usually vestigial and is absent in some cursorial genera (Figs. 6-3 and 6-14). There is no syndactyly. The foot posture is plantigrade in many species except the long-limbed jumping marsupial, the kultarr (*Sminthopsis laniger*), and the cursorial and carnivorous eastern quoll (*Dasyurus viverrinus;* Fig. 6-12). The marsupium is often absent; when present, it is often poorly developed. The tail is long and well furred, conspicuously tufted in some species, and never prehensile. The size of dasyurids ranges from that of a shrew (*Planigale*) to that of a small dog (*Sarcophilus*).

Dasyurids occupy a wide variety of terrestrial habitats; a few species are arboreal. There is a remarkably diverse array of marsupials in the family Dasyuridae. The smaller species fill the feeding niche occupied in Eurasia and North America by shrews (family Soricidae) and resemble those animals in the possession of long-snouted heads and unspecialized limbs. A group of rat-size dasyurids seems adapted to preying on insects and small vertebrates. The desert-dwelling genus *Sminthopsis* has long, slender limbs and a long, tufted tail and uses a rapid, bounding, quadrupedal gait. Another group, the quolls, consists of somewhat civet-like dasyurids that weigh about 0.5 to 3 kilograms and prey on a variety of small vertebrates. Quolls are agile and effective predators, and, although primarily terrestrial, they are capable climbers.

The largest dasyurid carnivore is the Tasmanian devil (*Sarcophilus laniarius;* Fig. 6-15B). The Tasmanian devil is a stocky, short-limbed dasyurid, weighing from 4.5 to 9.5 kilograms. Once widespread over Australia, it is now restricted to Tasmania. Devils occur in a variety of habitats but seem to prefer dry, sclerophyllous forests interspersed with grasslands. They are not territorial, and home ranges of several devils may overlap extensively. The species is a persistent scavenger but will also kill a wide variety of small vertebrates. They tend to forage alone, but groups of up to 22 have been reported feeding on large carcasses. As is typical of scavengers, they have strong jaws (they consume even bones) and can eat up to 40 percent of their body weight in a single night (Strahan, 1995). Tasmanian devils are highly vocal and disputes between neighbors can lead to clashes that result in serious wounds.

FAMILY MYRMECOBIIDAE. The most divergent dasyuromorph is the numbat, or banded anteater (*Myrmecobius fasciatus*), a small, long-snouted ani-

FIGURE 6-12   Four members of the family Dasyuridae: (A) a nulgara *(Dasyuroides byrnei)*; (B) a dunnart *(Sminthopsis crassicaudata)*; (C) a ningaui *(Ningaui sp.)*; (D) a quoll *(Dasyurus viverrinus)*. (A by J. Hudson; B and D by A. Robinson; C by D. Bos)

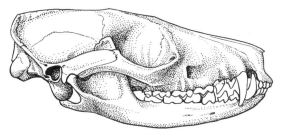

A. Australian Native "cat" *(Dasyurus viverrinus*, Dasyuridae)

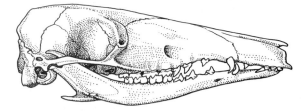

B. Bandicoot *(Perameles*, Peramelidae)

FIGURE 6-13   Skulls of dasyurid and peramelid marsupials: (A) Australian native cat or quoll, length of skull 72 millimeters, (B) Bandicoot, length of skull 81 millimeters.

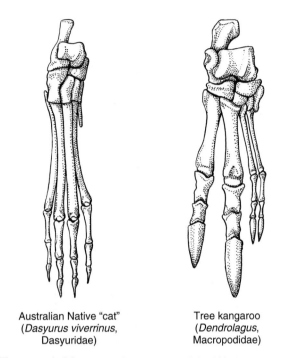

Australian Native "cat"
(*Dasyurus viverrinus*,
Dasyuridae)

Tree kangaroo
(*Dendrolagus*,
Macropodidae)

FIGURE 6-14 Feet of two marsupials: (A) a terrestrial species, (B) an arboreal species. *(After Marshall, 1972)*

A

B

FIGURE 6-15 (A) The thylacine *(Thylacinus cynocephalus*, Thylacinidae*),* which is now extinct. (B) The Tasmanian devil *(Sarcophilus laniarius,* Dasyuridae*).* *(A by E. H. Colbert; B by PhotoDisc, Inc.)*

mal that is the sole representative of the family Myrmecobiidae. The teeth are small and widely spaced in the long tooth row, and the long, protrusible tongue is used in capturing termites. This animal was formerly widespread in eucalyptus forests, in which fallen branches and logs provided lush populations of termites, but the commercial clearing of these forests has severely restricted the range of the numbat. Numbats are currently restricted to several isolated populations in southwestern Australia. In the late 1970s, numbat populations were thought to be fewer than 1000 individuals. Although numbats have several natural predators, the most significant predator is the introduced red fox. Recent efforts to control red foxes by the Australian government have resulted in an increase in numbat populations.

FAMILY THYLACINIDAE. Now extinct, the thylacine or Tasmanian "wolf" was doglike in size and general build (Fig. 6-15); it had long limbs and a **digitigrade** foot posture. Because of prominent stripes along the back and flanks, it was sometimes referred to as the Tasmanian "tiger." The thylacine was a pounce-pursuit predator that killed one- to

five-kilogram prey and filled a niche similar to that of some canids (Jones and Stoddart, 1998). The last thylacine (*Thylacinus cynocephalus*) died in captivity in 1936. Its demise in Australia probably resulted from competition with the dingo (*Canis lupus dingo*), which was introduced to mainland Australia some 3500 years ago (Strahan, 1995). Populations survived in Tasmania into the early 20th century because Tasmania became isolated from the Australian mainland between 8000 and 10,000 years ago, before introduction of the dingo. On Tasmania, these predators survived until the arrival of Europeans, whose flocks of sheep were thought to be easy prey for the thylacine. Bounties were placed on thylacines, and their populations went into steep decline, but it is argued that an epidemic of a distemper-like disease may have enhanced the eradication program.

ORDER NOTORYCTEMORPHIA

FAMILY NOTORYCTIDAE. This remarkable family is represented by a single species of marsupial "mole" that inhabits sandy soils in arid parts of northwestern and south-central Australia. Many of the diagnostic characters of these mouse-size animals are adaptations for fossorial life. The eyes are vestigial, covered by skin, lensless, and, as indicated by the specific name *Notoryctes typhlops* (*typhlops* means "blind" in Greek), nonfunctional. The ears lack pinnae. The nose bears a broad, cornified shield, and the nostrils are narrow slits. The dental formula is usually 4-3/3, 1/1, 2/3, 4/4 = 44 to 42, but the incisors vary in number. The incisors, canines, and all but the last upper premolar are unicuspid; the paracone and metacone of the upper molars form a prominent single cusp, and the lower molars lack a talonid. As an adaptation serving to brace the neck when the animal forces its way through the soil, the five posterior cervical vertebrae are fused. The forelimbs are robust, and the claws of digits three and four are remarkably enlarged and function together as a spade; the other digits are reduced. The central three digits of the hind feet have enlarged claws, the small first digit has a nail, and the fifth digit is vestigial. The rearward-opening marsupium is partially divided into two compartments, each with a single nipple. The fur is long and fine textured, varying in color from silvery white to yellowish red.

Notoryctids use their powerful forelimbs and armored rostrum to force their way through soft, sandy soil. When the animal forages near the surface, the soil is pushed behind it and no permanent burrow is formed. The food is predominantly invertebrate larvae.

ORDER PERAMELEMORPHIA

Members of the order Peramelemorphia include the bandicoots and bilbies. The families Peramelidae and Peroryctidae are characterized in general by an insectivoran dentition (Fig. 6-10C) and a trend toward specialization of the hindlimb for running or hopping. The marsupium is present and opens to the rear, and bandicoots alone among metatherians have a chorioallantoic placenta (p. 338). The order includes eight Recent genera, represented by 21 Recent species, mainly from Australia, Tasmania, and New Guinea. Some species of bandicoots have been extirpated or have become uncommon over parts of their former range, apparently owing to livestock grazing, to brush fires, and to the introduction of various eutherian mammals.

FAMILY PERAMELIDAE. This family includes all the nonspiny bandicoots (including the recently extinct pig-footed bandicoot, *Chaeropus ecaudatus*) and two species of bilby. The smaller bandicoots are the size of a rat, with the largest species weighing roughly 2 kilograms (Fig. 6-16). The dental formula is 4-5/3, 1/1, 3/3, 4/4 = 46 or 48. The incisors are small, and the molars are tritubercular or quadritubercular. The rostrum is slender (Fig. 6-13B) and the skull dorsoventrally flattened in cross section. The ears of some species resemble those of rabbits. Although often long, the tail is not prehensile. The fourth digit of the hind foot is always the largest, and the remaining digits are variously reduced (Fig. 6-17). The hind foot posture is usually digitigrade, and the hindlimbs are elongate. The opposable hallux, probably inherited by peramelids from an arboreal ancestral stock, is rudimentary or may be lost. The second and third digits of the hind foot are joined (syndactylous) as far as the distal phalanges by an interdigital membrane, and the muscles of these digits are partially fused, allowing them to act only in unison (F. Jones, 1924). An extreme degree of cursorial specialization occurred in the pig-footed bandicoot (*C. ecaudatus*): the forelimb was functionally **didactyl;** the second and third digits of the forelimb were large and had hooflike claws. The hind foot was functionally **monodactyl;** only the fourth toe was used during running.

The structure and function of the specialized peramelid hind foot are unique and have been described in detail by Marshall (1972) and Szalay (1994). In mammals, extreme reduction in the number of digits is usually associated with good running ability. Most highly cursorial ungulates have retained only the third digit (as in the horse) or digits three and four (as in some antelopes). A similar

A

B

FIGURE 6-16 Two peramelid marsupials: (A) New Guinea bandicoot *(Peroryctes raffrayanus),* (B) bilby *(Macrotis lagotis). (A by S. Grierson; B by A. Robinson)*

point of insertion for the extensors of the foot, muscles of great importance in locomotion.

Horses, pronghorns, and peramelemorphs provide beautiful examples of different structural means of solving a similar functional problem, in this case, refining running ability. In the horse only the third digit is retained, which is supported largely by the ectocuneiform, the navicular, and the astragalus bones (see Fig. 16-3). In the pronghorn, only two digits are retained and the cannon bone (the fused third and fourth metatarsals) is supported largely by the fused cuboid and navicular bones and the fused mesocuneiform and ectocuneiform bones; the calcaneum is no longer a weight-bearing element. In the most cursorial peramelid (*Chaeropus*), the fourth digit is greatly enlarged and is supported, as outlined above, by the cuboid, navicular, ectocuneiform, and astragalus bones (Fig. 6-17). Marshall (1972) points out, however, that the structure of the hindlimb of peramelids is not entirely modified for running, perhaps because of the burrowing tendencies of these animals. The fibula is large, and movement at the ankle joint is not restricted to a single plane, as it is in most cursorial mammals. (Cursorial adaptations are discussed in more detail on p. 261.)

Bandicoots and bilbies are largely insectivorous but also eat small vertebrates, a variety of invertebrates, and some vegetable material. Most inhabit relatively dry, open-country habitats. Some species take refuge in nests that they build of plant debris; both species of *Macrotis* dig burrows in which they hide during the day. Fossil peramelids are first known in the Miocene in Australia.

trend occurs in peramelemorphs. Probably partly due to an early development of syndactyly involving the second and third digits and the use of these digits for grooming, the general trend in the cursorial peramelids is toward the reduction of all digits but the fourth, with a great enlargement of this digit (Fig. 6-17). These specializations are accompanied by an alteration in the structure and function of the tarsal bones. The ectocuneiform bone makes broad contact with the proximal end of the fourth metatarsal and partially supports this digit, a character unique to peramelemorphs. The mesocuneiform is lost, and the weight of the body is borne mainly by the cuboid, ectocuneiform, navicular, and astragalus bones. The calcaneum does not serve a major weight-bearing function but, of course, serves as a

FAMILY PERORYCTIDAE. This monophyletic group (Groves and Flannery, 1990) probably originated and radiated in New Guinea, although they have no fossil record. Of the four genera and 11 species of peroryctid bandicoots, only 2 species are found outside of New Guinea. The rufous spiny bandicoot (*Echymipera rufescens*) is also found in the tip of Cape York in northeastern Australia, and *Rhynchomeles prattorum* is **endemic** to Seram Island, located between New Guinea and Sulawesi (Nowak and Paradiso, 1991). These bandicoots range in size from less than 100 grams (*Microperoryctes murina*) to the giant bandicoot (*Peroryctes broadbenti*), which can exceed 5 kilograms. All species are probably insectivorous and/or omnivorous. They

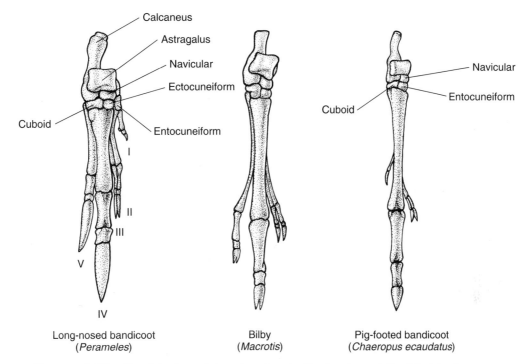

FIGURE 6-17   The right feet of three bandicoots. The least specialized foot is shown on the left and the most specialized is shown on the right. The digits are numbered in *Perameles. (After L. Marshall, 1972)*

occur in many habitats from subalpine grasslands to lowland rain forests; at middle elevations, up to six species may occur **sympatrically** in New Guinea (Flannery, 1995). Like peramelids, they possess a chorioallantoic placenta.

ORDER DIPROTODONTIA

FAMILY VOMBATIDAE. This family is represented by two living genera and three species. Known as wombats, these animals are completely herbivorous and show remarkable structural convergence toward rodents. Because of the efforts of humans, wombats have become scarce or absent over much of their former range and now are restricted to parts of eastern and southern Australia, Tasmania, and the islands between Australia and Tasmania.

Wombats are stocky animals with small eyes and rodent-like faces (Fig. 6-18); their body weight can exceed 36 kilograms (Strahan, 1995). The skull and dentition bear a striking resemblance to those of some rodents (Figs. 6-9D and 6-10D). The skull is flattened, the rostrum is relatively short, and the heavily built zygomatic arches flare strongly to the sides. The area of origin of the anterior part of the masseter muscle is marked by a conspicuous depression in the maxillary and jugal bones that is similar to the comparable depression in the maxillary and premaxillary bones of the beaver (*Castor*, Fig. 18-11B). The dental formula is 1/1, 0/0, 1/1, 4/4 = 24; all teeth are rootless and ever-growing. Only the anterior surfaces of the incisors bear enamel, and the incisors and the first premolars are separated by a wide diastema. The molars are **bilophodont** (Fig. 6-10D). As in rodents, the coronoid process of the dentary bone is reduced and the masseter muscle, rather than the temporalis, is the major muscle of mastication.

The limbs are short and powerful, and the foot posture is plantigrade. The forefeet have five toes; all digits have broad, long claws. The hallux is small and clawless, but the other digits have claws. Digits two and three of the hind feet are syndactylous. The tail is vestigial. The marsupium opens posteriorly and contains one pair of mammae.

This family is first known from the middle Miocene epoch, and both Recent genera have fossil species. The Pleistocene trend toward large size apparent in many other mammalian groups also occurred in the Vombatidae, as evidenced by the huge Pleistocene "wombat" *Phascolonus*.

Wombats are powerful burrowers. Their burrows are extensive networks of tunnels that are wide enough to admit a small person. Burrow entrances are often clustered to form **warrens** connected to other warrens by well-worn trails and marked by urine and feces. Although one warren system may house up to ten wombats, they are rarely found together in the same burrow. Young wombats learn to burrow in their mother's burrow system by digging small subsystems, but they abandon the maternal burrows about four months after leaving their mother's pouches. Wombats dig burrows in the open or beneath rock piles, as do marmots (*Marmota*), their North American rodent counterparts. Males have a well-developed dominance hierarchy, and females appear to be subordinate to all males.

The southern hairy-nosed wombat (*Lasiorhinus latifrons*) lives in semiarid areas and is able to go for long periods without drinking water. The species copes with the inhospitable environment and low quality diet, consisting primarily of herbs and grasses, by remaining in their burrows until the cooler evening hours. To further conserve energy, these wombats have a resting metabolic rate two thirds that of other metatherians. As a result, their high-fiber food may take over a week to pass through the gut (Strahan, 1995).

The common wombat (*Vombatus ursinus*) and the southern hairy-nosed wombat (*L. latifrons*) remain common in their limited ranges. The northern hairy-nosed wombat (*Lasiorhinus krefftii*) is endangered. Fewer than 70 individuals are restricted to an area less than 300 hectares (750 acres).

FAMILY PHASCOLARCTIDAE. The familiar koala (*Phascolarctos cinereus;* Fig. 6-19) is the sole member of this family. This highly specialized herbivore is restricted to some wooded parts of eastern Australia. The tufted ears, naked nose, and chunky, tail-less form make the koala one of the most distinctive of Australian marsupials. It is a fairly large marsupial; the adult ranges from 8 to 12 kilograms in weight. The skull is broad and sturdily built, and the dentary bones are deep and robust (Fig. 6-20A). The dental formula is 3/1, 1/0, 1/1, 4/4 = 30. The roughly quadrate molars have **crescentic** ridges (Fig. 6-20B), and there is a diastema in both the upper and lower tooth rows between the cheek teeth and the anterior teeth. Tree branches are grasped between the first two and the last three fingers of the hand and between the clawless first digit and the remaining digits of the foot; the long, curved claws help maintain purchase on smooth branches.

Koalas are fairly sedentary and feed on only a few species of smooth-barked eucalyptus trees. Dependence on eucalyptus trees, with their high concentration of toxic terpenes and phenolics, low protein content, and high levels of fiber, imparts considerable foraging costs. Koalas remain sedentary during most of the day, thereby reducing their energetic needs to a minimum. In addition, their powerful jaws grind leaves into a paste, extracting most of the cellular contents. Undigested fiber is retained in the caecum, where microbial fermentation extracts additional energy from the food. The liver aids in detoxification of plant secondary compounds (Lee and Martin, 1988).

Maturation of a koala takes considerable time. A single young is born and is carried in the pouch for six months, after which it rides on its mother's

FIGURE 6-18   A wombat (*Vombatus ursinus,* Vombatidae). *(Lone Pine Koala Sanctuary)*

FIGURE 6-19 The koala (*Phascolarctos cinereus*, Phascolarctidae). *(PhotoDisc, Inc.)*

back for a few more months. The young koala is dependent on its mother for one year, and sexual maturity is not reached until three or four years of age. Koalas are solitary and have individual home ranges that may overlap if population densities are high. Although they are not territorial, male koalas exhibit a dominance hierarchy and subordinate males are attacked when encountered. During the summer breeding season, males advertise their presence to females by making low, bellowing calls

that can be heard over half a kilometer away. Koalas lived in southwestern Australia during the late Pleistocene, but no longer occur there even though suitable habitat appears to be present.

FAMILY TARSIPEDIDAE. This family contains but one species, the highly specialized, slender-nosed honey possum, or noolbenger (*Tarsipes rostratus*). This remarkable animal's many specializations obscure its relationships to other marsupials, and its

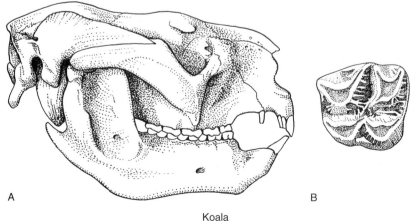

FIGURE 6-20 (A) Skull of the koala, length of skull 132 millimeters. (B) Occlusal view of the second molar, upper right tooth row, showing the crescentic areas of dentine, exposed by wear, and the complex pattern of furrows.

A   B

Koala
(*Phascolarctos cinereus*, Phascolarctidae)

taxonomic position has long been uncertain although it is certainly a diprotodont.

Tarsipes is small, only about 7 to 12 grams in weight, and has a long, prehensile tail. The pelage is marked by three longitudinal stripes on the back. The rostrum is long and fairly slim, and the dentary bones are extremely slender and delicate. The cheek teeth are small and degenerate, and only the upper canines and two medial lower incisors are well developed. The snout is long and slender, and the long tongue tapers to a fine point that has bristles at its tip (these specializations are similar to those of some nectar-feeding bats of the family Phyllostomidae). The muscles of the jaws are reduced and weak; those associated with the tongue are important in these obligate nectar feeders (Rosenberg and Richardson, 1995). All digits but the syndactylous second and third digits of the hind feet have expanded terminal pads resembling to some extent those of the primate *Tarsius*.

Honey possums occur in forested and shrubby areas of southwestern Australia. Like hummingbirds and nectar-feeding bats, honey possums feed on nectar, pollen, and, to some extent, small insects that live in flowers. The long, protrusible tongue is used to probe into flowers. *Tarsipes* can climb delicately over even the insecure footing of clusters of flowers at the ends of branches and often clings upside down to flowers while feeding. Although the animal is still common in some areas today, the expansion of agriculture in southwestern Australia is restricting the honey possum's range.

FAMILY ACROBATIDAE. This small family is represented by two living species, each in its own genus; the feathertail glider (*Acrobates pygmaeus*) from Australia and the feathertail possum (*Distoechurus pennatus*) from New Guinea. Both species are characterized by rows of long, stiff hairs along either side of the tail, from which they get the name "feathertail."

The feathertail glider is small, weighing approximately 10 to 15 grams. It has a membrane between the elbows and the knees, making it the smallest gliding mammal. The flattened, featherlike tail is used for steering; the prehensile tail is also used for gripping small branches. The eyes are large and anteriorly placed for improved binocular vision. Traction between the digits and the trunks

and branches of trees is increased by expanded pads at the tips of the fingers and toes; the surfaces of the pads have ridges that further increase the clinging ability of these animals. The molars are typical of insectivores, but the tongue is adapted for extracting nectar from flowers. Feathertail gliders are primarily nocturnal, and groups of up to 20 animals have been found nesting together (or torpid). Females exhibit **embryonic diapause,** the young remain in the pouch for over two months, and weaning may not be completed until 100 days. Because of the long developmental period, maternal investment is relatively high, and there are reports of females other than the mother helping with the care of young gliders.

Feathertail possums (*D. pennatus*) are united with the feathertail glider (*A. pygmaeus*) on the basis of three shared characters: the feather-like tail, an unusual middle ear anatomy, and the loss of the last molar (Flannery, 1995). Unlike the gliders, *D. pennatus* does not have a gliding membrane. The feathertail possum weighs between 40 and 50 grams, and females are larger than males.

FAMILY BURRAMYIDAE. The type genus of this family was known for many years only from Pleistocene fossil material; finally, in 1966, at a ski lodge on Mt. Hotham in Victoria, a representative of the genus was found alive. More recently, it has been found at other localities. This family contains two genera and five species of small, mouselike marsupials, called pygmy possums. They lack the feather-like tail of acrobatid possums.

These diminutive marsupials are from 120 to 295 millimeters in total length, are delicately built, and have large eyes and mouselike ears (Fig. 6-21). They weigh less than 40 grams and include the world's smallest possum, the 6 to 8 gram little pygmy possum (*Cercartetus lepidus*). Pygmy possums are highly arboreal and have strongly prehensile, nearly naked tails. Members of this family are restricted to wooded areas. They are apparently insectivorous-omnivorous, but the feeding habits of some members of the group are not known.

The mountain pygmy possum (*Burramys parvus*) inhabits cold alpine and subalpine regions and is known to hibernate for long periods. At high elevations (>1500 meters), where snow cover may last six months, mountain pygmy possums double their body weight in fat deposits before the winter

FIGURE 6-21   A pygmy possum (*Cercartetus concinnus*, Burramyidae). *(A. Robinson)*

snows. Hibernation lasts between two and seven months and is characterized by a body temperature around 2°C for periods of approximately three weeks, followed by short bouts at normal body temperature (Geiser and Broome, 1993). Because of its restricted mountaintop habitats, this unique mammal lives in several highly disjunct populations totalling less than 2600 individuals.

FAMILY PSEUDOCHEIRIDAE. Until recently, the ringtail possums and the greater glider were considered part of the Petauridae (Fig. 6-22), a family to which they are closely related. The pseudocheirids include five genera and 14 species. Ringtail possums have a strongly prehensile tail, but the greater glider's tail (*Petauroides volans*) is weakly prehensile and is used mainly as a steering rudder during gliding. All pseudocheirids have molar teeth that bear a series of sharp ridges used in grinding leaves. Microbial fermentation occurs in the large caecum and assists in the breakdown of plant material. The common ringtail (*Pseudocheirus peregrinus*) also practices **coprophagy** (reingestion of feces) to increase its ability to extract nutrients from its food. The largest gliding metatherian, the greater glider (*Petauroides volans*), can glide more than 100 meters and make nearly 90 degree turns. The unusual green ringtail possum (*Pseudochirops archeri*) has fur that appears green but is actually a combination of black, yellow, and white hairs. It has a highly specialized diet; its food is entirely leaves, chiefly those of fig trees (Proctor-Grey, 1984).

FAMILY PETAURIDAE. The striped possum, Leadbeater's possum, and five species of lesser glider, so named because of the membrane between their wrists and ankles, are currently recognized in the family Petauridae. Most members of this family are fairly small; weights range from about 100 grams to 700 grams. The skull is broad, and the molars have low, smooth cusps. The tail is long, bushy, and prehensile. All species have a dark dorsal stripe running from head to tail (Fig. 6-23). The gliders (*Petaurus*) have furred membranes that extend between the limbs and function as lifting surfaces for gliding. In these gliders, the claws are sharp and recurved, like those of a cat, and increase the ability of the animal to cling to the smooth trunks and large branches of trees. The petaurids are nocturnal, arboreal creatures that inhabit wooded areas and feed primarily on sap and nectar.

The gliders are strikingly similar to flying squirrels (*Glaucomys*) in gliding style and ability; some can glide more than 100 meters. Sugar gliders (*Petaurus breviceps*) live in family groups, and scent marking plays an important role in the social organization of the group. Each individual has a particular odor recognized by the others. The cohesion of the group is also aided by mutual scent marking, for all members of the group become permeated with the scent of the group's dominant males (Schultze-Westrum, 1965). Leadbeater's possum (*Gymnobelideus leadbeateri*) retains only a vestigial gliding membrane.

The four species of striped possums (*Dactylopsila*) lack a gliding membrane, and some may be

FIGURE 6-22 Ringtail possum (*Pseudocheirus forbesi*; Pseudocheiridae).*(S. Grierson)*

FIGURE 6-23 The gliding possum (*Petaurus breviceps*; Petauridae). *(S. Grierson)*

primarily terrestrial. They have a suite of unique characters associated with their specialized insectivorous diet. In *Dactylopsila,* the fourth digit of the hand is elongate and slender and its claw is recurved, similar to those of the primate *Daubentonia* (Fig. 6-24). In addition, the incisors are robust and function roughly as do those of rodents. Striped possums tear away tree bark with their incisors and extract insects from crevices and holes in the wood with the specialized fourth finger and the tongue. *Dactylopsila* wears a conspicuous, striped color pattern that is interesting because it is associated, as in skunks, with a powerful, musky scent.

FAMILY PHALANGERIDAE. There are six genera and 18 species of possums and cuscuses in the family Phalangeridae. Phalangerids are omnivorous and are known to eat a wide variety of plant material, as well as insects, young birds, and bird eggs. The brush-tail possum (*Trichosurus vulpecula*) is one of the most familiar Australian mammals, for it frequently maintains resident populations in suburban areas, where it often seeks shelter in roofs of houses and feeds on cultivated plants. Members of this family mostly inhabit wooded areas, but the adaptable brush-tail possum also occupies treeless areas, where it takes refuge in rocks or the burrows of other mammals. This animal is locally destructive to plantations of introduced pines. The brush-tail possum is solitary and has a sternal scent gland, considerably larger in males than females, which produces a musky smell that is used in scent marking objects within the animal's territory.

Phalangerids are of moderate size, ranging in weight from 1 to 6 kilograms. The skull is broad and has deep zygomatic arches (Fig. 6-9B). The molars are bilobed with rounded cusps (Fig. 6-10E, F). As adaptations to arboreal life, the hands and feet are large and have a powerful grasp, and the tail is prehensile. The cuscuses have short ears and woolly fur and resemble teddy bears (Fig. 6-25).

FAMILY POTOROIDAE. The family Potoroidae comprises several genera that depart from the familiar structural pattern of kangaroos and from the grazing or browsing habit. These animals are commonly known as potoroos, bettongs, and the musky rat-kangaroo. They seemingly represent a

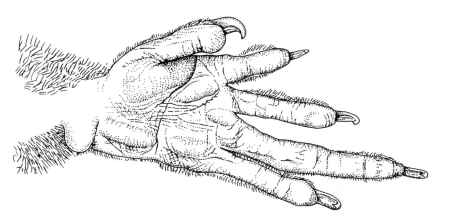

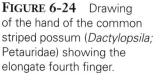

FIGURE 6-24 Drawing of the hand of the common striped possum (*Dactylopsila;* Petauridae) showing the elongate fourth finger.

conservative branch on the diprotodont tree. They retain a slightly prehensile tail and well developed upper canines but did not evolve the elaborate stomach of the macropodids. The musky rat-kangaroo (*Hypsiprymnodon moschatus*) is a muskrat-size inhabitant of rain forests and riparian situations. It has a tail of modest length and retains all the digits of the hind foot. The hindlimbs are not greatly elongate, and the animal uses quadrupedal rather than saltatorial locomotion. Animal material forms a large share of the omnivorous diet of this seemingly primitive diprotodont.

Potoroos and bettongs are small kangaroo-like animals weighing between 0.8 and 3.5 kilograms. The tail is sufficiently prehensile to carry nesting material, and the forelimbs are used in some species for digging the roots, tubers, and fungi that comprise their diets. Male rufous bettongs (*Aepyprymnus rufescens*) probably form loosely defended associations with several females whose ranges overlap their own. Many species dig burrows, but the recently extinct desert rat-kangaroo (*Caloprymnus campestris*) excavated a shallow depression lined with grasses under bushes. Of the nine Recent species of potoroids, two are probably extinct and two others are endangered. Competition with introduced rabbits and predation by dingos and introduced foxes and cats are likely responsible for these losses.

FAMILY MACROPODIDAE. Members of the familiar metatherian group Macropodidae, which includes 11 genera of kangaroos, euros, and wallabies (Fig. 6-26), are the ecological equivalents of such ungulates as antelope. The present distribu-

tion of the approximately 54 Recent species of macropodids includes New Guinea, the Bismarck Archipelago, the D'Entrecasteaux Group, Australia, and, by introduction, some islands near New Guinea and New Zealand. The family Macropodidae appears first in the late Oligocene or early Miocene of Australia. Wallabies and kangaroos are known from the late Tertiary, and unusually large macropodids occurred in the Pleistocene.

Living macropodids vary greatly in size and structure. The rock wallaby known as the monjon (*Petrogale burbidgei*) weighs only 900 to 1400 grams, whereas the great gray kangaroo (*Macropus giganteus*), the largest living marsupial, reaches 2 meters in height and approximately 90 kilograms in weight. The marsupium is usually large and opens anteriorly. The macropodid skull is moderately long and slender, and the rostrum is usually fairly long (Fig. 6-9C). The dental formula is 3/1, 1-0/0, 2/2, 4/4 = 34 or 32. The upper incisors have sharp crowns with their long axes oriented more or less front to back. The tips of the procumbent lower incisors are held against a leathery pad just behind the upper incisors when the animals gather vegetation. This specialized arrangement serves a cropping function similar to that of the lower incisors and the premaxillary pad in the artiodactylan ungulates that lack upper incisors. There is a broad diastema between the macropodid incisors and the premolars. The molars are quadritubercular and bilophodont (Fig. 6-10G). In many macropodids, the last molar does not erupt until well after the animal becomes adult. A unique situation occurs in the little rock wallaby (*Petrogale concinna*), in which nine molars

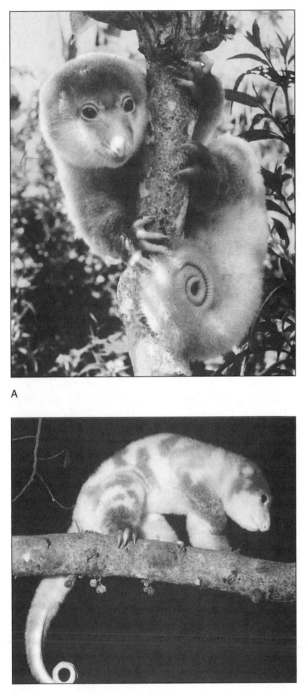

A

B

FIGURE 6-25 Two views of a New Guinea cuscus (*Spilocuscus maculatus,* Phalangeridae). Note the prehensile tail with the traction-producing ridges on the bare, distal part of the ventral surface. *(S. Grierson)*

may erupt in succession. Usually four or five molars are functional at one time, and replacement is from the rear as the molars are successively lost from the front.

Macropodids are highly specialized for jumping. The forelimbs are five-toed and usually small; they are used for slow movement on all fours or for food handling. The hindlimbs are elongate, especially the fourth metatarsal. The hallux is missing in all species. Digits two and three are small and syndactylous and used for grooming; the fourth is the largest digit; and the fifth is often robust (Fig. 6-27). The unusual pattern of digital reduction and the dominance of the fourth digit in the most highly cursorial Australian metatherians are perhaps due to the arboreal ancestry of these animals. In their ancestors, the foot was five-toed and the hallux opposable; the fourth was the longest remaining digit, and the foot was adapted to grasping branches. With specialization of the foot for running or hopping, the hallux was lost and the longest toe, the fourth, became the most important digit. In most macropodids, the foot is functionally two-toed during rapid locomotion, which is characteristically **bipedal.** In the macropodid tarsus, there is no contact between the ectocuneiform and the fourth metatarsal (in contrast to the arrangement in the Peramelidae; Fig. 6-17). Because the hindlimb posture of macropodids is basically plantigrade, the calcaneum is an important weight-bearing element of the tarsus (which it is not in the digitigrade Peramelidae). The macropodid tail is usually long and robust and functions in the more specialized species as a balancing organ and as the posterior "foot" of the tripod formed by the plantigrade hind feet and tail, on which the animal can sit when not in motion.

Macropodids are the only large-bodied mammals to utilize bipedal hopping for locomotion. Does hopping require more energy than quadrupedal running? When a typical quadrupedal mammal runs, the energetic cost increases linearly with increasing speed, regardless of whether the animal is metatherian or eutherian. Studies by Dawson and Taylor (1973), using kangaroos trained to hop on a treadmill, demonstrated that costs of bipedal hopping are initially high, but that once hopping has begun, it can be maintained at little or no additional cost over a wide range of speeds (p. 392). At moderate to high speeds, a

FIGURE 6-26 Three kinds of macropodid marsupials: (A) great gray kangaroo (*Macropus giganteus*), (B) pademelon (*Thylogale billardierii*), (C) red-necked wallaby (*Macropus rufogriseus*). *(D. Harrison)*

A

B

C

hopping kangaroo uses no more energy than a quadrupedal mammal. Part of the explanation for this comes from the energy stored in the huge elastic tendons attached to the hind foot.

The running ability of the larger kangaroos (*Macropus*) is impressive. Speeds on level terrain of roughly 65 to 70 kilometers per hour are attained, and leaps covering distances of 13.5 meters and height of 3.3 meters have been reported (Troughton, 1947). The solitary hare wallabies (*Lagorchestes*), which are roughly the size of a large

rabbit, also are renowned for their great speed. These animals have a jackrabbit style of escape. They hide beneath bushes or clumps of grass, burst out suddenly when frightened, and run away at high speed. The highly developed jumping ability of macropodids allows these animals to move easily for long distances between scattered sources of water or forage and to escape enemies by erratic leaps. These abilities, rather than the capacity for great speed, are perhaps of primary adaptive importance. Saltation may have been developed by

FIGURE 6-27 The right foot bones of some macropodid marsupials Least specialized on the left, and most specialized on the right. The digits are numbered in *Thyogale*.

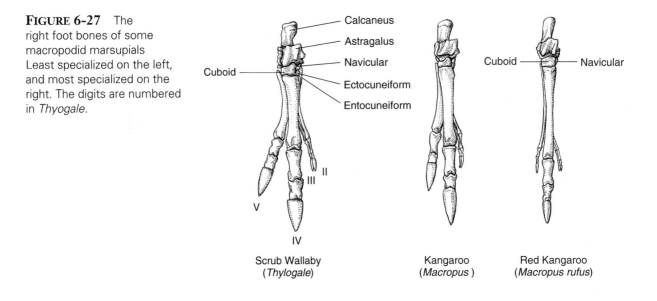

Scrub Wallaby
(*Thylogale*)

Kangaroo
(*Macropus*)

Red Kangaroo
(*Macropus rufus*)

small forms ancestral to kangaroos as a means of erratic escape in open areas. This **ricochetal** style of locomotion is known in a number of desert-dwelling rodents. The locomotion of rock wallabies (*Petrogale*) is adapted to the steep and rocky country they inhabit. According to F. W. Jones (1924), their movements are spectacular: "There seems to be no leap it will not take, no chink between boulders into which it will not hurl itself."

The tree kangaroos (*Dendrolagus*, Macropodinae) spend considerable time on the ground but frequently use their arboreal ability to escape from danger. This mode of life is reflected by the large and robust forelimbs with strong recurved claws, by the hindlimbs, which are not strongly elongate, and by the short, broad hind foot (Fig. 6-14). Saltation, typical of terrestrial kangaroos, has not been completely abandoned by tree kangaroos; not only are these animals agile climbers, but they also leap from one tree to another and from tree to ground. Their food is large fruit and leaves.

Macropodids resemble eutherian ungulates in their highly specialized limbs and cursoriality. These groups are also alike in being herbivorous and in having skulls and dentitions specialized for this mode of feeding. Even some specializations of the digestive system are similar between these two groups. Like eutherian ruminants, kangaroos have intestinal bacteria that digest the tough cell walls of plants. These (and probably other) macropodids can thus utilize the contents of the plant cells,

the byproducts of the bacterial digestion of cellulose, and the bacteria themselves (Hume, 1982; Dellow and Hume, 1982; Freudenberger et al., 1989). The ruminant ungulates also depend on bacteria to increase the efficiency of their utilization of vegetation (p. 264). Although both ungulates and macropodids are foregut fermenters (stomach chambers rather than a hindgut caecum contain the microbes responsible for fermentation), the kangaroo stomach is not truly multichambered. Instead, kangaroos have a long tubular stomach with three main sections (Fig. 6-28). The esophagus dumps food into an S-shaped blind sac called the sacciform forestomach, where much of the fermentation occurs. Over 60 percent of the organic matter is digested here, along with most of the sugars. Fiber is broken down in the next section of the stomach, the tubiform forestomach, which forms a coil. Here fiber is processed along the length of the coil and approximately one third of it is digested. The remaining fiber is digested in the colon. After passing through the tubiform forestomach, the undigested material passes into the third section, where acid and enzymes are secreted and final processing takes place prior to the food's entering the small intestine. In addition to the breakdown of fiber, the symbiotic microbes also manufacture vitamins, detoxify some plant compounds, and recycle nitrogenous compounds into proteins (Dawson, 1995).

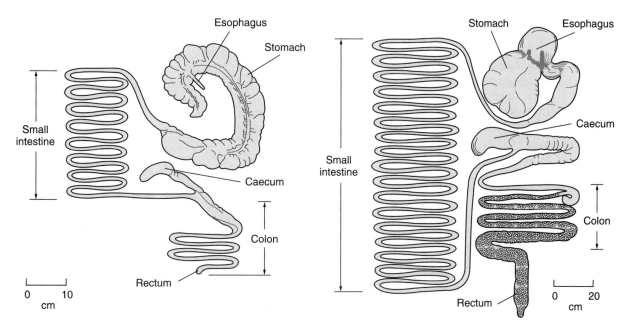

FIGURE 6-28 Comparison of the digestive systems of (A) a kangaroo (*Macropus giganteus*) and (B) a sheep (*Ovis aries*). *(After Dawson, 1995)*

WEB SITES

Australian Koala Foundation; information and links about koalas

> http://www.akfkoala.gil.com.au/

Basic information about koalas

> http://www.koala.net/

Information about tiger quolls

> http://www.npws.nsw.gov.au/wildlife/tigquoll.htm

Information about the extinct thylacine or Tasmanian "tiger"

> http://www.anca.gov.au/plants/threaten/information/species/animals/mammals/
> thylacine/thylacin.htm

Brief descriptions of many rare Australian mammals

> http://www.perthzoo.wa.gov.au/australia.html

Basic information on Tasmanian devils and other Tasmanian mammals

> http://www.parks.tas.gov.au/wildlife/mammals/devil.html

The Australian Mammal Society home page

> http://www.tesag.jcu.edu.au:80/mammal/

Many links to Australian mammal sites

> http://yoyo.cc.monash.edu.au/~tzvi/Echilinks.html

7 INTRODUCTION TO EUTHERIAN MAMMALS

The story of eutherian mammals begins with middle Cretaceous fossils from Khovboor, Mongolia (Kielan-Jaworowska and Dashzeveg, 1989). The record is sketchy, but improving, especially through finds in Asia. Most of the fossils are fragments of jaws with partial tooth rows or isolated teeth, and they represent only parts of Asia, western North America, and a single locality in France. The entire first half of the Cretaceous has yielded very little eutherian material, but a better record is available from the late Cretaceous. As the record of Cretaceous therians improves, it becomes more difficult to distinguish between eutherians and metatherians, based on tooth and jaw features alone (Cifelli, 1997; Luckett, 1993). The finding of a complete skeleton of an early therian could do much toward clarifying the early evolution of this group.

Despite the paucity of fossils, some conclusions regarding Cretaceous eutherians can be made. They were small, from the size of a shrew to the size of a marmot, and their structural diversity in the late Cretaceous reflects radiations during the 79 million years of Cretaceous time. Their skulls lacked an auditory bulla and were typically long, with a narrow braincase and long, narrow snout. The incisors varied in number from 5/4 to 3/3. Early Cretaceous eutherians had five premolars (Sigogneau-Russell et al., 1992), whereas late Cretaceous eutherians had four. There were three molars that were typically tribosphenic (bearing a protocone on the uppers that fit into a talonid basin surrounded by three cusps on the lowers) with high, sharp cusps (Fig. 7-1). The hands and feet of several genera are known, and these lack opposable first digits, indicating that Cretaceous eutherians were not basically arboreal.

Late Cretaceous eutherians probably played a variety of ecological roles. The diverse dentitions were seemingly adapted to insectivory, carnivory, and herbivory. It would be reasonable to speculate that Cretaceous eutherians functioned variously as climbers, jumpers, diggers, and those with generalized styles of locomotion. Interpretation of fossil postcranial material is difficult, for we know that some living mammals, such as squirrels (*Sciurus* sp., for example) practice several locomotor and dietary lifestyles.

The fossil record documents a late Cretaceous establishment of the evolutionary lines leading to the ungulates (Nessov and Kielan-Jaworowska, 1991; Archibald, 1996) and a latest Cretaceous establishment of the orders Insectivora, Carnivora, and Primates. This fossil record also suggests a very early Cenozoic (at least) differentiation of the remaining extant orders of eutherians. The tracing back of many eutherian evolutionary lines through the early Cenozoic and late Mesozoic awaits a more complete fossil record.

Lillegraven et al. (1987) provided an excellent review and intriguing discussion of the evolutionary development of the eco-physiological and reproductive strategies of eutherians and contrasted the evolutionary history of their reproduction with that of metatherians. Recent phylogenetic work on the major groups of eutherians has been led by Novacek (1989, 1992a, b; others listed in the references at the end of the book), who produced a phylogenetic tree for them and other major mammalian orders of the Cenozoic (Fig. 7-2). Molecular investigations into the phylogeny of eutherians include the review by Honeycutt and Adkins (1993), a controversial molecular time scale (Kumar and Hedges, 1998), and the promise of major taxonomic revisions to come (de Jong, 1998).

The accounts of eutherian mammals in the following chapters include brief comments about when the orders and families appeared and, in some cases, sketches of the evolutionary histories of taxa. The student should regard such comments as summaries of our current knowledge but should bear in mind that the fossil record is sadly incomplete for some groups. Mammals belonging to evolutionary lines that have been restricted to tropical areas, small and delicate species, and arboreal types had relatively little chance of leaving a fossil record. The first appearance of the Chiroptera (bats) in the early Eocene, for example, tells us only that by this time all the basic chiropteran adaptations had been perfected and that bats were part of the North American fauna. The history of these small, fragile, basically tropical creatures must extend far back, however, perhaps into latest Cretaceous times. We simply lack documentation of this history and doubtless of the early histories of many other groups, as well.

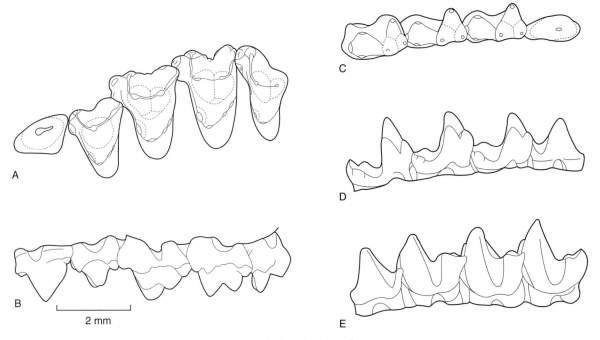

Prokennalestes trofimovi

FIGURE 7-1   (A) Occlusal and (B) labial view (view from the outside) of the left upper cheek teeth and (C) occlusal, (D) lingual (from the inside), and (E) labial views of the left lower cheek teeth of *Prokennalestes trofimovi*. The dentition of this eutherian is the most primitive of any yet known. *(From Kielan-Jaworowska and Dashzeveg, 1989)*

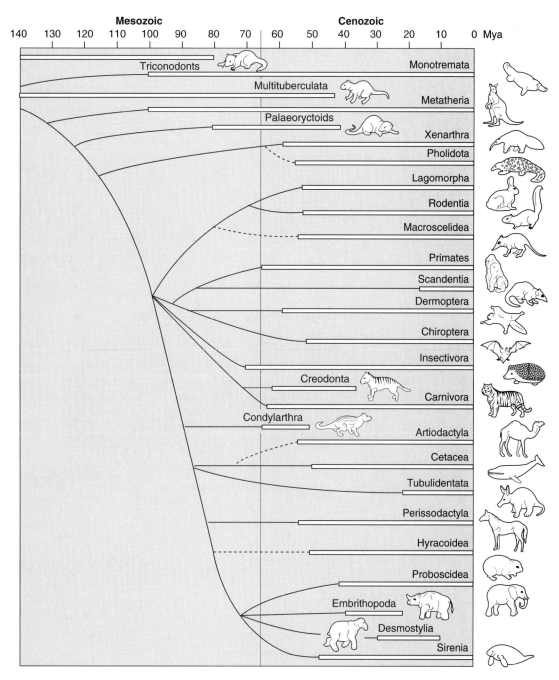

FIGURE 7-2 A phylogenetic tree showing relationships among the major mammalian orders. The solid horizontal bars indicate the known geologic age range of each order based on the fossil record. *Mya* = millions of years ago. *(From Novacek, 1992a)*

WEB SITES

An introduction to eutherian mammals

http://www.ucmp.berkeley.edu:80/mammal/eutheria/placental.html

8 INSECTIVORA

This chapter covers the order Insectivora, which includes the hedgehogs, moles, tenrecs, golden moles, solenodons, and shrews. As currently defined, the Insectivora includes six families and more than 400 extant species (Table 8-1). Although the Insectivora probably is a monophyletic group (Yates, 1984; Onuma et al., 1998), its members retain many characters that are primitive for

mammals. Because the most primitive eutherian mammals were insectivorous and because their descendants have often retained dentitions that remain adapted to an insect diet (although frequently highly specialized), the tendency has been to include some primitive types together with all modern descendants in the order Insectivora. This order has thus long been used as a conve-

nient catchall repository for taxa of doubtful affinities. As an example, until recently, tree shrews and elephant-shrews were also included in this order.

Taxonomic assignment of various equivocal fossil types has also been difficult and controversial. Members of the Cretaceous North American family Geolabididae, for example, have been regarded by some workers as early insectivorans (Carroll, 1988; Dawson and Krishtalka, 1984; Lillegraven et al., 1981). The early Tertiary Leptictidae of North America are thought to be the sister group to the Insectivora (Novacek, 1986). Kielan-Jaworowska et al. (1979) suggested that the insectivoran fossil record may begin with the poorly known *Batodon* from the late Cretaceous of North America. Other workers believe the Tertiary Apternodontidae are ancestral to extant insectivorans (VanValen, 1967). Insectivora as used here is approximately equivalent to the group called Lipotyphla by McKenna and Bell (1997). A modest fossil record coupled with a lack of shared derived characters makes the classification of "Insectivora" difficult, and the system used here is provisional at best (Fig. 8-1).

An insectivorous lifestyle was common to many Cretaceous eutherians, and, although their phylogenetic relationships are still a source of controversy, several major groups of Cretaceous mammals (Leptictidae, Palaeoryctidae, Geolabididae, Nyctitheriidae, etc.) have often been put in the order Insectivora. In any case, the lineages we are discussing are ancient: the roots of some of the modern families included here in the order Insectivora probably reach back into late Cretaceous or early Cenozoic times.

MORPHOLOGY OF THE INSECTIVORA

Although considerable morphological diversity is represented among the mammals grouped in the order Insectivora, for all Recent members the following characters are reasonably diagnostic. The tympanic bone is annular, no auditory bulla is present in some members, and the entotympanic bone is absent; the tympanic cavity is often partially covered by processes from adjacent bones; the cerebrum is smooth, and the olfactory bulbs are largely interorbital; the eyes and optic foram-

TABLE 8-1 Classification and Number of Species of Insectivora

Family	Genera	Species	Distribution
Solenodontidae	1	3[a]	Haiti and Cuba
Nesophontidae[b]	1	(8)	West Indies
Tenrecidae	10	24	Madagascar and central and western equatorial Africa
Chrysochloridae	7	18	Southern Africa
Erinaceidae	7	21	Africa, Eurasia, southeastern Asia and Borneo
Soricidae	23	312	Worldwide, except Australia and southern South America
Talpidae	17	42	North American, Europe, and Asia

[a]Includes 2 extant species and one extinct species.
[b]Includes 8 sub-Recent fossil species.

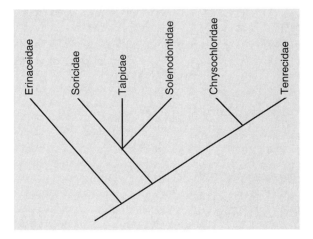

FIGURE 8-1   One phylogenetic hypothesis of the relationships of the extant Insectivora.

ina are usually small; the jugal is reduced or absent, and the zygomatic arch is incomplete in some groups; the orbitosphenoid bone is mainly anterior to the braincase; alisphenoid bone surrounds the foramen ovale. The teeth have sharp cusps, and usually the crown pattern of primitive placentals is recognizable; the anterior dentition is often modified by the enlargement and specialization of the incisors and the reduction of the canines. The intestinal caecum is absent; the limbs (except those of the fossorial groups) are usually unspecialized and are never adapted for saltation (jumping or leaping). Some genera have retained a cloaca.

ORDER INSECTIVORA

FAMILY SOLENODONTIDAE. Represented today by only one genus and two species, the solenodons are relict types that are unable to prosper in competition with other placentals recently introduced into their ranges. Solenodons occurred in sub-Recent and Recent times in Cuba, Puerto Rico, and Hispaniola but are now restricted to Haiti *(Solenodon paradoxus)* and to Cuba, where a declining and endangered population of *Solenodon cubanus* occurs. The introduction by humans of the house rat *(Rattus)*, the mongoose *(Herpestes)*, and dogs and cats into the West Indies, and the extensive clearing of land for agriculture combined to cause the rapid decline of the solenodons.

These animals are now listed as endangered species, and hopes for their survival under natural conditions are slender (Morgan and Woods, 1986). Solenodons may be derived from their close relatives, the extinct Apternodontidae of the Paleogene of North America.

Solenodons are roughly the size of a muskrat and have the form of an unusually large and big-footed shrew (Fig. 8-2). The five-toed feet and the moderately long tail are nearly hairless. The snout is long, slender, and highly flexible. The cartilaginous snout in *S. paradoxus* attaches to the cranium via a type of ball-and-socket joint (MacDonald, 1984). The eyes are small, and the pinnae are prominent. The zygomatic arch is incomplete, no auditory bulla is present, and the dorsal profile of the skull is nearly flat. The dentition is 3/3, 1/1, 3/3, 3/3 = 40. The first upper incisor is greatly enlarged and points backward slightly; the second lower incisor has a deep lingual groove that may function to transport the toxic saliva that empties from a duct at the base of this tooth. (*Solenodon* is from *solen* or "channel" and *dent* or "tooth"). The upper molars lack a W-shaped ectoloph and are basically tritubercular. A sharp, bladelike (trenchant) ridge is formed by a high crest at the outer edge of each molar.

Solenodons are generalized omnivorous feeders that prefer animal material. They are nocturnal and often find food by rooting with their snouts or by uncovering animals with their large claws. *Solenodon* produces a toxic saliva from its submaxillary salivary glands, which is carried by ducts to the base of the large and deeply channeled posterior surfaces of the second lower incisors and presumably enters a wound by capillary action. The saliva of this insectivoran is similar in effect to the saliva of the venomous shrews (p. 117). Solenodons prey primarily on invertebrates but may also scavenge vertebrate remains. Solenodons also have highly developed senses of touch, smell, and hearing. Vocalizations include high frequency "clicks," similar to those found in some shrews, that may be a crude form of **echolocation** (Nowak and Paradiso, 1991). Solenodons are relatively long-lived species (over 11 years in captivity) but have low reproductive rates. Being archaic creatures that seem to have little competitive ability, their distribution on islands is probably the key to their continued, if tenuous survival.

FIGURE 8-2   *Solenodon paradoxus* (Solenodontidae) is now restricted to Haiti, where populations of this large (800 gram) insectivoran are in decline. *(C. Wemmer)*

FAMILY NESOPHONTIDAE. This West Indian family is represented by eight fossil species in the genus *Nesophontes*. Of the eight species that survived the Pleistocene, only a few species survived into the 1900s, and all are now thought to be extinct (Morgan and Woods, 1986). Most of the information on this genus comes from skulls and skeletal remains from fossil sites and owl pellets found in 1930 (Nowak and Paradiso, 1991). The skulls of *Nesophontes* lack a jugal bone, zygomatic arch, and probably an auditory bulla. These mouse- to rat-sized animals probably resembled solenodons in having a long, flexible snout and small eyes. Little else is known except that they probably could not compete with the introduced *Rattus* and *Mus* (house mouse) brought to the West Indies by the Spanish. A 1985 expedition tried to locate surviving populations but was unsuccessful, and all species are presumed extinct (Woods et al., 1985).

FAMILY TENRECIDAE. The tenrecs are a distinctive group of primitive insectivorans that vary widely in morphology and habits. The family includes ten genera and 24 species and inhabits Madagascar, the Comoro Islands, and western central Africa. With the exception of the African subfamily Potamogalinae, the major evolution of the group occurred primarily on the island of Madagascar. A textbook example of an **adaptive radiation,** tenrecs bear a general resemblance to such diverse mammals as shrews *(Microgale)*, hedgehogs *(Echinops* and *Hemicentetes)*, otters *(Potamogale)*, desmans, *(Limnogale)*, and moles *(Oryzorictes)*. The

largest tenrec, *Tenrec ecaudatus,* has been recently introduced to the Comoro Islands, Mascarene Islands, and the Seychelles. The meager fossil record of tenrecs indicates little about their evolution. Fossils are known from the Miocene of Africa and the Pleistocene of Madagascar. Possibly the ancestral stock of the tenrecs on Madagascar dispersed there from Africa in the Miocene epoch, although the reverse is also possible.

Tenrecs vary from roughly the size of a shrew to the size of a cottontail rabbit. The snout is frequently long and slender. The jugal bone is absent, the eye is usually small, and the pinnae are conspicuous. The tympanic bone is annular, and the squamosal bone forms part of the roof of the tympanic cavity. The anterior dentition varies from species to species. The first upper premolars are never present, and the molars are 3/3 in all but *Tenrec* (4/3) and *Echinops* (2/2). The upper molars have crowns that are triangular in occlusal view, and only in one genus *(Potamogale)* is a W-shaped ectoloph present in these teeth. The urogenital canal and anus open into a common cloaca. The retractile penis rests in a fold ventral to the anus.

An unusually broad array of adaptive types occurs within the Tenrecidae. *Tenrec* roughly resembles a tailless, coarse-furred, long-snouted opossum and has spines interspersed with soft hairs (Fig. 8-3). It is omnivorous. *Echinops* (Fig. 8-4), *Hemicentetes* (Fig. 8-5), and *Setifer* are also spiny, and all three resemble hedgehogs (Erinaceidae) closely. In these two genera, the panniculus carnosus muscle is powerfully developed and enables the animals to erect the spines. It also contributes to the ability of these

FIGURE 8-3   A tenrec (*Tenrec ecaudatus,* Tenrecidae). *(J. Eisenberg and E. Gould)*

FIGURE 8-4   A Madagascar "hedgehog" (*Echinops telfairi,* Tenrecidae). *(J. Eisenberg and E. Gould)*

FIGURE 8-5   A streaked tenrec (*Hemicentetes semispinosus,* Tenrecidae). *(J. Eisenberg and E. Gould)*

animals to roll into a ball. The feet and head are tucked beneath the body during this protective movement, and "the sphincter muscles running around the body at the junction of the spiny dorsum and the hairy **venter** permit the spiny dorsal skin to be drawn together, thus enclosing the animal in an impregnable shield of spines" (Gould and Eisenberg, 1966). Gould and Eisenberg found that newborn *Echinops* and *Setifer* reacted to being disturbed by rolling into a ball. *Hemicentetes* has a group of 14 to 16 specialized quills on the middle of the back, the **stridulating organ,** that rub together when underlying dermal muscles are twitched to produce sounds in a variety of repetitive patterns. Differences in these sounds depend on differences in associated behavior of the animals (Gould, 1965) and may be used in **intraspecific** communication. Some tenrecs are known to become torpid under natural conditions during seasons of food shortage (Eisenberg and Gould, 1970). During torpor, the heart rate can drop to one beat every three minutes and the body temperature also drops substantially, depending on the ambient temperature.

The Malagasy shrew tenrecs of the genus *Microgale* have the most species (12) and are the least specialized tenrecs (Jenkins, 1992; Jenkins et al., 1997; MacPhee, 1987). Members of this genus lack spines and resemble shrews (Fig. 8-6). *Microgale longicaudata,* as its name implies, has 47 caudal vertebrae, the largest number of any mammal except certain pangolins. The tip of the tail in some species of *Microgale* is semiprehensile and is used when climbing thin branches as a fifth "hand" (Ryan, personal observation). The water tenrec, *Limnogale mergulus,* is semiaquatic. It lives adjacent to running water. Its forefeet are fringed with stiff hairs, and it has webbed hind feet to help it paddle through the water. Otherwise, little is known about the behavior or ecology of these small tenrecs.

The subfamily Potamogalinae, the otter shrews, includes animals that in many ways are the most remarkable members of the Tenrecidae. Members of this relict group live in western central Africa; they are the only living members of a primitive lineage and have probably survived because of their highly specialized, semiaquatic style of life. Although the giant otter shrew (*Potamogale velox*) has been known to scientists since 1860, the genus to which the dwarf otter shrews belong (*Micropotamogale*) was not described until 1954 (Heim de Balsac, 1954).

FIGURE 8-6   A shrew tenrec *(Microgale dobsoni)* from Madagascar *(J. Eisenberg)*

The giant otter shrew is large for an insectivoran, measuring 600 millimeters in length and weighing about 1 kilogram, and is highly specialized for the life of a miniature otter. The body is long and streamlined, the limbs are rather short and stocky, and the large tail is laterally compressed. Propulsion beneath the water is controlled by lateral movements of the flattened tail, and a number of features are associated with this locomotor style. The caudal vertebrae have high neural spines and transverse processes. These unusual caudal vertebrae provide attachment points for the powerful tail musculature, which is aided by the greatly enlarged gluteal muscles. The posterior parts of the gluteal muscles, which in quadrupedal mammals move the hindlimbs, attach to the muscles overlying roughly the first five caudal vertebrae and move the tail. In the sinuous motion of the back and tail, and even in overall body form, *Potamogale* resembles a large salamander. Otter shrews live in permanent streams and rivers and in coastal swamps, and although they rely partly on fish, they seem to prefer freshwater crabs to other food (Kingdon, 1984a; Vogel, 1983). The habits of potamogales remain poorly known and provide fascinating opportunities for the resourceful biologist.

FAMILY CHRYSOCHLORIDAE. Another type of insectivoran is typified by chrysochlorids, the golden moles. These animals resemble "true" moles (Talpidae), but even more closely resemble, in fossorial adaptations and in function, the marsupial "moles" (Notoryctidae). The seven genera and 18 species constituting the family Chrysochloridae occur widely in southern Africa, where

they occupy forested areas, savannas, and sandy deserts. The earliest fossil chrysochlorids from the Miocene of Kenya *(Prochrysochloris)* resemble Recent species, and these and Pleistocene fossil material give no firm evidence of the derivation of the group. Butler (1969) suggests that the Tenrecidae and the Chrysochloridae may be related. Bonner (1991 and 1995) and MacPhee and Novacek (1993) have argued that golden moles belong in a distinct suborder Chrysochloromorpha (or order Chrysochloridea of McKenna and Bell, 1997). Recent molecular data, however, suggest that golden moles may be more closely related to elephant-shrews or aardvarks (de Jong, 1998).

Golden moles have modes of life similar to those of the fossorial members of the Talpidae and possess some parallel adaptations, as well as some contrasting structural features. The ears of golden moles lack pinnae, and the small eyes are covered with skin and fur. The pointed snout has a leathery pad at its tip (Fig. 8-7). The zygomatic arches are formed by elongate processes of the maxilla, and the occipital area includes bones, the tabulars, not typically found in mammals. The skull is abruptly conical instead of being flattened and elongate as it is in many insectivorans. An auditory bulla is present that is formed largely by the tympanic bones; the malleus is enormously enlarged. The dental formula is usually 3/3, 1/1, 3/3, 3/3 = 40. The first upper incisor is enlarged, and the molars are basically tritubercular and lack the stylar cusps and W-shaped ectoloph typical of talpids. The permanent dentition of golden moles emerges fairly late in life. The forelimbs are powerfully built, and the forearm

FIGURE 8-7 The cape golden mole (*Chrysochloris asiatica;* Chrysochloridae). Note the leathery nose pad and the greatly enlarged claws of digits two and three. (*J. Jarvis*)

rests against a concavity in the rib cage. The fifth digit of the hand is absent, and digits two and three usually have a huge, picklike claw. A "third bone" is present in the forelimb, which is probably an ossified flexor tendon (Fig. 8-8; Gasc et al., 1986). The forelimbs are not rotated as are those of talpids but more or less retain the usual mammalian posture, with the **palmar** surfaces downward.

Golden moles are adept burrowers. Gasc and coworkers (1986) studied golden moles in the laboratory, using x-ray motion pictures, and found that the Namib golden mole burrows through loose sand by alternating a forelimb digging phase with a dorsoflexion of the head and shoulders that serves to "buttress" the sand above the animal and allow efficient tunneling through loose substrates (Fig. 8-9). Bateman (1959) also studied golden moles in the laboratory and found that a 60-gram golden mole could push up a 9-kilogram weight covering its cage; this amounts to exerting a force equal to 150 times the animal's weight. When the animal is close to the surface, a ridge marks the course of its progress. Both deep and shallow burrows are constructed; the depth of the burrow may depend on the amount of soil moisture. The roofs of shallow burrows in sandy soil frequently collapse, leaving a furrow in the sand as a trace of the former burrow. The diet of golden moles consists mostly of invertebrates; two desert-dwelling genera (*Cryptochloris* and *Eremitalpa*) also eat legless lizards. *Eremitalpa* lives in the dunes of the Namib Desert,

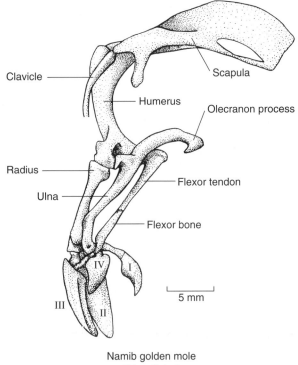

Namib golden mole
(*Eremitalpa*)

FIGURE 8-8 Lateral view of the left forelimb skeleton of a golden mole showing the enlarged olecranon process and addition of a third ossified element called the "flexor bone" in the forearm. The digits are labeled I through IV. (*After Gasc et al., 1986*)

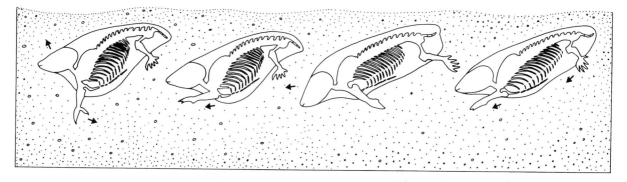

FIGURE 8-9   Schematic drawing of the burrowing behavior of the Namib golden mole *(Eremitalpa)*. Arrows indicate the direction of the thrust of the head and forefeet. *(After Gorman and Stone, 1990)*

where it forages mostly on the surface and makes occasional burrowing forays (Fielden et al., 1990). Termites are the primary food. Burrowing activity of *Amblysomus* varies with temperature, rainfall, and prey availability (see p. 389).

FAMILY ERINACEIDAE. Members of the family Erinaceidae, the hedgehogs, are morphologically primitive but remain successful even in areas highly modified by humans. The family is represented today by seven genera and 21 species; they occur in Africa, Eurasia, southeastern Asia, and on the island of Borneo. Erinaceids are first known from the Paleocene of North America (*Litolestes;* Novacek et al., 1985), and fossil material is known from the Eocene of Europe and Asia and the early Miocene of Africa. The family Adapisoricidae, containing primitive relatives of the hedgehogs, is represented from the late Paleocene to the Eocene in Europe.

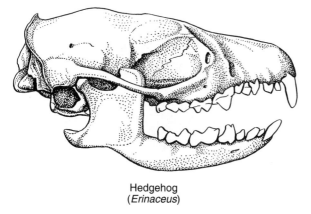

Hedgehog
(*Erinaceus*)

FIGURE 8-10   Skull of a hedgehog (length of skull 32 millimeters).

Erinaceids include two subfamilies; the Erinaceinae, or hedgehogs, and the Hylomyinae or moon rats (also called gymnures). They vary from the size of a mouse to the size of a small rabbit (1.4 kilograms). The eyes and pinnae are moderately large, and the snout is usually long (Corbert, 1988). The zygomatic arches are complete. The dental formula is 2-3/2-3, 1/1, 3-4/2-4, 3/3 = 34-44. The first upper and, in some species, the first lower incisors are enlarged, but the front teeth never reach the degree of specialization typical of shrews (Fig. 8-10). In hedgehogs, the upper molars have simple nonsectorial cusps, with the paracone and metacone near the outer edge; the hypocone completes the quadrate form of the upper tooth (Fig. 8-11A). Both the trigonid and talonid of the lower molars are well developed (Fig. 8-11B). The molars are thus better adapted to an omnivorous than to an insectivorous diet. The feet retain five digits in all but one genus, and the foot posture is plantigrade (walking on the soles of the hands and feet). An obvious specialization is the possession of spines in members of the subfamily Erinaceinae (Fig 8-12). In these animals, the panniculus carnosus muscle is greatly enlarged and controls the pulling of the skin around the body and the erection of the spines.

In various parts of their wide range, hedgehogs occupy deciduous woodlands, cultivated lands, and tropical and desert areas. They are omnivorous, but animal food seems to be preferred, and a wide variety of invertebrates are eaten (Reeve, 1994). Kingdon (1984a) reports that hedgehogs attack and kill small snakes and, during the attack, direct their spines forward, leaving only a small part of

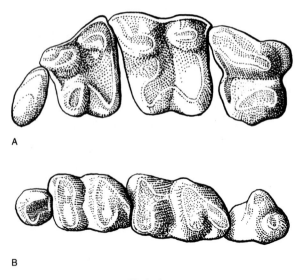

A

B

Hedgehog
(*Erinaceus*, Erinaceidae)

FIGURE 8-11   Cheek teeth of a hedgehog. Fourth premolar and three molars of the (A) upper right and (B) lower left tooth rows.

FIGURE 8-12   Drawing of the Algerian hedgehog (*Atelerix algirus*). *(After Reeve, 1994)*

their body exposed to the strikes of the snake. Hedgehogs seem remarkably resistant to snake venom. Some members of this family protect themselves by rolling into a tight ball with the spines erect.

Members of the subfamily Erinaceinae are probably **heterothermic.** (Heterothermic animals can regulate their body temperature physiologically, but temperature is not regulated precisely or at the same level at all times.) Hibernation occurs in the widespread genus *Erinaceus,* and entering a state of

torpor called **estivation** is practiced by the desert species *Hemiechinus aethiopicus.* A related species from India, *H. micropus,* has survived in captivity for periods of four to six weeks without food or water (Nowak and Paradiso, 1991), and this species and *Hemiechinus auritus* are known to have winter periods of dormancy in India. In Kenya, *Erinaceus albiventris* disappears and apparently hibernates through the long dry season from May to September or October. In this instance, the animals are probably responding primarily to food shortages, for temperatures remain moderate through this period.

The moon rat subfamily, Hylomyinae, includes three genera and six species restricted to southeastern Asia, the Malay Peninsula, Borneo, and the Philippines. Moon rats lack spines and instead have a thick coat of coarse hairs and a long, nearly naked tail (Fig. 8-13). They are largely nocturnal and solitary foragers (Gould, 1978), but little is known about their ecology or behavior. *Podogymnura truei* from the Philippines is listed as vulnerable to extinction primarily because of habitat loss.

FAMILY SORICIDAE. Members of this family, the shrews, are among the smallest and least conspicuous of mammals. In many areas, they are the most numerous insectivorans; they have the widest distribution of any of the Insectivora and are the most familiar. The family Soricidae is represented today by 23 genera and 312 species in two subfamilies (Soricinae and Crocidurinae). Shrews represent

FIGURE 8-13   Drawing of the greater moon rat (*Echinosorex gymnurus*) from Southeast Asia. *(After Reeve, 1994)*

A

B

FIGURE 8-14   (A) The common shrew *(Sorex araneus)* feeding on a worm. (B) The European pygmy shrew *(Sorex minutus). (A by D. Hosking, B by M. Andera)*

6.7 percent of all extant mammal species and occur throughout the world, except in the Australian area, most of South America, and the polar areas. *Cretasorex,* from the Upper Cretaceous of Asia (Nesov and Gureev, 1981), may be the first soricid to appear in the fossil record (but see McKenna and Bell, 1997). Other undoubted soricid fossils first appear in the Eocene of North America (Heterosoricinae; Krishtalka and Setoguchi, 1977) and Europe, and the Miocene of Africa and Asia. Because soricids are rare as fossils, their early evolution is obscure; they may have evolved from the fossil Nyctitheriidae (Dawson and Krishtalka, 1984).

Shrews are small: the smallest weigh 2.5 grams, and the largest weigh roughly 180 grams, the weight of a rat. The snout is long and slim; the eyes are small; and the pinnae are usually visible (Fig. 8-14). The feet are five-toed and unspecialized, except for fringes of stiff hairs on the digits in semiaquatic species and enlarged claws in semifossorial forms. The foot posture is plantigrade. The narrow and elongate skull usually has a flat dorsal profile (Fig. 8-15A); there is no zygomatic arch or tympanic bulla, and the tympanic bone is annular (Fig. 8-15B). The specialized dentition consists of 26 to 32 teeth; the dental formula of *Sorex* is 3/1, 1/1, 3/1, 3/3 = 32. In the subfamily Soricinae, the teeth are pigmented; the first upper incisor is large

and hooked and bears a notch and projection resembling those on the beak of a falcon (Fig. 8-15A). Behind the first upper incisor is a series of small unicuspid teeth (presumably incisors, a canine, and premolars); the fourth upper premolar is large and has a trenchant ridge; and the upper molars have W-shaped ectolophs (Fig. 8-16). Both the trigonid and talonid of the lower molars are well developed (Fig. 8-16), and the first lower incisor is greatly enlarged and procumbent (leaning forward).

Perhaps the most unusual skeleton belongs to the hero shrew *(Scutisorex somereni)* of West Africa (Fig. 8-17). The spinal column of this animal consists of vertebrae with numerous interlocking processes that presumably provide added strength to the vertebral column and additional attachment sites for the complex back muscles (Cullinane et al., 1998a, b). These shrews are thought to have incredible strength, but almost nothing is known of their ecology or the reasons for this unique vertebral system.

Because they are unusually small, shrews can exploit a unique mode of foraging. Many patrol for insects beneath logs, fallen leaves, and other plant debris and in the narrow spaces and crevices beneath rocks. Rodents' surface runways and burrows may also be used as feeding routes. Because of their often nocturnal style of foraging,

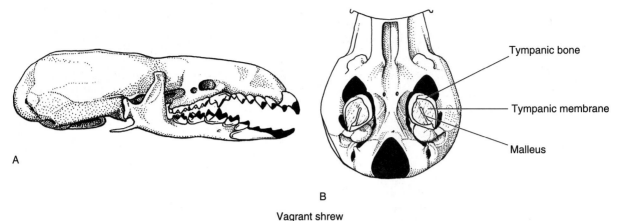

Vagrant shrew
(*Sorex vagrans*, Soricidae)

FIGURE 8-15   Skull of the vagrant shrew. (A) Side view, showing the pincer-like anterior incisors. The pigmented parts of the teeth are shown in black. (B) Ventral view of the basicranial region, showing the annular (ringlike) tympanic bone, the tympanic membrane, and the malleus (length of skull 17 millimeters).

FIGURE 8-16   Cheek teeth of insectivorans. (A) The fourth upper right premolar and first molar and (B) the first two lower left molars of the vagrant shrew. The pigmented parts of the teeth are shown in black. (C) The first and second upper right and (D) the comparable lower left molars of the eastern mole.

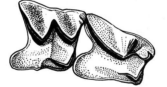

Vagrant shrew
(*Sorex vagrans*, Soricidae)

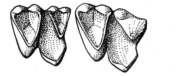

Eastern mole
(*Scalopus aquaticus*, Talpidae)

shrews are seldom observed, even in areas where they are common. Although shrews are typically associated with moist conditions, some species, such as the gray shrew *(Notiosorex crawfordi)* of the southwestern United States and the piebald shrew *(Diplomesodon pulchellum)* of southern Russia, inhabit desert areas. Aquatic adaptations in some species allow them to dive and swim and feed mainly on aquatic invertebrates. One of the most aquatic species is the Tibetan water shrew *(Nectogale elegans),* which inhabits mountain streams and feeds primarily on fish. In this species, the stream-

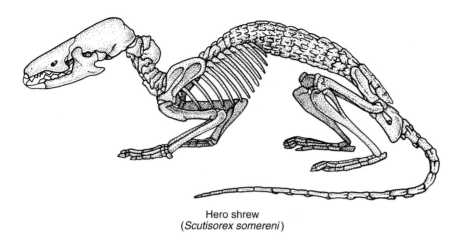

Hero shrew
(*Scutisorex somereni*)

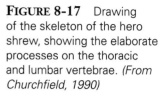

FIGURE 8-17 Drawing of the skeleton of the hero shrew, showing the elaborate processes on the thoracic and lumbar vertebrae. *(From Churchfield, 1990)*

lined shape is enhanced by the strong reduction of the pinnae, and the digits and feet have fringes of stiff hairs that greatly increase their effectiveness as paddles. The distal part of the tail is laterally compressed, and the edges bear lines of stiff hairs to aid in swimming.

Except for the duck-billed platypus (order Monotremata), the only mammals known to be venomous are shrews and solenodons. Over 350 years ago, there were reports on the symptoms that developed when humans were bitten by the short-tailed shrews of North America (*Blarina* species), and work in the present century has confirmed that short-tailed shrews have venomous saliva (Pournelle, 1968; Tomasi, 1978). It has also been demonstrated that the European water shrew (*Neomys fodiens*) and the Haitian solenodon (*Solenodon paradoxus*) are venomous (Pournelle, 1968; Pucek, 1968). The salivas of other insectivorans closely related to these two shrews have been studied, but are apparently not toxic. In some people, the bites of the musk shrew, *Suncus murinus,* cause minor aches and hypersensitivity and reddening of the skin, especially at the finger joints (G. L. Dryden, personal communication).

Both *N. fodiens* and *Blarina* species have similar adaptations for delivering venom, and the effects of the venoms are similar (Pournelle, 1968; Pucek, 1968; Tomasi, 1978). In these shrews, the first lower incisors have concave medial surfaces, forming a crude channel, and the ducts from the venom-producing submaxillary glands open near the base of these teeth. *Neomys fodiens* salivates copiously during attacks on prey, and saliva is seem-

ingly channeled to wounds via the two first lower incisors. Pearson (1942) showed that mice injected with extracts of the submaxillary glands of this shrew were strongly affected; the activity of the mice was reduced rapidly by what seemed to be a neurotoxin impairing normal function of the nervous system. Frogs bitten by *N. fodiens* were partially immobilized and, when forced to move, were uncoordinated. Laboratory mice injected with a homogenate of these salivary glands immediately developed paralysis of the hindlimbs.

What is the functional importance of venom to insectivorans? *Blarina* species can kill mice considerably larger than themselves, and Eadie (1952) and Getz et al. (1992) reported that meadow voles (*Microtus pennsylvanicus*) were an important fall and winter food of short-tailed shrews. Frogs and small fish are known to be preferred foods of *N. fodiens.* Both of these shrews attack prey from behind and direct bites at the neck and base of the skull, an area where neurotoxic venom might be readily introduced into the central nervous system. The adaptive importance to a very small predator of making its relatively large prey helpless seems to be great, and one wonders why more shrews are not venomous.

Shrews forage day and night and are not known to hibernate, which has led to their reputation as ravenous predators. Survival depends on their ability to consume nearly twice their body weight in food every day. The need for such large amounts of food can be explained by their small body size. In mammals, metabolic rate is inversely proportional to body size, resulting in a considerably

FIGURE 8-18 Drawing of a mother *Crocidura* and her young forming a caravan. *(After Churchfield, 1990)*

greater surface-area-to-volume ratio for smaller mammals. The net result is that small mammals have larger heat loss, and because metabolic heat is produced from the conversion of food into usable energy for cellular metabolism, smaller animals need to consume proportionally larger amounts of food (p. 368). In winter, demands for food increase precisely when invertebrate food supply declines, and shrews of the **Holarctic** genus *Sorex* are able to survive the winters of the north by doubling their **nonshivering thermogenesis** (a means of heat production) and reducing their body weight by up to 53 percent, while Old World tropical *Crocidura* undergo a daily torpor; Genoud, 1985; Merritt et al., 1994; Merritt, 1995). The high metabolic rate of shrews is accomplished by the highest blood oxygen content of any mammal measured, a heart rate between 900 and 1400 beats per minute in some species, and more than three times the number of red blood cells per cubic milliliter of blood found in humans (Grzimek, 1990).

Another interesting adaptation related to foraging is the ability of some shrews (*Sorex* and *Blarina*) to use a rudimentary form of echolocation to locate prey. These shrews emit 30 to 60 kilohertz pulses as they explore their environment (Gould et al., 1964; see Chapter 22).

Observations of shrew behavior in the wild are few, and what little is known comes from relatively few laboratory studies. One interesting behavior is the innate following behavior unique to some shrews of the subfamily Crocidurinae, which results in the formation of "caravans." When a female and her young are moving, the first young grabs the base of her tail with its mouth, the next young grasps the base of the tail of the first young in its mouth, and so on, forming a chain of young that, under some conditions, is dragged by the mother (Fig. 8-18).

A

B

FIGURE 8-19 Heads and forelimbs of (A) hairy-tailed mole *(Parascalops breweri)* and (B) star-nosed mole *(Condylura cristata)*. The unique "star" of finger-like structures is found in no other mammal. *(A by K. Catania and B from Catania and Kaas, © 1996 American Institute of Biological Sciences)*

FAMILY TALPIDAE. This family includes a group of small rat- or mouse-size animals usually referred to as moles. These predominantly burrowing insectivorans (17 genera and 42 species) occur in parts of North America, Europe, and Asia. The European fossil record of talpids begins in the Eocene (*Eotalpa;* Sigé et al., 1977); talpids are known first in the New World from the Oligocene. Apparently, the anatomical modifications typical of Recent fossorial genera were attained early, for the Recent European genus *Talpa* is first known from the Miocene. The fossil family Proscalopidae was once thought to be closely related to talpids, but recent discovery of postcranial material from the Miocene of Montana indicates that members of this fossil group did not use the same digging technique observed in modern moles and consequently may represent a distinct family of fossil insectivorans (Barnosky, 1982). Recent studies, using variations in proteins, suggest that talpids colonized the New World several times since the Miocene (Yates and Greenbaum, 1982).

The head and forelimbs of most talpids are modified for fossorial life (Fig. 8-19). The zygomatic arch is complete, the tympanic cavity is not fully enclosed by bone, and the eyes are small and often lie beneath the skin. The snout is long and slender, the ears usually lack pinnae, and the fur is characteristically lustrous and velvety. The dental formula is 2-3/1-3, 1/0-1, 3-4/3-4, 3/3 = 32-44. The first upper incisors are inclined backwards (Fig. 8-20), and the upper molars have W-shaped ectolophs (Fig. 8-16C and D). In the fossorial species, the forelimbs are more or less rotated from the usual orientation typical of terrestrial mammals in such a way that the digits point to the side, the palms face backward, and the elbows point upward (Fig. 8-21). In addition, the phalanges are short, the claws are long, and the clavicle and humerus are unusually short and robust. The scapula is long and slender and serves both to anchor the forelimb solidly against the axial skeleton and to provide advantageous attachments for some of the powerful muscles that pull the forelimb backward. The anteriormost segment of the sternum (the manubrium) is greatly enlarged and extends forward beneath the base of the skull (Fig. 8-21). These specializations increase the area for attachment of the large pectoralis muscles and move the shoulder joint forward, allowing the forepaws to remove or loosen soil beside the snout.

The clavicle is short and broad and provides a large secondary articular surface for the humerus. The double articulation of the shoulder joint, with articular contacts between the humerus and the scapula and clavicle, provides an unusually strong bracing for this joint during the powerful rotation of the humerus that accompanies the digging stroke of the forelimb. In some genera, the falciform bone is large and increases the breadth of the forepaw and braces the first digit (Fig. 8-21). Unlike golden moles (Chrysochloridae), talpids live in areas where it is possible to dig more or less permanent tunnel systems. As a result, they have evolved a different digging technique involving an alternating fore and aft scraping of their vertically oriented, spadelike hands against the tunnel walls (Skoczen, 1958; Yalden, 1966).

Fossorial talpids occur typically in moist and friable soils in forested, meadow, or streamside areas and feed largely on animal material. A species that occurs in the eastern United States *(Scalopus aquaticus),* however, locally penetrates the moderately dry sandhill prairies of eastern Colorado and Nebraska, where the characteristic ridges of soil made by the animals appear only during wet weather. In most areas, these ridges are a common evidence of the presence of moles and are made by the animals as they travel just beneath the surface by forcing their way through the soil (Yalden, 1966). Soil from deep burrows is deposited on the surface in more or less conical mole hills. Some species of mole appear to be territorial, although territories of males may overlap those of females (Fig. 8-22; Loy et al., 1994). Moles are not known to hibernate or estivate.

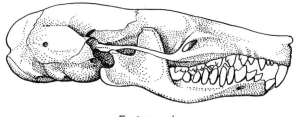

Eastern mole
(*Scalopus aquaticus,* Talpidae)

FIGURE 8-20 Skull of the eastern mole (length of skull 37 millimeters).

FIGURE 8-21   The pectoral girdle and forelimb of the eastern mole. (A) Side view of the pectoral girdle with forelimb bones removed. (B) Anterior view of part of the pectoral girdle and the forelimb, with the shoulder joint slightly disarticulated to show the head of the humerus and its secondary articular surface. The head of the humerus articulates with the glenoid fossa of the scapula, and a second articulation (involving considerably larger surfaces) occurs between the secondary articular surface of the humerus and an articular surface of the clavicle.

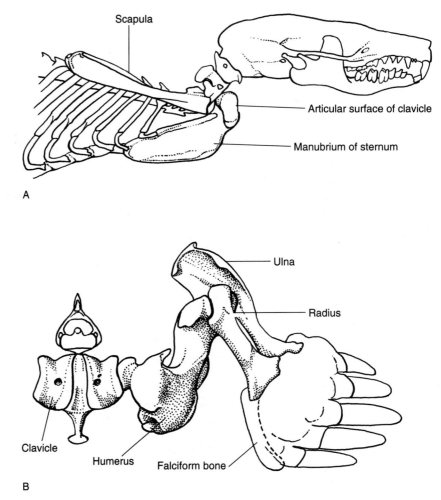

Eastern mole
(*Scalopus aquaticus*, Talpidae)

FIGURE 8-22   A three-dimensional plot of the home range of a mole. Each grid square represents 25m² of territory, and the height of the plot indicates the number of occurrences of the mole within that grid square. *(From Gorman and Stone, 1990)*

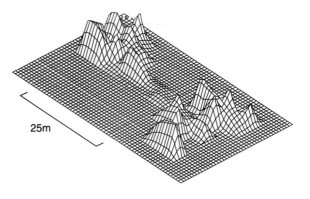

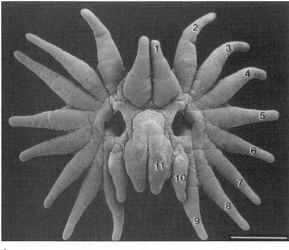

A

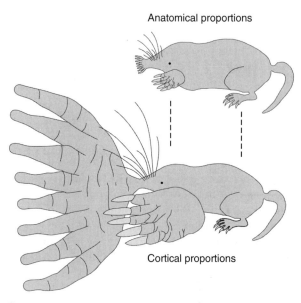

Anatomical proportions

Cortical proportions

C

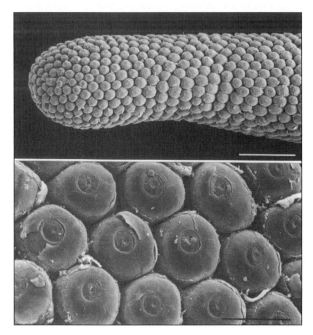

B

FIGURE 8-23   (A) Scanning electron micrograph of the nose of a star-nosed mole *(Condylura cristata)* showing the 22 fleshy appendages that ring the nostrils. Each appendage is covered with small epidermal papillae called Eimer's organs that are very sensitive to touch. Scale bar = 2 millimeters. (B) Close-up of the hundreds of Eimer's organs on each appendage. Scale bar = 250 microns. (C) Diagrammatic drawing of the body parts of a star-nosed mole in their normal anatomical proportions (above) and as they would be shown relative to their proportional representation in the somatosensory cortex (below). Note the relatively large somatosensory field for the nose of the mole. *(From Catania and Kaas, © 1996 American Institute of Biological Sciences)*

Moles forage along their tunnel systems and probe the walls of the tunnel with the aid of tens of thousands of touch receptors located on the snout in the **Eimer's organ** (Quilliam, 1966). These organs are also found in desmans (Richard, 1982). *Condylura cristata,* the star-nosed mole, possesses a ring of 22 fleshy "tentacles" on the tip of its snout (Fig. 8-23). This species is semiaquatic and uses these sensory "tentacles" when trying to locate prey. Recent experiments on the neurobiology of the nose of star-nosed moles demonstrated that the tentacles of the star contain an elaborate system of touch receptors (Eimer's organs). With approximately 25,000 Eimer's organs, the nose of *Condylura* contains five times as many touch receptors (mechanoreceptor) as the entire human hand (Catania and Kaas, 1996).

Thus, the nose of *Condylura* is used primarily for touch reception and not for olfaction.

The behavior of star-nosed moles suggests a second function of the sensory tentacles. Star-nosed moles frequently submerge all or part of their tentacles in the water before entering the water themselves. During this time, the proboscis slowly scans the underwater environment. This behavior, along with the density and complexity of the Eimer's organs, suggests that the tentacles may also serve as **electroreceptors** analogous to the function of the receptor pits in the duck-billed platypus (*Ornithorhyncus;* Grand et al., 1998). Intriguing as this idea may be, it has not yet been tested experimentally.

The evolution of Eimer's organs in talpids appears to be correlated with habitat preference (Catania, 1995). A gradient from wet to dry soils exists among three North American species, with the eastern mole *(S. aquaticus)* preferring drier soils, star-nosed moles *(C. cristata)* living in very wet habitats, and hairy-tailed moles *(P. breweri)* preferring intermediate soil types. Eimer's organs in eastern moles are degenerate, probably owing to the

thicker skin necessary to combat the considerable wear and tear of burrowing through dry, sandy soils. Conversely, Catania (1995) argued that the relatively wet, muddy soils inhabited by star-nosed moles allowed for the elaboration of the more delicate sensory structure containing thousands of sensitive mechanoreceptors.

The remarkable subfamily Desmaninae inhabits the Pyrenees Mountains and some mountains in Portugal, southeastern Europe, and parts of Russia. These animals are adapted to semiaquatic life. They live along the banks of lakes, ponds, or streams and feed largely on aquatic invertebrates. Their burrows open beneath the surface of the water and extend upward to a nest chamber above the water level. Desmans have webbed forefeet, and the greatly enlarged hind feet are webbed and bear a fringe of stiff hairs that increase their effectiveness as paddles. These animals also have flexible snouts that have an extremely highly developed sense of touch (Eimer's organ) and smell that enable the animals to detect underwater prey (Richard, 1982).

WEB SITES

Detailed scientific information on all families of Insectivora

http://www.oit.itd.umich.edu/~roger/bioweb/mirror/WebMirror-1.0/
Fall97.Deux/Chordata/Mammalia/Insectivora.shtml

The shrew site, with extensive information on shrews with photos, links, and large bibliography

http://members.vienna.at/shrew/index.html

Alphabetical list of hedgehog sites

http://hedgehoghollow.com

British Hedgehog Preservation Society homepage

http://www.argyll.demon.co.uk/bhps.html

Complete Internet hedgehog atlas with extensive list of links to other sites

http://www.ziplink.net/~amartine/edge/edgehog.html

9

XENARTHRA AND PHOLIDOTA

These two orders of mammals, Xenarthra (armadillos, sloths, and anteaters) and Pholidota (pangolins), are considered together in this chapter because many of their members share features associated with ant- and termite-eating such as elongated crania and tongues, large, recurved claws, and lowered metabolic rates. In the past, xe-narthrans and pangolins were frequently combined in an order Edentata. Xenarthrans underwent an impressive Tertiary radiation in South America, whereas extant pangolins are an Old World group that has conservatively maintained a single structural plan. Living xenarthrans exhibit a number of feeding habits, but insect-eating is most

popular, and the South American anteaters are highly specialized ant and termite eaters. Pangolins are also ant and termite eaters.

Whether there is a phylogenetic affinity between these orders is uncertain. Various morphological studies have suggested a close phylogenetic relationship between the two groups, but it has been difficult to separate features that are shared because of relatedness (phylogeny) or because of similar functional adaptations (convergence). Few comprehensive molecular studies have yet been done on pangolins and xenarthrans, but the results of several seem to indicate either a close relationship between pangolins and carnivorans or between pangolins and xenarthrans. The Xenarthra form a monophyletic lineage that is widely considered to be the most primitive living branch of the Eutheria, forming the sister group to all other eutherian mammals. The relationships of the various groups of xenarthrans to one another are fairly well understood on morphological grounds (Englemann, 1985; Patterson et al., 1989, 1992).

ORDER XENARTHRA

Although the xenarthrans are not represented by a large number of species today, including but 13 genera and 29 species, they are interesting animals because of their unique structure and unusual ecological roles, their large fossil types, and their remarkable Tertiary radiation in South America and Caribbean islands.

The living xenarthrans share a series of distinctive morphological features. Extra **zygapophysis**-like (xenarthrous) articulations brace at least the lumbar vertebrae (Fig. 9-1). The incisors and canines are absent; the cheek teeth, when present, lack enamel, and each has a single open root. The teeth in sloths erupt as simple cones with a harder outer dentine shell and a softer inner dentine core. They acquire wear features such as "cusps" and "basins" exclusively through wear (Naples, 1990). A septomaxilla, homologous with that of

the monotremes but found in no other living mammals, is present in the skull of many xenarthrans (Zeller et al., 1993). The tympanic bone is annular; the skull is usually long and rather cylindrical or conical in anteaters and armadillos, shorter and broader in sloths, and short and deep in glyptodonts. The scapula has a prominent spine, often has a secondary spine paralleling the acromion process, and the acromion and coracoid processes are unusually well developed. The clavicle is present. The ischium is variously expanded and specialized and usually forms an ischiocaudal, as well as an ischiosacral, symphysis (Fig. 9-2). The hind foot is typically five-toed, and the forefoot has two or three predominant toes with large claws. Major xenarthran structural trends are toward reduction and simplification of the dentition, specialization of the limbs for such functions as digging and climbing, and rigidity of the axial skeleton.

PALEONTOLOGY

Xenarthrans, in modern times strictly a New World group, are first known from the Paleocene record of South America. The earliest fossils from this region are from the late Paleocene and consist of **osteoscutes,** or bony plates, from the armor of armadillos (family Dasypodidae; Scillato-Yané, 1976). A possible anteater is known in the Eocene fossil record of Europe.

A diverse array of forms resulted from a mid-Tertiary radiation of South American armadillos: a large, Pliocene genus (*Macroeuphractus*) had enlarged, canine-like teeth and was probably a scavenger; the Miocene *Stegotherium* was most likely a termite eater; another Miocene form (*Peltephilus*) and its relatives had specialized scutes that formed pointed horns on the top of the snout; and pig-sized armadillos of the family Pampatheriidae lived in both North and South America in the Plio-Pleistocene.

The structural diversity of extinct South American armadillos indicates that they exploited a

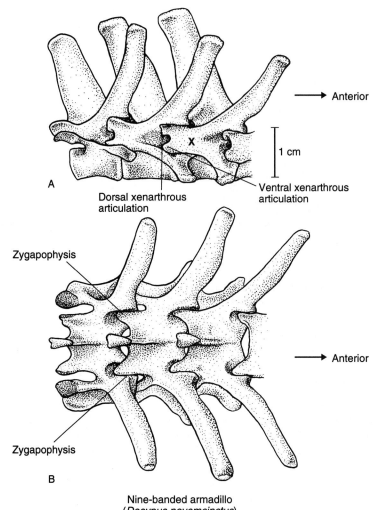

FIGURE 9-1 Three lumbar vertebrae of the nine-banded armadillo, showing the dorsal and ventral xenarthrous articulations upon the xenarthrous process *(x)*. (A) Right lateral view. (B) Dorsal view. The xenarthrous articulations supplement the normal articulations between zygapophyses, which are visible in the dorsal view. *(After Rose and Emry 1993)*

Anterior

1 cm

A

Dorsal xenarthrous articulation

Ventral xenarthrous articulation

Zygapophysis

Anterior

Zygapophysis

B

Nine-banded armadillo
(*Dasypus novemcinctus*)

variety of foods and together formed an important part of the South American Tertiary biota. The early dasypodids were armored with ossified dermal scutes, as are all modern species, and possibly arose from a slothlike ancestor.

From Paleocene until late Pliocene times, dispersal of mammals back and forth between North and South America was restricted. During this interval not only dasypodids but also other xenarthrans underwent a radiation in South America. Several now extinct evolutionary lines arose. The Glyptodontidae appeared in the middle Eocene epoch and represent one line that probably evolved from armadillo stock. These ponderous creatures, some of which were 3 meters long, had unusually deep skulls (Fig. 9-3A). Many of the

unique structural features of the glyptodonts are associated with their development of a nearly impregnable, turtle-like carapace composed of many fused polygonal scales. These are the most completely armored vertebrates known. The limbs are distinctive and highly specialized, and most of the thoracic, lumbar, and sacral vertebrae are fused into a massive arch that, together with the ilium, supports the carapace (Fig. 9-4). Patterson and Pascual (1972) suggested that the post-Miocene diversification of the glyptodonts was favored by the spread of pampas (grassland) in South America.

Additional evolutionary lines are represented by sloths of six or seven families, all but two of which are extinct. Sloths first appear in Eocene deposits in South America and Antarctica (Carlini et al.,

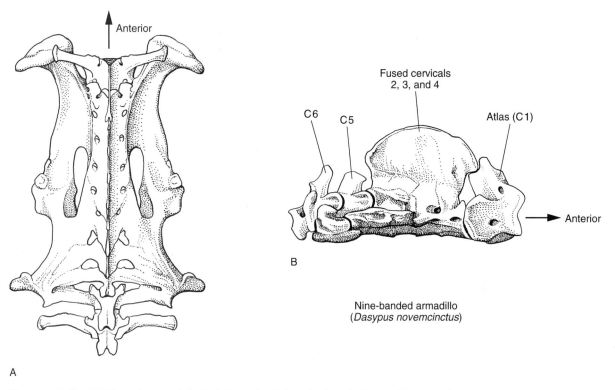

Anterior

Fused cervicals
2, 3, and 4

C6   C5   Atlas (C1)

Anterior

B

A

Nine-banded armadillo
(*Dasypus novemcinctus*)

FIGURE 9-2   (A) Dorsal view of the pelvic girdle of the nine-banded armadillo, showing the great degree of fusion of vertebrae with the ilium and ischium (anterior is at the top of the figure). (B) Lateral view of the cervical vertebrae (anterior is to the right). In this individual the axis and cervicals three and four are fused, and cervical five is partially fused.

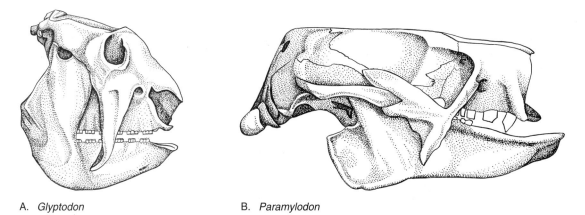

A. *Glyptodon*   B. *Paramylodon*

FIGURE 9-3   Skulls of extinct xenarthrans: (A) *Glyptodon,* length of skull 560 millimeters, (B) *Paramylodon,* length of skull 510 millimeters. *(After Romer, 1966 ©,Vertebrate Paleontology, University of Chicago Press)*

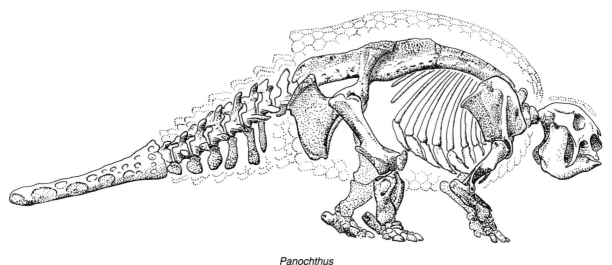

Panochthus

FIGURE 9-4 Skeleton of a glyptodont, *Panochthus,* from the Pleistocene of South America, with a superimposed outline of the carapace and head shield. The caudal armor, which in life probably bore horny spikes on the oval areas, is fused to the caudal vertebrae. Note the extensive fusion of vertebrae and massive limb bones.

1990). For a long time, extinct sloths have been called "ground sloths" because many types were very large and terrestrial. However, many of the smaller fossil species probably were arboreal or at least semiarboreal (White, 1997), and one large Pliocene species was probably semiaquatic, inhabiting shallow coastal waters (Muizon and McDonald, 1995). Moreover, one of the extant tree sloths *(Choloepus)* has recently been recognized as representing a family of "ground sloths," the Megalonychidae, long believed to be extinct (Webb, 1985). The other extant tree sloth, *Bradypus* (Bradypodidae) is phylogenetically related to megatherioid sloths (Megalonychidae and Megatheriidae). Only the extinct mylodontoid sloths, a separate lineage from the megatherioids, were strictly terrestrial. All sloths are herbivores with ever-growing teeth that lack enamel. The late Pleistocene megatheriid *Eremotherium* was common from Florida to Brazil; it reached the size of an elephant, lacked upper canine-like teeth, and walked on the outer edges of the unusually specialized and large hind feet.

The family Megalonychidae is closely related to the megatheriids but differs from that group in having the anteriormost cheek teeth modified into "canines." *Megalonyx,* a Pleistocene genus that reached the size of a cow, was widely distributed in North America. The remains of smaller species of

megalonychids have been found in the West Indies in association with human artifacts. Another family, the Mylodontidae, appeared in the Miocene and is characterized in part by the development of upper "canines" (Fig. 9-3B) and remarkably robust limbs. A degree of protection may have been afforded some members of this family by round dermal ossicles embedded in the presumably thick skin.

The glyptodonts, megatheriids, megalonychids, and mylodonts underwent much of their Tertiary evolution isolated from the North American mammalian fauna. When a land bridge between the Americas was re-established in the Pliocene, the glyptodonts and sloths were remarkably successful in invading North America. By some means of chance dispersal, megalonychid sloths reached North America before it was joined with South America, and one genus *(Megalonyx)* seemingly evolved in North America in the Miocene only to reinvade South America in the Pleistocene.

The plains-dwelling mylodont *Paramylodon* (sometimes known as *Glossotherium*) was widespread in North America and is the most common xenarthran in the Pleistocene deposits of Rancho La Brea in Los Angeles. This ground sloth had large claws on digits two and three (Fig. 9-5), an arrangement similar to that in the living armadillo (Fig. 9-6).

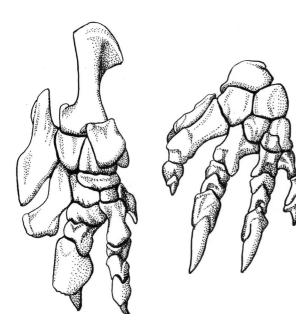

A. *Nothrotheriops*   B. *Paramylodon*

FIGURE 9-5 (A) The right pes of *Nothrotheriops*. (B) The right manus of *Paramylodon*. *(After Romer, 1966 ©,Vertebrate Paleontology, University of Chicago Press)*

The common megatheriid *Nothrotheriops* occurred in North America in the Pleistocene, and its remains from Gypsum Cave in Nevada include bones, skin, and hair (Fig. 9-7). *Nothrotheriops* probably walked on the sides of its highly modified hind feet, as did other large sloths (Fig. 9-5). The powerful claws of the forelimbs were probably used to grasp and tear down vegetation in preparation for ingestion. Dung of this animal is well preserved in dry caves in Nevada and Arizona. Dung from Rampart Cave in Arizona contains such plants as Mormon tea *(Ephedra)* and globe mallow *(Sphaeralcea)* (Hansen, 1978; Poinar et al., 1998). These plants remain common today in dry parts of the southwestern United States. This sloth persisted into the late Pleistocene in this area and probably did not disappear until 11,000 years ago. Thus ended a fascinating cycle of xenarthran evolution.

RECENT XENARTHRAN FAMILIES

FAMILY MYRMECOPHAGIDAE. Members of this family, the anteaters, are highly specialized ant and termite eaters (Fig. 9-8). They occur in tropical

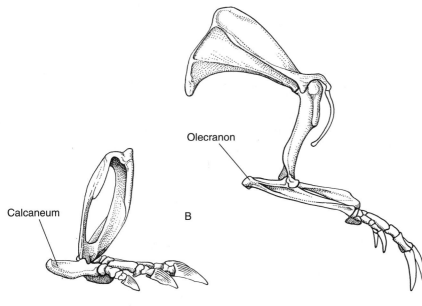

Olecranon

Calcaneum

B

A

Nine-banded armadillo
(Dasypus novemcinctus)

FIGURE 9-6 The right limbs of the nine-banded armadillo: (A) lower part of hindlimb (tibia, fibula, and pes); (B) forelimb. The flattening of the bones of the forearm and lower leg increases the surface area for attachment of flexor and extensor muscles, and the elongation of the olecranon and the calcaneum give added mechanical advantage to the muscles that insert on them.

FIGURE 9-7   How *Nothrotheriops* might have looked. This extinct megatheriid sloth survived until 11,000 years ago in the southwestern United States.

A

B

FIGURE 9-8   Living xenarthrans: (A) a southern tamandua (*Tamandua tetradactyla;* Myrmecophagidae) feeding on termites, (B) the giant anteater (*Myrmecophaga tridactyla*). (A by R. Warner; B by Corel Corp.)

forests of Central and South America and in South American savanna. There are four Recent species belonging to three genera.

The most obvious structural features of anteaters are associated with their ability to capture insects, to dig into or tear apart insect nests, and, except for the giant anteater, to climb. The skull is long and roughly cylindrical (Fig. 9-9A), the zygoma are incomplete, and the long rostrum contains complex, double-rolled turbinal bones. Teeth are absent, the dentary bone is long and delicate, and the **mandibular rami** are unfused. The jaw musculature is reduced, but the tongue musculature is greatly developed. The long, slender tongue is protrusible and covered with sticky saliva secreted by the enlarged and fused submaxillary and parotid salivary glands. The tongue muscles originate from the posterior end of the sternum (Fig. 9-10) rather than from the hyoid bones of the throat, as in most mammals. The soft palate is extremely long, extending posteriorly to the level of the fifth cervical vertebra, so that the oropharynx and nasopharynx extend posteriorly into the neck (Reiss, 1997). The forelimbs are powerfully built; the third digit is enlarged and bears a stout, recurved claw; the remaining digits are reduced. The giant anteater (*Myrmecophaga;* Fig. 9-8B) walks on its knuckles with its toes partly flexed, as did some extinct Pleistocene sloths; whereas the other anteaters (*Cyclopes* and *Tamandua*), which are fully or partly arboreal, walk on the side of the hand with the toes inward. The plantigrade foot has four or five clawed digits. Anteaters range in size from that of a squirrel (*Cyclopes*, 350 grams) to that of a large dog (*Myrmecophaga*, 25 kilograms). *Myrmecophaga* is covered with long, coarse fur, and its laterally compressed, nonprehensile tail has long hairs that hang downward. In the other

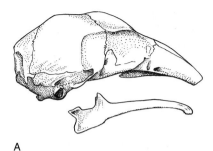

A

Two-toed anteater
(*Cyclopes didactylus*, Myrmecophagidae)

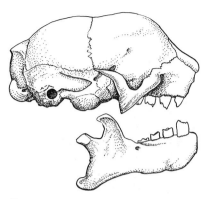

B

Three-toed sloth
(*Bradypus variegatus*, Bradypodidae)

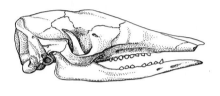

C

Nine-banded armadillo
(*Dasypus novemcinctus*, Dasypodidae)

FIGURE 9-9   Skulls of living xenarthrans: (A) two-toed anteater (length of skull 46 millimeters), (B) three-toed sloth (length of skull 76 millimeters), (C) nine-banded armadillo (length of skull 95 millimeters). *(A and B after Hall and Kelson, 1959 ©,* Mammals of North America, *reprinted with permission of John Wiley & Sons, Inc.)*

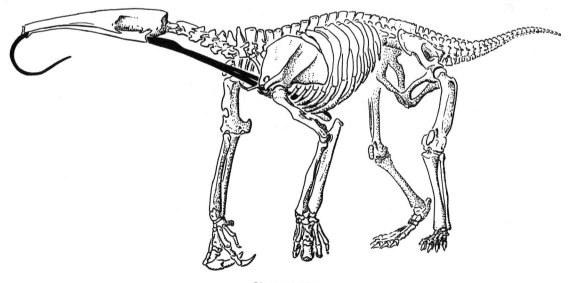

Giant anteater
(*Myrmecophaga tridactyla*, Myrmecophagidae)

FIGURE 9-10   Skeleton of the giant anteater. The tongue and tongue-retracting muscles are shown in black.

anteaters, the fur on the body and tail is shorter and the tail is prehensile.

Anteaters use the powerful forelimbs to expose ants and termites by tearing apart their nests; the insects are captured by the long tongue, swallowed whole, and possibly ground up by the thickened pyloric portion of the stomach. In the two genera of anteaters that climb trees, the claws of the manus are used to hook or to grasp as the animal travels along branches mainly by above-branch quadrupedal walking. *Cyclopes* is nocturnal and entirely arboreal, foraging for insects high in the trees. The two species of *Tamandua* are individualistically terrestrial and arboreal, nocturnal and diurnal. *Myrmecophaga* is entirely terrestrial and seems largely diurnal. *Cyclopes* and *Myrmecophaga* feed mainly on ants; *Tamandua* eats both ants and termites. The defensive behavior of these animals involves standing bipedally, bracing the body with the tail and hindlimbs, and slashing at the enemy with the forelimb claws.

FAMILY BRADYPODIDAE. Two living families of sloths are recognized. The members of Bradypodidae, called three-toed tree sloths or ais, are so highly modified for a specialized form of arboreal locomotion that they move somewhat awkwardly on the ground. They crawl along the ground slowly, using the claws of the forefeet and resting the body on the elbows. The hindlimbs are shorter and move as do those in other mammals (R. Timm, personal communication). The three Recent species of one genus *(Bradypus)* range from Central America (Honduras) through the northern half of South America to northern Argentina. These animals primarily inhabit tropical rain forests.

The adaptive zone of tree sloths is quite different from that of the anteaters and involves arboreal herbivory. The bradypodids differ strongly from the myrmecophagids, especially in skull characteristics. The tree sloth skull is short and fairly high, with a strongly reduced rostrum (Naples, 1982). The zygomatic arch is robust but incomplete, and its jugal portion bears a ventrally projecting jugal process similar to that present in many extinct xenarthrans (Fig. 9-9B). The premaxillary bones are greatly reduced, and the turbinal bones are complexly rolled, as in myrmecophagids. Five maxillary and four or five mandibular teeth are present. The persistently growing teeth are roughly cylindrical and have a central core of soft dentine surrounded successively by hard dentine and cementum.

A departure from the usual mammalian pattern of seven cervical vertebrae occurs in the bradypodids, in which eight or nine occur. **Xenarthrism** (Fig. 9-1) is developed in the lumbar and possibly thoracic vertebrae, and, as in some extinct ground sloths, the coracoid and acromion processes of the scapula are united. The three digits with long and laterally compressed claws are syndactylous (bound together). Three-toed tree sloths are of moderate size, weighing from 4 to 7 kilograms, and are covered with long, coarse hair. This fur provides a habitat for algae, which grow in transverse cracks on the surface of the hairs during the rainy season and tint the fur green. In addition, the adults of two genera of moths *(Bradypodicola* and *Cryptoses;* Pyralididae, Microlepidoptera*)* and a beetle hide in large numbers in the dense pelage. The tail is short.

These remarkably specialized animals eat leaves and descend to the ground only to defecate at five- to seven-day intervals. The stomach is chambered, and digestion is enhanced by fermentation aided by a **symbiotic** microbiota. Climbing is done in an upright position by embracing a branch or by hanging upside down and moving along hand-over-hand.

FAMILY MEGALONYCHIDAE. The sole surviving members of this once diverse family are two species of *Choloepus,* known as two-toed tree sloths or unaus (Fig. 9-11). They inhabit Neotropical rain forests from Nicaragua to Bolivia and southern Brazil. There are five upper and four or five lower teeth; the anteriormost teeth are canine-like and are kept sharp by abrasion between the posterior surfaces of the upper teeth and the anterior surfaces of the lower teeth (a process called **thegosis**). The number of cervical vertebrae is highly variable, from five to eight, the number differing from species to species and, in some cases, even from one individual to another of the same species. The forelimbs are considerably longer than the hindlimbs. As their name implies, there are two functional digits in the hand, with long grappling claws; the feet have three digits. Green algae and cyanobacteria grow on the hairs, camouflaging the

FIGURE 9-11   A two-toed tree sloth *(Choloepus hoffmanni),* an arboreal megalonychid. *(Corel Corp.)*

animal, as in *Bradypus,* but in longitudinal flutings rather than in transverse cracks. Like bradypodids, *Choloepus* are arboreal **folivores** (leaf eaters), and the two share many other convergently evolved aspects of their ecology, anatomy, and physiology.

FAMILY DASYPODIDAE. This family includes the armadillos, which differ from the South American anteaters and tree sloths in many ways but especially in having protective bony armor. Dasypodidae is the most diverse and widespread living family of xenarthrans. Some 20 Recent species, belonging to eight genera, collectively show a distribution from Kansas and much of the southeastern United States through Mexico and Central America into South America to near the southern end of Argentina. Armadillos occupy many ecological settings, from temperate and tropical forests to deserts.

The most obvious and unique structural feature of armadillos is the jointed armor. This consists of plates, bony scutes covered by horny epidermis, that occur in a variety of patterns but always include a head shield and protection for the neck and body (Fig. 9-12). Sparse hair usually occurs on the flexible skin between the plates and on the limbs and the ventral surface of the body. Individuals of some species can curl into a ball so that

their limbs and vulnerable ventral surfaces are largely protected by the armor. The largest species, the giant armadillo *(Priodontes maximus),* weighs up to 60 kilograms; the smallest, the pygmy armadillo *(Chlamyphorus truncatus),* is roughly the size of a small rat (120 grams).

The skull is often elongate and is dorsoventrally flattened; the zygomatic arch is complete, and the mandible is slim and elongate (Fig. 9-9C). With the exception of one species that has premaxillary teeth, the upper teeth are borne only on the maxillary bone. The teeth are **homodont** and nearly cylindrical and vary, both inter- and intraspecifically, from 7/7 to 18/19. Frequently, the teeth are partially lost with advancing age. The axial skeleton is fairly rigid and is partially braced against the carapace. The second and third cervical vertebrae, and, in four species, other cervical vertebrae as well, are fused (Fig. 9-2B); 8 to 13 sacral and caudal vertebrae form an extremely powerfully braced anchor for the pelvis (Fig. 9-2A); xenarthral articulations between thoracic and lumbar vertebrae (Fig. 9-1) produce a rigid vertebral column (Gaudin and Biewener, 1992). The skeleton braces, but does not come into contact with, the carapace.

In all armadillos, the limbs are powerfully built and the forefeet and hind feet bear large, heavy claws (Fig. 9-12). The feet are five-toed in all but one genus, and the foot posture is usually plantigrade. The tibia and fibula are fused proximally and distally and are highly modified for the origin

FIGURE 9-12   A naked-tail armadillo *(Cabassous centralis;* Dasypodidae). *(L. Ingles)*

of powerful muscles (Fig. 9-6A). Retia mirabilia occur in the limbs.

Armadillos are more generalized in their feeding habits and locomotion than other xenarthrans. Most species feed primarily on insects, but a variety of invertebrates, small vertebrates, and vegetable material are also eaten. All armadillos are at least partly adapted for digging, and some species are highly fossorial. Such a fossorial creature is the pygmy armadillo *(Chlamyphorus truncatus),* which utilizes a style of digging seemingly unique among mammals. The soil is dug away and pushed beneath the animal by the long claws of the forepaws, and the hind feet rake the soil behind the animal. The pelvic scute is then used to pack the soil behind the body. During the packing, the front limbs push the animal backward and the hindquarters vibrate rapidly from side to side (Rood, 1970). No permanent burrow is formed.

As indicated by their wide range, armadillos seem to be more resistant to cold than are tree sloths. In the northern parts of their range, however, armadillos may suffer 80 percent mortality during prolonged cold spells (Fitch et al., 1952). Nevertheless, armadillos have extended their range northward in the last 100 years (Humphrey, 1974; Taulman and Robbins, 1996).

Unusual reproductive strategies occur in the nine-banded armadillo, *Dasypus novemcinctus.* Females may experience an embryonic diapause sometimes lasting over two years; others give birth to litters in consecutive years without copulating with a male in between the first and second litters (Storrs et al., 1989). Often, the fertilized egg undergoes two divisions after which each of the four cells develop separately, resulting in identical quadruplets.

ORDER PHOLIDOTA

The pangolins, or scaly anteaters, members of the family Manidae, are represented today by a single genus *(Manis)* with seven species. Pangolins occur in tropical and subtropical parts of the southern half of Africa and in much of southeastern Asia. They are a monophyletic group, with their African and Asian branches forming two distinct radiations. Accordingly, some authors believe the separate branches should be afforded taxonomic

recognition as different genera. Their fossil record is poor but documents the Eocene occurrence of these animals in Europe, Asia, and Africa, and their Oligocene occurrence in North America (Rose and Emry, 1993).

At a glance, pangolins seem more reptilian than mammalian (Fig. 9-13). They are of moderate size, weighing approximately 5 to 35 kilograms. The skull is conical and lacks teeth; the dentary bones are slender and lack angular and coronoid processes. The tongue is extremely long and **vermiform.** Convergent with that of xenarthran anteaters, the tongue musculature of pangolins originates on a short posterior extension of the sternum (the xiphoid process) in Asian species, or on an enormously extended and bifid xiphoid process in African species. This extension passes into the posterior part of the abdominal cavity, curves upward, and ends near the kidneys. The tongue and its extrinsic musculature are therefore longer than the head and body together, allowing the tongue to be extremely protrusible (Chan, 1995).

The scales are the most distinctive feature of pangolins; they cover the dorsal surface of the body and the tail and are composed of agglutinated hair. The skin and scales account for a large share of the weight of these animals: Kingdon (1971) reports that these parts constitute one third to one half of the weight of the ground pangolin

FIGURE 9-13 A pangolin *(Manis crassicaudata,* Pholidota). *(Corel Corp.)*

(Manis temminckii). The manus and pes have long, recurved claws; the pes has five toes; and the manus is functionally tridactyl. The walls of the pyloric part of the stomach are thickened. This part of the stomach usually contains small pebbles (that may have been ingested accidentally) and seems to grind food as does the gizzard of a bird.

The food of pangolins is mostly termites, but ants and other insects are also eaten. The insects are located by smell, as in xenarthrans, and pangolins seem highly selective in their choice of food. Sweeney (1956) found that only rarely would a pangolin dig for the "wrong" species of ant or termite. Like xenarthran anteaters, some pangolins are nocturnal or diurnal; some are strictly terrestrial; some are semiarboreal; and two species (one in Java and one in Africa) are quite arboreal and have semiprehensile tails. Also like xenarthrans, pangolins have low metabolic rates. Pangolins roll up into a ball when disturbed, erect the scales, flail the tail, or move the sharp scales in a cutting motion. Some species spray foul-smelling fluid from the anal glands.

WEB SITES

Animal Diversity Web information about armadillos, anteaters, and sloths

http://www.oit.itd.umich.edu:80/bio108/Chordata/Mammalia/Xenarthra.shtml

10 DERMOPTERA AND CHIROPTERA

ORDER DERMOPTERA

Members of the Dermoptera are generally called colugos or, formerly, "flying lemurs" (a poor name choice, as they neither fly nor are lemurs). They have lemur-like faces and are able to glide long distances between trees. One family, Cynocephalidae,

with one genus (*Cynocephalus*) and two species (*C. variegatus* and *C. volans*), represents the order today. The two species are quite distinct and have often been considered to represent different genera (*Cynocephalus* and *Galeopithecus*). The distribution includes tropical forests from southern Myanmar (formerly Burma) and southern Indochina,

Malaya, Sumatra, Java, Borneo, and nearby islands to southern Mindanao and some of the other southern islands of the Philippine group.

In the past, numerous early Tertiary fossils from various parts of Europe and North America were considered to represent extinct dermopterans. None of these are true dermopterans related to the extant colugos, but they are convergent with them in dental features. The first true dermopteran fossil, *Dermotherium*, was reported recently from the late Eocene in Thailand (Ducrocq et al., 1992). The fossil, although only a jaw fragment, differs little from modern colugos despite being 34 million years old. Most phylogenetic hypotheses place the Dermoptera in a monophyletic group called Archonta, together with tree shrews, primates, and bats. A number of authors have considered the colugos to be most closely related to primates or to bats (Bailey et al., 1992; Novacek and Wyss, 1986; Simmons, 1993; Simmons and Quinn, 1994; Wible, 1993).

The two living species of *Cynocephalus* are modest in size, weighing roughly 1 to 1.75 kilograms, and have large eyes and faces that resemble those of Old World fruit bats and some lemurid primates (Fig. 10-1). The dorsal pelage color varies between the sexes and is very cryptic against tree bark; females are grayish and males brownish or chestnut, irregularly blotched with white. Although the molars have retained a basically three-cusped insectivore pattern, the broad cheek teeth have a shearing action that includes a large transverse component. This action, and the crenulated enamel of the molars, provide for efficient mastication of plant material (Rose and Simons, 1977). The anterior dentition is highly specialized: the lateral upper incisor (I2) is caniniform, and the first two lower incisors are broad and pectinate (comblike). The unusual lower incisors are used to groom the fur but may also be used to scrape leaves when the animal feeds (Rose et al., 1981). Unique among extant mammals, the canines are double-rooted. The dental formula is 2/3, 1/1, 2/2, 3/3 = 34.

A broad, furred membrane extends from the neck to near the ends of the fingers, between the limbs, and from the hind foot to the tip of the tail. The hands and feet retain five digits that bear needle-sharp, strongly curved claws for clutching branches. As in bats, the neural spines of the thoracic vertebrae are short, the sternum is slightly keeled, the ribs are broad, the radius is long, and the distal part of the ulna is strongly reduced (Fig. 10-2). A striking feature of colugos is their long, soft, and luxurious fur, similar to that of marsupial sugar gliders and flying squirrels. In addition to conferring excellent camouflage, the impressive fur may add an important aerodynamic component to gliding in that it appears to dampen turbulence (R. M. Timm, personal communication).

Colugos are **crepuscular** and seek refuge during the day in holes in trees. Several individuals may occupy the same den (Wharton, 1950). These animals invariably remain upside down while traveling along branches and feeding. Colugos are slow but skillful climbers, but they are unable to stand upright and are virtually helpless on the ground. They can glide distances well over 100 meters in traveling to and from feeding places. Their spectacular glides are facilitated in part by taking off from some of the tallest trees in the world's rain forests, with few lianas (vines) to block the way (Emmons and Gentry, 1983; Dudley and DeVries, 1990).

The diet includes leaves, buds, flowers, fruit, and sap from a variety of tree species. The enlarged tongue and specialized lower incisors are used in cowlike fashion in picking leaves (Winge, 1941). The great lengthening of the intestine typical of herbivorous mammals is well illustrated by colugos. *Cynocephalus*, which has a head plus body length of only about 410 millimeters, has an intestinal tract approaching 4 meters in length, nine times its head-and-body length (Wharton, 1950). The caecum, a blind diverticulum at the proximal end of the colon, is greatly enlarged (to about 48 centimeters in length) and is divided into

FIGURE 10-1   Colugo *(Cynocephalus volans)* with young clinging to a tree branch. *(P. Ward, Bruce Coleman Inc.)*

compartments. This chamber harbors microorganisms that help break down cellulose and other relatively indigestible carbohydrates. Caecal enlargement is usually associated with an herbivorous diet (as in many rodents).

The distribution of colugos is being restricted by the clearing of forests for agriculture, and in some regions the animals are hunted for their meat and their fur. Knowledge of the natural history of colugos remains rudimentary, and their study would make for difficult but fascinating field work, if it can be completed before their habitat is destroyed and they disappear.

ORDER CHIROPTERA

Accelerated research on bats in recent years has revealed fascinating aspects of the lives of these animals such as extraordinarily complex social behavior, involving polygamy by males and the use of an array of vocal communication signals. A coordinated assemblage of neuromuscular and behavioral adaptations allows bats to perceive in detail their prey and their environment by the use of sound. Bats are also unsurpassed by other mammals in the ability to survive through periods of stress or to conserve energy daily by drastic reductions in the metabolic rate and by tracking barometric pressure (Paige, 1995).

As many bat species become endangered, even as new genera and species are still being discovered (e.g., Kock and Storch, 1996), public awareness about bats and concern for their conservation has increased dramatically in recent years. Biologists and an increasing number of the general public are coming to realize not only that bats deserve respect as remarkably specialized products of at least 53 million years of evolution but also that they merit our protection for their importance in terrestrial ecosystems as efficient predators of insects and important pollinators of flowering plants.

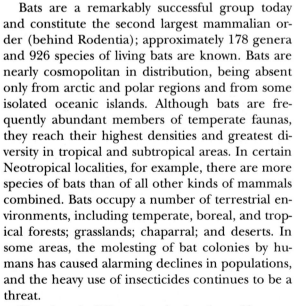

FIGURE 10-2 Skeleton of a colugo shown in a natural posture, hanging from a tree branch. In life, the terminal phalanges of the fingers and toes would bear very sharp, strongly curved claws for grasping.

Bats are a remarkably successful group today and constitute the second largest mammalian order (behind Rodentia); approximately 178 genera and 926 species of living bats are known. Bats are nearly cosmopolitan in distribution, being absent only from arctic and polar regions and from some isolated oceanic islands. Although bats are frequently abundant members of temperate faunas, they reach their highest densities and greatest diversity in tropical and subtropical areas. In certain Neotropical localities, for example, there are more species of bats than of all other kinds of mammals combined. Bats occupy a number of terrestrial environments, including temperate, boreal, and tropical forests; grasslands; chaparral; and deserts. In some areas, the molesting of bat colonies by humans has caused alarming declines in populations, and the heavy use of insecticides continues to be a threat.

Two sharply differentiated suborders of bats are recognized. The suborder Megachiroptera includes the family Pteropodidae, the Old World fruit bats, and the suborder Microchiroptera in-

cludes the other 17 families of bats. Microchiropterans are nearly cosmopolitan in distribution and are largely insectivorous. During recent years, a lively and exciting controversy flared about whether these two suborders are each other's closest relatives. One group of authors, lead mostly by J. D. Pettigrew, showed that certain Megachiroptera have a derived visual pathway in their brain that is similar to that in primates and different from that in Microchiroptera (Pettigrew, 1986, 1991a, 1991ab; Pettigrew et al., 1989; Pettigrew and Kirsch, 1995). These authors suggested that the complex visual pathway of megachiropterans and primates and other characters indicate that these two groups and Dermoptera are closely related, and that powered flight evolved independently in the two unrelated groups of bats. In contrast, other authors argued that the flight systems of the two groups of bats were inherited from a common ancestor and that the visual pathways of megachiropterans and primates evolved independently in these two groups (Baker et al., 1991; Wible and Novacek, 1988; Thewissen and Babcock, 1991; Simmons et

al., 1991; Simmons, 1993, 1994, 1995 and numerous references therein). This second group of authors generated further support for the monophyly of Mega- and Microchiroptera, based on
diverse other characters.

At the time this debate was peaking, preliminary
molecular data became available that supported
monophyly (Adkins and Honeycutt, 1993; Ammerman and Hillis, 1992; Bailey et al., 1992; Honeycutt
and Adkins, 1993; Mindell et al., 1991; Stanhope et
al., 1992). There is now a strong body of evidence
that the Microchiroptera and Megachiroptera are
monophyletic, but further molecular data are
needed from bats, primates, dermopterans, and
tree shrews to help resolve details of the relationships of these mammals. Recent studies also help to
clarify the phylogenetic relationships within the order Chiroptera and utilize diverse kinds of morphological and molecular data (Fig. 10-3) (Hoofer
and Van Den Bussche, 1998; Kirsch et al., 1998; Simmons and Geisler, 1998; Simmons, 1998).

Two functional contrasts between the megachiropterans and the microchiropterans are of particular importance. Megachiropterans are not known
to hibernate, although some nectar-feeding species
enter hypothermia with lowered rates of metabolism (McNab and Bonaccorso, 1995; Bonaccorso
and McNab, 1997). In contrast, many microchiropterans are heterothermic, and some hibernate
for long periods. In addition, whereas microchiropterans use sound that is associated with **echolocation** as a primary means of orientation and can
fly and capture insects, most megachiropterans use
vision. One exception is the megachiropteran
Rousettus, in which the ability to echolocate perhaps evolved independently. *Rousettus* uses clicks
made by the tongue as the basis for its acoustical
orientation (Novick, 1958a). All microchiropterans
use sound produced by the larynx. Often these
pulses are ultrasonic, that is, above the range of
human hearing. In most families the pulses are
emitted through the open mouth, but in those
with elaborate nose-leaves (Megadermatidae, Nycteridae, Rhinolophidae, and Phyllostomidae) the
sound pulses are emitted through the nostrils
while the mouth is kept closed. The differences
are reflected in underlying anatomy of the skull
and vocal apparatus (Freeman, 1984; Pedersen,
1998). Nose-leaves vary from simple to elaborate in
structure and are related in some yet unknown way

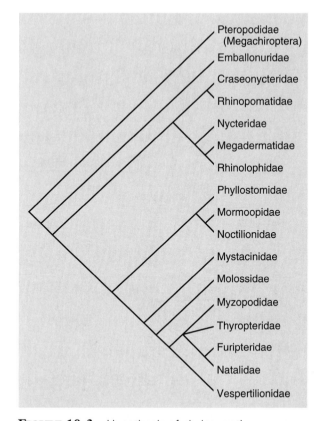

FIGURE 10-3   Hypothesis of phylogenetic
relationships of bats *(modified from Simmons, 1998,
to incorporate information from Hoofer and Van Den
Bussche, 1998; Kirsch et al., 1998; and Simmons and
Geisler, 1998).*

to echolocation and foraging style (Bogdanowicz
et al., 1997).

Echolocation, a means of perceiving the environment even in darkness or varied lighting conditions (see Chapter 22), and flight, a means of
great motility, have been two major keys to the success of bats. These abilities enable bats to occupy at
night many of the niches filled by birds during the
day. The abilities for echolocation and relevant
portions of the brain in Microchiroptera are incredibly complex and fascinating, possibly allowing
bats to "see" three-dimensional acoustic images
processed from information gleaned from echoes
(Simmons, 1995). In addition, the remarkably maneuverable flight of bats facilitates a mode of foraging for insects that birds have never exploited.
Heterothermy, allowing bats to hibernate or to operate at a lowered metabolic output during part of

the daily **torpor** cycle, has enabled these animals to occupy areas only seasonally productive of adequate food and to use an activity cycle involving only nocturnal or crepuscular foraging periods. The metabolic economy resulting from hibernation and from lowered metabolism during part of the daily cycle has affected the longevity of some bats. For their size, some microchiropteran bats are remarkably long-lived. *Myotis lucifugus,* a small bat weighing roughly 10 grams, may live as long as 35 years.

MORPHOLOGY

Many of the most important diagnostic features of bats are adaptations to flight. The bones of the arm and hand (with the exception of the thumb) are elongate and slender (Fig. 10-4). Flight membranes extend from the body and the hindlimbs to the arm and the fifth digit (**plagiopatagium**), between the fingers (**chiropatagium**), from the hindlimbs to the tail (**uropatagium**), and from the arm to the occipitopollicalis muscle (**propatagium**) (Fig. 10-5B). In some species, the uropatagium is present even when the tail is absent. The muscles bracing the wing membranes are often well developed and anchor a complex network of elastic fibers (Fig. 10-5A). Rigidity of the outstretched wing during flight is partly controlled by the specialized elbow and wrist joints, at which movement is limited to the anteroposterior plane.

In most microchiropteran species, the enlarged greater tuberosity of the humerus locks against the scapula at the top of the upstroke (Fig. 10-6), allowing the posterior division of the serratus ventralis thoracis muscle to tip the lateral border of the scapula downward to help power the downstroke of the wing (Vaughan, 1959; Fig. 10-7). The adductor and abductor muscles of the forelimb raise and lower the wings and are therefore the major muscles of locomotion. (In the contrasting arrangement found in terrestrial mammals, the flexors and extensors provide most of the power for locomotion.) The distal part of the ulna is reduced in bats, and the proximal section usually forms an important part of the articular surface of the elbow joint (Fig. 10-8). The clavicle is present and articulates proximally with the enlarged manubrium and distally with the enlarged acromion process and enlarged base of the coracoid process (see Fig. 10-7). The hindlimbs are either rotated to the side 90 degrees from the typical mammalian position and have a reptilian posture during quadrupedal locomotion or they are rotated 180 degrees, have a spider-like posture, and are used primarily to suspend the animal upside down from a horizontal support. The fibula is usually reduced, and support for the uropatagium, in the form of the **uropatagial spur** in Megachiroptera or **calcar** in Microchiroptera, is usually present (Fig. 10-9) (Schutt and Simmons, 1998).

The evolution of the muscular control pattern of the wing-beat cycle typical of microchiropteran bats has seemingly been strongly influenced by their use of echolocation. Highly maneuverable flight is essential for these bats because objects are perceived in detail only at fairly close range by echolocation (Aldridge and Rautenbach, 1987;

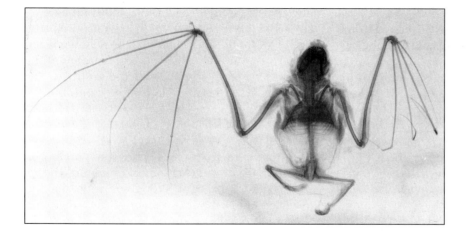

FIGURE 10-4   An x-ray photograph of the big fruit-eating bat (*Artibeus lituratus,* Phyllostomidae), showing the great elongation of the bones of the arm and hand. *(From Vaughan, 1970b)*

FIGURE 10-5 Ventral views of the wings of two bats, showing the parts of the wing and muscles and the elastic fibers that brace the membranes: (A) The big fruit-eating bat (*Artibeus lituratus,* Phyllostomidae); note the muscle strands that reinforce the plagiopatagium and the system of elastic fibers. This broad-winged bat does not remain on the wing for long periods. (B) The western mastiff bat (*Eumops perotis,* Molossidae). This narrow-winged bat is a fast and enduring flier. *(A from Vaughan, 1970a; B from Vaughan, 1970b)*

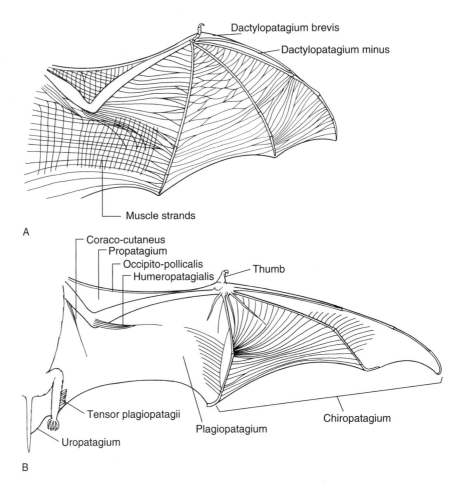

Fenton, 1994, 1995; Fenton et al., 1995). In contrast, birds use vision for more long-range perception of their environment and thus have relatively little need for extremely maneuverable flight. In both groups, similar trends toward rigidity of the axial skeleton and lightening of the wings occur, but many of the muscular and skeletal specializations that enable these animals to control their wings differ in the two groups. The pectoral girdle in birds is braced solidly by a tripod formed by the clavicula and coracoids anchored to the sternum and by the nearly bladelike scapula, which rests almost immovably against the rib cage. The pectoralis and supracoracoideus muscles, both of which originate on the sternum, supply nearly all the power for the wing beat (see Fig. 10-7).

In bats, nearly the reverse mechanical arrangement for flight occurs: the scapula is braced against the axial skeleton by the clavicle alone, movements of the clavicle during flight increase

flight efficiency (Hermanson, 1981), and the job of powering the wing beat is shared by many muscles (Vaughan, 1959; see Fig. 10-7). This division of labor is made possible partly because the scapula is free to rotate on its long axis. The pectoralis, the posterior division of the serratus ventralis thoracis, and the clavodeltoideus muscles control the downstroke of the wings; only the pectoralis originates on the sternum. The muscles of the deltoideus and trapezius groups and the supraspinatus and infraspinatus muscles largely power the upstroke. The subscapularis is responsible for fine control of the wings during the entire wing-beat cycle (Hermanson and Altenbach, 1981; Altenbach and Hermanson, 1987).

A morphological trend of critical importance to bats and all other flying animals is toward the reduction of wing weight. Propulsion is obtained in all flying animals by movement of the wings, and the kinetic energy produced by such movement de-

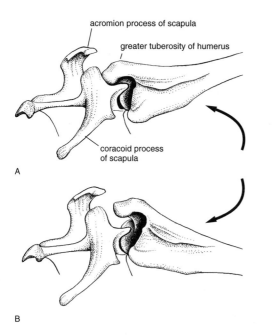

A

B

◄ **FIGURE 10-6** Anterior view of the left shoulder joint of a free-tailed bat *(Molossus ater)* at (A) the top of the upstroke and (B) during the downstroke. The greater tuberosity of the humerus locks against the scapula at the top of the upstroke, transferring the responsibility for stopping this stroke to the muscles binding the scapula to the axial skeleton. During the downstroke, the greater tuberosity of the humerus moves away from its locked position. This type of action and this type of shoulder joint also occur in the Vespertilionidae and other derived families of bats. *(From Vaughan 1970a)*

pends upon the speed and weight of the wing. The amplitude of a stroke and its speed are progressively greater toward the wing tip. Consequently, reduction of the weight of the distal parts of the wing results in a reduction of the kinetic energy developed during a wing stroke. A considerable advantage in metabolic economy is thus gained, for as less kinetic energy is developed during each stroke, less energy

is necessary to control the wings. In addition, light wings can be controlled with speed and precision during extremely rapid maneuvers when bats chase flying insects. Equally important, reduction in thickness of cortical bone in the humerus and radius of the bat wing is an adaptation to resist the large torsional stresses generated on these bones during flight (Swartz et al., 1992). Such torsion is not normally encountered during locomotion by terrestrial mammals, whose bones are relatively thick-walled, but bird and pterosaur forelimb bones are thin-walled like the humerus and radius of bats. Interestingly, the more distal elements in the wings (metacarpals and phalanges) of bats show the opposite trend, with greatly increased cortical thickness (Swartz et al., 1992).

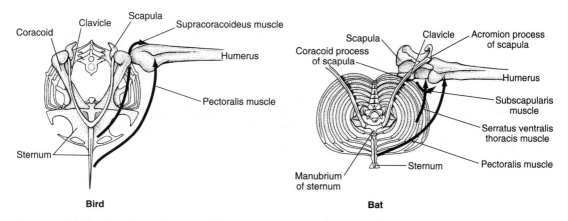

Bird   **Bat**

FIGURE 10-7 Anterior views of the thorax and part of the left forelimb of a bird and a bat, with some of the major muscles controlling the wing-beat cycle shown diagrammatically. In the bird, the supracoracoideus muscle raises the wing and the pectoralis muscle powers the downstroke; both muscles originate on the sternum. In the bat, the downstroke is primarily controlled by three muscles: the subscapularis, the serratus ventralis thoracis, and the pectoralis. Only the pectoralis originates on the sternum. Many muscles power the upstroke in bats. *(From Vaughan, 1970a)*

FIGURE 10-8 Lateral view of the right elbow of (A) a myotis (*Myotis volans*, Vespertilionidae) and (B) a free-tailed bat (*Molossus ater*, Molossidae). *(From Vaughan, 1970a)*

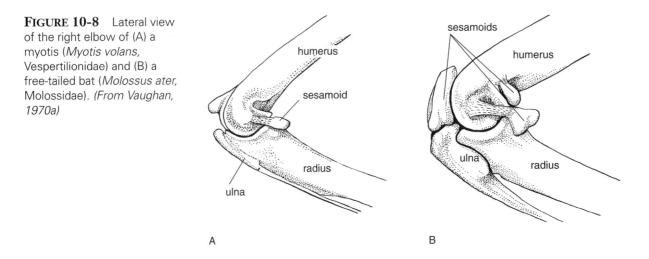

A   B

Reduction of the wing weight has been furthered in bats by many specializations. Movement at the elbow and wrist joints is limited to one plane, thus eliminating musculature involved in rotation and bracing at these joints. In addition, the work of extending and flexing the wings is transferred from distal muscles (of the forearm and hand) to large proximal muscles (pectoralis, biceps, and triceps), thereby allowing a reduction in the size of the distal musculature (Vaughan, 1959). Certain forearm muscles are made nearly inelastic by investing connective tissue. Because of this modification and specializations of their attachments,

these muscles "automatically" extend the chiropatagium with extension at the elbow joint or flex the chiropatagium with flexion at this joint (Vaughan and Bateman, 1970; Fig. 10-10).

The hindlimbs of bats are generally quite thin but are not drastically reduced in length because of their importance in supporting the trailing edge of the plagiopatagium and the lateral edge of the uropatagium. The thinness of the hind legs and the fact that bats generally hang upside down to roost probably evolved under selective pressures favoring reductions in weight and the advantage of a quick takeoff to escape predators. As an adaptation

A   B   C   D

FIGURE 10-9 A fishing bat *(Noctilio albiventris)*, showing several stages in the wing-beat cycle. A bone (the calcar) braces the uropatagium next to the ankle. (A) Top of the upstroke. (B) Midway through the downstroke. (C) End of the downstroke. (D) Midway through the upstroke. *(By J. Altenbach)*

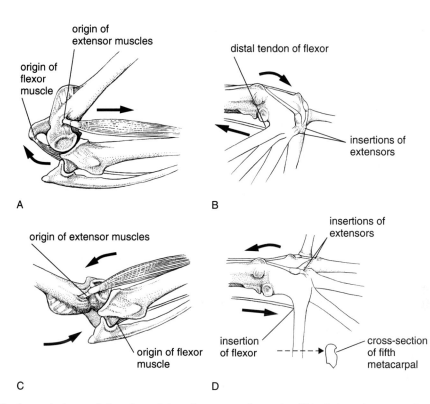

FIGURE 10-10 Lateral views of the elbow joint of the leaf-chinned bat *(Mormoops megalophylla),* showing the "automatic" flexion and extension of the fingers caused by certain forearm muscles in many advanced bats. (A) Flexion of the elbow joint moves the origins of the extensor muscles toward the wrist and the origin of one flexor muscle away from the wrist. (B) Because the flexor muscle is largely inelastic, with flexion at the elbow, the distal tendon of the flexor pulls on the fifth digit and tends to flex the fingers. (C) With extension at the elbow joint, the origin of the extensor muscles is moved away from the wrist and the origin of the flexor muscle moves toward the wrist. (D) This action pulls the extensor tendon toward the elbow and releases tension on the flexor tendon, thus extending the fingers. In D, the complex cross-sectional shape of the fifth digit is shown. *(From Vaughan and Bateman, 1970)*

to hanging, the delicate femur is suited to tensional stresses rather than compressional stresses associated with the femur of terrestrial mammals. In addition, many bats possess a tendon-locking mechanism on the tendons of the digital flexors that, when locked during hanging, reduces the muscular activity (Quinn and Baumel, 1993).

FLIGHT

The three modern groups of flying animals — insects, birds, and bats — are all highly successful. Viewing the terrestrial scene, there are more flying than nonflying species of animals, but each flying group has evolved a different type of wing: bird wings are formed of feathers braced by a simplified forelimb skeleton along the leading edge; insect wings are membranous sheets of chitin braced by intricate patterns of chitinous veins; and bat wings are sheets of skin braced by the five-digited forelimb and elastic connective tissue. Flight styles also differ. Birds usually depend on relatively fast and not especially maneuverable flight. Insects usually use extremely rapid wing beats, a variety of flight speeds, and often a remarkable ability to hover. Most bats, however, use slow, highly maneuverable flight. As might be expected, diverse and complex mechanical and aerodynamic problems are faced by these groups of fliers, and animal flight remains incompletely understood. Inasmuch as 20 percent of all mammal species are bats, it is important to understand the phenomenon of flight in bats. Competing hypotheses about the evolution of flight and echolocation in bats have been proposed or discussed recently by Fenton et al. (1995), Speakman (1993), Norberg (1994), Arita and Fenton (1997), and Simmons and Geisler (1998).

Most students have been introduced at least once to the basic aspects of aerodynamics; this topic can therefore be treated briefly. Because the wings of animals usually provide both the thrust and the lift necessary for sustained flight, whereas in aircraft the wings provide only the lift, flight in animals presents special problems.

Lift is generated when an airstream sweeps over a wing with an asymmetrical cross section. The profile of the cross section of a wing (the airfoil) varies widely from one species of flying animals to another, but characteristically in birds and bats it has an arched dorsal surface and a concave ventral surface (Fig. 10-11B). The tendency is for the parts of the airstream flowing over the opposite surfaces of the wing to arrive at the trailing edge simultaneously; this necessitates faster movement of air over the dorsal surface than over the ventral surface. The more rapidly the air moves over a surface, the less pressure it exerts, a relationship described by Bernoulli in 1738 and exploited by flying vertebrates for over 150 million years. The unequal forces on opposing wing surfaces creates lift, a force opposing the force of gravity on a flying animal.

Lift is also created when a surface is presented at an **angle of attack** to the airstream. (The angle of attack is the angle that the chord line of the airfoil makes with the plane of motion of the wing: Fig. 10-11A). Within limits, lift can be increased by raising the angle of attack. When, however, the angle of attack becomes so great that the air moving over the upper surface breaks away from the wing and forms turbulent eddies, the lift produced by the wing abruptly falls as the **drag** sharply rises and stalling occurs. (Drag is the force exerted by air on an object in motion and in a direction opposite that of the motion.)

Wing performance is influenced by a series of variables. Lift increases directly as the surface area of the wing, but so does drag; lift increases (within

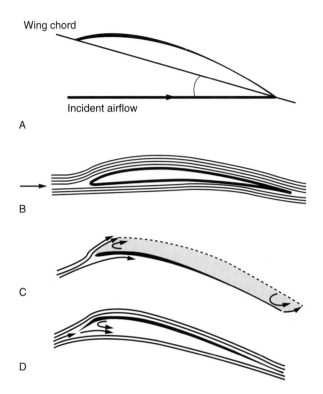

FIGURE 10-11   Cross sections of wings (airfoils) and air flow over the wings. (A) Thin airfoil, showing the angle of attack, the angle between the wing chord and the incident air flow. (B) Flow of air over a thick airfoil. (C) Turbulence and separated air flow over a wing at a high angle of attack. (D) The addition of a leading-edge flap keeps the air flowing smoothly over the surface. *(After Norberg, 1972)*

limits) as the **camber** of an airfoil increases (camber is the curvature, or arching, of an airfoil), but this also increases drag; lift increases as the square of the speed, as does drag. Intuitively, then, one might expect that some of the constraints forcing modifications of wing design on fast fliers are of relatively little importance in slow-flying bats. This seems to be true and leads us to a consideration of the unique structure and function of the chiropteran wing.

The wings of bats form very thin airfoils of high camber. Several important features enhance the performance of these wings in the low-speed flight typical of most bats (Norberg, 1969, 1972, 1981). Thin airfoils, essentially cambered membranes, are more effective in producing high lift at low speeds than are conventional airfoils with some thickness.

Of further importance is the ability of the bat to vary the camber of the wing in the interest of producing high lift at low speeds. Camber of the bat wing is largely under the control of the occipitopollicalis muscles, the flexors of the thumb, the inclination of the dactylopatagium minus (see Fig. 10-9), the fifth digit, and the hindlimbs.

Compared to birds, bats have low **wing loadings** (Norberg, 1981). (Wing loading is the ratio of body weight to wing area [W/S]. In general, the lower the wing loading, the slower an animal can fly and still maintain adequate lift to remain airborne). Most bats also have broad wings with a low **aspect ratio,** which is the relationship of the length of a wing to its mean breadth that for wings of irregular shape is expressed as the ratio of the span squared to the wing area (b^2/S). Some bats that fly rapidly and remain in flight for long periods have long, narrow, high-aspect-ratio wings (see Fig. 10-5B). Broad wings suffer some loss of lift owing to air spillage from the high-pressure area on the ventral surface to the low-pressure area on the dorsal surface of the wing tip. Wings that are strongly tapered toward the tip minimize this spillage and loss of lift and are typical of fast-flying bats.

To produce lift, an airfoil must move through the air, and this requires a means of propulsion. In animals, propulsion is created by movements of the wings, and photographs of the wing-beat cycle in bats in level flight indicate that the downstroke is the power stroke and the upstroke is largely a recovery stroke (see Fig. 10-9). During the downstroke, the wings are fully extended and the powerfully braced fifth digit and the hindlimbs maintain the plagiopatagium at a fairly constant angle of attack, but the air pressure against the membranes becomes progressively greater toward the wing tip as the speed of the wing increases. This increase in pressure, coupled with the elasticity of the membranes between the digits, causes the trailing edges of the chiropatagium to lag behind the well-braced leading edge. In effect, the wing tip is twisted into a propeller-like shape and serves a propeller-like function. As the wing tip sweeps rapidly downward, it tends to force air backward, resulting in forward thrust of the animal. The membrane between the third and fourth digits (dactylopatagium longus) is probably of primary importance in producing thrust. During the upstroke, or recovery stroke, the wing is partly

flexed, the stroke is directed upward and, to some extent, backward, and the force of the air stream partially aids the movement. Judging in part from the large muscles that power the downstroke and the relatively small muscles that control the upstroke, one would expect that the latter demands relatively little power and energy.

Some bats can fly very slowly, and some can hover; during these types of flight, the action of the wings is different from that used in level flight. When the nectar-feeding bat *Leptonycteris curasoae* hovers, the downstroke is directed largely forward and the upstroke is directed backward (Fig. 10-12). The posture of the wings during the downstroke is similar to that in level flight, but because the stroke is largely horizontal, vertical thrust is developed. The upstroke, however, is complicated by a reversal of the usual posture of the wing tip: the tip turns over in such a way that the dorsal surface of the chiropatagium faces downward, and the leading edge of the wing still leads in this stroke but is posterior to the trailing edge (Fig. 10-12). Toward the end of the upstroke, the reversed wing tip is flipped rapidly backward and produces considerable upward thrust; at the start of the downstroke, the wing tip swings into its normal posture. This powerful flip probably demands considerable energy, but the vertical thrust that it develops strongly augments the thrust resulting from the downstroke and enables the bat to remain nearly stationary in the air (Altenbach, 1977). Probably because of the high energy cost of hovering, it is generally used only briefly by bats.

During the early evolution of the bat wing, selection seemingly favored refinements in design that allowed the development of high lift at low speeds. Later, however, in the Eocene epoch, bats underwent an adaptive radiation involving, in part, exploitation of various styles of flight (Habersetzer and Storch, 1989; Habersetzer et al., 1994). The wings of some bats (members of the family Molossidae and some members of the family Emballonuridae, for example) developed characteristics advantageous during rapid flight. Because lift varies as the square of the speed of an airfoil, it would seem that rapid-flying bats could afford the luxury of higher wing loadings because of the greater lift developed per unit of wing area at higher speeds. Because drag also increases as the square of the speed, however, a reduction in wing surface area, angle of attack, and

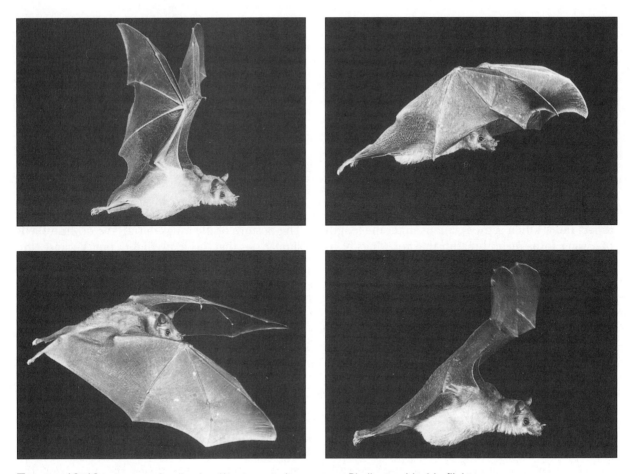

FIGURE 10-12   A nectar-feeding bat (*Leptonycteris curasoae,* Phyllostomidae) in flight, showing some positions of the wings during slow or hovering flight. *(J. Altenbach)*

camber during rapid flight would be highly advantageous. Wing design in rapid-flying bats is clearly the result of a series of evolutionary compromises, and not all these animals have wings that are alike; nonetheless, a number of bats have evolved roughly the same type of high-speed wing. (High speed is used here only in a relative sense, for probably few bats achieve speeds in level flight above 80 kilometers per hour.)

PALEONTOLOGY

Because of their small size, delicate structure, and greatest abundance in tropical areas (where fossils are more difficult to find), bats are relatively rare as fossils. Consequently, the evolution of bats is

poorly known. Ironically, however, the best bat fossils known also happen to be some of the earliest known (Simmons and Geisler, 1998). The earliest undoubted fossil bat, *Icaronycteris index* of the extinct family Icaronycteridae, is from early Eocene and possibly late Paleocene beds in Wyoming (Fig. 10-13). This species was described on the basis of one beautifully preserved specimen (Jepsen, 1966). Although this bat has several primitive features, such as claws on the first two digits of the hand and fairly short, broad wings, its basic limb structure is that of modern bats. The upper molars of *Icaronycteris* have the W-shaped ectoloph typical of most insectivorous bats, and this bat has been put in the suborder Microchiroptera. Different Eocene bats have been found to have moth scales (Micro- and

Macrolepidoptera), and parts of the chitinous exoskeletons of beetles (Coleoptera), cockroaches (Blattoidea), mosquitoes (Diptera: Culicidae), and caddis flies (Trichoptera) among their fossilized gut contents (Habersetzer et al., 1994). Eocene and early Oligocene deposits in Europe, Asia, and North America have yielded the earliest records of the modern microchiropteran families Emballonuridae, Megadermatidae, Nycteridae, Rhinolophidae, Natalidae, Molossidae, and Vespertilionidae (Simmons and Geisler, 1998; Sigé et al., 1994). Megachiropterans appear first in the fossil record in the late Eocene of Thailand (Ducrocq et al., 1992, 1993).

The Paleocene or Eocene appearance of a bat clearly well adapted for flight, the Eocene appearance of at least two dozen genera, including some belonging to modern families (Simmons and

Geisler, 1998), and the assignment by paleontologists of fossils from the late Eocene to the still-living genus *Rhinolophus* indicate an early origin of bats. This antiquity is further indicated by the fact that some genera were essentially as they are now in the Oligocene, a time when horses were three-toed and no bigger than sheep, and bears and antelope had not yet appeared. The wings of an Oligocene molossid bat of the genus *Tadarida,* for example, are nearly identical to those of present-day members of this family (Sigé, 1971). Although no unambiguous fossil evidence clarifies the matter, a late Cretaceous origin of bats seems probable, and molecular evidence is beginning to support this view.

SUBORDER MEGACHIROPTERA

Only one family represents this suborder today. The descriptions given here for the family Pteropodidae characterize the suborder Megachiroptera.

FAMILY PTEROPODIDAE. Most pteropodids are fruit eaters, others nectar and pollen feeders, and many species are called flying foxes because of their foxlike faces and large size. These bats are abundant and often conspicuous members of many tropical biotas in the Old World. This family is represented by 42 Recent genera and 166 Recent species. Pteropodids occur widely in tropical and subtropical regions from Africa and southern Eurasia to Australia and on many South Pacific islands eastward to Samoa and the Carolines. Members of this family are often large, up to 1.5 kilograms in weight and 1.2 meters in wing span, but some are small (13 grams in weight and 245 millimeters in wing span).

The Megachiroptera have the following features that set them apart from the Microchiroptera. The face is usually foxlike or lemur-like, with large eyes, usually a moderately long snout with a simple, unspecialized nose pad, and simple ears lacking a **tragus** (Fig. 10-14). (The tragus, a fleshy projection of the anterior border of the ear opening, may be seen in the Vespertilionidae; see Fig. 10-20C.) The orbits are large and are bordered posteriorly by well-developed postorbital processes that may meet to form a postorbital bar. The rostrum is never highly modified (Fig. 10-15). The dental formula is 1-2/0-2, 1/1, 3/3, 1-2/2-3 = 24-34. The molars are never **tuberculosectorial** with W-shaped ectolophs, as in most microchiropterans, but instead are low,

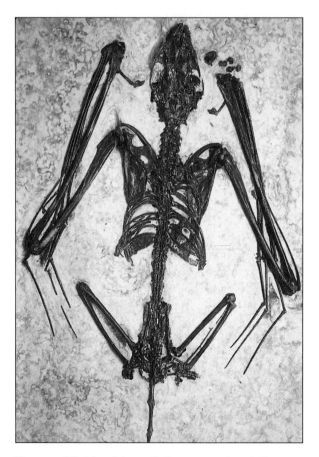

FIGURE 10-13   A beautifully preserved early Eocene bat (*Icaronycteris index,* Icaronycteridae). *(From Jepsen, 1970)*

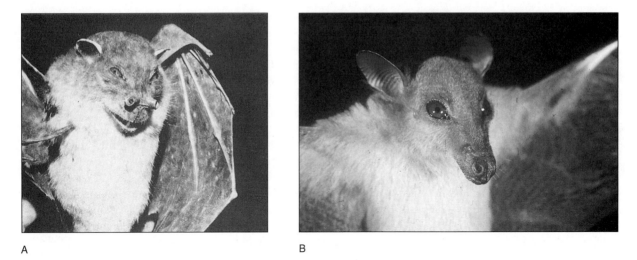

A   B

FIGURE 10-14 Faces of megachiropteran bats (Pteropodidae): (A) a tube-nosed bat
(*Nyctimene* sp.) from New Guinea, (B) a nectar-feeding pteropodid *(Macroglossus
minimus)* from Sabah, Borneo. *(A by M. Fenton; B by C. Francis)*

moderately flat-crowned, more or less quadrate, and
lacking in stylar cusps (Fig. 10-16). The teeth are
adapted basically to crushing fruit. The wing is primitive in having two clawed digits, except in the genera
Eonycteris and *Dobsonia* (in which only the thumb
has a claw); the greater tuberosity of the humerus
is not enlarged to make contact with the scapula
at the top of the upstroke (Fig. 10-17). The tail typically is short or rudimentary. The uropatagium is
supported posteriorly by a cartilaginous uropatagial spur that projects from the tendon of the gastrocnemius muscle, unlike the calcar of microchiropterans (Schutt and Simmons, 1998).

Broadly speaking, pteropodids eat two types of
food. One group are fruit eaters, whereas members of the other eat mostly nectar and pollen. Formerly the two types were considered to represent
two different subfamilies, the fruit eaters placed in
the subfamily Pteropodinae and the nectarivorous
bats (see Fig. 10-14) in the subfamily Macroglossinae. However, the nectar-feeding habit may have
evolved as many as five different times among
pteropodids, and the taxon Macroglossinae has
been abandoned (Kirsch et al., 1995). Few types of
pteropodids (some of the genus *Nyctimene*) eat insects. The fruit eaters have fairly robust or moderately reduced dentitions. The jaws in these species
are usually fairly long, or, in some species that presumably eat hard fruit, the jaws are shorter and the

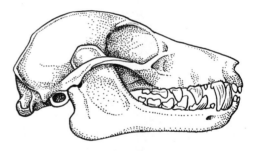

Pteropus sp., Pteropodidae

FIGURE 10-15 The skull of a megachiropteran bat,
length of skull 62 millimeters.

teeth and dentary bones are unusually robust. The
fruit bats often roost in trees in large colonies; in
the case of the Australian species *Pteropus scapula-
tus,* as many as 100,000 have been observed roosting together. Fruit bats occasionally travel long distances during their nocturnal foraging, and
Pteropus alecto regularly flies at least 70 kilometers
from roosting sites to feeding areas. The fruit
eaters usually are not particularly maneuverable
fliers but have a steady, direct style of flight.
Pteropodids are adroit at climbing about in vegetation, where the clawed first and second digits of
the wing come into play (Fig. 10-18).

Hypsignathus monstrosus, the hammer-head bat,
is unique among mammals in the fantastic degree

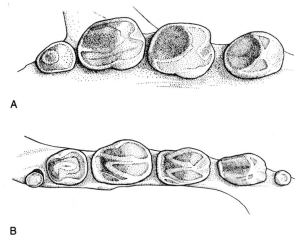

A

B

Pteropus sp.

FIGURE 10-16   The cheek teeth of a megachiropteran bat: (A) right upper tooth row, showing two molars and two premolars, (B) lower left tooth row, showing three premolars and three molars. *(From Vaughan, 1970b)*

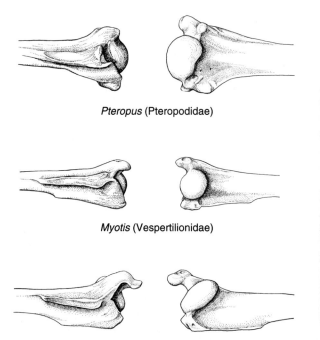

Pteropus (Pteropodidae)

Myotis (Vespertilionidae)

Molossus (Molossidae)

FIGURE 10-17   The proximal end of the right humerus in three bats. Anterior views are on the left and posterior views on the right. *(From Vaughan, 1970b)*

to which the vocal apparatus is specialized in the males. This large, frugivorous pteropodid, with a wing spread approaching 1 meter, occupies tropical forests in much of central Africa. Communal displays by males in courtship areas are important in the breeding cycle of *Hypsignathus*. The males on the courtship arena, which is called a **lek,** use a penetrating call, described by Kingdon (1984a) as "gutteral, explosive, and blaring," to attract females. The remarkable specializations of the vocal apparatus clearly evolved in association with the loud vocalizations employed during breeding displays.

Externally, the most striking feature of the male is the strange hammer-head appearance (Fig. 10-19). This is due in part to the enlarged and elevated nasal bones but is accentuated by a large pouch that encloses the rostrum and extends back over the cranium. These features enhance the resonance of the calls, and pharyngeal sacs in the throat are probably also resonators. Equally impressive are internal features attending the massive enlargement of the larynx. This structure, which contains huge vocal cords, has moved into the thorax, where it occupies most of the space filled by

FIGURE 10-18   A male Dyak bat *(Dyacopterus spadiceus)* hanging from a branch by its feet and thumb claws. This species is the only known mammal in which males produce milk from their nipples. *(C. Francis)*

the heart and lungs in other mammals. As a result of this migration, the large trachea lies against the diaphragm and curves sharply craniad to the lungs, which are also forced against the diaphragm. The thoracic cavity thus serves largely as a container for the huge larynx in male *Hypsignathus,* with a drastic sacrifice in lung capacity. Kingdon has called this animal a flying loud-speaker; this characterization seems especially apt when one considers the enlarged lips of the males, which can be formed into almost perfect megaphones. The anatomy of the larynx is described by Schneider et al. (1967), and early studies on the morphology of the vocal apparatus of this bat were carried out by Matschie (1899) and Lang and Chapin (1917).

The pteropodids that eat nectar and pollen have long, slender rostra, strongly reduced cheek teeth, and delicate dentary bones. The tongue is long and protrusible and has hairlike structures at its tip to which pollen and nectar adhere. Pollen, which adheres to the fur and is ingested when the bats groom themselves, is probably an essential source of protein to nectar-feeding bats. Some species roost in groups in caves, and some roost solitarily in vegetation. Flight is slow and maneuverable.

SUBORDER MICROCHIROPTERA

Recent members of this suborder are usually small, with body weights ranging from 2 to 196 grams. The eyes are often small, the rostrum is usually specialized, and the nose pad and lower lips may be modified in a variety of ways (Fig. 10-20). The ears have a tragus, a small flap of skin at the base of the pinna, in all but members of the family Rhinolophidae, are usually complex, and are frequently large. The postorbital process is usually small. Dentitions vary tremendously, but most microchiropterans (except some members of the Phyllostomidae) have tuberculosectorial molars; the upper molars have a W-shaped ectoloph with strongly developed stylar cusps, and in the lower molars the trigonid and talonid are roughly equal in size (Fig. 10-21A). In many insectivorous species and in some frugivorous members of the family Phyllostomidae, one or more premolars above and below are caniniform, and in some insectivorous species the premaxillae are separate (Fig. 10-22).

The flight apparatus of the microchiropterans is more derived than that of the megachiropterans. In microchiropterans, the second digit does not bear a claw and lacks a full complement of phalanges, and its tip is connected by a ligament to the joint between the first and second phalanges of the third digit. During flight, this connection allows the second digit to brace the third digit, which forms much of the leading edge of the distal part of the wing, against the force of the airstream (Norberg, 1969). The greater tuberosity of the humerus is usually enlarged and locks against a facet on the scapula at the top of the upstroke of the wings (see Fig. 10-6). The size of the tail and uropatagium varies. The uropatagium is supported posteriorly by a cartilaginous or bony calcar that articulates directly with the calcaneus of the ankle (see Fig. 10-9), unlike the uropatagial spur of megachiropterans (Schutt and Simmons, 1998). The shape of the wing varies according to foraging pattern and style of flight. In general, slow,

FIGURE 10-19   Two views of the head of a male hammer-head bat *(Hypsignathus monstrosus),* showing the highly modified lips and inflated rostrum. These specializations are associated with the ability to produce very loud, resonant sounds. *(R. Peterson)*

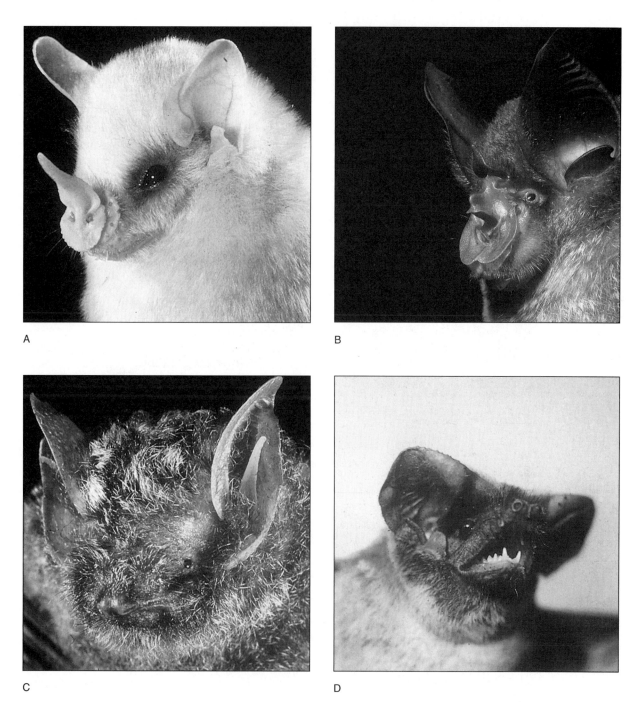

A

B

C

D

FIGURE 10-20   Faces of some microchiropteran bats: (A) Honduran white bat (*Ectophylla alba;* Phyllostomidae). (B) Old World leaf-nosed bat (*Rhinolophus malayanus,* Rhinolophidae) from Laos. (C) Groove-toothed bat (*Kerivoula [Phoniscus] atrox,* Vespertilionidae; this species has extremely long, flattened upper canines, the function of which is unknown) from Pasoh, peninsular Malaysia. (D) African free-tailed bat (*Tadarida pumila;* Molossidae) from Kenya. *(A by M. Tuttle, Bat Conservation International; B and C by C. Francis; D by R. Vaughan)*

maneuverable fliers have short, broad wings and rapid, enduring fliers have long, narrow wings (see Fig. 10-5).

Since their divergence from primitive insectivore stock, perhaps in the Cretaceous or early Paleocene, microchiropteran bats have undergone a remarkable adaptive radiation. There are 17 Recent families of microchiropterans (18 by some classification schemes); approximately 136 genera are now recognized. This large number of families and genera reflects the great structural diversity and widely contrasting modes of life within this suborder.

FAMILY EMBALLONURIDAE. This family contains a variety of bats that are frequently called sac-winged or sheath-tailed bats. They range in size from small (about 4 grams) to large (up to 105 grams). *Saccolaimus peli,* an African emballonurid, is among the largest of the insectivorous microchiropterans, with a wing spread of nearly 70 centimeters. The small emballonurids have a wing span of about 240 millimeters. Thirteen genera and about 47 species are currently recognized. The wide geographic range of emballonurids includes the Neotropics (much of southern Mexico, Central America, and northern South America),

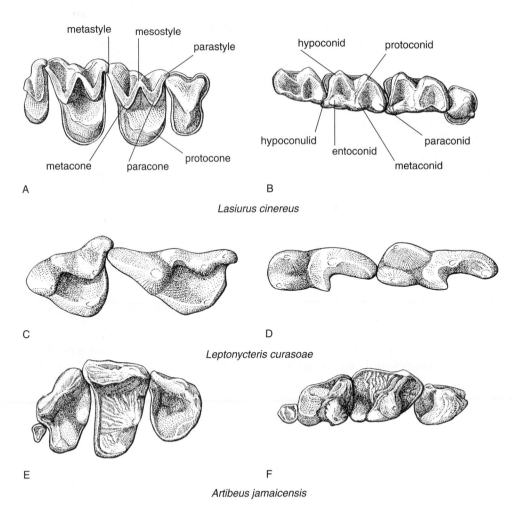

FIGURE 10-21   (A) Upper and (B) lower fourth premolar and three molars of an insect-eating vespertilionid. (C) Upper and (D) lower second and third molars of a nectar-eating phyllostomid. (E) Upper and (F) lower fourth premolar and three molars of a fruit-eating phyllostomid.

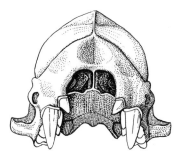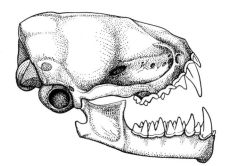

FIGURE 10-22 The skull of the hoary bat (Vespertilionidae): (A) anterior view, showing the emarginate front of the palate, (B) side view, showing the shortened rostrum typical of some insect-feeding bats. Length of skull 17 millimeters.

Hoary bat
(*Lasiurus cinereus*)

most of Africa, southern Asia, most of Australia, and the Pacific Islands east to Samoa. Emballonurids are known from the middle Eocene of Europe.

These small bats combine a number of primitive features with several noteworthy specializations. In the possession of postorbital processes and reduced premaxillaries that are not in contact with one another, emballonurids resemble pteropodids. In addition, the shoulder and elbow joints are primitive. In contrast to the rhinopomatids, emballonurids retain only the metacarpal in the second digit; the flexion of the proximal phalanges of the third digit onto the dorsal surface of the third metacarpal is a specialization also found in some advanced families of bats. Obvious external specializations include a glandular sac in the propatagium in some genera and the emergence of the tail from the dorsal surface of the uropatagium. The nose is simple; that is to say, it lacks leaflike structures or complex patterns of ridges and depressions. In addition to the more common gray and dark brown species of emballonurids, some species of two genera *(Rhynchonycteris* and *Saccopteryx)* have whitish stripes on the back, and members of the genus *Diclidurus* are entirely white.

These insectivorous bats inhabit tropical or subtropical areas, where they use a great variety of roosting sites. Emballonurids occupy houses, caves, culverts, rock fissures, hollow trees, vegetation, or the undersides of rocks and dead trees for daytime retreats and usually roost in colonies. They are often fairly tolerant of well-lighted situations. In East Africa, *Taphozous mauritianus* often roosts on the trunks of large trees such as baobab trees *(Adansonia digitata)*. In some areas, emballonurids proba-

bly forage mainly over water. Some members of the genus *Taphozous* have long, narrow wings, are swift and dashing fliers, and often forage in clearings and above the canopies of tropical forests. One emballonurid *(Taphozous mauritianus)* was recorded foraging at an altitude of 600 meters in Zimbabwe (Fenton and Griffin, 1997).

A distinctive feature of some emballonurids is a sac in the propatagium that may be glandular. Studies of one Neotropical species *(Saccopteryx bilineata)* has shown that this sac, especially well developed in males, is used in ritualized displays during the breeding season.

FAMILY CRASEONYCTERIDAE. This family is found only in Thailand. As far as is known, only one species *(Craseonycteris thonglongyai)* represents the family. The common name is bumble-bee bat or Kitti's hog-nosed bat.

Craseonycteris is delicately built and is one of the smallest living mammals; adults weigh about 2 grams. It has small eyes and large ears. The premaxillae are not fused to adjacent bones, a feature that may increase the mobility of the upper lip, and the much-reduced coronoid process of the dentary bone probably allows a wide gape of the jaws. The dental formula is 1/2, 1/1, 1/2, 3/3 = 28 and is of the usual insectivorous type with W-shaped ectolophs on the upper molars. The greater tuberosity of the humerus extends beyond the head of the humerus and may serve as a locking device; the second digit of the wing has only one very short phalanx; and the wing is broad. The pelvis and axial skeleton are highly specialized: the last three thoracic vertebrae and all but the last two lumbars are fused, and the sacral vertebrae are fused, whereas the pelvis is delicately built. The hindlimbs are slender, and the fibula is threadlike.

Bumble-bee bats are considered to be most closely related to the Rhinopomatidae, and they share a number of characteristics with most members of a clade known as the Yinochiroptera (including the families Rhinopomatidae, Megadermatidae, Nycteridae, and Rhinolophidae; Koopman, 1984, 1994; Simmons, 1998; Simmons and Geisler, 1998). For example, in certain members of all these families except Nycteridae, while the mother is roosting head-downward, the nursing young clings head-upward to her with its hind legs around the mother's neck and its mouth clinging to non–milk-producing pubic nipples (Duangkhae, 1990; Simmons, 1993). This tiny bat roosts by day in caves and eats small arthropods that are probably gleaned from leaves (Hill and Smith, 1981).

FAMILY RHINOPOMATIDAE. Members of this small family, containing but one genus with three species, occur in northern Africa and southern Asia east to Sumatra. These animals are called mouse-tail bats because of the long tail that is largely free from the uropatagium. No fossil representatives of the family are known.

These small bats (10 to 15 grams in body weight) have premaxillaries that resemble those of megachiropterans in being separate from one another, and their palatal portions are much reduced. The second digit of the hand, in contrast to the arrangement in all other microchiropterans, retains two well-developed phalanges. Perhaps the clearest indication of the primitiveness of these bats is the structure of the shoulder joint. In contrast to the situation in most microchiropterans, the greater tuberosity of the humerus is small and does not lock against the scapula at any point in the wing-beat cycle. Other rhinopomatid features include laterally expanded nasal chambers; no fusion of cervical, thoracic, or lumbar vertebrae; and a complete fibula.

The dentition is adapted to an insectivorous diet. The molars are tuberculosectorial; the upper molars have W-shaped ectolophs of the usual microchiropteran type. The dental formula is 1/2, 1/1, 1/2, 3/3 = 28. These are fairly small bats (the length of the head and body is up to 80 millime-

ters) with slender tails whose length approaches that of the head and body. The eyes are large, and the anterior bases of the large ears are joined by a fold of skin across the forehead. The nostrils are slitlike.

Mouse-tail bats are insectivorous and typically occupy hot, arid areas. They roost in a wide variety of situations, including fissures in rocks, houses, ruins, and caves; one species roosts in large colonies in some Egyptian pyramids. Although locally common, mouse-tail bats are outnumbered by other types of bats over much of their range and are not as important today as other microchiropteran families. Rhinopomatids perhaps hibernate in some areas. Large deposits of subcutaneous fat occur in the abdominal area and around the base of the tail in individuals from some localities. These bats tolerate body temperatures as low as 22°C and can spontaneously rewarm themselves (Kulzer, 1965).

FAMILY NYCTERIDAE. Members of this small family (12 species of one genus) are called slit-face bats. These bats occur in Madagascar, Africa, the western Arabian Peninsula, the Malay Peninsula, and parts of Indonesia, including Sumatra, Java, and Borneo. A fossil nycterid is known from the early Oligocene of Oman.

These fairly small- to moderate-sized bats weigh from about 6 to 35 grams and have wing spans ranging from 250 to 350 millimeters. They can be recognized by their large ears, very small eyes, distinctive "hollow" face, and a T- or Y-shaped cartilage at the end of the tail. The skull has a conspicuous interorbital concavity that is probably associated with the beaming of the sound pulses used in echolocation (Fig. 10-23). The extreme downward tilt of the rostrum relative to the basicranium strongly supports this suggestion (Freeman, 1984). The interorbital concavity is connected to the outside by a slit in the facial skin. The dental formula is 2/3, 1/1, 1/2, 3/3 = 32, and the molars are tuberculosectorial. Postcranially, these bats combine primitive and specialized features. The shoulder and elbow joints are fairly primitive, but the retention of only the metacarpal of the second digit of the hand and the reduction of the number of phalanges of the third digit to two are obvious specializations. The pectoral girdle

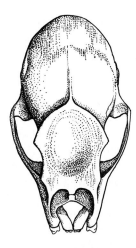 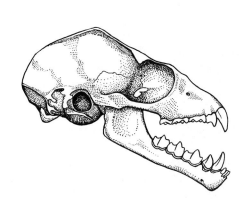

FIGURE 10-23   The skull of a slit-face bat (Nycteridae): dorsal view, showing the depression in the forehead; side view, showing the flattened profile. Length of skull 19 millimeters. *(After Hill and Carter, 1941)*

Slit-faced bat
(*Nycteris thebaica*)

is modified in the direction of enlargement and strengthening of the bracing of the sternum, a pattern parallel to the trend in birds. The sternum in nycterids is robust, and the mesosternum is strongly keeled. The manubrium is broad, the first rib is unusually strongly built, and the seventh cervical and first thoracic vertebrae are fused.

This general pattern also occurs in the family Megadermatidae and reaches its most extreme development in the family Rhinolophidae (see Fig. 10-27). Because the specializations of the pectoral girdle in these bats parallel to some extent the roughly similar modifications in birds, they could be associated with a progressive structural trend in bats. Actually, it is doubtful that this is the case. Some of the most advanced and successful families of bats have less birdlike pectoral girdles than those in the microchiropteran families listed above but have modifications of the shoulder and elbow joints and forelimb musculature that provide for efficient flight. Perhaps the nycterid-megadermatid-rhinolophid pectoral girdle is associated with a foraging style typified by short intervals of flight. In any case, this style of pectoral girdle seems to be a divergent type and does not represent a progressive morphological trend common to most "advanced" microchiropterans.

Slit-face bats inhabit tropical forests and savanna areas and seem to feed largely on arthropods picked from vegetation or from the ground.

Nycterids are amazingly delicate and maneuverable fliers. When foraging, they often seem to drift effortlessly around the trunks of large trees and near foliage. Flying insects form part of the diet, but orthopterans and flightless arachnids, such as spiders and scorpions, are also important food items. The large species *Nycteris grandis* of Africa eats a remarkably varied assortment of animals, including orthopteran and lepidopteran insects, fish, frogs, birds, and small bats (Fenton et al., 1990, 1993). These bats show a similar variability in their approach to foraging; some individuals hunt by short flights from a perch while others hunt while continuously on the wing.

Nycterids roost in a variety of situations, and some are even known to occupy burrows made by porcupines and aardvarks. In Kenya, in some remote safari camps, the pits dug for privies ("long-drops" in local parlance) are occasionally used as daytime retreats by *Nycteris thebaica*, to the consternation of the uninitiated users of these toilets.

FAMILY MEGADERMATIDAE.   This is a small family, consisting of but four genera and five species. These bats are known as false vampires, an inappropriate title as they neither resemble vampires nor feed on blood. They occur in tropical areas in East Africa, southeastern Asia including Indonesia, the Philippines, and Australia. The earliest fossil megadermatid is from the late Eocene of Europe.

FIGURE 10-24   The bat *Megaderma spasma*
(Megadermatidae) from Laos. Note the very large
nose leaf and huge ears. *(C. Francis)*

These are fairly large, broad-winged bats. The
largest species has a wing spread approaching 1
meter and weighs up to nearly 200 grams. Smaller
species have wing spreads of about 320 millimeters
and weigh about 25 grams. The ears are large and
are connected across the forehead by a ridge of
skin. The tragus is bifurcated. The snout bears a
conspicuous "nose leaf," and the eyes are large and
prominent (Fig. 10-24). The premaxillae and upper
incisors are absent; the upper canines project forward and have a large secondary cusp (Fig. 10-25B).
The molars are tuberculosectorial; the dental formula is 0/2, 1/1, 1-2/2, 3/3 = 26-28. In *Mega-*

derma, and to a still greater extent in *Macroderma,*
the W-shaped ectoloph of the upper molars is
modified by the partial loss of the commissures
connecting the mesostyle to the paracone and
metacone. This trend is toward the development
of an anteroposteriorly aligned cutting blade and
may be associated with the carnivorous habits of
these genera. The shoulder and elbow joints are
primitive; the second digit of the hand has one
phalanx, and the third has two phalanges. The
pectoral girdle has specializations similar to those
of the nycterids, but the strengthening of the pectoral girdle is carried further in megadermatids.
The manubrium of the sternum is broader in
megadermatids than in the nycterids and is fused
with the first rib and the last cervical and first thoracic vertebrae into a robust ring of bone. The
megadermatid sternum is moderately keeled. The
tail is very short or absent.

These bats occur in tropical forests and savannas, often near water, and eat a variety of foods. Of
the five species of megadermatids, three are
known to be carnivorous and two are mostly insectivorous. The Australian ghost bat *(Macroderma gigas),* an unusually large, pale megadermatid, feeds
on a variety of small vertebrates. In some areas, it
seems to feed largely on other bats. The ghost bat
and related species in southeastern Asia frequently
consume their prey while hanging from the ceilings of spacious covered porches or verandas of
large homes and detract from the gracious atmosphere by littering the floors with feet, tails, and
other discarded fragments of frogs, birds, lizards,
fish, bats, and rodents. Prakash (1959) observed
Megaderma lyra of India eating bats of the genera
Rhinopoma and *Taphozous,* a gecko, and a large insect. In the stomach contents of these bats, he
found bones of amphibians and fishes. The megadermatids may hunt partly by sight. One species,
the partially diurnal, insectivorous African *Lavia
frons,* uses a style of foraging similar to that of a fly-
catcher, hanging from a branch and making short
flights to capture passing insects.

Cardioderma cor of Africa has the most sedentary
style of foraging known for any insectivorous bat
(Vaughan, 1976). This entirely nocturnal bat hangs
from a perch in low vegetation when foraging and at
some seasons regularly gives loud, humanly audible
calls that seem to be territorial announcements. The
body revolves through approximately 360 degrees

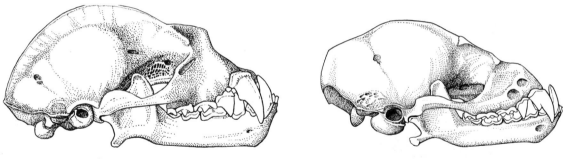

Giant leaf-nosed bat
(*Hipposideros commersoni*)

African false vampire bat
(*Cardioderma cor*)

FIGURE 10-25   Skulls of two African bats that eat large beetles: giant leaf-nosed
bat (Rhinolophidae), length of skull 32 millimeters; African false vampire bat
(Megadermatidae), length of skull 27 millimeters.

as the hanging bat meticulously "scans" the ground, listening for sounds made by terrestrial invertebrates, such as large beetles and centipedes. When prey is detected, the bat flies directly to the ground, snatches up the food, and returns to the same perch to consume it. When insects are abundant during the wet seasons, *Cardioderma* spends very little time in flight: on some nights it perches for periods averaging nearly 11 minutes and spends less than 1 percent of its foraging time in flight. In the dry season, however, when prey abundance declines, flights from perch to perch are more frequent and more time is spent in flight, although flights after prey average only 3 seconds. Considering all seasons, flights after prey average only 5 seconds. In *Cardioderma,* and perhaps in all megadermatids, the technique of searching for prey has departed markedly from that of most insectivorous bats, whereas the style of flight and the morphology of the forelimb have remained generalized.

Megadermatids roost in many types of places, from hollow trees, caves, and buildings in the case of most species, to sparse, occasionally sunlit vegetation in the case of *Lavia.* This bat has been found by Wickler and Uhrig (1969) to occupy fairly small foraging territories and to have several humanly audible calls during social interactions.

FAMILY RHINOLOPHIDAE. This is a large and successful Old World family, with 10 genera and approximately 130 species. As used here, the Rhinolophidae consists of two distinct subfamilies — the Rhinolophinae or horseshoe bats and the Hip-

posiderinae or Old World leaf-nosed bat — that are often considered separate families, Rhinolophidae and Hipposideridae. The term "horseshoe bats" refers to the complex, basically horseshoe-shaped, cutaneous ridges and depressions on the nose (Figs. 10-26, 10-20B). Some rhinolophids are quite small, with body weights of nearly 6 grams and wing spans of 250 millimeters, but *Hipposideros commersoni* of Africa, at the other extreme, is one of the largest insectivorous bats. Some individuals reach body weights of over 150 grams and wing spans of over 500 millimeters. The geographic distribution of rhinolophids includes much of the Old World from western Europe and Africa to Japan, the Philippines, Indonesia, Melanesia, and Australia. This is an ancient family, for some middle Eocene fossils from Europe have been assigned to the living genus *Rhinolophus.*

Because of the unique and complex face, rhinolophids are one of the most unmistakable groups of bats. The ears are usually large but lack a tragus, and the eyes are small and inconspicuous. The tail is of moderate length in some species but is small or rudimentary in others. The pectoral girdle is remarkable because it represents the extreme development of the trend (that occurs also in the Nycteridae and Megadermatidae) toward powerful bracing and enlargement of the sternum. In the most extreme manifestation of this trend, the seventh cervical vertebra, the first and second thoracic vertebrae, the first and most of the second rib, and the enormously enlarged and shield-like manubrium of the sternum are fused into a

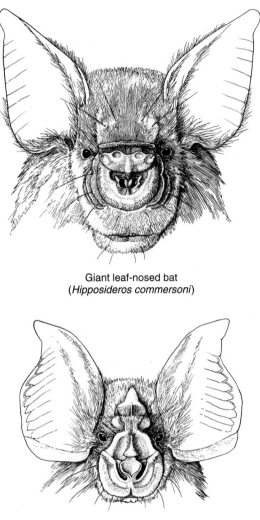

Giant leaf-nosed bat
(*Hipposideros commersoni*)

Rhinolophus landeri

FIGURE 10-26 Faces of bats of the family Rhinolophidae.

powerfully braced ring of bone (Fig. 10-27A, C). The shoulder joint has a moderately well-developed locking device. In some rhinolophids, all but the last two lumbar vertebrae are fused; a similar specialization occurs in the Natalidae (see Fig. 10-36). The pelvis is uniquely modified by enlargement of the anterior parts and by an accessory connection between the ischium and the pubis. These unusual pelvic specializations may be responses to the mechanical stresses imposed on the hindlimbs and pelvis by the repeated takeoffs and landings that occur during foraging in some of these bats. When they roost, these bats often hang

upside down, and the hindlimbs are rotated 180 degrees from the usual mammalian posture so that the plantar surfaces of the feet face forward. The extreme adaptations for strengthening the pectoral and pelvic girdles that are typical of rhinolophids occur to a comparable degree in no other family of bats.

Horseshoe bats are common in many areas, and in Germany the "Hufeisennase" is a familiar inhabitant of attics and church steeples. These bats have wide environmental tolerances. Various species inhabit temperate, subtropical, tropical, and desert regions. Rhinolophids hibernate in some parts of their range and characteristically rest or hibernate with the body enshrouded by the wing membranes. Several species in East Africa are migratory. The food is largely arthropods, and the style of foraging resembles that of the nycterids and some megadermatids. Horseshoe bats pick spiders and insects from vegetation or capture flying insects in midair, and *Rhinolophus ferrumequinum* was observed to alight on the ground and capture flightless arthropods (Southern, 1964). When foraging, *Hipposideros commersoni* of Africa hangs fairly high in trees, uses echolocation to detect large and straight-flying beetles at distances up to 20 meters, and makes brief and precise interception flights that last an average of 5.1 seconds (Vaughan, 1977). This bat returns to the perch to consume prey, which consists of very large beetles (up to 60 millimeters in length). Like *H. commersoni,* a number of rhinolophids make short foraging flights and do not remain continuously on the wing while foraging. Perhaps the wing membranes are important in some species in aiding in the capture of insects. Webster and Griffin (1962) demonstrated photographically that rhinolophids are able to capture insects in the chiropatagium.

In contrast to many bats that emit pulses from the open mouth in echolocation, rhinolophids keep the mouth closed during flight; the ultrasonic pulses used in echolocation are emitted through the nostrils and are beamed by the complex nasal apparatus (Möhres, 1953). A remarkable series of coordinated behaviors is associated with the highly specialized rhinolophid style of echolocation (see Chapter 22).

Most horseshoe bats are colonial, but some are solitary. Many kinds of roosting sites are used; caves, buildings, and hollow trees are generally

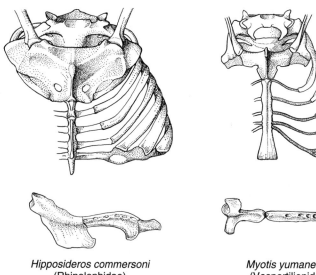

FIGURE 10-27 Ventral views of the thorax and lateral views of the sternum of a rhinolophid and a vespertilionid. Note the highly specialized sternum of *Hipposideros,* to which the first two ribs are fused. *(From Vaughan, 1970b)*

Hipposideros commersoni
(Rhinolophidae)

Myotis yumanensis
(Vespertilionidae)

preferred, but foliage and the burrows of large rodents are used by some species.

FAMILY MYSTACINIDAE. This rather aberrant family until recently consisted of two species in the genus *Mystacina,* restricted to New Zealand. One of the species is now extinct, not having been reported since 1965 (Hill and Daniel, 1985), and the surviving species is threatened. Late Pleistocene fossils of mystacinids are known in New Zealand, but extinct, plesiomorphic types were present in the Miocene in Australia (Hand et al., 1998).

The phylogenetic relationships between *Mystacina* and other bats has long defied explanation. In the past, based on various morphological features, it has been considered to have at least some affinities with practically every other microchiropteran family (Daniel, 1979). Recent studies using molecular evidence (Pierson et al., 1986; Kirsch et al., 1998) and hyoid morphology (Griffiths, 1997) suggest it is a noctilionoid, that is, a member of the radiation of endemic Neotropical families Noctilionidae, Mormoopidae, and Phyllostomidae. A phylogenetic analysis using a very broad data set placed mystacinids as the sister group to molossoid and vespertilionoid bats (Simmons, 1998). The recently discovered Tertiary fossils of mystacinids in Australia indicate that the family arose by at least the Miocene. These fossils and molecular data suggest that mystacinids colonized New Zealand from Australia (Hand et al., 1998; Kirsch et al., 1998).

Mystacina has an advanced locking shoulder joint, one phalanx in the second digit and two in the third, and the lack of fusion of presacral vertebrae. The skull is roughly like that of nataloids (natalids, thyropterids, furipterids, and myzopodids), with no anterior palatal emargination. The lower jaw symphysis is fused, an unusual characteristic for insectivorous bats but often found among nectar-feeding bats (Freeman, 1988, 1995). The teeth are tuberculosectorial, and the dental formula is 1/1, 1/1, 2/2, 3/3 = 28. The tongue is partly protrusible and bears a brush of fine papillae on its tip, also found in nectar-feeding bats (Freeman, 1995). The limbs resemble, in some ways, those of molossids: the wing membranes and uropatagium are tough and leathery, the first phalanx of the third digit folds back on the dorsal surface of the metacarpal, and the hind foot is unusually broad; the fibula is complete, and the hindlimb is robust. The tail is short and protrudes from the dorsal surface of the uropatagium. Unique among bats, each of the claws of the thumb and foot has a secondary talon at its ventral base.

These unusual bats have many distinctive aspects of their natural history that are probably related, in large part, to the fact that, until historic introductions of exotic species to New Zealand by humans, no other ground-dwelling small mammals occurred on the islands. In New Zealand, the two species of *Mystacina* tended to fill this niche (Daniel, 1990; Worthy et al., 1996). The limbs are well adapted to quadrupedal locomotion. The wing can be folded compactly, owing to the unique flexion pattern of the third digit, and during quadrupedal locomotion it is partially protected by the leathery proximal part of the plagiopatagium. The claws, and possibly the incisors, are used in excavating roosting tunnels in decaying kauri *(Agathis)* trees, in which several *Mystacina* will roost head-to-tail. These bats roost in a wide variety of places, including abandoned seabird burrows and holes in volcanic pumice.

Mystacina tuberculata forages primarily on the ground, where they have been observed burrowing through leaf litter. They feed on large ground-dwelling insects and other arthropods and on very small vertebrates, including nocturnal frogs and lizards (Worthy et al., 1996). However, they are not strictly predatory because their diet also includes fruit, nectar, and pollen (Daniel, 1979). They pollinate the endangered endemic plant *Dactylanthus* while crawling on the ground (Meyer-Rochow and Stringer, 1997). *Mystacina* enters torpor but apparently does not hibernate during the austral winter. They have been observed foraging at ambient temperatures as low as −2°C (Daniel, 1990). Another unusual aspect of their biology is that, unlike most other bats, they breed in leks.

FAMILY NOCTILIONIDAE. Although this family is not large in terms of numbers of species (it contains only two species in one genus), it is of particular interest because one species is highly specialized both structurally and behaviorally for eating fish. Noctilionid bats are often referred to as bull-dog bats or fishing bats. They inhabit the Neotropics from Sinaloa, Mexico, and the West Indies to northern Argentina. They are known as fossils from the middle Miocene of South America (Czaplewski, 1997).

Both in structure and in appearance, noctilionids are distinctive. They are fairly large (from roughly 30 to 60 grams in weight and up to 585 millimeters in wing spread), and the heavy lips, somewhat resembling those of a bulldog, pointed ears, and simple nose make the face unmistakable (Fig. 10-28). The dorsal pelage varies in color from orange to dull brown, and a whitish or yellowish stripe is usually present from the interscapular area to the base of the tail. The hindlimbs and feet are remarkably large, especially in *Noctilio leporinus,* and the feet have sharp, recurved claws. The premaxillae are complete, and in adults the two maxillae are fused together and are fused with the premaxillae, forming a strongly braced support for the enlarged upper medial incisors. The dental formula is 2/1, 1/1, 1/2, 3/3 = 28. The teeth are robust, and the molars are tuberculosectorial. *Noctilio* has unusually long canines and a low coronoid process for a bat; Freeman (1984) called it a "saber-toothed bat" because of several cranio-dental characteristics.

The seventh cervical vertebra is not fused to the first thoracic, the shoulder and elbow joints are primitive, and the second digit of the hand has a long metacarpal and a tiny vestigial phalanx. The pelvis is powerfully built, with the ischia strongly fused together and fused to the posterior part of the laterally compressed, keel-like sacrum. The

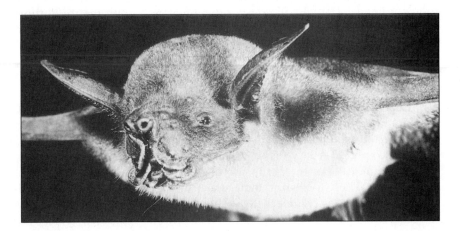

FIGURE 10-28 The face of the fishing bat *Noctilio albiventris* (Noctilionidae). *(N. Smythe and F. Bonaccorso)*

tibia and hind foot of *N. leporinus* have a series of unusual specializations, which will be considered below. The calcar in noctilionids is uniquely ossified, rather than cartilaginous as in other microchiropterans.

The two species of *Noctilio* primarily eat insects, but are quite flexible in foraging style and diet. *Noctilio leporinus* is well known for its habit of catching fish. The smaller species *Noctilio albiventris,* too, sometimes eats fish and occasionally even fruits, which are often taken from the surface of the water (Fleming et al., 1972; Hooper and Brown, 1968; Howell and Burch, 1974). *Noctilio leporinus* is able to catch flying insects in midair using the tail or wing membranes, but also catches fish, shrimp, terrestrial insects, and other invertebrates by using its feet (Brooke, 1994). The style of foraging of *N. leporinus* involves the use of the hind claws as gaffs (Bloedel, 1955; Brooke, 1994). This bat recognizes concentrations of small fish or single fish immediately beneath the surface of the water by detecting (by echolocation) the ripples or breaks in the surface that these fish create (Suthers, 1965, 1967). The bat skims low and drags its feet in the water, with the limbs rotated so that the hooklike claws are directed forward. (This involves rotation of the hindlimbs 180 degrees from the typical mammalian position.) When a small fish is "gaffed," it is brought quickly from the water and grasped by the teeth. From 30 to 40 small fish were captured in this fashion per night by one *N. leporinus* under laboratory conditions. Terrestrial invertebrates, such as large insects, scorpions, and small crabs, are gaffed from the surface of the ground in a similar manner (Brooke, 1994).

A series of modifications of the hindlimb are clearly advantageous in allowing this animal to pursue efficiently its specialized style of foraging. The long bony calcar, which is roughly as long as the tibia, the calcaneum, the digits and claws, and the distal part of the tibia are all strongly compressed so that they are streamlined with respect to their direction of movement when they are dragged through the water. During foraging sweeps, the short tail is raised and the bladelike calcar is pulled craniad and clamped against the flattened side of the tibia. In this way, the large uropatagium is brought clear of the water and the streamlined calcar and tibia knife through the water, producing a minimum of drag.

Noctilionids roost during the day in groups in hollow trees and rock fissures, caves, and occasionally buildings. *Noctilio leporinus* is seemingly most common in tropical lowland areas, frequently occurring along coasts, where it forages along rivers or streams, over mangrove-lined marshes and ponds, or over the sea. In western Mexico in the dry season, individuals often forage over small, disconnected ponds in nearly dry stream beds. Such ponds support large numbers of small fish.

FAMILY MORMOOPIDAE. The two genera and eight species that make up this family can appropriately be called leaf-chinned bats because in all species a conspicuous, leaflike flap of skin occurs on the lower lip. These bats are largely tropical in distribution and occur from the southwestern United States and the West Indies south to Brazil.

Leaf-chinned bats are fairly small, weighing between 7 and 20 grams, and have several distinctive external specializations. The snout and chin always have cutaneous flaps or ridges (that reach their most extreme form in *Mormoops*), but a nose leaf is not present. The ears are moderately large, have a tragus, and vary in shape but always have large ventral extensions that curve beneath the fairly small eyes. The tail is short and protrudes from the dorsal surface of the fairly large uropatagium. The rostrum is tilted more or less upward (this feature is most extremely developed in *Mormoops*), and the floor of the braincase is elevated. The coronoid process of the dentary bone is reduced, allowing the jaws to gape widely. The teeth are of the basic insectivorous type; the dental formula is 2/2, 1/1, 2/3, 3/3 = 34.

The second digit in the hand has one phalanx and the third, three, as in the Phyllostomidae, but the shoulder and elbow joints differ markedly from the phyllostomid pattern. The greater tuberosity of the humerus in mormoopids does not form a well-developed locking device with the scapula; the head of the humerus is more or less elliptical, perhaps favoring a specialized wing-beat cycle. The elbow joint is specialized in all species, and, in *Mormoops*, modifications of the distal end of the humerus and the forearm musculature provide for a highly efficient "automatic" flexion and extension of the hand. The musculature of the hand is reduced and simplified; this lightens the hand and probably favors maneuverability and endurance. The hindlimbs do not have the spider-like posture typical of phyllostomids but instead have a reptilian posture that allows the animal to crawl on the walls of caves with considerable agility.

Leaf-chinned bats are among the most abundant bats in many tropical localities, where they are seemingly the major chiropteran insectivores. They are most common in tropical forests but occur also in some desert areas. Some species appear early in the evening; their insect-catching maneuvers resemble those of their temperate zone counterparts, the vespertilionids. Leaf-chinned bats usually roost in caves or deserted mine shafts and may concentrate in large numbers. A colony of *Mormoops* observed by Villa-Ramirez (1966) in Nuevo León, Mexico, contained more than 50,000 bats, and a colony of four species of mormoopids in Sinaloa, Mexico, was estimated to contain 400,000 to 800,000 bats (Bateman and Vaughan, 1974). When the bats from the latter colony emerged in the evening, they swept down the nearby arroyos and trails in such numbers and at such speeds that one hesitated to move across their path. When they form large colonies, these bats seem to disperse many kilometers from their roosting site to forage at night and to remain continuously on the wing for several hours. Their impact on tropical ecosystems must be great, for the bats in the Sinaloan colony probably consume over 1400 kilograms of insects per night. It is not surprising that the bats must disperse over a wide area to forage.

FAMILY PHYLLOSTOMIDAE. This is the most diverse family of bats with respect to structural variation and contains more genera than any other chiropteran family. Forty-nine genera and 143 Recent species are included in the family. These Neotropical "leaf-nosed bats" are so named because of the conspicuous leaflike structure that is nearly always present on the nose (see Figs. 10-20A and 22-10A). These bats have exploited the widest variety of foods used by any family of bats. Some leaf-nosed bats have retained insectivorous feeding habits, but some are partly carnivorous and eat small vertebrates, including other bats, rodents, birds, frogs, and lizards. Some eat nectar and pollen, some are frugivorous, and some feed solely on the blood of other vertebrates. In recent years, leaves have been recognized as important items in the diet of some frugivorous species (Zortéa and Mendes, 1993;

Kunz and Diaz, 1995). Phyllostomids are the most important bats in the Neotropics and occur from the southwestern United States and the West Indies south to northern Argentina. These bats can be traced back to the middle Miocene of Colombia (Savage, 1951; Czaplewski, 1997).

The great structural variation that occurs in the phyllostomids is largely associated with an adaptive radiation into a wide variety of feeding niches. Phyllostomids form a monophyletic clade with the Mormoopidae and Noctilionidae (Pierson, 1986; Baker et al., 1989; Simmons, 1998), but the extensive functional specializations have made it difficult to determine the phylogenetic relationships within the family. Some phyllostomids are fairly small bats (*Choeroniscus* has a wing spread of roughly 220 millimeters and weighs 8 grams), and one species is the largest New World bat (*Vampyrum*, with a wing spread of over 1 meter and a body weight of up to 190 grams).

In most species, the nose leaf is conspicuous and spear-shaped, but in a few species the nose leaf is rudimentary or highly modified. The ears vary from extremely large to small, and a tragus is present. The tail and uropatagium are long in some species, with many stages of reduction and the absence of the tail and uropatagium in various species. Some species have a uropatagium but lack a tail; only *Sturnira* has completely lost the uropatagium. The wings are typically broad; the second digit has one phalanx and the third has three phalanges. The shoulder joint has a moderately well-developed locking device formed between the greater tuberosity of the humerus and the scapula, but the elbow joint and forearm musculature are primitive, and the forelimb is, for a bat, generalized. Probably all phyllostomids, whatever their feeding habits, remain on the wing only for short periods during foraging.

The forelimbs are not used only for flight but are important in many species in handling food, as well as in climbing over and clinging to vegetation (especially in the fruit-eating species). The importance of such use of the forelimbs has probably favored the retention in phyllostomids of limbs more generalized than those of many strictly insectivorous bats. The seventh cervical and first thoracic vertebrae are not fused, and in phyllostomids there is no fusion of elements to form the sturdy pectoral ring characteristic of rhinolophoid bats.

In some leaf-nosed bats, however, the sternum is strongly keeled. The ventral parts of the pelvis are lightly built in most species, but the ilia are robust and more or less fused to the sacral vertebrae. These vertebrae are fused into a solid mass that becomes laterally compressed posteriorly. The acetabulum is characteristically directed dorsolaterally; the hindlimbs are rotated 180 degrees from the usual mammalian orientation and have a spider-like posture. Because of this position of the hindlimbs, some phyllostomids are unable to walk on a horizontal surface and use the hindlimbs only for hanging upside down.

All the Recent leaf-nosed bats probably evolved from an ancestral type that had tuberculosectorial teeth adapted to a diet of insects. Of the eight Recent subfamilies recognized by Koopman (1993), however, only one has retained this dentition, and in some species there is no trace of the ancestral pattern. The noteworthy adaptive radiation of phyllostomids will be traced by considering the dentitions and foraging habits of each subfamily.

The subfamily Phyllostominae deviates least from the ancestral structural plan, and some species retain insectivorous feeding habits. This subfamily contains all the leaf-nosed bats with tuberculosectorial teeth of the ancestral type. However, in some species (*Chrotopterus* and *Vampyrum,* for example), the W-shaped ectoloph of the upper molars is distorted by the reduction of the stylar cusps and the closeness of the protocone, para-

cone, and metacone. Most members of this subfamily are insectivorous, and some species are known to pick insects either from vegetation or from the ground. However, a few of the largest phyllostomines resemble their Old World look-alikes and ecological counterparts, the megadermatids, in their partly carnivorous habits. The large phyllostomine species — *Phyllostomus hastatus, Trachops cirrhosus, Chrotopterus auritus,* and *Vampyrum spectrum*— are known to include small vertebrates in their diet. *Trachops* primarily eats insects, but in Panama it specializes in frogs (Fig. 10-29), which it locates using ears specially attuned to the low-frequency calls of the frogs (Tuttle and Ryan, 1981; Bruns et al., 1989). Beneath the roosts of *V. spectrum,* feathers and the tails of rodents and geckos give indications of feeding preferences (Vehrencamp et al., 1977). The means by which these carnivorous-omnivorous bats perceive small vertebrates is not known. They may well hear the faint sounds made as their prey moves, and the large eyes of these bats indicate that hunting may also involve vision. Bats of this type generally have large ears, however, suggesting highly discriminatory echolocation.

Nectar feeding is common among tropical vertebrates (as indicated by the presence of more than 300 species of hummingbirds in the American tropics) and has also been adopted by bats of the subfamilies Glossophaginae and Lonchophyllinae of Mexico and Central and South America and by bats of the subfamily Phyllonycterinae of the

FIGURE 10-29   A frog-eating bat *(Trachops)* about to capture a frog. *(By M. Tuttle, Bat Conservation International)*

West Indies. These bats feed on the nectar and pollen of a great variety of plants and have many structural features associated with this mode of life. The tongue is long and protrusible and has a brushlike tip; the rostrum is elongate, and the dentaries are slender (Fig. 10-30) and fused together (Freeman, 1995). The cheek teeth have largely lost the tuberculosectorial pattern (see Fig. 10-21C and D). The hairs of at least some of these nectar feeders have divergent scales that catch pollen as the bat feeds on nectar. This pollen is swallowed when the bats groom their fur and provides a protein supplement without which the animals could not survive.

In nectar-feeding bats, the wings are usually broad and the uropatagium is reduced. Some nectar feeders migrate long distances with the seasons, following the flowering of the plants that form their main food supply. Many can maneuver delicately through dense tropical vegetation or around spiny desert cacti and can hover. Flowers seem to be located by the sense of smell, and these bats feed by hovering briefly and thrusting the long tongue into the flowers. The pollination of many night-blooming Neotropical plants is accomplished by nectar-feeding phyllostomids, just as many plants in the Old World tropics are pollinated by nectar-feeding pteropodids. Selective forces determined by this method of pollination have probably been important in the evolution of flower structure and in the timing of pollen and nectar production in these plants. A dietary continuum from heavy reliance on insects (in *Glossophaga*) to dependence on nectar, pollen, and fruit (in *Leptonycteris*) occurs among "nectar-feeding" bats.

Recently, Ted Fleming and his colleagues have intensively studied the biology of nectar-feeding phyllostomids, especially *Leptonycteris* (Fleming et al., 1993; Fleming, 1995). Some of their studies have employed a method widely used by geochemists, plant physiologists, paleoecologists, and others: these scientists use stable isotopes of certain elements, especially carbon and nitrogen, to study various biological and physical processes. Fleming and his colleagues extracted stable isotopes from the tissues of these bats to determine indirectly the general type of plants on which the bats were feeding. Plants with different photosynthetic pathways (called C3, C4, and CAM pathways) store characteristic proportions of these isotopes in their tissues, which can be analyzed using a mass spectrometer. The poorly known Peruvian bat, *Platalina genovensium,* proved to be a specialist on CAM plants (columnar cacti of the genus *Weberbauerocereus*; Sahley and Baraybar, 1994). Biochemical studies have shown that some bats are migratory *(Leptonycteris)* and others are not *(Glossophaga),* and seasonal differences in diet among different populations of *Leptonycteris* have been detected (Fleming et al., 1993).

The members of three subfamilies of phyllostomids — Carolliinae, Brachyphyllinae, and Stenodermatinae — are frugivorous. The success of these groups and the richness of this food source in the Neotropics is indicated by the fact that within the family Phyllostomidae, the largest and most abundant group of Neotropical bats, nearly half of the

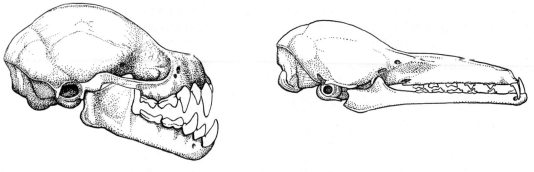

Artibeus phaeotis

Choeronycteris mexicana

FIGURE 10-30 Skulls of leaf-nosed bats (Phyllostomidae): a fruit eater *(Artibeus phaeotis)*, length of skull 19 millimeters; a nectar feeder *(Choeronycteris mexicana)*, length of skull 30 millimeters.

species (approximately 60 out of 143) are basically fruit eaters. Several variations on this fruit-eating theme can be recognized. Members of the subfamily Carolliinae have reduced molars with the original tuberculosectorial pattern largely obliterated. These bats apparently prefer ripe, soft fruit and are known to eat a great variety of it. Stenodermatines of the genus *Sturnira* are fairly small, often brightly colored bats that have robust molars with no trace of the basic tuberculosectorial pattern. Indeed, their molars strongly resemble those of New World monkeys (Cebidae). *Sturnira* eat small and often hard fruit, such as the fruits of low-growing species of nightshade (*Solanum* sp.).

Other stenodermatines have robust teeth that are highly adapted to crushing fruit. The upper molars have lost the stylar cusps, and the inner portion is much enlarged and marked by complex rugosities (Fig. 10-21E and F). The rostrum is short (Fig. 10-30), and the coronoid process of the dentary bone is fairly high in many species, conferring considerable mechanical advantage for powerful jaw action to the large temporal muscles. Large species of the stenodermatine genus *Artibeus* are remarkably abundant in some Neotropical areas, and their piercing calls are characteristic sounds of the tropical nights. Often many *Artibeus* of several species concentrate on a single fig tree (*Ficus* sp.; Handley et al., 1991; Kalko et al., 1996) with abundant fruit. In central Sinaloa, Mexico, two students and the senior author camped beneath such a fig tree — but only for one night. The activities of dozens of *A. lituratus, A. hirsutus,* and *A. jamaicensis* caused a nearly continuous rain of fruit and bat excrement throughout much of the night, and with sunrise came herds of aggressive local pigs to gather the night's fallout of figs. Stenodermatines often eat unripe and extremely hard fruit, and it is perhaps as an adaptation to this type of food that the robust teeth and powerful jaws evolved.

Interestingly, many different species of Stenodermatinae construct "tents" in which to roost by chewing the veins of leaves of different species of tropical plants (Timm, 1987; Timm and Lewis, 1991).

The subfamily Desmodontinae contains the vampire bats, the only mammals that feed solely on blood. Only three genera and three species constitute this group, but they are widely distributed from northern Mexico southward to northern Argentina, Uruguay, and central Chile.

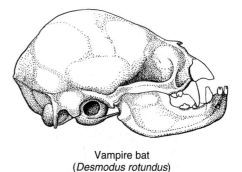

Vampire bat
(*Desmodus rotundus*)

FIGURE 10-31   The skull of a vampire bat (Phyllostomidae), length of skull 24 millimeters.

Vampire bats are fairly small. Mexican specimens of the common vampire bat *(Desmodus rotundus)* usually weigh about 30 grams and have an average wing span of 365 millimeters. The skull and dentition are highly specialized. The rostrum is short, the braincase is high (Fig. 10-31), and in all species the cheek teeth are reduced in both size and number. In *D. rotundus,* the most specialized species, the dental formula is 1/2, 1/1, 2/3, 0/0 = 20. The upper incisors are unusually large and are compressed and bladelike, as are the upper canines. These teeth have remarkably sharp cutting edges. The cheek teeth are tiny. Except for the canine, the lower teeth are small.

The flight of vampires is strong and direct and not highly maneuverable. However, the agility of vampires during quadrupedal locomotion is remarkable compared with most bats (Altenbach, 1979). The thumb in *Desmodus* is unusually long and sturdy and contributes an additional segment with three joints to the forelimb during quadrupedal locomotion. The hindlimbs are large and robust, and the fibula is not reduced. The proximal part of the femur and the tibia and fibula are flattened and ridged; these irregularities provide large surfaces for the attachment of the powerful hindlimb musculature. *Desmodus* can run rapidly and easily on the ground and can even jump short distances (Fig. 10-32). Their agility in terrestrial locomotion is reflected in the histochemistry of the primary flight muscle, the pectoralis. The histochemical composition of this muscle is complex, with four different fiber types similar to those in terrestrial mammals and unlike the pectoralis of other bats (Hermanson et al., 1993). *Diphylla* feeds entirely on the blood of birds; in order to do this it

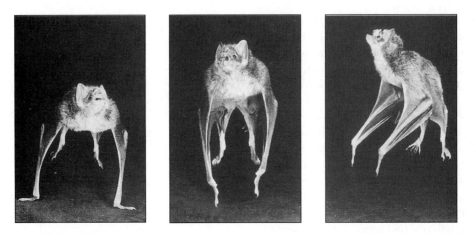

FIGURE 10-32 A vampire bat *(Desmodus rotundus)* leaping. Note the use of the long, robust thumbs. *(J. Altenbach)*

must climb and feed arboreally. During arboreal locomotion, *Diphylla* climbs head upward, gripping branches with opposed thumbs and hind feet, and uses the short, stout calcar on its hindlimbs like an opposable "sixth digit" to facilitate grasping (Schutt and Altenbach, 1997).

Vampire bats can detect temperature differences, an ability probably of importance when they are seeking a place to bite. Temperature stimuli are probably detected in three pits in the hairless skin surrounding the nose (Kurten and Schmidt, 1982). The surface temperature of this skin is up to 9°C lower than that of nearby parts of the face.

The feeding habits of vampire bats are of particular interest. Vampires begin foraging after dark and have one foraging period per night. *Diaemus* and *Diphylla* prefer the blood of birds, but *Desmodus* is more generalized, feeding on both mammals and birds. Indeed, *Desmodus* is a major pest of livestock and poultry, such as chickens and turkeys kept in poorly enclosed pens in Latin America, and there are reports of it feeding on penguins and sea lions along the coast of South America. *Desmodus* alights on the ground near its chosen host, often a cow, horse, or mule, and climbs up the foreleg to the shoulder or neck. The bat uses its upper incisors and canines to make an incision several millimeters wide from which it "laps" blood with its tongue. Vampire bats occasionally feed on the feet of cattle, at which time their ability to jump quickly may enable them to avoid injury when the host animal moves its feet.

In *Desmodus*, the ingestion of blood is facilitated by an **anticoagulant** in the saliva that retards clotting. It has been estimated that each bat takes a meal of blood each night that amounts to over 50 percent of the fasting weight of the bat; a vampire bat weighing 34 grams, then, takes roughly 18 grams of blood per night (Wimsatt, 1969a). Because of this nightly drain of blood from cattle in certain localities and because vampire bats transmit rabies and other diseases, they have great economic impact in many Neotropical areas. Occasionally vampire bats feed on humans. Female *Desmodus* that have fed will regurgitate blood to feed their young and closely related roost mates, a form of altruistic behavior (Wilkinson, 1987). Related bats and those that share the same roost will also share the same feeding sites on their prey.

FAMILY MOLOSSIDAE. Members of this family, the free-tailed bats, are important components of tropical and subtropical chiropteran faunas throughout much of the world. The family comprises 12 genera and 80 species, which includes *Tomopeas,* a genus formerly placed in the family Vespertilionidae but recently shown to belong with the Molossidae (Sudman et al., 1994). Molossids occupy the warmer parts of the Old World, from southern Europe and southern Asia southward, as well as Australia and the Fiji Islands. In the New World, they occasionally occur as far north as Canada, but the main range begins in the southern and southwestern United States and the West Indies and extends southward through all but the southern half of Chile and Argentina. Fossils of molossids are known from as early as the middle Eocene of Canada (Legendre, 1985).

Structurally, this is a peripheral group of bats; the most extreme manifestations of many of the typically chiropteran adaptations for flight occur in the Molossidae. These bats weigh from 8 to 196 grams, and wing spans range from 240 to at least 516 millimeters. The greater tuberosity of the humerus is large (see Figs. 10-6, 10-17), and the locking device between it and the scapula is highly developed. The origins of the extensor carpi radialis longus and brevis and flexor carpi ulnaris muscles are well away from the center of rotation of the elbow joint and probably act more effectively than in any other bats as "automatic" extensors and flexors of the hand. The wing is typically long and narrow (Fig. 10-5B), with the fifth digit no longer than the radius, and the membranes are leathery because they are reinforced by numerous bundles of elastic fibers. In many species, there are structural refinements that favor high-speed flight. In many molossids, for example, the radius and forearm muscles are flattened and the arrangement of specialized hairs (Fig. 10-33) is such that the forearm is streamlined with respect to the airstream during flight (Vaughan and Bateman, 1980). Adding rigidity to the outstretched wing during flight, movement at the wrist and elbow joint is strictly limited to one plane. The muscles that brace the fifth digit and maintain an advantageous plagiopatagium attack angle during the downstroke of the wings are large and unusually highly specialized. Except for fusion of the last cervical and first thoracic vertebrae, the presacral vertebrae are unfused. The body of the sternum is not keeled.

The general appearance of molossids is distinctive. The tail extends well beyond the posterior border of the uropatagium when the bats are not in flight, and the fur is usually short and velvety. (In one genus, *Cheiromeles*, hair is virtually absent.) Typically, the ears are broad, project to the side, and are like short winglets. As viewed from the side, the pinnae are arched and resemble an airfoil of high camber. The ears are frequently braced by thickened borders and are connected by a fold of skin across the forehead. Because of their unique design, the ears in most species do not di-

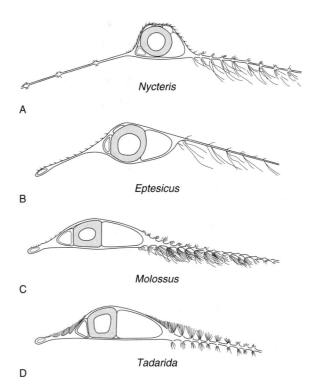

FIGURE 10-33 Cross-sectional views of the forearms and wing membranes of four species of bats, showing in C and D the pronounced streamlining of the wing by hair tracts and the flattening of the forearm in molossids. Two slow fliers: (A) *Nycteris thebaica* and (B) *Eptesicus fuscus.* Two fast fliers: (C) *Molossus ater* and (D) *Tadarida condylura. (Reprinted from Vaughan and Bateman, 1980, p. 74, by permission of Texas Tech Press)*

rectly face the force of the airstream during flight, an adaptation probably of considerable importance to these fast-flying bats. The muzzle is broad and truncate, and the thick lips are wrinkled in some species (Fig. 10-20D). In *Eumops* the lips are finely wrinkled and almost prehensile when used in feeding (Freeman, 1981).

The molossid skull is broad, the teeth are tuberculosectorial, and the dental formula is 1/1-3, 1/1, 1-2/2, 3/3 = 26-32. Several characteristically molossid features are associated with the well-developed quadrupedal locomotion typical of these bats. The first phalanges of digits three and four flex against the posterodorsal surfaces of their respective metacarpals, allowing the chiropatagium to be folded into a compact bundle, no longer than the forearm, that is manageable when the animals run.

The feet are broad and have sensory hairs along the outer edges of the first and fifth toes. The fibula is not reduced, and the short hindlimbs are stoutly built. Within the structural limits of the basic chiropteran plan, these bats have seemingly made the best of two types of locomotion. The highly specialized wings are clearly adapted to fast, efficient flight, and the primitive hindlimbs have not lost their ability to serve in rapid quadrupedal locomotion. In *Cheiromeles* the big toe is opposable and has a nail instead of a claw.

These insectivorous bats are remarkable for their speedy and enduring flight. Whereas most bats fly fairly close to the ground or to vegetation when foraging, many molossids fly high and may move long distances during their nightly foraging. Some populations of Brazilian free-tailed bats *(Tadarida brasiliensis)*, the bats that occur in great numbers in Carlsbad Caverns and other large caves in the southwestern United States, fly at least 90 kilometers to their foraging areas each night (Davis et al., 1962). These bats were observed in Texas with radar and helicopters, and dispersal flights were tracked (Williams et al., 1973). Dispersing bats were recorded at elevations of over 3000 meters, and masses of bats moved at an average speed of 40 kilometers per hour. The western mastiff bat *(Eumops perotis)* forages over broad areas and in southern California may on occasion fly more than 650 meters above the ground (Vaughan, 1959). Because of the temperature inversions that frequently prevail for many nights in this area, in the winter these high-flying bats may be surrounded by air warmer than that at the ground and may be catching insects that are flying in the warm air strata. Using bat detectors on helium-filled balloons, Fenton and Griffin (1997) recorded several species of African molossids foraging 600 meters above the ground. Some molossids remain in flight for much of the night; foraging periods of at least six hours have been recorded for some species.

Some molossids make spectacular dives when returning to their roosts. The western mastiff bat, for example, often makes repeated high-speed dives and half loops past the roosting site. It returns to its roost in a cliff by diving toward the cliff base, pulling sharply upward at the last instant, and entering the crevice with momentum to spare. Several other molossids are known to return to

their roosting places by similar maneuvers. Because the wings of many molossids are narrow and have relatively small surface areas relative to the weights of the bats, these animals must attain considerable speed before they can sustain level flight. As a result, some species roost high above the ground in cliffs, buildings, or palm trees, in situations where they can dive steeply downward for some distance in order to gain appropriate flight speed. These species are unable to take flight from the ground.

Most molossids inhabit warm areas. Migration, therefore, is not generally characteristic of these bats. The Brazilian free-tailed bat, however, is known to make extensive migrations from the United States to as far south as southern Mexico (Villa-Ramirez and Cockrum, 1962). The tremendous deposits of **guano** in some large caves inhabited by molossids attest to the effect that large colonies of these bats must have on insect populations in some areas.

FAMILY VESPERTILIONIDAE. This is the largest family of bats in terms of numbers of species and the most widely distributed. Thirty-five genera and approximately 320 species are included in this family, and in temperate parts of the world these are usually by far the most common bats. In the New World, vespertilionids occur from the tree line in Alaska and Canada southward throughout the United States, Mexico, and Central and South America. All of the Old World is inhabited north to the tree line in northern Europe and Asia. Most islands, with the exception of some that are remote from large land masses, support vespertilionids. As can be inferred from their geographic distribution, these bats occupy a wide variety of habitats, from boreal coniferous forests to barren sandy deserts. In the Neotropics, however, they are greatly outnumbered by bats of other families, particularly by leaf-nosed bats (Phyllostomidae).

Perhaps because of the diversity of habits and structure represented within the Vespertilionidae, no common name for this group is in general use; they are usually called simply "vespertilionid bats." The genus *Myotis* is remarkable for its broad geographic distribution, which includes roughly the entire area occupied by the Vespertilionidae, and for its long fossil record, which begins in the early Oligocene of Europe.

Vespertilionids are rather plain-looking bats that lack the distinctive facial features characteristic of many families. A nose leaf is rarely present (only in the subfamily Nyctophilinae, best developed in *Pharotis,* which is now extinct; Flannery, 1995), and complex flaps or pads do not occur on the lower lips. The eyes are usually small. The ears are of moderate or large size, and the tragus is present but differs in shape markedly from one species to another. These bats are usually small, weighing from 4 to 83 grams, with wing spans ranging from 200 to 400 millimeters. The wings are typically broad, and the uropatagium is large and encloses the tail. The shoulder joint is of a derived type and provides for a locking of the large greater tuberosity of the humerus (see Fig. 10-17) against the scapula at the top of the upstroke of the wing. The elbow joint is also derived, and the spinous process of the medial epicondyle, which is well developed in many species, enables certain forearm muscles to "automatically" extend and flex the hand (see Fig. 10-10). The shaft of the ulna is vestigial, but the proximal portion forms an essential part of the elbow joint. The second digit of the hand has two bony phalanges, and the third digit has three. The fibula is rudimentary. The manubrium of the sternum has a keel, but the body of the sternum has at best a slight ridge. In all species the presacral vertebrae are unfused.

The teeth are tuberculosectorial, and the W-shaped ectoloph of the upper molars is always well developed. The dental formula is 1-2/2-3, 1/1, 1-3/2-3, 3/3 = 28-38. The skull lacks postorbital processes; the palatal parts of the premaxillaries are missing, and the front of the palate has an open gap (see Fig. 10-22). In general, vespertilionids are mostly small, plain bats that are characterized by refinements of echolocation and of the flight apparatus that enable them to forage and maneuver efficiently.

Most vespertilionids are insectivorous, and in their ability to capture flying insects they are unsurpassed. Most children in Europe and North America gain their first experience with bats by watching vespertilionid bats, silhouetted against the twilight sky, making abrupt turns and sudden dives while pursuing insects. The most commonly used vespertilionid foraging technique is probably the most demanding: it involves the pursuit and capture of flying insects, which means that the bats

must remain on the wing throughout most of their foraging periods. The insects, detected by echolocation, are usually followed in their erratic flight by a series of intricate maneuvers by the bat and are either captured in the mouth or, in some species of bats, trapped by a wingtip or by the uropatagium (Webster and Griffin, 1962). This type of foraging demands highly maneuverable flight, and this is the type of flight to which vespertilionids seem best adapted.

Styles of foraging vary from one type of vespertilionid to another. The tree-roosting bats *(Lasiurus)* remain on the wing throughout their foraging, while some other bats alight to eat large prey. Some vespertilionids capture insects from the surface of the water, and several species of *Myotis* also capture fish or crustaceans from the water, probably by gaffing the prey with the large claws of the enlarged feet. Still other kinds of vespertilionids glean insects from the surfaces of vegetation or the ground. The pallid bat *(Antrozous pallidus),* a common species in the southwestern United States, feeds on such large terrestrial arthropods as scorpions, Jerusalem crickets *(Stenopelmatus),* and sphinx moths (Sphingidae). After localizing these prey items mostly by listening for the sounds they make, the bats snatch them from leaves or pounce on them on the ground. Occasionally, pallid bats kill and eat small vertebrates, such as horned lizards *(Phrynosoma).* This bat pollinates cactus and *Agave* flowers in southwestern North American deserts while capturing and eating insects that are in the flowers (Fleming, 1995). The feeding habits of one of the tube-nosed bats, *Harpiocephalus harpia,* are poorly known, but the species seems to have evolved some morphological features that are convergent with certain fruit-eating bats, such as a broad rostrum and strongly modified teeth (Fig. 10-34).

A wide variety of roosting places are used by vespertilionids. They adapt well to urban life and frequently roost during the day in attics, in spaces between rafters, or behind shutters or loose boards. Crevices in rocks, spaces beneath rocks or behind loose bark, caves, mines, holes in trees, and foliage are also utilized. Often these bats are colonial, frequently with maternity colonies of females with young occupying one roost and adult males using another. Many species, however, such as the foliage-roosting bats, roost singly or in small groups.

FIGURE 10-34   The murinine bat *Harpiocephalus harpia* (Vespertilionidae: Murininae) from Laos. This poorly known vespertilionid seems to possess some cranio-dental features that are convergent with certain fruit-eating bats. *(C. Francis)*

Some species rest for part of the night beneath bridges, in grottoes in cliffs, or in porches or buildings, often in places never used as daytime retreats.

In temperate regions, many vespertilionids hibernate. Although the hibernation sites of many species are not known, some well-known species hibernate in caves and mines or buildings, and some species migrate fairly long distances to reach favorite hibernacula.

FAMILY MYZOPODIDAE. The only species representing this family is *Myzopoda aurita*, the sucker-footed bat, a species restricted to Madagascar. Fossils of this family are known from the early Pleistocene in East Africa.

The structure of its shoulder joint indicates that this bat is probably related to the Natalidae, Furipteridae, and Thyropteridae. The lumbar vertebrae are not fused as they are in natalids. The dental formula is 2/3, 1/1, 3/3, 3/3 = 38. The ears are very large, and the ear opening is partly covered by an unusual mushroom-shaped structure found in no other bat. The claw of the thumb is rudimentary, and the thumb bears a sucker disk. Only the metacarpal of the second digit is bony; the third digit has three ossified phalanges. The foot bears a sucker disk on its sole, and, as in thy-

ropterids, each digit has only two phalanges. However, unlike thyropterids, the disks of *Myzopoda* seem to function by gluing instead of active suction (Thewissen and Etnier, 1995). In *Myzopoda,* the metatarsals are fused and all the toes fit tightly against one another. *Myzopoda* appears to be rare, and its life history is poorly known. It is insectivorous, feeding on small moths. One individual was captured roosting in a traveler's palm *(Ravenala madagascariensis),* and a captive specimen readily roosted in *Ravenala* fronds by using its thumb and foot disks, orienting itself head-upward, and bracing itself with its stiff tail (Fig. 10-35; Göpfert and Wasserthal, 1995).

FAMILY NATALIDAE. This small Neotropical family includes a single genus with five species. These bats are commonly referred to as funnel-eared bats and occur from Baja California, northern Mexico, and the West Indies southward to Colombia, Venezuela, and Brazil. The living genus *Natalus* is known as a fossil from the Pleistocene, but extinct nataloids are known from the early Eocene of Wyoming (Beard et al., 1992) and Eocene-Oligocene of France (Sigé, 1974).

These small bats weigh from roughly 5 to 10 grams and have slender, delicate-looking limbs,

FIGURE 10-35   Golden bat, or sucker-footed bat, *Myzopoda aurita* from Madagascar. The animal is clinging by its sucker discs to a frond of the palm *Ravenala. (L. Wasserthal)*

broad wing membranes, and a large uropatagium that encloses the long tail. The funnel-shaped ears with a tragus, the simple nose lacking any sort of nose leaf, and the long, soft pelage that is frequently yellowish or reddish are characteristic. The skull has a long, wide rostrum with complete premaxillaries, and the braincase is high. The teeth are tuberculosectorial; the dental formula is 2/3, 1/1, 3/3, 3/3 = 38. The humeroscapular locking device is well developed, reduction of the phalanges of the hand is well advanced (the second digit lacks a phalanx and the third has two), and the manubrium of the sternum is unusually broad and has a well-developed keel. Some of the most distinctive natalid features, however, are those of the axial skeleton that reduce its flexibility: the thoracic vertebrae are anteroposteriorly compressed and fit tightly together; the ribs are broad, and the narrow intercostal spaces are largely spanned by sheets of bone. All except the last two lumbar vertebrae are fused into a solid, laterally compressed, dorsally and ventrally keeled mass (Fig. 10-36); and the sacral vertebrae are mostly fused. As a result of these specializations, the strongly arched thoracolumbar section of the vertebral column is nearly rigid, with movement between this and the sacral section of the column allowed only by the "joint" formed by the last two lumbar vertebrae. These specializations seem to brace and cushion the vertebral column against shock transmitted to it by the hindlimbs when this bat alights on the ceilings of caves.

Funnel-eared bats are insectivorous, and their foraging flight is slow, delicate, and maneuverable. Individuals released in dense vegetation are amazingly adroit at flying slowly through small openings between the interlacing branches of trees and shrubs. These bats inhabit tropical and semitropical lowlands and foothills and typically roost in groups in warm, moist, and deep caves or mines. These are handsome little bats; groups of *Natalus stramineus* scattered over the ceiling of a cave look like bright orange jewels in the beam of a flashlight.

FAMILY FURIPTERIDAE. This small family contains but two genera, each with one species. These bats occur from Costa Rica south to southern Brazil and northern Chile, and in Trinidad. Furipterids, known as smoky bats because of their gray pelage, are seemingly closely related to the Natalidae, Thyropteridae, and Myzopodidae. All these groups share certain structural similarities. The extant genus *Furipterus* is known as a fossil in the late Pleistocene of Brazil.

Externally, furipterids resemble natalids in the structure of the ears and in the slender build. The shoulder joint and the fused lumbar vertebrae are also similar in these families. Furipterids differ from natalids in minor features of the skull and dentition, such as partially cartilaginous premaxillaries and reduced canines. The furipterid dental formula is 2/3, 1/1, 2/3, 3/3 = 36. The thumb of smoky bats is greatly reduced.

These bats apparently are not common, and their habits are poorly known. They are insectivorous and have been found in caves and buildings. Most of the area inhabited by smoky bats is tropical, but *Amorphochilus* occurs in arid coastal sections of northwestern South America.

FAMILY THYROPTERIDAE. This family includes three small Neotropical species of bats that are known as disk-winged bats because of the remarkable sucker disks that occur on the thumbs and feet (Fig. 10-37). These animals are the only bats and the only mammals that have true suction cups. Disk-winged bats occur in southern Mexico, Central America, and South America as far as Peru and southern Brazil. Fossil thyropterids have been recorded in the middle Miocene of Colombia.

In general appearance and in many skeletal details, these small (weight about 4 grams; wing span 225 millimeters), delicately formed bats resemble natalids, but the lumbar vertebrae are not

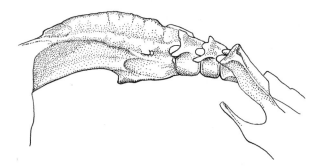

FIGURE 10-36 Lateral view of the left side of the fused lumbar vertebrae of a funnel-eared bat *(Natalus stramineus,* Natalidae). *(From Vaughan, 1970b)*

FIGURE 10-37 A disk-winged bat, *Thyroptera* sp., from La Selva, Costa Rica, clinging by its suction discs to a smooth-surfaced leaf. *(L. Wasserthal)*

fused as in the latter. The skulls of natalids and thyropterids are similar, and the dental formulas are the same. The thumb is reduced but retains a small claw, and its first phalanx has a sucker disk. The second digit is short, being represented by only a rudimentary metacarpal, and, as a result, the membrane between digits two and three is unusually small. The third digit has three bony phalanges. The digits of the feet have only two phalanges each, the third and fourth digits are fused, and the metatarsals bear suction disks, which have a complex structure that allows them to act as suction cups. The bats can cling to smooth surfaces and can even climb a vertical glass surface. A fibrocartilaginous framework braces each disk; the rim of the disk consists of 60 to 80 chambers, each supplied by a sudoriferous (sweat) gland, which improves the tightness of contact with the substrate by ensuring that the face of the disk is constantly moistened. The disk itself lacks muscles, but specialized forearm muscles produce suction by cupping the middle of the face of the disk and release suction by lifting a section of the disk rim (Wimsatt and Villa-Ramirez, 1970).

Disk-winged bats are insectivorous and restricted to tropical forests. Their roosting habits are highly specialized. They roost only in the young, slightly unfurled leaves of certain tropical plants that are partially or completely shaded by larger trees. Such a roosting site is provided by the "platanillo" (*Heliconia* sp.), which resembles the banana plant. When a young leaf of this plant is beginning to unroll, it forms a tube roughly 1.3 meters long and 25 millimeters in diameter with a small opening at its tip. Several disk-winged bats may occupy such a tubular leaf in a head-to-tail row, heads upward, with the sucker disks anchoring them to the slippery surface of the smooth leaf. Because the leaf soon unfurls, it is suitable for occupancy for only about 24 hours, and the bats move periodically to new and more suitable leaves. Findley and Wilson (1974) found that these bats usually roost in social groups of six or seven, that the bats of a given group always roost together, and that each group occupies an exclusive area within which it roosts in the daytime.

WEB SITES

Animal Diversity Web information about bats

http://www.oit.itd.umich.edu:80/bio108/Chordata/Mammalia/Chiroptera.shtml

Bat Conservation International home page

http://www.batcon.org/

Bat Conservation Trust home page, United Kingdom

http://www.bats.org.uk/

Information about bat detectors and echolocation

http://home.earthlink.net/~infocentr/

Bat echolocation and bioacoustics information

http://www.ozemail.com.au/~jollys/

On Australian Megachiroptera

http://online.anu.edu.au/srmes/wildlife/batatlas/at_foxes.html

Phylogenetic tree of bats and information about characteristics

http://research.amnh.org/tol/chiroptera/chiroptera.html

Many bat-related links along with plans for building bat houses

http://www.nyx.net/~jbuzbee/bat_house.html

11 PRIMATES AND SCANDENTIA

PRIMATES

The order Primates is of particular interest not only because it includes our closest relatives but also because its members display a fascinating breadth of structural and behavioral adaptations to their environments. The primate radiation can be viewed as an exploitation of arboreal herbivory, arboreal locomotion, manual dexterity, stereoscopic vision, and complex social behavior and communication. The ancestors of humans are among the relatively few primates that adopted a largely terrestrial mode of life.

Primates other than humans have been most successful in tropical and subtropical areas, where

today they pursue mostly arboreal modes of life. Some primates, such as baboons and chimpanzees, have become partly or mostly terrestrial, but only humans have become fully bipedal. Approximately 60 genera and 233 species of primates live today, of which 16 genera and about 90 species occur in the New World.

Although primates are largely herbivorous, many are omnivorous and seem to be primarily opportunistic feeders. The molars of primates are largely bunodont and brachydont and possess the quadrate form typical of molars of many herbivores or generalized feeders (Kay, 1975). Early in the evolution of primates, a hypocone was added to the upper molar and the paraconid of the lower molar disappeared, leaving a basically four-cusped pattern (Fig. 11-1). The rostrum of primates is proportionately shorter than in most mammals, resulting in a shorter tooth row. This is probably related to the importance of stereoscopic vision.

Primates constitute one of the oldest eutherian orders, possibly dating from the latest Cretaceous Period of North America. Primates underwent an early radiation from the early Paleocene to the late Eocene in both Europe and North America (Fig. 11-2). Traditionally, living primates have been grouped according to two major classification schemes (Fig. 11-3). One such scheme divides primates into the Prosimii, including tarsiers, lemurs, lorids, and their fossil relatives; and the Anthropoidea, which includes monkeys, apes, and humans. A second classification scheme groups primates according to the structure of the rhinarium. The more primitive families among living primates (Cheirogaleidae, Lemuridae, Megaladapidae, Indridae, Daubentoniidae, Loridae, and Galagonidae) are included here in the Suborder Strepsirhini, whereas the more advanced families (Tarsiidae, Callitrichidae, Cebidae, Cercopithecidae, Hylobatidae, and Hominidae) are placed in the Suborder Haplorhini. This classification largely follows that of Thorington and Anderson (1984) and Andrews (1988), and is consistent with that of Kay et al. (1997b).

The exact origin of primates is controversial because there are few shared derived characters uniting early primates with members of the other mammalian orders (Novacek et al., 1983). The earliest fossil evidence of primates is a lower molar from the early Paleocene of North America. The primate fossil record becomes more abundant and diverse in the middle Paleocene and Eocene deposits of North America and Europe. The Paleocene and Eocene Plesiadapiformes are thought by most authorities to be the ancestors of modern primates (Carroll, 1988; Benton, 1990). However, Andrews (1988) has argued for removing plesiadapiforms from primates altogether, and Cartmill (1974) advocated allying them with Insectivora. Two additional fossil families, Adapidae and Omomyidae, arose in the late Paleocene or early Eocene. Adapids have many characteristics in common with strepsirhines, especially lemurs; however, many of these features are shared primitive characters. Eocene omomyids were small and resembled the living tarsiers (Carroll, 1988).

There is little doubt that primates are a monophyletic group. Martin (1986) and Andrews (1988) defined primates based on the presence of 28 shared derived characters mostly absent from tree shrews (Table 11-1). These characters are associated with the style of locomotion, the presence of stereoscopic vision, increase in size and complexity of the brain, patterns of reproduction with associated morphological features, modifications of the auditory complex, and dentition. The evolution of primates is still the subject of heated debate, and a consensus on phylogeny does not yet exist (MacPhee, 1993, and articles therein).

A great deal of literature on primates has appeared during the past two decades, and at least six journals are devoted exclusively to primatology. So vast and ever-growing has the literature on primates become that the Regional Primate Research Center of the University of Washington maintains a computerized primate bibliographic service (see web sites). Excellent treatments of primate evolution, ecology, and behavior can be found in several

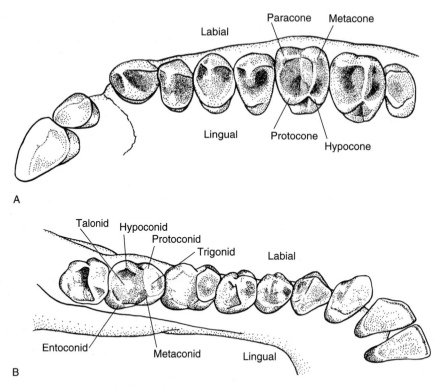

FIGURE 11-1 A diagrammatic representation of the maxillary and mandibular tooth row of the platyrrhine primate *Aotus*: (A) left upper tooth row, (B) left lower tooth row. Lingual refers to the side of the tooth row adjacent to the tongue. *(After Ford, 1986)*

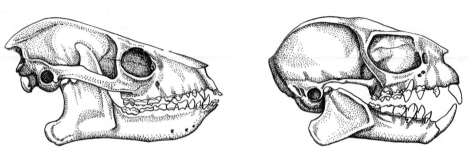

Notharctus, Adapidae   *Tetonius*, Anaptomorphidae

FIGURE 11-2 Skulls of a fossil Eocene lemuroid (A), with a skull length of 75 millimeters, and an Eocene tarsier-like primate (B), length of skull 46 millimeters. *(After Romer, © 1966, Vertebrate Paleontology, University of Chicago Press)*

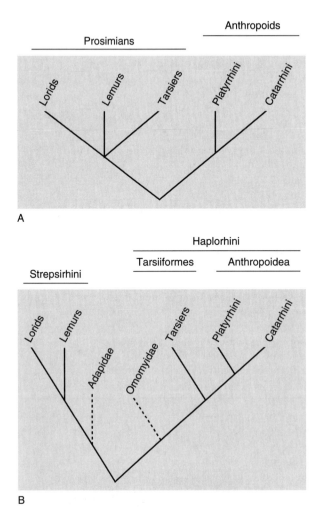

FIGURE 11-3 Two contrasting primate phylogenies: (A) the classical prosimian-anthropoid split, (B) a more recent phylogeny in which the tarsiers are allied with the anthropoids. The two extinct families Adapidae and Omomyidae are represented by dashed lines.

texts, including Martin (1990), Fleagle (1988), Ciochon and Chiarelli (1980), Luckett and Szalay (1975), Mittermeier and Coimbra-Filho (1981), and Szalay and Delson (1979).

SUBORDER STREPSIRHINI

The extant strepsirhine primates include the five families of lemurs endemic to Madagascar and the Loridae and Galagonidae. The suborder derives its name from the features of the nose, with its naked **rhinarium,** unfused nasal prominences, and slitlike nostrils. The strepsirhine nose is of little use, however, in defining this assemblage of primates because these features are shared by many nonprimate mammals (Andrews, 1988). Strepsirhine primates are better defined by presence of a tooth comb, composed of lower incisors and canine teeth, and a grooming claw on the second digit of the foot. The tooth comb (secondarily lost in *Daubentonia*) projects forward from the lower jaw and is used during grooming. Tree shrews have a similar tooth comb, but it does not include the canine teeth (Rose et al., 1981). The extant dermopteran *Cynocephalus* also has elaborate tooth combs with up to 15 tines, but these "combs" evolved independently and are used to crop leaves and not for grooming. The lemurs are distinguished from the lorids and galagos by a suite of dental characters (Schwartz and Tattersall, 1987). The strepsirhine fossil record begins in the early Eocene of Africa and Asia, too late to provide information concerning the strepsirhine/haplorhine divergence.

FAMILY CHEIROGALEIDAE. Of the five families of primates commonly called lemurs that inhabit the island of Madagascar, the cheirogaleids include only four genera and seven species commonly known as mouse and dwarf lemurs (Fig. 11-4). The family Cheirogaleidae lacks a fossil record. In the past, cheirogaleids were considered a subfamily of the Lemuridae (Simpson, 1945). However, cheirogaleids are now considered a valid family (Cartmill, 1975). Cheirogaleids, like other Malagasy primates, possess a "tooth comb" comprising the incisors and canines (Thorington and Anderson, 1984). These small lemurs are nocturnal and weigh less than 500 grams. Mouse lemurs *(Microcebus)* and the hairy-eared dwarf lemur *(Allocebus)* are among the smallest living primates. They are highly arboreal and move by means of quadrupedal walking or by a series of bipedal leaps between branches or trees. Mouse lemurs feed primarily on fruits, flower nectar, leaves, insects, and even an occasional vertebrate, including tree frogs and chameleons (Hladik et al., 1980). *Microcebus* tends to forage relatively close to ground level (less than 10 meters) and inhabits a wide variety of forest types, from desert forests of southern Madagascar to the montane rain forests along the eastern escarpment (Tattersall, 1982; Schmid and Kappeler, 1994).

TABLE 11-1 Shared Derived Features of the Order Primates

Locomotor characters	Grasping hands and feet with opposable thumbs and toes
	Hallux containing a nail
	Nails present on all or most digits (may be lost secondarily)
	Elongation of the calcaneus bone
	Hindlimb dominance during locomotion
	Center of gravity shifted towards hindlimbs
Stereoscopic vision	Some degree of forward rotation of the orbits and narrowing of the interorbital distance
	Enlargement of the orbital cavity
	Exposure of the ethmoid bone on the inner orbital wall
	Stereoscopic vision in which approximately half of the retinal axons project to the ipsilateral side of the brain
Brain characters	Increased fetal brain size compared to fetal body weight, retained in neonates
	Sylvian sulcus and triradiate calcarine sulcus both present on brain
Reproductive features	Descent of testes early in life
	Urogenital sinus absent in females
	Long gestation times relative to body size, resulting in small litters
	Sexual maturity relatively late in life
	Long lifespans (typically)
Auditory characters	Auditory bulla bony; the floor is made up of the petrosal bone
	Extension of the ectotympanic into the auditory meatus
Dentition	Loss of one incisor and one premolar from the ancestral eutherian condition

The dwarf lemurs, *Cheirogaleus,* have a more pointed snout and a longer body. They are arboreal quadrupeds and rarely resort to bipedal leaping. *Cheirogaleus* species tend to be more **frugivorous** than other lemurs. In the drier habitats of southern and western Madagascar, dwarf lemurs may estivate (undergo a period of reduced metabolic activity similar to hibernation) for up to six months, relying on fat reserves stored in their tails. Their body weight may drop nearly 50 percent during this period (Hladik et al., 1980). *Phaner furcifer,* the fork-marked lemur, is the largest member of the family Cheirogaleidae. Their name derives from the dark rings around their eyes that join to form a single stripe down the middle of the back. These lemurs forage almost exclusively on gums and sap secreted from wounds in the outer bark of certain species of trees. They use their procumbent incisors to gouge into the inner gum-producing layers of the tree, and their large caecum contains microbes used to break down the gum (Charles-Dominique and Petter, 1980). The tiny

FIGURE 11-4 The mouse lemur (*Microcebus murinus,* Cheirogaleidae) from the eastern rain forest of Madagascar *(J. Visser, American Society of Mammalogists).*

A

B

FIGURE 11-5   (A) A ring-tailed lemur (*Lemur catta,* Lemuridae) and (B) a sifaka (*Propithecus verreauxi,* Indridae) from the island of Madagascar. *(A by J. Ryan and B by Corel Corp.)*

hairy-eared dwarf lemur *(Allocebus trichotis)* was, until very recently, considered extinct, but was rediscovered in northern Madagascar in 1989. Studies using DNA hybridization suggest a close relationship between *Allocebus* and *Microcebus* (Rumpler et al., 1994). Little is known about these very rare lemurs, but the similarities in their dentition with *Phaner* suggests a diet consisting largely of tree gums. The discovery of a new species of mouse lemur *(Microcebus ravelobensis)* from western Madagascar in 1997 indicates just how much remains to be learned about the endemic mammals of Madagascar (Zimmerman et al., 1998).

FAMILY LEMURIDAE. Lemurids inhabit Madagascar and the nearby Comoro Islands. Among the ten Recent species belonging to four genera, some are arboreal, some are semiarboreal, and one species, *Lemur catta,* is largely terrestrial (Fig. 11-5A). Only one fossil species is known, *Pachylemur insignis,* which is similar in dentition to the ruffed lemur (Seligsohn and Szalay, 1974).

In contrast to most primates, the cranium of lemurs is elongate and the rostrum is usually of moderate length, giving the face a foxlike appearance. The dental formula is 0-2/2, 1/1, 3/3, 3/3 = 32-36. The lower incisors also form a tooth comb and the upper incisors are reduced. Between those of the two sides is a broad diastema. The pollex and hallux are more or less enlarged and are opposable in all genera (Fig. 11-6). The pelage is woolly, the tail is long and heavily furred, the limbs are usually slim, and the tarsal bones are not greatly elongated as in unrelated tarsiers (Fig. 11-7). Conspicuous color patterns occur in some species.

Lemurs are variously herbivorous-frugivorous and, depending on species, are primarily diurnal. They are agile climbers, and the hands are used both for climbing and for handling food. Three species of *Hapalemur,* the bamboo lemurs, inhabit restricted areas of the southeastern rain forest; in some areas, all three species occur sympatrically. As their common name implies, they feed almost entirely on bamboo shoots and leaves. Until 1986, only *H. griseus* and the much larger *H. simus* were known, but two groups of scientists independently discovered a third species, *H. aureus,* living in isolated patches of primary rain forest (Meier, et al., 1987; Meier and Rumpler, 1987). Even more remarkable was the discovery that they are able to coexist in the same block of forest because they partition their food resources in such a way that each species feeds on a different species of bamboo or eats a different part of the bamboo plant.

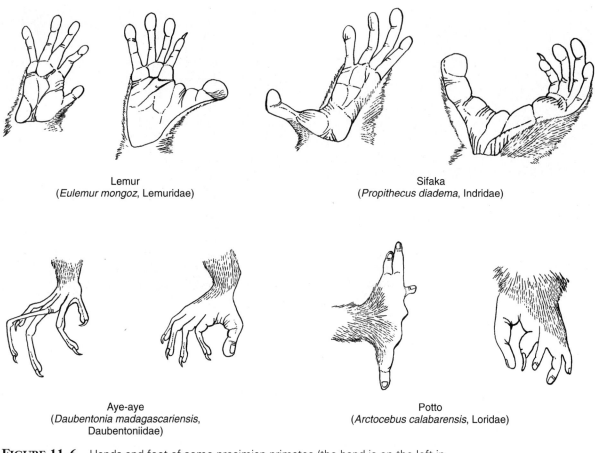

Lemur
(*Eulemur mongoz*, Lemuridae)

Sifaka
(*Propithecus diadema*, Indridae)

Aye-aye
(*Daubentonia madagascariensis*,
Daubentoniidae)

Potto
(*Arctocebus calabarensis*, Loridae)

FIGURE 11-6   Hands and feet of some prosimian primates (the hand is on the left in each pair).

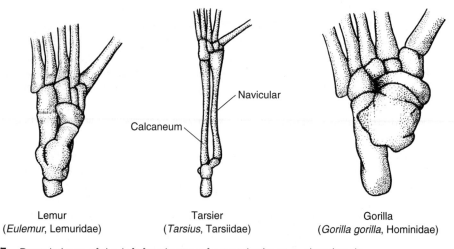

Navicular

Calcaneum

Lemur
(*Eulemur*, Lemuridae)

Tarsier
(*Tarsius*, Tarsiidae)

Gorilla
(*Gorilla gorilla*, Hominidae)

FIGURE 11-7   Dorsal views of the left feet bones of several primates, showing the tarsal bones. Note the remarkable elongation of the calcaneum and the navicular in the tarsier. *(After Clark, 1971)*

Hapalemur griseus eats primarily young bamboo leaves and the tender base of mature leaves, while *H. simus* eats mature leaves and the densely lignified main stems of the bamboo plant (Santini-Palka, 1994).

Ring-tailed lemurs *(Lemur catta)* are largely terrestrial and live in large social groups of up to 50 individuals. These social groups spend a great deal of time traveling each day in search of food. Interestingly, most lemur societies are female dominance hierarchies (females are dominant to males; Kappeler and Ganzhorn, 1993; Tattersall, 1982).

FAMILY MEGALADAPIDAE. The sportive lemurs, genus *Lepilemur,* were previously grouped with the Lemuridae but are distinct in lacking upper incisors and possessing an unusual jaw articulation and are now placed in a separate family, Megaladapidae (Tattersall and Schwartz, 1974). The fossil *Megaladapis* weighed up to 80 kilograms and is one of the largest fossil lemurs. Living sportive lemurs are nocturnal folivores and rely on bacteria in their digestive tracts, especially in the large caecum, to digest plant cellulose. Leaves are poor in nutritional content, and sportive lemurs conserve energy by ingesting their feces.

FAMILY INDRIDAE. The Malagasy indrids include the woolly lemur *(Avahi laniger;* Fig. 11-8), three species of sifaka *(Propithecus;* Fig. 11-5B), and the largest living lemur, *Indri indri.* Sub-Recent indrid fossils from Madagascar include a terrestrial lemur, *Archeoindris,* with limb dimensions similar to the extinct ground sloths of the New World and an estimated weight of 200 kilograms (Carlton, 1936; Mittermeier et al., 1994).

Indrids are fairly large, measuring up to 900 millimeters in head-an-body length and weighing about 10 kilograms. They have a shortened rostrum and a monkeylike face. They have a reduced dental formula of 2/2, 1/0, 2/2, 3/3 = 30, with only four incisors making up the tooth comb. The hands and feet are highly modified for grasping branches during climbing (Fig. 11-6). The limbs are relatively long and the tail is approximately equal to the head-and-body length, except in *Indri,* which has a short stub of a tail. The pelage is mottled brown in the nocturnal *Avahi,* either black and white or orange-brown and white in *Propithecus,* and black and white in *Indri.*

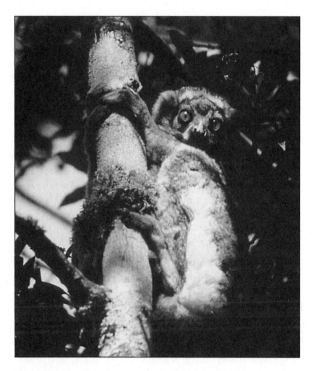

FIGURE 11-8   A nocturnal avahi *(Avahi laniger,* Indridae) in Madagascar. *(J. Ryan)*

These primates are largely herbivorous. Their leaf-eating habits resemble those of the Neotropical howler monkey *(Alouatta;* Cebidae) and the African colobus monkey *(Colobus;* Cercopithecidae). Indrids are typically fairly slow, deliberate climbers but can move rapidly through the canopy by using a series of bipedal leaps. The hindlimbs are long relative to the front limbs; when traveling on the ground, these primarily arboreal animals proceed by a series of hops. The hands are used for climbing and for handling food, but manual dexterity seems limited, and food is often picked up in the mouth. The nocturnal *Avahi* lives in **monogamous** pairs, forages through the rain forests for leaves, and sleeps huddled in small family groups during the day. Sifakas and indri live in small family groups and forage during the day. The specialized larynx enables *Indri* to produce loud, resonant calls. These howls are given with greatest frequency in the morning and evening, as are the calls of the howler monkey, and may help maintain territorial boundaries between neighboring bands. All indrids are vocal to some extent.

FAMILY DAUBENTONIIDAE. This family is represented by only one highly specialized living species, the aye-aye *(Daubentonia madagascariensis).* This secretive nocturnal animal occurs locally in northern and eastern Madagascar, where it is restricted to dense forests and stands of bamboo. The fossil record of Daubentoniidae consists of sub-Recent fossils from Madagascar of an extinct species that was slightly larger than the surviving aye-aye.

Aye-ayes weigh approximately 2 kilograms. They have prominent ears and a long, bushy tail. The skull and dentition are remarkably specialized and depart strongly from the usual primate plan (Fig. 11-9). The skull is short and moderately high. The orbit is prominent and faces largely forward; the postorbital bar and zygomatic arch are robust, and the rostrum is short and deep. The dentition differs from the basic primate type both in the extensive loss of teeth and in the strong specialization of the teeth that are retained. The dental formula is 1/1, 0-1/0, 1/0, 3/3 = 18 or 20. The canine is often absent, and the cheek teeth have flattened crowns with no clear cusp pattern. The laterally compressed incisors are greatly enlarged and wear to a sharply beveled edge because only the anterior surfaces are covered with enamel (as in rodents); they are ever-growing. Because of the shape of the teeth and the presence of a diastema between the incisors and the cheek teeth, *D. madagascariensis* was first described as a rodent. The hand is unique among primates; the digits are clawed, all but the nonopposable pollex are long and slender, and the third digit is remarkably slender (Oxnard, 1981; Fig. 11-6). In the hind foot, the hallux is opposable and bears a nail, but the other digits are clawed.

Aye-ayes are arboreal, nocturnal, and mainly insectivorous. Their foraging technique is noteworthy. The elongate third finger is used to tap on wood harboring wood-boring insects; the aye-aye then listens carefully for insects within the wood, and the remarkable third digit is used for removing adult and larval insects from holes or fissures in the wood (Erickson, 1991). When necessary, the powerful incisors tear away wood to enable the third digit to reach insects in deep burrows. Interestingly, this strange mode of foraging is shared (convergently) by metatherians of the family Petauridae from Australia and New Guinea (Strahan, 1995; Flannery, 1995). In *Dactylopsila,* the most specialized of these metatherians, the front incisors are modified and the manus is specialized along lines parallel to those in the hand of *Daubentonia,* except that the fourth rather than the third digit is the probing finger (see Fig. 6-24). As is the case with many extremely specialized mammals that occupy limited geographic areas, the future of the aye-aye and many other species of Malagasy primates seems dim, and its survival depends on preservation of natural forests in northern and eastern Madagascar.

FAMILY LORIDAE. The lorids are more widely distributed than are the primitive primates of Madagascar and can be common at some localities. Lorids occur in Africa south of the Sahara, in India, Sri Lanka, and southeastern Asia, and in the East Indies. The fossil record of lorids is scanty, but

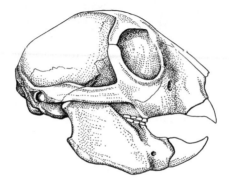
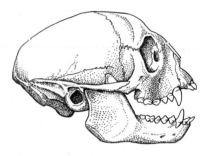

FIGURE 11-9 Skulls of an aye-aye (length of skull 110 millimeters) and a marmoset (length of skull 51 millimeters). *(Marmoset after Hall and Kelson, Mammals of North America, © 1959 by permission of John Wiley & Sons, Inc.)*

Aye-aye
(*Daubentonia madagascariensis*, Daubentoniidae)

Marmoset
(*Saguinus geoffroyi*, Callitrichidae)

suggests that these animals evolved in the Old World and have never occurred elsewhere. A Miocene lorid is known from Africa. There are four living genera and six species.

The eyes face forward in the lorids (Fig. 11-10), rather than more or less to the side as in lemurs, and the rostrum is short. Lorids are arboreal, and their locomotion usually involves methodical hand-over-hand climbing. Lorids vary from the size of a rat to the size of a large squirrel. The braincase is globular, the facial part of the skull is often short and ventrally placed, and the anteriorly directed orbits are separated by a thin interorbital septum. The dental formula is 1-2/2, 1/1, 3/3, 3/3 = 34 or 36. The upper incisors are small, the lower canine is incisiform, and the molars are basically quadritubercular. Tails are either absent or short, less than 15 percent of head-and-body length.

FIGURE 11-10 A potto (*Perodicticus potto*, Loridae). *(J. Eisenberg, American Society of Mammalogists)*

The manus and pes are specialized in a variety of ways for clutching branches. In the genus *Arctocebus,* a pincer-like hand has been developed by the reduction of digits two and three and a change in the posture of the remaining digits; the first digit of the pes is opposable and frequently greatly enlarged (Fig. 11-6). Circulatory adaptations in the appendages provide for an increased blood supply to the digital flexor muscles, which are used in gripping branches during extended periods of contraction. Similar circulatory modifications, involving the formation of a rete mirabile, are also important in this and many other mammals in conserving body heat (p. 371).

These nocturnal primates are insectivorous and carnivorous, and prey is usually captured by the hands after a stealthy approach. Pottos (*Perodicticus* and *Arctocebus)* are slow climbers that prefer continuous canopy. *Perodicticus* is generally found higher in the canopy, and restricts its movements to larger branches, while *Arctocebus* forages in the understory, below 5 meters, where it climbs on very small branches and vines. Their diet depends on the region, but generally *Perodicticus* prefers fruits and gums, and *Arctocebus* is primarily insectivorous (Fleagle, 1988; Oates, 1984; Walker, 1969). Asian lorids also exhibit two distinct body types; the slender loris *(Loris)* of southern India and Sri Lanka is slightly built relative to the much stockier slow loris of southeastern Asia *(Nycticebus)*.

FAMILY GALAGONIDAE. Galagos reach the size of a large squirrel, have very large eyes and expressive ears that resemble those of some bats, and possess a remarkable ability to make prodigious arboreal leaps. Eleven species of galagos in four genera are recognized by Groves (1993). The tail is long and well furred and is used as a balancing organ during leaping. Galagos have unusually long hindlimbs with powerful thigh muscles. The skull has a long rostrum (relative to those of other primates), and the dental formula is 2/2, 1/1, 3/3, 3/3 = 36. Specialized lower incisors and canines are procumbent (Fig. 11-11) and form a tooth comb (Fig. 11-12) used in grooming the fur and in feeding on tree resin (gum). The specialized hands and feet are well adapted to grasping: both the thumb and hallux are large and opposable, the fourth digits are unusually long, and the distalmost pads of the digits have well-developed traction

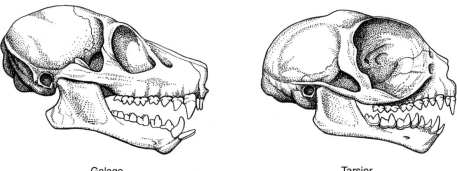

Galago
(*Galago* sp., Galagonidae)

Tarsier
(*Tarsius spectrum*, Tarsiidae)

FIGURE 11-11   Skulls of a galago (length of skull 65 millimeters) and a tarsier (length of skull 36 millimeters). *(After Clark, 1971)*

FIGURE 11-12   Teeth of a galago (strepsirhine) and a tarsier (haplorhine). *(From Clark, 1971)*

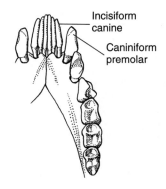

Incisiform canine

Caniniform premolar

Lower teeth

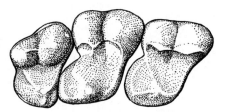

Upper right molars

Galago
(*Galago* sp., Galagonidae)

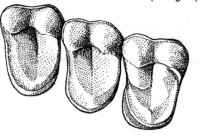

Upper right molars

First and second lower left molars

Tarsier
(*Tarsius spectrum*, Tarsiidae)

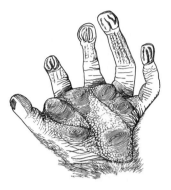

FIGURE 11-13   Traction patterns on the palm of the hand of the greater galago *(Otolemur crassicaudatus).*

ridges (Fig. 11-13) that are important during climbing. The second digit of the hind foot is short and bears a claw used for grooming, but all other digits bear flattened nails. The long foot segment of the hindlimb is associated with leaping ability; the elongation involves the tarsus but is not as extreme as that in *Tarsius* (Fig. 11-7).

Galagos are common in many sections of Africa, and the arboreal leaps of *Galago senegalensis* are a fascinating part of the twilight scene in some parts of East Africa. By day this species (Fig. 11-14) usually takes refuge in family groups in holes in trees such as the baobab. The evening dispersal from these retreats often involves the use of favorite pathways through the trees. Leaps of up to 7 meters between adjacent trees have been reported (Kingdon, 1971).

Galagos have an extremely varied diet but seem to prefer insects. When insects are abundant during the rainy season, galagos depend primarily on this food source. A precisely timed leap terminating in a quick grab with one hand is a common style of capturing insects. Kingdon reports that *G. senegalensis* feeds also on seeds and small vertebrates and that it takes nectar from the large flowers of baobab trees (Kingdon, 1971, 1997). Of particular interest is the habit of eating the resin of trees such as acacias. In some areas with long dry seasons and periodic food shortages, these gums, which contain polymers of pentose sugars, are the key to survival (Charles-Dominique, 1971; Martin and Bearder, 1979).

Galagos have a variety of vocalizations that serve as warnings, as communication signals between mother and young, and perhaps as appeasement during intraspecific encounters. *G. senegalensis* has a "vocabulary" of about ten basic sounds (Andersson, 1969). The loud, raucous calls of the greater galago *(Otolemur crassicaudatus)* are an impressive addition to the chorus of night sounds in wooded parts of East Africa.

SUBORDER HAPLORHINI

The haplorhine primates include the fossil families Omomyidae and Eosimiidae, the extant and fossil tarsiers (family Tarsiidae), and the New and Old World anthropoid primates (Kay et al., 1997b; Andrews, 1988). Haplorhine primates are phylogenet-

FIGURE 11-14
Bushbabies, or lesser galagos *(Galago senegalensis):* (A) a group of five that slept together on a *Commiphora* branch during the day, (B) an individual preparing to leap from a *Commiphora* branch. *(A by T. Huels; B by R. Bowker)*

A   B

ically united by shared dental and cranial features, as well as synapomorphies of the soft anatomy, molecular characters, and the presence of a **hemochorial placenta** and a **fovea centralis** in the retina associated with improved visual acuity (Porter et al., 1994; Kay and Williams, 1994; Kay et al., 1997b).

The position of the tarsiers has been controversial (Fig. 11-3), but most primatologists now agree that they are a sister group (closest genealogical relative) of the Anthropoidea, based on dental characters (including the lack of a tooth comb), a postorbital septum (a thin wall of bone separating the orbit from the temporal fossa), and several features of the auditory region (Kay et al., 1997b).

The suborder Anthropoidea includes the two extant New World families and three Old World families. New World anthropoids are a monophyletic group, usually called the Platyrrhini, that includes the marmosets and tamarins (Callitrichidae) and the New World monkeys (Cebidae). The Catarrhini are an Old World assemblage including macaques, baboons, and mangabeys (Cercopithecidae), along with the gibbons (Hylobatidae) and the apes (Hominidae), including humans. Whether ancestral anthropoids arose in Africa or Asia is not clear, but Kay and coworkers (1997b) provide evidence that the earliest anthropoids were small, primarily insectivorous, arboreal quadrupeds that were active during the day. They hypothesize that tarsiers diverged from this ancestral plan by becoming nocturnal, more carnivorous, and switching to a bipedal leaping style of arboreal locomotion. Anthropoids, on the other hand, shifted to an herbivorous diet, increased in body size, and evolved a more quadrupedal form of arboreal locomotion.

FAMILY TARSIIDAE. The Tarsiidae family is represented today by five species in a single genus, *Tarsius*, and occurs in jungles and secondary growth in Borneo, southern Sumatra, some East Indian islands including Sulawesi, and some of the Philippine Islands. Tarsiids are known from the late middle Eocene of China and are thought to be closely related to either eosimiids (Beard et al., 1994, 1996) or to omomyids (Kay et al., 1997b).

The tarsier is roughly the size of a small rat and, with its large head, huge eyes, long limbs, and long tail, has a distinctive appearance (Fig. 11-15). The most conspicuous cranial features are the enormous orbits, which face forward and have expanded rims and a thin interorbital septum (Fig. 11-11). The eye of the tarsier is apparently adapted entirely to night vision, for it lacks cones in the retina. However, tarsiers also lack a reflecting tapetum lucidum in the retina, suggesting they evolved from a diurnal ancestor. Tarsiers and anthropoids are the only mammals that possess a fovea centralis for improved visual acuity. The dental formula 2/1, 1/1, 3/3, 3/3 = 34 is unique among primates. The medial upper incisors are enlarged, the premolars are simple, the crowns of the upper molars are roughly triangular, and the lower molars have large talonids (Fig. 11-12). The neck is short, a characteristic of many saltatorial (jumping) vertebrates. All but the clawed second and third pedal digits have flat nails, and all digits have disklike pads (Fig. 11-16A). The limbs, especially the hind ones, are elongate; the tibia and fibula are fused.

FIGURE 11-15   Tarsier (*Tarsius bancanus,* Tarsiidae). *(M. Roberts, American Society of Mammalogists)*

Tarsiers share with other haplorhines a hemochorial placenta (capillaries of the chorion burrow into the uterine wall, making direct contact with the maternal blood), unlike the **epitheliochorial placenta** of lemurs, where the capillaries are separated from maternal blood by the uterine lining (Fleagle, 1988).

The trend toward jumping ability that is apparent in galagos is developed to an extreme degree in the family Tarsiidae. As in all highly specialized jumpers, the hind foot is elongate, but in the tarsier the elongation is unique. It involves two tarsal bones (hence the name *Tarsius*) rather than metatarsals, as in such jumpers as jerboas and kangaroo rats. In *Tarsius,* the calcaneum and navicular are greatly elongate (Fig. 11-7), whereas the metatarsals are not unusually long in relation to the phalanges (Fig. 11-16A). An important functional end is achieved by this unusual system of foot elongation: the elongation that has occurred in the tarsus has not sacrificed the dexterity and grasping ability of the digits themselves (the metatarsals and phalanges). Elongation of the metatarsals would have caused a reduction of dexterity. In elephant-shrews and kangaroo rats, dexterity and gripping ability of the hind foot are not important and a more "direct" means of elongation—lengthening of the already somewhat elongate metatarsals—occurred.

Tarsiers are primarily arboreal and nocturnal and feed largely on insects, which they pounce upon and grasp with the hands. Fogden observed tarsiers quietly watching and waiting on a low perch and leaping down to the ground to capture insects (Fogden, 1974). Although more highly adapted to leaping than any other primate, tarsiers can walk and climb quadrupedally, hop or run on their hind legs on the ground, and slide down branches (Sprankel, 1965). Tarsiers and some species of galagos share the ability to leap long distances with great precision, and in both of these types of primates the landing from a leap is largely bipedal. In association with jumping ability, much of the weight of the tarsier is concentrated in the hindlimbs, which together constitute 21 percent of the total weight of the animal; the musculature of the thighs

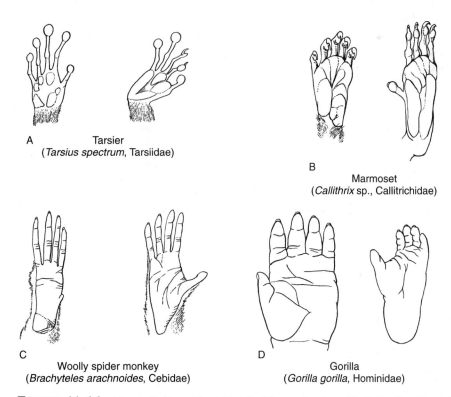

A   Tarsier
(*Tarsius spectrum*, Tarsiidae)

B
Marmoset
(*Callithrix* sp., Callitrichidae)

C
Woolly spider monkey
(*Brachyteles arachnoides*, Cebidae)

D
Gorilla
(*Gorilla gorilla*, Hominidae)

FIGURE 11-16   Hands (left) and feet (right) of four primates. *(A, B, D after Clark, 1971)*

alone equals 12 percent of the body weight, largely owing to great enlargement of the quadriceps femoris, a powerful extensor of the shank (Grand and Lorenz, 1968). Tarsiers usually live in pairs, but single animals are most frequently observed, suggesting that only large, dominant males associate with females (Fogden, 1974). Studies of captive animals have demonstrated a considerable degree of variation in social structure among the species of tarsiers. *Tarsius spectrum* live in small family groups and are quite vocal in defending their territories, while Bornean tarsiers *(T. bancanus)* live a largely solitary life (Crompton and Andau, 1987). Tarsiers have an unusually long gestation period (up to six months) for their size and give birth to a single offspring that can weigh 30 percent of the mother's weight.

FAMILY CALLITRICHIDAE. This New World family includes four genera and 26 species commonly known as tamarins and marmosets (Rylands, 1993). Callitrichids are the smallest of the New World primates, weighing between 100 and 750 grams, depending on the species. The genus *Callimico,* or Goeldi's monkey, is currently assigned to this family (Groves, 1993; Fleagle et al., 1997), based on a number of shared derived characters (Thorington and Anderson, 1984). Considerable debate remains about the evolutionary history of this family. Hershkovitz (1977) considered callitrichids to be a relatively primitive group, while several other researchers (Rosenberger, 1984; Sussman and Kinzey, 1984) argued that they represent derived forms specialized to a unique insectivorous and resin-feeding niche.

Callitrichids differ from other New World primates in having triangular molars, in lacking the third molar (2/2, 1/1, 3/3, 2/2 = 32 except in *Callimico* where a third molar is retained), in having chisel-shaped medial incisors, and in having claws on all digits but the hallux, which bears a nail (Fig. 11-16B). Several species of these small primates have heads adorned with manes or conspicuous tufts of fur.

Marmosets and tamarins are **omnivorous.** Their diet consists mostly of fruit, tree resins (Garber, 1984), and insects; lizards and small birds and their eggs may be important foods for some species. Marmosets and tamarins are social and typically live in groups characterized by one breed-

ing female and several adult males, plus offspring. Group size in the wild remains relatively constant despite frequent **immigration** and **emigration** of group members (Garber, 1984; Terborgh and Goldizen, 1985). In most marmosets and tamarins, females give birth to twins and the young are cared for by the mother and one or more of the adult males (Terborgh, 1983). However, in *Callimico* females give birth to one young, and male group members share in parental care (Hershkovitz, 1977; Heltne et al., 1981; Pook and Pook, 1981). The recent discoveries of several new species of callitrichids in South America demonstrate that much remains to be learned in tropical forests threatened by deforestation (Van Roosmalen et al., 1998; Mittermeier et al., 1992; Lorini and Persson, 1990).

FAMILY CEBIDAE. The New World monkeys belonging to this family include 11 Recent genera and 58 species. Cebids range from southern Mexico through Central America to southern Brazil. They first appear in the late Oligocene record of Bolivia *(Branisella).* According to one hypothesis, primitive primates ancestral to the cebids may have entered South America from Central America on logs or debris that floated across the stretch of water that separated these land masses or by island-hopping along an early Antillean archipelago during much of the Tertiary Period (Gingerich, 1980; Simpson, 1945; Marshall and Sempere, 1993). Alternatively, New World primates may have arrived by chance rafting from Africa, when Africa and South America were much closer together (Ciochon and Chiarelli, 1980). Once established in the New World, cebids became diverse and ecologically successful components of the South and Central American fauna (Fleagle et al., 1997; Tejedor, 1998).

Cebids have elongate limbs, the digits bear curved nails, and the pollex is not opposable and is small or absent in some species (Fig. 11-16C). The hallux is strongly opposable. The tail is typically long and is prehensile in many species (Fig. 11-17). The size ranges from about 275 grams to 9 kilograms. The skull is more or less globular, with a high braincase and short rostrum. The orbits face forward, and the nostrils are separated by a broad internarial pad and face to the side. The dental formula is 2/2, 1/1, 3/3, 3/3 = 36, and the

FIGURE 11-17   New World woolly monkeys (*Lagothrix* sp.; Cebidae). Note the prehensile tail used to grasp branches. *(Corel Corp.)*

lateral pair and medial pair of cusps of the quadrate molars are separated by a central depression (Fig. 11-18). Brightly colored and bare patches of skin on the rump **(ischial callosities)** do not occur in cebids, as they often do in Old World monkeys (family Cercopithecidae). Lack of color markings in New World primates may be the result of their lack of **trichromatic** color vision (color vision based on three types of cone cells that absorb light in the blue, green, and red parts of the visible spectrum). Most cebids that have been studied have a single pigment gene located on the X chromosome, and multiple alleles of this gene allow several types of dichromatic color vision in these primates (heterozygous females may have trichomatic color vision). The howler monkey, *Alouatta,* is perhaps unique among platyrrhine monkeys in that they have multiple pigment genes and, like catarrhine primates, have trichromatic color vision (Jacobs et al., 1996)

Cebids typically occur in tropical forests, and most are diurnal. They are primarily vegetarians; fruit is often preferred, but a wide variety of plant and animal material is eaten. The night monkey *(Aotus trivirgatus),* an inhabitant of Central and South America, is nocturnal, but its eyes are not highly specialized for night vision; the eyes retain both rods and cones in the retina and lack a tapetum lucidum (Fleagle, 1988). Cebids are active, intelligent animals and adroit climbers; some species move with amazing speed through the trees. For dazzling arboreal ability, the Neotropical spider monkey *(Ateles)* is probably surpassed only by the Old World gibbons (Hylobatidae).

Most cebids are vocal to some extent, and several species have loud, penetrating calls. Outstanding among these are the howler monkeys, in which the **hyoid apparatus** is enlarged into a resonating chamber. The males emit loud roaring sounds that carry for long distances through the tropical rain forest. These sounds seem important in maintaining the cohesiveness of the troop, which may include up to 40 individuals. The territories of troops are probably announced, and partly maintained, by the loud vocalizations. Most cebids are gregarious, the most common social aggregation consisting of a family group. Some species form unusually large troops; the squirrel monkey *(Saimiri)* occurs in bands of up to 100 animals. Night monkeys *(Aotus)* live in monogamous family groups, while sakis *(Pithecia)* and spider monkeys in the genus *Ateles* live in more fluid social groups, often referred to as a fission-fusion social system because members of the larger group often split into smaller temporary bands (McFarland, 1986).

FAMILY CERCOPITHECIDAE. These Old World monkeys, belonging to the catarrhine group, are the most successful primates in terms of number of species (81 Recent species in 18 genera). They occupy a wide range, including Gibraltar, northeastern Africa, Africa south of the Sahara, the southern Arabian Peninsula, much of southeastern Asia east to Japan, Indonesia east to Timor, and the Philippine Islands. Among nonhominid primates, cercopithecids have the greatest tolerance for cold climates; some occupy high forests in Tibet, and others live in northern Honshu, Japan, where winter snows occur.

FIGURE 11-18 Cheek teeth of anthropoid primates. Note the cross lophs on the teeth of the mangabey and the baboon, and the extra posterior cusp (hypoconulid) on the third lower molar of the baboon. *(D from Clark, 1971)*

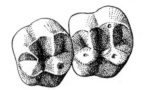

Second and third upper right molars

Orange-crown mangabey
(*Cercocebus torquatus*, Cercopithecidae)

First and second upper right molars

Saki monkey
(*Pithecia monachus*, Cebidae)

Hypoconulid

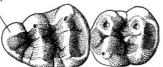

Second and third lower molars

Baboon
(*Papio hamadryas*, Cercopithecidae)

Upper right molars

Orangutan
(*Pongo pygmaeus*, Hominidae)

Harrison (1987) lists 21 cranial, postcranial, and dental characters that define a monophyletic cercopithecid clade (monophyletic group). Cercopithecids first appear as fossils in the early to middle Miocene of western Kenya, but older relatives in Asia and Africa date back to the Eocene. Late Eocene parapithecid and oligopithecid fossils appear morphologically similar to living platyrrhine primates (Kay et al., 1997b). *Propliopithecus* (Pliopithecidae) from the Oliogocene of Egypt and several Miocene primates from Europe are probably the sister group of extant catarrhines. Just as in many other groups, some Pleistocene cercopithecids reached large sizes; an extinct South African baboon reached the size of a gorilla.

In weight, cercopithecids range from 1.5 to over 50 kilograms, and some species are stocky in build. The nostrils are close together and face downward, a condition termed **catarrhine** (Fig. 11-19). The skull is often robust and heavily ridged, and, relative to that of cebids, the rostrum is long (particularly in the baboons; Fig. 11-20). The dental formula is 2/2, 1/1, 2/2, 3/3 = 32, as in the hylobatids and hominids. The medial upper incisors are often broad and roughly spoon-shaped; the upper canines are usually large and, in some species, tusklike. When the jaws are closed, the lower canine rests in a diastema between the upper canine and the last incisor. The first lower premolar is enlarged and forms a shearing blade that rides against the sharp posterior edge of the upper canine (Fig. 11-20). Most of the molars have four cusps, the outer pair connected to the inner pair by two transverse ridges, producing a bilophodont tooth. The last lower molar has an additional posterior cusp, the hypoconulid (Fig. 11-18).

All of the digits have nails, and the pollex and hallux are opposable, except in the strongly arboreal, leaf-eating genus *Colobus*, in which the pollex is vestigial or absent. The tail is vestigial in some species but long in others. Ischial callosities are well developed in many species, and the bare rump skin is frequently bright red. The conspicuous patch is used in conjunction with ritualized postures as a means of communication between members of a social group. Bare facial skin may also be red but is bright blue and red in the mandrill *(Mandrillus sphinx)*. These patches of skin are more brightly colored in the male than in the female in most species. The olfactory epithelium is greatly reduced in cercopithecids, and apparently their sense of smell is rudimentary. The facial muscles are well developed and produce a wide variety of facial expressions. Some cercopithecids are

A

B

C

FIGURE 11-19   (A) Japanese macaques (*Macaca fuscata;* Cercopithecidae) in northern Honshu, Japan. (B) A mother gibbon (*Hylobates* sp.; Hylobatidae) with her young. (C) A gorilla (*Gorilla gorilla;* Hominidae) from East Africa. *(A by C. Koford, B by PhotoDisc, Inc.; C by Corel Corp.)*

brightly or conspicuously marked (Kingdon, 1989c, 1997). For example, the variegated langur of Indochina *(Pygathrix nemaeus)* has a bright yellow face, a chestnut strip beneath the ears, black and chestnut limbs, a gray body, and a white rump and tail.

Although most cercopithecids are probably largely omnivorous, some are adapted to an herbivorous diet. Members of the subfamily Colobinae (the arboreal langurs and colobus monkeys) are herbivorous and frugivorous, and some species seem to feed primarily on leaves. The baboons *(Papio, Theropithecus)* are the most successful terrestrial cercopithecids, and one species, in areas where suitable trees are not available, assembles on cliffs in large groups of up to 750 individuals (Kummer, 1968). Baboons, alone among cercopithecids, have adapted locally to extremely harsh desert conditions. The remarkably complicated social behavior of baboons and of certain other Old World monkeys is reasonably well known and is discussed on page 460 (Altman and Altman, 1970; Dunbar, 1984; Eisenberg et al., 1972; Lindberg, 1980; Struhsaker and Leland, 1979; van Noordwijk, 1985).

Interesting contrasts between the behavior of the baboons and that of the equally terrestrial patas monkey *(Erythrocebus patas)* are discussed by

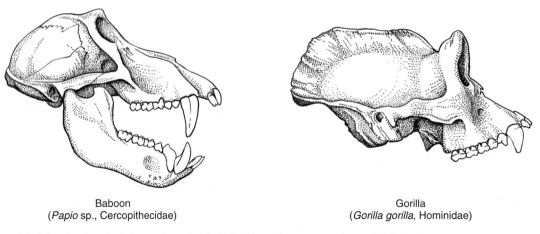

Baboon
(*Papio* sp., Cercopithecidae)

Gorilla
(*Gorilla gorilla*, Hominidae)

FIGURE 11-20   Skulls of a baboon (length of skull 200 millimeters) and a gorilla (length 320 millimeters).

Hall (1968). Savanna baboons are highly vocal, live in fairly large troops controlled by several dominant males, and are prone to noisy, rough, aggressive interactions. In contrast, the patas monkey usually maintains "adaptive silence" but has a repertoire of soft calls, lives in small troops, each with a single adult male that serves as a sentry, is rarely aggressive, and never fights. The patas monkey has a slim greyhound-like build and is the fastest runner of all primates, having been timed at a speed of 55 kilometers per hour. Adaptations for speed in this animal include elongation of the limbs, carpals, and tarsals; shortening of the digits; reduction of the pollex and hallux; and the development of palmar and plantar pads. This remarkably cursorial primate has a quiet mode of life. It usually attempts to escape detection and depends on its speed to escape danger. In these respects, the patas monkey is the primate counterpart of the small antelopes. The noisy baboon troop, however, frequently depends on its aggressive dominant males to confront and discourage a predator, and terrestrial locomotion is relatively unimportant as a means of escaping enemies.

Sexual dimorphism is pronounced in both the baboons and the patas monkey, as it is in many primates. The male baboon of southern and eastern Africa weighs roughly 33 kilograms, the female 16.5 kilograms. Similarly, the male patas monkey averages 13 kilograms, and the female 6.5 kilograms (Hall, 1968). Probably all cercopithecids are basically social, and vocalizations and facial expressions play central roles in social interactions. The

life span of these monkeys is long: a Chacma baboon (*Papio hamadryas*) lived in captivity for 45 years, and life spans of 20 to 25 years in the wild may be common (Nowak and Paradiso, 1991).

Baboons are known to kill and eat small mammals opportunistically, including young gazelles, hares, and even vervet monkeys (Strum, 1981). These predatory events are relatively uncommon, and the bulk of their diet consists of fruits, roots, tubers, seeds, and leaves. Colobine monkeys (*Colobus* and *Procolobus*), in contrast, rely on a multi-chambered stomach containing colonies of cellulose digesting bacteria, allowing them to forage exclusively on leaves (Bauchop, 1978).

FAMILY HYLOBATIDAE. This family includes 11 species of Southeast Asian gibbons in one genus, *Hylobates*. Sometimes called lesser apes, gibbons share several features with apes (family Hominidae) and with cercopithecid monkeys. The hylobatid fossil record is scanty, and fossils are known only from the Miocene to the Recent of southeastern Asia (Thorington and Anderson, 1984). There is no consensus concerning the fossil ancestry of modern gibbons (Szalay and Delson, 1979). Andrews and Martin (1987) listed 11 morphological characters shared by extant gibbons and apes, and molecular data also support a close relationship between hylobatids and hominids (Andrews and Martin, 1987).

Gibbons are small, weighing 4 to 11 kilograms, and lack the sexual dimorphism in size found in some other primates. They retain the same dental formula as cercopithecids and have bunodont molars and prominent canines. The most notable feature of gibbon morphology is their limb proportions (Fig. 11-19B). They have extremely elongate forelimbs, modified for **brachiation** (arm-over-arm swinging), with forelimbs over twice the length of the body (Fig. 11-21B). Hindlimbs are also elongate and are nearly one and a half times the body length. The digits on the hand and feet are long and slender, and there is a deep cleft between the first and second digits. The pollex and hallux are strongly opposable. Ischial callosities are present, but gibbons lack an external tail and cheek pouches.

Gibbons occur throughout tropical regions of southeastern Asia from Bangladesh to Vietnam and south to Malaysia, Sumatra, Borneo, Java, and neighboring islands. They prefer tropical forests with well developed canopy structure that facilitates brachiation and feed principally on fruit, leaves, and flowers (Chiver, 1974; Fleagle, 1976; Leighton, 1986). Gibbons live in small family groups that defend territories by loud vocaliza-

tions. Once believed to be monogamous, recent evidence suggests that individual gibbons occasionally migrate between groups consisting of small, non-nuclear families (Gibbons, 1998).

FAMILY HOMINIDAE. The family Hominidae has until recently included only one living member, *Homo sapiens,* with the great apes being assigned to a separate family, Pongidae. However, recent data indicate a close relationship between humans and chimpanzees and necessitated a redefinition of the Hominidae (Grove, 1989). With the exception of humans, which now occur worldwide, the great apes are found in isolated habitats in equatorial Africa *(Pan* and *Gorilla)* and in Borneo and Sumatra *(Pongo).* The fossil record of hominids dates from the late Oligocene of eastern Africa and includes *Kenyapithecus* and *Dryopithecus,* which are ancestral to African forms, and *Sivapithecus* from Asia, the likely ancestor of orangutans (Carroll, 1988; Ward and Pilbeam, 1983). The fossil record of the genus *Homo* began in Africa approximately between 2 and 2.5 million years ago, and the genus

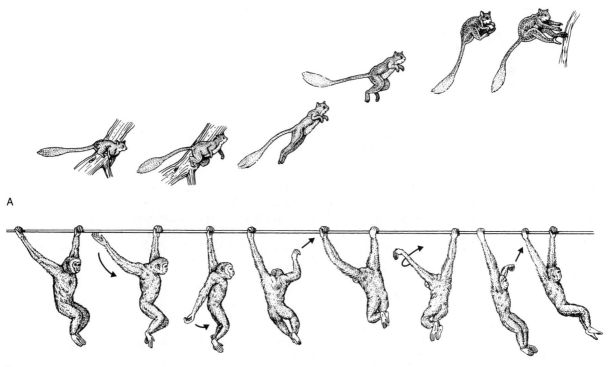

FIGURE 11-21   Modes of locomotion in two primates: (A) the bipedal leap and landing of a galago (*Galago senegalensis,* Galagonidae), (B) brachiation in a gibbon (Hylobatidae). *(After Rogers, 1986)*

probably arose from a primitive *Australopithecus* (Johanson and Edey, 1981; Kimbel, et al., 1984). Excellent treatments of the evolution of hominids, and the humans in particular, can be found in Ciochon and Corruccini (1983), Delson (1985), and Sibley and Ahlquist (1984; 1987).

The great apes (excluding *Homo*) vary from 48 to 270 kilograms in weight and have robust bodies and powerful arms. The hands and feet are similar to those of humans, but the hallux is opposable (Fig. 11-16D). The elongate skull is typically robust and in older animals is marked by bony crests and ridges (Fig. 11-20). The dental formula is 2/2, 1/1, 2/2, 3/3 = 32, as in cercopithecids and humans. The incisors are broad, and the premaxillae and anterior parts of the dentary bones are broadened to accommodate them. The canines are large and stoutly built. The upper molars are quadrangular and usually four-cusped, and the lower molars have an additional posterior cusp (hypoconulid). In contrast to cercopithecids, a trend toward elongation of the molars does not occur in great apes, and the molars lack well-defined cross ridges (Fig. 11-18). The tooth rows are parallel, and the mandibular symphysis is braced by a bony shelf (the "simian shelf").

The forelimbs are longer than the hindlimbs, and the hands are longer than the feet; all digits bear nails. Hominids have no tails. The thorax is wide, and the scapula has an elongate vertebral border. Adaptations allowing advantageous muscle attachments during erect or semierect stances include lengthening of the pelvis and enlargement and lateral flaring of the ilium. Regarding structural details, locomotor ability, molecular genetics, brain size, and level of intelligence, the great apes are closer to humans than are any other mammals (Diamond, 1993).

The great apes are largely vegetarian, but some chimpanzees *(Pan)* are occasionally carnivorous. For example, chimpanzees occasionally catch and eat colobus monkeys. Arboreal locomotion in chimpanzees and *Pongo* (orangutan) involves brachiation. Although the gorilla *(Gorilla)* and chimpanzee are able climbers, both are capable of a bipedal stance and limited bipedal locomotion. The behavior of hominids and cercopithecids has been studied intensively in recent years (Dixon, 1981; deWaal, 1982; Fossey, 1983; Goodall, 1986; Ghiglier, 1987; Wrangham, 1987; Schwartz, 1988).

Owing primarily to the efforts of humans, some of the great apes are dangerously close to extinction. Destruction of habitat and killing of animals has led to serious reduction of some populations. Populations of wild orangutans *(Pongo pygmaeus)* are small, and these are restricted to parts of the islands of Sumatra and Borneo (Fig. 11-22). As a sad stroke of irony, an important drain on the declining populations of orangutans and chimpanzees

FIGURE 11-22 The face of a male orangutan (*Pongo pygmaeus*, Hominidae). (*PhotoDisc, Inc.*)

resulted from their capture and exportation to European and American zoos, institutions dedicated to the preservation of vanishing species. Similarly, large numbers of rhesus and other monkeys were taken from the wild for use in medical research before breeding colonies were established solely for that purpose.

Humans *(Homo sapiens)* are the only living member of the genus *Homo*. In humans, the skull has a greatly inflated cranium, housing a large cerebrum, and the rostral part of the skull is virtually absent. The foramen magnum is beneath the skull, a feature associated with an upright stance. The dentition is not as robust as in their fellow hominids: the incisors are less broad, the canines typically rise but slightly above adjacent teeth, and the cheek teeth are less heavily built. The premolars are usually bicuspid. The upper molars have four cusps; the first lower molar has five cusps, the second has four, and the third has five. The dental formula $2/2, 1/1, 2/2, 3/3 = 32$ occurs in most individuals, but one or more of the posteriormost molars (the "wisdom teeth") may not appear. The tooth rows are not parallel as they are in great apes, nor is the simian shelf present in the mandible. The pollex, but not the hallux, is opposable. With a change in posture and use of the forelimbs, the thorax has become broad and the scapulae have come to lie dorsal (or posterior) to the rib cage, as in bats, rather than lateral to the rib cage, as in most other mammals. As in many primates, human males are significantly larger than the females. Recent molecular data indicate that humans are more closely related to chimpanzees than they are to either orangutans or gorillas (Miyamoto et al., 1987).

SCANDENTIA

The order Scandentia is represented by one family, Tupaiidae, which consists of five genera and 19 extant species, all of which are referred to as tree shrews. Members of this family resemble small long-snouted squirrels and occur from India through Burma to the islands of Sumatran Borneo and the Philippines (Fig. 11-23). The first definitive fossil tupaiids appear in middle Eocene deposits from Asia, but these fossils are morphologically similar to extant genera, and add little to our understanding of the phylogenetic position of tree shrews relative to other mammalian orders. Most modern systematists exclude tree shrews from the Primates and place them in their own order, but there is strong disagreement about their phylogenetic relationships. Morphologists link them with Primates (Kay et al., 1997b; Wible and Covert, 1987) but molecular geneticists have variously allied them with the Dermoptera, Lagomorpha, or a Dermoptera-Macroscelidea clade (Bailey et al., 1992; Adkins and Honeycutt, 1993; Stanhope et al., 1993).

The dental formula is $2/3, 1/1, 3/3, 3/3 = 38$; the upper incisors are caniniform, and the upper canine is reduced. Tree shrews have a tooth comb consisting of the middle four lower incisors, but, unlike lemurs, the comb does not include the canines (Rose et al., 1981). The upper molars have **trenchant,** W-shaped ectolophs, and the lower molars retain the basic insectivore pattern (Figs. 11-24). There is no loss of digits, and all digits have strongly recurved claws. The long tail is usually heavily furred. One unique feature of tree shrews

FIGURE 11-23 A tree shrew (*Tupaia longipes,* Tupaiidae). *(M. Sorenson)*

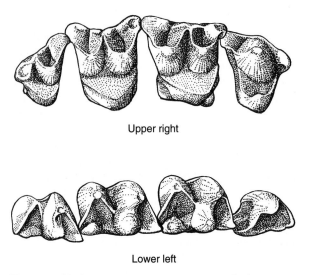

FIGURE 11-24   Fourth premolars and molars of a tree shrew (*Tupaia* sp., Tupaiidae).

is a prominent hole in the zygomatic arch (Fig. 11-25). Tree shrews have well-developed postorbital processes that join the zygoma, a feature long used as evidence of a close relationship with primates. Extant tree shrews, however, lack virtually all of the characteristics that define the order Primates (Table 11-1), and the superficial similarities with primates may stem from the shared arboreal and diurnal niche.

Tree shrews occupy deciduous forests and forage both in the trees and on the ground. They are opportunistic feeders and utilize a variety of foods. Although tree shrews were long considered to be primarily insectivorous, Emmons (1991) demonstrated that four species in Malaysia are mostly frugivorous. Her studies reveal that tree shrews eat a wide variety of fruit, which passes rapidly (in 13 to 29 minutes) through their simple digestive tracts. In their rapid food transit times and simple gut morphology, tupaiids resemble frugivorous bats and may be important seed dispersers of some tropical forest trees.

The feather-tailed tree shrew *(Ptilocercus lowei)* of the Malay Peninsula is arboreal and nocturnal, but other tree shrews are diurnal. Tree shrews are characteristically quick moving and highly vocal. The mountain tree shrew *(Tupaia montana)* lives in social groups in which a rigid dominance hierarchy is apparent, whereas in parts of Borneo other species occupying lowland areas do not form social groups (Sorenson and Conaway, 1968). The common tree shrew *(Tupaia glis)* is territorial, and small family groups, made up of a male and female and their offspring, patrol the hectare-sized territory regularly, frequently scent-marking the boundaries. Parental care appears to be rudimentary, and mothers visit the highly **altricial** young in the nest only to suckle them every other day (Martin, 1968; Emmons and Biun, 1991). The milk is rich in protein, however, and the short suckling bouts are enough to sustain the developing young. This form of maternal behavior, called "absentee parental care," probably serves as an anti-predator strategy because the nestlings remain silent and mothers approach the nest so as not to leave a conspicuous odor trail (Emmons and Biun, 1991).

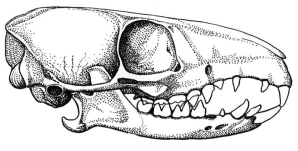

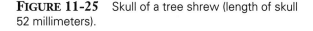

Tree shrew
(*Tupaia* sp., Tupaiidae)

FIGURE 11-25   Skull of a tree shrew (length of skull 52 millimeters).

WEB SITES

Wisconsin Regional Primate Center home page with many links

http://www.primate.wisc.edu/pin/

Leakey Foundation home page on human origins

http://www.leakeyfoundation.org/

Duke University's prosimian site

http://www.duke.edu/web/primate/index.html

International Primatological Society home page

http://www.primate.wisc.edu/pin/ips.html

American Society of Primatologists home page

http://www.asp.org

The Douc Langur Project home page

http://www-rohan.sdsu.edu/faculty/lippold1/

The Gorilla Foundation home page

http://www.gorilla.org/

Jane Goodall Institute's Center for Primate Studies

http://www.cbs.umn.edu/chimp/

Orangutan Foundation International home page

http://www.ns.net/orangutan/index1.html

IUCN tree shrew status survey and conservation action plan

http://members.vienna.at/shrew/itsesAP95-cover.html

University of Washington Regional Primate Center home page

http://www.rprc.washington.edu

12 CARNIVORA

Predation in mammals is an ancient means of procuring food. Primitive carnivorous mammals (creodonts) appeared in the early Paleocene before most of the Recent mammalian orders. Mammalian carnivores probably evolved in response to the food source offered by an expanding array of herbivorous mammals and underwent adaptive radiation as herbivores diversified.

The classification of carnivorans remains a subject of considerable debate (Benton, 1988; Dragoo and Honeycutt, 1997; Hunt and Tedford, 1993; Wayne et al., 1989; Wozencraft, 1989; Wyss and

Flynn, 1993). Living members of the Carnivora include terrestrial carnivores (such as dogs, cats, weasels, and bears) as well as aquatic carnivores (seals, sea lions, and walruses). No consensus on the classification of the Carnivora has been reached. However, most mammalogists agree that the Carnivora is composed of two clades: the Feliformia, which includes felids, hyaenids, viverrids, and herpestids, and the Caniformia, including the canids, ursids, procyonids, mustelids, odobenids, otariids, and phocids (Wozencraft, 1993; Wyss and Flynn, 1993). According to this phylogeny, the seals, sea lions, and walruses (Pinnipedia) are most closely related to the ursids (Fig. 12-1).

Wyss and Flynn (1993) list several morphological characteristics that diagnose extant carnivorans: (1) an expanded braincase in which the fronto-parietal suture is located posteriorly relative to the postorbital constriction, (2) a fully or partially ossified entotympanic that is firmly fused to the skull, (3) fused scaphoid and lunar bones in the carpals, and (4) the loss of the third trochanter on the femur. Molecular data summarized by Wayne et al. (1989) confirms the monophyly of the Carnivora with its two clades, Feliformia and Caniformia.

Most Recent carnivorans are predaceous and have a remarkable sense of smell. Cursorial ability may be limited, as in the Ursidae and Procyonidae, or may be strongly developed, as in the cheetah and some canids. The braincase is large. The orbit is usually confluent with the temporal fossa. The turbinal bones are usually large, and their complex form provides a large surface area for olfactory epithelium. There are usually 3/3 incisors (3/2 in the sea otter, *Enhydra lutris*), and the canines are large and usually conical. The cheek teeth vary from 4/4 premolars and 2/3 molars in long-faced carnivores, such as the Canidae and Ursidae, to 2/2 premolars and 1/1 molars, as in some cats. In most species the fourth upper premolar and the first lower molar are **carnassials** (specialized shearing blades). The teeth are rooted. The condyle of

the dentary bone and the glenoid fossa of the squamosal bone are elongated transversely and allow no rotary jaw action and only limited transverse movement.

Cursorial adaptations evident in the carpus include the fusion of the scaphoid and lunar bones and the loss of the centrale (Fig. 12-2). The foot posture is plantigrade, as in ursids and procyonids, or digitigrade, as in canids, hyaenids, and felids. Little reduction from the ancestral number of five digits has occurred. The greatest reduction occurs in the hyenas and in the African hunting dog (*Lycaon pictus*), in which the manus (forefoot) and pes (hind foot) have four toes.

The oldest carnivorous mammals, order Creodonta, appeared in the early Paleocene. They

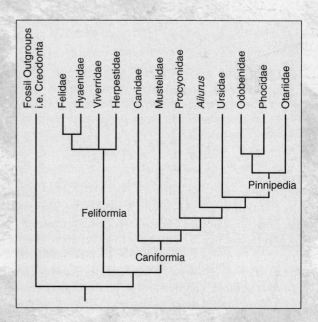

FIGURE 12-1 Consensus phylogeny of the Carnivora based on 64 morphological characters. The Carnivora include two major clades; feliforms and caniforms. The pinnipeds are a monophyletic group within the caniform clade. *(After Wyss and Flynn, ©1993, Springer-Verlag, New York, Inc.)*

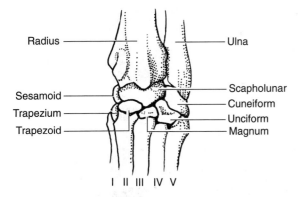

FIGURE 12-2   Anterior view of the left carpus of the gray fox *(Urocyon cinereoargenteus).* Roman numerals at the bottom indicate the digit number.

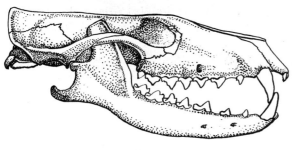

Sinopa

FIGURE 12-3   The skull of *Sinopa,* an Eocene creodont in which the carnassials are M2 and m3; length of skull is approximately 150 millimeters. *(After Romer, © 1966,* Vertebrate Paleontology, *University of Chicago Press)*

were the typical carnivores of the Paleocene and Eocene and persisted in Old World tropical **refugia** into the Miocene. Some creodonts appear to have retained the insectivorous food habits of their ancestors, whereas some were carnivorous and some were omnivorous. Two Eocene genera had saber-like upper canines and probably killed relatively large prey. The creodont skull (Fig. 12-3) differs from that of Carnivora in lacking an ossified auditory bulla and in having either no carnassial pair or having M1 and m2 or M2 and m3 forming the carnassials (instead of P4 and m1 as in Carnivora). The feet of creodonts were usually five-toed and plantigrade, and the limbs were often short. The scaphoid and lunar of the carpus were

not fused, and the centrale was present. The distal phalanges were fissured and, in some species, bore flattened, rather than clawlike, nails. As currently recognized, the creodonts include two families, the Oxyaenidae and Hyaenodontidae.

The earliest members of the Carnivora are represented by two primitive families (Viverravidae and Miacidae) of Paleocene and Eocene age. Modern carnivorans probably arose from one or both of these two families, but few fossils representing modern carnivoran families are known until the Oligocene, by which time many of the modern families were already formed (Carroll, 1988; Heinrich and Rose, 1995; Hunt and Tedford,1993; Wyss and Flynn, 1993). Miacids were small and perhaps mostly arboreal carnivores. They had the modern carnassial arrangement (P4 over m1) but lacked ossified bullae. In contrast to the creodonts, the distal phalanges of miacids were not fissured.

FELIFORMIA

The feliform clade includes four families of extant carnivorans: cats (Felidae), hyenas and the aardwolf (Hyaenidae), civets and genets (Viverridae), and mongooses (Herpestidae). A number of morphological and biochemical features diagnose this clade (Wayne et al., 1989; Wyss and Flynn, 1993).

FAMILY FELIDAE. Of all the carnivorans, the cats are the most proficient predators. Some species regularly kill prey as large as themselves and may occasionally overcome prey several times their own weight (as in the case of the African lion and the giraffe). Throughout the history of the order Carnivora, the cats have been the carnivores most highly specialized morphologically for predation.

Included in the family of living cats are 18 genera and about 36 species. All cats, from the pampered tabby to the tiger, bear a strong family resemblance (Fig. 12-4). This family occurs nearly worldwide, with the exception of Antarctica, Australia, Madagascar, and various isolated islands.

In members of the family Felidae, the rostrum is short, an adaptation furthering a powerful bite, and the orbits in most species are large (Fig. 12-5). The number of teeth is reduced. The typical dental formula is 3/3, 1/1, 3/2, 1/1 = 30, and the anteriormost upper premolar is strongly reduced or

A

FIGURE 12-4 (A) Caracal *(Caracal caracal)*, in Augrabies National Park, South Africa. This small cat (up to 19 kilograms) occurs widely in Africa, the Middle East, and as far east as India. (B) A Central and South American ocelot *(Leopardus pardalis)*. *(A by T. Vaughan, B by N. Vandermey, Feline Conservation Center, Rosamond, CA).*

B

lost (as in *Lynx*). The carnassials are well developed and have specializations that enhance their shearing ability (Fig. 12-6). The foot posture is digitigrade. The forelimbs are strongly built, and the manus can be rotated so that the soles face upward; the claws are sharp and recurved and are completely retractile, except in the cheetah *(Acinonyx)*. These features of the forelimbs allow cats to clutch and grapple with prey. Some species are spotted or striped. These color patterns enable the animals to conceal themselves effectively (Fig. 12-4). The weight of cats varies from about that of a domestic cat (3 kilograms) to that of the tiger *(Panthera tigris)* at 275 kilograms.

The groundwork for an understanding of the phylogeny of cats was laid by Matthew (1910), and new fossil material described in the last decade has improved our knowledge of the felid fossil record, but until it is more complete, any felid phylogeny must be provisional.

The earliest catlike mammals, which appeared in Eurasia in the Eocene, were already strongly

African lion
(Panthera leo, Felidae)

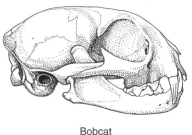

Bobcat
(Lynx rufus, Felidae)

FIGURE 12-5 Skulls of an African lion (length of skull 366 millimeters) and a bobcat (length of skull 120 millimeters).

FIGURE 12-6 Occlusal views of the cheek teeth of the bobcat. Note the lack of crushing molars. The cusps of the carnassials (P4 and m1) are labeled. Only sectorial teeth are present; the parastyle of P4 increases the length of its shearing blade, and the loss of the talonid of m1 makes this tooth entirely bladelike.

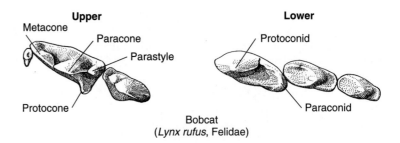

Upper

Metacone
Paracone
Parastyle
Protocone

Lower

Protoconid
Paraconid

Bobcat
(*Lynx rufus*, Felidae)

differentiated from other carnivores. All had retractile claws, sectorial carnassials, reduced cheek teeth anterior and posterior to the carnassials, and some had saber-like upper canines (Fig. 12-7). Recently, these fossil "cats," known as false sabertooths, were divided into two subfamilies (Nimravinae and Barbourofelinae), based on characters of the auditory region, dentition, and skull (Bryant, 1991; Neff, 1983). They were placed in the family Nimravidae separate from Felidae. The phylogenetic position of the Nimravidae, with respect to the rest of the Carnivora, remains a subject of considerable debate. For example, Martin (1980) included all members of the early sabertooth radiation in the Nimravidae and advocated a close relationship between nimravids and the Felidae (Martin, 1989). However, the Nimravidae have also been allied with the canids (Flynn and Galiano, 1982), or separated from both the canids and felids and placed in a more basal position as a sister group to the Carnivora (Neff, 1983). Despite the disagreement over the placement of the Nimravi-

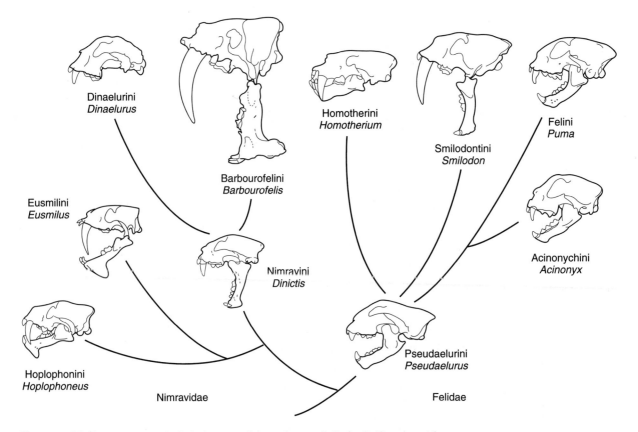

Dinaelurini
Dinaelurus

Barbourofelini
Barbourofelis

Homotherini
Homotherium

Smilodontini
Smilodon

Felini
Puma

Eusmilini
Eusmilus

Nimravini
Dinictis

Acinonychini
Acinonyx

Hoplophonini
Hoplophoneus

Nimravidae

Pseudaelurini
Pseudaelurus

Felidae

FIGURE 12-7 A hypothetical phylogeny of the sabertooth "cats." *(Reprinted from L. Martin, 1980, p. 151, by permission of the Nebraska Academy of Sciences)*

dae, this radiation culminated in the Miocene with the most highly specialized of all sabertooths, *Barbourofelis fricki.* The nimravids are not known after the Miocene. Features that distinguish nimravids from felids include the lack of an auditory bulla in most species and the absence of a cruciate sulcus (a conspicuous, deep groove) on the brain.

The family to which all living cats belong, Felidae, appeared in the early Oligocene *(Proailurus).* The basal members of this family had relatively short upper canines, but a Miocene and Pliocene radiation resulted in both sabertooth types and cats with short upper canines (Fig. 12-7). The Pleistocene felid *Smilodon,* although not as extreme in sabertooth specialization as *Barbourofelis,* was an imposing predator that survived until the end of the Pleistocene and may have coexisted with humans (Miller, 1969). All living felids have short, conical upper canines.

The extinction of the sabertooth cats is difficult to explain. Saber-like upper canines evolved independently four times among mammals (in Eocene creodonts, nimravids, felids, and the Mio-Pliocene marsupial family Thylacosmilidae), each time, perhaps, in response to the abundance of large herbivores. Probably, in each case, the extinction of the sabertooth carnivores was linked to the decline or extinction of their preferred prey.

Cats usually catch prey by a stealthy stalk followed by a brief burst of speed (p. 436). They are typically sight hunters, and some species spend considerable time watching for prey and waiting for it to move into striking distance. Many kinds of animals are eaten, from fish, mollusks, and small rodents to ungulates as large as bison and cape buffaloes. The fishing cat *(Prionailurus planiceps)* of southeastern Asia is unusual in that it forages near streams and rivers for fish and frogs (Nowak and Paradiso, 1991).

Felids have well-developed senses of smell, sight, and hearing. Many species hunt nocturnally, despite having good color vision. In nocturnal species a tapetum lucidum located in the retina helps intensify existing light. Most cats are agile climbers, and leopards *(Panthera pardus)* often drag their kills up into the branches of trees to protect the carcass from scavengers. Cheetahs *(Acinonyx jubatus)* are poor climbers, but their ability to sprint is unsurpassed (Fig. 12-8). Cheetahs make short dashes after prey and can attain speeds in excess of 90 kilometers per hour (about 60 miles per hour). Unlike most cats, cheetahs have relatively long limbs and a slender build. Cheetahs can accelerate from a standstill to nearly 90 kilometers per hour in only a few seconds. In most chases, however, the average speed is less than 65 kilometers per hour and does not cover more than 300 meters. Prey are usually killed by a suffocating bite to the ventral surface of the neck after being knocked off their feet by the cheetah (Estes, 1991). Despite their speed, chases result in kills only 40 percent of the time.

One fundamental difference between most large cats, such as members of the genus *Panthera,* and most smaller cats *(Felis, Prionailurus,* etc.) is in the structure of their hyoid, which in the larger cats is partially replaced with a flexible cartilage, allowing them to roar instead of purr. Many of the larger cats, particularly those with spotted or striped coats, are hunted because their fur is valuable in the fur trade. Others, most notably the tiger, are hunted for their body parts, which are

FIGURE 12-8 The cheetah *(Acinonyx jubatus),* the fastest cursorial mammal. Note how the extension of the spine contributes to an increase in stride length. *(PhotoDisc, Inc.)*

sold at high prices to supply the ever-increasing demands for home remedies in the Asian traditional medicine markets. Habitat loss has also contributed to the decline of many cat populations worldwide. Except for the domestic cat, the entire family Felidae is now listed in Appendix II of CITES (Convention on International Trade in Endangered Species), except for some species listed in Appendix I. In Australia and many islands, the spread of introduced feral cats is exacerbating the decline of native mammals.

FAMILY HYAENIDAE. Many carnivorans will eat **carrion** if the opportunity arises, but most members of the family Hyaenidae have become more or less specialized for carrion feeding. This is a small family, with but four Recent species in four genera. The distribution includes Africa, Turkey, and the Middle East to parts of India. The Hyaenidae, probably derived from viverrid stock, appeared in Eurasia and Africa in the late Miocene and, except for *Chasmaporthetes* (which probably crossed the Bering Strait land bridge and is known from the Pliocene-Pleistocene of North America), has been an entirely Old World family.

Except for the aardwolf (*Proteles cristatus*), hyenas are characterized by rather heavy builds, forelimbs longer than hindlimbs, strongly built skulls, and powerful dentitions (Fig. 12-9). The carnassials are well developed, and all of the cheek teeth have heavily built crowns adapted to crushing bone. The dental formula is 3/3, 1/1, 4/3, 0-1/1 = 32-34. The feet are digitigrade, and both

forepaws and the hind paws have four toes that bear blunt, nonretractile claws. The pelage is either spotted (*Crocuta*) or variously striped (*Hyaena*). Hyenas weigh up to 80 kilograms.

The aardwolf is more lightly built than the hyenas and has a more delicate skull and smaller teeth (Figs. 12-9 and 12-10A). All teeth except the canines are small, and the cheek teeth are simple and conical. The dental formula is generally 3/3, 1/1, 3/2-1, 1/2-1 = 28-32, but frequently some of these teeth are lost (as in the skull shown in Fig. 12-9). The forefeet have five toes, and the hind feet have four. The animal is striped and has a mane of long hair from neck to rump. The tail is quite bushy. When the animal is threatened and adopts a defensive posture, the hair of the mane and tail is erected, but the mouth remains closed. *Proteles* has abandoned the open-mouthed threat used by most carnivores—which in its case would merely advertise the weakness of the dentition—in favor of extensive erection of the long hair. The aardwolf also releases fluid from the well-developed anal glands when attacked.

Hyenas in some areas specialize in scavenging on the kills of lions and other large carnivores and are able to drive cheetahs (*Acinonyx*) from their kills. They may also forage in villages at night for edible refuse. In the ability to crush large bones, they are unsurpassed, but studies by Kruuk (1972) and Mills (1989) have shown that spotted hyenas are also powerful predators (Fig. 12-10B). Often hunting in packs of up to 30 animals, these nocturnal hunters can bring down animals as large as

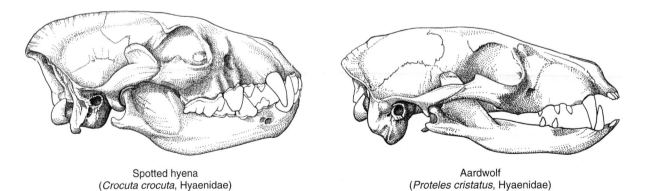

Spotted hyena
(*Crocuta crocuta*, Hyaenidae)

Aardwolf
(*Proteles cristatus*, Hyaenidae)

FIGURE 12-9   Skulls of a spotted hyena (length of skull 248 millimeters) and an aardwolf (length of skull 148 millimeters).

A   B

FIGURE 12-10 (A) Aardwolf *(Proteles cristatus)* foraging for termites. Note the proximity of the apparently unconcerned Thompson's gazelle. (B) Spotted hyena *(Crocuta crocuta)* tearing apart a topi. *(A from Kruuk and Sands, 1972,* African Journal of Ecology, Blackwell Science, Ltd.; *B from Corel Corp.)*

zebras. In the Kalahari Desert of southern Africa, spotted hyenas kill 72 percent of the food they consume, and nearly half of those kills are of large ungulates such as gemsbok (*Oryx gazella*) and wildebeest (*Connochaetes taurinus;* Mills, 1989). Indeed, in the Ngorongoro Crater in Tanzania, a reversal of the usual pattern of interactions between lions and hyenas has occurred: spotted hyenas are better able than the lions to make regular kills, and lions live by driving hyenas from their kills and eating the carrion (Ewer, 1968). In contrast, Kalahari brown hyenas, *Parahyaena brunnea,* are primarily scavengers, killing and scavenging mainly small animals, such as springhares, with kills making up only 5.8 percent of the diet in the Kalahari (Mills, 1989). The differences in diet between these two species of hyenas may explain the differences in their hunting behavior; brown hyenas are mostly solitary scavengers, while spotted hyenas usually hunt in groups of at least three individuals.

The aardwolf eats mostly termites, to which it is attracted largely by the sounds they make (Kruuk and Sands, 1972). Its unusually large auditory bullae may be associated with an enhanced sense of hearing. In contrast to many termite feeders, the aardwolf does not dig for termites but laps them from the surface of the ground.

Spotted hyenas live in "clans" of up to 80 individuals, and they scent mark the borders of their territories, which may exceed 30 square kilometers (Nowak and Paradiso, 1991). These loosely defined clans are usually made up of smaller hunting groups spread out across the territory. Hyena clans in Kruger National Park spent one third of their activity time on territorial patrols (Henschel and Skinner, 1990). Fierce fighting can break out among adjacent clans when territorial borders are violated (p. 463). Care of the young in spotted hyenas is the sole responsibility of the mother, although all young from the same clan may live in a communal den. Young spotted hyenas suckle from their mothers for up to 18 months because food is not brought back to the den by the mother (MacDonald, 1984). In contrast, brown hyena mothers will suckle any infant in the clan and adults frequently bring food back to the den for the young to consume (Skinner et al., 1980).

FAMILY VIVERRIDAE. The family Viverridae includes the Old World civets and genets. Twenty genera and 34 species are recognized, but the taxonomy of this family is still uncertain because of a lack of clearly derived features defining it. Viverrids inhabit much of the Old World, but the center of their distribution is in tropical and southern temperate areas, and they are absent from northern Europe and all but southern Asia, as well as from New Guinea and Australia. This is an old group: it appeared in Europe in the early Oligocene, but the first African records are from the Miocene. Viverrids did not reach Madagascar until the Recent (Holocene). Viverrids may have a long, unknown history in Africa and Asia (Martin, 1989).

Viverrids are small to medium sized, short-legged, long-tailed carnivores. They vary in size from about 600 grams in the Oriental linsang *(Prionodon pardicolor)* to roughly 20 kilograms in the African civet *(Civettictis civetta;* Fig. 12-11). The viverrid skull has a moderately long rostrum (Fig. 12-12). The premolars are large, and the carnassials are usually trenchant. The upper molars are tritubercular and are wider than they are long; the lower molars have well-developed talonids. The

A

B

FIGURE 12-11 (A) African civet *(Civettictis civetta).* (B) Large-spotted genets *(Genetta tigrina),* agile climbers that often feed on birds. *(A by R. Bowker, and B by T. Huels)*

dental formula is generally 3/3, 1/1, 3-4/3-4, 2/2 = 36-40. The five toes on each foot include a much-reduced pollex or hallux. The foot posture is plantigrade or digitigrade, and the claws are semiretractile and covered by a fleshy sheath in some genera. The tail is typically long and bushy. Some species are banded, others are spotted, and still others are striped. Most species have well-developed perineal scent glands (near the anus) used in intraspecific and interspecific communication.

Palm civets are agile climbers and, with the aid of traction pads on the hind feet and hooklike claws on its medial toes, they can crawl headfirst down a nearly smooth tree trunk. Several viverrids, including the aquatic genet *(Osbornictis piscivora)* and the otter civet *(Cynogale bennettii),* show adaptations for a semiaquatic lifestyle. Aquatic genets, for example, have naked palms on the forefeet that allow a better grip on slippery fish. As the name implies, otter civets have an otter-like face, complete with nostrils that open on the top of the rhinarium and close by special flaps that prevent water from entering the nostrils. Long stiff vibrissae (whiskers) that are located on the snout and under the ears serve as tactile hairs in the water. The feet of otter civets are only slightly webbed, and the animals are probably relatively slow swimmers, but they are good climbers and often seek the safety of trees (Nowak and Paradiso, 1991).

Viverrids are primarily carnivorous and eat small vertebrates or insects. Most are nocturnal ambush predators, and many are excellent climbers. The palm civets *(Nandinia)* are omnivorous and feed primarily on fruit or other plant material. The semiaquatic genera *Osbornictis* and *Cynogale* feed primarily on fish and frogs. The aquatic genet *(Osbornictis)* is represented in museums by only a few specimens and is exceedingly rare (Hart and Timm, 1978).

Viverrids display considerable diversity in form and function. Some are agile climbers and forage for prey in trees and on the ground. The rare Malagasy falanouc *(Eupleres goudotii)* has a reduced dentition of short conical teeth used to grasp earthworms, snails, and other invertebrates. Another Malagasy species, *Cryptoprocta ferox,* resembles a small mountain lion. It is nocturnal and arboreal with sharp retractile claws. Its diet includes many small vertebrates, and it is thought to be a principal predator of small lemurs. The binturong

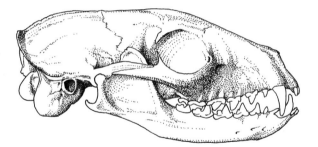

White-tailed mongoose
(*Ichneumia albicauda*, Herpestidae)

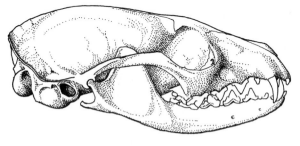

Large-spotted genet
(*Genetta tigrina*, Viverridae)

FIGURE 12-12 The skulls of a white-tailed mongoose (length of skull 106 millimeters) and a large-spotted genet (length of skull 87 millimeters). Note the large auditory bulla of the mongoose.

(Arctictis binturong) of Asia and Indochina is a slow deliberate climber and has a bushy tail that is prehensile at its tip. Binturongs are highly vocal and live in small groups consisting of adults and their offspring.

The oriental civet of India *(Viverra civettina)* is considered critically endangered, and the southeast Asian otter civet, the falanouc of Madagascar, and the central African crested genet *(Genetta servalina)* are considered endangered. The principal reason for the decline in these populations appears to be the continued destruction of appropriate habitat.

FAMILY HERPESTIDAE. Members of the family Herpestidae, commonly known as mongooses, include 37 Recent species in 18 genera. The fossil record of mongooses dates back to the late Oligocene of Europe (Martin, 1989). They are strictly an Old World group with a poor fossil record. Long considered a subfamily of the Viverridae, the herpestids are now considered a distinct lineage (Hunt 1987; Neff, 1983; Wozencraft, 1984, 1989). Herpestids inhabit most of Africa (except the Sahara Desert), southeastern Europe, the Middle East, India, Sri Lanka, and much of southeastern Asia to the Philippine Islands and Borneo. Several genera are restricted to Madagascar and are thought to have arrived on that island in the Recent. Mongooses have been introduced to several of the Caribbean and Hawaiian Islands. Herpestids occupy habitats ranging from deserts to tropical forests.

Mongooses are small, typically long-bodied and long-tailed carnivores (Fig. 12-13). They range in size from about 270 grams (the dwarf mongoose, *Helogale parvula*) to 5 kilograms (the white-tailed mongoose, *Ichneumia albicauda*. The postorbital part of the skull is relatively long (Fig. 12-12), and the complex structure of the large auditory bullae, featuring an expanded ectotympanic and a circular external auditory meatal tube, is unique among carnivores (van der Klaauw, 1931). The dental formula is 3/3, 1/1, 3-4/3-4, 2/2 = 36-40. Usually all but the first premolar are large; the upper molars are tritubercular and are broader than long (Fig. 12-14). The third upper premolar has an internal cusp in most mongooses (Fig. 12-14). The upper carnassial is usually trenchant and has a large inner, protocone-bearing lobe. In most herpestids, the front and hind feet have five toes. The feet are digitigrade to semiplantigrade, and the forefeet bear long, protracted claws in many species. One African species, the marsh mongoose *(Atilax paludinosus)*, has long-toed feet similar to those of a raccoon. The marsh mongoose, like the raccoon, forages by feeding along the bottom of a pool with its dextrous "hands." The herpestid anal scent glands, situated adjacent to the anus in an anal pouch, are highly specialized sacs that secrete carboxylic acid, a byproduct of bacterial metabolism (Gorman et al., 1974). Herpestids have facial glands that are used in scent marking.

Complex social life, characterized by highly structured social behavior and the lavish use of scent marking, has evolved in roughly half the genera of mongooses. Social mongooses are diurnal, live in clans, and are primarily insectivorous

A

B

FIGURE 12-13   (A) Yellow mongoose *(Cynictis penicillata),* in Kalahari Gemsbok National Park. This animal is diurnal, social (up to 20 in a colony) and often lives in association with colonies of cape ground squirrels. (B) The Malagasy broad-striped mongoose *(Galidictis fasciata;* Herpestidae) is nocturnal and solitary. *(A by T. Vaughan, and B by J. Ryan)*

(p. 465). Sociality has enabled such species (for example, the dwarf mongoose, the banded mongoose, *Mungos mungo,* the meerkat, *Suricata suricatta)* to forage by day in open situations where threats from aerial and terrestrial predators abound. By concerted group action, these species can deter some predators.

Although herpestids eat a wide range of food, from various invertebrates and small vertebrates to eggs and even fruit, insects are the mainstay. Of the 18 genera of mongooses, 15 are basically insectivorous. Adult and larval beetles *(Coleoptera),* termites *(Isoptera),* and grasshoppers and their kin *(Orthoptera)* are most important. One species, Meller's mongoose *(Rhynchogale melleri),* is seemingly a termite specialist. Because in Africa some termites inhabit ungulate dung, as do the larvae of the diverse and numerous dung beetles, the abundance of African ungulates may influence population levels of mongooses.

A number of herpestids break the shells or exoskeletons of such items as eggs or crabs by throwing them backward, between their hind legs, against a hard object. Only among the mongooses and their ecological counterparts, the spotted skunks *(Spilogale),* is this behavior known.

CANIFORMIA

The remaining carnivorans form the caniform clade, which includes the families Canidae, Procyonidae, Mustelidae, Ursidae, and three families of aquatic carnivorans (Fig. 12-1). There is general agreement that the pinnipeds form a monophyletic clade allied to the bears (Ursidae).

FAMILY CANIDAE. Fourteen genera and about 34 Recent species make up the family Canidae. Canids — wolves (Fig. 12-15), jackals, foxes (Fig.

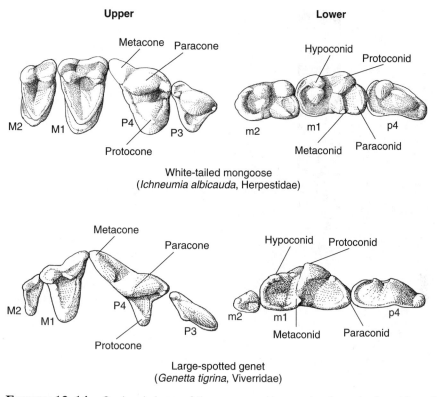

FIGURE 12-14   Occlusal views of the upper and lower cheek teeth of a white-tailed mongoose and a large-spotted genet. Note the large inner lobes (bearing the protocone) of P4 (carnassials) of both species and the large inner lobe of P3 of the mongoose.

12-16A and B), and dogs — occupy a great array of environments from arctic to tropical. Prior to their domestication and dispersal with humans, canids occurred nearly worldwide, except on most oceanic islands. The dingo (*Canis*) was probably brought to Australia by early humans between 11,000 and 3,500 years ago (Menkhorst, 1995). The fossil record of canids dates from the Oligocene of North America *(Hesperocyon)*. The African and South American fossil record is restricted to the Pleistocene.

Canids are broadly adapted carnivores, which is reflected in their morphology. The canid skull typically has a long rostrum that houses a large nasal chamber with complex turbinal bones, a feature associated with a remarkable sense of smell (Fig. 12-17). Most canids have a nearly complete eutherian complement of teeth (3/3, 1/1, 4/4, 2/3 =

42). The canines are generally long and strongly built, and the carnassials retain the shearing blades (Fig. 12-18). The postcarnassial teeth have crushing surfaces, indicating a more flexible diet than that of the more strictly carnivorous cat family (Felidae). In cursorial species, the limbs are long, and rotation at the joints distal to the shoulder and hip joints is reduced in the interest of cursorial ability. The clavicle is absent. The feet are digitigrade, and the well-developed but blunt claws are nonretractile. The forefoot usually has five toes, and the hind foot has four. The weight of canids ranges from 1 to 75 kilograms.

Some canids forage tirelessly over large areas, and lengthy pursuit is frequently part of the hunting technique. The coyote (*Canis latrans*), probably one of the swiftest canids, can run at speeds of about 65 kilometers per hour. The fact

FIGURE 12-15   An Alaskan wolf. *(From* The Wolves of Mount McKinley, *by A. Murie, 1944)*

A

B

that coyotes in many areas depend partly on jackrabbits for food is an impressive testimonial to this carnivore's speed. Canids often hunt in open country. Wolves *(Canis lupus)* and the African hunting dog *(L. pictus;* Fig. 12-16C) seem to rely more on endurance than on speed when hunting. These canids, and the eastern Asian dholes *(Cuon alpinus),* habitually hunt in packs and kill larger prey than could be overcome by a solitary hunter. The gray fox *(Urocyon cinereoargenteus)* does not generally forage in open areas but is amazingly agile and can run rapidly through the maze of stems beneath a canopy of chaparral. The foods of canids include vertebrates, arthropods, mollusks, carrion, and many types of plant material. Blackbacked

▶**FIGURE 12-16**   (A) A bat-eared fox *(Otocyon megalotis)* showing the large forward-facing ears. (B) The San Joaquin kit fox *(Vulpes velox).* (C) An African hunting dog *(Lycaon pictus). (A by T. Vaughan; B by Corel Corp.; C by PhotoDisc, Inc.)*

C

jackals *(Canis mesomelas)* once were a problem in parts of South Africa because of their extensive feeding on pineapples (Ewer, 1968). Coyotes in parts of the western United States feed heavily on such cultivated crops as melons and such uncultivated plant material as juniper berries and prickly-pear cactus fruit. The average canid is clearly an opportunist; this may in large part account for the great success of this family.

Canid hunting behavior is correlated loosely with body size, whereby small canids, such as arctic foxes *(Alopex lagopus),* bat-eared foxes *(Otocyon megalotis),* and kit foxes *(Vulpes velox)* are solitary hunters; medium-sized canids (coyotes) hunt alone or in small family groups; and large canids (wolves, African hunting dogs, and dholes) hunt in large cooperative groups (Moehlman, 1989). One exception is the large maned wolf *(Chrysocyon brachyurus),* which is a solitary hunter. Before recent declines, African hunting dogs formed large packs of up to 50 dogs. Members of these packs hunt cooperatively, and, although the young are usually the progeny of only one dominant breeding pair, all adult pack members provision the young (Estes, 1991). Hunting success varies with prey species, season, and presence of potential competitors (Fuller and Kat, 1993).

FAMILY MUSTELIDAE. The family Mustelidae is large, with 25 genera and some 65 Recent species, including the weasels, badgers, skunks, and otters (Fig. 12-19). For consistency, we follow Wozencraft (1993) in retaining the skunks in the family Mustelidae, although we recognize that a separate

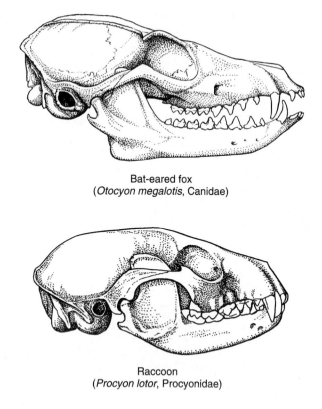

Bat-eared fox
(*Otocyon megalotis,* Canidae)

Raccoon
(*Procyon lotor,* Procyonidae)

FIGURE 12-17   Skulls of a bat-eared fox (length of skull 111 millimeters), and a raccoon (length of skull 115 millimeters).

family Mephitidae (including skunks and the stink badger) has merit (Dragoo and Honeycutt, 1997). Mustelids occupy virtually every type of terrestrial habitat, from arctic tundra to tropical rain forests,

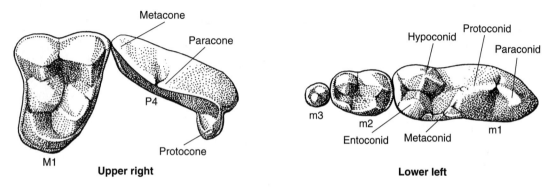

FIGURE 12-18   Occlusal view of selected cheek teeth of a coyote (*Canis latrans,* Canidae). The cusps of the carnassials (P4 and m1) are labeled. Notice that the cheek teeth both shear and crush.

A

B

FIGURE 12-19 (A) Striped skunk *(Mephitis mephitis)*. (B) A river otter *(Lontra canadensis)*. *(A by T. Vaughan; B from PhotoDisc, Inc.)*

and live in rivers, lakes, and oceans. The distribution is nearly cosmopolitan, but they do not inhabit Madagascar, Australia, or oceanic islands. Mustelids appear in the fossil records of Europe and Asia in the early Oligocene.

Mustelids are typically fairly small, long-bodied carnivores with short limbs and a pushed-in face. The skull generally has a long braincase and a short rostrum (Figs. 12-20), and the postglenoid process partially encloses the glenoid fossa so that in some species the condyle of the dentary bone is difficult to disengage from the fossa. Obviously, little lateral and no rotary jaw action is possible. The dentition is quite variable but is generally 3/3, 1/1, 3/3, 1/2 = 34. The carnassials are trenchant in

many species (Fig. 12-21) but have been modified into crushing teeth in others. In the sea otter *(E. lutris)*, for example, none of the cheek teeth are trenchant, the carnassials have rounded cusps adapted to crushing, and the postcarnassial teeth, M1 and m2, are broader than they are long (Fig. 12-21C and D). The first upper molar is frequently hourglass-shaped in occlusal view (Fig. 12-21) or may be expanded into a large crushing tooth, as in skunks.

The limbs are usually short, the five-toed feet are either plantigrade or digitigrade, and the claws are never completely retractile. Anal scent glands are usually well developed; they are extraordinarily large in skunks and are used for defense. The tail is generally long, and the pelage may be conspicuously marked, as in skunks and badgers. Some mustelids have beautiful, glossy fur that has considerable value in the fur trade. In size, mustelids range from the smallest member of the order Carnivora, a circumboreal weasel *(Mustela nivalis)* that weighs 35 to 250 grams, to the wolverine *(Gulo)* and the sea otter at 32 and 45 kilograms, respectively.

Mustelids, although basically carnivorous, pursue many styles of feeding. Most aggressively search for prey in burrows, crevices, or dense cover, and many are able killers. A male long-tailed weasel *(M. frenata)* can kill young cottontail rabbits *(Sylvilagus audubonii)* roughly twice its own weight. The weasel typically kills by repeatedly biting the back of the rabbit's skull. Some mustelids, such as the beautiful and graceful marten *(Martes americana)*, are swift and agile climbers who often catch arboreal squirrels. Otters are semiaquatic, or almost completely aquatic in the case of the sea otter, and feed on a wide variety of vertebrates and invertebrates. Sea otters are unique among mustelids in that they use "tools," in the form of rocks brought up from the bottom of the ocean, to crack open the hard shells of crabs and mollusks (Fig. 12-22). The skunks, with no claim to remarkable agility or killing ability, seem to feed on whatever animal material is readily available, which during the summer is generally insects.

Mustelids are unusual in several features of their reproduction. Apparently, all mustelids must prolong copulation to induce the female to **ovulate** (release an ovum for fertilization). As a result, copulation may last for several hours in some

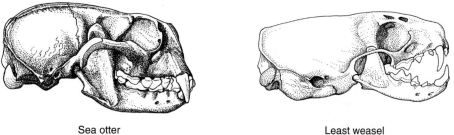

Sea otter
(*Enhydra lutris*, Mustelidae)

Least weasel
(*Mustela nivalis*, Mustelidae)

FIGURE 12-20   The skull of the sea otter (length of skull 152 millimeters), with heavy cheek teeth adapted to crushing marine invertebrates, and a least weasel (length of skull 31 millimeters).

species. Once fertilization has taken place, the embryo may or may not immediately implant in the lining of the uterus. **Delayed implantation** occurs is some, but not all, mustelids (p. 348). For example, in the Old World badger *(Meles meles)*, implantation may be delayed for up to ten months, and implantation of the **blastocyst** takes place in response to environmental cues, such as temperature and day length (Canivenc and Bonnin, 1979; Mead, 1989).

FAMILY PROCYONIDAE. The family Procyonidae includes the raccoons *(Procyon)*, ringtails *(Bassariscus astutus)*, and their relatives. As with the bears, omnivorous feeding habits have become predominant in procyonids. Six genera and 18 species are known. The Procyonidae, as currently recognized, does not include the lesser panda *(Ailurus)*. Procyonids occupy much of the temperate and tropical parts of the New World, from southern Canada through much of South America. Pro-

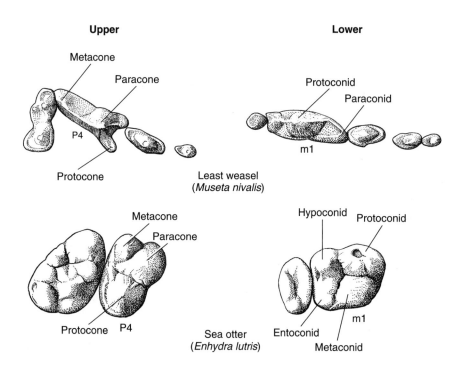

FIGURE 12-21   Occlusal views of the cheek teeth of two carnivores. The crushing teeth, M1 and m2, of the least weasel are reduced, and the shearing function of the cheek teeth is of major importance. The teeth of the sea otter have rounded cusps and are adapted to crushing; the carnassials retain no shearing function.

FIGURE 12-22   California sea otter *(Enhydra lutris)* feeding while floating on its back. *(L. Consiglieri, NOAA Corps)*

cyonids chiefly inhabit forested areas, but the range of one species of ringtails includes arid desert mountains and foothills. Procyonids are known from the late Eocene of North America, early Oligocene of Europe, and from the late Miocene to Recent in Asia. They reached South America from North America at the end of the Miocene (Dawson and Krishtalka, 1984).

The structural and functional departure of procyonids from the carnivorous norm has included adaptations favoring both omnivorous feeding habits and climbing ability (Fig. 12-23). Associated with the omnivorous trend has been a specialization of the cheek teeth. The premolars are not reduced, as in the bears, but the shearing action of the carnassials is nearly lost. Instead, the carnassials are low-cusped crushing teeth; a hypocone was retained in the upper, and in the lower the talonid was enlarged and broadened (Fig. 12-24). In contrast to the elongate upper molars of bears, those of procyonids are usually broader than they are long. The coati *(Nasua narica)* has flattened, bladelike canines that are formidable defensive weapons. The dental formula is usually 3/3, 1/1, 4/4, 2/2 = 40. There are five toes on each foot; the foot posture is usually plantigrade; and the claws are nonretractile or semiretractile. The limbs are fairly long. The toes are separate, and the forefoot has considerable dexterity in some

species and is used in handling food. Tracks left by the human-like hand of the raccoon are familiar to many people. The tail is long, generally marked by dark rings, and is prehensile in the arboreal kinkajou *(Potos flavus)*. Procyonids are of modest size, weighing from less than 1 to about 20 kilograms.

The familiar raccoon often takes advantage of cultivated crops. Corn is a staple food item for raccoons living in the middle and western United States, and they eat grapes, figs, and melons in parts of California (Grinnell et al., 1937). In addition, they prey on a variety of small vertebrates and some invertebrates. Some tropical procyonids are largely vegetarians. Hall and Dalquest (1963) reported that in Veracruz the coati eats corn, bananas, and fruit of the coyol palm and that kinkajous eat mostly fruit. The ringtails, on the other hand, are known to feed mostly on small rodents in some areas. Procyonids reach their greatest diversity and greatest densities in the Neotropics, where they are largely arboreal; in tropical forests several species may occur together. In such areas, the nocturnal, quavering cries of kinkajous can be heard regularly. Coatis are social animals and assemble in female-young tribes of from 5 to 20 or so animals. These animals are highly vocal and have a varied repertoire of communication calls.

FAMILY URSIDAE. Except for the lesser panda *(Ailurus fulgens)*, the bears are notable for their large size and their departure from a strictly carnivorous mode of life (Fig. 12-25). The family Ursidae contains six genera and nine Recent species. Morphological and molecular evidence suggests that the giant panda, *Ailuropoda melanoleuca,* is a bear, and it is included here in the Ursidae (Chorn and Hoffmann, 1978; Goldman et al., 1989; O'Brien et al., 1985; Wozencraft, 1989). The taxonomic position of the lesser panda is more problematic, and it has been allied with the procyonids, based on molecular and chromosomal evidence (Goldman et al., 1989). However, Todd and Pressmann (1968) and Zhang and Shi (1991) provide evidence for a closer relationship between the lesser panda and bears. Furthermore, Flynn et al. (1988) could find no shared derived characters to unite *Ailurus* with the procyonids. The most recent evidence seems to support a closer association between bears and *Ail-*

A

B

FIGURE 12-23   (A) A raccoon *(Procyon lotor)* from North America and (B) a kinkajou *(Potos flavus)* from South America. The arboreal kinkajou has a prehensile tail and is able to rotate its feet to grasp branches. *(A by PhotoDisc, Inc., and B by Corel Corp.)*

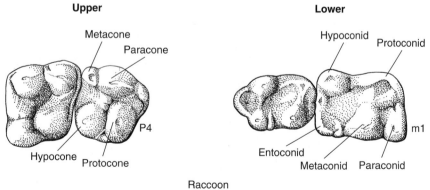

Raccoon
(*Procyon lotor*, Procyonidae)

FIGURE 12-24   Occlusal view of selected cheek teeth of the raccoon. Note that the upper carnassial has a hypocone and that all the teeth are adapted to crushing.

urus (Decker and Wozencraft, 1991; Wozencraft, 1989).

The distribution of ursids includes most of North America and Eurasia, the Malay Peninsula, the South American Andes, the Atlas Mountains of extreme northwestern Africa, and, for the lesser panda and giant panda, Nepal, Myanmar (formerly Burma), and parts of China. Bears inhabit diverse habitats, from drifting ice in the Arctic to the tropics but are most important in boreal and temperate areas.

Bears first appeared in Europe, Asia, and North America in the early Miocene. They probably reached South America and northwestern Africa in the early Pleistocene and late Miocene, respectively (McKenna and Bell, 1997).

The bear skull retains the long rostrum typical of the canids, but the orbits are generally smaller, and the dentition is very different (Fig. 12-26). The postcarnassial teeth are greatly enlarged, and the occlusal surfaces are "wrinkled" and adapted to crushing. On the other hand, the first three premolars are usually rudimentary or may be lost, and a diastema usually occurs between premolars. The upper carnassial is roughly triangular because of the posterior migration of the protocone and is much smaller than the neighboring molars (Fig.

12-26B); both upper and lower carnassials no longer have a shearing function. The dental formula is usually 3/3, 1/1, 4/4, 2/3 = 42, but premolars may be lost with advancing age. The limbs, especially the forelimbs, are strongly built; the plantigrade feet have long, nonretractile claws. There are five toes on each foot. The ears are small, and the tail is extremely short. In size, bears range from that of a large dog to over 800 kilograms in the polar bear *(Ursus maritimus)*. The nocturnal lesser panda has shorter limbs, a long ringed tail, and facial markings reminiscent of raccoons.

The abandonment of cursorial ability in favor of power in the limbs and the loss of the shearing function of the cheek teeth in favor of a crushing battery have accompanied the adoption of omnivorous feeding habits. The strong forelimbs can aid in the search for food by rolling stones or tearing apart logs, and the crushing surfaces of the molars can cope with many kinds of food, from insects and small vertebrates to berries, grass, and pine nuts. Carrion is also avidly sought. The polar bear has a more restricted diet, consisting mainly of seals, and the giant panda eats mostly bamboo shoots (Schaller et al., 1985). In areas with cold winters, some bear species retreat for much of the

FIGURE 12-25 A Kodiak brown bear *(Ursus arctos)*, also known as the grizzly bear, fishes for salmon in Alaska. *(Corel Corp.)*

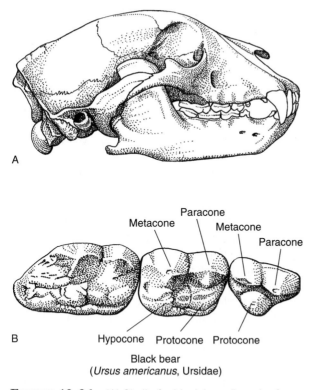

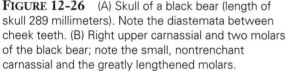

FIGURE 12-26   (A) Skull of a black bear (length of skull 289 millimeters). Note the diastemata between cheek teeth. (B) Right upper carnassial and two molars of the black bear; note the small, nontrenchant carnassial and the greatly lengthened molars.

Black bear
(*Ursus americanus*, Ursidae)

winter to caves or other retreats protected from drastic temperature fluctuations. Here they become lethargic or dormant, and they live off fat reserves accumulated during the fall. Whether this is "true" hibernation is a matter of semantics. Some ecophysiologists do not consider this hibernation because the bear's body temperature does not drop substantially and bears can easily arouse. However, Folk et al. (1976) maintain that bears do hibernate because their pulse rate drops by nearly half.

PINNIPEDIA

The aquatic members of the Carnivora include the earless seals (Phocidae), the eared seals (Otariidae), and the walrus (Odobenidae). Based on fea-

tures of the skull, Tedford (1976) proposed that the pinnipeds (aquatic carnivorans) are not a monophyletic group and that otariids and odobenids are derived from ursids, while the phocids share a common ancestor with the mustelids. However, Wyss (1987) provides evidence from morphology and biochemical data to support the hypothesis that pinnipeds are monophyletic (Fig. 12-1). According to this hypothesis and basicranial evidence (Hunt and Barnes, 1994), the pinnipeds form a distinct group most closely related to ursids.

Many of the distinctive morphological features of aquatic carnivorans are adaptations to marine life. They are larger than land carnivores, ranging from 90 to 3600 kilograms in body weight. Large size saves energy in cold environments because of the favorable surface area-to-volume ratio of large animals (p.367). According to Scheffer (1958), large size in pinnipeds is primarily an adaptation to a cold environment. The body is insulated by thick layers of blubber and, in otariids, by thick insulating fur as well. The pinnae are either small or absent, the external genitalia and mammary nipples are withdrawn beneath the body surface, the tail is rudimentary, and only the parts of the limbs distal to the elbow and knee protrude from the body surface. As a result, the torpedo-shaped body has smooth contours and creates little drag during swimming. The slitlike nostrils are normally closed and opened by voluntary effort.

The skull is partially telescoped, with the supraoccipital partially overlapping the parietals; the rostrum is usually shortened, and the orbits are usually large and encroach on the narrow interorbital area. Either one or two pairs of lower incisors are present. The canines are conical. The cheek teeth are homodont (none modified as carnassials), two-rooted, and usually simple and conical (Fig. 12-27); they vary in total number from 12 to 24. In some pinnipeds, cheek teeth are characteristically lost with advancing age. The limbs and girdles are highly specialized. The clavicle is absent, and the humerus, radius, and ulna are short and heavily built (Fig. 12-28). The pollex is the longest and most robust of the five digits and forms the leading edge of the winglike fore flipper. The pelvic girdle is small and nearly parallel to the vertebral column in phocids (Fig. 12-28). The femur is broad and flattened. The first and fifth are the

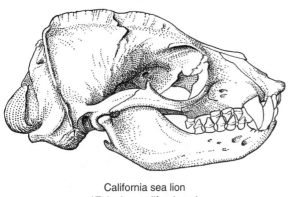

Male (290 mm)

California sea lion
(*Zalophus californianus*)

FIGURE 12-27 Skull of a male California sea lion (length of skull 290 millimeters).

longest digits of the pes, and both the manus and pes are fully webbed.

The reduction of the vertebral zygapophyses and the absence of the clavicle allow the vertebral column and the forelimbs considerable flexibility and freedom of movement; these features may favor rapid maneuvering during the pursuit of prey. Although terrestrial locomotion is characteristically slow and laborious in most species, the importance of terrestrial locomotion when the animals haul themselves out on rocks or ice, or are on breeding grounds, has probably limited the extent to which the limbs of aquatic carnivores have become specialized for swimming.

Aquatic carnivorans are extremely capable divers, with some species surpassing most cetaceans in this skill. The diving performance of the eared seals (Otariidae) is probably similar to that of many dolphins. The most spectacular diving occurs among earless seals (Phocidae). The elephant seal (*Mirounga angustirostris*) of the North Atlantic reaches depths of a least 1500 meters and can stay beneath the surface for up to 90 minutes (Le Boeuf et al., 1993; Stewart, 1997). The Weddell seal (*Leptonychotes weddellii*) commonly reaches depths of 300 to 400 meters and is known to dive to 600 meters (Kooyman, 1981). Dives often last more than 40 minutes, and a dive of 1.17 hours was recorded (Kooyman et al., 1971). This duration surpasses that recorded for any marine mammal except the sperm whale (*Physeter catodon*), bow-

head whale (*Balaena mysticetus*), and bottle-nosed whale (*Hyperoodon ampullatus*). Distances traveled during dives by Weddell seals are also remarkable. Kooyman (1968) found that during a single dive they can swim 5 kilometers from their breathing holes in the ice and return.

An integrated array of specializations is associated with the diving ability of seals (King, 1983; Kooyman, 1968, 1975, 1981; Kooyman et al., 1976, 1980). The lungs of aquatic carnivores tend to be larger than those of terrestrial mammals of comparable size, a character important primarily in contributing to buoyancy and allowing the animals to rest at sea. The respiratory airways, even the terminal segments supplying the alveoli, are made rigid by cartilage and muscle. This feature, shared by cetaceans, allows for free passage of air from the alveoli when the lungs collapse under the great pressures to which the body is subjected during deep dives. In addition, this rigidity contributes to the animal's ability to expire air from the lungs extremely rapidly and provides for very quick exchanges of large gas volumes. Another adaptation for prolonged diving is the large blood volume of seals. Weddell seals usually have approximately 14 percent of their body weight in the form of blood, compared to only around 7 percent for terrestrial animals of similar size. During extended dives, there is a redistribution of blood: blood flow to the major muscle masses is restricted to maintain adequate oxygen supply to the brain and heart. It has been estimated that the metabolic rate of a seal during a long dive is less than 20 percent of the basal rate.

FAMILY ODOBENIDAE. The family Odobenidae contains only one species, *Odobenus rosmarus*, the walrus (Fig. 12-29). This species occurs near shorelines in arctic waters of the Atlantic and Pacific Oceans but may stray southward along the coastlines. Odobenids first appeared in the early Miocene and, like otariids, may have had a North Pacific origin.

The walrus is a large pinniped (up to 1270 kilograms) with a robust build, a nearly hairless skin, and no external ears. The hind flippers can be brought beneath the body and are used for terrestrial locomotion, which is ponderous and slow. In both sexes, the upper canines are modified into

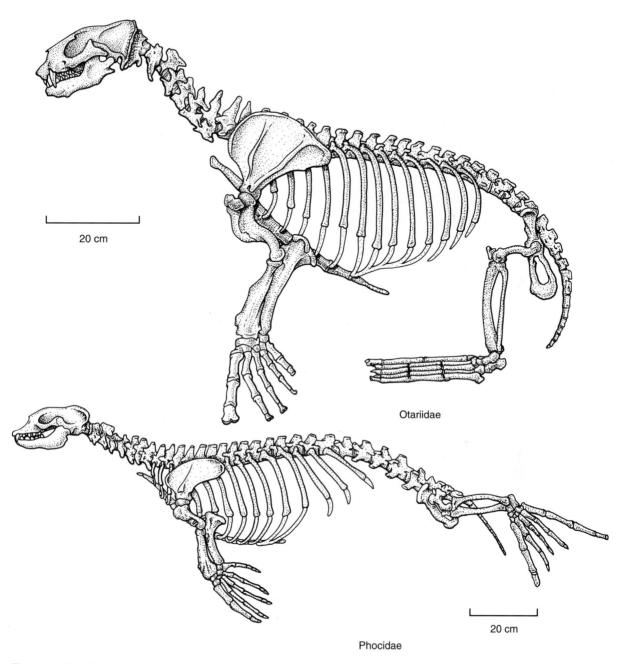

Otariidae

Phocidae

20 cm

20 cm

FIGURE 12-28   Drawings of the skeletons of an otariid seal and a phocid seal. *(From King, 1983).*

long tusks (Fig. 12-30), which in the adult lack enamel. There are no lower incisors in adults, and 12 cheek teeth are usually present. The dental formula is 1-2/0, 1/1, 3-4/3-4, 0/0 = 18-24. On the huge mastoid processes attach the powerful neck muscles that pull the head downward.

Walruses feed on mollusks, which they take from the sea floor by means of their lips, opening the shells with tongue suction. Their huge tusks are apparently not used for digging (Fay, 1981). Walruses are **gregarious** and **polygynous** and frequently assemble in large groups of more than 1000 individu-

FIGURE 12-29   A walrus (*Odobenus rosmarus;* Odobenidae). *(L. Consiglieri, NOAA Corps)*

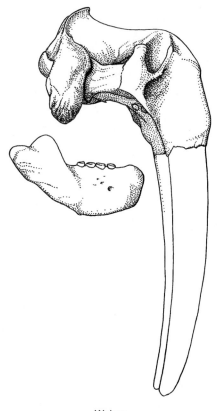

Walrus
(*Odobenus rosmarus*)

FIGURE 12-30   The skull of a walrus (length of skull approximately 355 millimeters). The tusks are enlarged upper canines.

als. They are migratory to some extent, moving southward in winter. Walruses make a variety of loud noises when out of the water and make a church-bell sound and rasps and clicks underwater (Schevill et al., 1963, 1966). The rasps and clicks made during swimming suggest their use in echolocation, but this has not been confirmed.

FAMILY PHOCIDAE. Earless seals include 19 species and ten genera, making the Phocidae the most species-rich family of aquatic carnivorans. They occur along most northern (above 30 degrees northern latitude) and most southern (below 50 degrees southern latitude) coastlines and in some intermediate areas. The Caspian seal *(Phoca caspica)* is restricted to the Caspian Sea, and the Baikal seal *(P. siberica)* lives only in freshwater Lake Baikal in eastern Russia. Phocids appear in the early Miocene and probably arose in the Atlantic Ocean (Muizon, 1982).

The earless seals are more highly specialized for aquatic life than are the other aquatic carnivorans. As the vernacular name implies, there is no external ear. The hind flippers are useless on land but, as a result of lateral undulatory movements, are the primary propulsive organ in the water (Fig. 12-31). The fore flippers are short and well furred. The structure of the cheek teeth is highly variable but is usually fairly simple. In the crab-eater seal *(Lobodon carcinophagus),* however, the cheek teeth have complex cusps (Fig. 12-32). The pelage of most phocids is spotted, banded, or mottled. These seals frequently have extremely heavy layers

of subcutaneous blubber that give the body smooth contours and, in some cases, a nearly perfect **fusiform** shape. Most species weigh from about 80 to 450 kilograms, but male elephant seals *(Mirounga)* occasionally weigh as much as 3600 kilograms.

Many phocids are monogamous and form small, loose groups in which no social hierarchy is evident, but some, such as the elephant seal, are gregarious and polygynous and have a dominance hierarchy. The monogamous species are quiet, whereas the polygynous species are highly vocal (Evans and Bastian, 1969). The sole function of the **proboscis** of the male elephant seal is the production of vocal threats (Bartholomew and Collias, 1962).

The usual foods of phocids are fish, cephalopods, and other mollusks. Large prey may be taken: Sterling (1969) reports a fish weighing 29.5

kilograms removed from the stomach of a Weddell seal, and the powerful leopard seal *(Hydrurga leptonyx)* eats penguins and small seals. Two species of phocids are filter feeders and use the complex cheek teeth to filter crustaceans and other plankton from the water. So abundant is the filter-feeding Antarctic crab-eater seal that it constitutes a major share of the world population of phocid seals. Dehnhardt et al. (1998) report that harbor seals *(P. vitulina)* use their vibrissae as hydrodynamic receptors to detect the water movements of nearby prey.

Many Weddell seals of the Antarctic are year-round residents as far south as 79 degrees, where broad areas are covered all year with stationary sea ice and the seals must depend on scattered breathing holes. The seals use the upper canines and incisors to ream these holes open by using violent side-to-side thrashing of the head (Kooyman, 1975). The ice is about 1 meter thick away from the holes, and the survival of seals depends on their ability at the end of a dive to find their way back to the original breathing hole or to locate a new one. These animals must be skilled navigators in darkness, for many spend the winter where the sun does not appear above the horizon for up to 3.5 months.

FAMILY OTARIIDAE. The family Otariidae, containing seven genera and 14 Recent species, includes the eared seals and the sea lions (Fig. 12-33).

FIGURE 12-31   Leopard seal (*Hydrurga leptonyx;* Phocidae) in Hughes Bay, Antarctica. (*P. Boveng, NOAA, NMFS, National Marine Mammal Laboratory)*

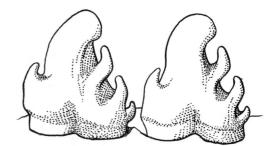

FIGURE 12-32   Medial view of two right lower cheek teeth of the crab-eater seal. These complex teeth enable this animal to depend on filter-feeding.

These animals inhabit many of the coastlines of the Pacific Ocean and parts of the South Atlantic and Indian Oceans, including the coasts of southern Australia and New Zealand. They are common along the Pacific coast of North America. The earliest otariids are known from the middle Miocene (*Pithanotaria* and *Thalassoleon;* Carroll, 1988), and may have originated in the food-rich kelp reefs of the North Pacific (Repenning, 1976; Scheffer, 1958).

Otariids differ from phocids in being less highly modified for aquatic life and better able to move on land. The hind flippers can be brought beneath the body and used in terrestrial locomotion (Fig. 12-28). Well-developed nails occur on the three middle digits. A small external ear is present. Males are much larger than females; in the northern fur seal *(Callorhinus ursinus),* males weigh 4.5 times as much as females. Considerable sexual dimorphism in the shape of the skull occurs in some species. In males, the skull becomes larger and more heavily ridged with advancing age. The dental formula is 3/2, 1/1, 4/4, 1-3/1 = 34-38. The body is covered with uniformly dark fur. Weights of otariids range from roughly 60 to 1000 kilograms.

These seals are generally highly vocal and utter a great variety of sounds. They tend to be gregarious all year round and are social during the breeding season, when they assemble in large breeding rookeries (p. 454). Propulsion in the water is accomplished by powerful downward and backward strokes of the forelimbs; speeds up to 27 kilometers per hour have been recorded (Scheffer, 1958). Otariids eat mostly squid and small fish that occur in schools, and they maneuver rapidly in pursuit of prey.

FIGURE 12-33 Group of Steller's sea lions (*Eumetopias jubatus;* Otariidae) on the rocky coast of Alaska. *(S. Mizroch, NOAA, NMFS, National Marine Mammal Laboratory)*

WEB SITES

General

Descriptions of many species of carnivores

http://www.oit.itd.umich.edu/bio108/Chordata/Mammalia/Carnivora.shtml

Craighead Environmental Research Institute home page

http://montana.avicom.net/ceri/

Carnivore Preservation Trust home page

http://ils.unc.edu/~mcphk/cpt/

Felidae

Big Cats Online—Excellent resource for information about all kinds of big cats

http://dialspace.dial.pipex.com/agarman/

Feline Conservation Center

http://www.cathouse-fcc.org/

UC Berkeley's page about fossil felids

http://www.ucmp.berkeley.edu/mammal/carnivora/sabretooth.html

Hornocker Wildlife Institute—Conservation of large cats

http://www.uidaho.edu/rsrch/hwi/

IUCN World Conservation Union—links to endangered species information

http://www.IUCN.org/themes/themes.html

Tiger Information Center—Tiger conservation information

http://www.5tigers.org/

Hyaenidae
General site about hyenas

http://www.csulb.edu/~persepha/hyena.html

Canidae
List of links to other canid sites

http://www.birminghamzoo.com/ao/dogs.htm

Site dedicated to wolves

http://www.wolfsden.org/

International Wolf Center home page

http://www.wolf.org/

Ursidae
The International Association for Bear Research and Management

http://weber.u.washington.edu/~hammill/iba/iba.html

Mustelidae
Detailed information about mustelids

http://www.oit.itd.umich.edu/bio108/Chordata/Mammalia/Carnivora/Mustelidae.shtml

Detailed information about mustelids

http://sciweb.onysd.wednet.edu/sciweb/zoology/mammalia/weasels.html

Pinnipedia
Marine Mammal Society home page

http://pegasus.cc.ucf.edu/~smm/home.html

Sea World site about harbor seals

http://www.seaworld.org/HarborSeal/hsintro.html

Seal Conservation Society home page

http://www.greenchannel.com/tec/pinniped.htm

13 CETACEA

Cetaceans (order Cetacea) are notable for being the mammals most fully adapted to aquatic life. Baleen whales (suborder Mysticeti), the largest living or fossil mammals known, are mainly plankton feeders (although some eat fish) and thus draw from the tremendous productivity near the bottom of the marine food web. Some of the large toothed whales (suborder Odontoceti), in contrast, exploit the top of the marine food web. Sperm whales, for example, eat giant squid, large sharks, and bony fish. The food of killer whales includes fish, seals, porpoises, and baleen whales. Remarkable swimming and diving ability, the capability of many (perhaps all odontocetes) to echolocate, considerable

intelligence, and complex social behavior have all contributed to the great success of the cetaceans.

Cetaceans have long intrigued and inspired humans. Leaping dolphins have been the embodiment of beauty and exuberance to seafarers for centuries: 4000 years ago Minoan artists included graceful drawings of dolphins in their frescoes at the palace of Knossos on Crete. It has remained for modern humans to deplete the cetacean populations that once seemed limitless. All species of cetaceans are now listed in appendix I or II of the Convention on International Trade in Endangered Species (CITES).

MORPHOLOGY

All cetaceans are completely aquatic, and their structure reflects this mode of life. The body is fusiform (cigar-shaped), nearly hairless, insulated by thick blubber, and lacks sebaceous glands. Most vertebrae have high neural spines (Fig. 13-1), and the cervical vertebrae are highly compressed (Fig. 13-2). The clavicle is absent, the forelimbs (flippers) are paddle-shaped, and no external digits or claws are present. Little movement is possible between the joints distal to the shoulder. The proximal segments of the forelimb are short, whereas the digits frequently are unusually long because of the development of more phalanges per digit than the basic eutherian number (Fig. 13-3). The hindlimbs are vestigial, do not attach to the axial skeleton, and are not visible externally. The **flukes** (tail fins) are horizontally oriented. To aid in streamlining the body, the single pair of mammae lies flat along the abdomen and the teats are enclosed within slits adjacent to the urogenital opening. In males, the testes remain abdominal and the penis is retractile (Rice, 1984).

The skull is typically highly modified as a result of the posterior migration of the external nares (Fig. 13-4). The premaxillary and maxillary bones form most of the roof of the skull, and the occipitals form the back. The nasals and parietals are telescoped between these bones and form only a minor part of the skull roof (Fig. 13-5), and large frontal bones are mostly covered by the maxillaries and premaxillaries. The tympanoperiotic bone (the bone that houses the middle and inner ear) is not braced against adjacent bones of the skull in most cetaceans and is partly insulated from the rest of the cranium by surrounding air sinuses (Fig. 13-6). Members of the families Ziphiidae and Physeteridae have a pneumatic, bony strut that braces the tympanoperiotic bone against the skull.

Cetaceans have well-developed retia mirabilia (systems of countercurrent capillaries, singular rete mirabile) in the body, fins, basicranial regions, and even in the tongue in some, that function in temperature regulation.

Experimental work on the bottlenose dolphin (*Tursiops truncatus*) by Herman et al. (1975) showed that visual acuity is similar above and below the water. The above-water acuity may be due to the "pinhole camera" effect of the pupil. Mechanisms for fine adjustment of lens shape or displacement are seemingly absent in cetaceans, but in bright light the central part of the pupil closes completely, leaving two tiny apertures that yield great depth of field, as does a pinhole camera. This allows the dolphin to receive a sharp image from distant objects when ambient light levels are relatively high (Dawson et al., 1979; Dawson, 1980).

PALEONTOLOGY

The earliest known cetaceans, *Pakicetus, Ambulocetus,* and several other extinct genera, are known from the early to middle Eocene of Pakistan, where their fossils are found in association with various land mammals in river sediments deposited at the border of a shallow sea (Gingerich et al., 1983; Thewissen, 1994; Williams, 1998). These and other primitive whales belong to the extinct suborder Archaeoceti. They show morphological features

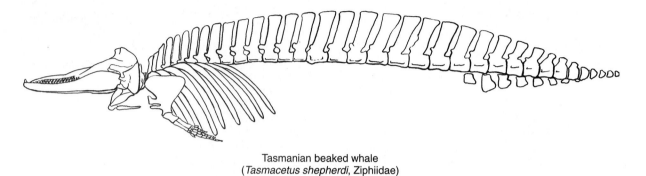

Tasmanian beaked whale
(*Tasmacetus shepherdi*, Ziphiidae)

FIGURE 13-1   The skeleton of the Tasmanian beaked whale. Total body length is approximately 6.6 meters.

that clearly indicate they were phylogenetically derived from mesonychians, an extinct, scavenging or carnivorous group of primitive ungulates. The closest living relatives of cetaceans are the Artiodactyla, as confirmed by morphological and molecular evidence (Messenger and McGuire, 1998; Milinkovitch, 1992; Milinkovitch et al., 1994; Novacek, 1993b). At least two genera, *Ambulocetus* and

Rhodocetus, are known to have had legs and show other primitive features of the terrestrial mesonychians, as well as derived features of archaeocete whales (fossil whales), including adaptations of the ear (Gingerich et al., 1994; Milinkovitch and Thewissen, 1997; Thewissen and Hussain, 1993; Thewissen et al., 1994; Thewissen, 1994, 1998).

In fact, fossils of *Ambulocetus, Pakicetus, Rhodocetus,* and several other genera provide remarkable examples of a major evolutionary transition and even offer evidence (from isotope studies) for changes in osmoregulatory abilities as the animals moved from fresh to salt water (Thewissen et al., 1996; Roe et al., 1998). Interestingly, these Eocene whales originated in shallow seas that were the remnants of the ancient Tethys Sea, which previously had separated the northern supercontinent Laurasia and the southern supercontinent Gondwana (including India, which at about the time was beginning its slow collision with Asia; see Fig. 25-5). Some archaeocetes attained large size. *Basilosaurus,* a late Eocene type from Egypt, was elongate and slender. Its body was roughly 17 me-

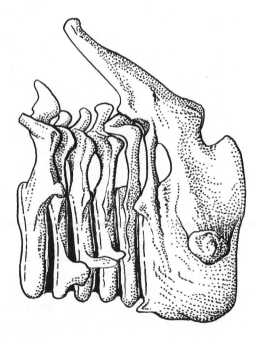

FIGURE 13-2   Cervical vertebrae of a dolphin (*Delphinus delphis,* Delphinidae). Only the axis and atlas are fused, whereas most of the series are fused in some cetaceans.

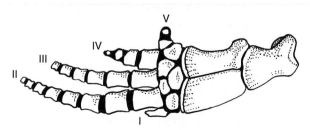

FIGURE 13-3   Dorsal view of the right forelimb of the bottlenose dolphin (*Tursiops truncatus,* Delphinidae).

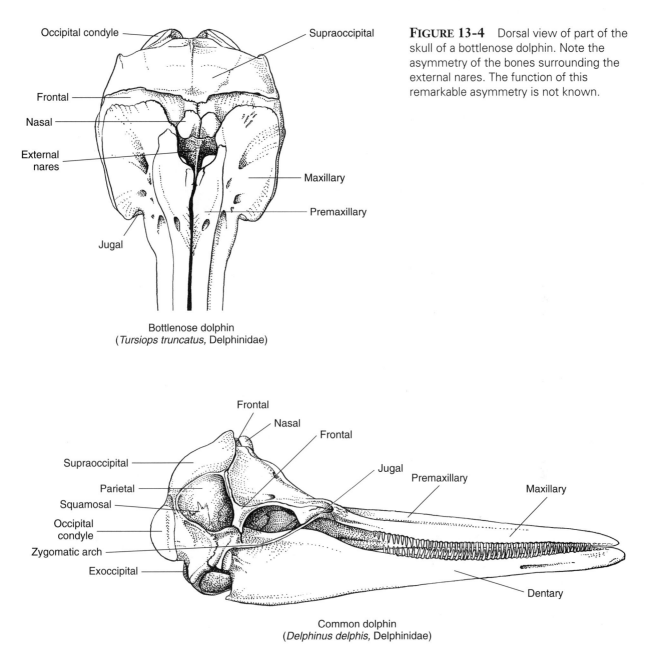

Occipital condyle
Supraoccipital
Frontal
Nasal
External
nares
Jugal
Maxillary
Premaxillary

Bottlenose dolphin
(*Tursiops truncatus*, Delphinidae)

FIGURE 13-4   Dorsal view of part of the skull of a bottlenose dolphin. Note the asymmetry of the bones surrounding the external nares. The function of this remarkable asymmetry is not known.

Frontal
Nasal
Frontal
Supraoccipital
Jugal
Premaxillary
Maxillary
Parietal
Squamosal
Occipital
condyle
Zygomatic arch
Exoccipital
Dentary

Common dolphin
(*Delphinus delphis*, Delphinidae)

FIGURE 13-5   Skull of a common dolphin. Note the highly telescoped skull with the maxillary and frontal bones roofing the small temporal fossa. The frontal bone is barely exposed on the skull roof. The length of the skull is 475 millimeters.

ters long with a skull 1.5 meters long. In primitive cetaceans (Fig. 13-7), the external nares had not migrated so far toward the back of the skull as they have in modern cetaceans.

Although it may seem counterintuitive that a cow is more closely related to a whale than to a horse, evidence from molecular, morphological, and paleontological studies strongly agree that cetaceans and artiodactyls had a common ancestry (Berta, 1994; Milinkovitch and Thewissen, 1997; Montgelard et al., 1997; Shimamura et al., 1997; Thewissen, 1994). Cetaceans in general appear to

FIGURE 13-6   Ventral view of part of the skull of the grampus, showing the large air sinuses (stippled) anterior to the tympanoperiotic bone and partly surrounding it. *(After Purves, 1966)*

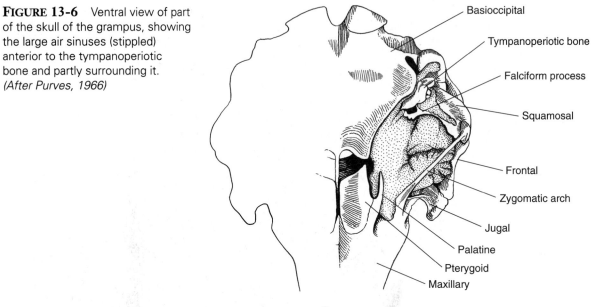

Basioccipital

Tympanoperiotic bone

Falciform process

Squamosal

Frontal

Zygomatic arch

Jugal

Palatine

Pterygoid

Maxillary

Grampus
(*Grampus griseus*, Delphinidae)

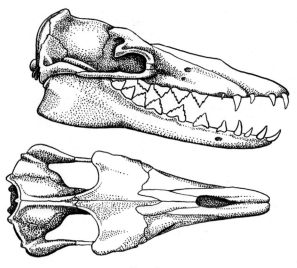

Dorudon

FIGURE 13-7   Skull of a primitive cetacean *(Dorudon)*, a fossil archaeocete from the Eocene. Length of skull approximately 600 millimeters. Note the lack of telescoping of the skull and the heterodont dentition. *(After Romer, ©1966, Vertebrate Paleontology, University of Chicago Press)*

form a monophyletic group, but considerable debate remains over the relationships within the order; controversies over the phylogeny of odontocetes have been especially intense (Árnason and Gullberg, 1994; Cerchio and Tucker, 1998; Messenger and McGuire, 1998; Milinkovitch et al., 1993; Milinkovitch and Thewissen, 1997).

Fordyce and Barnes (1994) gave a concise history of important events in cetacean evolution (Fig. 13-8). The earliest records of the suborder Mysticeti, the baleen whales, are from the middle Eocene record of Antarctica. Late Oligocene species from the North Pacific and North Atlantic were roughly intermediate between primitive toothed forms and true baleen whales (Barnes, 1976; Barnes and Sanders, 1996; Emlong, 1966). The suborder Odontoceti, the toothed whales, appeared in the Oligocene record of Australia, Europe, and New Zealand. By the Miocene, cetaceans had undergone considerable radiation (Fig. 13-9); over half of the known cetacean genera appeared in this epoch. Advanced odontocetes with highly telescoped skulls, homodont dentitions, and many

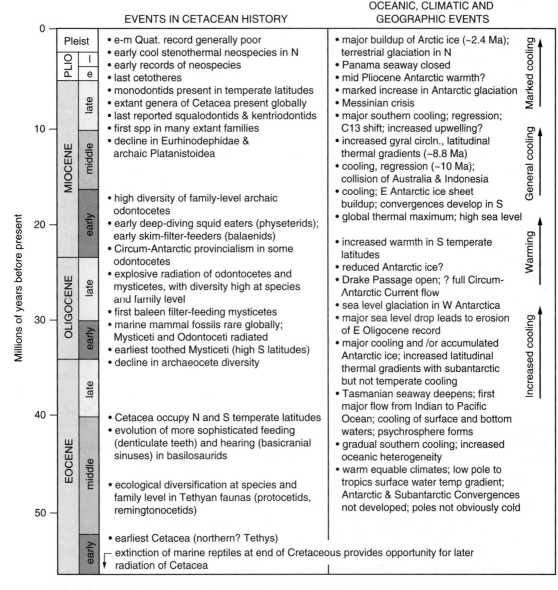

EVENTS IN CETACEAN HISTORY

OCEANIC, CLIMATIC AND
GEOGRAPHIC EVENTS

FIGURE 13-8   Major climatic events and the evolution of cetaceans beginning in the early Eocene. *(From Fordyce and Barnes, with permission from the* Annual Reviews of Earth and Planetary Sciences, *Vol. 22, ©1994, by Annual Reviews)*

more teeth than the primitive eutherian complement, are known from the Miocene (Fig. 13-10). During the Pliocene, a bizarre odontocete related to the narwhals and belugas evolved a skull morphology convergent with that of walruses (Muizon, 1993).

CETACEAN ADAPTATIONS

The vulnerable point of cetaceans is their need to breathe atmospheric air, but, unlike most other mammals, many cetaceans are able to alternate between periods of **eupnea** (normal breathing) and

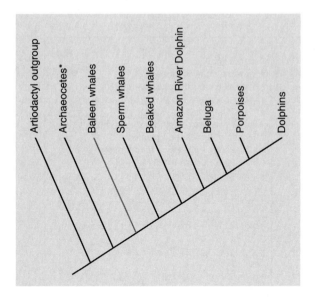

FIGURE 13-9 Phylogeny of the Cetacea based on the combined analysis of morphological and DNA sequence characters described in Messenger and McGuire (1998). Asterisk denotes the fossil Archaeocete whales for which no DNA data were available.

long periods of **apnea** (cessation of breathing). Some whales, such as sperm whales *(Physeter catodon)* remain submerged for over 70 minutes. Most delphinids can hold their breath 4 to 5 minutes, but they often surface to breathe several times a minute. The ability of cetaceans to remain active during periods of apnea probably depends on many adaptations. Rapid gas exchange is enhanced by two layers of capillaries in the interalveolar septa. During expiration, most of the air can be exhausted from the lungs, and up to 12 percent of the oxygen from inhaled air is utilized (the corresponding figure for terrestrial mammals is only 4 percent). Cetaceans have at least twice as many erythrocytes per volume of blood as terrestrial mammals do and about two to nine times as much myoglobin (a molecule able to store oxygen and release it to tissue) in the muscles. During deep dives, the heart rate drops to roughly half the surface rate and vascular specializations allow blood to bypass certain muscle masses, while maintaining flow to the brain. Important physiological adaptations to prolonged submersion include tolerance to high levels of lactic acid and a relative insensi-

tivity to carbon dioxide. Good discussions of cetacean adaptations to deep diving are given by Elsner (1969), Irving (1966), Kooyman and Anderson (1969), and Lenfant (1969).

Most small odontocetes are seemingly shallow divers, but some large odontocetes and some mysticetes can perform deep dives (Table 13-1). During deep dives, cetaceans are subjected to tremendous pressures, for with every 10-meter increase in depth an additional 1 atmosphere of pressure is exerted on the body. A cetacean swimming at a depth of only 200 meters, probably a common depth for many species, is subjected to 20 atmospheres of pressure, or 294 pounds per square inch. Any gases that remain within the body cavities are therefore subjected to great pressure, resulting in a decrease in their volume and an in-

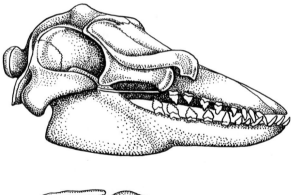

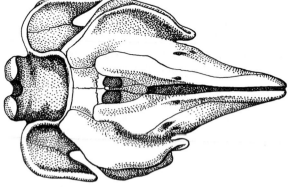

Prosqualodon

FIGURE 13-10 Skull of *Prosqualodon,* a fossil porpoise from the Miocene. Length of skull approximately 450 millimeters. The skull is highly telescoped, and the maxillary and frontal bones form a roof over the temporal fossa. *(After Romer, © 1966, Vertebrate Paleontology, University of Chicago Press)*

TABLE 13-1   Depths at Which Cetaceans Have Been Recorded

Species	Depth (m)	Method of Observation
Balaenopteridae	500	Harpooned and collided
Fin whale, *Balaenoptera physalus*		with bottom
	355	Depth manometer on harpoon
Physeteridae	900	Entangled in deep-sea cable
Sperm whale, *Physeter catodon*	1134	Entangled in deep-sea cable
	520	Echo-sounder
Delphinidae	30	Attached to depth recorder
Rough-toothed dolphin,		
Steno bredanensis		
	366	Inferred from feeding behavior
North Pacific pilot whale,		
Globicephala macrorhynchus		
Bottlenose dolphin,	92	Visual observations from
Tursiops truncatus		underwater craft
	185	Vocalizations near underwater craft
	170	Trained to activate buzzer

Data from Kooyman and Andersen (1969), who cite the sources of the observations.

crease in the amount of gases that dissolve in the body solvents, such as blood. One serious result of similar activities in humans is a condition known as "the bends," or decompression sickness. When humans use equipment that allows them to breathe underwater and undergo prolonged exposure to high pressures during diving, greater than normal amounts of gases are dissolved in the tissues and the blood. If decompression is too rapid, these gases cannot be carried to the lungs rapidly enough to be removed from the body; instead they quickly leave the solution and appear as bubbles in the tissues. Intravascular bubbles may occlude capillaries and result in injury to tissues or even death.

Cetaceans are not known to have these problems, probably because their lungs collapse in the first 100 meters of a dive and they have evolved the following anatomical specializations for deep diving: (1) A large proportion of the ribs lack an attachment to the sternum or other ribs, which allows rapid lung collapse. (2) The lungs are dor-

sally situated above the oblique diaphragm. (3) In deep divers, the lungs are small and the volume of the nonvascular air passages is relatively large. (4) The trachea is short and often of large diameter, and the cartilaginous rings bracing the trachea are nearly complete, have small intermittent breaks, or are fused (Slijper, 1962). (5) The bronchioles are reduced in length, and the entire system of bronchioles, to the very origin of the alveolar ducts, is braced by muscles and cartilaginous rings. (6) The lungs, especially the walls of the alveolar ducts and the septa, contain unusually high concentrations of elastic fibers. (7) Finally, in some of the small odontocetes, a series of myoelastic sphincters occur in the terminal sections of the bronchioles.

The adaptive importance of all these features is not yet completely understood. Clearly, the specializations of the ribs and the placement of the lungs relative to the diaphragm permit the lungs to collapse and air to move from space where gas exchange occurs to space where it does not occur.

Bracing the respiratory passages may facilitate alveolar collapse and rapid inhalation and exhalation at the surface. The alveoli fully collapse during deep dives, forcing air into respiratory passages.

Cetaceans are fast swimmers. Powerful dorsoventral movements of the tail provide propulsion, and the flippers are used for steering. Dolphins can swim up to 36 kilometers per hour, and speeds of 55 and 27 kilometers per hour have been reported for killer whales and a pilot whale, respectively (Lang, 1966). Gawn (1948) reports the huge blue whale's speed as 37 kilometers per hour for 10 minutes.

This remarkable swimming performance of cetaceans has proved difficult to explain. Recent studies have demonstrated that their speed is due not to muscles that are vastly more powerful than those of other mammals, but to specializations that greatly reduce resistance (drag) as the animals swim. The amount of resistance depends on the type of water flow over the body surface. If the flow is smooth, parallel to the surface, it is said to be laminar. When such smooth flow is interrupted by water movements that are not consistently parallel to the body surface, however, turbulent flow occurs. All other conditions being equal, laminar flow creates much less resistance than does turbulent flow. If the bodies of small dolphins were subjected to turbulent flow, swimming at 38 kilometers per hour would require their muscles to be five times as powerful as those of humans (Lang, 1966). Assuming flow to be nearly laminar, however, this speed is approximately that expected if their power output were that of a well-trained human athlete. Scientists for many years have attempted without success to design bodies shaped so that air or water flow over the surface is laminar or nearly so. What is the cetacean solution to this problem?

Several factors seemingly contribute importantly to reducing resistance as a porpoise swims rapidly (Hertel, 1969). The body is hairless, and no obstructions except the streamlined appendages break the extremely smooth surface. In addition, the body form of dolphins is approximately parabolic; this form creates even less resistance than the rounded (elliptical) head end and tapered body of such rapid swimmers as a trout. In addition to being streamlined, dolphins have a compliant spongy layer in their outer skin that acts to reduce local pressure fluctuations that cause

turbulence, and therefore reduces drag (McNeill Alexander, 1968). Body size also plays an important role in swimming speed. Power output is directly proportional to muscle mass, and muscle mass generally increases with overall body mass. Simply stated, larger animals generate more power. At the same time, drag as a function of body mass decreases with body mass (at a given speed). In other words, the most important components of drag, surface area and cross-sectional area, increase with the square of a linear dimension, such as body length, while body mass, which determines power output, increases with the cube of body length. Thus, a larger aquatic animal will always be able to swim faster than a smaller animal of identical body shape (Eckert et al., 1988).

Large body size is also important in thermoregulation. Many cetaceans spend considerable time in polar waters, where water temperatures often approach −2°C (the freezing point of sea water). Smaller cetaceans use a great deal of metabolic energy to maintain a body temperature of approximately 35°C. Heat loss is minimized by retia mirabilia, a system of countercurrent heat exchangers in the flippers and tail flukes. In addition, a thick layer of blubber under the skin acts as a layer of insulation. Larger whales have a low surface area to volume ratio and, because of this, have little problem keeping warm in polar seas. Gray whales, which filter feed in near-freezing Arctic waters using their huge (1.6 meter long), uninsulated tongues, even have a countercurrent heat exchanger in the tongue to conserve heat (Heyning and Mead, 1997).

A poorly understood adaptation is the ability of cetaceans to use the Earth's geomagnetic fields for navigation. Evidence from geomagnetic anomaly maps and records of cetacean strandings point to the use of some sort of geomagnetic detection system in whales. Klinowska (1990) and Kirschvink (1990) hypothesize that cetaceans use the flux density of the earth's magnetic field as a map. The evidence comes from the fact that cetaceans generally migrate parallel to the geomagnetic fields along the sea bottom. Furthermore, this geomagnetic field exhibits regular fluctuations that allow cetaceans to mark their progress along the geomagnetic map. Although they do not use the geomagnetic information like a compass, cetaceans can apparently use the information for local and long-distance migrations. Further support for this theory comes from data on strandings of live whales along coastal beaches. Dates of mass strandings coincide with unusually large fluctuations in the normal geomagnetic flux (Kirschvink et al., 1986; Klinowska, 1985a,b), suggesting that whales blunder onto beaches while trying to navigate along contours of geomagnetic minima (Kirschvink, 1990; Walker et al., 1986; Walker et al., 1992).

SUBORDER MYSTICETI

Suborder Mysticeti contains the huge baleen whales, which inhabit all oceans. There are 11 Recent species, grouped in six genera and four families. Before intensive whaling decimated their populations, the mysticetes were perhaps more important than the

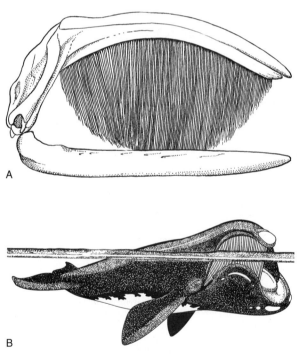

Right whale
(*Eubalaena glacialis*, Balaenidae)

FIGURE 13-11 (A) Skull of the Atlantic right whale; length of skull roughly 4 meters. Note the baleen plates attached to the maxilla. (B) Right whale feeding on plankton at the surface of the sea. *(B reprinted from Pivorunas, 1979, vol. 67, by permission of American Scientist and Sigma Xi, The Scientific Research Society)*

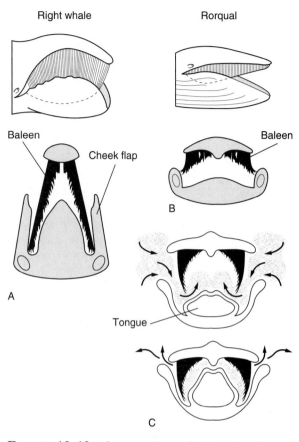

FIGURE 13-12 Cross sections of the heads of two kinds of whales, showing contrasting arrangements of baleen: (A) right whale; (B) rorqual. (C) The pattern of water flow into and out of the mouth during feeding. *(A and B reprinted from Pivorunas, 1979, vol. 67, by permission of American Scientist and Sigma Xi, The Scientific Research Society)*

odontocetes in terms of biomass, although there are fewer species of mysticetes.

Although all baleen whales are filter feeders, they utilize three rather distinct feeding styles (Pivorunas, 1979). The first is typical of the right whales (Balaenidae). These are large-headed animals with long **baleen plates** and conspicuous lips (cheek flaps; Fig. 13-11). Right whales generally graze on small plankton, usually copepods less than 1 centimeter in length, that concentrate in layers at or near the water surface. The whales swim slowly through these concentrations; the water and plankton flow in the front of the mouth, and the water passes along the side of the tongue and out through the baleen, which traps the plankton (Fig. 13-12). The trapped plankton is then removed from the baleen plates by the action of the tongue. This sifting of the water can involve skimming a surface layer of plankton or moving through plankton swarms at greater depths.

The second method is that of the rorquals (Balaenopteridae). These whales have huge mouths and heads, relatively short baleen (Fig. 13-12B), and extensive furrowing of the blubber of the throat and anterior abdomen that forms a pouch that is highly distensible (Lambertsen et al., 1995; Fig. 13-13). Rorquals engulf food that occurs in dense swarms, usually krill or fish. One of the rorquals, the blue whale *(Balaenoptera musculus)*, has been observed to swim up to its prey and engulf it, along with huge amounts of water (up to 6400 kilograms). The food and water are contained briefly in the capacious ventral pouch. The pouch is then contracted, the water passes through

the baleen, and the food trapped by the baleen is swallowed. During feeding, the mandible is lowered until it forms a 90-degree angle with the plane of movement through the water. Recently, Lamertsen and coworkers (1995) described a frontomandibular stay apparatus consisting of a strong fibrous tendon that prevents hyperextension of the mandible as the whale moves through the water (Fig. 13-14). A number of variations on the basic feeding pattern are used. On occasion, for example, the humpback whale *(Megaptera novaeangliae)* blows a ring of bubbles, a bubble "net," near the surface and, by rising upward within the net, engulfs animals that hesitate to pass through the net and are trapped at the surface (Fig. 13-15).

The third feeding style is that of the gray whale *(Eschrichtius robustus)*, which feeds by scooping or suctioning material from the bottom and filtering out bottom-dwelling organisms. Gray whales dive to the bottom and, turning on their sides, plow

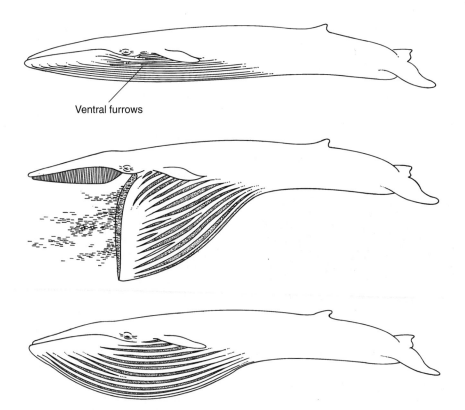

Ventral furrows

FIGURE 13-13   The style of feeding of rorquals and the use of the furrowed, expandable pouch. *(Reprinted from Pivorunas, 1979, vol. 67, by permission of American Scientist and Sigma Xi, The Scientific Research Society)*

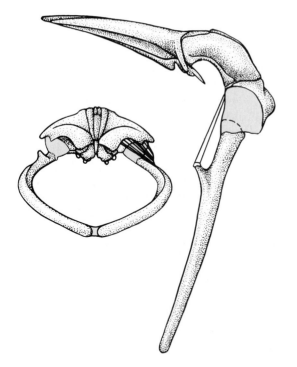

FIGURE 13-14   Drawing of the frontomandibular stay apparatus of Balaenopteridae. The stay apparatus is a fibrous part of the temporalis muscle that mechanically links the frontal bone of the skull with the coronoid process of the lower jaw, thereby limiting hyperdepression of the lower jaw during feeding. Lateral view of the skull shows the maximum gape. Anterior view of the skull shows a moderate gape angle. In both figures the shaded areas represent connective tissues covering the temporomandibular joint and the mandibular symphysis. (From Lambertsen et al., 1995)

FAMILY BALAENIDAE. Family Balaenidae, the right whales (Fig. 13-11B), were killed in such great numbers during the height of whaling activities that they are rare today and are protected by international treaty. The family includes three Recent species of two genera that inhabit most marine waters except tropical and south polar seas. Balaenids are known as fossils from the early Miocene of Argentina to the Recent (McLeod et al., 1993).

These are large, robust whales that reach about 18 meters in length and over 67,000 kilograms in weight. The head and tongue are huge; the head amounts to nearly one third of the total length. There are more than 350 long baleen plates on each side of the upper jaw; these plates fold on the floor of the mouth when the jaws are closed. No furrows are present on the skin of the throat or

their open mouths through the muddy sediments, leaving shallow trenches in the bottom. They filter out the crustaceans, polychaete worms, and other bottom-dwelling organisms on their baleen plates (MacDonald, 1984).

FIGURE 13-15   Two humpback whales (*Megaptera novaeangliae*, Balaenopteridae) breaching the surface during feeding. Note that a flipper is visible in the foreground. (S. Mizroch, NOAA, NMFS, National Marine Mammal Laboratory)

chest. The cervical vertebrae are fused, and the skull is telescoped to the extent that the nasal bones are small and the frontal bones are barely exposed on the top of the skull. The rostrum is arched to accommodate the long baleen plates (Fig. 13-11). The flippers are short and rounded, and the dorsal fin is usually absent.

Right whales feed largely on copepods and are most common near coastlines or near pack ice. The bowhead whale *(Balaena mysticetus)* never ventures far south of the Arctic Circle (Burns et al., 1993). Eskimos claim it can break through ice nearly 1 meter thick to reach air (Rice, 1984). The southern right whale *(Eubalaena australis)* of the southern oceans makes long annual migrations from temperate or tropical waters to spend the austral summer in Antarctic waters. The northern right whale *(E. glacialis)* remains in North Atlantic or North Pacific waters.

FAMILY BALAENOPTERIDAE. The family Balaenopteridae, usually known as rorquals, includes six Recent species and two genera. The fossil record of this family dates back to the middle Miocene of the western South Pacific (Bearlin, 1988). The distribution includes all oceans. These whales vary in size from the fairly small (for whales) Minke whale *(Balaenoptera acutorostrata)*, at 11 meters, to the extremely large, 31-meter and 160,000-kilogram, blue whale *(B. musculus)*. In some species of rorquals, the body is slender and streamlined, but it is chunky in others. The baleen plates are short and broad, and the skin of the throat and chest is marked by numerous longitudinal furrows (Figs. 13-13). The nasals are small, and the frontals are either not exposed or only barely exposed on the skull roof.

Some of these whales feed in cold waters near the edges of the ice where up-welling, nutrient-rich water results in great growths of plankton in summer. Planktonic crustaceans and small schooling fish are eaten (Tershy, 1992). During the northern winter, the Northern Hemisphere populations move southward toward the equatorial areas, and during the southern winter, southern populations move northward. Wintering adults do not feed but live off stored blubber. Breeding occurs in the wintering areas, but because the southern and northern winters are six months out of phase, no interbreeding between populations occurs. The humpback whale, an animal given to spectacular leaps (Fig. 13-16), makes remarkably melodious and varied underwater sounds that have been recorded and discussed by Payne (1970) and Payne and McVay (1971).

FIGURE 13-16 Humpback whale (*Megaptera novaeangliae;* Balaenopteridae) breaching. *(W. Lawton, NOAA, NMFS, National Marine Mammal Laboratory)*

Excessive commercial exploitation has resulted in a tremendous decline in the populations of such rorquals as fin whales *(B. physalus)*, humpbacks, and blue whales. Over broad areas, blue and humpback whales are so scarce as to be "commercially extinct" (Rice, 1984), and some species of smaller rorquals are also in decline due to overhunting. In the 120 years prior to their protection, it has been estimated that 338,000 blue whales were killed (Baskin, 1993); at up to 175 tons each, this equals 59 million tons of blue whales! Although blue whales are possibly beginning to increase since being legally protected in 1968, it has been difficult to estimate their numbers, a problem exacerbated by the intentional cover-up and severe under-reporting of numbers taken by some whaling countries (Yablokov, 1994).

FAMILY ESCHRICHTIIDAE. Family Eschrichtiidae is represented today only by the gray whale *(Eschrichtius robustus)*, which occupies parts of the North Pacific. The fossil record for this species dates back only as far as the late Pleistocene (McLeod et al., 1993).

The gray whale is fairly large, weighing up to 31,500 kilograms and measuring 15 meters in length, and has a slender body with no dorsal fin. It has a relatively small head, with a narrow, arched rostrum. The baleen plates are short, and the telescoping of the skull is not extreme. The nasal bones are large, and the frontals are broadly visible on the roof of the skull. The throat usually has two longitudinal furrows in the skin.

Gray whales migrate extremely long distances but probably no farther than blue whales. The round trip distance of the migration is from 10,000 to 22,000 kilometers. Gray whales occupy parts of the North Pacific (the Bering, Chukchi, and Okhotsk seas) in the summer. Here they feed largely on bottom-dwelling crustaceans (amphipods), which they take by stirring up the sediments with their snouts. In late autumn, they migrate southward along the coastlines. The western Pacific population winters along the coast of Korea, and the eastern Pacific gray whales winter along the coast of Baja California and Sonora. Young are born in shallow coastal lagoons in the wintering areas. Many people each year watch migrating gray whales from a vantage point at Cabrillo National Monument, near San Diego, and from numerous whale-watching vessels.

The future of the gray whale today seems reasonably bright. Driven nearly to extinction by whaling activities between 1850 and 1925, they are now protected by the International Convention for the Regulation of Whaling, and their numbers have increased greatly in recent years.

FAMILY NEOBALAENIDAE. Family Neobalaenidae consists of one living species, the pygmy right whale *(Caperea marginata)*, which inhabits the marine waters of the Southern Hemisphere. It is unusual in that it is small, usually between 5 to 6 meters in length, and has more ribs (34) than any other whale. It is also slender, like some rorquals, has narrow flippers, and a small dorsal fin. Pygmy right whales feed mainly on copepods. These whales are not thought to make long-distance migrations but may instead form small, localized populations. Unlike the true right whales, pygmy right whales do not breach or "lob-tail" (slap the water surface with the tail fluke). Little is known about these whales, but they are thought to swim and feed primarily near the surface (Mitchell, 1975; Ross et al., 1975).

SUBORDER ODONTOCETI

The odontocetes — the toothed whales, porpoises, and dolphins — form the largest suborder of cetaceans in terms of abundance and species diversity, and they are the most widely distributed. Odontocetes include 67 Recent species within 35 genera and six families; they occur in all oceans and seas connected to oceans. Some members of three families inhabit some rivers and lakes in North America, South America, Asia, and Africa. Odontocetes are readily observed; they frequently forage close to shore, often make spectacular leaps, and roll repeatedly out of the water. Some ride the bow waves of ships much as humans ride shore waves.

Just as the mysticetes are characterized by a unique filter-feeding mode of life, odontocetes are characterized by their ability to echolocate. An echolocating animal emits acoustic signals and detects objects by receiving and interpreting echoes of these signals. Much has been learned about cetacean echolocation since the original discovery by McBride in 1947 that dolphins *(Tursiops truncatus)* could avoid nets made invisible by turbid water (Au, 1993; McBride, 1956). Odontocetes emit a great variety of sounds, broadly grouped into two categories: narrow-band continuous tones, such as whistles and squeaks used for intraspecific communication, and broad-band clicks used for echolocation (Au, 1993; Herman and Tavolga, 1980).

In addition to the use of sound for prey detection, odontocetes may acoustically stun their prey. Bel'kovich and Yablokov (1963) were the first to suggest that an intense, high-frequency sound of the sort that could be made by even a small dolphin could produce a shock sufficient to stun prey. Berzin (1971), who carefully studied sperm whales, was impressed by the fact that large squid and sharks found in the stomachs of sperm whales bore no teeth marks and that even sperm whales with deformed or injured jaws, incapable of grasping prey, had normally full stomachs. Berzin proposed that sperm whales acoustically stun their prey and that this reduced the importance of jaws for feeding. More recently, Hult (1982) observed disorientation among schooling fish and hypothesized that this resulted from the high-intensity sounds made by approaching bottlenose dolphins.

Primitive cetaceans, from the early Eocene, generally had laterally compressed, serrated teeth that probably functioned, as do the similar teeth of the crab-eater seal (see Fig. 12-32) today, to allow water to escape but to trap small prey when the jaws are closed. As odontocetes radiated in the Miocene, these serrated teeth were progressively replaced by many simple conical teeth borne on slim, elongate jaws. Such pincer-type jaws probably enabled Miocene odontocetes to capture small, underwater prey by quick snaps and thrusts of the jaws. Many living odontocetes retain such teeth and beaklike jaws. In living odontocetes, engulfment of prey is caused by a rapid retraction of the piston-like tongue, with the resultant rush of water and prey into the mouth. In many living odontocetes, the beak pincers are lost or reduced, perhaps in association with increasing perfection of the ability to stun prey acoustically.

Two additional important and obvious evolutionary trends in odontocetes may also be related to this ability. The first is a trend among advanced odontocetes (such as the bottlenose dolphin) toward the focusing of sound energy into a concentrated beam. Such a specialization furthers long-range echolocation ability, and prey debilitation could be its byproduct. A second suggestive trend involves the loss of teeth, a reduction in their size, or the development of specialized teeth unsuitable for capturing prey (such as the long narwhal tusk). A specialized means of debilitating prey would open the door to such dental changes.

Experimental trials with trained bottlenose dolphins showed that they can produce sounds of intensities equal to the lethal threshold of some marine fish and close to those that have been observed to kill moderately large squid. These intense sounds were not harmful to the dolphins but were some

Figure 13-17   A pair of killer whales (*Orcinus orca;* Delphinidae) swimming in Puget Sound, Washington. *(S. Mizroch, NOAA, NMFS, National Marine Mammal Laboratory)*

five orders of magnitude above levels usually reported for wild or captive dolphins.

Although acoustic debilitation of prey by odontocetes has not been proved, it seems probable and could have evolved by a series of entirely plausible steps. Highly specialized sound-producing structures evolved in response to sociality and the importance of communication. Sound production later became used for echolocating obstacles and prey. With an increasing ability to produce intense and focused sounds for long-range echolocation, a remarkable new ability appeared: prey subjected to these sounds could be stunned. This innovation would free the jaws and dentition from many selective pressures associated with killing or grappling with prey and could allow the shortening of beaks, reduction of dentition, or development of teeth primarily adapted to ritualized combat, as in the case of the male narwhal.

FAMILY DELPHINIDAE. Family Delphinidae, the dolphins, is by far the largest and most diverse group of cetaceans. Because some species come close to shore and roll and jump conspicuously, they are the most frequently observed cetaceans. The word "dolphin" is commonly applied to any small cetacean having a beaklike snout, whereas the term "porpoise" generally refers to small cetaceans with a blunt snout and a less streamlined body form. These terms can be misleading, however, as not all delphinids have a beaklike snout. Thirty-two Recent species representing 17 genera are known. Delphinids inhabit all oceans and some large rivers and estuaries in southern Asia, Africa, and South America. Fossil delphinids appear first in the Oligocene record of Europe.

Small delphinids are roughly 1.5 meters in length and 50 kilograms in weight, but the killer whale *(Orcinus orca)* reaches 9.5 meters and at least 7000 kilograms (Fig. 13-17). The facial depression of the skull is large, and the frontal and maxillary bones roof over the reduced temporal fossa (Fig. 13-5).

The "melon," a lens-shaped fatty deposit that lies in the facial depression, is well developed and gives many delphinids a forehead that bulges prominently behind the beaklike snout. Some delphinids, such as the killer whale, lack a beak and have a rounded profile. The number of teeth varies from 65/58 to 0/2. From two to six cervical vertebrae are fused (Fig. 13-2). Males are typically

FIGURE 13-18 Upper body of a beluga *(Delphinapterus leucas;* Monodontidae). *(R. Angliss, NOAA, NMFS, National Marine Mammal Laboratory).*

larger than females, and in some species there is considerable sexual dimorphism in the shape of the flippers and dorsal fin. Coloration is varied: some species are uniformly black or gray; some have striking patterns of black and white; and still others have yellowish or tan stripes or spots.

Delphinids characteristically feed by making shallow dives and surfacing several times a minute. They are rapid swimmers, and some species regularly leap from the water during feeding and traveling. In the Gulf of California, bottlenose dolphins have been observed leaping completely out of the water and catching mullet in midair. Pacific striped dolphins *(Lagenorhynchus obliquidens)* have been trained at Marineland of the Pacific to leap over a wire 4.8 meters above the water. Most small delphinids eat fish and squid, but the killer whale is known to take a great variety of items, including large bony fish, sharks, seabirds, sea otters, seals and sea lions, porpoises, dolphins, and whales.

Delphinids are typically highly gregarious, and assemblages of approximately 100,000 individuals have been observed. Some groups of delphinids kept in large tanks establish a dominance hierarchy, with an adult male having the highest position (Bateson and Gilbert, 1966). Most dolphins spend their entire lives in schools (Norris and Dohl, 1980; Pryor and Norris, 1991). Schooling behavior enhances the effectiveness of food searching, prey capture, and predator avoidance and may increase reproductive synchrony and efficiency. Dolphins have highly evolved systems of communication and social behavior (Herman and Tavolga, 1980). They display apparently **altruistic behavior** and, apart from humans and gorillas, are the only other animals known to assist a member of another species in distress. Repeated observations have been made of one species of delphinid supporting at the surface a distressed individual of another species. Many recent studies have indicated that cetaceans are remarkably intelligent, inventive, and capable of "higher-order" learning (Pack and Herman, 1995). Herman (1980) compares the learning ability of odontocetes with that of primates: "Both taxa have advanced capabilities for classifying, remembering, and discovering relationships among events. . . ." In cetaceans, these capabilities are based on auditory perception, whereas primates use mostly visual perception. Remarkably inventive behavior was observed by Hoese (1971), who watched two bottlenose dolphins cooperatively pushing waves onto a muddy shore and stranding small fish. The dolphins rushed up the bank, snatched the fish from the mud, and then slid back into the water.

FAMILY MONODONTIDAE. Family Monodontidae contains two species: the narwhal *(Monodon monoceros),* remarkable for its long, straight, forward-directed tusk; and the beluga *(Delphinapterus leucas),* also called the white whale (Fig. 13-18). These two species occur in the Arctic Ocean and the Bering and Okhotsk seas, in Hudson Bay, in the St. Lawrence River in Canada, and in some large rivers in Siberia and Alaska. The earliest monodontids are from the early Miocene of Europe.

These are small to medium-sized cetaceans; belugas reach about 6 meters in length and 2000 kilograms in weight, and narwhals, without the tusk, are of similar length. In both species, the facial depression in the skull is large and the maxillary and frontal bones roof over the reduced temporal fossa; the zygomatic process of the squamosal bone is strongly reduced; and the cervical vertebrae are not fused. The beluga has 11/11 teeth, and the narwhal has 1/0. The single upper tooth of the narwhal (usually the left) forms a straight, spirally grooved tusk up to 2.7 meters long; the corresponding tooth in the other upper jaw is rudimentary.

The gregarious belugas and narwhals are characteristic of northern seas, where in winter they assemble in open-water areas. In summer, belugas move far up large rivers. They feed largely on fish, both benthic (bottom-dwelling) kinds and those that live at intermediate depths, and squid. Narwhals are seemingly largely pelagic (open-sea dwellers). Male narwhals fence with their long tusks, and a tusk occasionally becomes imbedded and broken off in the head of one of the combatants (Silverman and Dunbar, 1980). Both belugas and narwhals are quite vocal, and the trilling sounds made by belugas account for their common name of "sea canary."

FAMILY PHOCOENIDAE. Members of family Phocoenidae are generally called porpoises. Six Recent species of four genera are recognized. They occur widely in coastal waters of all oceans and connected seas of the Northern Hemisphere, as well as in some coastal waters of South America and some rivers in southeastern Asia. The earliest fossil records of phocoenids are from the late Miocene of North America and Asia.

Phocoenids are small, from about 1.5 to 2.1 meters in length and 90 to 118 kilograms in weight, and have fairly short jaws and no beak. The dorsal fin is either low or absent. The skull resembles that of the Delphinidae but has conspicuous prominences anterior to the nares. The teeth of most phocoenids are distinctive in being laterally compressed and spadelike; the crowns have two or three weakly developed cusps. *Phocoenoides,* however, has conical teeth. The number of teeth varies from 15/15 to 30/30. From three to seven cervical vertebrae are fused.

Some phocoenids (*Phocoena* and *Neophocaena*) inhabit inshore waters, such as bays and estuaries, whereas the swift white-flank porpoises (*Phocoenoides*) generally inhabit deeper water. Schools of at least 100 phocoenids may assemble, and crescentic formations associated with feeding have been noted (Fink, 1959). A variety of food is taken, including cuttlefish and squid, crustaceans, and fish.

FAMILY PLATANISTIDAE. Family Platanistidae, the long-snouted river dolphins, is remarkable because its members live largely in rivers. The distribution includes some large river systems in India and Pakistan (*Platanista gangetica* and *P. minor*), the Amazon and Orinoco river systems of South America (*Pontoporia blainvillei* and *Inia geoffrensis*), and some large rivers in China *(Lipotes vexillifer)*. In the past, each of these genera has been placed in its own family (Rice, 1984). The fossil record dates back to Miocene times.

These are small cetaceans, from 1.5 to 2.9 meters in length and 40 to 125 kilograms in weight. The jaws are unusually long and narrow and bear numerous teeth (from about 26/26 to 61/61); the forehead rises abruptly and is rounded, giving the head an almost birdlike aspect. In *Platanista,* there are large crests that are probably of sesamoid origin. The large temporal fossa is not roofed by the maxillary and frontal bones. None of the cervical vertebrae are fused. The eyes of all members of the family are reduced, and presumably food and obstacles are detected by echolocation. *Platanista,* which usually swims on its side, lacks eye lenses and can perhaps detect only light and dark. The eyes of the Chinese river dolphin *(Lipotes)* are greatly reduced, and vision is presumably poor. The eyes of the other river dolphins are small but presumably functional.

These cetaceans often inhabit rivers that are made nearly opaque by suspended sediments. Under these conditions echolocation may completely supplant vision. A variety of fish and crustaceans are eaten, some of which are captured by probing muddy river bottoms. The Amazon dolphins *(Inia)* feed largely on fish and, during the rainy seasons, may move deep into flooded tropical forests (Humbolt and Bonpland, 1852). River dolphins are seemingly not as social as many other cetaceans. Only 14 percent of Layne's (1958) observations of Amazon dolphins were of groups containing more than four individuals. Layne made observations that suggest fairly acute vision above water. Individuals approaching a narrow channel used their eyes above water, presumably to scan the banks for danger.

FAMILY PHYSETERIDAE. The sperm whales occur in all but Arctic oceans, and the giant sperm whale *(Physeter catodon)* of Moby Dick fame has long been an important species to the whaling industry. There are two genera and three species of living sperm whales, and numerous fossil forms are known from as far back as the early Miocene.

Physeter is large, attaining a length of over 18 meters and a weight in excess of 53,000 kilograms. The pygmy sperm whales *(Kogia)* are relatively small, reaching about 4 meters in length and 320 kilograms in weight. The head is huge in *Physeter,* accounting for about one third of the total length. In both genera, the rostrum is truncate, broad, and flat. The facial region of *Physeter* contains a massive **spermaceti organ,** which contains great quantities of oil. This organ plays an important role in thermoregulation. The blowhole is toward the end of the left side of the snout, and the nasal passages are highly specialized to exchange heat. The upper jaw (except in occasional individuals) lacks functional teeth; the lower jaw has 25 functional teeth on each side in *Physeter* and from 8 to 16 in *Kogia.* All of the cervicals are fused in *Kogia,* and all but the atlas are fused in *Physeter.*

The habits of *Kogia* are not well known, but those of *Physeter* are better understood, probably because humans have persistently hunted this animal for many years. *Physeter* is social and assembles in groups with occasionally as many as 1000 individuals. Schools of females with their calves, together with male and female subadults, are overseen by one or more large adult males, whereas younger males congregate in "bachelor schools." Some adult males are solitary. Sperm whales generally forage in the open sea at depths where little or no light penetrates (the use of echolocation by *Physeter* is discussed on p.425). Dives to depths of 1000 meters are probably usual, and dives of 1130 meters have been recorded (Heezen, 1957). *Physeter* feeds largely on deep-water squid, including giant squid; sharks and skates; and such bony fish as tuna and barracuda. Males commonly migrate north in summer and are occasionally seen in the Bering Sea, but females remain in temperate and tropical waters. *Kogia* occurs in small schools and feeds largely on cephalopods such as squid and cuttlefish. Giant sperm whales were one of the most important species to the whaling industry until the early 1980s, when hunting of these animals was prohibited by the International Whaling Commission.

FAMILY ZIPHIIDAE. The beaked whales are widely distributed — they occupy all oceans — but are rather poorly known; some species have never been seen alive. Nineteen Recent species of six genera are recognized. The earliest fossil record of ziphiids is from the early Miocene of Australia.

These are medium-sized cetaceans with fairly slender bodies. The length varies from 4 to over 12 meters, and the weight reaches 11,500 kilograms. The snout is usually long and narrow, and in some species the forehead bulges prominently. One species *(Tasmacetus shepherdi)* has a large number of teeth; in the others, the dentition is strongly reduced. Only two lower teeth on each side occur in the two species of *Berardius;* in all remaining ziphiids, there is only a single functional tooth, a lower one, on each side (Fig. 13-19). In some species, the lower jaw extends beyond the upper jaw and the teeth are visible outside the mouth. Two to seven cervical vertebrae are fused. The stomach is divided into 4 to 14 chambers, an unusual morphology for a mammal that eats largely squid and fish.

Beaked whales are deep divers able to remain submerged for long periods. The North Atlantic bottlenose whale *(Hyperoodon ampullatus)* can dive for periods well over 1 hour. Some species forage in the open ocean far from land. Most beaked whales are highly social and travel in schools in which all members surface and dive in synchrony. The teeth in those species with reduced dentition

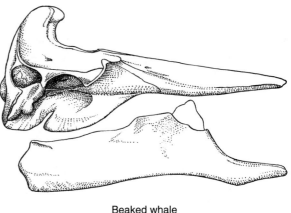

Beaked whale
(*Mesoplodon* sp., Ziphiidae)

FIGURE 13-19 Skull of a beaked whale (Ziphiidae). Note the single large tooth in the dentary bone. The length of the skull is about 590 millimeters.

may be used primarily during intraspecific social interactions and may be of little use during feeding. The primary food is squid, but deep-sea fish are also taken. The North Atlantic bottlenose whale is known to make annual migrations, and other species are probably migratory as well.

WEB SITES

The Dolphin Study Group — research about dolphin behavior

http://dsg.sbs.nus.edu.sg/

National Marine Fisheries Service — guide to the marine mammal protection act

http://kingfish.ssp.nmfs.gov/

Cousteau Society — information on marine environments

http://www.math.clemson.edu/~rsimms/cousteau/

The Cetacean Research Unit — research and educational information about whales

http://www.cetacean.org/

American Cetacean Society home page

http://www.redshift.com/~estarr/acs/

Cetacean Society International home page

http://elfi.com/csihome.html

A listserve newsgroup dedicated to marine mammals

listserv@uvvm.uvic.ca

Whale Net — educational resources about whales

http://whale.wheelock.edu/

International Whaling Commission home page

http://ourworld.compuserve.com/homepages/iwcoffice/

The Marine Mammal Center

http://www.tmmc.org/

14 PROBOSCIDEA, SIRENIA, AND HYRACOIDEA

If general appearance were used as the single criterion for evaluating phylogenetic relationships, the rodent-like hyraxes, massive elephants, and aquatic sirenians would be judged to be three very distantly related groups of mammals. In this case, however, appearances are deceptive, for the fossil record suggests that these groups evolved in Africa from a common ancestral stock related to the ungulates (hoofed mammals). Because of this presumed near-ungulate ancestry, hyraxes, elephants, and sirenians have long been called Paenungulates. Indeed, recent eutherian phylogenies support a clade containing sirenians and proboscideans, with hyracoids as their sister group (Novacek et al., 1988; Novacek, 1986; 1993). However, not all hypotheses agree on this point. Several recent phylogenies unite hyracoids with perissodactyls (Gaudin et al., 1996; Fischer and Tassy, 1993; Tassy and Shoshani, 1988).

ORDER PROBOSCIDEA: ELEPHANTS

Throughout much of the Cenozoic, some of the largest and most spectacular herbivores were proboscideans; in the late Tertiary Period, a varied array of these animals occurred widely in North America, Europe, and Africa (Fig. 14-1). The diversity of proboscideans was reduced in the Pleistocene Epoch, and today only two species represent this remarkable group. Because elephants now often threaten the interests of humans and because of the great value of their tusks (trade in elephant ivory is now restricted by international treaty), they are being extirpated over wide areas as human populations increase. Today elephants occur only in Africa south of the Sahara Desert *(Loxodonta)* and in parts of southeastern Asia *(Elephas)*. Regrettably, we may be witnessing the final stages in the history of one of the most interesting mammalian orders.

The fossil record of proboscideans begins in the late Paleocene or early Eocene of North Africa with *Phosphatherium escuilliei* and *Numidotherium* (Gheerbrant et al., 1996, Mahboubi et al., 1984). By the late Eocene, African proboscideans were reasonably diverse and already exhibited the typical columnar limbs of today's elephants. These Eocene forms are in the genus *Palaeomastodon* (Fig. 14-1; Carroll, 1988). Proboscideans expanded out of Africa in the Miocene, when Africa and Asia became contiguous, and apparently reached North America in the middle Miocene. These rather primitive proboscideans had brachydont teeth with few ridges; most or all the cheek teeth were in place at one time, and both an upper and a lower set of tusks usually were present (Fig. 14-2). The extant family Elephantidae appears in the early Miocene and is characterized by molar teeth that erupt and are replaced in sequence.

Both living species belong to the family Elephantidae, which is represented first by the Miocene African genera *Stegotetrabelodon* and *Stegodibelodon*. Mammoths *(Mammuthus)* evolved in Africa, radiated outward into Eurasia and North America, and at least three species of mammoth coexisted in North America with early human populations in the Pleistocene. One Pleistocene species *(Mammuthus primigenius)* was in some ways more specialized than the living elephants. It had a remarkably foreshortened skull (Fig. 14-3) and long, upwardly curved tusks that sometimes crossed; the last molar had up to 30 laminae, more than occur in the living elephants. Entire frozen woolly mammoths have been found in Siberia and Alaska, and many graceful drawings made by paleolithic peoples on the walls of caves depict these animals.

The two living proboscideans — the African elephant, *Loxodonta africana,* and the Indian elephant, *Elephas maximus* — are the largest land mammals, reaching weights of 6000 kilograms (Fig. 14-4). In Africa, there are thought to be two subspecies of elephant: the larger plains elephant *(L. africana*

FIGURE 14-1   One example of the proboscidean family tree from the Eocene to present. *(Courtesy of* Natural History, *based on artwork by U. Kikutani and G. Marchant)*

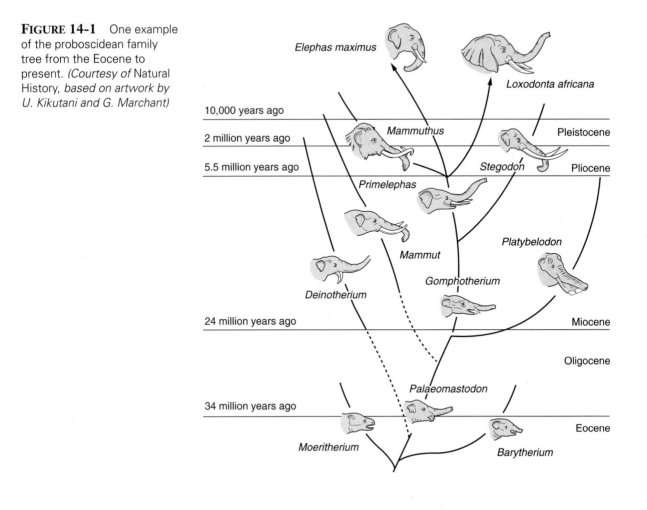

FIGURE 14-2   The skull of *Gomphotherium,* a Miocene proboscidean. Length of skull and tusks roughly 1 meter. *(After Romer, © 1966,* Vertebrate Paleontology, *University of Chicago Press)*

FIGURE 14-3   The skull of *Mammuthus,* a Pleistocene elephantid. Length of skull and tusks roughly 2.8 meters. *(After Romer, © 1966,* Vertebrate Paleontology, *University of Chicago Press)*

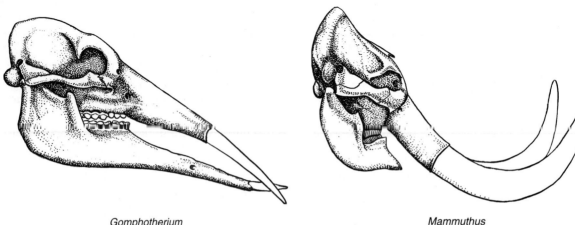

A   B

FIGURE 14-4   (A) A group of African elephants *(Loxodonta africana)* at a waterhole in Etosha National Park. (B) The Indian elephant *(Elephas maximus)* has smaller ears and a flatter forehead. *(A by T. Vaughan, and B by Corel Corp.)*

africana) and the smaller forest elephant *(L. africana cyclotis)*. In addition to its smaller size, the forest elephant also has five front nails and four on the foot, while the plains elephant has four in front and three in back. They have a long proboscis (trunk) with one or two finger-like structures at its tip, large ears, and **graviportal** limbs.

Asian and African elephants differ in that Asian elephants have much smaller ears, 19 pairs of ribs instead of 21 pairs, a flattened forehead, the top of the head is dome shaped and is the highest point on the animal, and there is a single finger-like process at the tip of the trunk. In African elephants, the shoulders are generally the highest point and there are two finger-like processes at the tip of the trunk.

In both species, the limb bones are heavy and the proximal segments of the limbs are relatively long; the ulna and fibula are unspecialized; and the bones of the five-toed manus and pes are short and robust with an unusual, spreading, digitigrade posture (Fig. 14-5). A heel pad of dense connective tissue braces the toes and largely supports the weight of the animal. An adaptation allowing the efficient support of great weight, the long axis of the pelvic girdle is nearly at right angles to the vertebral column, and the acetabulum faces ventrally. In addition, when the weight of the body is sup-

ported by the limbs, there is little angulation between limb segments; that is to say, each segment is roughly in line with other segments. The gait is unusual. As described by Howell (1944), an elephant "relies exclusively upon the walk or its more speedy equivalent, the running walk, which permits it to keep at least two feet always on the ground. Not only does the weight make it advisable that this be

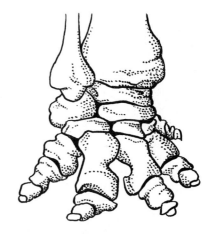

FIGURE 14-5   The right hind foot of *Mammut,* a late Tertiary and Pleistocene proboscidean. *(After Romer, © 1966,* Vertebrate Paleontology, *University of Chicago Press)*

distributed among each of the four feet when the animal is in motion, but the bulk doubtless requires that the equilibrial stresses be shifted as gradually as possible to each foot, rather than more abruptly, as in the trot or gallop."

The skull is usually short and high, perhaps in response to a need for great mechanical advantage for the muscles that attach to the lambdoidal crest and raise the front of the head and the tusks. The skull contains numerous large air cells, particularly in the cranial roof.

The highly specialized dentition consists of the tusks (each a second upper incisor) and six cheek teeth in each half of each jaw. The pattern of cheek tooth replacement is remarkable. The cheek teeth erupt in sequence from front to rear, but only a single tooth or one tooth and a fragment of another are functional in each half of each jaw at one time. As a tooth becomes seriously worn, it is replaced by the next posterior tooth. The first three cheek teeth erupt during an animal's youth. The fourth erupts at 4 to 5 years of age, the fifth tooth at age 12 to 13, and the final tooth erupts around 25 years of age and lasts for the remaining 50 or so years (Benton, 1990; Laursen and Bekoff, 1978). The hypsodont cheek teeth are formed of thin laminae, each consisting of an enamel band surrounding dentine, with cementum filling the spaces between the ridges (Fig. 14-6). The last molar, the tooth that must serve for much of the animal's adult life, has the greatest number of laminae. The premolars are considerably smaller, simpler, and less durable than the molars. In old elephants, some of the anterior laminae of the third molar may be lost while the remainder of the tooth is still functional. The unique pattern of tooth replacement and the complex occlusal surface provide for an enduring dentition that lasts for up to about 70 years (Laursen and Bekoff, 1978).

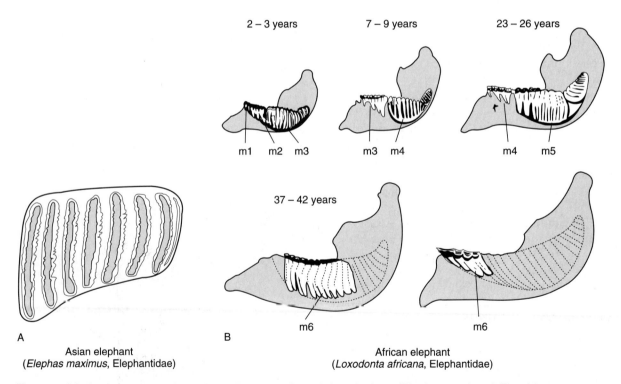

2 – 3 years   7 – 9 years   23 – 26 years

m1   m2   m3   m3   m4   m4   m5

37 – 42 years

m6   m6

A

B

Asian elephant
(*Elephas maximus*, Elephantidae)

African elephant
(*Loxodonta africana*, Elephantidae)

FIGURE 14-6   (A) The occlusal surface of a molar of the Asian elephant *(Elephas maximus)*. The ridges of the lamellae are enamel, the gray areas are dentine, and the white areas are cementum. (B) The progression of molar toothwear in the African elephant *(Loxodonta africana)* from birth to approximately 55 years of age. Some researchers consider the first three cheek teeth in elephants to represent deciduous premolars and m4–m6 (as numbered here) to represent the permanent m1–m3. *(After Kingdon, 1989b)*

Elephants occupy forests, semi-open or dense scrub, savanna, and even semidesert regions in Namibia, but are restricted to areas near water. They feed on a variety of trees, shrubs, grasses, and aquatic plants and characteristically strongly influence their environments (p.496; Laws, 1970). Each individual eats over 200 kilograms of forage daily. Their great size and strength enable them to "ride down" fairly large trees in order to feed on the leaves.

Female elephants are highly social. African elephants live largely in **matriarchal** kinship groups from which adult males are excluded; these groups are held together by close social ties between adult females and between mothers and their offspring. Adolescent males leave the maternal herd and become solitary or form loose bachelor herds. Male elephants of both species experience episodes of **musth,** periods of increased aggression and heightened sexual activity coupled with elevated serum testosterone levels (Poole and Moss, 1981; Rasmussen et al., 1996).

Vocal, as well as behavioral, communications play an important role in maintaining group cohesion. Vocalizations are diverse and include the familiar trumpeting sounds made when animals are threatened. Poole et al. (1988) discovered that African elephants are capable of communicating over vast distances using **infrasound.** These very low-frequency sounds (14 to 24 Hertz) are not audible to humans but can be sensed as a series of air pressure vibrations not unlike those produced by the lowest notes of large pipe organs.

The greatest danger to elephants is encroachment on their habitat by humans and poaching for the ivory in their tusks. In Luangwa National Park in Zimbabwe, poaching of elephants for their tusks has increased selection for a genetic condition that results in tuskless elephants. Tuskless female elephants have increased in frequency to 38.2 percent of the population in 1989, from only 10 percent two decades ago (Jachmann et al., 1995). In Africa, national parks, where settlement by people is prohibited, offer some hope for the survival of wild populations. However, elephants are still killed when they leave the parks and destroy adjoining crops or kill villagers.

ORDER SIRENIA: SIRENIANS

The sirenians, or seacows — dugongs and manatees — are the only completely aquatic mammals that are herbivorous and are one of the most anomalous mammalian orders. There are four living species of two genera (*Dugong,* Dugongidae; *Trichechus,* Trichechidae). Sirenians occur in coastal waters from eastern Africa to the Ryukyu Islands of southern Japan, the Indo-Australian Archipelago, the western Pacific and Indian Oceans; in tropical western Africa; on the Atlantic coastlines of the Western Hemisphere from 30°N to 10°S, in the Caribbean region, and the Amazon and Orinoco drainages in South America; and formerly in the Bering Sea (Rathbun, 1984). Sirenians probably share a common ancestry with the proboscideans and are known from Eocene deposits from such scattered points as Europe, Africa, and the West Indies. Nearly all fossil sirenians were tropical and marine.

Living sirenians are large, reaching weights in excess of 1500 kilograms (O'Shea, 1994). They are nearly hairless, except for bristles on the snout, and have thick, rough, or finely wrinkled skin (Fig. 14-7). The nostrils are **valvular,** the nasal opening extends posterior to the anterior borders of the orbits, and the nasals are reduced or absent. The skull is highly specialized, and the dentary is deep (Fig. 14-8). The tympanic bone is semicircular, and the external auditory meatus is small. The periotic

FIGURE 14-7 Female manatee *(Trichechus manatus latirostris)* and two offspring at Blue Spring, Florida. Twins are rare in sirenians. *(T. O'Shea, USGS, Biological Resources Division)*

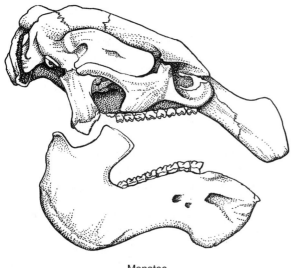

Manatee
(*Trichechus manatus*, Trichechidae)

FIGURE 14-8   The skull of a manatee; length of skull 360 millimeters. *(After Hall and Kelson, © 1959, Mammals of North America, by permission of John Wiley & Sons, Inc.)*

bone has no bony attachments to the skull but is attached by ligaments. The lungs are unlobed, unusually long, oriented horizontally, and separated from the massive gut by a long, horizontal dia-

phragm. The orientation of the lungs and dense, heavy bone allow the animal to use minor adjustments in lung volume to maintain a horizontal attitude while feeding at various depths. The heavy bone may counterbalance the added buoyancy from gas production in the gut. Postcranially, sirenians somewhat resemble cetaceans. The five-toed manus is enclosed by skin and forms a flipper-like structure, the pelvis is vestigial, and the tail is a horizontal fluke. There is no clavicle, and the scapula is narrow and bladelike as in cetaceans.

The teeth of dugongs *(Dugong)* are large and columnar, lacking enamel, and cementum-covered. They have open roots, and the occlusal surfaces are wrinkled and bunodont. In manatees *(Trichechus),* by contrast, there is an indefinite large number of enamel-covered, cementumless teeth, each with two cross ridges and closed roots. As teeth at the front of the tooth row wear out, they are replaced by the posterior teeth pushing forward. Five to eight teeth in each side of the jaw are functional at one time. Horny plates cover the front of the palate and the adjacent surface of the mandible in all genera. The skull of *Trichechus* is modified by elongation of the nasal cavity, and these animals, virtually alone among mammals, have only six cervical vertebrae. Some differences between the two families of sirenians are shown in Table 14-1. Manatees

TABLE 14-1   Comparison of Characteristics of Two Sirenian Families

Dugongidae	Trichechidae

Functional dentition:
  1/0, 0/0, 0/0, 2-3/2-3 =10-14
	No functional incisors;
	  indeterminate number of cheek teeth

Cheek teeth columnar,
  no enamel, cementum-covered;
  roots single
	Teeth with cross ridges,
	  covered with enamel,
	  cementum absent, roots double,
	  continuous tooth replacement

Premaxillaries large;
  nasals absent; nasal cavity short
	Premaxillaries small;
	  nasals present; nasal cavity long

Slender neural spines and ribs	Robust neural spines and ribs
Flippers lacking nails	Flippers with nails in two of the three species
Tail notched, as in whales	Tail not notched but spoon-shaped

evolved from dugongid ancestral stock in South America and completely replaced the dugongs in the New World.

Sirenians are heavy-bodied, slow-moving animals that inhabit coastal seas, large rivers, and lakes and graze while submerged for periods of up to about 15 minutes. Due to a low nutrient diet, manatees have very slow metabolisms, resulting in the production of relatively little body heat for animals of their size (Gallivan and Best, 1986). Water is an excellent conductor of heat, and, because manatees have low metabolic rates and their blubber does not provide the insulation seen in other marine mammals, they tend to lose heat rapidly in cooler waters. For this reason, they tend to be restricted to tropical regions (Irvine, 1983). Some individuals inhabiting the coasts of Florida move into rivers, springs, and industrial warm-water effluents in winter to avoid cold water.

Sirenians are known to make a variety of sounds underwater (Anderson and Barclay, 1995). O'Shea (1994), observing free-ranging West Indian manatees *(Trichechus manatus),* found that these calls form a series of complex, graded signals used to maintain contact between individuals and to communicate basic behavioral information. The underwater calls of *Dugong dugon* may also be used to maintain exclusive "activity zones" that are patrolled and defended against intruders (Anderson, 1997). Recently, Anderson (1997) presented evidence that *D. dugon* may form **leks,** aggregations of males competing for access to females.

Humans have been responsible for the great range reductions of sirenians and exterminated the Steller sea cow *(Hydrodamalis)* in about 1769, only 27 years after its discovery (Walker, 1968). This gigantic sirenian was nearly 8 meters in length, probably reached over 6000 kilograms in weight, and inhabited shallow parts of the Bering Sea, where it fed on seaweed. Present serious declines in the populations of dugongs and manatees in some areas are due to persistent hunting by humans. In Florida, where they are stringently protected from hunters, most manatees bear scars inflicted by boat propellers, and collisions with boats and other human-caused factors are responsible for a large proportion of the observed mortality (O'Shea et al., 1995). In addi-

tion, **epizootic** infections associated with dinoflagellate blooms have lead to recent die-offs of manatees along the eastern coast of the United States (O'Shea et al., 1991). Because of the rapidly increasing human population in Florida and the low reproductive rate of manatees, serious population decline of manatees due to habitat loss and accidental deaths seems likely in the near future.

ORDER HYRACOIDEA: HYRAXES

Members of this unusual order are small, rodent-like creatures commonly called hyraxes or dassies, and their appearance gives little indication of their relationship to the ungulates (Fig. 14-9). This is a small order with a single Recent family, Procaviidae, and one extinct family, Pliohyracidae. Recent members include three genera with six species. Hyraxes, which today occupy nearly all of Africa (except the arid northwestern part) and parts of the Middle East, appeared first in lower Eocene beds of Egypt. Some early members of the Pliohyracidae reached the size of a tapir. The distribution of the structurally more conservative surviving family extended north of its present limits during the Pliocene, when a gigantic procaviid occurred in western Europe as far north as France.

The relationships of the hyracoids are uncertain, but they probably descended from an early "ungulate" stock (Condylarthra). Sudre (1979) and Novacek (1992a,b) linked hyraxes with the elephants and sirenians, while other recent work (Fischer, 1989; Fischer and Tassy, 1993) suggests a closer relationship between hyraxes and perissodactyls.

The roughly rabbit-size procaviids of today have a short skull with a deep lower jaw (Fig. 14-10). The dental formula is 1/2, 0/0, 4/4, 3/3 = 34. The incisors are specialized: the pointed, ever-growing uppers are broadly separated, and the flattened posterior surfaces lack enamel; the lowers are chisel-shaped and generally tricuspid. Behind the incisors is a broad diastema, and the cheek teeth are either brachydont or hypsodont. The molars resemble those of a rhinoceros: the uppers have an ectoloph and two cross lophs, and the lowers have a pair of V-shaped lophs. The body is

FIGURE 14-9   Rock hyrax *(Procavia capensis)* basking in the sun at Augrabies National Park, South Africa. *(T. Vaughan)*

fairly compact, and the tail is tiny. The forefoot has four toes and the hind foot has three. The feet are **mesaxonic** (the plane of symmetry goes through the third digit). The digits are joined to the bases of the last phalanges, and, except for the clawed second digit of the pes, all digits bear flattened nails (Fig. 14-11). The plantigrade feet have specialized elastic pads on the soles that are kept moist by abundant skin glands; in addition, the soles may be "cupped" by specialized muscles and provide remarkable traction. Although the clavicle is absent, as in cursorial mammals, the centrale of the carpus is present, a feature decidedly not characteristic of runners. The stomach is simple, but digestion is aided by microbiota in the pair of caeca of the colon and in the single ilio-colic caecum.

Hyraxes are mainly herbivorous and are nimble climbers and jumpers. They occur in a variety of habitats, from forests and scrub country to rock outcrops and lava beds in grassland, and at elevations up to 5000 meters. The bush hyrax *(Heterohyrax)* and rock hyrax *(Procavia)* live in cliffs, ledges, and talus and are adroit at climbing rapidly over steep rock faces and in trees. The three species of *Dendrohyrax* are highly arboreal, spending most of their time foraging in trees.

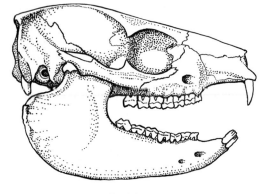

Bush hyrax
(Heterohyrax sp., Procaviidae)

FIGURE 14-10   Skull of a Bush hyrax; skull length is 98 millimeters. *(After Hatt, 1936)*

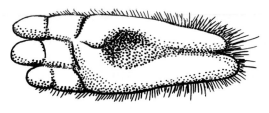

Tree hyrax
(Dendrohyrax dorsalis, Procaviidae)

FIGURE 14-11   The sole of the right hind foot of a tree hyrax.

The terrestrial species are diurnal and often form polygynous family groups, with one territorial adult male and several genetically related females and their young (Happold, 1987; Hoeck et al., 1982). Hyraxes are relatively long-lived for small mammals, some individuals surviving more than ten years in the wild (Hoeck, 1982). In contrast, tree hyraxes are nocturnal, mostly solitary, and maintain territories in the canopy by loud barking calls (Rahm, 1969).

The body temperature of the bush hyrax is quite variable (Bartholomew and Rainy, 1971). This animal is known to make wide use of behavioral thermoregulation (p.385).

WEB SITES

University of Michigan's animal diversity web site

http://www.oit.itd.umich.edu/bio108/Chordata/Mammalia/Proboscidea.shtml

Save the Manatee Club — good general information site

http://www.savethemanatee.org/

Florida manatee information, including telemetry

http://www.dep.state.fl.us/psm/webpages/manatee.htm

Elephant Net — Malaysian elephant tracking project site

http://www.asiaconnect.com.my/elephantnet/

Biosis links to various elephant sites

http://www.york.biosis.org/zrdocs/zoolinfo/mam_prob.htm

15 TUBULIDENTATA

The order Tubulidentata, the aardvarks, includes but one family (Orycteropodidae). The one Recent species (*Orycteropus afer;* Fig. 15-1) inhabits much of Africa south of the Sahara Desert. The earliest record of tubulidentates is from the late Eocene–early Oligocene of Europe, and most of the group's history occurred in Africa and Europe. An extinct member of *Orycteropus* occupied parts of Europe and Asia as recently as the late Pliocene.

In the past, tubulidentates often were considered closely related to ungulates. Recent phylogenetic studies using morphological data cast doubt on a relationship with ungulates but could not otherwise clarify the group's affinities (Thewissen, 1985; Novacek and Wyss, 1986). Recent data from mitochondrial and nuclear genes, however, support a relationship among aardvarks, elephant-shrews, paenungulates (hyraxes, sirenians, and proboscideans), and golden moles (Chrysochloridae). These ecologically divergent adaptive types probably originated in Africa; the molecular evidence implies that they may all have arisen there from a common ancestor that existed in the Cretaceous Period, when Africa was isolated from other continents (Springer et al., 1997).

Aardvarks are powerful diggers that feed on ants and termites. They weigh up to 65 kilograms, and the thick, sparsely haired skin provides protection from insect bites. The skull is elongate, and the dentary bone is long and slender (Fig. 15-2). Incisors and canines are lacking in the dentition of adults (but are present in the deciduous dentition); the cheek teeth are 2-3/2 premolars and 3/3 molars. Each tooth is rootless and consists of as many as 1500 hexagonal prisms of dentine, each surrounding a slender, tubular pulp cavity. The columnar teeth lack enamel but are surrounded by cementum. The anteriormost teeth erupt first and are often lost before the posterior molars are fully erupted. The slender tongue is protrusible.

Olfaction is used in finding insects; the olfactory centers of the brain are unusually well developed, and the turbinal bones are remarkably large and complex. The nostrils are highly specialized in a fashion not found in any other mammal. Fleshy tentacles, which presumably have a tactile or other sensory function, occur on the nasal septum (Fig. 15-3), and dense hair surrounds the nostrils, which can be sealed when the aardvark digs. The pinnae are large and may also be used in locating insects. The pollex is absent, and the hind feet are five-toed; the robust claws are flattened and blunt, intermediate between hoof and nail.

Many of the adaptations of aardvarks are specializations for feeding on termites and ants. Some of these, like the elongation of the snout, the long sticky tongue, and the simplification of the dentition, are features that are convergent with those of other ant-eating mammals like the pangolins (order Pholidota) and xenarthran anteaters (order Xenarthra). In addition to these structural features, aardvarks are physiologically convergent with other ant-eating mammals; they exhibit a low basal metabolic rate and a low body temperature of about 34.5°C (McNab, 1984).

FIGURE 15-1   The aardvark (*Orycteropus afer*). (C. Berlinky, © Educational Images, Ltd.)

The powerful forelimbs are used in burrowing and in dismantling rock-hard termite and ant nests, and the hindlimbs thrust accumulated soil from burrows. Not only do aardvarks have phenomenal strength but they can dig astonishingly fast; a 1-meter section of burrow the diameter of the animal can be dug in 5 minutes in appropriate soils. Burrows dug by aardvarks are extensive and numerous in some areas and are used as retreats by a variety of mammals, such as warthogs, hyenas, porcupines, jackals, bat-eared foxes, hares, bats, ground squirrels, and civets, as well as monitor lizards and owls. In Botswana, Smithers (1971) recorded 17 species of mammals, 2 species of reptiles, and a bird using aardvark burrows. One bird, the ant-eating chat *(Myrmecocicla formicivora)*, nests in occupied burrows. In parts of Africa, the abundance of warthogs depends on the availability of abandoned aardvark burrows (Melton, 1976).

Although the foot posture is digitigrade, aardvarks are slow runners and can be outrun by a human. They are almost entirely nocturnal so are rarely seen by people, even those living in the same area. As a result, their disappearance over much of their historic range has gone largely unnoticed.

During their nightly foraging, aardvarks may travel 10 to 30 kilometers. Often their foraging strategy is to move along in a zigzag path about 30 meters wide, sniffing and probing the ground and frequently digging small or large excavations. Although ants and termites are the main foods eaten, other insects such as beetles and grasshoppers are occasionally taken. In some areas they drink regularly, but in parts of southern Africa they occupy areas that lack surface water in the dry season. During the driest months of the year, aardvarks dig up and eat the moist, fleshy fruits of an unusual species of cucurbit plant (*Cucumis humifructus;* Cucurbitaceae), known as the "aardvark cucumber." The association between the aardvark

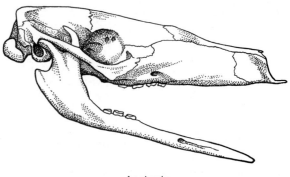

Aardvark
(Orycteropus afer)

FIGURE 15-2   The skull of the aardvark (length of skull 240 millimeters). *(After Hatt, 1934)*

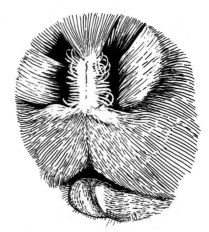

FIGURE 15-3   The complex nose of an aardvark *(Orycteropus afer)*. Note the fleshy tentacles on the nasal septum and the dense tracts of hair that can seal the nostrils. *(After Kingdon, 1971)*

and this plant may be symbiotic. The geographic distributions of the two species (mammal and plant) are similar, and the cucumber is the only member of its family that produces an underground fruit. Aardvarks pass the ingested seeds intact in their feces, burying and thus replanting and fertilizing them. Aardvarks are the only known agent of dissemination of the seeds; presumably, they acquire water and nutrients from the fruits (Patterson, 1975).

WEB SITES

Animal Diversity Web information about aardvarks

http://www.oit.itd.umich.edu:80/bio108/Chordata/Mammalia/Tubulidentata.shtml

16 PERISSODACTYLA

Since Eocene times, some of the most specialized and spectacular cursorial mammals have been perissodactyls (horses, rhinos, and tapirs). In the early Tertiary, these were the most abundant ungulates, but their diversity declined in the Oligocene, perhaps due to changing environments. With the diversification and "modernization" of the artio-

dactyls (pigs, camels, antelopes, and their relatives) in the Miocene, the fortunes of perissodactyls further declined. The surviving perissodactylan fauna (consisting of six genera and 17 species) is but an insignificant remnant of this once important and diverse group and is vastly overshadowed by an impressive living artiodactylan assemblage (consisting

of 220 species). Perissodactyls occur today largely in southern areas — Africa, parts of central and southern Asia, and tropical parts of southern North America and northern South America.

The term **ungulate** has no taxonomic status here but refers to all hoofed mammals, both the Perissodactyla and the Artiodactyla (and many other extinct groups). Ungulates typically are herbivorous and are adapted to rapid cursorial locomotion. Among the ungulates are some of the most graceful and handsome mammals and some that are in serious danger of extinction. A key to understanding the biology of ungulates is knowledge of their cursorial and feeding adaptations.

CURSORIAL SPECIALIZATION

Exceptional running ability has evolved independently in a number of mammalian groups. It provides a means of escaping predators (as in ungulates, rabbits, and some rodents) or of capturing prey (as in carnivores). A whole series of specializations for cursoriality are unique to mammals and involve the integration of locomotor and respiratory functions. Galloping mammals synchronize their breathing and stride cycles, using the inertia of their intestines as a "visceral piston" to help ventilate the lungs as the liver "sloshes" back and forth against the diaphragm with each running stride (Bramble and Jenkins, 1993, 1998). As an aid to this process, a phenomenon known as "pneumatic stabilization" involves tracheal valving to shunt air from side to side between lungs and help to control inhalation and exhalation (Simons, 1996). Thus pressurized, the lungs not only function in respiration but also help to stabilize the shoulder and chest wall on alternating sides as the leading forelimb strikes the ground (Bramble and Jenkins, 1998).

Cursorial adaptations appear in the early Cenozoic history of mammals. The earliest known perissodactyl, *Hyracotherium*, of the early Eocene Epoch, was not highly specialized (Radinsky, 1966). The early Eocene genus *Diacodexis,* one of the earliest known artiodactyls, had slim, elongate limbs and was highly cursorial (Carroll, 1988; Rose, 1982, 1996). The refinement of cursorial adaptations in ungulates was favored by the Miocene expansion of grasslands. Not only was speed the primary means of escaping predators in this open country but long daily or seasonal movements to seek water or nutritious food probably became an important part of the ungulate mode of life.

Running speed is determined basically by two factors: stride length and stride rate (the number of strides per unit of time). Cursorial specializations lengthen the stride or increase its rate. Perhaps the most universal cursorial adaptation that lengthens the stride is lengthening of the limbs. In generalized mammals, or in many powerful diggers, the limbs are fairly short and the segments are all roughly the same length (Fig. 16-1). In cursorial species, however, the limbs are long and, in the most specialized runners, the metacarpals and metatarsals are elongate and the manus and pes are the longest segments. Loss or reduction of the clavicle contributes further to the length of the stride. This occurs in carnivorans, leporids, and ungulates. With the loss of the clavicle, the scapula and shoulder joint are freed from a bony connection with the sternum, and the scapula is free to rotate to some extent about a pivot point approximately at its center.

Substantial lengthening of the stride also results from an inchworm-like **flexion** and **extension** of the spine (Fig. 16-2). In small or moderate-size runners, the flexors and extensors of the vertebral column are powerfully developed; the vertebral column extends as the forelimbs reach forward and the hindlimbs are driving against the ground, and it flexes when the front feet move backward while braced against the ground as the hindlimbs swing forward. Hildebrand estimated that such movements of the vertebral column could propel the cheetah at nearly 10 kilometers per hour if the animal had no legs!

The speed of limb movements and thus stride rate are increased by a combination of structural

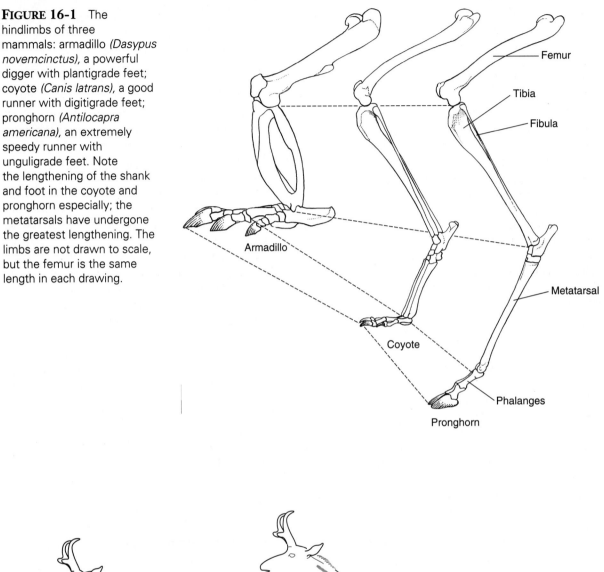

FIGURE 16-1 The hindlimbs of three mammals: armadillo *(Dasypus novemcinctus)*, a powerful digger with plantigrade feet; coyote *(Canis latrans)*, a good runner with digitigrade feet; pronghorn *(Antilocapra americana)*, an extremely speedy runner with unguligrade feet. Note the lengthening of the shank and foot in the coyote and pronghorn especially; the metatarsals have undergone the greatest lengthening. The limbs are not drawn to scale, but the femur is the same length in each drawing.

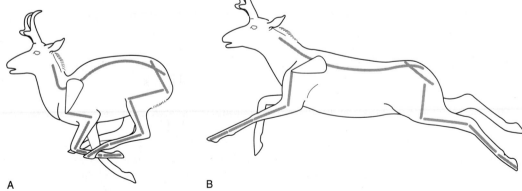

FIGURE 16-2   Two positions of a running pronghorn, showing the flexion and extension of the vertebral column and the changing position of the scapula. (A) The forelimbs have just left the ground; the hindlimbs are reaching forward and will touch the ground as the forelimbs swing forward. (B) The animal is bounding ahead after the limbs have driven against the ground; the forelimbs are reaching forward.

modifications. The total speed of the foot, which drives against the ground and propels the animal, depends on the speed of movement of the distal joints of the limb. If another movable joint is added to the distal limb, the speed of the limb will be increased by the speed of movement at the new joint. The greater the number of joints that move in the same direction simultaneously, the greater the speed of the limb. One way to add a joint to the distal limb is to change the foot posture such that only the hoof-bearing tips of the digits contact the ground. This lifting of the heel from the ground allows another limb joint, that between the metapodials and the phalanges, to contribute to limb speed. Not surprisingly, nearly all cursorial mammals have abandoned a plantigrade foot posture in favor of one that is a digitigrade or **unguligrade.** In addition, the movable scapula and the flexion of the vertebral column (in some cursorial species), which help to increase stride length, also contribute their motion to total foot speed.

Specializations of the musculature also add importantly to limb speed and, hence, to running speed. A trend in many cursorial mammals is toward a lengthening of the tendons of some limb muscles (in association with the elongation of the distal segments of the limbs) and, in some cases, a migration of the insertion points of these muscles toward the body. Generally, the nearer the insertion point of a muscle approaches the joint it spans and at which it causes motion, the greater the advantage for speed.

Speedy limb movements are further facilitated by a reduction of the mass of the distal parts of the limbs and the resultant reduction in the amount of inertia that must be overcome at the end of one limb movement and the start of another. Because as the distal part of the limb moves more rapidly than the proximal part during a stride, reduction of mass of the distal parts is especially advantageous. Several specializations commonly serve this end. The most obvious is the loss of certain digits. Also, the heaviest muscles are mostly in the proximal segment of the limb, thus keeping the center of gravity of the limb near the body. The combined effect of these modifications that reduce and redistribute weight is to favor rapid limb movement and to reduce the outlay of energy associated with that movement. For excellent discussions of cursorial adaptations in mammals, see Gambaryan (1974), Hildebrand (1959, 1960, 1965, 1985, 1987), and Kardong (1998).

In the ungulate ankle joint, the calcaneum is pushed aside, so to speak. In mammals in which no drastic reduction of digit number has occurred, the distal surface of the astragalus articulates with the navicular, the calcaneum articulates with the cuboid (see Fig. 2-26), and the weight of the body is transferred through the digits, the distal carpals, both the astragalus and the calcaneum, and the tibia and fibula. In ungulates, a different arrangement occurs in association with the reduction of digit number: the astragalus rests more or less directly on the distal tarsal bones, which may be highly modified by fusion and loss of elements (Fig. 16-3, see Fig. 17-3), and the weight of the body is borne by the central digits (or digit, in the case of equids), the distal tarsals, and the astragalus. The astragalus thus becomes the main weight-bearing bone of the two proximal tarsals.

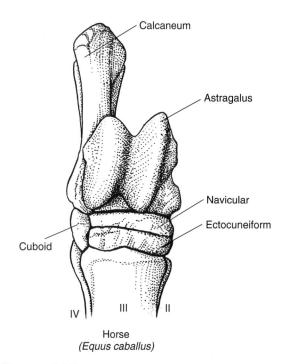

FIGURE 16-3   The tarsus of the horse *(Equus caballus).* The metatarsals are numbered.

The calcaneum remains important as a point of insertion for extensors of the foot, but it no longer is a major weight-bearing bone of the tarsus. A similar bypassing of the calcaneum occurs in the cursorial peramelid marsupials (see Fig. 6-17).

Two distinctive ungulate specializations involve connective tissue. The nuchal ligament is a heavy band of elastin (an elastic protein found in vertebrates) that is anchored posteriorly to the tops of the neural spines of some of the anteriormost thoracic vertebrae and attaches anteriorly high on the occipital part of the skull. This ligament, especially robust in large, heavy-headed ungulates such as the horse and moose, helps support the head so that the burden on the muscles that lift the head is greatly lightened. The elasticity of the ligament allows the animal to lower its head when it is eating or drinking. A second specialized ligament, the springing ligament, occurs in the front and hind feet of ungulates; it evolved from muscles that flexed the digits (Camp and Smith, 1942; Hildebrand, 1985). In the hind foot of the pronghorn *(Antilocapra)*, for example, the springing ligament arises from the proximal third of the back of the **cannon bone** and inserts distally on the sides of the first phalanges of digits three and four (see Fig. 17-4). When the foot supports the weight of the body, the phalanges are extended, thereby stretching the springing ligament. As the foot begins to be relieved of the weight of the body toward the end of the propulsion stroke of the stride, however, the elastic ligament begins to rebound; when the foot is leaving the ground, the phalanges snap toward the flexed position. The familiar backward flip of the horse's foot just as it leaves the ground is controlled by the springing ligament. This flip gives a final increase in speed and thrust to the stride and increases the ungulate's speed without the use of muscular effort.

FEEDING SPECIALIZATION

The herbivorous diet characteristic of most ungulates has favored the development of cheek teeth with large and complex occlusal surfaces that finely section plant material as an aid to digestion. Premolars tend to become molariform and thus to increase the extent of the grinding battery, and the anterior dentition becomes variously modified.

In advanced types, there is a diastema between the anterior dentition and the cheek teeth.

Diet puts unusual demands on the digestive systems of ungulates. Vegetation yields far less fat and protein than meat, is more difficult to digest, and is often protected by defensive secondary compounds. In addition, an herbivore must break down the cell wall, a fairly rigid structure formed largely of cellulose, not so much for the energy it yields as to gain access to the proteins, carbohydrates, and lipids within the cells. This process is difficult, however, for mammals lack enzymes that digest cellulose. All herbivores must therefore have an alimentary canal that can digest cellulose by means other than the herbivore's own enzymatic action. Both perissodactyls (horses, rhinoceroses, and tapirs) and ruminant artiodactyls (camels, deer, antelope, sheep, goats, and cattle) utilize a fermentation process that breaks down cellulose with the cellulolytic enzymes of microorganisms living in the alimentary canal. **Microbial fermentation** is a relatively slow process requiring specialized fermentation chambers. Microbial digestion of cellulose in a modified stomach is called **gastric fermentation** and is typical of artiodactyls (p. 273).

In perissodactyls, protein is digested and absorbed in the relatively small and simple stomach, whereas microbial fermentation takes place in the large intestine and enlarged caecum (**intestinal fermentation** also occurs in rodents, lagomorphs, elephants, and hyraxes). The horse caecum is an expanded portion of the ascending colon with a blind sac at one end. As a result of microbial fermentation, cellulose is broken down and sugars and fatty acids are released and absorbed across the intestinal walls.

Perissodactyls and **ruminant** artiodactyls have thus evolved contrasting nutritional strategies. Ruminant artiodactyls can satisfy their nutritional needs with relatively unnutritious food; in areas with marked seasonality, the only areas where ruminants are abundant, they can remain widespread through times when forage is nutritionally poor. During comparable times, perissodactyls must seek sites that support the greatest quantity of vegetation and the most nutritious vegetation, and their range is correspondingly restricted. When ruminants compete with perissodactyls for nutritious food, however, ruminants are the losers. Ruminants are less efficient in utilizing nutritious food because proteins are used inefficiently by the ruminant system of microbial fermentation (Reid, 1970; Smith, 1975). Ruminants have reduced their efficiency in transforming forage into animal biomass while enhancing their ability to survive on low-quality food (Kinnear et al., 1979).

An extremely interesting African grazing succession, strongly influenced by differences between digestive efficiencies and food requirements, has been described by Bell (1971). He studied primarily the most abundant ungulates: the zebra *(Equus burchellii)*, a nonruminant, and the wildebeest *(Connochaetes taurinus)* and Thompson's gazelle *(Gazella thompsonii)*, both ruminants. He found that, in the Serengeti Plains, the zebra was the first of these ungulates to be forced by food shortages to move from the preferred short-grass area down into the longer, coarser grasses of the lowlands. After the zebras' feeding and trampling in the lowlands had removed the coarse upper parts of the grass and had made the lower, more nutritious parts more readily available, the wildebeest, a more selective feeder, moved in. By this time, the zebras were becoming less able to get sufficient quantities of forage and were moving to new tall-grass pastures. A similar replacement of wildebeest by Thompson's gazelles occurred after the wildebeest had removed still more grass and had made available to the small, highly selective gazelles the fruits and leaves of low-growing forbs. Competition between these abundant ungulates is minimized by this grazing pattern, and the activities of the early members of the grazing succession were highly advantageous to the later, more selective members. Bell's study clearly illustrates that differences in the digestive systems of ungulates have pronounced effects on food preferences, migratory patterns, and many basic interactions within a grazing ecosystem.

PERISSODACTYL EVOLUTION

Perissodactyls evolved in Asia from herbivorous condylarths of the family Phenacodontidae (Fig. 16-4). The extinct order Condylarthra (which is probably not monophyletic) includes a diverse group of ancient ungulates that existed from the early Paleocene to the Oligocene, and was probably the basal stock for many mammalian orders,

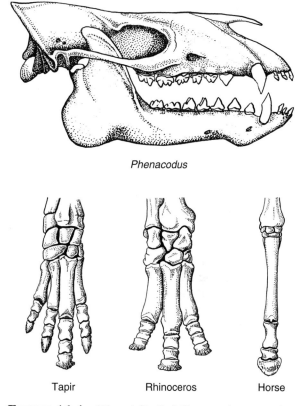

FIGURE 16-4 (Above) Skull of *Phenacodus*, an early Eocene primitive ungulate (order Condylarthra); length of skull 125 millimeters. *(After Romer, © 1966, Vertebrate Paleontology, University of Chicago Press).* (Below) Front feet of three perissodactyls: a tapir; a rhinoceros; and a horse. *(After Howell, © 1944, Speed in Animals, University of Chicago Press)*

including the Recent orders Artiodactyla, Perissodactyla, Cetacea, Tubulidentata, Hyracoidea, Proboscidea, and Sirenia. Perissodactyls appeared in the early Eocene in North America and underwent rapid diversification. Thirteen of the 14 families appeared in the Eocene, but, in addition to the living families Tapiridae, Rhinocerotidae, and Equidae, only the anomalous extinct family Chalicotheriidae survived into the Pleistocene.

The features of several important perissodactylan families illustrate the considerable structural and functional diversity within the group. The dentition and cranial morphology developed in response to herbivorous feeding habits. Living perisodactyls have elongate skulls, owing to an enlargement of the facial region to accommodate a full series of large cheek teeth (often hypso-dont); some have a complete complement of 44 teeth. The teeth are usually lophodont and are either hypsodont in grazing types (all Equidae and *Ceratotherium* of the Rhinocerotidae) or brachydont in browsers (all Tapiridae and *Rhinoceros* and *Dicerorhinus* of the Rhinocerotidae). Many postcranial specializations further cursorial ability. The clavicle is absent, and usually the manus has three or four digits and the pes, three digits. In the equids, however, only one functional digit is retained on each foot (Fig. 16-4). The feet are **mesaxonic;** that is, the plane of symmetry of the foot passes through the third digit, whereas this plane passes between digits three and four in the **paraxonic** foot of artiodactyls.

FAMILY EQUIDAE. Horses, the most highly cursorial and graceful perissodactyls, now occur wild only in Africa, the Middle East, and parts of western and central Asia. In addition, feral populations of domestic horses and burros live in various places. There is but one genus with eight living species.

Wild horses in general are not as large as domesticated breeds. The average weight of a female zebra *(Equus burchellii)* is given by Bell (1971) as 219 kilograms, but some domestic breeds of horses weigh over 1000 kilograms (Fig. 16-5). The skull has a fairly level profile, and the rostrum is long and deep (Fig. 16-6); the dental formula is 3/3, 0-1/0-1, 3-4/3, 3/3 = 36-42. The cheek teeth are hypsodont and have complex patterns on the occlusal surfaces (Fig. 16-7). The limbs are of a highly cursorial type: only the third digit is functional, all but the proximal joints largely restrict movement to one plane, and the foot is greatly elongate. In the tarsus, the main weight-bearing bones are the ectocuneiform, navicular, and astragalus; the calcaneum is mostly posterior to the astragalus (Fig. 16-3).

The evolution of horses is well documented by an excellent and largely New World fossil record and is discussed by MacFadden (1992) and Radinsky (1984). Equids are first represented by *Hyracotherium* from the early Eocene record of Europe and North America. This primitive type had a generalized skull with 44 teeth (Fig. 16-8). The upper and lower molars were brachydont and basically four-cusped. The upper molars bore a protoconule and a metaconule, and the paraconid of the lower

FIGURE 16-5 Mountain zebra *(Equus zebra zebra)* in Mountain Zebra National Park, South Africa. *(T. Vaughan)*

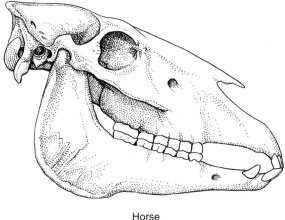

Horse
(*Equus caballus,* Equidae)

FIGURE 16-6 The skull of a horse (length of skull
530 millimeters).

molars was reduced. The premolars were not molariform. The limb structure reflected considerable
running ability: the limbs were fairly long and slender; the front foot had four toes and the hind foot
had three toes, but the animal was functionally tridactyl. Each digit terminated in a small hoof, and
the foot posture was unguligrade. *Hyracotherium*
species ranged in size from 25 to 50 centimeters at
the shoulder and presumably browsed on low-growing vegetation in forested or semiforested areas.

Side branches from the main stem of equid evolution developed at various times. During part of
the late Miocene, North American savannas supported 12 species of horses. The main evolutionary
line can be traced through such intermediate genera as *Merychippus, Dinohippus,* and *Pliohippus* to
the Pleistocene and Recent *Equus. Merychippus,* a
pony-size, early Miocene type, was functionally tri-

dactyl but retained short lateral digits. The dentary
bone was deep, the face was long, and the orbit
was fully enclosed. The cheek teeth were high-
crowned and covered with cementum and had an
occlusal pattern similar to that of *Equus* (Fig. 16-7).
Pliohippus occurred in the late Miocene; it had the
skull features of its progenitor, *Merychippus,* but
was more progressive in having higher crowned
teeth and lateral digits usually reduced to splint-
like vestiges. *Equus,* as well as an extinct evolutionary side branch of short-legged South American
horses (typified by *Hippidion*), evolved from *Pliohippus. Equus,* the genus to which all living horses
belong, differs from *Pliohippus* in greater size and
in a more complex crown pattern of the cheek
teeth. Major evolutionary trends of the Equidae include increase in size, lengthening of legs and feet,
reduction in the size of the lateral toes and emphasis on the middle toe, molarification of premolars, increase in height of the cheek teeth crowns,
lengthening of the facial part of the skull to accommodate the large cheek teeth, and deepening
of the maxillary and dentary bones to accommodate the high-crowned teeth (MacFadden, 1992;
Radinsky, 1984; Simpson, 1951). In addition, the
profile of the angular border of the dentary bone
swept progressively farther forward and the origin
of the masseter muscles migrated forward. These
adaptations increased the force the masseter muscles could exert on the dentary bone.

Cenozoic changes in climate and in the flora of
North America may have had a critical influence
on the evolution of horses. Especially important
was the Miocene development of grasslands and savannas over much rolling or nearly level land in
the present Great Plains, the Great Basin, and the
southwestern deserts of the United States. Many of

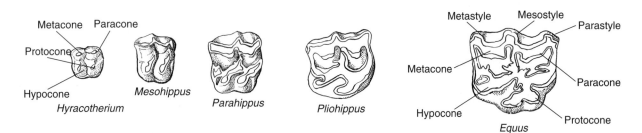

FIGURE 16-7 Right upper molars of four fossil equids and the extant *Equus.* These
teeth illustrate stages in the evolution of the equid molars. *(After Romer, © 1966,*
Vertebrate Paleontology, *University of Chicago Press)*

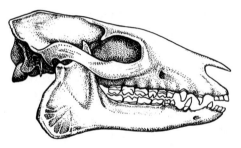

Hyracotherium, Equidae

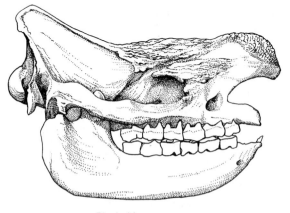

Black rhinoceros
(*Diceros bicornis,* Rhinocerotidae)

FIGURE 16-8   Skull of *Hyracotherium,* first known equid (length of skull 134 millimeters), and of a black rhinoceros (length of skull 692 millimeters). (Hyracotherium *after Romer,* © *1966,* Vertebrate Paleontology, *University of Chicago Press*)

the most progressive equid skull and dental features probably arose in response to the shift to a grazing habit. Grass, at least at certain times of the year, has low nutritional value and must be eaten in large quantities to sustain life. High-crowned, persistently growing teeth evolved to cope with grasses made highly abrasive by silica in the leaves and by particles of soil deposited on leaves by wind and the splash effect of rain. Also of great adaptive value were the highly cursorial limbs with single-toed feet, which facilitated rapid and efficient locomotion on the firm, level footing of the grasslands. Cursorial ability was perhaps as advantageous for traveling between widely scattered concentrations of food and distant water holes in semiarid regions as for escaping from predators.

For some unknown reason, horses disappeared from the New World, their place of origin and the primary center of their evolution, before historic times. Their decline began toward the end of the Miocene Epoch, when savannas were being replaced by cooler and drier steppe conditions. Although wild horses now occupy only Africa and parts of Asia, within historic times they occurred throughout much of Eurasia. Wild equids inhabit grasslands in areas ranging from tropical to subarctic in climate. Equids are polygynous animals (one male controls access to more than one breeding female) that often form large herds consisting of extended family groups or "clans." Within these groups, a social hierarchy exists, usually led by a stallion. Status within the group is maintained by complex behavioral and vocal communication.

FAMILY TAPIRIDAE. Tapirs occupy tropical parts of the New World and the Malayan area. The family includes one living genus and four species. Structurally, tapirs are notably conservative and share many features with the common ancestors of all perissodactyls. "True" tapirs are known first from the Oligocene, but possible ancestral types occurred in the Eocene of North America.

Tapirs have a stocky build and weigh up to about 320 kilograms. The limbs are short, and both the ulna and fibula are large and separate from the radius and tibia, respectively. The front feet have four toes (Fig. 16-4) and vestiges of the fifth (the pollex), and the hind feet have three toes. Tapirs retain a full placental complement of 44 teeth. Three premolars are molariform, and the brachydont cheek teeth retain a simple pattern of cross lophs. The short proboscis (Fig. 16-9) and reduced nasals are among the few specializations of tapirs.

These animals have probably always occupied moist forests, where their primitive feet serve well on the soft soil and their teeth are adequate for masticating plant material that is not highly abrasive. Tapirs today inhabit mostly tropical areas and are usually found near water. They are rapid swimmers and often take refuge from predators in the water. Tapirs are solitary and nocturnal, and their presence is frequently made known chiefly by their systems of well-worn trails between feeding areas, resting places, and water. Their food is largely succulent plant material, including fruit.

FIGURE 16-9   A Brazilian tapir *(Tapirus terrestris)* crossing a river. The upper lip is elongate, forming a down-curved proboscis. *(Corel Corp.)*

FAMILY RHINOCEROTIDAE. Family Rhinocerotidae is represented today by four genera and five species and is restricted to parts of tropical and subtropical Africa and southeastern Asia. These ponderous creatures — the armored tanks of the mammal world — are surviving members of the spectacular late Tertiary and Pleistocene ungulate fauna. Although now a declining group, rhinoceroses have an illustrious past.

The fossil record of the rhinoceroses and their relatives is remarkably complex and parallels that of the horses in documenting the early and middle Tertiary success and the late Tertiary decline of the group. Two genera that illustrate well the diversity of early Tertiary rhinocerotoids are *Hyracodon* and *Indricotherium. Hyracodon* (Hyracodontidae), a small North American Eocene–Oligocene "running rhinoceros," had slender legs and tridactyl feet and was probably similar in cursorial ability to Oligocene horses. Hyracodonts became extinct in early Miocene. A contemporary of *Hyracodon* in the Oligocene was the Eurasian form *Indricotherium* (Hyracodontidae), the largest known land mammal. This giant was nearly 5 meters high at the shoulder, and probably weighed up to 15 to 20 tons and the skull (small in proportion to the great size of the rest of the animal) was 1.2 meters long. The neck was long, and they may have browsed on high vegetation in giraffe-like fashion. The limbs were long and graviportal, but the tridactyl feet were unique in that the central digit was greatly enlarged and terminated in a broad hoof, whereas the lateral digits were much smaller than in any other rhinocerotoid. Rhinoceroses died out in the New World in the earliest Pliocene but remained common and diverse in Eurasia through the Pleistocene. The Pleistocene woolly rhinoceros *(Coelodonta)* was apparently adapted to cold climates. Entire preserved specimens of this animal have been found in an oil seep in Poland.

All Recent rhinoceroses are large, stout-bodied herbivores with fairly short, graviportal limbs (Fig. 16-10). Weights range up to about 2800 kilograms. The front foot has three or four toes (Fig. 16-4), and the hind foot is tridactyl. The nasal bones are thickened and enlarged, often extend beyond the premaxillary bones, and support a horn. Where there are two horns, the posterior one is on the frontals. The horns are of dermal origin and lack a bony core. The occipital part of the skull is unusually high and yields good mechanical advantage for neck muscles that insert on the lambdoidal crest and raise the heavy head (Fig. 16-8). The incisors and canines are absent in some rhinoceroses and are reduced in number in others; the dental formula is 0-1/0-2, 0/0-1, 3-4/3-4, 3/3 = 24-36. The cheek teeth have a primitive pattern of cross lophs far simpler than that of equids (Fig. 16-11).

Rhinoceroses inhabit grasslands, semideserts, savannas, brushlands, forests, and marshes in tropical and subtropical areas. Some species are usually solitary *(Diceros)*, whereas others occur in family groups *(Ceratotherium)* or even in assemblages including up to 24 animals (Heppes, 1958).

FIGURE 16-10   The white rhinoceros *(Ceratotherium simum)* of South Africa. This species had been decimated throughout its range by the beginning of the twentieth century. Due to strenuous conservation efforts in South Africa, the species is now well represented in a number of reserves and parks in southern Africa. The northern subspecies *(C. s. cottoni)* of East Africa, however, is critically endangered. *(T. Vaughan)*

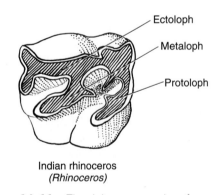

Ectoloph

Metaloph

Protoloph

Indian rhinoceros
(Rhinoceros)

FIGURE 16-11   The right upper molar of a rhinoceros.

Rhinoceroses are territorial and practice scent marking by establishing dunghills along well-worn trails. A variety of plant material is eaten; white rhinos have a square lip for cropping grasses, and others have a pointed prehensile upper lip adapted to browsing.

Adults are nearly invulnerable to predation, except by humans, but young rhinoceroses are occa-

sionally killed by lions, spotted hyenas, or tigers. The Asian rhinoceroses *(Rhinoceros* and *Dicerorhinus)* and the African black rhinoceros *(Diceros)* are facing possible extinction. Because of the supposed medicinal properties of the horn and other parts, rhinoceroses have been hunted persistently for at least 1000 years. There are thought to be fewer than 400 Sumatran rhinos *(Dicerorhinus)* surviving in highly fragmented populations in Indonesia and Malaysia. Due to strict protection from Indian and Nepalese governments, Indian rhino *(Rhinoceros)* populations have recovered from less than 200 to over 2000 today. The African black rhinoceros continues to decline over broad areas despite efforts to protect it from poaching. It has suffered an estimated 85 percent decline in the past two decades, but recently its numbers have stabilized, and a thriving population lives in Etosha National Park, Namibia. Regrettably, the future of rhinoceroses seems extremely dim, and all five species are currently listed as endangered.

WEB SITES

Tapir Specialists Group IUCN World Conservation Union (includes extensive bibliography)

http://www.tapirback.com/tapirgal/iucn-ssc/tsg/

The Tapir Gallery — extensive information about tapirs with pictures and links

http://www.tapirback.com/tapirgal/default.htm

Fossil Horses in Cyberspace — information on the evolution of horses

http://www.flmnh.ufl.edu/natsci/vertpaleo/fhc/fhc.htm

Horse evolution site

http://www.talkorigins.org/faqs/horses.html

Animal Diversity Web — general information about perissodactyls

http://www.oit.itd.umich.edu/bio108/Chordata/Mammalia/Perissodactyla.shtml

International Rhino Foundation — extensive information about rhinos and their conservation status

http://www.rhinos-irf.org/

UC Berkeley site with good background information about perissodactyls

http://www.ucmp.berkeley.edu/mammal/mesaxonia/perissodactyla.html

17 ARTIODACTYLA

Today artiodactyls (pigs, camels, deer, antelope, cattle, and their kind) far overshadow perissodactyls in diversity and abundance. Although the perissodactyls appeared in the late Paleocene and reached their greatest diversity in the Eocene, the first artiodactyl (the rabbit-size *Diacodexis*) is from the early Eocene (Rose, 1996) and the major artio-

dactyl radiation occurred in the Miocene. Since the Miocene, the perissodactyls have steadily declined, but the artiodactyls have remained diverse and successful. Of the 36 families present in the Cenozoic, 10 families (and 81 genera) survive to the present. In contrast, the perissodactyls, represented by 14 Cenozoic families, are reduced today

to 3 families and only 5 genera. It is tempting to relate the decline of the perissodactyls to the rise of the artiodactyls and to regard the latter as the more effective competitors. Although many structural differences between perissodactyls and artiodactyls are apparent, the functional advantages conferred by many of the features of the latter are not easily recognized.

MORPHOLOGY

In ruminant artiodactyls, the enlarged and multichambered stomach harbors microorganisms that break down cellulose (Langer, 1988). Initially, food passes into the large saclike **rumen** that serves as a holding chamber and a microbial fermentation vat (see Fig. 2-12). Large, undigested food particles float on top of the fluid in the rumen. During **regurgitation,** these larger plant fibers are forced into the esophagus by a combination of contractions of the diaphragm and peristaltic contractions in the esophagus itself, finally reaching the mouth where the fibers are rechewed; the animals "chew their cud." Following the remastication of the cud, it is swallowed a second time. The well-digested, finer particles of food are drawn from the **reticulum** into the **omasum** by the negative pressure caused by a relaxation of the omasum walls. Contractions of the omasum then forces the slurry into the **abomasum.** The abomasum is glandular and equivalent to the true stomach of other mammals. Gastric fermentation (also called foregut fermentation) has evolved independently in sloths, some leaf-eating primates, peccaries, camels, hippopotami, many rodents, hyraxes, and kangaroos.

The complex system of recycling and reconstituting food in the stomach enhances the nutritional yield of poor-quality food in four ways. First, gastric fermentation releases proteins, carbohydrates, and lipids early in the digestive process so they can be efficiently absorbed by the intestines. Second, remastication results in more complete breakdown of cell walls. Third, nitrogen, a byproduct of cellulose digestion, is used by the microbes to make their own proteins. The ruminant periodically flushes these microbes into its intestines and thereby digests the proteins formed by the microbes. Finally, ruminants can quickly gather large volumes of plant material into the huge stomach and digest it later. As might be expected, ruminants can subsist on remarkably low-quality food.

However, there are disadvantages to gastric fermentation. The ruminant's food is digested slowly and its rate of passage through the gut is slow. Food takes 70 to 100 hours to pass through the gut of a cow, compared to only 30 to 45 hours for a nonruminant horse. Thus, when food is abundant, nonruminants, such as horses, consume greater quantities of food each day, process the easily digestible portion, and excrete the rest.

Because small herbivores have relatively high metabolic rates, they must digest their food more rapidly to meet their higher metabolic demands. Consequently, most small herbivores (weighing less than 5 kilograms) are intestinal fermenters that rely on rapid transit of abundant, high-quality vegetation. Large herbivores, however, have lower metabolic rates and relatively high-volume digestive systems. Thus, large herbivores can afford the slower ruminant digestive system because their metabolic rates are lower. If the animal is large enough and its metabolic rate is low enough, intestinal fermentation may rival the efficiency of gastric fermentation. This explains why the largest terrestrial herbivores, elephants and rhinoceroses, are intestinal fermenters.

In the order Artiodactyla, the structure of the foot is especially diagnostic. The foot is paraxonic; that is, the plane of symmetry passes between digits three and four. The weight of the animal is borne primarily by these digits: the first digit is always absent in living members, and the lateral digits (two and five) are always more or less reduced in size. Four complete and functional digits occur in the families Suidae, Hippopotamidae, and Tragulidae and in the forelimb of the Tayassuidae (the hindlimb has the medial digit suppressed). Two complete toes, with the lateral digits absent (Camel-idae, some Bovidae, Antilocapridae, and Giraffi-dae) or with incomplete remnants of the lateral digits (Cervidae and some Bovidae) occur in the more cursorial families. The cannon bone (fused third and fourth metapodials) is present in the families Camelidae, Cervidae, Giraffidae, Antilo-capridae, and Bovidae. Typically, the terminal pha-langes are encased in pointed hoofs. The limbs have springing ligaments (Figs. 17-1 and 17-2), and the astragalus has a "double pulley" arrangement of articular surfaces (Fig. 17-1 and Fig. 17-3) that completely restricts lateral movement. The proxi-mal articulation (with the tibia) and the distal ar-ticulation (with the navicular and cuboid, which are fused in many advanced types) of the astra-galus are critical in allowing great latitude of foot and digit flexion and extension, as is the extension of the articular surfaces and keels on the distal ends of the cannon bones to the anterior surfaces

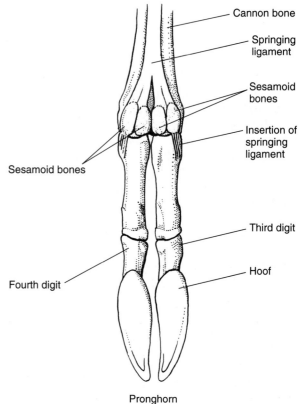

Pronghorn
(Antilocapra americana)

FIGURE 17-2 Posterior view of the left hind foot of an artiodactyl *(Antilocapra americana)*, showing the position of the springing ligament.

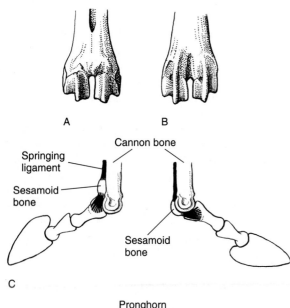

Pronghorn
(Antilocapra americana)

FIGURE 17-1 Left hind foot of the pronghorn *(Antilocapra americana)*. (A) Anterior view of the distal end of the cannon bone. (B) Posterior view of this bone. (C) Position of the phalanges: when the foot is supporting the weight of the body and the springing ligament (shown in black) is stretched and when the foot leaves the ground and the springing ligament flexes the phalanges.

(Fig. 17-1). The distinctive artiodactyl astragalus is regarded by some as a key to the success of the group. Perhaps more important, however, is the re-markable efficiency of the ruminant digestive sys-tem, developed in the most successful subgroup of artiodactyls.

The limbs of artiodactyls, especially the distal segments, are usually elongate and fairly slim. The femur lacks a third trochanter. Whereas this prominence serves as a point of insertion of gluteal muscles in perissodactyls, in artiodactyls these muscles insert more distally, on the tibia. The distal parts of the ulna and the fibula are usu-ally reduced and may fuse with the radius and tibia, respectively; this fusion is associated with the restriction of limb movement to one plane. The clavicle is seldom present. The intrinsic muscles of the feet (those that both originate and insert on

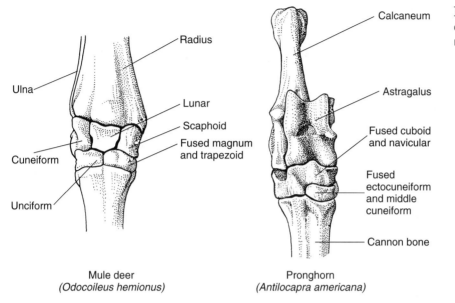

FIGURE 17-3   The right carpus of a mule deer and the right tarsus of a pronghorn.

Mule deer
(*Odocoileus hemionus*)

Pronghorn
(*Antilocapra americana*)

the feet) are usually absent, being replaced by specialized tendons and ligaments.

The skull usually has a long preorbital section, and a postorbital bar or process is always present. Horns, always of bone or with a bony core, are most often borne on the frontal bones, which are enlarged at the expense of the parietal bones. The teeth are brachydont or hypsodont and vary from 30 to 44 in number. The crown pattern is bunodont or, more often, selenodont. The premolars are not fully molariform, in contrast to the perissodactyl situation, and considerable specialization of the anterior dentition occurs in advanced artiodactyls.

The system of classification used here, that of Grubb (1993), recognizes the living suborders Suiformes, Tylopoda, and Ruminantia, all of which are first known from the Eocene (Scott and Janis, 1993; Simpson, 1984). The extinct Dichobunidae contain the most primitive known artiodactyls from the early Eocene (Carroll, 1988).

SUBORDER SUIFORMES

In members of the suborder Suiformes, which includes pigs, peccaries, and hippopotami, the molars are bunodont, the canines (and in the hippopotami, the incisors also) are tusklike (Figs. 17-4 and 17-5), and the feet usually retain four toes with complete and separate digits. The skull contrasts

with those of other artiodactyls in having a posterior extension of the squamosal bone that meets the exoccipital bone and conceals the mastoid bone. Pigs have a nonruminant type stomach with as many as three chambers but employ caecal fermentation. Hippopotami (and possibly peccaries), in contrast, have multichambered stomachs and gastric fermentation (Langer, 1988).

FAMILY SUIDAE. Swine are omnivorous and lack many structural modifications typical of more specialized artiodactyls. The Suidae is an Old World family whose present distribution includes much

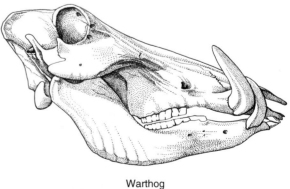

Warthog
(*Phacochoerus aethiopicus*)

FIGURE 17-4   Skull of the African warthog (length of skull 376 millimeters).

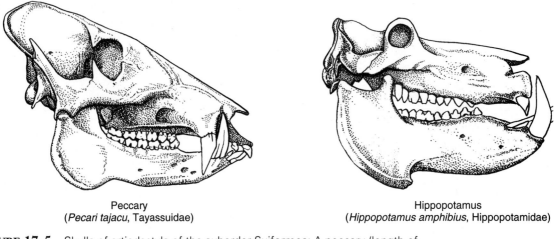

Peccary
(*Pecari tajacu*, Tayassuidae)

Hippopotamus
(*Hippopotamus amphibius*, Hippopotamidae)

FIGURE 17-5   Skulls of artiodactyls of the suborder Suiformes: A peccary (length of skull 225 millimeters), and a hippopotamus (length of skull 600 millimeters). *(Hippo after Romer, ©1966,* Vertebrate Paleontology, *University of Chicago Press)*

of Eurasia and Africa south of the Sahara. There are 16 Recent species in five genera. Suids appeared in the Oligocene. The entelodonts (Entelodontidae), an early branch on the swine evolutionary line, were huge piglike creatures with skulls up to 1 meter in length.

Most suids resemble the domestic swine *(Sus)*. Adults may weigh as much as 275 kilograms and typically have thick, sparsely haired skin. The skull is long and low and usually has a high occipital area and a concave or flat profile (Fig. 17-4). The large canines are ever-growing, and the upper canines form conspicuous tusks that protrude from the lips and curve upward. In *Babyrousa,* an Indonesian suid, the upper canines are rotated upwards to protrude from the top of the snout. Suid molars are bunodont, and the

last molars are often elongate, with many cusps and a complexly wrinkled crown surface (Fig. 17-6). The dental formula is variable even within a species; the total number of teeth ranges from 34 to 44. The limbs are usually fairly short, and the four-toed feet never have cannon bones (Fig. 17-7).

Swine inhabit chiefly forested or brushy areas, but the warthog *(Phacochoerus)* favors savanna or open grassland and is entirely herbivorous. Most suids are gregarious, and some assemble in groups of up to 50 individuals. Most species eat a broad array of plant food and carrion, and, given the opportunity, kill and eat such animals as small rodents and snakes. By comparison with ruminant artiodactyls, cursorial ability in suids is modest. Warthogs (Fig. 17-8) are fairly swift, however, and

Swine
(*Sus scrofa*, Suidae)

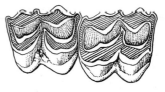

Elk
(*Cervus elaphus*, Cervidae)

FIGURE 17-6   Second and third right upper molars of the swine and the comparable teeth of an elk, with the enamel ridges unshaded, the enamel-lined depressions stippled, and the dentine cross-hatched. The molars of *Sus* are bunodont; those of *Cervus* are selenodont.

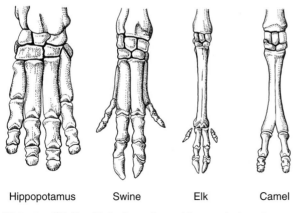

FIGURE 17-7   Right front feet of four artiodactyls: A hippopotamus *(Hippopotamus)*, swine *(Sus)*, elk (Cervus), and camel *(Camelus)*. *(After Howell, ©1944, Speed in Animals, University of Chicago Press)*

FIGURE 17-8   A pair of warthogs *(Phacochoerus aethiopicus)* showing the prominent tusks. *(Corel Corp.)*

escape their predators by speed and by taking refuge in burrows.

FAMILY TAYASSUIDAE. Tayassuids, usually called javelinas or peccaries (Fig. 17-9), are restricted to the New World, where they occur from the southwestern United States to central Argentina. There are but three Recent species of three genera *(Pecari, Tayassu,* and *Catagonus). Catagonus* was first known from Pleistocene fossils and was for many years thought to be extinct, but Wetzel et al. (1975) reported a surviving population in the biologically poorly known Gran Chaco area of Paraguay. The fossil record of tayassuids begins in the Eocene. Pre-

sumably, peccaries evolved from Old World suids, but they are not known from the Old World after the Miocene. They are more progressive in limb structure than are suids and are less carnivorous.

Peccaries are much smaller than suids; the weight of peccaries ranges up to about 30 kilograms. The skull has a nearly straight dorsal profile, and the zygomatic arches are unusually robust (Fig. 17-5). The canines are long and are directed slightly outward; the upper canines never turn upward, however, and have sharp medial and lateral edges. These opposing canines slide against one another, and the anterior surface of the upper and posterior surface of the lower are planed flat by this contact. These interlocking canines form "occlusal guides" that, together with the hingelike jaw joint, stabilize the jaw joint against forces generated when hard nuts or seeds are cracked by the rear teeth (Kiltie, 1981). The molars are roughly square and have four cusps; they have thick enamel and lack the complex wrinkled and multicusped pattern typical of suids. The dental formula is 2/3, 1/1, 3/3, 3/3 = 38.

The feet of peccaries are slender and appear delicate, and the side toes are small relative to those of suids and usually do not reach the ground. Of the three living genera of peccaries, *Catagonus* is the most cursorial. All three genera have four toes on the front foot. Although in *Catagonus* digits two and five of the hind foot are vestigial (they lack phalanges and hoofs), digit two is complete and digit five is vestigial in *Tayassu.* In

FIGURE 17-9   Collared peccary (*Pecari tajacu;* Tayassuidae). *(T. Vaughan)*

all genera, the medial metatarsals are partly fused. An additional difference is the more elongate distal segments of the limbs in *Catagonus* (Wetzel, 1977). Modern peccaries are not as cursorial as was *Mylohyus*, an extinct Pleistocene species of North America in which the side toes of the forefoot were very strongly reduced and the didactyl hind foot had a fully developed cannon bone.

Peccaries occupy diverse habitats, from deserts and oak-covered foothills in Arizona to dense tropical forests and thorn scrub in southern Mexico, Central America, and South America. They are highly social, each social group containing up to 12 individuals. Peccaries are omnivorous but seem to rely more heavily on plant material than do suids. *Catagonus* lives in small groups that range over 2 kilometers per day in search of ground cacti (Mayer and Brandt, 1982; Taber et al., 1993). The presence of peccaries is often indicated by shallow excavations where roots have been exposed beneath bushes or patches of prickly-pear cactus. Despite their chunky build, peccaries are rapid and extremely agile runners.

FAMILY HIPPOPOTAMIDAE. Family Hippopotamidae is represented today by the genera *Hippopotamus* and *Hexaprotodon* (the pigmy hippopotamus), each with one species. The group first appeared in the early Miocene in East Africa. By late Miocene, the family occurred widely in the southern parts of the Old World and was in southern Asia by the Pleistocene. Although subfossils of two species are known from Madagascar, hippopotami now occur only in Africa; in North Africa they are restricted to the Nile River drainage, but they occur widely in the southern two thirds of the continent.

Hippopotami are bulky creatures with huge heads and short limbs (Fig. 17-10). They are large; *Hippopotamus* weighs up to 3600 kilograms and *Hexaprotodon* about 180 kilograms. Some of the distinctive features of these animals probably evolved in association with their amphibious mode of life. Specialized skin glands secrete a pink, oily substance that protects the sparsely haired body from the sun. The highly specialized skull has elevated orbits and enlarged and tusklike canines and incisors (Fig. 17-5). The bunodont molars are basically four cusped; the dental formula is 2-3/1-3, 1/1, 4/4, 3/3 = 38-44. The limbs are robust, and the feet are four-toed (Fig. 17-7). The foot posture

FIGURE 17-10   A pair of hippopotami *(Hippopotamus amphibius)* head for shore in Kenya. *(Corel Corp.)*

is semidigitigrade; only the distal phalanx of each toe touches the ground. The broad foot is braced by a sturdy "heel" pad of connective tissue, and the central digits are nearly horizontal.

Hippopotami are restricted to the vicinity of water. *Hippopotamus* is gregarious, and groups spend much of the day in the water. When bodies of water are scarce during the dry season, hippopotami often concentrate in stagnant ponds, which they churn into muddy morasses. They are good swimmers and divers and, when submerged, are able to walk on the bottom of rivers or lakes, using a slow-motion gait. Hippos are known to communicate by producing a wide variety of sounds; some of their underwater sounds resemble the sonar clicks of dolphins (p. 423). At night, hippopotami may move far inland to feed on vegetation. *Hexaprotodon* is solitary or occurs in pairs and inhabits forested areas. Instead of seeking shelter in the water when disturbed, as is characteristic of *Hippopotamus*, *Hexaprotodon* seeks refuge in dense vegetation.

SUBORDER TYLOPODA

FAMILY CAMELIDAE. Camelids are primitive ruminants that are restricted to arid and semiarid regions. *Camelus*, with two species, occupies the Old World, and wild populations persist in the Gobi Desert of Asia. *Lama*, with three species, occurs in South America from Peru through Bolivia, Chile,

Argentina, and Tierra del Fuego. The genus *Vicugna* contains one species restricted to the Andes Mountains south of Peru (Fig. 17-11). Camels probably arose in the Old World and migrated in the middle Eocene to North America, where their fossil record begins.

Of special interest, as an example of a reversal of a well-established evolutionary trend, is the development of the camelid foot. By the Oligocene, the camel's foot was already highly specialized. It was nearly unguligrade in posture and didactyl, and the distalmost phalanges probably bore hoofs. The distinctive distal divergence of the metapodials (Fig. 17-7), however, was already recognizable. In the Miocene, the central metapodials fused to form a cannon bone, but, at this same time, a retrograde trend toward the secondary development of a digitigrade foot posture began, and from Pliocene times onward camels were digitigrade. They are the only extant fully digitigrade ungulates. Because semiarid conditions developed and became widespread in the Miocene, it is tempting to relate the changes in the camelid foot posture to changing soil conditions. In any case, the camelid foot clearly provides effective support on soft, sandy soil, into which the feet of "conventional" unguligrade artiodactyls sink deeply. Taking advantage in the Pleistocene of land bridges between North America and Asia, and between

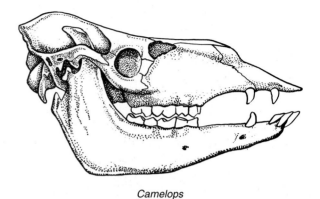

Camelops

FIGURE 17-12   Skull of an extinct Pleistocene New World camel *(Camelops);* length of skull 565 millimeters. *(After Romer, © 1966,* Vertebrate Paleontology, *University of Chicago Press)*

North and South America, camels spread to the Old World and to South America.

Although highly specialized in foot structure, camelids are the most primitive living ruminants. They are large mammals, ranging in weight from about 60 to 650 kilograms, and have long necks and long limbs. The dentition has advanced less toward herbivorous specialization than has that of the Ruminantia. In camelids, only the lateral upper incisor is present, but it is caniniform (Fig. 17-12); the lower canines are retained and are little modified.

FIGURE 17-11 The vicuña *(Vicugna vicugna),* a camelid that inhabits the central Andes of South America. *(C. Koford)*

The lower incisors are inclined forward and occlude with a hardened section of the gums on the premaxillary bones. A broad diastema is present, and the premolars are reduced in number (to 3/1-2 in *Camelus* and 2/1 in *Lama*). As in other ruminants, the limbs are long and the ulna and fibula are reduced; the trapezium is absent in the carpus, and the mesocuneiform and ectocuneiform are fused in the tarsus. The digitigrade feet are didactyl, but the cannon bone is distinctive in that the distal ends of the metapodials remain separate and flare outward (Fig. 17-7). The toes are separate, and each is supported by a broad cutaneous pad that largely encases the second phalanx and greatly increases the surface area of the foot. The short ungual phalanges do not bear hoofs but have nails on the dorsal surfaces.

Camels are remarkably well adapted to arid areas. They have the ability to go without drinking free water for months (p. 395). They conserve water by producing dry feces and little urine. They have the ability to allow body temperature to rise 6°C during the heat of the day. Camels have specialized nasal cavities that reduce water loss via evaporation. In addition, camels can store fat in their humps for use during food shortages (Fig. 17-13; MacDonald, 1984). They are grazers and can survive in regions with only sparse vegetation. The guanaco *(Lama guanicoe)* and vicuña *(Vicugna*

vicugna) of South America are gregarious and live in small social groups dominated by an adult male. Guanacos are fairly speedy runners but are especially adroit at moving rapidly over extremely rough terrain. Camelids are highly vocal. The dromedary (Arabian) camel *(Camelus dromedarius)* makes an assortment of snarling sounds, and guanacos give a yammering call.

SUBORDER RUMINANTIA

Suborder Ruminantia includes giraffes, deer, antelope, sheep, goats, and cattle. Members of this most advanced artiodactylan suborder have been in the past, and remain, the dominant artiodactyls. In general, these animals are committed strictly to an herbivorous diet and to highly cursorial locomotion. Ruminants chew their cud; the stomach has three or four chambers (see Fig. 2-12) and supports microorganisms that have cellulolytic enzymes. Ruminants have selenodont molars (Fig. 17-6), and the anterior dentition is variously specialized by loss or reduction of the upper incisors, by the development of incisiform lower canines, and commonly by the loss of upper canines. The skull differs from those of members of the suborder Suiformes in the exposure of the mastoid bone between the squamosal and exoccipital bones. Antlers or horns, often large and complex structures, are present in

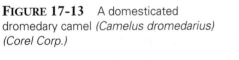

FIGURE 17-13 A domesticated dromedary camel *(Camelus dromedarius)* *(Corel Corp.)*

the most progressive families. In the limbs, there is a pronounced trend toward elongation of the distal segments, fusion of the carpals and tarsals, and perfection of the two-toed foot. A diagnostic feature of the ruminants is the fusion of the navicular and cuboid bones, over which the astragalus is nearly centered (Fig. 17-3).

The advanced artiodactyls (Cervidae, Giraffidae, Antilocapridae, and Bovidae) share a series of progressive features. The upper incisors are absent, the upper canines are usually absent, the lower canines are incisiform, and the cheek teeth are selenodont. The dental formula is typically 0/3, 0/1, 3/3, 3/3 = 32. The cannon bone is present in fore and hindlimbs, its distal articular surfaces are extensive, and the lateral digits are always incomplete (Fig. 17-7) and are often lacking. Movement of the foot is strongly limited to a single plane by the tongue-in-groove contacts between the astragalus and the bones with which it articulates and by the specialized articular surfaces of the joint between the cannon bone and the first phalanges (Fig. 17-2). Some fusion of carpal elements always occurs and further restricts movement to one plane. The navicular and cuboid bones are always fused (Fig. 17-3), and a variety of fusion patterns occur in the other elements. The four-chambered stomach is of a ruminant type. Although all ruminants but the tragulids share this basic structural plan, each family has distinctive features, usually related to diet and degree of cursorial ability.

FAMILY TRAGULIDAE. Family Tragulidae, which contains the chevrotains (mouse deer), has only three living genera with four species, but is of interest because these animals probably resemble in many ways the ancestors of the more advanced ruminants. Chevrotains are small, delicate creatures, weighing from 2.3 to 4.6 kilograms, that occur in tropical forests in central Africa *(Hyemoschus)* and in parts of southeastern Asia *(Tragulus)*. The tragulid fossil record begins in the Miocene of Eurasia and Africa.

Although apparently related to more advanced ruminants, chevrotains combine a unique complex of features. The tragulid skull never bears antlers, but, seemingly in compensation, the upper canines are tusklike and are used by males in intraspecific combat. Otherwise, the dentition resembles that of higher ruminants: the upper incisors are lost, the lower canine is incisiform, and the cheek teeth are selenodont. Unique to tragulids is an ossified plate, derived from an **aponeurosis** (a membranous sheet of tendon) to which the sacral vertebrae attach. The limb structure is a mosaic of primitive and advanced features. Although the limbs are long and slender and the carpus is highly specialized in having the navicular, cuboid, and ectocuneiform bones fused, the lateral digits are complete, a condition never present in the more advanced ruminants. In addition, although a cannon bone occurs in the hindlimb, the metacarpals of the central digits are separate in the African tragulid and are partly fused in the Asian form, whereas the cannon bone is represented by fully fused metapodials in all other ruminants.

Tragulids are secretive, chiefly solitary, nocturnal creatures that inhabit forests and underbrush and thick growth along water courses. They escape predators by darting along diminutive trails into dense vegetation or into water in the case of the water chevrotain *(Hyemoschus aquaticus)*. Their food consists of grass, the leaves of shrubs and forbs, and some fruit. They have a four-chambered stomach that is not as specialized as in other ruminants.

FAMILY GIRAFFIDAE. Family Giraffidae is represented today by but two monotypic genera, *Giraffa* and *Okapia*. The family occurs in much of Africa south of the Sahara.

The robust cheek teeth of giraffids are brachydont and are marked with rugosities. Short, bony horns, covered with furred skin, are borne on the front part of the parietals, and a medial thickening of the bone in the area where the nasals and the frontals join is conspicuous (Fig. 17-14). In some populations from north of the equator in East Africa, this thickening produces a median horn. Horns occur in both sexes and are never shed. The lateral digits of the elongate limbs are entirely gone, and the tarsus is highly specialized. Distal to the astragalus and calcaneum, only two tarsal bones are present. One is formed by the fusion of the navicular and cuboid bones and the other by the fusion of the three cuneiform bones. The okapi lacks the extreme elongation of the neck and legs that is typical of giraffes but has an even

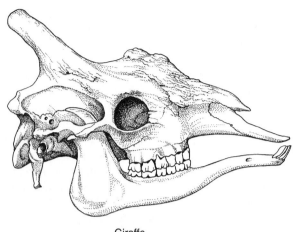

Giraffe
(Giraffa camelopardalis)

FIGURE 17-14 Skull of a male giraffe (length of skull 708 millimeters). The heavy deposits of bone on the frontal and nasal bones seemingly protect the skull when the head is used as a weapon in fights between males.

more specialized tarsus in which all bones distal to the astragalus and calcaneum are fused. The fossil record of the giraffids begins in the Miocene Epoch of Africa. The okapi, not known to zoologists until 1900, is remarkable in its close resemblance to primitive late Miocene and early Pliocene giraffids long known to paleontologists (Palaeotraginae).

Giraffes occur in savannas, semideserts, and lightly wooded areas, where their exceptional height enables them to browse on branches of leguminous trees up to 6 meters above the ground. Some of the species of *Acacia* that these animals feed on bear long thorns, but giraffes adroitly use their long tongue and prehensile upper lip to gather leaves from even the most thorny acacias. Acacia leaves are high in protein and are the most preferred and important food of giraffes (Sauer et al., 1982). Despite their considerable weight (to nearly 1820 kilograms in males), giraffes can gallop for short distances up to 60 kilometers per hour (Fig. 17-15; MacDonald, 1984). Relative to those of lighter, shorter limbed artiodactyls, the limbs of giraffes are flexed little during each stride, producing a stiff-legged gait. When walking, giraffes swing both legs from the same side forward at nearly the same time, but when galloping

the hind legs swing forward together. When the animal is walking or galloping, its center of gravity is partly controlled by fore-and-aft movements of the head and neck (Dagg, 1962). Ritualized fighting by males involves powerful blows by the head against the opponent's head, neck, and body.

The okapi lives in dense tropical forests of central Africa. Okapis eat leaves and fruit. They are diurnal and usually solitary but will form small family groups.

FAMILY MOSCHIDAE. The musk deer were traditionally placed in the family Cervidae but are now placed in a separate family (Groves and Grubb, 1987; Janis and Scott, 1987; Webb and Taylor, 1980). This small family contains one genus and four species that occupy an area including north-

FIGURE 17-15 Two giraffes *(Giraffa camelopardalis)* running across the African savanna. *(Corel Corp.)*

ern Afghanistan, Pakistan, India, the Himalayan plateau of southern China, Nepal, Tibet, and North Vietnam. The earliest musk deer are known from the Oligocene of Europe (Dawson and Krishtalka, 1984). Musk deer have coarse fur, and their hindlimbs are longer than the forelimbs. They lack antlers, but have saber-like upper canines. As their name implies, males have a musk gland in the abdomen that secretes a waxy substance that is used as a base for expensive perfumes. Due to extensive hunting, musk deer populations have declined rapidly, and all species are listed as threatened or vulnerable by the IUCN (International Union for the Conservation of Nature and Natural Resources). The Chinese have succeeded in breeding musk deer in captivity for the production of musk, and this may take the pressure off some wild populations (MacDonald, 1984).

FAMILY CERVIDAE. Members of family Cervidae, which includes the muntjacs, deer, elk, caribou, and moose, occur throughout most of the New World and in Europe, Asia, and northwestern Africa; they have been introduced widely elsewhere. Grubb (1993) recognizes 16 genera and 43 species, not including the recently discovered giant muntjac *(Megamuntiacus vuquangensis)* from the Annamite Mountains on the border between Laos and Vietnam (Schaller and Vrba, 1996; Fig. 17-16). Cervids appeared in the early Miocene in Asia, reached North America in the Miocene, and entered South America in the Pleistocene over the Panama land bridge (Carroll, 1988).

Antlers are the most widely recognized characteristic of members of the family Cervidae (Fig. 17-17). Antlers attain spectacularly large size in some species and vary widely from one species to another. People have long been fascinated by their complexity, variety, and symmetry. All but two of the 43 species of cervids have antlers, and they occur only in males, except in caribou *(Rangifer;* Fig. 17-18). In some antlered cervids, the upper canines are retained but reduced (as in the elk, *Cervus elaphus)*. Two cervids with short antlers have enlarged canines *(Muntiacus* and *Elaphodus),* and in one deer there are no antlers *(Hydropotes),* but the canines are enlarged sabers. Antlers usually arise from a short base on the frontals (the pedicel, Fig. 17-17) and are entirely bony. Of particular interest is the annual cycle of rapid growth of the

FIGURE 17-16   A giant muntjac *(Megamuntiacus vuquangensis)* with antlers in velvet. *(G. Schaller, from Schaller and Vrba, 1996).*

antlers, their use during the breeding season in ritualized social interactions, and their subsequent loss.

This annual cycle has been thoroughly studied in the white-tail deer *(Odocoileus virginianus)* of North America (Wislocki, 1942; Wislocki et al., 1947; Goss, 1983). The cycle is primarily under the control of testicular and pituitary hormones. In the Northern Hemisphere, secretions from the pituitary, activated by increasing daylength in the spring, initiate antler growth in April or May. Some time later, pituitary gonadotropin stimulates growth of the testes. The growing antlers are covered by "velvet," a fur-covered skin that carries blood vessels and nerves. In the fall, androgen from the enlarging testes inhibits the action of the pituitary antler-growth hormone, leading to the drying and loss of the velvet. Then the animals rub and thrash their antlers against vegetation, and, as

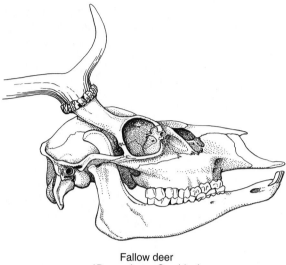

Fallow deer
(*Dama dama,* Cervidae)

FIGURE 17-17   Skull of a male fallow deer (length of skull 265 millimeters). The bony antlers are shed yearly.

the velvet is removed, the antlers are stained by resins and take on a brown, polished look. In the fall and early winter, androgens maintain the connection between the dead bone of the antlers and the live frontal bones, and, during the fall breeding season, the antlers are used in clashes between males competing for females. In winter, pituitary stimulation of the testes declines as daylength is reduced, and androgen secretion diminishes. This results in decalcification in the **pedicel,** weakness at the point of connection between the antler and the pedicel, and shedding of the antlers. For several months in late winter, before reinitiation of antler growth, the males are antlerless.

The cheek teeth of cervids are brachydont, reflecting a browsing habit. Body sizes of cervids range widely: the pudu deer *(Pudu)* weighs from 7 to 10 kilograms, whereas the moose *(Alces)* weighs up to roughly 800 kilograms. The feet are always four-toed, but the lateral toes are often greatly reduced. Distal to the astragalus and calcaneum, the tarsus is usually composed of three bones: the fused navicular and cuboid, the fused ectocuneiform and mesocuneiform, and the internal cuneiform (Fig. 17-3).

Cervids occur from the Arctic to the tropics. Many are well adapted to boreal regions and occupy mountainous or subarctic areas with severely cold winters. Effective insulation is provided in

many cervids by the long, hollow hairs of the pelage. Some species are gregarious for much of the year and may assemble in large herds during the winter and during migratory movements.

FAMILY ANTILOCAPRIDAE. Family Antilocapridae is represented today by one species, the pronghorn *(Antilocapra americana;* Fig. 17-19), which occupies open country from central Canada to north central Mexico. The fossil record of these animals is entirely North American and begins with *Merycodus* in the early Miocene. The horns of this antilocaprid were cervid-like in form, that is, they were forked and had a basal burr. In contrast to cervids, however, the cores were never shed, and both sexes had horns. In addition, the unusually prominent orbits were situated high and far back in the

FIGURE 17-18   A male caribou *(Rangifer tarandus)* with antlers in velvet. *(Corel Corp.)*

FIGURE 17-19   The pronghorn *(Antilocapra americana)*, one of the fastest cursorial mammals. The long, slender limbs are typical of the more cursorial ungulates. *(G. Bear)*

skull, as in *Antilocapra*. The horns of fossil pronghorns were generally more complex than those of *Antilocapra*, but the limbs and teeth of fossil forms were similar to those of the present species. The pronghorn fauna was at one time more diverse than it is now. From the middle Pleistocene Tacubaya Formation in central Mexico, which contains numerous fossils of mammals that lived in a small area at one time, Mooser and Dalquest (1975) list four species of extinct pronghorns, ranging in size from a tiny species to one at least as large as the living species.

The pronghorn is unique in being the only mammal that sheds its horn sheaths annually. The sheaths are of keratinized skin. The old sheath, beneath which a new sheath is beginning to develop, is shed annually in early winter, and the new sheath is fully grown by July (O'Gara et al., 1971). Whereas the mature sheath is forked, the bony core is a single, laterally compressed blade. Both sexes have horns, but those of the females are small and inconspicuous. The dental formula is that typical of pecorans, and although the animals are largely browsers on low shrubs and forbs, the cheek teeth are high-crowned. The abrasive soil particles adhering to the low-growing vegetation they eat probably make high-crowned teeth advantageous. Perhaps as an adaptation allowing a pronghorn to watch for danger while its head is close to the ground, the orbits are unusually far back in the skull (Fig. 17-20). The legs are long and slender, and all vestiges of lateral digits are gone. The tarsus distal to the astragalus and calcaneum consists of only three bones (Fig. 17-3).

Pronghorns inhabit open prairies and deserts that support at least fair densities of low grasses, shrubs, and forbs. The numbers of pronghorn were seriously reduced during the pioneering period of the western United States. Today pronghorns are common in a number of the western states (O'Gara, 1978), but they have been all but eliminated from much of their original range in states such as Kansas. They are among the fastest of cursorial mammals. At full speed on level footing, they can attain a speed of at least 85 kilometers per hour (McLean, 1944; Einarsen, 1948; Kitchen, 1974).

Although it may yield top speed to the cheetah, in high-speed endurance running, the pronghorn is probably unsurpassed by any other mammal. Pronghorns can sustain high speeds (about 65 kilometers per hour) over distances greater than 10 kilometers (McKean and Walker, 1974). Hence, they can cover 11 kilometers in only 10 minutes. This performance demands a rate of oxygen uptake over three times that predicted for a similar-size cursor. This aerobic ability is not facilitated by unusual locomotor efficiency (energy cost per unit of distance covered) or unusually large muscles, but by an integrated suite of muscular and respiratory adaptations that provide for extremely rapid and prolonged oxygen uptake (Lindstedt et al., 1991). Compared to other similar-size mammals, pronghorns have enlarged airways to the lungs (trachea and bronchi), greater lung surface areas, higher concentrations of hemoglobin, and greater densities of muscle capillaries and mitochondria.

FAMILY BOVIDAE. The family Bovidae, which includes African and Asian antelope, bison, sheep, goats, and cattle, is the most important and most diverse living group of ungulates (Table 17-1). The family includes 45 genera and 137 species, and wild species occur throughout Africa, in much of Europe and Asia, and in most of North America.

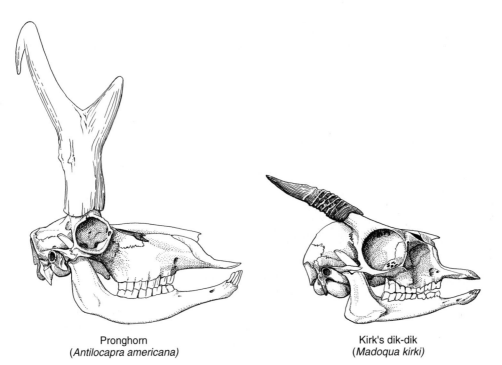

Pronghorn
(*Antilocapra americana*)

Kirk's dik-dik
(*Madoqua kirki*)

FIGURE 17-20   Skulls of a pronghorn (length of skull 292 millimeters) and a Kirk's dik-dik (length of skull 108 millimeters). The receding nasal bones of the dik-dik are an adaptation allowing mobility of the short proboscis.

Two new bovid species, the saola (*Pseudoryx nghetinhensis*) and another species known only from its horns (*Pseudonovibos spiralis*), were recently discovered in the forests of Laos and Vietnam (Peter and Feiler, 1994; Schaller and Rabinowitz, 1995; Robichaud, 1998). The domestication of bovids began in Asia roughly 8000 years ago (MacDonald, 1984), and domesticated bovids are nearly as cosmopolitan in distribution as are humans.

The systematic relationships of genera and species within the Bovidae remain controversial (Allard et al., 1992; Essop et al., 1997), and several authors even disagree about the monophyly of the family (Allard et al., 1992; Gatesy et al., 1992). Bovids seemingly derived from traguloid ancestry in the Old World and first appeared in the Oligocene of Asia (*Eotragus;* Ginsburg and Heintz, 1968). Africa was the center of the early and rapid radiation that gave rise to a diverse bovid fauna,

judging from the many kinds of bovids known from the Miocene. More Miocene than Recent genera of bovids are known (78 versus 49).

Toward the end of the Pleistocene, most bovids were driven from Europe by the southward advance of the cold climate. A few reached the New World in the Pleistocene via the Bering Strait land bridge. Because this boreal avenue of dispersal was under the influence of cold climates, it functioned as a "filter bridge" (Simpson, 1965b), and only animals adapted to the cold dispersed across it.

As a consequence, the New World received from Asia such bovids as the bighorn sheep (*Ovis*) (Fig. 17-21), the mountain "goat" (*Oreamnos;* Fig. 17-22), the musk ox (*Ovibos*), and the bison (*Bison*). Bovids less able to withstand boreal conditions—the Old World antelopes and the gazelles are prime examples—were forced from the northern parts of Europe and Asia in the Pleistocene back to their pre-

TABLE 17-1 Classification of Bovidae and Distribution of Recent Genera

Genera (number of species)	Common Name and Continent(s)
Subfamily Aepycerotinae	
Aepyceros (1)	Impala (Africa)
Subfamily Alcelaphinae	
Alcelaphus (1)	Hartebeest (Africa)
Connochaetes (2)	Wildebeest (Africa)
Damaliscus (3)	Hartebeest, topi, blesbok (Africa)
Sigmoceros (1)	Hartebeest (Africa)
Subfamily Antilopinae	
Ammodorcas (1)	Dibatag (Africa)
Antidorcas (1)	Springbok (Africa)
Antilope (1)	Blackbuck (Asia)
Dorcatragus (1)	Beira antelope (Africa)
Gazella (16)	Gazelle (Africa)
Litocranius (1)	Gerenuk (Africa)
Madoqua (4)	Dik-dik (Africa)
Neotragus (3)	Royal, pygmy, and suni antelope (Africa)
Oreotragus (1)	Klipspringer (Africa)
Ourebia (1)	Oribi (Africa)
Pantholops (1)	Chiru (Asia)
Procapra (3)	Black-tailed gazelle (Asia)
Rhaphicerus (3)	Steinbuck, grysbuck (Africa)
Saiga (1)	Saiga (Asia, Europe)
Subfamily Bovinae	
Bison (2)	Bison (Europe, North America)
Bos (5)	Cattle (worldwide)
Boselaphus (1)	Nilgai (Asia)
Bubalus (5)	Asiatic buffalo (Asia)
Syncerus (1)	African buffalo (Africa)
Taurotragus (2)	Eland (Africa)
Tetracerus (1)	Four-horned antelope (Asia)
Tragelaphus (7)	Bushbuck, nyala, kudu, bongo (Africa)
Subfamily Caprinae	
Ammotragus (1)	Barbary sheep (North Africa)
Budorcas (1)	Takin (Asia)
Capra (9)	Ibex, goat (Asia, Europe, North America)
Hemitragus (3)	Tahr (Asia)
Naemorhedus (6)	Goral (Asia)
Oreamnos (1)	Rocky mountain "goat" (North America)
Ovibos (1)	Musk ox (North America, Greenland)
Ovis (6)	Mouflon, argali, bighorn sheep, sheep (Asia, Europe, North America, North Africa)
Pseudois (2)	Nahur (Asia)
Rupicapra (2)	Chamois (Southwest Asia)
Subfamily Cephalophinae	
Cephalophus (18)	Duiker (Africa)
Sylvicapra (1)	Bush duiker (Africa)

(continued)

TABLE 17-1 *(continued)*

Genera (number of species)	Common Name and Continent(s)
Subfamily Hippotraginae	
Addax (1)	Addax (Africa)
Hippotragus (3)	Roan, sable (Africa)
Oryx (3)	Oryx (Africa)
Subfamily Peleinae	
Pelea (1)	Rhebuck (Africa)
Subfamily Reduncinae	
Kobus (5)	Waterbuck, kob, lechwe (Africa)
Redunca (3)	Reedbuck (Africa)

From Grubb, 1993.

sent strongholds in Africa and Asia and hence did not disperse across the Bering bridge to North America. An exception is the saiga antelope *(Saiga),* which now inhabits arid parts of Asia but which occurred in Alaska in the Pleistocene. *Bison* were extremely abundant members of grassland faunas in the Recent in North and Central America, where they occurred as far south as El Salvador. Some structural divergence occurred in Pleistocene bison; in some areas, several species may have occupied common ground. Some Pleistocene bison were considerably larger than the present *Bison bison.* Specimens of the Pleistocene species *Bison latifrons* from North America indicate that this animal was over 2 meters high at the shoulder and had horns that in larger individuals spanned more than 2 meters.

Bovids characteristically inhabit grasslands, and their advanced dentition and limbs probably de-

FIGURE 17-21 Male Rocky Mountain bighorn sheep *(Ovis canadensis). (G. Bear)*

FIGURE 17-22 A mountain goat *(Oreamnos americanus). (PhotoDisc, Inc.)*

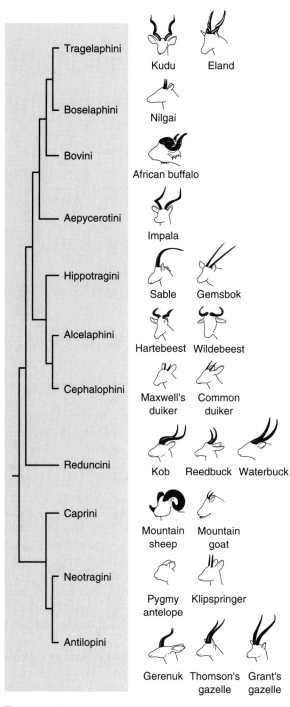

FIGURE 17-23   Phylogeny of the Bovidae based on mitochondrial rRNA data of Allard et al., 1992 showing typical horn morphology for certain members of each tribe. *(From Lundrigan, 1996)*

A

B

FIGURE 17-24   Some members of the diverse bovid fauna of Africa: (A) greater kudu *(Tragelaphus strepsiceros)*, (B) red hartebeest *(Alcelaphus buselaphus). (T. Vaughan)*

veloped in association with grazing habits. The cheek teeth are high-crowned, and the upper canines are reduced or absent. **Preorbital vacuities** in the skull are present in some bovids and absent in others. The lateral digits are reduced or totally absent, the ulna is reduced distally and is fused with the radius, and only a distal nodule remains as a vestige of the fibula. Horns, formed of a bony core and a keratinized sheath, are present in males of all wild species, and females often bear horns also

FIGURE 17-25 Gemsbok *(Oryx gazella). (T. Vaughan)*

(Geist, 1966). The entire horns (including both sheath and core) are never shed and in some species grow throughout the life of the animal. Bovid horns are never branched but are often large and occur in a variety of forms (Fig. 17-23). Males of the Indian four-horned antelope *(Tetracerus quadricornis)* are unique in having four short, dagger-like horns. The horns are frequently used in fights between males during the breeding season, but, in many species, the horns are used in ritualized tests of body strength in such a way as to minimize injury (p. 441). Some bovids, such as oryx and gemsbok *(Oryx),* can use their horns as awesome defensive weapons, respected even by lions.

Horns occur in both sexes of many antelope, but the females of some species are hornless; horns are usually present in both sexes of the larger species but absent in the females of the smaller species. Females have horns in 75 percent of the genera in which the average female weight is more than 40 kilograms, whereas females nearly always lack horns in genera in which females weigh less than 25 kilograms. Further, when both sexes have horns, those of the males are thicker at the base, more complex in shape, and adapted to withstanding the forces encountered during intraspe-

cific combat (Lundrigan, 1996). The horns of the female are straighter and thinner, better adapted to stabbing, and thus more effective defensive weapons. The relationship in females between body weight and the possession of horns is probably correlated with antipredator behavior: larger species depend on direct defense, but small species flee or use concealment. Large antelope offer especially effective antipredator defense because they are larger than most of their predators and are much larger than the predators of their young.

The last great strongholds of bovids are the grasslands and savannas of East Africa, but a diversity of bovids still occurs in parts of central and southern Africa (Fig. 17-24 and Fig. 17-25). Seemingly, every conceivable bovid niche has been occupied. Some antelope, such as the Bohor reedbuck *(Redunca redunca)* and the lechwe *(Kobus leche),* inhabit river borders and swampy ground, while at the other extreme, the oryx *(Oryx)* and springbok *(Antidorcas marsupialis)* live in arid plains and deserts, where they have limited access to drinking water. The protection afforded game in some parts of Africa should allow the survival of many species of this diverse and handsome group.

WEB SITES

General artiodactyl information

http://www.ipl.org/exhibit/dino/artiodactylstext.html

General information about ungulates

http://blackbox1.wittenberg.edu/academics/biol/courses/mammals/perart.htm

Details of the biology of many artiodactyl species from the Animal Diversity web

http://www.oit.itd.umich.edu/biol08/Chordata/Mammalia/Artiodactyla.shtml

Artiodactyl information with details of many species

http://www.pathcom.com/~dhuffman/artiodactyla.html

Pronghorn information

http://yellowstone.minns.com/jh/Wildlife/Antelope.htm

The Giraffe Project — lots of information about giraffes

http://www.whidbey.com/giraffe/

Basic information about camels

http://www.arab.net/camels/welcome.html

National Zoo's page on pygmy hippos

http://web3.si.edu/organiza/museums/zoo/zooview/exhibits/elehouse/hippo/wildp.htm

Deer Facts home page with information and links about deer

http://www.mich.com/~serenget/facts.htm

18 RODENTIA

Because plants are the most abundant food source for terrestrial mammals, it is not surprising that most members of the largest mammalian order, Rodentia, are herbivorous. Approximately 43 percent of all mammals are rodents, with some 29 living families, 443 genera, and roughly 2004 species (Wilson and Reeder, 1993). Rodents have been, and remain today, spectacularly successful: they are nearly cosmopolitan in distribution, exploit a broad spectrum of foods, are important members of most terrestrial faunas, and often reach extremely high population densities. Repeated rodent radiations have occurred at many times in many places. As a result of convergent evolution,

many rodents have lifestyles and morphological features similar to those of members of other orders. Among rodents that radiated in South America, for example, are species that resemble rabbits or small antelopes, and one species fills the ecological role of a miniature hippopotamus. Rodents are an unusually complex group with respect to morphological diversity, lines of descent, and parallel evolution of similar features in different evolutionary lines. Parallelism is regarded by Wood (1935) as "the evolutionary motto of the rodents in general." Because of these complexities, disagreement among zoologists as to relationships among rodent taxa has been the rule (Luckett and Hartenberger, 1985).

The terms "sciuromorph," "myomorph," and "hystricomorph" have been used repeatedly to designate major divisions within the order Rodentia and to refer to basic patterns in the arrangement of the masseter muscles, the skull, and the zygomatic arch. There has been little agreement, however, regarding the use of these terms in the taxonomic scheme of rodents. The classification used here (Table 4-1) is primarily that of Wilson and Reeder (1993), who follow Carleton (1984) in recognizing two main divisions, the suborders Sciurognathi and Hystricognathi.

Rodents are fascinating partly because of the very features that make them difficult to classify.

Their complex patterns of evolution, different morphological solutions to similar basic functional problems, intricate systems of resource allocation, and finely tuned adaptations to such extreme environments as those in arctic and desert areas make rodents a rewarding group to study.

MORPHOLOGY

Rodents range in size from about 5 grams to 50 kilograms, and Recent members of the order Rodentia share a series of distinctive cranial features (Table 18-1). The upper and lower jaws each bear a single pair of persistently growing incisors, a feature developed early in the evolution of rodents and one that committed them to an essentially herbivorous mode of feeding, while still permitting the exploitation of such abundant foods as insects. Because only the anterior surfaces are covered with enamel, the incisors assume a characteristic beveled tip as a result of wear. The occlusal surfaces of the cheek teeth are often complex and allow for effective sectioning and grinding of plant material. In some rodents, the cheek teeth are ever-growing. The dental formula never exceeds 1/1, 0/0, 2/1, 3/3 = 22, and a diastema is always present between the incisors and the premolars. The incisors and canines are always 1/1, 0/0.

TABLE 18-1 Uniquely Derived Characters Used to Diagnose a Monophyletic Rodentia

- One pair of upper and lower incisors; each tooth is enlarged, sharply beveled, and ever-growing
- Broad diastema (space) between incisors and premolars of both upper and lower jaws resulting from the loss of canines and some cheek teeth
- Incisor enamel restricted to the outside surface only
- Paraconid lost on lower cheek teeth
- Orbital cavity lying just dorsal to cheek teeth
- Ramus of the zygoma lies anterior to the first cheek tooth
- Glenoid fossa is an anterior-posterior trough allowing fore and aft movement of the mandible

---

From Hartenberger, 1985.

The glenoid fossa of the squamosal bone is elongate and allows anteroposterior and transverse jaw action. The mandibular symphysis has sufficient give in many species to enable the transverse mandibular muscles to pull the ventral borders of the rami together and spread the tips of the incisors. The masseter muscles are large and complexly subdivided, provide most of the power for mastication and gnawing, and, in all but one species *(Aplodontia rufa),* have at least one division that originates on the rostrum. The temporal muscles are usually smaller than the masseters, and their point of insertion, the coronoid process, is usually reduced. In general, rodents have undergone little postcranial specialization. There are notable exceptions, however, among saltatorial, fossorial, and gliding rodents.

PALEONTOLOGY

Rodents are an old group, dating back to the late Paleocene of Asia and North America. The earliest fossils are jaws and scattered teeth, about 56 million years old, representing the primitive families Alagomyidae and Paramyidae (Meng et al., 1994; Dawson and Beard, 1996). Early Eocene ctenodactyloid rodents from Asia seem equally primitive in zygomasseteric structure (the structure of the masseter muscles and the parts of the skull from which they originate) but are probably more primitive dentally (Dawson et al., 1984). The structure of the skull of primitive rodents (Fig. 18-1) indicates that the temporalis muscle was large and the masseter muscles were not highly specialized and originated entirely from the zygomatic arch (Fig. 18-2). The primitive dental formula is 1/1, 0/0, 2/1, 3/3 = 22, and the cheek teeth were brachydont.

Rodents may have diverged early into two major groups. This hypothetical dichotomy is sometimes recognized systematically by the suborders Sciurognathi and Hystricognathi, names that refer to contrasting types of mandibles. In the **sciurognathous** type (Fig. 18-3A), the angular process of the dentary bone originates in the vertical plane that passes through the alveolus of the incisor and is ventral to the alveolus. In the **hystricognathous** type (Fig. 18-3B and C), the angular process is not in the same plane as the horizontal ramus (Korth, 1994). In contrast to the sciurognathous dentary, the hystricognathous dentary tends to have a more strongly reduced coronoid process and its lower border is generally marked by a more prominent projection at the base of the incisor root (Fig. 18-3C). These characters, however, vary widely within each suborder. It is possible that the hystricognathous condition evolved in parallel in the extinct middle Eocene rodent *Prolapsus* (Sciuravidae, Sciurognathi) (Wilson and Runkel, 1991).

The early success of rodents is indicated by their Eocene abundance in Eurasia and North America and by their rapid radiation. In the Eocene, when they were seemingly replacing the

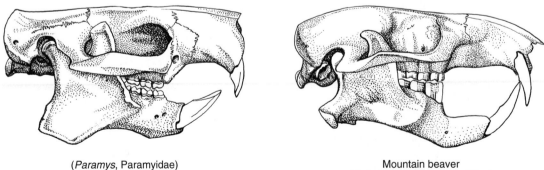

(*Paramys*, Paramyidae)

Mountain beaver
(*Aplodontia rufa*, Aplodontidae)

FIGURE 18-1   Skulls of a primitive late Paleocene and early Eocene sciuromorphous rodent (Paramyidae; length of skull 89 millimeters) and a mountain beaver (Aplodontidae; length of skull 68 millimeters). (Paramys *after Romer,* © *1966,* Vertebrate Paleontology, *University of Chicago Press)*

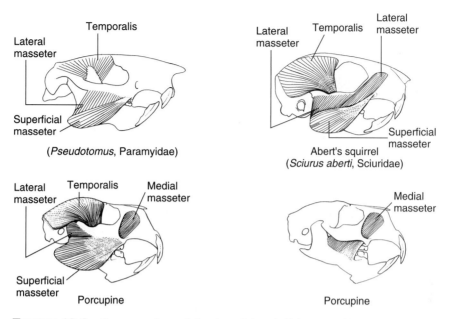

FIGURE 18-2 Patterns of specialization of the skull, jaws, and jaw musculature of rodents. The jaw muscles are restored in *Pseudotomus* (Paramyidae), a primitive Eocene rodent. Note that the masseter muscles originate entirely on the zygomatic arch. In Abert's squirrel (*Sciurus aberti,* Sciuridae), the anterior part of the lateral masseter originates on the rostrum and the zygomatic plate. In the porcupine (*Erethizon dorsatum,* Erethizontidae), the anterior part of the medial masseter originates largely on the rostrum and passes through the greatly enlarged infraorbital foramen. The temporalis muscle, typically reduced in size in hystricomorphous rodents, is unusually large in porcupines. (*Pseudotomus after Wood, 1965*)

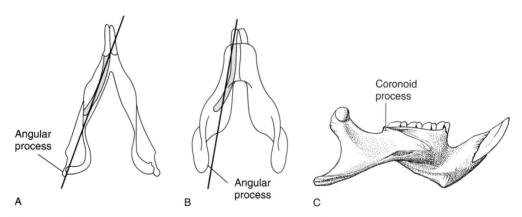

FIGURE 18-3 The sciurognathous and hystricognathous dentary bones. (A) Ventral view of the sciurognathous pattern of *Sciurus niger* with angular process in line with the ramus of the dentary bone (line). (B) Ventral view of an hystricognathous dentary of the porcupine *(Erethizon dorsatum),* showing the angular process located off the plane of the dentary. (C) Lateral view of the hystricognathous dentary of the nutria (*Myocastor coypus,* Myocastoridae). Note the strongly reduced coronoid process. Compare this dentary to that of *Aplodontia,* shown in Fig. 18-1.

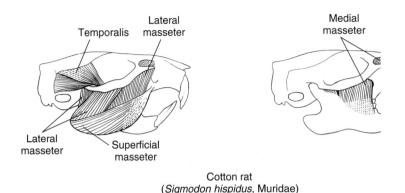

Cotton rat
(*Sigmodon hispidus,* Muridae)

FIGURE 18-4   Zygomasseteric pattern in a myomorphous rodent, the cotton rat (*Sigmodon hispidus,* Muridae). The superficial masseter originates on the rostrum, and the anterior part of the lateral masseter originates on the anterior extension of the zygomatic arch. The superficial muscles have been removed on the right to expose the medial masseter, which originates partly on the ro*strum* and passes through the slightly enlarged infraorbital foramen. *(After Rinker, 1954)*

ancient and formerly highly successful multituberculates, rodents were abandoning the primitive zygomasseteric arrangement. The time between the late Eocene and middle Oligocene was one of accelerated rodent evolution (Korth, 1994; Wilson, 1972), and it was then that the major patterns of jaw muscle specialization typical of modern rodents (Figs. 18-2 and 18-4) were established. Over half of the living families of rodents appeared by the end of the Oligocene. The Miocene spread of grasslands or savannas in both the Old World and the New World provided new adaptive zones for rodents. The evolution of the jerboas (Dipodidae) in the Old World and the kangaroo rats and pocket mice (Heteromyidae) in the New World was probably decisively affected by increasing aridity in the late Miocene and the appearance of deserts in the Pliocene.

Because of the overwhelming diversity and degree of evolutionary parallelism among rodents, phylogenetic relationships are difficult to divulge and are controversial. Numerous authors using both morphological and molecular data are actively investigating the relationship among rodents with the result that some broad patterns are discernible as follows: (1) the Rodentia is a monophyletic order (Luckett and Hartenberger, 1993; Philippe, 1997; Sullivan and Swofford, 1997; but see D'Erchia et al., 1996 and Reyes et al., 1998). (2) Lagomorphs are the most likely living sister group of the Rodentia (Luckett and Hartenberger, 1985). (3) Alagomyids including *Alagomys* and *Tribosphenomys* are the most primitive rodents morphologically and represent the probable sister group to all other rodents (Dashzeveg, 1990; Dawson and Beard, 1996). (4) Ctenodactyloids also represent an early, primitive line of rodent evolution. (5) Similarities among early geomyoids (represented today by the Heteromyidae and Geomyidae), the Muridae, and Dipodidae indicate that these groups share a common ancestral stock. (6) The Sciuridae and Aplodontidae (and possibly the Castoridae) are each other's closest living relatives. (7) The relationships of the Anomaluridae, Pedetidae, Castoridae, and Myoxidae to other families are unresolved. (8) The Hystricognathi originated in Asia and includes a number of closely related groups. (Fig. 18-5).

JAW MUSCLE AND SKULL SPECIALIZATIONS

Despite their diversity and success in adapting to contrasting environments and lifestyles, rodents have rather consistently followed certain basic trends in the evolution of the jaw muscles, the bones from which these muscles take origin or insertion, and the teeth. Even in the early stages of

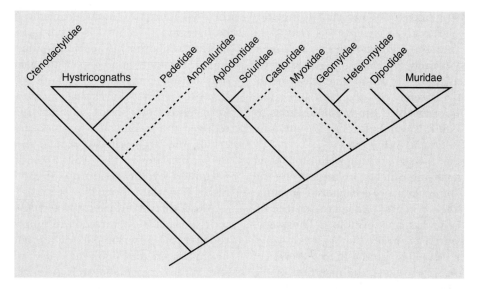

FIGURE 18-5   One possible phylogeny of the order Rodentia based on several morphological and molecular character sets summarized by Luckett and Hartenberger (1985). In this phylogeny, ctenodactylids are considered the sister group to all other rodents but are most closely allied with hystricognaths. Sciurognath rodents are represented by the clade that includes Aplodontidae through Muridae. Dashed lines indicate unresolved relationships.

their evolution, selective pressures apparently favored forward migration of the jaw muscles.

Since their appearance in late Paleocene times, rodents have had a dentition featuring a division of labor between incisors and cheek teeth. This dentition serves chiefly two functions. The incisors serve as chisels with which food is gnawed, vegetation is clipped, or, in some fossorial forms, soil and rocks are dug away. These teeth are subject to heavy wear and became ever-growing early in the evolution of rodents. The cheek teeth, separated from the incisors by a broad diastema, perform a different function, that of mastication of food. A complicated jaw action allows the lower cheek teeth to move transversely or anteroposteriorly against the upper teeth, producing a crushing and grinding action. Not only are the gnawing and grinding functions performed by different teeth, but they must be performed separately. When the cheek teeth are in position for grinding, the tips of the incisors do not meet; the lower jaw must therefore be moved forward for the incisors to be in position for gnawing.

This division of labor between incisors and cheek teeth clearly "guided" the evolution of the rodent jaw musculature. During gnawing, muscles that attach far forward on the jaw and skull are advantageous because they confer great power on the jaw action through increased mechanical advantage. Furthermore, because forward movement of the lower jaw is a prerequisite for gnawing, selection probably favored the attachment of some of the jaw muscles far forward on the rostrum; contraction of these muscles caused a forward shift of the jaw. During the grinding of food by the cheek teeth, jaw muscles with mechanical advantage for power are also important, and the complex jaw action associated with grinding demands jaw musculature with precise control over anterior, posterior, and transverse jaw movement. The jaw musculature and skull specializations to be considered can best be put in functional perspective if the importance of complex grinding movements and powerful forward movements of the rodent jaw are kept in mind.

Not every rodent can readily be classified as **sciuromorphous, hystricomorphous,** or **myomorphous,** and, as mentioned above, experts do not always agree on how the types evolved. The myomorphous pattern, for example, may have evolved through an hystricomorphous ancestry (Klingener, 1964). The broad trend in rodents was away from the primitive condition, in which the masseter muscles originated entirely on the zygomatic arch, toward the placement of the origin of at least one division of the masseter on the rostrum. The primitive condition, termed **protrogomorphous** by A. E. Wood (1965), is retained today by only one rodent (*Aplodontia rufa*, Aplodontidae).

Presumably as a result of competition among the rapidly diversifying rodents, in late Eocene the skull and jaw musculature were altered in some phylogenetic lines in a way that increased the effectiveness of gnawing and grinding. These specializations involved primarily the lateral and medial masseter muscles and their areas of attachment. In some rodents, the insertion of the anterior part of the lateral masseter shifted onto the anterior surface of the zygomatic arch and the adjacent part of the rostrum (Fig. 18-2). This pattern is termed sciuromorphous and occurs in the Sciuridae and Castoridae. In these families, the temporalis muscle is relatively large and the coronoid process is moderately well developed.

A second pattern of zygomasseteric specialization involves the shift of the origin of the medial masseter from the zygomatic arch to an extensive area on the side of the rostrum. This muscle passes through the often greatly enlarged infraorbital foramen, and the arrangement is termed hystricomorphous (Fig. 18-2). It occurs in the suborder Sciurognathi in the families Dipodidae, Ctenodactylidae, and Pedetidae and in most families in the suborder Hystricognathi. Although included in the Hystricognathi, members of the African family Bathyergidae are not hystricomorphous as adults: in this family the infraorbital foramen is reduced and transmits a small slip of the medial masseter only in the embryo (Maier and Schrenk, 1987). Lavocat (1973) believes that ancestral bathyergids were hystricomorphous but that they secondarily reduced the infraorbital foramen. In living bathyergids, the large anterior part of the medial masseter originates in the orbit, a condition perhaps permitted by the reduction of the eye in these highly fossorial rodents.

Most rodents utilize a third type of zygomasseteric specialization, termed myomorphous. In such rodents, the anterior part of the lateral masseter originates on the highly modified anterior extension of the zygomatic arch (the zygomatic plate and zygomatic spine; Fig. 18-4), and the anterior part of the medial masseter originates on the rostrum and passes through the somewhat enlarged infraorbital foramen. The temporalis is typically reduced, and the coronoid process ranges from well developed to vestigial (Fig. 18-6). Many rodents belonging to the suborder Sciurognathi, including all members of the huge family Muridae (1325 species), are myomorphous.

SUBORDER SCIUROGNATHI

FAMILY APLODONTIDAE. Family Aplodontidae is interesting primarily because of the unique, primitive morphological features that characterize its

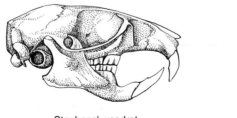

Stephens' woodrat
(*Neotoma stephensi*, Sigmodontinae)

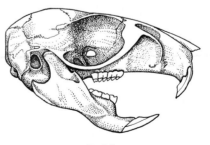

Gerbil
(*Tatera humpatensis*, Gerbillinae)

FIGURE 18-6   Skulls of myomorphous rodents, including a Stephens' woodrat (Muridae, Sigmodontinae; length of skull 42 millimeters) and a gerbil (Muridae, Gerbillinae; length of skull 38 millimeters). *(Gerbil after Hill and Carter, 1941)*

one extant member, *Aplodontia rufa,* the mountain "beaver," an animal restricted to parts of California and the Pacific Northwest. Mountain beavers are roughly the size of a small rabbit and have a robust, short-legged form. *Aplodontia* is the only living rodent in which the masseters have an entirely zygomatic origin. The skull is flat, and the coronoid process of the dentary bone is large (Fig. 18-1). The cheek teeth are ever-growing (a specialized feature) and have a unique crown pattern (Fig. 18-7). The dental formula is 1/1, 0/0, 2/1, 3/3 = 22.

The earliest records of aplodontids are from the late Eocene of western North America. Derived from aplodontid ancestry, the highly specialized family Mylagaulidae appeared in late Oligocene. These extinct marmot-size rodents were fossorial and one or two forms are notable for having prominent nasal "horns." The aplodontids spread later to Europe and Asia but since the middle Pliocene have lived only in the moist, forested parts of the Pacific slope of North America, where their relict distribution today is from central California to southern British Columbia. Widespread late Tertiary aridity may have restricted the aplodontids to their present range.

Aplodontia favors moist areas supporting lush growths of forbs and often builds its burrows next to streams. The tunnel systems have multiple exits, and this rodent seldom forages far from its burrows. The diet includes a variety of forbs and the buds, twigs, and bark of such **riparian** plants as willow *(Salix)* and dogwood *(Cornus).* On occasion, *Aplodontia* builds "hay piles" of cut sections of forbs (Grinnell and Storer, 1924). During winter months, mountain beavers may climb shrubs or small trees to cut off branches and twigs.

FAMILY SCIURIDAE. The successful and widespread family Sciuridae includes 273 Recent species representing 50 genera. Squirrels, chipmunks, marmots (Fig. 18-8), and prairie dogs belong to this family. Sciurids probably evolved from ischyromyid ancestors. Sciurids appeared first in the late Eocene of North America. Ground squirrels and tree squirrels can be distinguished from each other by the end of the Oligocene. Sciurids remain widespread today, absent only from the Australian region, Madagascar, the polar regions, southern South America, and certain Old World desert areas.

Sciurids are distinctive structurally. The skull is usually arched in profile (Fig. 18-9), and the front

FIGURE 18-8   Cape ground squirrel (*Xerus inauris,* Sciuridae) in the Kalahari Desert of South Africa. *(T. Vaughan)*

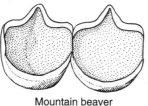

Mountain beaver
(*Aplodontia rufa,* Aplodontidae)

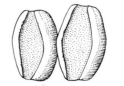

Merriam's kangaroo rat
(*Dipodomys merriami,* Heteromyidae)

FIGURE 18-7   Crowns of first two right upper molars of a mountain beaver (note the simplified and unique crown pattern) and a Merriam's kangaroo rat (note the highly simplified crown pattern). The labial border of the tooth is above; anterior is to the right. The unshaded part is enamel, and the stippled part is dentine.

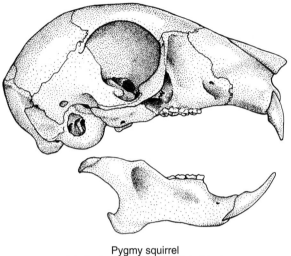

Pygmy squirrel
(*Exilisciurus whiteheadi*, Sciuridae)

FIGURE 18-9 Lateral view of the skull and mandible of a pygmy squirrel (length of skull 28 millimeters). *(After Heaney, 1985)*

of the zygomatic arch is flattened where the anterior part of the lateral masseter rests against it. The dental formula is 1/1, 0/0, 1-2/1, 3/3 = 20-22. The cheek teeth are rooted and usually have a crown pattern that features transverse ridges. Sciurids have relatively unspecialized bodies: a long tail is usually retained, and the limbs seldom have a loss of digits or reduction of freedom of movement at the elbow, wrist, and ankle joints. Several semifossorial types, including ground squirrels *(Spermophilus)*, prairie dogs *(Cynomys)*, and marmots *(Marmota)*, have variously departed from this plan in the direction of greater power in the forelimbs and, in some cases, reduction of the tail.

Sciurids are basically diurnal herbivores, but a great variety of food is utilized. Tree squirrels occasionally eat young birds and eggs; chipmunks *(Tamias)* and the antelope ground squirrels *(Ammospermophilus)* are seasonally partly insectivorous in some areas. Sciurids are tolerant of a great range of environmental conditions. Some, such as marmots, prairie dogs, chipmunks, and some ground squirrels, hibernate during cold parts of the year.

Styles of locomotion vary among sciurids; the most specialized style occurs in the "flying" squirrels. This group of 14 genera, constituting the subfamily Pteromyinae, is characterized by gliding surfaces formed by broad folds of skin between the forefoot and hind foot (Johnson-Murray, 1977; Thorington et al., 1998). These animals usually glide distances of only 10 to 20 meters, but some can glide considerable distances. The giant flying squirrel *(Petaurista)* of southeastern Asia, for instance, can glide up to 450 meters and can bank in midair (Nowak and Paradiso, 1991).

FAMILY CASTORIDAE. The family Castoridae is represented today by only two species, *Castor canadensis* of the United States and Canada and *Castor fiber* of northern Europe and northern Asia. Beavers had an important role in the history of the United States, for much of the early exploration of some of the major river systems in the western United States was done by trappers in quest of valuable beaver pelts.

The fossil record of beavers begins in the late Eocene of Asia, and beavers reached North America and Europe in the early Oligocene. Several lines of descent developed in the Tertiary. One line developed fossorial adaptations; the fossil remains of the North American Miocene beaver *Palaeocastor* have been found in the spectacular corkscrew-shape burrows apparently dug by these animals. Another evolutionary line led to the bear-size giant beaver *(Castoroides)* of the North American Pleistocene. Throughout their history, castorids have been restricted to the Northern Hemisphere.

Extant beavers are semiaquatic, and some of their distinctive structural features are adaptations to this mode of life (Fig. 18-10). The animals are

FIGURE 18-10 A beaver *(Castor canadensis)* repairs damage to a dam. *(Corel Corp.)*

large, reaching over 30 kilograms in weight. Their large size is associated with a surface-to-volume ratio that is more advantageous in terms of heat conservation than that of smaller rodents. In addition, the body is insulated by fine underfur protected by long guard hairs. These are important adaptations in animals that frequently swim and dive for long periods in icy water. The large hind feet are webbed, the small eyes have **nictitating membranes** (membranes that arise at the inner angle of the eye and can be drawn across the eyeball), and the nostrils and ear openings are valvular and can be closed during submersion.

Because of two structural specializations, beavers can open their mouths when gnawing under water and while swimming can carry branches in the submerged open mouth without danger of taking water into the lungs (Cole, 1970). The epiglottis is internarial (it lies above the soft palate), an arrangement allowing efficient transfer of air from the nasal passages to the trachea but not allowing mouth breathing or panting. Also, the mid-dorsal surface of the posterior part of the tongue is elevated and fits tightly against the palate and, except when the animal is swallowing, blocks the passage to the pharynx (Cole, 1970).

The tail is broad, flat, and largely hairless. The skull is robust. The zygomasseteric structure is specialized in that the rostrum is marked by a conspicuous lateral depression from which a large part of the lateral masseter muscle originates (Fig. 18-11). The jugal is broad dorsoventrally, and the external auditory meatus is long and surrounded by a tubular extension of the auditory bulla. The dental formula is 1/1, 0/0, 1/1, 3/3 = 20. The premolars are molariform, and the complex crown pattern features transverse enamel folds (Fig. 18-12).

Beavers are always found along waterways. Although they are most typical of regions supporting coniferous or deciduous forests, they also live in some hot desert regions, as, for example, along the lower Colorado River of Arizona and California. In the southwestern and middle-western United States, beavers dig burrows in the river banks, but in northern and mountainous regions they build lodges of sticks and mud in ponds formed behind their dams. Beavers remain active beneath the ice throughout the winter, feeding on the cambium of aspen and willow branches that they have stuck in the mud bottoms of the ponds. Numerous mounds of mud marked with **castoreum** (a urine-based secretion of the castor glands) and scattered within the territory serve to advertise the family unit's territorial boundaries (Schulte, 1998).

Beavers are remarkable in their ability to modify their environment by felling trees and building

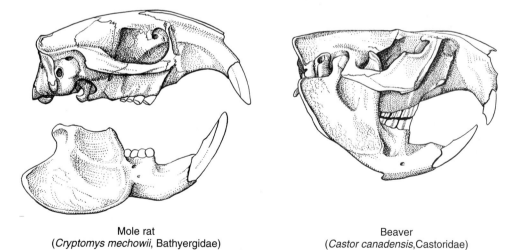

Mole rat
(*Cryptomys mechowii*, Bathyergidae)

Beaver
(*Castor canadensis*, Castoridae)

FIGURE 18-11 Skulls of a mole-rat (note the large, procumbent incisors, used for digging; length of skull 57 millimeters) and a beaver (note the depression in the side of the rostrum from which the anterior part of the lateral masseter originates; length of skull 139 millimeters). *(Mole-rat after Hill and Carter, 1941)*

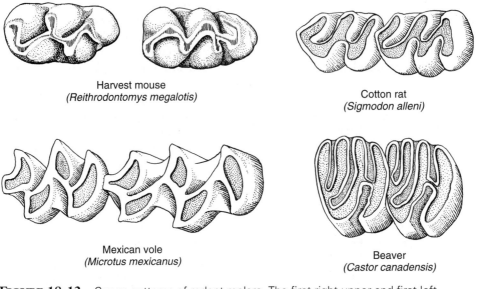

Harvest mouse
(*Reithrodontomys megalotis*)

Cotton rat
(*Sigmodon alleni*)

Mexican vole
(*Microtus mexicanus*)

Beaver
(*Castor canadensis*)

FIGURE 18-12   Crown patterns of rodent molars. The first right upper and first left lower molar of a harvest mouse (Muridae). The first two right upper molars of a cotton rat (Muridae) and a Mexican vole (Muridae). The first two right upper molars of a beaver (Castoridae). Unshaded areas on occlusal surfaces are enamel; stippled areas within the enamel folds are dentine.

dams. Many high valleys in the Rocky Mountains have been transformed by beaver dams from a series of meadows through which a narrow, willow-lined stream meandered to a terraced series of broad ponds bordered by extensive willow thickets and soil saturated with water. A valley suitable for grazing by cattle before occupancy by beavers may be more suitable afterward for moose, trout, and waterfowl.

FAMILY GEOMYIDAE. Members of family Geomyidae, the pocket gophers, are the most highly fossorial North American rodents. They are distributed from Saskatchewan to northern Colombia. The family includes roughly 35 Recent species in five genera. Pocket gophers appeared first in the lower Oligocene of North America. Although they are not restricted to semiarid habitats today, many of their most characteristic specializations probably evolved in response to the soil conditions and floral assemblages of the semiarid and plains environments that developed in the Miocene. Geomyids, along with their sister group the Heteromyidae, form a monophyletic clade and share external, fur-lined cheek pouches (Genoways and Brown, 1993; Ryan, 1989; Wahlert, 1985).

The most obvious structural characteristics of pocket gophers were developed in response to fossorial life. These animals are fairly small, weighing from 100 to 900 grams. They have small pinnae, small eyes, and a short tail. The head is large and broad, and the body is stout. External fur-lined cheek pouches are used for carrying food. The dorsal profile of the geomyid skull is usually nearly straight, the zygomatic arches flare widely, and, in the larger species, the skull is angular and features prominent ridges for muscle attachment. The rostrum is broad, robust, and marked laterally by depressions from which the lateral masseter muscles take origin. The large incisors often protrude forward, in some species beyond the anteriormost parts of the nasals and premaxillae; the lips close behind the incisors, which are therefore outside the mouth. The dental formula is 1/1, 0/0, 1/1, 3/3 = 20. The cheek teeth are ever-growing and have a highly simplified crown pattern. There is no loss of digits. The forelimbs are powerfully built and bear large, curved claws; the toes of the forefoot have fringes of hair that presumably increase the effectiveness of this foot during digging (Fig. 18-13).

Pocket gophers occupy friable soils in environments ranging from tropical to boreal, eat a variety

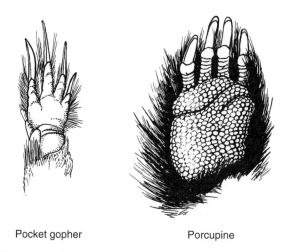

Pocket gopher   Porcupine

FIGURE 18-13   Ventral views of the left manus of
the pocket gopher (*Thomomys bottae,* Geomyidae).
Note the fringes of hairs on the toes. A right manus of
the porcupine (*Erethizon dorsatum,* Erethizontidae).
Note the pattern of tubercles on the pads, a design
that increases traction.

of above- and underground parts of forbs, grasses,
shrubs, and trees, and strongly affect their environment. Pocket gopher burrows provide channels
that allow deep penetration of water and decrease
surface erosion during periods of snow melt in
mountainous areas. The disturbance of the soil
and the mounds of soil created by pocket gophers
strongly influence vegetation by favoring pioneer
plants. In some mountain meadows, roughly 20
percent of the ground surface is covered with
mounds, and in such meadows in Utah pocket gophers may eat more than 30 percent of the annual
underground productivity of forbs (Andersen and
MacMahon, 1981). Because their preference for al-
falfa and some other cultivated plants results in
great crop damage, large amounts of money have
been spent by farmers and federal agencies to control pocket gophers on cultivated land in the western United States.

FAMILY HETEROMYIDAE.   Members of family Heteromyidae are the North American rodents most
fully adapted to desert life. Heteromyids are restricted to the New World, where they range from
southern Canada through the western United
States to Ecuador, Colombia, and Venezuela. Although they occupy areas ranging from temperate
to tropical, they reach their greatest diversity and

density in arid and semiarid regions. This family
contains 59 Recent species in six genera.

Heteromyids first appeared in the Oligocene
of North America. The kangaroo rats (*Dipodomys;*
Fig. 18-14A) are known from the late Miocene,
when the deserts and semiarid brushlands that heteromyids now frequently inhabit became widespread in western North America. Large eyes,
small ears, and handsome facial markings are typical of kangaroo rats. Certain diagnostic characteristics of kangaroo rats, such as the greatly enlarged
auditory bullae (Fig. 18-15) and features of the
hindlimbs that favor saltation, probably evolved
under constraints associated with desert or semi-

A

B

FIGURE 18-14   (A) A bannertail kangaroo rat
(*Dipodomys spectabilis,* Heteromyidae) from the
desert southwest of the United States. (B) An Old
World jerboa (*Jaculus jaculus,* Dipodidae). *(A is
by J. McDonald, and B is by M. Andera)*

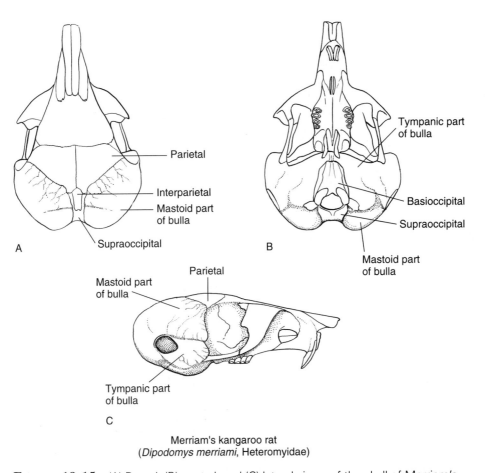

A

Parietal

Interparietal

Mastoid part
of bulla

Supraoccipital

Tympanic part
of bulla

Basioccipital

Supraoccipital

Mastoid part
of bulla

B

Parietal

Mastoid part
of bulla

Tympanic part
of bulla

C

Merriam's kangaroo rat
(*Dipodomys merriami*, Heteromyidae)

FIGURE 18-15   (A) Dorsal, (B) ventral, and (C) lateral views of the skull of Merriam's kangaroo rat (Heteromyidae). Note the great enlargement of the auditory bulla, the chamber surrounding the middle ear (length of skull 45 millimeters). *(After Grinnell, 1922)*

desert conditions (Genoways and Brown, 1993; Grinnell, 1922; Nikolai and Bramble, 1983).

Heteromyids are specialized for jumping. Such adaptations are most strongly developed in the kangaroo rats *(Dipodomys)* and kangaroo mice *(Microdipodops)*. In these heteromyids, the forelimbs are small, the neck is short, and the tail is long and serves as a balancing organ. The hindlimbs are elongate, and the thigh musculature is powerful (Howell, 1932; Ryan, 1989). The hind foot is elongate, but except for the almost complete loss of the first digit in some kangaroo rats (Fig. 18-16E), there is no loss of digits. The cervical vertebrae are largely fused in *Microdipodops,* and in *Dipodomys* they are strongly compressed and partly fused, producing a short, rigid neck (Fig. 18-16C). These species are mostly bipedal when moving rapidly; when frightened, they move by a series of erratic hops.

As in the pocket gophers, fur-lined cheek pouches are present in heteromyids (see Ryan, 1986 for a discussion of the evolution of cheek pouches in rodents). The skull is delicately built, with thin, semitransparent bones; the zygomatic arch is slender. The auditory bullae are usually large and in some genera are enormous, being formed largely by the mastoid and tympanic bones (Fig. 18-15). The enlargement of the bullae in heteromyids (and convergently in jerboas, Dipodidae) greatly increases auditory sensitivity. The nasals are slender and usually extend well forward of the slender upper incisors. The dental formula is 1/1, 0/0, 1/1, 3/3 = 20. The ever-growing cheek teeth have a strongly simplified crown pattern (Fig. 18-7) resembling that of pocket gophers, to which heteromyids are closely related (Ryan, 1989; Wahlert, 1985).

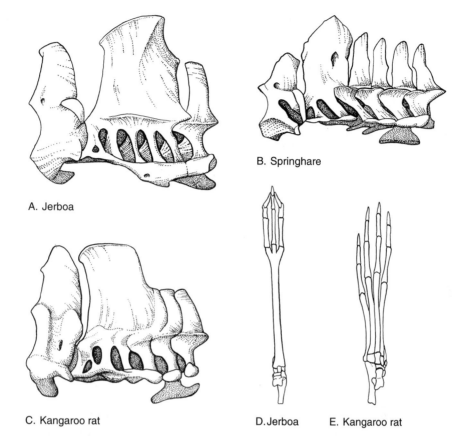

A. Jerboa

B. Springhare

C. Kangaroo rat

D. Jerboa   E. Kangaroo rat

FIGURE 18-16   The cervical vertebrae and hind feet of several saltatorial rodents. Cervical vertebrae of (A) a jerboa (*Jaculus* sp., Dipodidae), (B) the springhare (*Pedetes capensis,* Pedetidae), and (C) Heermann's kangaroo rat (*Dipodomys heermanni,* Heteromyidae). (D) Dorsal view of hind foot of a jerboa (*Jaculus* sp., Dipodidae); note the reduction of digits and the cannon bone formed by metatarsals two, three, and four. (E) Left hind foot of the desert kangaroo rat (*Dipodomys deserti,* Heteromyidae); note the near loss of the first digit and the elongation of the foot. *(A, B, and C after Hatt, 1932; D after Howell, © 1944,* Speed in Animals, *University of Chicago Press; E after Grinnell, 1922)*

Small annual plants in deserts are able to make the most of irregular moisture by germinating, growing, and flowering rapidly and by producing abundant seeds that remain dormant for various periods. This enormously abundant seed crop is the major food source of heteromyids. Perhaps the most remarkable heteromyid adaptation is the ability to survive for long periods on a diet of dry seeds with no free water (MacMillen, 1983a, b). This capability does not occur in all heteromyids, nor is it developed to the same degree in all species adapted to dry climates. In the species of kangaroo rats and pocket mice (*Chaetodipus* and *Perognathus*)

that occupy the desert, however, this ability is well developed.

FAMILY MYOXIDAE. Family Myoxidae includes the dormice, a group (eight genera with 26 Recent species) of squirrel-like Old World rodents known first from the middle Eocene of France. Myoxids are an old and distinctive group that was one of the first to branch from the primitive and extinct family Ischyromyidae. Dormice are entirely Old World in distribution, occurring in much of Africa south of the Sahara, England, Europe from southern Scandinavia southward, Asia Minor, southwestern

Russia, southern India, southern China, and Japan. The desert dormouse or dzhalman (*Selevinia betpakdalaensis*), is restricted to the Betpak-dala Desert of Russia (and until recently was placed in its own family).

Dormice are small (up to 325 millimeters in length), and most genera have bushy or well-furred tails (Fig. 18-17). The skull has a smooth, rounded braincase, a short rostrum, and large orbits. The dental formula is 1/1, 0/0, 1/1, 3/3 = 20. The crowns of the brachydont molars have parallel cross ridges of enamel, or, in some cases, the ridges are reduced and the crowns have basins. The infraorbital foramen is somewhat enlarged and transmits part of the medial masseter muscle. The limbs and digits are fairly short, and the sharp claws are used in climbing. The manus has four toes, and the pes has five. The desert dormouse eats primarily insects and spiders and, as is typical of many desert mammals, has greatly enlarged auditory bullae. Its dental formula is 1/1, 0/0, 0/0, 3/3 = 16; the cheek teeth are small and short-crowned, and the much-simplified crown pattern features smooth, concave surfaces.

Myoxids are swift and agile climbers that occupy trees and shrubs, rock piles, or rock outcrops. These rodents are omnivorous but are able little predators capable of killing small birds and large insects. In temperate areas, the animals hibernate in winter and may have an above-ground activity season of only four to six months. In years of low autumn food availability, male *Myoxus glis* enter hibernation with minimal fat reserves and may forego reproducing in the spring in order to replenish their energy stores (Bieber, 1998). A unique feature of myoxids is their ability to lose and regenerate the tail (Mohr, 1941).

FAMILY DIPODIDAE. Family Dipodidae includes the jerboas (strongly saltatorial), the jumping mice (moderately saltatorial), and the birch mice (non-saltatorial). Jerboas occur in arid and semiarid areas in northern Africa, Arabia, and Asia Minor and in southern Russia eastward to Mongolia and northeastern China. The Nearctic jumping mice occupy Alaska and much of Canada and in the United States live (mostly in boreal habitats) as far south as New Mexico and Georgia. Palearctic jumping mice and birch mice are found from Scandinavia into central Europe and in Russia, Mongolia, and China. The family first appeared in the early Eocene of Asia and is represented today by 15 genera and 51 species.

Jerboas have a compact body, large head, reduced forelimbs, and elongate hind limbs—features associated with saltatorial locomotion (Fig. 18-18A and 18-14B). The tail is long and usually tufted, and, as in the New World kangaroo rats (Heteromyidae), the tuft is frequently conspicuously black and white. The posterior part of the skull is broad (Fig. 18-19A), owing mostly to the enlargement of the auditory bullae, which are huge in some species. The rostrum is usually short, the orbits are large, and through the enlarged infraorbital canal passes most of the anterior part of the medial masseter, which originates largely on the side of the rostrum. The zygomatic plate is narrow and below the infraorbital canal. The dental formula is 1/1, 0/0, 0-1/0, 3/3 = 16-18, the cheek teeth are hypsodont, and the crown pattern usually features re-entrant enamel folds.

The hindlimbs are elongate in all genera, but varying stages of specialization for saltation are represented (Stein, 1990). In members of the subfamily Cardiocraniinae, the toes vary in number from three to five and the metatarsals are not fused. At the other extreme are such genera as *Dipus* and *Jaculus* (subfamily Dipodinae), which represent the greatest degree of hindlimb specialization for saltation in rodents. In these genera, only three toes (digits two, three, and four) remain and

FIGURE 18-17 A forest dormouse (*Dryomys nitedula*, Myoxidae). *(M. Andera)*

A   B

FIGURE 18-18   (A) Jerboa (*Allactaga elater,* Dipodidae) showing the elongate tail and enlarged hind feet. (B) A birch mouse (*Sicista betulina,* Dipodidae). *(M. Andera)*

the elongate metatarsals are fused into a cannon bone (Fig. 18-16D). An additional specialization that occurs in some species is a brush of stiff hairs on the ventral surface of the phalanges. The ears of jerboas vary from short and rounded to long and rabbit-like.

Jumping mice and birch mice are small and graceful, weighing approximately 10 to 25 grams, with a long tail and, in all genera but *Sicista,* elongate hindlimbs (Fig. 18-18B). The coloration in most species is striking: the belly is white and the dorsum is bright yellowish or reddish brown. Much of the anterior part of the medial masseter muscle

originates on the side of the rostrum and passes through the enlarged infraorbital foramen (Klingener, 1964). Although all living dipodids have an hystricomorphous zygomasseteric arrangement, they are regarded as related to the myomorphous Muridae. Both Wilson (1949) and Klingener (1964) concluded that the murid myomorphous masseter was derived from a dipodoid-like ancestor. The dental formula is 1/1, 0/0, 0-1/0, 3/3 = 16-18; the cheek teeth are brachydont or semihypsodont and have quadritubercular crown patterns with re-entrant enamel folds. The hindlimbs in jumping mice are elongate and somewhat adapted

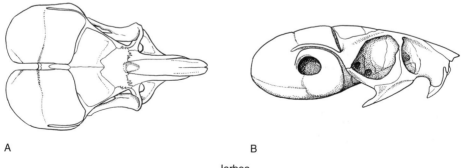

A   B

Jerboa
(*Salpingotus kozlovi,* Dipodidae)

FIGURE 18-19   (A) Dorsal and (B) lateral views of the skull of a jerboa (*Salpingotus kozlovi,* Dipodidae), length of skull 27 millimeters. Note the greatly enlarged auditory bullae and the general resemblance between this skull and that of the kangaroo rat (Fig. 18-15). *(After Allen, 1940)*

for hopping, but unlike the case with more specialized saltatorial rodents, all digits are retained. As an additional contrast with jerboas and other specialized saltators, the cervical vertebrae of jumping mice are unfused.

Jerboas occupy arid areas and lead lives that resemble in some ways those of members of the New World family Heteromyidae (Fig. 18-14; Mares, 1980). They live in burrows that are frequently plugged during the day, a habit that favors water conservation by keeping the humidity in the burrow as high as possible. They are nocturnal, and many species sift seeds from sand or loose soil with the forefeet, although some species depend largely on insects for food. Unlike kangaroo rats, jerboas hibernate during the winter in fairly deep burrows. Locomotion in jerboas is chiefly bipedal, but when they are moving slowly, the forefeet may be used to some extent. When frightened, jerboas move rapidly in a series of long leaps, each of which may cover 3 meters. Such a rapid and, more important, erratic mode of escape from predation is especially effective in the barren terrain jerboas occupy (Schröpfer et al., 1985).

Jumping mice and birch mice usually inhabit boreal forests. Some species occur typically in coniferous forests, while others appear in birch stands or in mixed deciduous forests. Usually, jumping mice favor moist situations, and *Zapus princeps* of the western United States is most abundant in dense cover adjacent to streams or in wet meadows. These mice hibernate in the winter and emerge during or after snow melt. Food consists of a variety of seeds, fungus *(Endogone)*, and other plant material, but insect larvae and other animal material made up approximately half of the food of *Z. hudsonius* in New York (Whitaker, 1963) and roughly one third of the diet of *Z. princeps* in Colorado (Vaughan and Weil, 1980).

FAMILY MURIDAE. Family Muridae is huge, including some 66 percent of the living species of rodents (roughly 1325 species and 281 genera) and is nearly worldwide in distribution; its members occupy environments ranging from high arctic tundra to tropical forests to desert sand dunes. Table 18-2 lists the subfamilies of murids, their common names, and their general distributions. Excellent

TABLE 18-2   The Subfamilies of the Family Muridae

Subfamily (no. of species)	Common Name	Distribution
Arvicolinae (143)	Voles, lemmings, muskrat	Holarctic
Calomyscinae (6)	Mouse hamsters	Middle East and SW Asia
Cricetinae (18)	Hamsters	Palearctic
Cricetomyinae (6)	Pouched rats and mice	Africa south of Sahara
Dendromurinae (23)	Climbing rats, forest mice	Africa south of Sahara
Gerbillinae (110)	Gerbils, jirds, sand rats	Africa, southern Asia
Lophiomyinae (1)	Maned rat	Eastern Africa
Murinae (529)	Old World rats and mice	Nearly worldwide
Myospalacinae (7)	Zokors	Siberia, northern China
Mystromyinae (1)	White-tailed rat	South Africa
Nesomyinae (14)[a]	Malagasy rats and mice	Madagascar
Otomyinae (12)	Vlei rats, Karroo rats	Parts of Africa
Petromyscinae (5)	Rock mice, swamp mouse	Parts of Africa
Platacanthomyinae (3)	Spiny mouse, blind tree mouse	India, southeastern Asia
Rhizomyinae (15)	Mole rats, bamboo rats	East Africa, south Asia
Sigmodontinae (422)[a]	New World rats and mice	New World
Spalacinae (8)	Blind mole-rats	Mediterranean region

From Musser and Carleton 1993.
[a]Does not include recently discovered species.

coverage of murid rodents is given by Carleton and Musser (1984).

Most murids retain a "standard" mouselike form, with a long tail, generalized limb structure, and no loss of digits. Murids range in size from about 10 grams in weight, as in the pygmy mouse *(Baiomys)*, to two kilograms, as in the New Guinean rat *Mallomys*. The skull varies widely in shape (Figs. 18-4, 18-20, 18-21), but the infraorbital foramen is always above the zygomatic plate and is enlarged dorsally for the transmission of part of the medial masseter, which originates on the side of the rostrum. Through the narrowed ventral part of this foramen pass blood vessels and a branch of the trigeminal nerve. The maxillary root of the zygomatic arch is platelike and provides surface for the origin of part of the lateral masseter. This myomorphous zygomasseteric arrangement was perhaps derived through a hystricomorphous ancestry. The dental formula is generally 1/1, 0/0, 0/0, 3/3 = 16; in some species the molars are reduced in number. Molars range from brachydont to hypsodont and ever-growing. The basic cusp pattern involves transverse crests (Fig. 18-22), but crests are absent in some species and molar crown patterns vary widely (Fig. 18-12).

Within the array of murid species, a variety of modes of life and morphological and behavioral

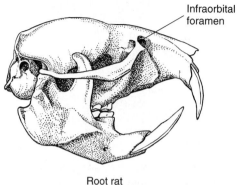

Root rat
(Tachyoryctes splendens)

FIGURE 18-21   Skull of an African root rat (Muridae, Rhizomyinae). Note the procumbent incisors, which are used for digging, and the dorsal position of the infraorbital foramen (length of skull 41 millimeters).

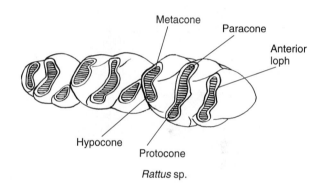

Rattus sp.

FIGURE 18-22   Occlusal surfaces of the right upper molars of a murid rodent *(Rattus).* With wear, the cross lophs become lakes of dentine (cross-hatched areas) rimmed with enamel.

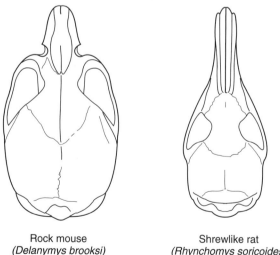

Rock mouse
(Delanymys brooksi)

Shrewlike rat
(Rhynchomys soricoides)

FIGURE 18-20   Extremes in skull shape in rodents of the family Muridae. The rock mouse is a rock-dwelling omnivore, while the shrewlike rat is a rare species that apparently feeds on invertebrates.

specializations are represented. The following paragraphs discuss several subfamilies that display some of this broad structural and functional variety.

The Murinae (Fig. 18-23) is the largest murid subfamily (529 species), occurs nearly worldwide, and includes a wide diversity of types adapted to terrestrial, fossorial, largely aquatic, or arboreal life. Some murines live in close association with humans in situations ranging from isolated farms to the world's largest cities. As a result of introductions by humans, these animals have become nearly cosmopolitan in distribution and are probably the rodents most familiar to us. Murines that are not commensal with humans occur in much of

A

B

FIGURE 18-23 (A) A striped mouse (*Rhabdomys pumilo*, Muridae, Murinae) from Karroo National Park, South Africa. (B) A field vole (*Microtus agrestis*, Muridae, Arvicolinae). *(A is by T. Vaughan, B is by M. Andera)*

southeastern Asia, Europe, Africa, Australia, Tasmania, and Micronesia. Tropical and subtropical areas are centers of murine abundance, but these animals have occupied a wide variety of habitats, and some genera are highly adapted to specialized modes of life. Murines range in size from about 10 grams to about 2 kilograms. Although the tail is usually more or less naked and scaly, it is occasionally heavily furred and bushy. The molars are rooted or ever-growing and usually have crowns with crests or chevrons (Fig. 18-12); great simplification of the crown pattern occasionally occurs. The dental formula is usually 1/1, 0/0, 0/0, 3/3 = 16. In some murines, the reduction of the cheek

teeth has become extreme. The greatest reduction occurs in *Mayermys*, a rare mouse from New Guinea, in which only one molar is retained on each side of each jaw. The feet retain all of the digits, but the pollex is rudimentary.

Murines appear fairly late in the fossil record (middle Miocene), but the subfamily has been remarkably plastic from an evolutionary point of view (Carleton and Musser, 1984). In both Africa and the Australian faunal region, murines have undergone impressive radiations and are variously amphibious, terrestrial, semifossorial, arboreal, and saltatorial. The water rats *(Crossomys)* have greatly reduced ears, large, webbed hind feet, and nearly waterproof fur. These animals live along waterways in New Guinea. At the other extreme is the hopping mouse *(Notomys alexis)*, a saltatorial inhabitant of extremely arid Australian deserts. This rodent needs no drinking water and has the greatest ability to concentrate urine as a means of conserving water of any animal in which water metabolism has been studied (MacMillen and Lee, 1969). Murines feed on a variety of plant material and on invertebrate and vertebrate animals. In association with the great diversity of feeding habits of murines, the skull form varies widely within the subfamily (Fig. 18-20), with a shrewlike elongation of the rostrum occurring in some insectivorous genera.

Extremely high population densities have been recorded for feral populations of some murines, often commensal with humans. A 35-acre area near Berkeley, California, which had only an occasional house mouse for a number of years after population studies began in 1948, supported 7000 *Mus* in June 1961 (Pearson, 1964). Among other factors, the high reproductive rate of *Mus* contributes to its ability to reach high densities quickly. Murines that live with humans have great economic impact. Not only do they spread such serious diseases as bubonic plague and typhus, but the damage they do to stored grains and other foods is so severe that in many countries *Rattus* and *Mus* compete effectively and devastatingly with humans for food.

The Sigmodontinae is the second largest murid subfamily (422 species) and occupies South America and most of North America. *Peromyscus*-like rodents *(Copemys)* entered North America from Eurasia in the early Miocene, but underwent little radiation until late Miocene and Pliocene, when such sigmodontine genera as *Peromyscus, Neotoma, Onychomys,* and *Reithrodontomys* appeared. Neotropical sigmodontines may have been derived from a separate invasion from the Old World in the late Miocene *(Calomys)*. They also underwent rapid radiation in the late Miocene and Pliocene; by the early Pliocene, the extant genus *Sigmodon* appeared. Sigmodontines possibly entered South America when the Panamanian land bridge formed in late Pliocene but could have entered earlier by **waif dispersal** across the seaway separating the Americas (Engel et al., 1998). Experts disagree as to whether the present diversity of South American sigmodontines resulted from a tropical North American radiation before access to South America was gained or whether this radiation occurred in South America after the land bridge formed (Baskin, 1978; Patterson and Pascual, 1972; Reig, 1980). In any case, of the 79 living sigmodontine genera, 86 percent live in South America or no farther north than tropical Central America.

Today sigmodontines occupy habitats ranging from subarctic to tropical and are variously terrestrial, amphibious, fossorial, or arboreal. In many parts of the western United States, the large stick-and-debris houses of woodrats *(Neotoma)* are conspicuous features, and white-footed mice *(Peromyscus)* seem to occupy every conceivable North American terrestrial habitat.

The subfamily Arvicolinae includes the voles, lemmings, and muskrat, a group of 26 genera and 143 species of rodents distributed throughout the Northern Hemisphere (Fig. 18-23B). These rodents frequently have short tails, ear openings that are partially guarded by fur, and a chunky, short-legged appearance. The cheek teeth often feature complex crown patterns (Fig. 18-12) adapted to masticating forbs and grasses. The voles and lemmings have high reproductive rates and undergo remarkable population fluctuations in some areas (p. 518).

The subfamily Gerbillinae, the gerbils, includes 14 genera and 110 species and is a group of rodents that resembles jerboas (Dipodidae) and kangaroo rats (Heteromyidae) in being semifossorial, more or less saltatorial, and in inhabiting mainly desert regions. Gerbils now occur in arid parts of Asia, in the Middle East, and in Africa. The hindlimbs are large, the central three digits are larger than the lateral ones, and the tail is often long and

functions as a balancing organ (Fig. 18-23). The skull does not depart strongly from the general murid plan (Fig. 18-6). In their ability to hop and in their choice of habitats, gerbils resemble heteromyid rodents. Gerbils maintain water balance in hot, arid conditions partly by eating food with a high water content and partly by concentrating urine, as do heteromyids.

The subfamily Spalacinae contains animals usually called blind mole-rats. There are eight Recent species in two genera *(Spalax* and *Nannospalax)*. These rodents first appear in the Pliocene and occur today in the eastern Mediterranean region, southeastern Europe, and into the Ukraine and southwestern Russia. They are fossorial and the eyes are small and covered with skin. The eye muscles and optic nerve are degenerate. The ears are absent, and the tail is vestigial. Unlike many fossorial rodents, *Spalax* digs primarily with its large incisors and bulldozes soil with its blunt head. As adaptations to this style of digging, the neck and jaw muscles are powerful, the incisors are robust, and the nose is protected by a broad, horny pad. The feet, surprisingly, are not unusually large; the claws have been described as blunt, round nubbins. These nocturnal rodents burrow in both alluvial and stony soils and eat both above- and underground parts of plants. *Spalax* often lives in burrows in water-saturated or snow-covered soil; these burrows have extremely high levels of carbon dioxide and extremely low levels of oxygen. Whereas in most mammals these conditions interfere seriously with heart action, *Spalax* can raise its heart rate and maintain a stable pulse at low oxygen levels (Arieli and Ar, 1981a). As an adaptation to improve oxygen delivery to tissues under these conditions, the capillary density in the heart and skeletal muscles of *Spalax* is nearly twice as high as in the laboratory rat (Arieli and Ar, 1981b).

Members of the subfamily Rhizomyinae, called root rats, occur in southeastern Asia and East Africa and are first known from the early Miocene of southern Asia. The subfamily includes three genera, represented by 15 species. These rodents range from a total length of 200 to 500 millimeters and have short, robust limbs and a compact body. In two genera, the incisors are procumbent (Fig. 18-21). Root rats live primarily in areas with at least 500 millimeters of annual precipitation and occupy habitats ranging from dense bamboo thickets in Asia to subalpine slopes at 4000 meters on Mount Kenya in East Africa. The digging behavior of the African root rat, *Tachyoryctes,* has been described by Jarvis and Sale (1971). This animal burrows by slicing away the soil with powerful upward sweeps of the protruding lower incisors. The dislodged soil is moved behind the animal by synchronous thrusts with the hindlimbs. When the burrow becomes blocked with freshly dug soil, *Tachyoryctes* turns and pushes the load to the surface with the side of its head and one forefoot. The conspicuous mounds (up to 6 meters in diameter) associated with the activities of *Tachyoryctes* on Mount Kenya resemble the Mima mounds in some parts of the western United States that are formed by pocket gophers (see Fig. 24-25). *Tachyoryctes* is solitary and aggressive and eats a variety of below- and above-ground parts of plants.

An obscure, but interesting, murid subfamily is the Nesomyinae of Madagascar. This probably monophyletic group of rodents appeared in the Miocene in Africa, dispersed to Madagascar, and subsequently radiated in isolation from their African ancestors. Until recently, seven genera and only ten species of nesomyine rodents were known (Musser and Carleton, 1993). Detailed surveys of the island in the past several years, coupled with taxonomic revisions of several genera, indicate that the nesomyine fauna is much more diverse (Carleton and Schmidt, 1990; Carleton, 1994; Carleton and Goodman, 1996, 1998; Musser and Carleton, 1993). In fact, recent surveys of the previously unexplored high mountain forests in Madagascar have resulted in the addition of two new genera and species of nesomyine rodents: *Monticolomys koopmani* and *Voalavo gymnocaudus* (Carleton and Goodman, 1996, 1998).

FAMILY ANOMALURIDAE. Family Anomaluridae, composed of seven Recent species of three genera, includes the scaly-tailed squirrels. These animals occupy forested, tropical parts of western and central Africa. They resemble the flying squirrels (Sciuridae) in some structural features and in gliding ability. The relationship of anomalurids to other rodents groups is not well understood.

Increased surface area for gliding is provided in anomalurids by a fold of skin that extends between the wrist and the hind foot and is supported and extended during gliding by a long cartilaginous rod, roughly the length of the forearm, that originates on the posterior part of the elbow (Fig. 18-24; Kingdon, 1984b). (In gliding sciurids, in contrast, a short cartilaginous brace arises from the wrist.) In anomalurids, folds of skin similar to the uropatagium of bats extend between the ankles and the tail a short distance distal to its base (Johnson-Murray, 1987). The fur over most of the membrane is fine and soft, but a tract of stiff hairs occurs along the outer edge of the membrane behind the cartilaginous elbow strut. This tract may improve the efficiency of gliding by controlling the flow of the boundary layer of air sweeping over the membrane (Johnson-Murray, 1987). Relative to those of most rodents, the limbs of anomalurids are unusually long and lightly built; they provide for a wide spreading of the gliding membranes. One genus *(Zenkerella)* does not have a gliding membrane. The anomalurid tail is usually tufted and has a bare ventral area near its base that has two rows of keeled scales (Fig. 18-24). These scales seemingly keep the animals from losing traction when they cling to the trunks of trees. The feet are strong and bear sharp, recurved claws. In the anomalurid skull, the infraorbital canal is enlarged and transmits part of the medial masseter muscle. The dental formula is 1/1, 0/0, 1/1, 3/3 = 20, and the cheek teeth are rooted.

Anomalurids are handsome animals beautifully adapted to an entirely arboreal life. Their diet includes leaves, sap, and insects (Julliot et al., 1998). Derby's anomalure *(Anomalurus derbianus)* is a graceful and highly maneuverable glider, capable of glides of over 100 meters and of midair turns. Groups of anomalurids often take shelter during the day in cavities in trees. Group size varies from 6 or 8 animals to colonies of over 100 anomalurids of several species. Rosevear (1969) reports four species occupying the same hole in a tree. On occasion, these rodents share a hollow tree with dormice *(Graphiurus)* and with several species of bats (Julliot et al., 1998).

FAMILY PEDETIDAE. This family is represented by one distinctive species, *Pedetes capensis,* the springhare (Matthee and Robinson, 1997, have argued for the recognition of a second species). Its distribution includes East Africa and southern Africa, where it inhabits sandy soils in semiarid regions. Sparsely vegetated areas or places where the vege-

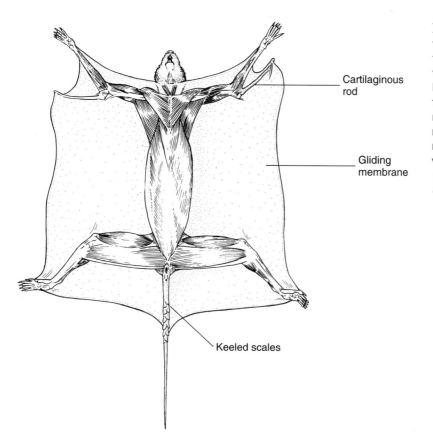

Cartilaginous rod

Gliding membrane

Keeled scales

FIGURE 18-24 Drawing of the ventral surface of a scaly-tailed flying squirrel (Anomaluridae) with the skin and fur removed to highlight the gliding membrane and the cartilaginous rod supporting the membrane near the elbow. As their name implies, the tail has several rows of keeled scales on the ventral surface. *(After Kingdon, 1984b, © East African Mammals, University of Chicago Press)*

tation has been heavily grazed by ungulates are preferred. *Megapedetes,* a lower Miocene species from eastern Africa, is the earliest member of this family. The phylogenetic position of this family within the Rodentia is controversial because pedetids share both sciurognathous and hystricognathus features. Attempts to resolve pedetid relationships using molecular data suggest that *Pedetes* belongs to a clade containing the Heteromyidae, Geomyidae, and Muridae (Matthee and Robinson, 1997b).

The springhare is saltatorial and roughly the size of a large rabbit, weighing up to about 4 kilograms. The eyes are extremely large, suggesting perhaps a reliance on vision for detecting predators. The forelimbs are short but robust and bear long claws that are used in digging. The hindlimbs are long and powerfully built, the fibula is reduced and fused distally to the tibia, and the feet have only four toes. The long tail is heavily furred throughout its length. Through the enormous infraorbital foramen (Fig. 18-25) passes the large an-

terior division of the medial masseter. As in a number of saltatorial rodents, the cervical vertebrae are partly fused (Fig. 18-16B). The dental formula is 1/1, 0/0, 1/1, 3/3 = 20, and the cheek teeth are ever-growing with a simplified crown pattern. A tragus fits against the ear opening and keeps out sand and debris when the animal digs.

Springhares dig fairly elaborate burrows. Because of their restriction to friable soils, these animals are not evenly distributed and appear to occur in colonies. When frightened, springhares can make tremendous bipedal leaps of over 6 meters, but when foraging and moving slowly, they are quadrupedal. A variety of plant material is eaten, including bulbs, seeds, and leaves. Water balance may be maintained during some seasons by eating succulent vegetation or insects. The springhare has an unusually low reproductive rate for a rodent. There are only two pectoral mammae, and typically the female bears only one young. Newborn young are large—roughly one third the size of the adult (Coe, 1967; Hediger, 1950)—and well

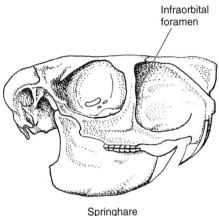

Infraorbital
foramen

Springhare
(*Pedetes capensis*, Pedetidae)

FIGURE 18-25   The skull of the springhare
(Pedetidae). Note the enormously enlarged infraorbital
foramen. *(R. Vaughan)*

developed and remain in the maternal burrows
until they weigh at least half as much as adults.

FAMILY CTENODACTYLIDAE. Members of family
Ctenodactylidae, commonly called "gundis," inhabit arid parts of northern Africa from Senegal,
Chad, Niger, and Mali on the west to Somaliland
on the east. There are four Recent genera and
five Recent species. The earliest known ctenodactyloids are from the early Oligocene of Asia.
Ctenodactylids share many derived characters with
the Hystricognathi but ctenodactylids retain many
primitive features and are considered a possible
sister group to both sciurognath and hystricognath rodents (Luckett and Hartenberger, 1985).

These are small (about 175 grams), compact,
short-tailed rodents with long, soft fur. The ears
are round and short and in some species are protected from wind-blown debris by a fringe of hair
around the inner margin of the pinnae. The infraorbital canal is enlarged, and through it passes
part of the medial masseter muscle. The skull is
flattened, and the auditory bullae and external auditory meatus are enlarged. The cheek teeth are
ever-growing, the crown pattern is simple, and the
premolars are nonmolariform. The dental formula
is 1/1, 0/0, 1-2/1-2, 3/3 = 20-24. The limbs are
short; the manus and pes each have four digits.

These herbivorous rodents occur in arid and
semiarid areas, where they are restricted to rocky

situations. They are diurnal and crepuscular and
scurry into jumbles of rock or fissures when threatened. Gundis spend much time basking and do
not emerge on cold or stormy days, habits suggesting that they conserve energy by the use of behavioral thermoregulation.

SUBORDER HYSTRICOGNATHI

Although the African hystricognaths (Bathyergidae, Hystricidae, Petromuridae, Thryonomyidae)
and those of South America (termed caviomorphs)
share many morphological characters (Table 18-3)
and are regarded by most authorities to be closely
related, their origins and relationships remain a
source of controversy and speculation. The controversy centers in part on the geographic origins of
these groups. One school of thought (Wood, 1980)
holds that the South American caviomorphs are
derived from early members of the Tertiary Frani-
morpha, a basal stock from North America. Lavocat
(1976, 1980), on the other hand, believes that Africa
was the source for South American caviomorphs.
Recent discovery of a new fossil rodent from Chile
pushes the origin of caviomorph-like rodents in
South American back to the Eocene-Oligocene
boundary. It is now generally agreed that the
South American hystricognaths (caviomorphs)
arose from an African ancestry (Wyss et al., 1993;
Martin, 1994).

Competition from other orders of South American mammals was apparently not intense, for the
caviomorphs rapidly radiated; among fossils from
the late Oligocene (Deseadan) 7 of the 13 living
caviomorph families can be distinguished. The
Oligo-Miocene climatic changes in South America
that were accompanied by an expansion of grasslands strongly affected the fortunes of the caviomorphs. The Octodontidae increased its range in
the Pliocene, whereas the range of the Echimyidae
shrank. The Dasyproctidae became less common
in the Pliocene, and the Chinchillidae became
less diverse. From South America, the hutias
(Capromyidae) probably reached the Lesser Antilles in the Oligocene by rafting and had varying
degrees of success there. The New World porcupines have been extremely successful in moving
northward from their ancestral home in South
America. They are now widespread in the North

TABLE 18-3   Shared Derived Characters in Hystricognathous Rodents

- Mandible hystricognathous
- Infraorbital foramen enlarged and hystricomorphous
- Deepened pterygoid fossa opening into the orbit
- Enlarged slip of the superficial masseter muscle attached to the medial mandible
- Malleus and incus closely appressed or fused
- Internal carotid artery and canal absent
- Stapedial artery absent
- Molars four or five crested

After Luckett and Hartenberger, 1985.

American tropics, and one species is widely distributed in North American coniferous forests.

FAMILY BATHYERGIDAE. Family Bathyergidae contains the African mole-rats, a group of highly specialized fossorial rodents. The family includes five genera with nine living species. Mole-rats occupy much of Africa from Ghana, Sudan, Ethiopia, and Somaliland southward. The earliest fossil records of bathyergids are from the early Miocene of Africa. Bathyergids are allied with the Old World hystricognaths (Luckett, 1980b; Maier and Schrenk, 1987; Sarich and Cronin, 1980; Sherman et al., 1991).

Bathyergids are from 120 to 330 millimeters in total length, and they possess a number of unique structural features associated with their fossorial life. The eyes are small in all species, vision apparently is poorly developed, and the visual centers in the brain are reduced (Jarvis and Bennett, 1991). The ears lack or nearly lack pinnae. The skull is robust, the powerful incisors are procumbent in all species (Fig. 18-11), and the roots of the upper incisors usually extend behind the molars. The lips close tightly behind the incisors so that dirt does not enter the mouth when the animal is burrowing. The cheek teeth are hypsodont but rooted and typically have a simplified crown pattern. The dental formula is variable (1/1, 0/0, 2-3/2-3, 0-3/0-3 = 12-28). In *Heliophobius* there are six cheek teeth, but not all are functional simultaneously. The zygomasseteric structure is distinctive. The infraorbital foramen transmits little or no muscle. The masseter muscles, however, are highly specialized: the large anterior part of the medial masseter

originates from the upper part of the medial wall of the orbit, and the superficial part of the lateral masseter originates partly on the anterior face of the zygoma. The mandibular fossa and angular part of the dentary bone are greatly enlarged and provide an extensive area for the insertion of the masseter muscles (Fig. 18-11). The limbs are fairly robust in all species but are apparently used for digging only in *Bathyergus*. The hind feet are broad, and the animals back up against a load of soil and push it from the burrow with the hind feet. The tail is short and is used as a tactile organ. The pelage is normal in most species, but in *Heterocephalus glaber,* the skin is nearly naked with only a sparse sprinkling of long hairs (Fig. 18-26).

The African mole-rats are herbivorous and eat plant underground storage organs **(geophytes),** such as tubers, corms and bulbs, which the mole-rats find by burrowing. They seldom appear above ground and typically occupy soft loamy or sandy soils in desert and savanna areas. The huge incisors are the major digging tools in all species except *Bathyergus suillus*. Three of the five bathyergid genera are solitary, but *Heterocephalus* and *Cryptomys* are colonial and their burrow systems are far more extensive than those of solitary species. Burrowing *Heterocephalus* workers form "digging chains" with an organized division of labor (p. 439). *Heterocephalus* and *Cryptomys* are of tremendous interest because they are the only known **eusocial** mammals (p.473).

FAMILY HYSTRICIDAE. The Old World porcupines are a widely distributed group of rodents (three genera with 11 Recent species) that resem-

FIGURE 18-26   The naked mole-rat (*Heterocephalus glaber*, Bathyergidae) of East Africa. The lips close behind the incisors, and a fold of skin guards the nostrils. These adaptations keep soil from being inhaled and ingested when the incisors are involved in digging. (J. Jarvis)

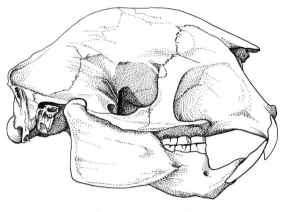

African crested porcupine
(*Hystrix cristata*, Hystricidae)

FIGURE 18-27   Skull of the African crested porcupine (Hystricidae). Note the greatly inflated rostrum and frontal part of the skull.

ble the New World porcupines (Erethizontidae) in having quills for protection. Hystricids occur throughout Africa (except Madagascar), in Sicily, Italy, Albania, and northern Greece (the Romans perhaps introduced porcupines into this area), in southern Asia and South China, and in Borneo, southern Sulawesi, Flores, and the Philippines. Hystricids first appear in the Miocene of Europe.

These large and imposing rodents weigh up to 27 kilograms and have a stocky build. The occipital region of the skull is unusually strongly built and provides attachment for powerful neck muscles. The zygomasseteric arrangement is hystricomorphous, with the large anterior part of the medial masseter originating on the deep rostrum. The nasoturbinal, lacrimal, and frontal bones are highly pneumatic (filled with air cavities) in some African species (Fig. 18-27). The dental formula is 1/1, 0/0, 1/1, 3/3 = 20. The hypsodont cheek teeth have re-entrant enamel folds that, with wear, become islands on the occlusal surfaces. Some of the hairs are stiff, sharp spines that reach at least 40 centimeters in length; in some species, open-ended, hollow spines make a noise when rattled that appears to have a warning function. One genus (*Trichys,* of Borneo, the Malay Peninsula,

and Sumatra) lacks stiff spines. The large, plantigrade feet of hystricids are five-toed, and the soles are smooth.

Hystricids are herbivorous, but, in contrast to New World porcupines, are terrestrial rather than partly arboreal and often dig fairly extensive burrows that are used as dens. The quills of some species are conspicuously marked with black and white bands. The two African species of *Hystrix* have an impressive intimidation display, involving fanning and "whirring" the quills, stamping the feet, and rushing rump-first toward the adversary (Kingdon, 1984b). Aside from humans, the larger cats are the main predators of hystricids.

FAMILY PETROMURIDAE. The relict family Petromuridae includes but a single species, *Petromus typicus,* the dassie rat (Fig. 18-28). This animal lives in parts of southwestern Africa, including the extremely dry Namib Desert. The family is known by fossils only in the Pleistocene of Africa.

The dassie rat is a small rodent (about 200 grams) with a squirrel-like appearance. The infraorbital foramen is enlarged and transmits part of the medial masseter. The rooted, hypsodont cheek teeth have a simplified crown pattern; the dental formula is 1/1, 0/0, 1/1, 3/3 = 20. Structurally, these animals are most remarkable for specializations enabling them to seek shelter in narrow crevices. Such specializations include a strongly flattened

FIGURE 18-28   A dassie rat (*Petromus typicus;* Petromuridae). *(M. Griffin)*

skull, flexible ribs that allow the body to be dorsoventrally flattened without injury, and mammae situated laterally, at the level of the scapulae, where the young can suckle while the female is wedged in a rock crevice.

Dassie rats are diurnal and feed largely on leaves and flowers. They are restricted to dry, rocky areas, where they take shelter in crevices. They bask in the early morning sun on cool mornings, a well-known energy-conservation behavior of another leaf eater, the hyrax (Hyracoidea). The daily energy expenditure of the dassie rat is much lower than the average for mammals its size (Withers et al., 1979), perhaps partly as a result of the use of behavioral thermoregulation.

FAMILY THRYONOMYIDAE. One genus with only two species makes up the small family Thryonomyidae, the members of which are known as cane rats. These animals are broadly distributed in Africa south of the Sahara. The earliest records of cane rats are from the late Eocene of Africa. Fossil cane rats from the Oligocene of the Mediterranean region, and an extinct species of *Thryonomys* from the central Sahara, indicate that the range of cane rats was once far greater than it is now.

Cane rats are large rodents, from 4 to 6 kilograms in weight, with a coarse, grizzled pelage. The snout is blunt, and the ears and tail are short. The robust skull has prominent ridges, a heavily built occipital region, and a large infraorbital fora-

men. The cheek teeth are hypsodont, and the large upper incisors are marked by three longitudinal grooves. The dental formula is 1/1, 0/0, 1/1, 3/3 = 20. The fifth digit of the forefoot is small, and the claws are strong and adapted to digging.

Cane rats are capable swimmers and divers and are largely restricted to the vicinity of water, where they take shelter in matted vegetation or in burrows. Males indulge in ritualized snout-to-snout pushing contests, and the blunt shape of the snout seems to enable the animals to avoid damage during these bouts. They are herbivorous and do considerable local damage to crops, particularly cassava and sugar cane. Cane rats are prized for food in many parts of Africa. The animals are often taken by snares, during organized drives using dogs, or are driven from their hiding places and captured when hunters set fire to reeds.

FAMILY ERETHIZONTIDAE. To this small family belong the New World porcupines, a group including four Recent genera with 12 living species. These animals occur from the Arctic south through much of the forested part of the United States into Sonora, Mexico, in the case of *Erethizon,* and from southern Mexico through much of the northern half of South America in the case of the other genera. Porcupines are of interest to most people because of their remarkable coat of quills and because the animals often have little fear of humans and can be observed easily.

New World porcupines are large, heavily built rodents, weighing up to 16 kilograms; all species have quills on at least part of the body (Fig. 18-29). The stiff quills are usually conspicuously marked by dark- and light-colored bands, and the sharp tips have small, proximally directed barbs. These barbs make the quills difficult to remove from flesh and aid in their penetration, which progresses at the rate of 1 millimeter or more per hour. The skull is robust, the rostrum is deep, and the greatly enlarged infraorbital foramen is nearly circular in some species (Fig. 18-30) and accommodates the highly developed medial masseter. The dental formula is 1/1, 0/0, 1/1, 3/3 = 20; the rooted cheek teeth have occlusal patterns dominated by re-entrant enamel folds (Fig. 18-31). New World porcupines have some arboreal adaptations that are lacking in their more terrestrial Old World counterparts. The feet of erethizontids have broad

FIGURE 18-29 A North American porcupine (*Erethizon dorsatum*, Erethizontidae) in winter pelage. *(PhotoDisc, Inc.)*

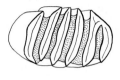

Porcupine   Capybara

FIGURE 18-31 Crowns of hystricognath molars. The upper right molars one and two of the porcupine (*Erethizon dorsatum*, Erethizontidae) and the third lower left molar of the capybara (*Hydrochaeris hydrochaeris*, Hydrochaeridae). The cross-hatched areas on the porcupine teeth are dentine; the stippled areas on the capybara tooth are cementum. *(Capybara after Ellerman, 1940)*

ginning in the Oligocene, this family underwent its early evolution in South America, becoming established in North America only after the emergence of the previously inundated Isthmus of Panama in the Pliocene.

New World porcupines eat a variety of plant material. Cambium is a staple winter food for *Erethizon*, and in many timberline areas trees missing large sections of bark and cambium give evidence of long-term winter occupancy by porcupines. In summer, porcupines eat a variety of plants and often "graze" on sedges *(Carex)* along the borders of meadows. Most species of porcupines are able climbers, and *Coendou* spends most of its life in trees. *Erethizon* is inoffensive and at times almost oblivious to humans. When in danger, however, the animal directs its long dorsal hairs forward, exposing the quills; it erects the quills and arches its back. The tail is flailed against an attacker as a last

soles marked by a pattern of tubercles that increase traction (Fig. 18-13); in some species, the hallux is replaced by a large, movable pad. The toes bear long, curved claws, and the limbs are functionally four-toed. In *Coendou*, the long tail is prehensile and curls dorsally to grasp a branch. Be-

FIGURE 18-30 Skulls of two hystricognath rodents. Note the great enlargement of the zygomatic arch in the paca (length of skull 150 millimeters) and the large infraorbital foramen in the porcupine (length of skull 115 millimeters). *(Paca after Hall and Kelson, ©1959, Mammals of North America, by permission of John Wiley & Sons, Inc.)*

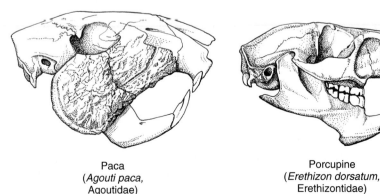

Paca
(*Agouti paca*,
Agoutidae)

Porcupine
(*Erethizon dorsatum*,
Erethizontidae)

resort. Surprisingly, *Erethizon* is killed by a variety of carnivores. Some mountain lions learn to flip porcupines on their backs and kill them by attacking the unprotected belly. Occasionally, however, dead or dying carnivores are found with masses of quills penetrating the mouth and face, indicating that learning to prey on porcupines may be a fatal undertaking. *Erethizon* characteristically takes shelter in rock piles, beneath overhanging rocks, or in hollow logs, but does not dig burrows as do Old World porcupines.

FAMILY CHINCHILLIDAE. One member of family Chinchillidae, the chinchilla *(Chinchilla)*, is somewhat familiar to many because of the publicity given to chinchilla fur farming some years ago. The family also includes the viscachas *(Lagidium* and *Lagostomus)*. Three genera with six Recent species represent the family, which occurs in roughly the southern half of South America in the high country of Peru and Bolivia and throughout much of Argentina to near its southern tip. The fossil record of this group is entirely South American and extends from the late Oligocene to the Recent.

Chinchillids are densely furred and of moderately large size (1 to 9 kilograms), with a long, well-furred tail. Mountain viscachas *(Lagidium)* and chinchillas have fairly large ears and a somewhat rabbit-like appearance, whereas the plains viscacha *(Lagostomus)* has short ears. The cheek teeth are ever-growing, and the occlusal surfaces are formed by transverse enamel laminae with intervening cementum. The dental formula is 1/1, 0/0, 1/1, 3/3 = 20. There are some cursorial adaptations, but the clavicle is retained. The forelimbs are fairly short and tetradactyl; the hindlimbs are long, however, and the elongate feet have four *(Chinchilla* and *Lagidium)* or three *(Lagostomus)* toes.

Chinchillids are herbivorous and occupy a variety of habitats, including open plains (pampas), brushlands, and barren, rocky slopes at elevations ranging from 800 to 6000 meters. The mountain viscachas and chinchillas are diurnal and seek shelter in burrows or rock crevices. Although adept at moving rapidly over rocks and broken terrain, they seem not to depend on speed in the open to escape enemies. The plains viscacha *(Lagostomus)*, in contrast, occurs in open pampas areas with little cover, where colonies live in extensive burrow sys-

tems marked by low mounds of earth and accumulations of such debris as bones, livestock droppings, and plant fragments. The habit of collecting items is displayed even by captive animals. Colonies may occupy large areas; one such area measured 20 by 300 meters, and this colony was known to have been in existence for at least 70 years (Weir, 1974). In *Lagostomus*, cursorial ability is highly developed: these animals are able to make long leaps and to evade a pursuer by abrupt turns. They have considerable endurance and can run at speeds up to 40 kilometers per hour.

FAMILY DINOMYIDAE. Family Dinomyidae includes a single South American species, *Dinomys branickii*, the pacarana. This rare animal inhabits the foothills of the Andes and adjacent remote valleys in Peru, Colombia, Ecuador, and Bolivia. An extinct member of this family reached spectacular size: the Pliocene dinomyid *Telicomys* was the size of a rhinoceros. Although dinomyids appear to be near extinction today, they were more successful and diverse in the past with up to 58 species, the oldest of which appears in the Oligocene of South America (Mones, 1981).

The pacarana weighs up to 15 kilograms; the dark brown pelage is marked by longitudinal white stripes and spots. Pacaranas lack the cursorial adaptations of the Caviidae and Hydrochaeridae. Instead, the broad tetradactyl feet of pacaranas have long, stout claws seemingly adapted to digging, and the foot posture is plantigrade. The clavicle is complete, another departure from the conventional cursorial plan. The unusually hypsodont cheek teeth consist of a series of transverse plates. The dental formula is 1/1, 0/0, 1/1, 3/3 = 20.

This unusual rodent is herbivorous, slow-moving, and docile in captivity. It is probably nocturnal and communicates by foot-stamping and an elaborate series of vocalizations (Collins and Eisenberg, 1972). It is extremely rare in the wild and is currently listed as endangered.

FAMILY CAVIIDAE. Family Caviidae is small, including just five genera and 14 species, contains the familiar guinea pig *(Cavia)*, as well as several similar types, and the Patagonian mara *(Dolichotis)*, an animal remarkable in having many cursorial adaptations. Caviids occur nearly throughout

South America, except in Chile and parts of eastern Brazil, and first appeared in the Oligocene of South America.

The guinea pig–like caviids (subfamily Caviinae) are chunky and moderately short-limbed and weigh from 400 to 700 grams. *Dolichotis* (Dolichotinae), in contrast, has long, slender legs and feet and weighs up to approximately 16 kilograms. All caviids have ever-growing cheek teeth with occlusal patterns consisting basically of two prisms. The dental formula is 1/1, 0/0, 1/1, 3/3 = 20. The dentary bone has a conspicuous lateral groove into which insert the temporal muscle and the anterior part of the medial masseter. Although only *Dolichotis* is strongly cursorial, all caviids have certain features typical of cursorial mammals: the clavicle is vestigial, the tibia and fibula are partly fused, and the digits are reduced to four on the manus and three on the pes. Members of the subfamily Caviinae, despite these cursorial adaptations, have a plantigrade foot posture and scuttle about in mouselike fashion. Locomotion in *Dolichotis*, however, is rapid and involves long bounds. The foot posture of *Dolichotis* during running is digitigrade, and specialized pads beneath the digits (Fig. 18-32D) cushion the impact when the feet strike the ground. With its deep, somewhat laterally compressed skull and large ears, *Dolichotis* resembles a large rabbit.

Caviids are herbivorous, and some have complex systems of social behavior and vocal communication (Lacher, 1981; Rood, 1970b, 1972). They occupy habitats ranging from grassland and open pampas to brushy and rocky areas and forest edges. They are nocturnal, diurnal, or **crepuscular** (active at dawn and dusk) and often live in large colonies with distinct social hierarchies. The mara is an inhabitant of open, arid regions. It is diurnal and forms large, loosely knit groups on occasion, but the basic social unit is a monogamous pair that mates for life.

FAMILY HYDROCHAERIDAE. Family Hydrochaeridae contains the largest living rodent, the capybara *(Hydrochaeris)*. The single species occupies Panama and roughly the northern half of South America east of the Andes. Extinct species of capybara are known from the Pliocene and Pleistocene in North America and the Miocene to Recent in South America. An extinct Pliocene type *(Protohydrochaerus)* probably weighed over 200 kilograms.

Capybaras are large (up to 50 kilograms in weight), robust, rather short-limbed rodents with a coarse pelage (Fig. 18-33). The head is large and has a deep rostrum and truncate snout. The skull and dentary bone are similar to those of members of the Caviidae, but the paroccipital processes are unusually long. The teeth are ever-growing. Both upper and lower third molars are much larger than any other cheek tooth in their respective rows and are formed by transverse lamellae united by cementum (Fig. 18-31). The dental formula is 1/1, 0/0, 1/1, 3/3 = 20. The tail is vestigial, and the same cursorial features listed for the caviids occur in the hydrochaerids. The digits are partly webbed and unusually strongly built, adaptations that prob-

FIGURE 18-32 Ventral views of the left hind foot of some South American rodents. (A) Chinchilla (*Chinchilla* sp., Chinchillidae), (B) degu (*Octodon* sp., Octodontidae), (C) *Dasyprocta* sp. (Dasyproctidae), (D) Patagonian mara (*Dolichotis* sp., Caviidae). *(After Howell, Speed in Animals, © 1944, University of Chicago Press)*

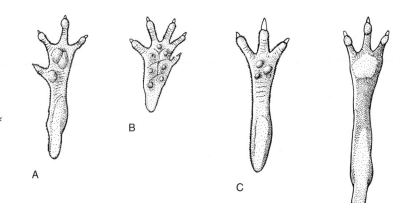

FIGURE 18-33   Two capybaras (*Hydrochaeris hydrochaeris,* Hydrochaeridae) in a streambed. *(A. Brouwer)*

ably allow the support of the considerable body weight on marshy ground.

Capybaras are semiaquatic and occur along the borders of marshes or the banks of streams, where they forage on succulent herbage (Quintana et al., 1998). They are largely crepuscular, and, although they can run fairly rapidly, usually seek shelter in the water. They swim and dive well and can remain submerged beneath water plants with only the nostrils above water. In both form and function, a capybara resembles a miniature hippopotamus. In parts of Venezuela, capybaras are raised commercially as food for human consumption.

FAMILY DASYPROCTIDAE. Members of family Dasyproctidae and their relatives, called agoutis (Fig. 18-34A), occur in the Neotropics from southern Mexico south to Ecuador, Bolivia, Paraguay, and northeastern Argentina. Two genera and 13 species (some of which are of doubtful validity) are included in this family. The earliest fossils of rodents in South America (Eocene–Oligocene boundary) are of a dasyproctid from Chile.

These rodents are medium sized, up to 2 kilograms in weight. The tail is short. The skull is robust, the incisors are fairly thin, and the crowns of the hypsodont cheek teeth are flat and bear five crests. The dental formula is 1/1, 0/0, 1/1, 3/3 = 20. Although these rodents are compactly built, the limbs are slim and have many cursorial adaptations. The forefeet are tetradactyl, and the plane of symmetry passes between digits three and four (as in the Artiodactyla). The hind feet have three toes, and the plane of symmetry passes through digit three (Fig. 18-32C), as in the Perissodactyla. The clavicle is vestigial, and the claws are hooflike.

These herbivorous rodents typically inhabit tropical forests, where they are largely diurnal. *Dasyprocta punctata* is territorial and scatter-hoards fruit and nuts in times of plenty to be used during lean times. These animals are probably important seed dispersers. A male *D. punctata* sprays a female with urine during courtship; the female then goes into a "frenzy dance" and allows the male to copulate (Smythe, 1978). Some species take refuge in

A   B

FIGURE 18-34   (A) An agouti (*Dasyprocta* sp., Dasyproctidae) and (B) a paca, *Agouti paca* (Agoutidae). *(L. Ingles)*

burrows that they dig in the banks of arroyos, beneath roots, or among boulders. Agoutis are rapid and agile runners and usually travel along well-worn trails. Agoutis have a habit of remaining still when approached by a predator and then bursting from cover and running away after the fashion of a small antelope.

FAMILY AGOUTIDAE. Family Agoutidae includes two species of a single genus (*Agouti*). These rodents live in tropical forests from central Mexico to southern Brazil and are known from the Oligocene to Recent.

Often called pacas, these rodents are large, weighing up to about 12 kilograms, nearly tailless, and have a conspicuous pattern of white spots and stripes on the body (Fig. 18-34B). They have an exceptionally ungraceful form, with short legs and a blunt head. There are four digits on the forefeet and five digits on the hind feet. The cheek teeth are high-crowned, and the dental formula is 1/1, 0/0, 1/1, 3/3 = 20. Resonating chambers are formed by concavities in the maxillaries and by greatly broadened zygomatic arches (Fig. 18-30); air is forced through associated pouches, producing a resonant, rumbling sound. The massively enlarged zygoma are unique to these animals.

These terrestrial rodents live in tropical forests along streams and rivers, where they dig burrows in banks. The diet consists of a variety of plant material, including fallen fruit. They are nocturnal and not particularly swift on land but are good swimmers and often escape enemies by fleeing into the water. Intensive hunting for their highly prized meat, coupled with the fact that pacas seldom bear more than one young, has lead to dramatic declines in their populations in many areas.

FAMILY CTENOMYIDAE. Members of family Ctenomyidae, called tuco-tucos, are fossorial and superficially resemble pocket gophers (Geomyidae). They occupy much of the southern two thirds of South America, from Peru to Tierra del Fuego, including the Andes Mountains to elevations of 4000 meters. There is a single genus (*Ctenomys*) with 38 species. Ctenomyids first appear in the late Pleistocene of South America.

These rodents range in size from 100 to 700 grams and are unusual in having simplified cheek teeth that are roughly kidney-shaped; the dental formula is 1/1, 0/0, 1/1, 3/3 = 20, and the third molar is vestigial. The skull is broad and dorsoventrally flattened. Among South American hystricognathous rodents, only in the Ctenomyidae and in one species of the Octodontidae (*Spalacopus cyanus*) are fossorial adaptations strongly developed. The head of the tuco-tuco is large and broad, and the stout incisors protrude permanently from the lips. The eyes and ears are small, the neck is short and powerfully built, the forelimbs are powerful, the manus has long claws, and the tail is short and stout. In contrast to pocket gophers, tuco-tucos have greatly enlarged hind feet with powerful claws, and they lack external cheek pouches. Fringes of hair on the toes of the fore and hind feet in tuco-tucos are presumably an aid to the animals when they are moving soil.

Tuco-tucos are herbivorous and eat such underground parts of plants as roots, tubers, and rhizomes. They dig extensive burrow systems in open, often barren areas, and most species live in "colonies" composed of many solitary individuals, each with its burrow systems spaced widely apart from those of its neighbors. An animal typically occupies a given burrow system permanently but periodically seeks adjacent foraging areas by digging new burrows. Tuco-tucos occasionally make short forays from their burrows to gather leaves and stems (Pearson, 1959). In contrast to pocket gophers (Geomyidae), tuco-tucos are quite vocal and give distinctive cries from burrow entrances. Tuco-tucos share **karyotypic** similarities with the Octodontidae (George and Weir, 1972). The rapid speciation of *Ctenomys* is probably due to the isolation afforded by their fossorial lifestyle.

FAMILY OCTODONTIDAE. Although mostly burrow dwellers, octodontids are ratlike in general appearance, and most species lack the fossorial specializations typical of the Ctenomyidae (Fig. 18-35). Octodontids, variously called degus, cururos, or rock rats, have a restricted range near the west coast of South America from southwestern Peru south to northern Argentina and northern Chile. There are six genera and nine species. The earliest fossil records are from the lower Oligocene of South America. These small rodents (200 to 300 grams) derive their family name from the "eight-shaped" crown pattern of the cheek teeth; the dental formula is 1/1, 0/0, 1/1, 3/3 = 20. Most

FIGURE 18-35 A rock rat or degu (*Octodon degus,* Octodontidae) *(M. Andera)*

species have large ears, large eyes, and long vibrissae. The forefeet have four digits, and the hind feet have five; the tail varies from long to rather short.

Octodontids occupy a variety of habitats, from grassy areas to high Andean forests to dry cactus and acacia slopes. *Octodon* is an able climber that takes shelter in rocks or the burrows of other animals, and *Octodontomys* lives in burrows and in rock crevices or caves and feeds on acacia pods and cactus. Neither of these genera is specialized for fossorial life, nor is *Octomys*. Of the remaining genera, *Aconaemys* is somewhat modified for fossorial life and *Spalacopus* is strongly so. In Chile, *Spalacopus cyanus* occupies sandy coastal areas where it occurs in colonies, all members of which occupy a common burrow system. The animals feed entirely below ground, and the tubers and underground stems of huilli, a species of lily *(Leucoryne ixiodes),* form the bulk of the diet. *Spalacopus* is nomadic, an exceptional mode of life for a rodent. When a colony exhausts the supply of huilli roots at one place, the animals abandon this foraging site and move to a nearby undisturbed area (Reig, 1970). *Spalacopus* is unusually vocal for a rodent and gives distinctive calls at burrow openings. It uses its forelimbs and teeth to loosen soil and its large hind feet to throw dirt from the mouth of the burrow. The ranges of the tuco-tucos (Ctenomyidae) and the similarly adapted *Spalacopus* do not overlap. The restricted diet and nomadic nature of *Spalacopus* have not allowed the rapid speciation found in the more sedentary Ctenomyidae (Reig, 1970).

FAMILY ABROCOMIDAE. Members of family Abrocomidae, the "chinchilla" rats, occur in parts of west central South America. Their range includes southern Peru, Bolivia, and northwestern Argentina and Chile. The family is represented today by one genus *(Abrocoma)* with three species. This family appeared first in the South American Pliocene.

Chinchilla rats look roughly like large woodrats *(Neotoma;* Muridae), reach over 400 millimeters in total length, and resemble octodontids in many ways. The pelage is long and dense. The skull has a long, narrow rostrum, and the bullae are enlarged. The cheek teeth are ever-growing; the upper teeth have an internal and an external enamel fold, while the lowers have two internal folds. The dental formula is 1/1, 0/0, 1/1, 3/3 = 20. The limbs are short and have short, weak nails. The pollex is absent. *Abrocoma bennetti* is unusual among rodents in having 17 pairs of ribs.

These herbivorous rodents are poorly known. They are seemingly colonial, climb well, and usually seek shelter beneath or among rocks or in tunnels under rocks or bushes. *A. cinerea* lives in cold, bleak, rocky areas in the Andes at elevations between 3700 and 5000 meters. The extremely long digestive tract (2.5 meters) and voluminous caecum suggest a diet of leaves.

FAMILY ECHIMYIDAE. Members of the important Neotropical family Echimyidae, which includes a variety of roughly rat-size rodents, are called spiny rats. Most of the living species have flattened, spinelike hairs with sharp points and slender basal portions. Approximately 78 living species of 20 genera are recognized (four new species were described by da Silva, 1998). Spiny rats are widely distributed in the Neotropics, occurring from Nicaragua southward through the northern half of South America to Paraguay and southeastern Brazil.

Echimyids have prominent eyes and ears. Some species are fairly large for rodents, weighing over 600 grams. The tail, which in some genera is longer than the head and body, is lost readily, a feature perhaps of value in aiding escape from predators. The point of weakness is at the centrum of the fifth caudal vertebra. Among 637 *Proechimys* taken in Panama, 18 percent were tailless (Fleming, 1970). The cheek teeth are rooted, and the occlusal surfaces in most species are marked by transverse re-entrant folds. The dental formula is 1/1, 0/0, 1/1, 3/3 = 20. The feet are not highly specialized in most genera. In the arboreal tree rats *(Echimys)*, however, the digits are elongate and partially syndactylous. When an animal is climbing, the first two digits grasp one side of a branch in opposition to the remaining digits, which grasp the other side. Tree rats *(Echimys)* inhabit the canopy zone and, as a result their ecology and behavior, are poorly known.

Echimyids are an old group, appearing first in the late Oligocene of South America. Two extinct genera are known from skeletal material found in Indian kitchen middens in Cuba and Haiti. These genera seemingly became extinct fairly recently. In the case of the genus from Haiti *(Brotomys)*, extinction may have resulted from the introduction of predators by Europeans.

As far as is known, spiny rats are completely herbivorous. They are variously semifossorial, terrestrial, or arboreal. In Panama, fruit was the primary food found in the stomachs of many *Proechimys semispinosus* (Fleming, 1970). *Kannabateomys* apparently prefers the young shoots of bamboo and inhabits dense bamboo thickets near waterways (Emmons, 1990). Several arboreal species take shelter in tree holes. The bamboo rat *(Dactylomys dactylinus)* of South America is completely nocturnal and arboreal, eats leaves and buds, and gives explosive calls that presumably play a role in territoriality (Emmons, 1981). *Proechimys* is terrestrial and is among the most abundant lowland rainforest rodents (Eisenberg, 1989; Emmons, 1990).

FAMILY CAPROMYIDAE. Members of family Capromyidae are known locally as hutias, zagouties, cavies, or coneys and are restricted to the West Indies. Of the 20 species in eight genera listed by Woods (1993), 6 species are now extinct and 2 others are probably extinct (Woods et al., 1985). The 12 living species occupy the Bahamas, Cuba, Isle of Pines, Hispaniola, Puerto Rico, and Jamaica. The oldest fossils are those of early Miocene age. These mostly herbivorous rodents weigh up to about 7 kilograms and look like unusually large rats. They closely resemble the nutria (Myocastoridae) structurally and are often included in the same family. Hutias have thick fur, hypsodont and flat-crowned molars. They are primarily herbivorous but are

known to take small lizards. Although most species are solitary, *Geocapromys ingrahami* lives in colonies.

These rodents, adapted to the insular conditions of the West Indies before the coming of Europeans, were unable to cope with predation by the introduced mongoose *(Herpestes)* or by humans and their dogs. Of the 35 Recent species of capromyids, 23 are extinct; the remaining ones are restricted to steep or inaccessible areas (Woods, 1982; Woods et al., 1985). One living species *(Mesocapromys nana)* was first described from bones found in a cave but was later found alive. (Members of the West Indian family Heptaxodontidae have gone extinct in Recent or sub-Recent times and are not considered here.)

FAMILY MYOCASTORIDAE. The nutria *(Myocastor coypus)*, the only living member of family Myocastoridae, is familiar to many people in North America, Europe, and Asia because this South American rodent has been introduced widely and has thrived in certain areas. In some places, it has become a serious pest because of its destruction of aquatic vegetation and crops and its disruption of irrigation systems. This animal is native to southern South America, from Paraguay and southern Brazil southward, but now also occurs in some 15 states in the United States, as well as in some countries in Europe and East Africa. The family is also represented by nine extinct genera ranging from the early Miocene to the Recent in South America.

The nutria is large, up to roughly 8 kilograms, and looks like a rat-tail beaver *(Castor;* Castoridae). The skull is heavily ridged and has a deep rostrum. The zygomasseteric structure is hystricomorphous; in association with the reduction of the temporal muscles, the coronoid process of the dentary bone has nearly disappeared and is represented by a small knob. The hypsodont cheek teeth well illustrate changes in crown pattern that occur with increasing age and wear (Fig. 18-36). The dental formula is 1/1, 0/0, 1/1, 3/3 = 20. The feet have heavy claws, and a web joins all but the fifth toe of the pes.

Nutrias resemble beavers in some of their habits. They dig burrows in banks, use cleared trails through vegetation, are extremely destructive to plants, and are capable swimmers and divers. Because of their dense, fine underfur, they have been raised in some fur farms in the United States,

Light wear   Heavy wear

Nutria
(*Myocastor coypus,* Myocastoridae)

FIGURE 18-36   First and second upper right molars of the nutria. Note the tremendous changes in the crown pattern due to wear. Stippled areas on the occlusal surfaces surrounded by enamel (unshaded) are dentine.

and are trapped for their fur in some states. In the trapping season of 1975 to 1976, the nutria sold to the fur trade in Louisiana yielded $8 million (Woods, 1984). Most biologists strongly oppose the indiscriminate introductions of such non-native animals (referred to as "exotics") as the nutria. The activities of non-native species often result in the alteration of the vegetation, with the resultant disappearance of native species and the destruction, perhaps irretrievably, of the original biotic community.

WEB SITES

The squirrel page with links to other squirrel sites

http://www.io.com/~hmiller/squirrel.html

General information about lemmings

http://stud1.tuwien.ac.at/~e8826423/LemmZool.html

Many links to rodent sites

http://netvet.wustl.edu/rodents.htm

General information about capybaras

http://www.rebsig.com/capybara

Gerbil home page with lots of information and links

http://users.bart.nl/~fredveen/gerbiluk.htm

19 LAGOMORPHA AND MACROSCELIDEA

The taxonomic position of the Lagomorpha (pikas and rabbits) with respect to the other mammalian orders has been vigorously debated over the past century. Lagomorphs were traditionally united with rodents in the cohort Glires (Gregory, 1910; Simpson, 1945). However, such groups as pri-

mates, insectivorans, and artiodactyls have also been considered sister groups of the lagomorphs (Wood, 1957). Cranial characteristics (Novacek, 1985; Gaudin et al., 1996), placental membranes (Luckett, 1985), and new fossil evidence (Li and Ting, 1985; Li et al., 1987) support a close relationship

between rodents and lagomorphs. Unfortunately, molecular studies not only fail to corroborate this relationship, they lack consensus as to the position of lagomorphs within the Mammalia (Sarich, 1985; Shoshani, 1986; Li et al., 1990; Miyamoto and Goodman, 1986; Springer and Kirsch, 1993). Until further evidence is available, we support the placement of lagomorphs as a sister group of rodents.

Macroscelidids (elephant-shrews) were traditionally associated with tupaiids and insectivores (Simpson, 1945; Anderson and Jones, 1984), a view based primarily on shared primitive characters. Others consider macroscelidids to be a sister group of Glires (rodents and lagomorphs; McKenna, 1975; Novacek and Wyss, 1986; Novacek, 1992b; Wible and Covert, 1987; Gaudin et al., 1996; Szalay, 1977), hence the inclusion of these two orders in the same chapter. Recent studies of both molecular and morphological characters suggest that macroscelids are an ancient African radiation with no close living relatives and that they may belong to a clade consisting of elephants, hyraxes, and aardvarks called the "paenungulata" (De Jong et al., 1993; Woodall and FitzGibbon, 1995; Balter, 1997; Madsen et al., 1997).

ORDER LAGOMORPHA

Although lagomorphs — the rabbits (Leporidae) and pikas (Ochotonidae) — are not a diverse group, including but 13 genera with 80 Recent species, they are important members of many terrestrial communities and are nearly worldwide in distribution. Considering large land masses only, lagomorphs were absent only from Antarctica, the Australian region, and southern South America before recent introductions by humans. Lagomorphs occupy diverse terrestrial habitats from the Arctic to the tropics. In some temperate and boreal regions, rabbits and hares are subject to striking population cycles marked by periods of great abundance alternating with periods of extreme scarcity (p. 519). In such regions, the population cycles of some carnivores are influenced strongly by changes in rabbit population densities.

Many important diagnostic features of Recent lagomorphs are related to their herbivorous habits and, in the case of leporids, to their cursorial locomotion. Lagomorphs have a **fenestrated** skull (having areas of thin, lattice-like bone), a feature highly developed in some leporids (Fig. 19-1). The anterior dentition resembles that of a rodent, but whereas rodents have 1/1 incisors, rabbits and pikas have 2/1 incisors; the second incisor is small and peglike and lies immediately posterior to the first (Fig. 19-1). As in rodents, the lagomorph incisors are ever-growing. A long post-incisor diastema is present in lagomorphs, and the canines are absent. The cheek teeth are hypsodont and rootless, and the crown pattern features transverse ridges and basins (Figs. 19-2A and 19-3B). The distance between the two upper tooth rows is greater than that between the two lower rows, restricting occlusion of upper and lower cheek teeth on only one side at a time and requiring a lateral or oblique jaw action. The masseter muscle is large, and the pterygoideus muscles are well developed and help control transverse jaw movements. The temporalis is small, and the coronoid process, its point of insertion, is rudimentary.

The skull of leporids is unique among mammals in having a clearly defined joint at which slight movement occurs. This joint fully encircles the skull just anterior to the occipital and otic bones. This unusual cranial specialization appears first in Miocene leporids and is a mechanism that absorbs shock to the skull while the animal is bounding at high speeds (Bramble, 1989). The clavicle is either well developed (Ochotonidae) or rudimentary (Leporidae), and the elbow joint limits movement to a single anteroposterior plane (Fig. 19-2B and C). The tibia and fibula are fused distally; the front foot has five digits, and the hind foot has four or five digits. The soles of the feet, except for the distalmost toe pads in *Ochotona*, are covered with hair.

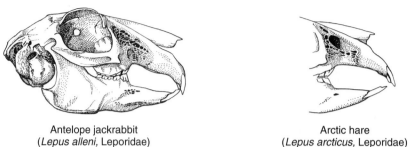

Antelope jackrabbit
(*Lepus alleni*, Leporidae)

Arctic hare
(*Lepus arcticus*, Leporidae)

FIGURE 19-1 Skull of the antelope jackrabbit *(Lepus alleni);* note the highly fenestrated maxillary and occipital bones. On the right is the anterior part of the skull of the arctic hare *(L. arcticus),* showing the procumbent incisors and the receding nasals, specializations associated with this animal's habit of using the incisors to scrape away ice and snow to reach food. *(Arctic hare after Hall and Kelson, © 1959,* Mammals of North America, *by permission of John Wiley & Sons, Inc.)*

The foot posture is digitigrade during running but plantigrade during slow movement. The tail is short and in pikas is not externally evident.

The first fossil record of mammals with lagomorph-like characters is from the Paleocene of China. Although rodents and rabbits were long regarded as unrelated, recent evidence indicates that these groups share a common ancestry within the early Paleocene order Anagalida (Li, 1977; Li et al., 1987). The family Leporidae probably originated in Asia in the Eocene but underwent most of its early (Oligocene and Miocene) evolution in North America. Leporids diversified in the Old World in the Miocene, and the advanced subfamily Leporinae arose there. The pikas appeared first in the Eocene of Eurasia and spread in the Oligocene to Europe and North America. The Recent genus *Ochotona* is known from the late Miocene. In contrast to the leporids, which have remained widespread since the Pliocene, the ochotonids reached their greatest diversity and widest distribution in the Miocene, when they occupied Europe, Asia, Africa, and North America (Dawson, 1967); they have declined since. In North America, pikas are

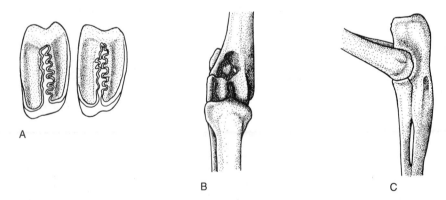

FIGURE 19-2 Leporid structural features, exemplified by the antelope jackrabbit (*Lepus alleni*). (A) Occlusal view of upper right premolars three and four. (B) Anterior view of right elbow joint; movement is limited to the anteroposterior plane by this "tongue and groove" articulation. When the forearm is fully extended, a process on the olecranon of the ulna locks into the conspicuous hole in the humerus and braces the joint. (C) Medial view of right elbow, showing the tight fit between the articular surface of the humerus and those of the radius and ulna. The radius and ulna are partially fused.

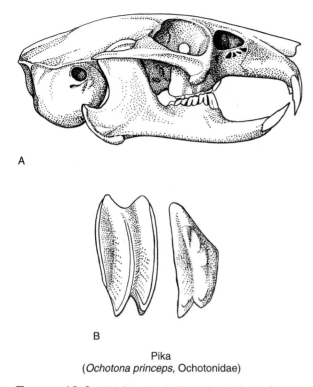

A

B

Pika
(*Ochotona princeps*, Ochotonidae)

FIGURE 19-3   (A) Skull and (B) occlusal view of upper right premolars three and four of the pika.

now of local occurrence in high mountains north of Mexico. They occur more widely and are more diverse in the Old World, where they inhabit eastern Europe and much of northern and central Asia.

The environmental tolerance, or perhaps the competitive success, of ochotonids has seemingly changed in the New World since the Miocene. For example, ochotonids occupied riparian and marshy communities in the Great Plains, Rocky Mountains, and Great Basin regions of North America (Wilson, 1960; Dalquest et al., 1996). Wetland situations in these areas no longer support pikas. The factors influencing the striking post-Miocene decline of the ochotonids are unknown.

Why, although they are an old and thriving group, have lagomorphs not undergone a greater adaptive radiation? Perhaps their conservatism is related to the limitations of their functional position as "miniature ungulates." Competition with members of the larger and more diverse order Artiodactyla, a group highly adapted to an herbivorous diet and cursorial locomotion, may have limited lagomorphs to the exploitation of but a single, limited adaptive zone, although this zone was oc-

cupied with great success over broad areas. Of interest in this regard are the scarcity and local occurrence of lagomorphs in many parts of East Africa, where there is an extremely rich ungulate fauna.

FAMILY OCHOTONIDAE. The pikas are represented today by two genera with roughly 26 species. Pikas are less specialized with regard to cursorial adaptations than are rabbits and usually venture only short distances from shelter. Pikas occur in the mountains of the western United States and south central Alaska and over a wide area in the Old World, including eastern Europe and much of Asia southward to northern Iran, Pakistan, India, and Burma.

Pikas are smaller than rabbits, weighing about 100 to 150 grams. They have short, rounded ears, short limbs, and no externally visible tail (Fig. 19-4). The ear opening is guarded by large valvular flaps of skin that may provide protection during severe weather. The skull is strongly constricted between the orbits and lacks a supraorbital process; the rostrum is short and narrow. The skull is less strongly arched in ochotonids than in leporids (Fig. 19-3A), and the angle between the basicranial and palatal axes is smaller. The maxilla has a large fenestra. The dental formula is 2/1, 0/0, 3/2, 2/3 = 26. The third lower premolar has more than one re-entrant angle, and the re-entrant enamel ridges of the upper cheek teeth are straight (Fig. 19-3B). The anal and genital openings are enclosed by a common sphincter, and males have no scrotum.

FIGURE 19-4   A pika (*Ochotona princeps*) at a lookout point on a rock. *(© PhotoDisc, Inc.)*

In North America, pikas usually inhabit talus slopes in boreal or alpine situations and occur from near sea level in Alaska to the treeless tops of some of the highest peaks in the Rocky Mountains and Sierra Nevada–Cascade chain. When frightened, pikas seek shelter in the labyrinth of chambers and crevices between rocks and seldom forage far from such shelter. Large "hay piles" are built each summer in the shelter of large, usually flat-bottomed boulders; these provide food when snow covers the ground. In Eurasia, pikas occupy an extensive geographic range and a wide variety of habitats, including talus, forests, rock-strewn terrain, and open plains and desert-steppe areas. Unusually large hay piles, weighing up to 20 kilograms, are made by pikas inhabiting dry areas in southern Russia (Formozov, 1966).

FAMILY LEPORIDAE. The rabbits and hares are a remarkably successful group in terms of ability to occupy a variety of environments over broad areas; they are now nearly cosmopolitan. Their distribution before introduction by humans included most of the New and Old Worlds, and rabbits have been introduced into New Zealand, Australia, parts of southern South America, and various oceanic islands in both the Atlantic and the Pacific Ocean. Eleven Recent genera represented by 54 Recent species are known.

Several major leporid evolutionary structural trends are recognized by Dawson (1958). The cheek teeth have become hypsodont, some of the premolars have become molariform, and the primitive crown pattern has been modified into a simple arrangement in which most traces of the primitive cusp pattern have been lost. These changes resemble those in some groups of strictly herbivorous rodents. The skull has become arched, and the angle between the basicranial and palatal axes has increased. The changes are associated with a posture involving a greater angle between the long axis of the skull and the cervical vertebrae than that typical of primitive leporids. Trends in limb structure leading to increased cursorial ability include elongation of the limbs and articulation specializations that limit movement to one plane.

The leporid skull (Fig. 19-1) is arched in profile, and the rostral portion is fairly broad. The maxillae, and often the squamosal, occipital, and parietal bones, are highly fenestrated, and a prominent supraorbital process is always present. The auditory bullae are globular, and the external auditory meatus is tubular. The dental formula is usually 2/1, 0/0, 3/2, 3/3 = 28; the re-entrant enamel ridges of the upper cheek teeth are usually crenulated (Fig. 19-2A). The clavicle is rudimentary and does not serve as a brace between the scapula and the sternum. The limbs, especially the hindlimbs, are more or less elongate; movement at the elbow joint is limited to the anteroposterior plane (Fig. 19-2B, C). The tail is short. The ears have a characteristic shape: the proximal part is tubular, and the lower part of the opening is well above the skull (Fig. 19-5). The testes become scrotal during the mating season. In some species that inhabit regions

FIGURE 19-5 A snowshoe rabbit *(Lepus americanus)* in its summer pelage. *(© PhotoDisc, Inc.)*

with snowy winters, the animals molt into a white winter pelage in the fall (Fig. 19-6) and into a brown summer pelage in the spring. Wild leporids weigh from 0.3 to 5 kilograms.

Leporids inhabit a tremendous array of habitats, from arctic tundra and treeless and barren situations on high mountain peaks to coniferous, deciduous, and tropical forests, open grassland, savanna, and deserts. Some species, such as *Sylvilagus palustris* and *S. aquaticus* of the southeastern United States, are excellent swimmers and lead semi-aquatic lives. Leporids are entirely herbivorous and eat a wide variety of grasses, forbs, and shrubs. Several species are known to reingest fecal pellets and are thought to obtain essential nutrients (proteins and some vitamins) from material as it passes through the alimentary canal a second time.

FIGURE 19-6 A white-tail jackrabbit *(Lepus townsendii)* in its partially white winter pelage. This animal is from Colorado, but in more northerly parts of its range, *L. townsendii* is entirely white in winter. *(G. Bear)*

Habitat preference and cursorial ability differ markedly from species to species and are strongly interrelated. Broadly speaking, species with less cursorial ability, such as *Brachylagus idahoensis* and *S. bachmani* of the western United States, scamper short distances to the safety of burrows or dense vegetation when disturbed. Cottontails, such as *S. floridanus* of the eastern and *S. audubonii* of the western United States, are intermediate in cursorial ability and typically inhabit areas with scattered brush, rocks, or other cover and do not run long distances to reach a hiding place. Representing the extreme in cursorial specialization among lagomorphs are some members of the genus *Lepus*, such as the New World jackrabbits (*L. californicus, L. townsendii,* and *L. alleni* and their relatives) and some hares of the Old World (such as *L. capensis* of Africa). These animals, which have greatly elongate hindlimbs, have adopted a bounding gait and occupy areas with limited shelter, such as deserts, grasslands, or meadows, where they take shelter in shallow depressions in the grass, called "forms." Instead of taking cover at the approach of danger, they depend for escape on their running ability. Jackrabbits and other similarly adapted members of the genus *Lepus* are extremely rapid runners for their size; some attain speeds up to 70 kilometers per hour. This speed allows them to occupy areas with little cover, where they can outrun predators. The arctic hare *(L. arcticus)* of the North American Arctic often uses bipedal locomotion and can stand and jump using only its hind legs.

Rabbits are strong competitors and remarkably adaptable. In some parts of Australia, the extinction or near extinction of certain metatherians is perhaps due primarily to competition from introduced European rabbits *(Oryctolagus)*. In addition, these prolific rabbits have caused great damage to crops and rangeland and at various times have been a primary agricultural pest in many parts of Australia as well as in New Zealand, where they were also introduced. Leporids have adapted to a wide range of environmental conditions. Along the arctic coasts of Greenland, *L. arcticus* uses its protruding incisors (Fig. 19-1) to scrape through snow and ice to reach plants during the long arctic winters. Far to the south, in the deserts of northern Mexico, jackrabbits *(L. alleni)* maintain their water balance through hot, dry periods by eating cactus and yucca.

ORDER MACROSCELIDEA

The relationships between the elephant-shrews and other groups of mammals are unclear, due in part to a poor fossil record. Elephant-shrews have often been regarded as members of the heterogeneous and inclusive order Insectivora but differ from the hedgehogs, shrews, moles, and tenrecs in having complete auditory bullae, entotympanic bones, a large jugal bone and complete zygoma, and relatively small olfactory lobes. Elephant-shrews are now considered a separate order that comprises a single family (Macroscelididae).

Elephant-shrews are distinctive African mammals adapted to rapid cursorial locomotion. Their present distribution includes northeastern Africa (one species from Morocco, Algeria, Tunisia, and western Libya) and sub-Saharan Africa; they are absent from the Sahara region and western Africa. Areas of particular diversity include eastern and southwestern Africa. There are four genera and 15 living species. Fossils are known only from Africa, where their sparse fossil record dates back to the Eocene of Tunisia (Corbert and Hanks, 1968; Corbert, 1971; Patterson, 1965). *Rhynchocyon* is known from the Miocene, and two other extant genera *(Macroscelides* and *Elephantulus)* are present in the Pliocene-Pleistocene of South Africa and Tanzania.

Elephant-shrews vary in size from the diminutive *Macroscelides proboscideus* (45 grams) to *Rhynchocyon chrysopygus* (approximately 550 grams). They have large eyes and ears and remarkably long snouts (Figs. 19-7 and 19-8). The limbs are long and slender, and the tail is moderately long. The forelimbs and hindlimbs have four or five digits; in *Rhynchocyon* the forefeet are functionally tridactyl, with the first digit absent and the fifth much reduced, whereas the hind feet have four digits. This more extreme loss of digits in the forelimb than in the hindlimb is unusual in mammals. The large orbits of the skull are never bordered by a complete postorbital bar (Fig. 19-8). The dental formula is 1-3/3, 1/1, 4/4, 2/2-3 = 36-42; the last upper premolar is the largest molariform tooth, and the upper canine is double-rooted. The upper molars are quadrate and have four major cusps.

Many habitats are occupied by elephant-shrews, including open plains, savannas, deserts, thornbush, and tropical forests. *Rhynchocyon,* the giant elephant-shrew, is a forest dweller and, in some

A

B

FIGURE 19-7 Two species of elephant-shrews (Macroscelididae): (A) rufous elephant-shrew *(Elephantulus rufescens);* (B) golden-rumped elephant-shrew *(Rhynchocyon chrysopygus). (G. Rathbun)*

places, such as the coast of Kenya, occupies relict strips of forest. *R. chrysopygus* (Fig. 19-7B) has been studied by Rathbun (1979) and FitzGibbon (1997) in the coastal forests of Kenya, where this diurnal animal feeds on a wide variety of invertebrates, many of which it digs from the leaf litter and soil with the long claws of the front feet. This species is strikingly colored: a dark chestnut brown, with a slightly purplish cast, covers the back and flanks, against which the bright yellow rump patch stands out in sharp contrast. These elephant-shrews typi-

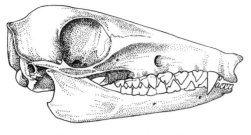

Golden-rumped elephant-shrew
(*Rhynchocyon chrysopygus,* Macroscelididae)

FIGURE 19-8   Skull of a golden-rumped elephant-shrew (length of skull 67 millimeters).

cally live in pairs occupying stable territories; scent marking by both sexes is an important territorial behavior (Rathbun, 1979; FitzGibbon, 1995, 1997). A resident pair chases conspecific intruders, with males chasing males and females chasing females. Rathbun (1978) postulates that the conspicuous yellow rump of *R. chrysopygus* functions as a target during aggressive encounters. He has found that the skin beneath this patch is thicker than skin over the rest of the body and that scars and cuts are concentrated beneath the patch.

Studies by Rathbun (1979) on the crepuscular rufous elephant-shrew have shown that the territories of this brush dweller contain intricate patterns of trails and that 24 percent of the daylight behavior of a territorial male is devoted to cleaning these trails by using the front feet to meticulously sweep away leaf litter. This male contribution to the common territory of a female and male may help explain why facultative monogamy evolved in this species (Rathbun, 1992). This species spends its life above ground, never seeking refuge in burrows, and usually forages within 1 meter of trails. These elephant-shrews escape predators by bounding along the trails at amazing speed. Scent marking by feces and sternal glands is concentrated on territorial borders. Rathbun found that the bulk of the diet is termites and ants. In coastal Kenya, where the four-toed elephant-shrew (*Petrodromus tetradactylus*) and the golden-rumped elephant-shrew (*R. chrysopygus*) are sympatric, niche separation is maintained by a combination of different activity patterns, diet preferences, and habitat separation (FitzGibbon, 1995). Elephant-shrews in some areas regularly bask, probably as a means of reducing energy expended on thermoregulation.

WEB SITES

General information about elephant-shrews

http://www.uq.edu.au/~anpwooda/pages/el-shrws.html

IUCN lagomorph specialist group — information on rabbit conservation

http://www.ualberta.ca/~dhik/lsg/

NetVet links to many other web resources on rabbits

http://netvet.wustl.edu/rabbits.htm

20 REPRODUCTION

Because of its primary importance to all life, reproduction is tied to virtually every structural, physiological, and behavioral adaptation of an individual or a species. The unique mammalian pattern of reproduction must be of great antiquity: mammary glands, nourishment of the newborn or newly hatched with milk, and a close mother-young bond probably evolved together with the diphyodont dentition in the late Triassic, though the bearing of live young (viviparity) may have appeared later.

Of all the vertebrate classes, the reproductive pattern typical of mammals departs most from that of other vertebrates. Primitive ancestral vertebrates

presumably were egg-layers, and this style of repro-duction, or some fairly modest variation on this theme, is typical of all classes of vertebrates but the Mammalia. In all mammals except prototherians, young remain within the uterus during their em-bryonic and fetal life, and it is here that embryonic tissues and organs differentiate and the fetus grows. Nourishment and protection for the in-trauterine young are provided by the mother, and, under most conditions, fetal survival rates are high. After birth, all young mammals are nour-ished by milk from the mother, and parental care, or in most cases maternal care, lasts until the young are reasonably capable of caring for them-selves. The young of some mammals stay with their parents through an additional period in which they learn complex foraging and social behavior. In sharp contrast, in most nonmammalian verte-brates (birds are an exception), the young have lit-tle or no parental care after hatching or, in the case of **ovoviviparous** animals (those that produce eggs that are incubated and hatch within the par-ent's body), after birth.

In mammals, the combined effect of high fetal survivorship and extended **postpartum** care is an increase in reproductive efficiency in terms of en-ergy expenditure per young reaching reproductive maturity. Most lower vertebrates lay great numbers of eggs at tremendous metabolic cost, and the suc-cess of the species depends on the survival of an extremely small percentage of young. For any given young of most lower vertebrates, survival is unlikely. In mammals, on the contrary, relatively few young are produced but the likelihood for sur-vival of any given young is fairly high.

MONOTREME EGGS

The three living monotremes are of great interest because of their combination of reptilian and mammalian characteristics. One characteristic that truly sets monotremes apart from other mammals is that females are oviparous (lay eggs). The repro-

ductive tract of female monotremes (see Fig. 5-3) resembles that of reptiles. Relative to placental mammals, the ovaries of monotremes are much larger because of the greater amounts of yolk in each egg. Mature **oocytes** (approximately 3 to 4 millimeters in diameter) are shed into the **in-fundibulum,** where they are fertilized before en-tering the oviduct. Although monotremes retain the paired oviducts of their reptilian ancestors, only the left oviduct is functional. During passage of the fertilized egg down the fallopian tube, a mu-coid coat (consisting of glycoprotein) is secreted around the egg (Hughes et al., 1975). Farther down the fallopian tubes, in the tubal gland re-gion, a basal layer of ovokeratin is secreted, form-ing the shell (Griffiths, 1978). After the egg passes into the anterior uterus, a second shell membrane is added (Hill, 1941). In the uterus, development proceeds via **meroblastic** cleavage (the cleavage furrow does not pass through the yolk), forming a **blastodisc** on top of the yolk. Eventually, the blas-todisc completely envelops the yolk, and embryo-genesis continues. Uterine secretions provide addi-tional nutrition at this stage. Once the embryo reaches about 10 to 12 millimeters in diameter, the third, more porous, layer of the shell is applied, and the eggs are laid soon afterwards.

Very little is known about the manner in which eggs are laid and their subsequent incubation by the female, but the following comes from a variety of observations of the echidna (*Tachyglossus aculea-tus*), summarized by Griffiths (1978). Apparently the egg moves through the cloaca directly to the pouch, where further incubation occurs. The fe-male curls up and, in this position, the cloaca nearly enters the pouch. Presumably, this curled posture allows the egg to be shed directly into the pouch. The pouch forms a shallow depression with swollen lips that forms the incubation chamber. Newly laid eggs are covered with a sticky coating that keeps the egg in the pouch when the female must move about. Approximately ten days after the egg is laid, the tiny young uses its **egg tooth** (another characteristic shared with reptiles) to

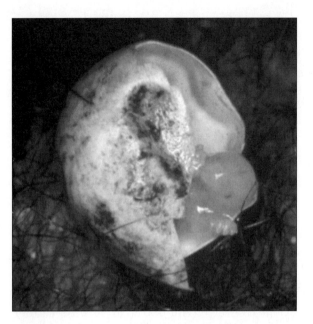

FIGURE 20-1 Young echidna (*Tachyglossus* sp.) emerging from the egg. *(P. Rismiller and M. McKelvey, © Rismac)*

emerge from the shell (Fig. 20-1). After hatching, the young finds its way to a mammary lobule inside the pouch and begins to suckle the rich milk. Young *Tachyglossus* remain in the pouch for approximately 55 days. When their spines begin to emerge, they are dropped from the pouch by the mother. Evicted young can continue to suckle from their mother for several more months (Griffiths, 1978).

It is a mistake to assume that because monotremes lay eggs their reproduction is similar to that of reptiles. Monotreme eggs are much smaller at ovulation, contain considerably less yolk, and, for most of the intrauterine development, growth of the embryo depends on absorption of **endometrial gland** secretions across the yolk sac. Furthermore, monotreme young hatch at a very early stage of development (as do metatherian young) and depend on lactation for continued development.

THE MAMMALIAN PLACENTA

One of the most distinctive and important structures associated with reproduction in therian mammals is the placenta. Differences among the

major placental types have been used in distinguishing some of the higher taxonomic categories of mammals (subclasses and infraclasses), and some primary contrasts between reproductive patterns in mammals relate to placental differences.

Fertilization normally takes place in the oviducts shortly after ovulation. The fertilized egg, or **zygote,** begins a series of cell divisions as it is passed down the oviduct by **peristaltic** muscular contractions. By the time the blastocyst reaches the uterus, it is a hollow ball of approximately 64 cells (called a **morula**). The journey usually takes three to four days from ovulation to arrival in the uterus. The inner uterine lining, or **endometrium,** has been conditioned for the arrival of the blastocyst by the combined action of FSH and LH hormones from the anterior pituitary, and progesterone from the **corpus luteum.** At implantation, the embryo, enclosed within a sphere of trophoblast cells, adheres to the surface of the endometrium, triggering a round of proliferation in the **trophoblast** cells. These cells apparently secrete enzymes that erode the lining of the endometrium, permitting the trophoblast (and the enclosed embryo) to invade the highly vascular lining of the uterus. The trophoblast grows rapidly at this stage, extending thousands of **chorionic villi** into the surrounding endometrium and eventually establishing the placenta (Fig. 20-2).

A functional connection between embryo and uterus is necessary in animals in which the fetus develops within the uterus and in which nutrients for the fetus come directly from the uterus rather than from yolk stored in the ovum. This connecting structure, the placenta, allows for nutritional, respiratory, and excretory interchange of material by diffusion between the embryonic and maternal circulatory systems. The placenta achieves intimate contact between embryonic and uterine tissues. The placenta also functions as a barrier that excludes bacteria and many large molecules from the embryonic circulation. In addition, the eutherian placenta produces certain food materials and functions as an endocrine organ by producing a variety of hormones. Amino acids supplied by the maternal circulation, for example, provide the placenta with the raw materials for protein synthesis for fetal growth. In some mammal species, the placenta produces progesterone and estrogen to help maintain pregnancy. Hormones, such as placental lacto-

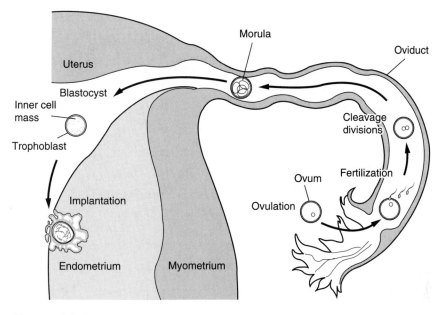

FIGURE 20-2   Drawing of early embryonic development in a human from ovulation to implantation of the trophoblast into the endometrium of the uterus. Fertilization occurs in the upper oviduct, and it takes three to four days for the embryo to reach the uterus.

gen, are also secreted to promote mammary development (Talamantes, 1975). Mammals are not unique in having a placenta, for certain fishes and reptiles establish placenta-like connections that allow diffusion of materials between the vascularized oviduct and the embryo. Among mammals, the major types of placentae differ sharply in structure and in the efficiency with which they facilitate the nourishment of the embryo.

CHORIOVITELLINE PLACENTA

This, the most primitive type of mammalian placenta, occurs to varying degrees in all metatherians. In a **choriovitelline placenta,** the yolk sac is greatly enlarged to form a placenta. After the shell membrane degenerates, there is considerable variation in the degree to which the blastocyst invades the endometrium. In some cases, the blastocyst does not actually implant itself deep in the **uterine mucosa,** as it does in eutherians, but merely sinks into a shallow depression in the mucosa. The contact is strengthened by the wrinkling of the blastocyst wall that lies against the uterus; this wrinkling increases the absorptive surface area of the blasto-

cyst. The embryo is nourished largely by "uterine milk," a nutritive substance secreted by the uterine mucosa and absorbed by the blastocyst. The embryo also derives nourishment from limited diffusion of substances between the maternal blood in the eroded depression in the mucosa and the blood vessels within the large yolk sac of the blastocyst. In contrast, the yolk sacs of older embryos of opossums (*Didelphis virginiana*) are fused to each other and to the endometrium such that they cannot be pulled apart (New et al., 1977). In this species, scanning electron micrographs reveal abundant microvilli that are thought to increase absorption from the endometrium (Krause and Cutts, 1984).

Metatherian placentae come in four basic types, based on the degree of association between fetal and maternal tissues and the structure of the allantois relative to the chorion (Hughes, 1984; Tyndale-Biscoe and Renfree, 1987). The eastern gray kangaroo (*Macropus giganteus*) exhibits the first and most common of these, in which a highly vascularized yolk sac makes an intimate connection with the endometrium of the uterus (Fig. 20-3). Similar placentae have been described in some members

▶ **FIGURE 20-3** Drawing of the arrangements of fetal membranes in metatherian and eutherian mammals. The kangaroo has a relatively simple yolk sac (choriovitelline) placenta. The bandicoot (Peramelidae) has a combined chorioallantoic and choriovitelline placenta. The chorioallantoic placenta of a typical primate shows the penetration of the endometrium by thousands of chorionic villi. *(Redrawn from Dawson, 1995)*

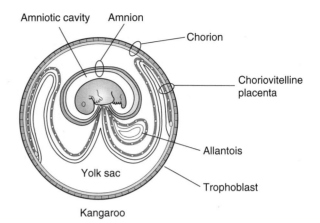

Kangaroo

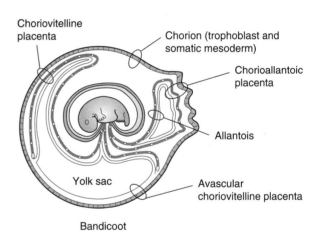

Bandicoot

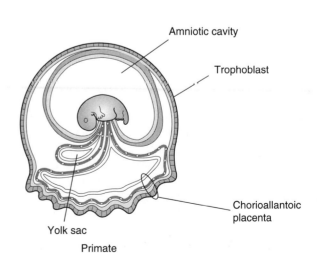

Primate

of the Didelphimorphia and other Diprotodontia. In *Dasyurus viverrinus* (Dasyuromorphia), a second placental type occurs when a subtle variation occurs in which the allantois and the chorion become apposed early in development, but the allantois degenerates as development proceeds. In the third type, occurring in some diprotodonts such as the koala *(Phascolarctos cinereus)* and the common wombat *(Vombatus ursinus),* the allantois is apposed to the chorion, forming a rudimentary respiratory exchange surface. The yolk sac retains its role as the primary surface for absorption of nutrients and gas exchange. Finally, a true **chorioallantoic placenta** is formed in addition to the choriovitelline placenta in members of the Peramelemorphia (Fig. 20-3).

Although similar to the eutherian chorioallantoic placenta in basic structure, the peramelid placenta achieves less effective transfer of substances between the fetal and maternal circulations. In peramelids, the allantois is fairly large and becomes highly vascularized; the blastocyst rests against the endometrium on the side where the allantois contacts the chorion. At the point of contact with the blastocyst, the uterus is highly vascularized and the part of the chorion against the vascularized endometrium is more or less lost. Because of the close apposition of the maternal bloodstream and the allantois, exchange of materials occurs across the allantoic membranes. Because the peramelid allantois lacks villi and only its corrugations increase its absorptive surface, a limited surface area is available for exchange of material between the maternal and fetal bloodstreams. Supplementary nutrition is supplied by uterine milk absorbed across the choriovitelline surfaces. Peramelids have one of the shortest gestation periods of any marsupial, yet their offspring are born relatively well developed (Table 20-1), suggesting that the choriovitelline placenta is less efficient

TABLE 20-1 Breeding Cycles for Several Metatherian Families

Family	Breeding season	Polyestrus or monestrus	Litter size	Gestation period (days)	Suckling period (days)
Didelphidae	March–October	Either	2–25	13	70–80
Dasyuridae	April–December	Monestrus	3–12	8–34	49–150
Peramelidae	March–June	Polyestrus	1–7	15	59
Phalangeridae	All year	1 litter/2 years or polyestrus	1–6	17–35	42–165
Macropodidae	All year	Either	1–2	24–43	64–270

Data on suckling period from Sharman, 1970.

than the chorioallantoic placenta (Padykula and Taylor, 1982).

CHORIOALLANTOIC PLACENTA

In eutherian mammals, the chorioallantoic placenta reaches its most advanced condition with regard to facilitating rapid diffusion of materials between the fetal and uterine bloodstreams. In eutherians, the blastocyst first adheres to the uterus and then sinks into the endometrium. As implantation proceeds, chorionic villi grow rapidly and push farther into the endometrium as local breakdown of uterine tissue occurs. The resulting tissue "debris," often called **embryotroph,** is absorbed by the blastocyst and nourishes the embryo until the villi are fully developed and the embryonic vascular system becomes functional. In response to the presence of the blastocyst, the uterus becomes highly vascularized at the site of implantation. When the eutherian placenta is fully formed, the complex and highly vascularized villi provide a remarkably large surface area through which rapid interchange of materials between the maternal and fetal circulations can occur (Fig. 20-3). The extent to which the villi increase the surface area available for diffusion is difficult to imagine but is suggested by the fact that the total length of the villi in the human placenta is roughly 48 kilometers (Bodemer, 1968).

Among eutherians, the degree to which the maternal and fetal bloodstreams are separated in the placenta varies widely (Fig. 20-4). Lemurs, some ungulates (suids and equids), and cetaceans have an **epitheliochorial placenta,** in which the epithelium of the chorion is in contact with the uterine epithelium and the villi rest in pockets in the endometrium. Under these structural conditions, oxygen, nutrients, and **immunoglobins** (in many species) must pass through the walls of the uterine blood vessels and through layers of connective tissue and epithelium before entering the fetal bloodstream. In ruminant artiodactyls, the uterine epithelium is eroded locally and there is contact between the chorionic ectoderm and the vascular uterine connective tissue. This is a **syndesmochorial placenta.** In carnivorans, erosion of the endometrium is carried further and the epithelium of the chorion is in contact with the endothelial lining of the uterine capillaries. This is called an **endotheliochorial placenta.** Destruction of the endometrium in some mammals may involve even the endothelium of the uterine blood vessels, allowing blood sinuses to develop in the endometrium; the chorionic villi may then be in direct contact with maternal blood (Fig. 20-4). This **hemochorial placenta** occurs in some insectivorans, bats, higher primates, and some rodents. In rabbits and some rodents, the destruction of placental tissue is so extreme that only the endothelial lining of the blood vessels in the villi separates the fetal blood from the surrounding maternal blood sinuses (Arey, 1974). In this case, a **hemoendothelial placenta** results.

The rate at which substances move from the maternal to the fetal bloodstream in the placenta is, of course, increased when the number of interposed membrane barriers is reduced. Because of

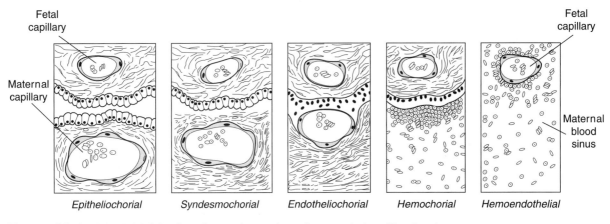

FIGURE 20-4 Drawing of the five placental types based on proximity of fetal and maternal circulations. Maternal and fetal blood supply is separated by six layers in the epitheliochorial placenta. At the opposite extreme, the fetal blood supply is separated from the maternal blood sinus by only a single layer of cells making up the capillary walls. *(After Arey, 1966)*

the difference between the number of such barriers in the human and the pig, for example, sodium is transferred 250 times more efficiently by the human placenta than by the pig placenta (Flexner et al., 1948). The remarkable absorptive ability of the allantoic placentae of such mammals as insectivorans, bats, primates, rabbits, and rodents is due largely to the great surface area afforded by the complex system of villi, to the extensive erosion of uterine mucosa and the resulting development of blood sinuses into which the villi extend, and to the loss of nearly all of the membranes separating uterine from fetal blood.

The shape of the placenta is governed by the distribution of villi over the chorion. Several different distributions of villi occur in mammals. The lemurs, some artiodactyls, and perissodactyls have a **diffuse placenta,** which has a large surface area because the villi occur over the entire chorion. Ruminant artiodactyls have a **cotyledonary placenta** consisting of more or less evenly spaced groups of villi scattered over the mostly avillous chorion. Carnivorans have a **zonary placenta,** in which a continuous band of villi encircles the equator of the chorion. Insectivorans, bats, some primates, rabbits, and rodents have a **discoidal placenta,** in which villi occupy one or two disk-shaped areas on the chorion.

At birth, the fetal contribution to the placenta is always expelled as part of the "afterbirth," but the maternal part may or may not be lost at this time. In mammals with an epitheliochorial placenta, the villi pull out of the uterine pits in which they fit, none of the endometrium is pulled away, and no bleeding occurs at birth. This placenta is termed **nondeciduous.** In mammals with placentae allowing more intimate approximation of uterine and fetal bloodstreams, there is extensive erosion of the uterine tissue and extensive intermingling of uterine and chorionic tissue, and the uterine part of the placenta is torn away at birth, resulting in some bleeding. This type of placenta is **deciduous.** The hemorrhaging after birth is soon stopped by the collapse of the uterus, by contractions of the **myometrium** (the smooth muscle layer of the uterus), which tend to constrict the blood vessels, and by clotting of blood. The loss of the placenta suddenly removes a major source of progesterone, an inhibitor of prolactin that is critical for the release of milk.

THE ESTRUS CYCLE, PREGNANCY, AND PARTURITION

In mammals, reproduction is characterized by a series of cyclic events that are under nervous and hormonal control. As with many complex functions of the vertebrate body, the regulation of the reproductive cycle is maintained by environmental

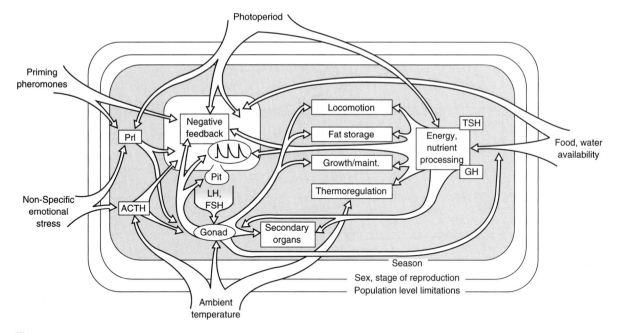

FIGURE 20-5   Diagram of the multiple interactions between environmental cues and neuroendocrine responses that mediate reproductive cycles in mammals. Important environmental factors are located outside the shaded box, and pathways are indicated by arrows. *(From Bronson, © 1989, Mammalian Reproductive Biology, University of Chicago Press)*

and social cues and by reciprocal controls between endocrine organs and their secretions (Fig. 20-5). The events characterizing the stages in the mammalian reproductive cycle are well known, but details of the hormonal regulation of these events are not completely understood. The ovarian cycle results in the development of ova, their release from the ovary, and their passage into the uterus; the uterine cycle involves a series of cyclic changes in the uterus.

The mammalian **estrus** cycle includes four phases: proestrus, ovulation, metestrus, and diestrus (Fig. 20-6). **Proestrus** is characterized by the growth of the follicle (the ovum, or egg cell, surrounded by specialized cells of the ovary) and its secretion of estrogen. Ovulation signals entry into estrus and the development of the corpus luteum (a glandular structure) from the ruptured follicle. The corpus luteum secretes the hormones estrogen and progesterone. Declining progesterone secretion leads to regression of the corpus luteum during **metestrus.** A period of variable

length called **diestrus** occurs before the next proestrus. In **monestrus** mammals, diestrus lasts most of the year and females are receptive to males for only a short period during the year. In contrast, **polyestrus** mammals have a short diestrus and therefore are receptive on a regular schedule throughout the year, or part of the year (Fig. 20-7).

The estrus cycle is controlled largely by the anterior pituitary hormones, FSH (follicle-stimulating hormone) and LH (luteinizing hormone), and the ovarian hormones, estrogen and progesterone (Fig. 20-6). Hormonal control of the estrus cycle is best understood in humans, where diestrus is more or less equivalent to the period of **menstruation.** Menstruation is typified by low levels of all four hormones, a thin uterine lining, and early stages of follicle development. Low estrogen levels in the ovary trigger the hypothalamus to produce FSH-releasing factor, which causes the anterior pituitary to release FSH, resulting in the maturation of ovarian follicles. As the follicle matures, it releases estrogen, which in turn causes a cessation of FSH

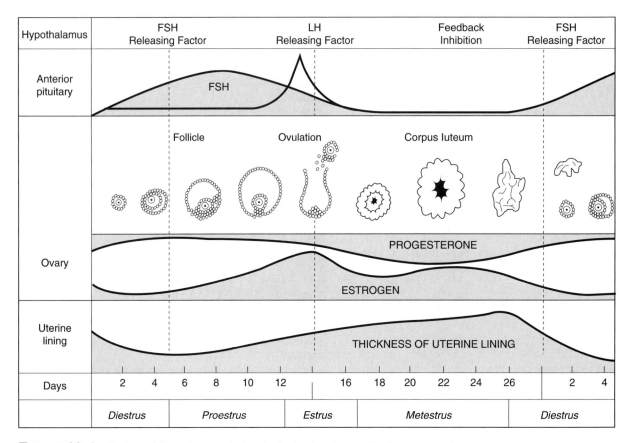

FIGURE 20-6   Timing of the estrus cycle in a typical polyestrus eutherian mammal, showing the levels of important regulatory hormones, state of the uterus, and the transition from the follicle to the corpus luteum. *(After Withers, 1992)*

production by the hypothalamus-pituitary axis. Rising estrogen levels stimulate the hypothalamus to produce LH-releasing factor, which turns on the secretion of LH from the anterior pituitary and also causes the uterine lining to thicken and become more vascularized.

At this point, the cycle enters estrus and the mature follicle ruptures, releasing the ovum. After the release of the ovum, the remaining follicle cells enlarge to form the corpus luteum in response to LH. The corpus luteum secretes estrogen and progesterone, which sensitizes the uterus for implantation and forms part of a negative feedback loop that stops FSH and LH production. If the ovum is not fertilized, the corpus luteum regresses, resulting in the sloughing off of the thickened uterine lining (the menstrual flow), in humans, Old World primates, and elephant-shrews. The hy-

pothalamus and pituitary respond by resuming production of FSH, and another ovarian cycle is initiated.

This pattern, involving **spontaneous ovulation,** occurs widely among mammals, but some deviations are common. **Induced ovulation** occurs in several carnivores (including *Procyon*), some rodents, some insectivorans, and most lagomorphs. The follicles develop, but ovulation does not occur until after **copulation.** In rabbits, the follicles do not develop fully before copulation and a long estrus may occur; ripening of the follicles and ovulation is initiated by copulation. In some rodents, pheromonal and tactile stimuli from the male may initiate ovulation.

The ovarian and uterine cycles in Old World primates are different, to some extent, from those in other mammals. This primate cycle is called the

Phyllostomus hastatus   *Artibeus jamaicensis*

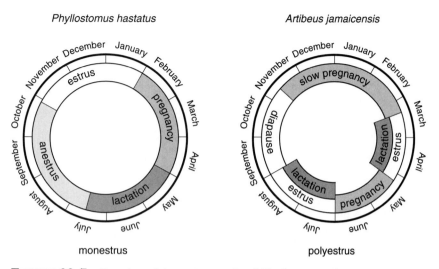

monestrus   polyestrus

FIGURE 20-7   Drawing of the estrus cycle of *Phyllostomus hastatus*
(Phyllostomidae), exhibiting a typical pattern for a monestrus bat, and the estrus
cycle of *Artibeus jamaicensis* (Phyllostomidae), a polyestrus bat species.

menstrual cycle rather than an estrus cycle. In humans and in Old World primates, considerable bleeding is typical of the time of endometrial breakdown, ovulation occurs at regular intervals throughout the year, and females may be receptive to copulation over an extended period. Even in such advanced primates as the gorilla *(Gorilla gorilla)*, however, copulation is cyclic and occurs rarely except when the vulva of the female is swollen in association with ovulation (Nadler, 1975).

When copulation occurs, the sperm reach the oviducts in a matter of minutes, in some species, and fertilization of the ova usually occurs within 24 hours of ovulation. The zygotes move down the oviducts, aided by peristaltic contractions of the muscles of the oviducts and movement of cilia, and usually reach the uterus and implant within a few days.

In eutherians, the delicate hormonal control of pregnancy is exerted by interactions between hormones produced by the pituitary, the ovary, and the uterus. In the early part of pregnancy, chorionic gonadotropin is of critical importance in preserving the corpora lutea and in preventing regression of the endometrium. This hormone is produced first by the trophoblast during its implantation in the endometrium and then by the chorion, which develops from the trophoblast.

During early pregnancy, the corpus luteum, because of its production of progesterone, is important in maintaining pregnancy by keeping the endometrium in a thickened and highly vascularized condition and by preventing coordinated contractions of the myometrium that might expel the embryo. In some species, progesterone sensitizes the endometrium and increases the efficiency of blastocyst implantation.

The maintenance of pregnancy in many mammals is not entirely under the control of progesterone from the corpus luteum. Instead, as pregnancy continues, hormones are produced progressively more by the placenta and less by the ovary. In humans, the placenta produces estrogens and is probably the most important source of progesterone after the third month. The third month is a time of great risk of miscarriage, when corpus luteum progesterone declines and placental progesterone commences. During the latter stages of pregnancy, the placenta seems to be a nearly independent endocrine gland that, in humans at least, takes over the functions of the pituitary gland and the corpus luteum. As mentioned above, the placenta now produces a form of prolactin called "placental lactogen," which induces mammary growth.

An important hormone of pregnancy, but one whose function is important mainly at **parturition** (delivery of the infant), is relaxin. This hormone, which causes relaxation of the pelvic ligaments and the pubic symphysis in preparation for parturition, is known to occur in a variety of mammals and may be universal among mammals. The concentration of relaxin in the bloodstream increases toward the end of pregnancy. Relaxin may be produced by the uterus or by the placenta and in humans is known to be produced by the ovaries during pregnancy (Guyton, 1976). Passage of the very large fetus through the birth canal of the Brazilian free-tail bat *(Tadarida brasiliensis),* and perhaps other bats, is facilitated by an elastic interpubic ligament that can stretch to 15 times its original length (Crelin, 1969).

Birth is accomplished by rhythmic and powerful contractions of the uterine myometrium, aided by the abdominal muscles. Continued contractions force the placenta from the uterus and the vagina. These contractions are seemingly under the control of interacting hormones. Oxytocin, produced by the hypothalamus and stored in the posterior lobe of the pituitary, occurs in increasingly higher concentrations in the maternal bloodstream towards the end of pregnancy; oxytocin can initiate contractions of the uterus and also induces maternal behavior. Apparently the reduced concentrations of progesterone late in pregnancy are insufficient to block the effects of the increasing levels of oxytocin, with resulting contractions of the myometrium and parturition. The amount of estrogen also increases late in pregnancy, and may sensitize the uterus shortly before parturition, allowing oxytocin to initiate uterine contractions.

The newborn mammal is nourished by milk produced by the mother's mammary glands. Under the influence of estrogen, progesterone, insulin, and placental lactogen from the placenta, these glands undergo considerable growth during pregnancy. Milk production is stimulated by suckling of the young and regulated by prolactin, produced by the anterior lobe of the pituitary. Prolactin is secreted in progressively larger amounts during the latter part of pregnancy and after parturition; it is inhibited by progesterone. When the inhibition of prolactin by progesterone is removed after birth, milk secretion can begin. Oxytocin, released under the stimulus of suckling, causes contraction of the myofibrils surrounding the milk-containing alveoli of the mammary glands,

inducing a release of milk (milk-letdown) for the young. Milk production is partly under neural control and continues only as long as the suckling stimulus persists.

THE METATHERIAN-EUTHERIAN DICHOTOMY

The recognition of a metatherian-eutherian dichotomy is based on a number of biological differences; primary among these are the contrasting reproductive patterns. Metatherians bear virtually embryonic young after a brief gestation period. In contrast, many eutherians bear anatomically complete, highly **precocious** young after a relatively long gestation period (Fig. 20-8). Why this difference?

One hypothesis invokes the extremely effective vertebrate **immune-response** system that serves to destroy invading foreign **antigenic** materials. During prolonged gestation in vertebrates lacking substantial stored energy reserves (such as yolk), an intimate contact between mother and fetus must ensure efficient physiological exchange, especially throughout the period of **organogenesis** and rapid growth. In this critical period, however, because the paternal antigens of the fetus may be recognized as foreign by the maternal system, the fetus risks destruction during its extended "parasitism" of the mother. The mechanisms tending to avoid this immune response are not understood for most vertebrates, but for mammals some information is available (Anderson, 1972; Hughes, 1974).

Although eutherian mammals lack shell membranes, the early stages of the zygote are separated from maternal tissues by the **zona pellucida** (a noncellular layer surrounding the zygote) and do not elicit an immune response. Later embryological stages are protected by the trophoblast and by the decidua (the uterine mucosa contacting the trophoblast) and its noncellular external secretion (Wynn, 1971). In humans, chorionic gonadotropin blocks the action of maternal lymphocytes, protects the surface of the trophoblast, and allows the fetus to be accepted (Adcock et al., 1973). Throughout the gestation, despite the close apposition of the uterine tissues and the fetus, at least one layer of trophoblast constantly provides a barrier between fetal and maternal tissues.

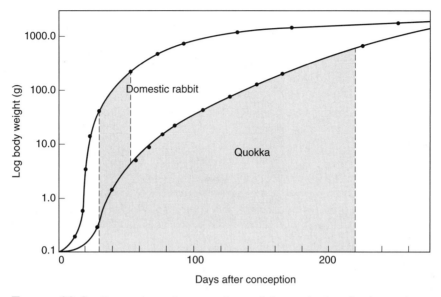

FIGURE 20-8   Comparison of postnatal growth in a eutherian, the domestic rabbit (*Oryctolagus cuniculus,* Leporidae) and a metatherian, the quokka (*Setonix brachyurus,* Macropodidae). The shaded region under each growth curve represents the time from birth to weaning.

In those metatherians that have been studied, the eggshell membranes are retained through the early two-thirds of the gestation period. Although these membranes permit the passage of nutrients and probably of enzymes and antibodies, they interpose a barrier between the antigen-bearing parts of the embryo and the lymphocytes in the uterine fluid that initiate the immunological reaction. Immune-response rejection of the early fetus is thus avoided. Late in the gestation period, however, after the shedding of the shell membranes has occurred and the fetal and maternal tissues are in close contact, the stage is set for an immunological attack. The apparently precipitous birth of the rudimentary young after the short gestation period was hypothesized to be an adaptation enabling metatherians to avoid this attack (Moors, 1974; Lillegraven, 1975). This argument is based on an assumption that metatherians lack a trophoblast (Tyndale-Biscoe and Renfree, 1987). Taylor and Padykula (1978) have shown, however, that metatherians have a trophoblast. Furthermore, studies of the Tammar wallaby, *Macropus eugenii,* designed to compromise pregnancy by exposing females to repeated skin grafts of a particular male, failed to cause an immunosuppressive response (Rodger et al., 1985). These studies strongly suggest that the immunosuppressive function of the trophoblast occurs in metatherians and likely evolved prior to the evolutionary separation of the Metatheria and Eutheria (Rodger et al., 1985; Tyndale-Biscoe and Renfree, 1987).

Lillegraven (1975) suggests that the common ancestor of metatherians and eutherians was viviparous and bore virtually embryonic young; he hypothesizes that viviparity did not evolve independently in metatherians and eutherians. These groups of mammals seemingly diverged in the early Cretaceous, and compelling paleontological and anatomical evidence indicates an origin from a common ancestral stock. Temporary closures of the eyes and ears in newborn metatherians guard against desiccation, and partial closure of the mouth ensures secure attachment to the nipple and immovable jaws in which the dentary-squamosal jaw joint (absent at birth) can develop. Among eutherian mammals, the young of many Insectivora (arguably the least derived mammals) are born in a highly altricial condition (Fig. 20-9). Although not as poorly developed at birth as monotreme and marsupial young, they also share the closure of the eyes, ears, and mouth at similar stages as those of metatherians. In eutherians that bear precocious young, these closures have no

FIGURE 20-9 A six-day-old shrew (*Sorex araneus,* Soricidae) showing an altricial state of development. *(M. Andrea)*

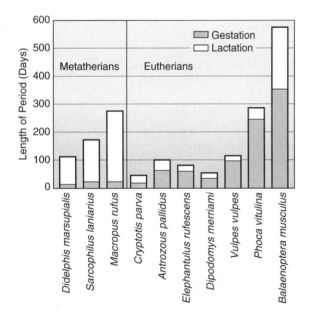

FIGURE 20-10 A comparison of the gestation and lactation periods in metatherian and placental mammals. Note that metatherians invest more heavily in the lactation period. Placental mammals show a wide range of variation but generally invest in a longer gestation period. Data are from Dawson, 1995 and Eisenberg, 1981.

function, yet they still occur briefly during intrauterine development and perhaps represent a developmental stage inherited from a common ancestor with metatherians. This ancestor may have borne rudimentary young after a brief gestation period, just as metatherians do today. From these similar beginnings, Tyndale-Biscoe and Renfree (1987) propose that eutherians evolved reproductive characters associated with longer gestation, while metatherians evolved along a separate trajectory with an emphasis on lactation for early growth of the young (Fig. 20-10 and Table 20-2).

MAJOR REPRODUCTIVE PATTERNS

In eutherian mammals, the length of time from fertilization to implantation is considerably shorter than the period between implantation and birth. Typically, fertilization occurs shortly after ovulation, and the development of the embryo from fertilization to birth is an uninterrupted process. Perhaps in response to specialized activity cycles, some mammals have abandoned this usual pattern of continuous development. One departure involves a delay of ovulation and fertilization until long after copulation **(delayed fertilization);** another is typified by normal fertilization and early cell cleavages but arrested embryonic development at the blastocyst stage **(delayed implantation);** and another involves a long delay in the development of the blastocyst after it has implanted **(delayed development).**

DELAYED FERTILIZATION

This pattern of development occurs in a number of bats inhabiting northern temperate regions. As early as 1879, Fries recognized that the males of some species of the families Rhinolophidae and

TABLE 20-2 Comparison of Reproductive Characters in Small- and Large- Bodied Metatherian and Eutherian Mammals

	Small mammals		Large mammals	
Character	Metatherian	Eutherian	Metatherian	Eutherian
Activity	Nocturnal	Nocturnal	Nocturnal, crepuscular	Diurnal
Breeding	Polyestrus	Polyestrus	Polyestrus	Polyestrus, monestrus
Ovulation	Polyovular	Polyovular	Monovular	Monovular, polyovular
Gestation	Very short	Short to very short	Short to very short	Long
Development at birth	Very altricial	Altricial	Very altricial	Precocial
Thermoregulation at birth	Poikilothermic	Poikilothermic	Poikilothermic	Homeothermic
Jaw development at birth	Incomplete	Incomplete	Incomplete	Complete
Brain growth	Postnatal	Postnatal	Postnatal	Prenatal
External protection	Pouch	Nest	Pouch	No nest
Lactation	Long, > gestation	Short, > gestation	Very long	Short, < gestation
Lifespan	1–2 years	1 year or less	> 1 year	> 1 year

After Tyndale-Biscoe and Renfree, 1987.

Vespertilionidae could store viable sperm through the winter, long after **spermatogenesis** had ceased; later studies detailed the reproductive cycles of the females of these species (Guthrie, 1933; Hartman, 1933; Wimsatt, 1944, 1945). These remarkable reproductive tactics are seemingly adaptations to continuous or periodic winter dormancy and occur in a number of New World and Old World species in the genera *Rhinolophus, Myotis, Pipistrellus, Eptesicus, Nycticeius, Lasiurus, Plecotus, Miniopterus,* and *Antrozous.* Delayed fertilization may be the typical pattern in all but the tropical members of the family Vespertilionidae. Papers by Wimsatt (1944, 1945), Racey (1982), Uchida and Mori (1987), and by Bernard and Cumming (1997) describe delayed fertilization as it occurs in Microchiroptera, and the following remarks are based largely on those studies.

The reproductive cycle of the little brown bat (*Myotis lucifugus*) follows a timetable similar to that of many temperate-zone vespertilionids. The testes descend into the scrotum in the spring. This descent is caused mostly by increased production of testosterone, which is cyclical in bats. The testes begin to enlarge in the spring, and spermatogenesis peaks in the spring and ends by September. Accessory reproductive organs remain enlarged throughout the winter. A female may be inseminated repeatedly in the fall and winter, and males frequently copulate with hibernating females, although usually all females are inseminated by the end of November. In males of this species, mating occurs when circulating testosterone levels are at their lowest (Fig. 20-11; Gustafson and Damassa, 1985). In the females, a single follicle enlarges in the autumn but remains in the ovary throughout the winter. The most typical vespertilionid pattern is for copulation to occur before hibernation. The sperm are stored in the uterus, where they remain motile for at least 198 days in *Nyctalus noctula*

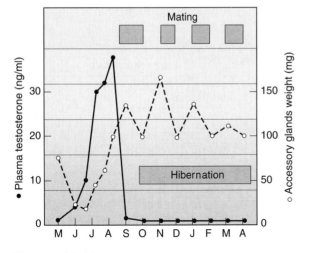

FIGURE 20-11   Graph of the reproductive cycle in male little brown bats *(Myotis lucifugus)*. Circulating testosterone levels peak in September and fall to very low levels during the mating season, which begins just before hibernation. *(Redrawn from Gustafson and Damassa, 1985)*

(Racey, 1973). Sperm remain viable for such long periods because the sperm heads are embedded in the microvilli of the uterus where secretions from the uterine lining nourish them (Uchida and Mori, 1987). The gestation period in this species is highly variable, probably due to regional differences in ambient temperatures and hence to the different body-temperature routines that occur in bats of widely separated colonies. Periodic torpor or low body temperature after the beginning of gestation slows the development of the embryo.

Several features of this unique reproductive cycle are especially noteworthy. The development of the male reproductive organs is out of phase; that is, the testes have regressed when the caudal epididymides and accessory organs are most enlarged and when breeding activity is at its peak. Males retain viable sperm in the caudal epididymides long after spermatogenesis has ceased. Females do not ovulate until long after they have been inseminated, but are able to store viable sperm for several months. Because of differing metabolic routines in different individuals, the rate of development of the embryo is highly variable.

Delayed fertilization is seemingly a highly advantageous adaptation in mammals with long periods of dormancy. Spermatogenesis, enlargement of reproductive organs, and copulation require considerable energy. In species that practice delayed fertilization, these activities occur in the late summer and autumn, when males are in excellent condition and have abundant food, rather than in spring, when the animals are in their poorest condition and when food (insects) may not yet be abundant. Ovulation and zygote formation occur almost immediately upon emergence from dormancy, rather than being delayed until after males attain breeding condition and copulation occurs. The female can therefore channel more energy into nourishment of the embryo than would be available if copulation occurred after hibernation. Perhaps the major advantage is that the time of parturition is hastened; thus, young have the longest possible time to develop before the next period of winter dormancy.

DELAYED IMPLANTATION

This deviation from the normal reproductive pattern occurs in a variety of mammals representing the orders Chiroptera, Xenarthra, Carnivora, and Artiodactyla (Tables 20-3 and 20-4). These mammals obviously do not share a direct common ancestor; in addition, they occupy a wide variety of habitats and pursue different modes of life. Delayed implantation in each group, therefore, has probably evolved independently in response to different selective pressures. Delayed implantation is either obligate, and constitutes a consistent part of the reproductive cycle, or facultative, and provides for a delay of implantation on occasions when an animal is nursing a large litter. A good discussion of delayed implantation is given by Daniel (1970) and Mead (1989).

In mammals with obligate delayed implantation, ovulation, fertilization, and early cleavages up to the blastocyst stage occur normally, but further development of the blastocyst is arrested and it does not implant in the uterine endometrium. The blastocyst remains dormant in the uterus for periods from 12 days to 11 months. Little blastocyst growth occurs during dormancy, which begins generally when the embryo consists of approximately 100 to 400 cells. The western spotted skunk *(Spilogale putorius)* of western North America, studied by Mead (1968), follows a reproductive pattern fairly typical of mammals with delayed implantation. Males become fertile in the summer, and copula-

TABLE 20-3 Periods During Which Blastocysts Remain Dormant in Some Mammals with Obligate Delayed Implantation or Delayed Development

Species	Dormancy of blastocyst (months)
Order Chiroptera	
Equatorial fruit bat	3+
(*Eidolon helvum*)	
Jamaican fruit bat	2.5
(*Artibeus jamaicensis*)	
Order Xenarthra	
Nine-banded armadillo	3.5–4.5
(*Dasypus novemcinctus*)	
Order Carnivora	
Grizzly bear (*Ursus arctos*)	6+
Polar bear (*U. maritimus*)	8
River otter	9–11
(*Lontra canadensis*)	
Harbor seal (*Phoca vitulina*)	2–3
Gray seal	5–6
(*Halichoerus grypus*)	
Walrus (*Odobenus rosmarus*)	3–4
Order Artiodactyla	
Roe deer	4–5
(*Capreolus capreolus*)	

Data mostly from J.C. Daniel, 1970, data on *A. jamaicensis* from Fleming, 1971.

tion and fertilization of the ova occur in September. The zygote undergoes normal cleavage but stops at the blastocyst stage; the blastocysts float freely in the uterus for 180 to 200 days. After implantation, the gestation period is about 30 days, and the young are usually born in May. During dormancy, each blastocyst is covered by a thick and durable zona pellucida. This general pattern of delayed implantation occurs in a number of carnivorans, but the timing of the cycle varies from species to species (Table 20-4). One of the fascinating features of delayed implantation in the western spotted skunk, is that the pattern of spermatogenesis and mating differs dramatically across the continental divide in the Rocky Mountains (Mead, 1968a, b). Litters are born in May and June after only a 50- to 65-day gestation on the eastern side of the divide; west of the divide mating occurs in the early fall, and blastocysts remain in the uterus over the winter months, to implant in time for litters to be born in May.

All pinnipeds probably practice delayed implantation, which in these animals enhances the survival of young by providing for the optimal timing of births (Boyd, 1991). The timing of life-history events relative to the reproductive cycle in the northern fur seal (*Callorhinus ursinus*) was described by York and Scheffer (1997). These seals are widely dispersed in the North Pacific for most of the year but assemble each summer at breeding colonies (rookeries) on islands off the coasts of California, Alaska, and Russia. In the Pribilof Islands, females arrive from mid-June through early August. They bear a single young within two days of arrival and mate within the next week. Cell division in the blastocyst accelerates just before implantation, which occurs mostly in November, 123 days after copulation and 237 days before birth. This schedule allows males and females to disperse widely and feed in separate areas for most of the year and to assemble at rookeries to give birth and to breed.

Facultative delayed implantation occurs in some species in which the female is inseminated soon after the birth of a litter. This type of delay is known in some metatherians, some insectivorans, and many rodents. In rodents that have postpartum estrus, implantation of blastocysts is delayed when the female is suckling her first litter.

Our understanding of the factors controlling normal blastocyst development or dormancy in eutherian mammals is incomplete. Present evidence suggests that estrogen causes the uterine endometrium to form proteins essential for rapid growth of the blastocyst and that a deficiency of these proteins results in blastocyst dormancy (Daniel, 1970; Heideman and Powell, 1998). Experimentally administered doses of estrogen or progesterone, or both, have been used in an attempt to reinitiate growth of a dormant blastocyst in mammals with obligate delayed implantation. These procedures have not been successful in renewing growth of the blastocyst, indicating that in such animals some blocking of the action of estrogen in the endometrium must occur (Daniel, 1970). McLaren (1970) has proposed that, during lactation in mice (*Mus*), implantation is delayed by

TABLE 20-4 Reproductive Cycles of Some North American Mammals with Delayed Implantation

Species	Breeding season	Time of implantation	Length of delay period (months)	Time of parturition	Litter size	Gestation period (months)
Long-tailed weasel (*Mustela frenata*)	July	March	8	April–May	6-10	9
Ermine (*Mustela erminea*)	June–July	March	8.5–9	April–May	6–10	9.5–10
Mink (*Mustela vison*)	Feb–March	March	0–1	April–May	3–8	1.3–2.3
Marten (*Martes americana*)	July–August	March–April	8	May	2–3	9
Fisher (*Martes pennanti*)	March–April	Feb–March	11	March–April	2–4	11.5–12
Wolverine (*Gulo gulo*)	Spring?–Summer	Jan–Feb	5+	March–April	2–4	8–9+
Badger (*Taxidea taxus*)	July–Aug	Feb	6	March–April	2–3	8
Western spotted skunk (*Spilogale putorius*)	Sept	April	6–7	May–June	4–7	8
Black bear (*Ursus americanus*)	June	Nov	6	Jan–Feb	1–4	7
Northern fur seal (*Callorhinus ursinus*)	Late July	Nov–Dec	3.5–4.5	Late July	1	12

From P.L. Wright.

an initial inability of the blastocyst to "hatch" from the zona pellucida, which must be shed before implantation can occur.

Delayed implantation is an important part of the reproductive cycles of many metatherians. In most macropodids for which embryonic diapause is known, the mother undergoes postpartum estrus. A female with a newborn in the pouch may copulate and produce a fertilized zygote. In this case, the blastocysts cease growing at the 70- to 100-cell stage. The metatherian blastocyst is surrounded by protective membranes consisting of an albumin layer and a shell membrane. When the young leaves the pouch, development of the corpus luteum and growth of the blastocyst resume, the blastocyst implants, and rapid growth of the embryo resumes. The resumption of blastocyst growth may be due to the secretion of progesterone. Embryonic diapause is common in the red kangaroo (*Macropus rufus*), a large herbivore living in arid regions of Australia. When conditions are good, a female *M. rufus* may have a blastocyst in di-

apause, a neonate suckling in the pouch, and a **joey** at her side (Fig. 20-12). When drought occurs, the joey and the neonate may both die, in which

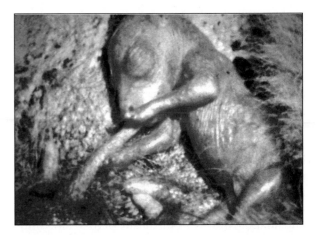

FIGURE 20-12 Newborn red kangaroo *(Macropus rufus)* attached to a nipple inside the mother's marsupium. *(C. Berlinky, © Educational Images, Ltd.).*

case the corpus luteum reinitiates the development of the blastocyst (Tyndale-Biscoe, 1984). Thus, this pattern of delayed implantation prevents births from occurring during periods of drought.

In metatherians, the young suckle after they leave the pouch for a period roughly comparable to the suckling period in eutherian mammals of similar size, but the intrauterine period for the metatherian fetus is often short (Table 20-1). As a result, a newborn young may be attached to a nipple and suckling while a much older young is returning periodically to suckle from a separate nipple. During double suckling in kangaroos, a remarkable thing occurs: separate mammary glands concurrently produce vastly different milks. The gland supporting the pouch young produces milk containing little fat, and the gland supporting the advanced young produces milk with three times as much fat. When the newborn young suckles, it stimulates the secretion of mesotocin (the metatherian equivalent of oxytocin) which, in turn, induces the release of a small amount of milk from that nipple and none from the other nipple. The mammary gland supplying milk to the older young produces milk that is richer in protein and fat than the milk supplied to the newborn. In both metatherians and eutherians, the composition of the milk changes during nursing. In metatherians, the milk secreted early in the suckling period contains little or no fat, whereas milk secreted later in the period may contain as much as 20 percent fat (Fig. 20-13).

DELAYED DEVELOPMENT

The California leaf-nosed bat *(Macrotus californicus)* does not hibernate. This phyllostomid, like the vespertilionid *M. lucifugus,* mates in the fall, and females do not give birth until the following summer. In *M. californicus,* however, fertilization also occurs in the fall shortly after mating. Development proceeds until the embryo reaches the blastocyst stage, at which time the blastocyst implants in the uterine lining. Once implanted, embryonic development ceases for about four months, to resume again in the spring (Bradshaw, 1962).

Fleming (1971) showed that, in the Jamaican fruit bat *(Artibeus jamaicensis),* development of the blastocyst is delayed. A blastocyst conceived after

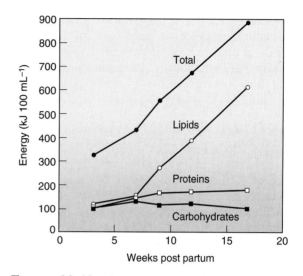

FIGURE 20-13 Changes in the composition and energy of the milk through the lactation period of the eastern quoll *(Dasyurus viverrinus). (After Green et al., 1997)*

the birth of young in July or August soon implants in the uterus, as in most mammals, but then becomes dormant, and further development is delayed until mid-November. This delayed development allows the young resulting from late summer matings to be born in early spring, when fruit is abundant.

Heideman (1989), Heideman et al. (1993), and Heideman and Powell (1998) demonstrated that delayed development also occurs in several species of Old World fruit bats, including the pteropodids *Otopteropus cartilagonodus, Haplonycteris fisheri,* and *Ptenochirus jagori.* In all reported cases of delayed development in bats, the delay occurs at a very early stage: before gastrulation of the blastocyst. Post-implantation delays in tropical species, including phyllostomids and pteropodids, are not the result of lowered body temperatures as they are in some temperate zone vespertilionids (Racey, 1982; Heideman and Powell, 1998). Heideman and his colleagues have suggested that delayed development in tropical bats may have evolved to allow females some flexibility in responding to poor environmental conditions or to synchronize the birthing period so that all young are born within a short time period (which may be important in species with nursery colonies). The exact mechanism producing these delays, however, is not known. Heideman and Powell (1998) hypothesize that some maternal signal, such as inhibitory or stimulatory molecules, triggers the inhibition of one or many genes in the embryo that are required for gastrulation and the formation of extra-embryonic membranes.

CONTROL OF THE TIMING OF REPRODUCTION

A major factor affecting reproductive success in mammals is the precise timing of reproduction to coincide with favorable environmental conditions. This timing is mediated by interactions between environmental, behavioral, and physiological stimuli (Fig. 20-5). These complex interactions defy simple explanation but have recently become better understood (Bronson, 1989; Jameson, 1988).

Many environmental factors have some effect on the timing of reproduction in mammals, but **photoperiod** (the period of light during the daily light-dark cycle), temperature, energy, and nutrition are probably of prime importance. Seasonal breeding in many temperate-zone mammals is partly regulated by photoperiod (Elliott, 1976; Kenagy and Bartholomew, 1981; Reiter, 1983; Bronson, 1989). In mammals, the pineal gland is photosensitive and can transmit information about the photoperiod to the endocrine system. The pineal secretes the hormone melatonin. When the photoperiod is short (short daylength), melatonin levels increase, causing a depression of gonadal activity in the Syrian hamster *(Mesocricetus auratus)*. The white-footed mouse *(Peromyscus leucopus)* is found over a great range of latitudes, and, as expected, the number of hours of daylight it needs to preserve testicular function varies with latitude (Lynch et al., 1981). Laboratory studies of Tammar wallabies *(M. eugenii)* show that their parturition times are determined by daylength (Sadlier and Tyndale-Biscoe, 1977). Further evidence for a photoperiod cue in these wallabies comes from experiments in which the pineal was denervated, which abolished the pattern of delayed development of the blastocyst (Renfree et al., 1981). However, some opportunistic species that respond to unpredictable food abundance are relatively unaffected by photoperiod. Lemmings, which live in the Arctic, are unresponsive to changes in photoperiod (Hasler et al., 1976).

The effects of temperature on reproduction are often more difficult to interpret. The testes of male pocket mice *(Chaetodipus formosus)* exposed experimentally to high temperatures were small relative to those of mice exposed to low temperatures; at high temperatures (35° C), reproduction in house mice was depressed (Pennycuik, 1969). This inverse relationship is unusual, however, for low temperatures usually inhibit testicular growth (Clarke and Kennedy, 1967; Kenagy, 1981). Temperature may have little effect on gonadal activity, but changes in ambient temperature can deplete energy stores, resulting in changes in testicular activity.

Two prime factors affecting the timing of reproduction are energy and nutrition. Gestation and lactation demand large amounts of energy, and it is axiomatic that, in mammals faced with seasonal variations in food availability, breeding coincides with the time (or times) of food abundance. Unpredictable food resources may cause irregular reproduction: In Nevada deserts, when the annual seed crop fails to appear because precipitation has been scant, Merriam's kangaroo rats *(Dipodomys merriami)* do not breed (Beatley, 1969). In the Great Basin Desert of southwestern Idaho, the Townsend ground squirrel *(Spermophilus townsendii)* has a single pulse of reproduction in early spring when green forage is available but will suspend reproduction in response to inadequate food supply. Availability of free water may be a more important factor controlling reproductive timing in desert environments. Addition of supplemental free water available to rodents in the Namib Desert caused a marginal improvement in reproductive success (Christian, 1979).

The nutritional quality of the food supply also has an effect on reproductive performance. Lack of sufficient protein or a particular vitamin can negatively effect **gametogenesis.** Specific factors, such as plant secondary compounds, have been known to influence reproduction in such herbivores as voles. Both male and female montane voles *(Microtus montanus)* become reproductive rapidly when they ingest 6-methoxybenzoxazolinone (6-MBOA), a compound derived from young, actively growing plants (Berger et al., 1981; Sanders et al., 1981). This chemical cue allows the voles to initiate reproduction when the growing season has begun—when survival of the young of this short-lived rodent would be high—and synchronizes breeding with abundant plant growth in a fluctuating environment. Although still subject to experimental verification, there is evidence that 6-MBOA is one cue for reproduction in a number of vertebrates and may affect population dynamics in arvicoline rodents. Other plant compounds have the opposite effect. At the end of the growing season in August, salt grass (*Distichlis* sp.) becomes dry and brown and accumulates high levels of the phenol 4-vinylguaiacol, which is known to suppress reproduction in female montane voles (*M. montanus*; Berger et al., 1977). A wide variety of **phytoestrogens** (plant compounds that mimic animal estrogens) occur in nature, and the role they play in regulating mammal reproduction is likely to be complex.

Behavioral and physiological regulation of reproduction is clearly of great importance, and the importance of **pheromones** has been established by laboratory studies. Mammals deposit urine at various places in the course of their activities, and this urine serves as an individual's olfactory "signature" or "fingerprint" (Caroom and Bronson, 1971; Jones and Nowell, 1973a, b) that provides information on the species, sex, reproductive condition, and social status. Such olfactory cues can regulate reproduction by triggering endocrine responses or modifying behavior. In association with tactile stimuli, chemicals in male urine called "priming pheromones" regulate the reproductive maturation and timing of ovulation in female house mice *(Mus musculus)* and deer mice *(Peromyscus maniculatus)* by inducing a series of hormonal responses (Bronson, 1971; Bronson and Maruniak, 1975; Bronson, 1989). Males are also sensitive to urinary cues: the male endocrine system responds to female urine by secreting hormones that Bronson (1979) hypothesizes increase the effect that the male's urine has on the female's reproductive system. Among prepubertal females, female urine suppresses reproductive maturation and overrides the effect of male urine for a period during development. In the presence of a male, female **puberty** can occur at 25 days of age but may not occur until 50 days in the absence of a male. Tactile stimuli are important at various stages of the reproductive cycle, and domination of one mouse by another can drastically reduce the secretion of sex hormones in the subordinant.

Bronson (1989), in his detailed summary of the proximate and ultimate factors controlling reproduction, points to what he calls the "three dimensions of complexity" in mammalian reproductive patterns. First, "mammals have evolved in ways that allow them to surmount environmental complexity." By this, he means that mammals can respond appropriately to environmental complexity and that no single pattern of responses exists in mammals. The second dimension of complexity "relates to the fact that the actions and interactions of environmental factors can vary greatly with the stage of the mammal's life cycle during which they are preceived." Responses to a particular set of factors may change dramatically after a female becomes pregnant, for example. The third dimension corresponds to "the ongoing interplay between a mammal's need to reproduce and its need to survive." In other words, reproductive processes are not independent of other demands on the organism. Bronson (1989) illustrates this point by describing reproduction as a process that competes for calories with thermoregulation. Calories allocated to reproduction are given low priority in nonpregnant females, but the balance shifts when the female becomes pregnant.

INFANTICIDE AND TERMINATION OF PREGNANCY

Male-induced termination of pregnancy is a remarkable phenomenon mediated by social, olfactory, and endocrine factors (Bruce, 1966; Hausfater and Hrdy, 1984). Such termination is called the **Bruce effect** and is initiated in the laboratory by the replacement of the original male by an unfamiliar male or exposure of a pregnant female to the odor of an unfamiliar male. In response to a new male, females of several species of voles will abort their fetuses, enter estrus, and breed with the new male within a few days (Stehn and Jannett, 1981). The appearance of the new male induces estrus regardless of the reproductive condition of the female. The Bruce effect is thought to result from a drop in prolactin secretion in the female induced by nervous inputs to the brain from the olfactory lobes. Male-induced abortion is to be expected in many arvicolines and could influence the population dynamics of wild populations.

Infanticide by males is an additional factor affecting the timing of reproduction in some mammals. Under some conditions, male langur monkeys *(Semnopithecus entellus)* kill young during a troop takeover (Hrdy, 1977), and infanticide is recorded among several other primates (Rudran, 1973, 1979; Goodall, 1977; Struhsaker, 1977). In a four-year study of a troop of Namib Desert baboons living under harsh conditions, only 3 of 22 infants survived more than six months (Brain, 1992, Brain and Bohrmann, 1992). Infant deaths were due to unusually heavy tick (*Rhipicephalus* sp.) infestations around the nose and mouth that resulted in an inability to suckle and to infant kidnapping by high-ranking, nonlactating females (called "aunts"). Hrdy (1976) also details infant-kidnapping by females, which was referred to as "aunting-to-death" in other primate species.

Infanticide in male collared lemmings *(Dicrostonyx groenlandicus)* is directed toward unfamiliar young, but males show little or no infanticidal behavior toward their own young (Mallory and Brooks, 1978). Young Asian rhinoceroses *(Rhinoceros unicornis)* that are unprotected by their mothers are commonly killed by adult males (Dinerstein et al., 1988). Male infanticide is especially well understood in lions (p. 464). A new male or group of males taking over a pride often kill the cubs to hasten the onset of estrus in the females; this alteration in the timing of reproduction seemingly increases the reproductive fitness of the infanticidal male.

REPRODUCTIVE CYCLES AND LIFE-HISTORY STRATEGIES

Underlying the tremendous variation in mammalian reproductive cycles is a broad pattern. Mammals can be segregated into two groups, those that bear altricial young (helpless, naked young in which the eye and ear openings are covered by membranes and locomotion and thermoregulation are undeveloped) and those that bear precocial young (well-developed, fur-bearing young in which the eyes and ears are functional and thermoregulation and locomotion are well developed). Each of these patterns is typically associated with a different life-history strategy.

Mammals with altricial young live under unstable conditions with seasonal or unpredictable food abundance. They are small and subject to heavy predation pressure. Litters are large (often seven young or more), the young are born in a nest, and the gestation and suckling periods are short. The young grow rapidly and reach sexual maturity early. Life spans are short, the brain size is relatively small, the mother-young bond is brief, and social behavior is simple. Estrus is short and frequently triggered in part by male-female interactions. Under favorable environmental conditions, breeding may occur repeatedly throughout the year. Their high reproductive rates allow these mammals to be reproductive opportunists and take advantage of even brief periods of food abundance. With such opportunism goes high population turnover and population densities that are unstable seasonally and from year to year. An array of mammals, including tree shrews, insectivorans, many kinds of rodents, and small carnivorans, fit this pattern. Examples of altricial mammals are the European lemming *(Lemmus lemmus)* and the montane vole. Females of both these species can breed at a remarkably early age (15 and 21 days, respectively; Kalela, 1961), and the gestation period is only some 21 days. The polyestrus females may have several litters during the summer growing season with postpartum estrus, but seldom will the females survive to bear young during a second reproductive season. In extreme cases, reproduction may continue nearly all year in some arvicoline rodents (Baker and Ranson, 1933; Greenwald, 1956). Populations of lemmings and voles are famous for their dramatic fluctuations (p. 518).

Mammals with precocial young, among which are many ungulates, cetaceans, primates, hyraxes, and some hystricognath rodents, typically live in relatively stable environments with a more predictable food base. These mammals are often large, can reach sexual maturity late, and some are not subject to intense predation pressure. The estrus cycle is long, ovulation is usually spontaneous, and gestation is prolonged. Usually a single young is not born in a nest but commonly accompanies (or clings to) the female virtually from birth. Lactation is also prolonged, with an enduring mother-young social bond developing in many species. The brain of the precocial young mammal is usually large, social behavior is complex, and an individual may spend its entire life as a member of a social group. These mammals have a low reproductive rate (exceptions include *Lepus* spp.), but the survival rate of young is high because of the extended period of maternal care. Population stability, a low reproductive rate, low population turnover, and dependence on a stable environment make these animals vulnerable to habitat alteration by humans. An example of a mammal with precocial young is the South American woolly monkey *(Lagothrix lagotricha)*. This 5.4-kilogram primate mates first at eight years of age, has a long gestation (255 days), one large (1 kilogram) young, and young are born at widely spaced intervals (1.5 to 2 years). Lactation lasts from 9 to 12 months, the mother-young bond is tight, and social behavior is complex.

The length of the gestation period in eutherian mammals is, in general, positively correlated with body weight and with the degree of development of the newborn. The larger the mammal, the longer the gestation period; for mammals of equal weight, the species with the heaviest neonate has the longest gestation (Huggett and Widdas, 1951). Elephant-shrews, xenarthrans, hystricognathous rodents, cetaceans, some pinnipeds, and primates depart sharply from this trend, however. These departures are the basis for the following examination of the relationships between specific aspects of life-history strategies and reproductive patterns.

Elephant-shrews have unusually long gestation periods and bear highly precocial young. The mouse-size (45 grams) *Macroscelides proboscideus,* for example, has a gestation period nearly two weeks longer (76 versus 63 days) than the wolf *(Canis lupus),* an animal over 600 times larger. Most species of elephant-shrews do not use burrows or nests. They rest on the surface of the ground, remain perpetually alert for predators, and escape by rapid bounds along well-known trails. Survival of young elephant-shrews depends on their precociously developed sensory and locomotor abilities and, thus, indirectly on their long gestation period.

Xenarthrans, which have a very specialized lifestyle (see Chapter 9), have a long gestation period, but there is no consistent progression toward longer gestation with increasing body size. McNab (1979, 1980) has found that xenarthrans have unusually low metabolic rates and has related this to their **myrmecophagous** (ant-eating) or folivorous (leaf-eating) habits. The case of the two-toed sloth *(Choloepus hoffmanni)* is of particular interest. This modest-size (9 kilograms) animal has a gestation period of 332 days, some 3.5 months longer than

the American elk or wapiti *(Cervus elaphus),* which is 22 times larger (200 kilograms). Both of these mammals are herbivorous and bear a single, precocious young at long intervals (12 months for the elk and 18 months for the sloth). The elk is highly mobile and ranges widely, feeding selectively on a wide variety of plants. It has relatively modest protein and energy needs per unit of weight because of its large size. In contrast, the arboreal sloth is sedentary, occupies a small (1.96 hectares) home range, and feeds on the leaves of but a few tree species. Not only are the leaves often low in energy and protein but they also contain defensive secondary compounds (such as tannins and terpenoids) that retard digestion. This sloth became adapted to a dependable and ubiquitous, but energetically marginal, food by adopting a very low metabolic rate, heavy insulation, and a countercurrent heat exchange system in the limbs. The reproductive pattern is also the result of selection pressure to reduce energy needs: the extremely long gestation period avoids the rapid mobilization of energy, allows for the development of a precocial young able to cling to its mother, and reduces the energetically costly lactation period.

Baleen whales stand out as remarkable exceptions to the general body-weight, gestation-period trend. These whales, the largest animals of all time, have gestation periods similar to or shorter than those of camels or horses. Thus, the blue whale *(Balaenoptera musculus),* 250 times heavier than the camel *(Camelus bactrianus),* has a shorter gestation period (360 versus 406 days). The reason for this amazing rate of fetal growth is unknown. One possibility is that, by being extremely efficient harvesters of plankton, the most abundant marine food source, and by spending summers in boreal or austral waters where plankton reaches peak productivity, female whales may be able to invest great amounts of energy in the fetus and still store the reservoir of blubber necessary for migration and early lactation (young are usually born in tropical waters). Another possibility is that carrying a fetus through more than one migratory cycle and continuing pregnancy through tropical fasting periods is unfeasible energetically; the very rapid development of a fetus may be more economical.

The Australian sea lion *(Neophoca cinerea)* is unique in having an extended, nonannual, and nonseasonal breeding cycle. The pupping period is five months; lactation is prolonged (15 to 18 months); the females stay near the breeding grounds all year; and breeding events shift forward 13.8 days earlier every 18 months (Higgins, 1993). The only other pinniped with a nonannual cycle is the walrus *(Odobenus rosmarus),* in which individual females breed about every 24 months, while breeding occurs annually in the population but does not involve all reproductive females (Fay, 1981).

The primates as a group have long gestation periods, and those of the small primitive primates are most remarkable. Although most primate young are precocial, they depend on a long period of maternal and sometime paternal care and often indoctrination into a complex social system. This is reflected by the typically long lactation period. The evolution of an early-primate reproductive pattern involving long gestation and precocial young may have been critical in setting the stage for the highly social lives of higher primates.

AVOIDANCE OF INBREEDING

Seemingly, no mammalian species is immune to the deleterious effects of **inbreeding** (Lacy, 1997). Inbreeding (breeding between close relatives) increases the probability that an individual will have two identical **alleles** at a **locus** due to inheritance from an ancestor shared by the individual's parents. Observed effects of inbreeding include heightened mortality, susceptibility to disease, and increased frequency of developmental defects, as well as reduced **fecundity,** lowered growth rates, inability to withstand stress, and lowered competitive ability (Wright, 1977; Ralls et al., 1988). Reduced ability to adapt to environmental change is often cited as an especially adverse result of depletion of **heterozygosity.** Inbreeding in mammals is avoided or reduced in many ways, including dispersal by one or both sexes from the **natal** area before sexual maturity in some species, in others by male dispersal from natal social groups, and by recognition of **kin** and avoidance of breeding with them. In *Microtus,* the degree of inbreeding avoidance differs among species, perhaps being highest in monogamous species (Berger et al., 1997). Avoidance in *Microtus* may be based entirely on kin recognition due to familiarity of parents and siblings before weaning (Ferkin, 1989, 1990). Delayed reproductive development was observed by Rissman and Johnson (1985) in male California voles

that were raised in bedding impregnated with "family" odor. This was probably a response to a unique family scent. These authors regarded this delay as a response that avoided inbreeding. Similarly, young white-footed mice *(P. leucopus)* were reproductively inhibited by air-borne chemicals present in their mother's urine (Terman, 1992). In this species, movement away from the pre-weaning social environment may trigger reproduction (Wolff et al., 1988). Belding ground squirrels *(Spermophilus beldingi),* in contrast, can recognize kin without prior contact (Holmes and Sherman, 1983). Clearly, avoidance of inbreeding in mammals is mediated by a variety of physiological and behavioral mechanisms.

LITTER SIZE AND REPRODUCTIVE SEASONS

Because the metabolic cost of raising large, well-nourished litters is paid by a lowering of future reproduction, litter size represents the best reproductive investment for the environmental situation in which any population is living (Williams, 1967). This best investment may differ within a species from area to area (Spencer and Steinhoff, 1968). Within a species, large litters occur at far northern latitudes and high elevations, where severe winters and brief growing seasons limit the number of litters per year. The general pattern is for the mammals of a boreal community to have a few large litters each year, for those of a less severe temperate area to have smaller but more frequent litters, and for those in tropical communities to have many small litters each year. Within any area, however, strategies typically differ from one species to another. Thus, in Panama, most rodents breed throughout the year, but some species breed seasonally (Fleming, 1970). In a subalpine community in Colorado, some rodents have several litters each summer and some have only one (Vaughan, 1969).

In the view of Pianka (1976), optimal reproductive tactics involve maximizing an individual's reproductive fitness (the sum of all present plus future offspring) at every age. Reproductive effort, therefore, should vary inversely with residual reproductive value (expectation of offspring). The oldfield mouse *(Peromyscus polionotus)* follows these tactics (Dapson, 1979). In one population of this species, the mice lived about two years, reproduced at a moderate rate, and had few differences in reproductive pattern among the **age cohorts** (segments of the population). A contrasting strategy was followed by another population of this same species. One cohort, made up of younger mice that had a chance of surviving to reproduce the following year, had a moderate reproductive effort, whereas another cohort, consisting of older animals with low residual reproductive value (animals that would not survive to reproduce the following season), had litters in rapid succession and had double the number of young per female.

ADAPTIVE SEX RATIOS

Many researchers have observed skewed **sex ratios** in newborn young of deer species that have polygynous mating systems. These observations have prompted several hypotheses regarding adaptive variation in sex ratios. Perhaps the most thoroughly tested hypothesis is that of Trivers and Willard (1973), who proposed that females have the ability to vary the sex ratios of offspring, and that among **polygamous** species, in which reproductive success is more variable in males than in females, maternal investment should be biased toward males. Thus, well-nourished mothers should wean more sons than daughters, as male embryos usually require more energy from the mother. Conversely, poorly nourished mothers should invest more heavily in female offspring, because reproduction is less seriously prejudiced by poor condition in females than in males. When the fitness of parents benefits more from investment in one sex rather than in the other, parents should invest more heavily in the more "profitable" sex (Frank, 1990). Male-biased mortality *in utero* or after birth is known to increase in several species of mammals in response to poor nutrition of the mother (Robinette et al., 1955; McClure, 1981; Clutton-Brock et al., 1985). There is also evidence for adjustment of sex ratios in such species as the opossum (*Didelphis marsupialis;* Austad and Sunquist, 1986), bushy-tailed woodrat (*Neotoma cinerea;* Moses et al., 1995), eastern woodrat (*N. floridana;* McClure, 1981), elk (Smith et al., 1996), and mule deer (*Odocoileus hemionus;* Kucera, 1991). In these species, heavier, well-nourished mothers had male-

biased sex ratios of their offspring. Seemingly, not all mammals function as the Trivers-Willard hypothesis predicts, however, for Sikes (1996) found that food-stressed female grasshopper mice *(Onychomys leucogaster)* either tried to raise all their young or killed and ate all of them. More remains to be learned about the causes and occurrence of biased sex ratios among mammals.

SEASONAL TIMING OF REPRODUCTION

The timing of reproduction is a vital factor influencing reproductive success. During gestation, and especially during lactation, the mother's energy needs increase tremendously. During the period when young are weaned and are becoming independent, their survival depends on adequate food. All phases of reproduction may be within the seasonal period of food abundance if it is long. If this period is short, however, gestation may occur at a stressful time, with lactation and weaning (or just weaning) occurring when food is most abundant.

Even in tropical regions, there is typically seasonality in food abundance, and mammalian reproductive patterns are responsive to this. In many species of Neotropical bats, the weaning of young coincides with peak food abundance just after the start of the rainy season (Wilson, 1979; Bernard and Cumming, 1997), but among phyllostomid bats there are three common reproductive patterns: aseasonal polyestry, bimodal polyestry, and monestry (Fig. 20-7). Similarly, there are several reproductive patterns among African bats (Table 20-5). The timing of reproduction in some species varies geographically. In Gabon, Africa, colonies of the bat *Hipposideros caffer* usually give birth in October in southern latitudes and in March north of the equator. The majority of tropical rainforest mammals breed seasonally (Bourliere, 1973). This is the case for rodents in Africa (Dubost, 1968; Rahm 1970; Delany, 1971) and the typical pattern for monkeys of the genus *Cercopithecus* (Bourliere, 1973).

In boreal and montane areas, brief growing seasons and long, severe winters compress breeding seasons and select for optimal allocation of energy between reproduction and hibernation. Long-term field studies by Michener and her coworkers

TABLE 20-5   Reproductive Patterns in Selected Paleotropical (African) Bats

Polyestrus: year long asynchronous breeding	Polyestrus: asynchronous breeding for part of year	Diestrus: Two synchronous breeding periods	Monestrus: One synchronous breeding period
Rousettus aegyptiacus (P)	*Epomophorus wahlbergi* (P)	*Rousettus aegyptiacus* (P)	*Eidolon helvum*[a] (P)
R. lanosus (P)		*Myonycteris torquata* (P)	*Myonycteris torquata* (P)
Epomophorus labiatus (P)		*Taphozous mauritianus* (E)	*Rhinopoma hardwickei* (RM)
Epomops franqueti (P)		*Nycteris hispida* (N)	*Coelura afra* (E)
		N. thebaica (N)	*Taphozous nudiventris* (E)
		Cardioderma cor (MG)	*Hipposideros commersoni* (R)
		Rhinolophus landeri (R)	*H. cyclops* (R)
		Mops condylurus (M)	*H. caffer* (R)
		Lavia frons (MG)	*Pipistrellus nanus* (V)
			Eptesicus somalicus (V)
			Chaerephon pumila (M)

[a]This species has delayed fertilization; copulation and parturition seem not to be synchronous within a population.
Abbreviations: E, Emballonuridae; M, Molossidae; MG, Megadermatidae; N, Nycteridae; P, Pteropodidae; R, Rhinolophidae; RM, Rhinopomatidae; V, Vespertilionidae.
Data mostly from Kingdon, 1974; some data taken by T.J. O'Shea and T.A. Vaughan in Kenya.

(Michener and Michener, 1977; Michener and Locklear, 1990a, 1990b; Michener, 1998) have revealed that the extremely tight activity schedules for the Richardson's ground squirrel *(Spermophilus richardsonii)* differ between the sexes. Typically, males awake from hibernation in early February but remain in their burrows for a week, during which they eat seeds stored the previous summer, renew fat deposits, and begin spermatogenesis. Males emerge from their burrows in late February, two to three weeks before the appearance of females. Females emerge soon after awakening, over a two- or three-week period in early March. Each female enters estrus within four days after emergence and is usually receptive to males for only a part of a single afternoon.

Males eat little while competing intensely for females, lose weight, sustain injuries from fighting, and suffer high mortality. In poor physical condition, surviving males eat, replenish fat deposits, store seeds in their burrows, and enter hibernation in late June. Females bear young in mid-April, and young emerge from burrows a month later. Adult females then spend several weeks laying on fat deposits and enter hibernation in early July. After emergence, juvenile ground squirrels gain weight

rapidly and begin hibernation from about 7 weeks (for females) to approximately 15 weeks (for males) after the adults (Fig. 20-14). Adults of both sexes are active for only about 110 days a year. Males spend this time preparing for mating (by storing fat), mating, and preparing for hibernation and next year's mating (by storing body fat and caching seeds). Females spend about half their active time in pregnancy (23 days) and in nursing young (30 days) and half in eating and storing fat for hibernation. Thus, in a given activity season, the fitness of a male depends primarily on his ability to gain body mass and competitive condition in the brief time (about five weeks) between awakening from hibernation and the availability of estrus females. A female's fitness, however, depends on her ability to assimilate sufficient energy after emergence for pregnancy and lactation. (This pattern, typical of mammals, involves males competing for females and females competing for energy to invest in young.)

The reproductive cycles of desert rodents are timed to take advantage of seasonal bursts in the growth of ephemeral forbs (Beatley, 1976). Reichman and Van De Graaff (1975) found that, in Arizona, both sexes of Merriam's kangaroo rat ate

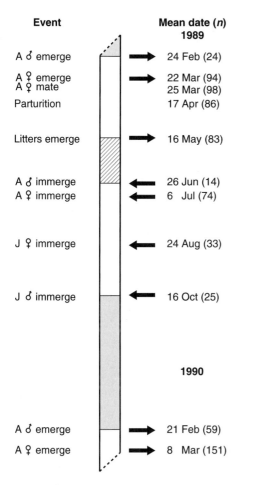

Event		Mean date (n)
		1989
A ♂ emerge	→	24 Feb (24)
A ♀ emerge	→	22 Mar (94)
A ♀ mate	→	25 Mar (98)
Parturition		17 Apr (86)
Litters emerge	→	16 May (83)
A ♂ immerge	←	26 Jun (14)
A ♀ immerge	←	6 Jul (74)
J ♀ immerge	←	24 Aug (33)
J ♂ immerge	←	16 Oct (25)
		1990
A ♂ emerge	→	21 Feb (59)
A ♀ emerge	→	8 Mar (151)

FIGURE 20-14 The annual cycle of Richardson's ground squirrels in Alberta, Canada, as indicated by major events in parts of 1989 and 1990: emergence from hibernation for adult males (A ♂) and adult females (A ♀); mating, parturition, emergence of litters from natal burrows, and immergence (entry into hibernation burrows) for adults and juvenile males (J ♂) and juvenile females (J ♀). Arrows pointing right designate mean emergence dates; arrows pointing left designate mean immergence dates. The cross-hatched segment (16 May to 26 June) is the only period when all age and sex classes are active above ground at the same time. Figures in parentheses are sample sizes. *(After Michener, 1998)*

more green plant material during the semiannual periods of plant growth than during the rest of the year and that there were surges in kangaroo rat reproduction immediately following these periods. Two Arizona sites only 147 kilometers apart differed markedly in rainfall: one site received three times more rain in the autumn than the other and

had vastly more green vegetation. At the wetter site, 90 percent of the adult kangaroo rats were reproductively active, whereas only 14 percent of the males and none of the females sampled at the drier site were in reproductive condition (Van De Graaff and Balda, 1973).

REPRODUCTIVE COSTS FOR MALES

Because of the high energetic costs of pregnancy and lactation in mammals, reproduction is often regarded as far more of an energetic hardship for females than for males. For some polygynous, **harem**-forming artiodactyls, however, this assumption may be false. Studies of Polish red deer *(Cervus elaphus)* by Bobek et al. (1990) indicated that reproduction was at least as costly for males as for females. Adult males lost an average of 102 kilograms of body mass (much of which was fat) during the 30-day rutting season. The estimated daily energy loss was an impressive 63.1×10^3 kJ per day. The tremendous energy reserve of males was accumulated during the 135 days of spring and summer preceding the rut. Males devoted 25 percent of their annual energy budget to reproduction, whereas the comparable figure for females was 18 percent. Red deer that hold harems fight more often and feed less than other males, and the energy invested in reproduction increases with increasing harem size (Clutton-Brock et al., 1982). Males that try to assemble excessively large harems suffer drastic weight losses during the rut. These animals are therefore especially vulnerable to predation and starvation the following winter and may not survive to breed again. Males of some other species may also invest heavily in reproduction. Michener (1998) judged mating effort to be more costly for male Richardson's ground squirrels than was parenting for females, if proximate factors such as depletion of fat, loss of body mass, and frequency of injury were considered.

LACTATION AND POSTNATAL GROWTH

The period of lactation is one of the most critical times in the life of a female mammal. The survival and vigor of her young, as well as her own fitness, depend on a female's ability to meet the energetic demands of lactation, which requires far more energy than does pregnancy (Table 20-6). This is due to the postnatal thermoregulatory and activity costs of developing young. In a laboratory study, food intake by pregnant cotton rats *(Sigmodon hispidus)* increased 25 percent over the intake of nonreproductive females; during lactation, the increase was 66 percent (Randolph et al., 1977). This disparity is actually greater than it seems, for about two thirds of the energy accumulated by a pregnant female was not invested in young but was stored and used during lactation, when the mother could not assimilate energy fast enough to support her rapidly growing young. Even with ample food, lactating cotton rats lost about 11 percent of their body weight (Mattingly and McClure, 1985).

Lactation is especially costly for metatherians because of their long lactation periods. Female eastern quolls *(Dasyurus viverrinus)* provided 11, 33, and 55 percent of their total digestible energy intake to litters of one, three, and five young, respectively. Consequently, mothers with large litters lost weight during the lactation period even though young from large litters were smaller at weaning than those from smaller litters (Green et al., 1997).

In yearling, female Richardson's ground squirrels, the energetic costs of gestation are low enough to allow these animals to complete the growth of the skeleton to adult size during pregnancy (Dobson and Michener, 1995). Further, for both yearling and older females, over half the weight gained during pregnancy is in body mass. Thus, females are heavier postpartum than just prior to pregnancy. During lactation, in sharp contrast, females do not gain weight, and they deplete some of their fat reserves (Michener and Locklear, 1990a).

Female ungulates of several species respond to the cost of gestation and lactation and poor nutrition by periodic infertility. In northern Alaska, for example, female caribou *(Rangifer tarandus)* have a reproductive pause once every four years; this may increase reproductive performance over a female's lifetime (Cameron, 1994).

Although lactation alone is energetically costly, many species of rodents, rabbits, and otariid seals

TABLE 20-6 Increase (in Percentage) in Dietary Energy Used by Pregnant or Lactating Animals Relative to Energy Used by Nonreproductive Individuals

Species	Pregnancy	Lactation	Reference
Homo sapiens (humans)	+10%	+50%	Nat. Res. Council Food and Nutrition Board (1968)
Ovis aries (sheep)	+81%	+146%	NRC (1974)
Rattus norvegicus (Norway rat)	+46%	+152%	NRC (1972)
Mus musculus (house mouse)	+33%	+111.9%	Myrcha et al. (1969)
Microtus arvalis (European vole)	+32%	+133%	Migula (1969)
Clethrionomys glareolus (bank vole)	+24%	+92%	Kaczmarski (1966)
Sigmodon hispidus (cotton rat)	+25%	+66%	Randolph et al. (1977)

After Randolph et al., 1977

have a postpartum estrus, and pregnancy and lactation are concurrent (Asdell, 1964). The energetics of concurrent pregnancy and lactation in laboratory colonies of cotton rats and eastern woodrats (*Neotoma floridana*) was studied by Oswald and McClure (1990). Compared to females that were only lactating, females of both species that were concurrently pregnant and lactating had higher resting metabolic rates (26 percent higher for cotton rats, 14 percent for woodrats). In contrast to concurrently pregnant and lactating cotton rats, woodrats in this condition increased food intake, delayed implantation of the second litter, and had smaller second litters. The length of delay of implantation was positively correlated with the number of suckling young, an effect also observed in some other rodents. The woodrat's period of lactation was twice as long as that of the cotton rat (24 versus 14 days). For species with prolonged lactation, delayed implantation postpones the progressively increasing costs of pregnancy until after the litter in the nest is weaned (Oswald and McClure, 1990). Probably as a result of energetic stress, concurrent pregnancy and lactation in wild rabbits (*Oryctolagus cuniculus*) in Wales frequently resulted in the death and resorption of the embryos in the uterus (Asdell, 1964). Although concurrent pregnancy and lactation are an energetic strain that may prejudice a female's post-breeding survival, selection seemingly favors this strategy in some mammals (such as some rodents, shrews, and rabbits) that live only long enough to breed during a single season.

Female mammals meet the cost of reproduction in several ways. Probably all species increase food intake during reproduction, and this increase can be drastic. In addition, some females may save energy during reproduction by decreasing activity. Energy stored in the body during pregnancy is used by some species during lactation. Mother white-footed mice (*P. leucopus*) metabolized body fat during lactation to augment food intake; when food was restricted, however, they sacrificed their litters rather than threaten their own survival (Miller, 1975). With restricted food, lactating cotton rats drew heavily on their own energy stores, losing 5.2 grams of body weight per day (Mattingly and McClure, 1985).

Some mammals make dietary shifts in an attempt to compensate for the heavy demands of lactation (Brody, 1945). A particularly interesting example involves spotted dolphins (*Stenella attenuata*) in the eastern Pacific. The diet of the general population of these dolphins, including pregnant females, was squid, but lactating females switched to a diet of flying fish (Bernard and Holn, 1989). Analysis of these foods showed that the fish yielded more energy than did squid (per 100 grams of muscle: 420 kilojoules versus 310 kilojoules) and higher levels of protein (21.2 percent versus 17.6 percent).

In general, small mammals have relatively higher postnatal growth rates than do large mammals. That is to say, small mammals attain adult weight, or a given percentage thereof, at an earlier age than do large mammals. Because of the higher metabolic rate of the small mammal, it can presumably mobilize energy for reproduction rapidly. Even among mammals of similar size, however, both metabolic rates and postnatal growth rates vary.

Low growth rates in certain small mammals may be associated with unusually low metabolic rates (as is probably the case with the tenrec, *Hemicentetes semispinosus*) or with an adaptively long mother-young association. Consider the following case discussed by Eisenberg (1981). Relative to body weight of the adult, postnatal growth is more rapid in vespertilionid bats than in phyllostomid bats (Kleiman and Davis, 1978), and the former apparently have richer milk. The Neotropical phyllostomids can afford a long suckling period because the young need not prepare for hibernation.

Postnatal growth rates reflect the life history of a mammal. High rates have evolved under such demanding conditions as stressful environments and short seasons for preparing for hibernation. The need for high growth rates in young bats that must store fat for a long hibernation has been mentioned. Equally demanding conditions have selected for extremely rapid growth in some marine mammals whose young must rapidly prepare for life at sea. This is true for the northern and southern elephant seals (*Mirounga* spp.). These are huge animals; males reach weights of over 3000 kilograms. An average pup weighs about 35 kilograms at birth, and this weight doubles by about 11 days of age. Pups often triple their weights by 28 days of age, when they are weaned (Condy, 1978). The young Weddell seal (*Leptonychotes weddellii*) doubles its weight within two weeks after birth (Bertram, 1940).

The rapid growth of pinnipeds is facilitated by the high-energy milk these animals produce (up to 53 percent fat). In some species, the suckling period is long, up to 1.5 years in the walrus and one year in some California sea lions (*Zalophus californianus*; Peterson and Bartholomew, 1967). The rapid growth allowed by high-calorie milk and a long suckling period prepares the young for facing the stresses of winter at sea.

Because pinnipeds forage at sea but must haul themselves onto land or ice floes to give birth and nurse their young, a different suite of maternal strategies has evolved in otariid and phocid seals. Female elephant seals (Phocidae) typically give birth to a single pup per year, and the duration of maternal care varies from four days to three years. Males do not contribute to parental care, therefore, the lifetime reproductive success of those males able to gain access to harems greatly exceeds that of females. A harem bull may mate with over 100 females in a single season (LaBoeuf and Reiter, 1988), while a successful female has a rela-

tively low lifetime reproductive output. Successful female northern elephant seals wean an average of ten pups in a lifetime.

Pinnipeds exhibit three basic maternal strategies with respect to lactation and maternal care of pups: aquatic nursing, foraging cycle, and fasting cycle strategies (Boness and Bowen, 1996). Walruses are unique among pinnipeds in exhibiting an aquatic nursing strategy (Fig. 20-15). Females accumulate blubber prior to parturition and fast for the first few days after giving birth. Young follow their mothers out to sea and nurse in the open water (Miller and Boness, 1983).

A foraging cycle strategy is typified by many otariids, where females rely on stored blubber to support lactation for the first two weeks postpartum. After this initial fasting period, females make foraging trips to sea (from 3 to 5 days for Antarctic fur seals to 5 to 14 days for northern fur seals) before returning to the beach to nurse their pups. The milk is high in fat, and lactation can last from four months to three years.

The fasting strategy, typical of phocids, requires females to arrive on the birthing beaches with

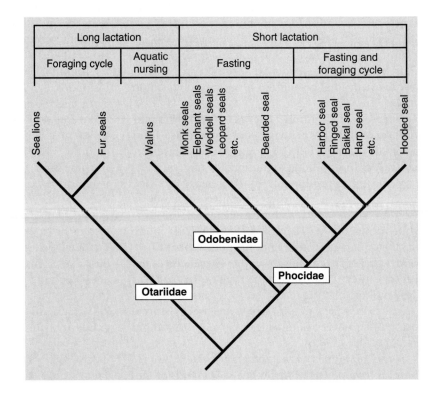

FIGURE 20-15 Lactation and feeding strategies of pinnipeds. Phocid seals typically have short lactation periods relative to otariids and the walrus. *(After Boness and Bowen, 1996, Bioscience, American Institute of Biological Sciences)*

TABLE 20-7   Characteristics of Maternal Support and Lactation in Pinnipeds

Species	Female mass (kg)	Body fat (%)	Lactation period (d)	Milk fat (%)	Daily energy (MJ/d)
Phocidae					
Mirounga leonina	515	23	24	47	184
M. angustirostris	504	40	27	54	98
Cystophora cristata	179	40	4	61	250
Phoca vitulina	84	25	24	50	31
Otariidae					
Eumetopias jubatus	273	–	330	24	21
Zalophus californianus	88	–	300	44	10
Callorhinus ursinus	37	–	125	42	6

After Boness and Bowen, 1996.

large stores of blubber because they must fast throughout the lactation period. Consequently, lactation periods are short (from 4 to 50 days) and the milk is extremely high in fat. Hooded seal pups, for example, take in 250 megajoules of energy per day over a four-day lactation period, compared to only approximately 10 megajoules per day for the California sea lion, an otariid (Table 20-7; Oftedal et al., 1987; 1993).

Until recently, it has been assumed that otariids used a foraging cycle strategy and that phocids used the fasting strategy. Recent advances in radio telemetry and isotope labeling have revealed that several small phocids, including harbor seals *(Phoca vitulina)*, use the foraging cycle strategy but retain a short lactation period (Bowen et al., 1992; Boness et al., 1994). Boness and Bowen (1996) concluded that the energetic constraints of small body size allowed the evolution of a maternal foraging cycle strategy in harbor seals. According to their hypothesis, small-bodied seals cannot store sufficient energy in the form of blubber to sustain their 24-day lactation period and must replenish these stores with short foraging bouts, especially toward the end of the lactation period.

WEB SITES

Society for the Study of Reproduction
   http://www.ssr.org/~ssr/ssr/ssr.htm

Reproduction, Fertility, and Development **journal**
   http://www.publish.csiro.au/journals/rfd/

Australian Society for Reproductive Biology
   http://Numbat.murdoch.edu.au/spermatology/asrb.html

21 ASPECTS OF PHYSIOLOGY

Some barriers to mammalian distribution are easily recognized. Bodies of water, ice fields, or mountains may be absolute barriers to dispersal, depending on the environmental tolerances of the specific mammals. Equally limiting, however, are environmental temperatures and the availability of water. The distributions of some mammals—

Neotropical sloths, for example—might be described most precisely by reference to the extremes of temperature and to the seasonal patterns of temperature change that can be tolerated. Air temperatures from −50 to 50°C may be encountered at various times and places on the earth, but, at best, mammals can only survive body tempera-

tures of approximately 0° to 45°C and can be normally active only within the narrow range of body temperatures between approximately 30° and 42°C. Just as some mammals are adapted to a few food sources or to a restricted type of habitat, some can live only within a narrow range of temperatures. In mammals, interspecific differences in the ability to withstand temperature extremes or scarcity of water occur even among closely related species, and it is not surprising that no one species is adapted to facing the full range of environmental extremes known for mammals as a group. Knowledge of various aspects of mammalian physiology is essential to understanding how mammals adapt to the great array of ecological settings they occupy.

ENDOTHERMY: BENEFITS AND COSTS

Most animals are **ectothermic** (their body temperature is regulated behaviorally by heat gained from the environment rather than by metabolic heat), but mammals and birds are **endothermic** (their body temperature is controlled largely by a combination of metabolic activity and physiological regulation of heat exchange with the environment, with behavioral thermoregulation of lesser importance). Most endotherms maintain a high and fairly constant body temperature throughout life, but for this they pay an extremely high cost in energy use. When not under thermal stress, a mammal expends five to ten times more energy for maintenance than a reptile of equal size and equal body temperature for body temperatures from 38° to 40°C. At lower temperatures, the cost of maintaining a high body temperature rises abruptly: a mammal uses 33 times more energy than a reptile at 20°C and 100 times more at 10°C. A foraging mouse uses 20 to 30 times more energy than a foraging lizard of equal weight. In small mammals, such as most rodents, 80 to 90 percent of the total energy budget is spent on thermoregulation. The

costs of endothermy are clearly high. What are the benefits?

A variety of benefits have been recognized. Endotherms can be active under an imposing array of temperature extremes, ranging from intense desert heat to extreme arctic cold. Moreover, they are freed from dependence on the warmer sunlit part of the daily light-dark cycle and from becoming inactive during cold seasons. Most mammals have responded to this freedom by being nocturnal, and many are active through all seasons. High body temperatures are supported by high oxygen-transport capabilities and high rates of enzymatic action. The primary advantage of endothermy is in greatly enhancing the ability to sustain high levels of activity (Bennett and Ruben, 1979). The maximum capacity of endotherms to use aerobic metabolism to produce power surpasses that of ectotherms by a factor of roughly 10. An example given by Bennett and Ruben of the relative ability of an ectotherm and an endotherm to increase oxygen consumption at high activity levels compares a 1-kilogram lizard and a 1-kilogram mammal. Resting and maximal rates of oxygen transport are 2 and 9 milliliters of oxygen per minute in the lizard and 9 and 54 milliliters of oxygen per minute in the mammal. In mammals (or in cynodont therapsids antecedent to mammals), endothermy probably evolved in the late Triassic under selective pressure for sustained activity and reduced body size; as a costly byproduct of these capabilities, the resting metabolic rate was raised (McNab, 1978).

Although core body temperatures are held relatively constant within groups, there is considerable variation in mammalian body temperature (T_b) between groups. Monotremes and xenarthrans have relatively low body temperatures, between 29° and 33°C. With the exception of shrews, insectivorans and metatherians also have lower T_b than many other groups of mammals (Fig. 21-1).

Each endotherm has a **thermal neutral zone** within which little or no energy is expended on temperature regulation. This zone is "the range of temperatures over which a **homeotherm** can vary its

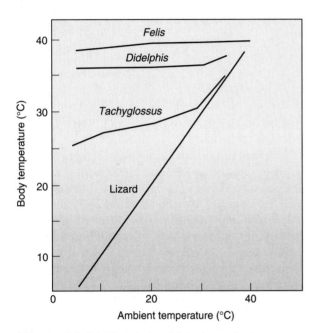

FIGURE 21-1 The relationship between body temperature and ambient temperature in various mammals. The cat *(Felis)* and the opossum *(Didephis)* are homeotherms capable of maintaining a constant body temperature over a wide range of ambient temperatures. The echidna *(Tachyglossus)*, a monotreme, is somewhat homeothermic, and the lizard is a strict heterotherm deriving virtually all of its body heat from the environment. *(After Marshall and Hughes, 1980)*

thermal conductance (C) in an energetically inexpensive manner and on a short time scale" and maintain a constant body temperature (Gordon et al., 1977). Within this zone, the fluffing or compressing of the fur, local vascular changes, or shifts in posture suffice to maintain thermal homeostasis.

One consequence of maintaining a constant body temperature is that the level of metabolic activity required to maintain T_b increases with extremes in ambient temperature (Fig. 21-2). Outside the thermal neutral zone, metabolic costs go up. To see why this is so, consider the relationship between the amount of metabolic heat (VO_2) needed to offset heat lost to the environment and ambient temperature (T_a).

$$VO_2 = C(T_b - T_a)$$

Because T_b is held constant in mammals, and C is determined by the degree of insulation (that is,

the better the insulation, the lower the C is), metabolic heat production (VO_2) varies linearly with ambient temperature (T_a) outside of thermal neutrality. At the lower limit of the thermal neutral zone is the **lower critical temperature,** the point below which the balance between metabolic heat production and heat lost to the environment cannot be maintained by metabolically inexpensive variations in C. Below the lower critical temperature, oxidative metabolism must be increased to keep the body temperature constant. The rate at which oxidative metabolism must be increased (denoted by the slope of the line) is determined by the C. The better the insulation, the lower the slope and the less metabolic heat that must be produced to keep the body temperature constant for an animal of given body size (Fig. 21-2). Thermal conductance (C) also depends on body size because body size is an important determinant of metabolic rate. Obviously, if a constant body temperature is to be maintained over a wide range of ambient temperatures, adjustments of both thermal conductance (through changes in insulation) and heat production (through metabolic changes) are necessary.

The **upper critical temperature** is the point above which a constant body temperature can be maintained only by increasing the metabolic work being done above the resting level in order to dissipate heat. This temperature is far less variable than the lower critical temperature but is of great importance to desert mammals, which usually do not have access to drinking water and must strictly minimize water loss. Many mammals faced with temperatures above the upper critical temperature dissipate heat by **evaporative cooling,** which involves considerable water loss (physiological regulation). Because such loss in desert species is extremely disadvantageous, these animals try to avoid temperatures above the upper critical temperature, often by spending the day in cooler shelters and being active on the surface only at night (behavioral regulation).

Selection in mammals has favored body temperatures and metabolic rates that save energy or facilitate the exploitation of a particular environment. There are many variations on the endothermic theme among mammals, and the survival of a species depends just as much on matching its thermoregulation pattern to its lifestyle and envi-

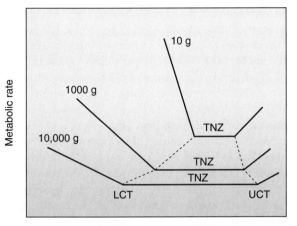

FIGURE 21-2   Relationship between metabolic rate and ambient temperature for mammals of various body sizes ranging from 10 grams to 10 kilograms. Metabolic rate stays relatively constant within the thermal neutral zone (TNZ). Above the upper critical temperature (UCT), the animal must expend additional energy to dissipate heat. Below the lower critical temperature (LCT), metabolic rate increases to keep the animal warm. Larger mammals have greater thermal inertia and require less metabolic energy to keep body temperature constant below the LCT.

ronment as it does on foraging efficiency or predator avoidance. Many large mammals have sufficient body mass, and therefore enough **thermal inertia,** to maintain their body temperature within narrow limits. For small mammals, however, in which the ratio of mass to surface area favors rapid heat exchange, wider fluctuations of body temperature are common. Indeed, for small mammals, a regular pattern of body temperature change through the daily temperature cycle may be the rule rather than the exception (Kenagy and Vleck, 1982).

COPING WITH COLD

There are essentially four mammalian strategies for surviving in cold environments. Mammals inhabiting cold climates can: (1) over many generations, evolve large body size, resulting in a more favorable surface-area-to-volume ratio and consequently reduced heat loss; (2) decrease their rate of heat loss through increased insulation or behav-

ioral thermoregulation; (3) increase their rate of metabolic heat production; or (4) abandon their normal body temperature and allow it to decline to a level closer to ambient temperatures **(hypothermia).** Obviously, the "best" strategy will depend on the environmental conditions and physiological constraints of the individual at any given time.

CONSEQUENCE OF BODY SIZE

Boreal mammals are generally larger than their ecological counterparts and close relatives in warmer areas (referred to as Bergmann's Rule). For example, the collared lemming (*Dicrostonyx groenlandicus*) of the Greenland tundra, a giant among nonaquatic arvicoline rodents, is more than twice as large (76 grams versus 32 grams) as its related eastern North American counterpart, the meadow vole (*Microtus pennsylvanicus*). Other cold-climate mammals that outsize their warm-climate relatives are the Arctic hare, the Alaskan wolf, and the Alaskan moose. The uniformly large size of marine mammals (which will be discussed later) is an adaptation to living in cold water.

Large size is a common and effective adaptation to cold. In general, large size favors heat conservation and small size favors heat dissipation: the larger the animal, the greater the volume or mass relative to surface area, and the smaller the animal, the greater the surface area relative to mass. Surface area is proportional to the square of body length, and volume is proportional to the cube of length. The surface-area-to-volume ratio, then, varies as the two-thirds power of the weight. The empirical relationship between body temperature (T_b), lower critical temperature (T_{lc}), and body weight (W) in mammals is represented by the expression $T_b - T_{lc} = 4W^{0.25}$. Because body temperature is usually held constant in mammals, as weight decreases, T_{lc} approaches T_b (Gordon et al., 1977; Fig. 21-2). In contrast, when foraging, small nocturnal mammals usually face temperatures below their lower critical temperature. This temperature for a 20-gram mouse is approximately 29°C, a temperature considerably higher than that usually encountered by nocturnal mammals.

Basal metabolic rate (the minimum necessary for simply maintaining life in a resting organism in thermal neutrality), lower critical temperature, and thermal conductance all vary inversely with body size and are intimately interrelated. Mass-specific metabolic rate, as measured by oxygen consumption per gram of body weight per hour, rises so precipitously with decreasing body weight that the smallest mammals (Etruscan shrew, *Suncus etruscus,* and Kitti's hog-nosed bat, *Craseonycteris thonglongyai*), weighing about 2 grams, probably represent the lower limit of mammalian body size (neonates are considered in the following discussion). The mass specific metabolic rate of a mammal the size of the Etruscan shrew, for example, is approximately 141 J/g/hr (joules/gram/hour) compared to only 1.6 J/g/hr for an elephant (Withers, 1992). Rates of oxygen consumption differ markedly even among small mammals (Fig. 21-3): the tiny masked shrew consumes oxygen at a rate over four times that of the larger deer mouse. Carrying the comparison further, the mouse consumes oxygen ten times faster than the horse (Krebs, 1950).

The smallest mammals are often **neonates** (newborns). Until relatively recently, studies of neonatal thermoregulation involved measurements made on individual young removed from the nest. Under these highly artificial conditions, altricial neonates appeared to be devoid of physiological regulation of body temperature and were erroneously referred to as ectothermic (Case, 1978). As Hill (1992) points out, in nature, neonates huddle in groups in well-insulated nests and are often warmed by the close proximity of their parents and siblings. Measurements of neonatal and weanling thermoregulation under these more natural conditions revealed that even altricial young begin to thermoregulate at an early age. Thus, neonates rely on shared body warmth from litter mates and parents for much of their early development. Hill (1992) hypothesized that neonates suspend their own thermoregulation facultatively when the mother is present but begin endogenous **thermogenesis** when she is out foraging.

Basal metabolic rates (BMR) are typically measured under controlled laboratory conditions far removed from those experienced by the animal in the field. Applying BMR data from the laboratory to free-living mammals responding to a wide variety of environmental and behavioral conditions, is difficult. Field metabolic rates (FMR) can now be measured with accuracy using the doubly labeled water technique. FMR is the total energy cost (including BMR, thermoregulation, locomotion, feeding, digestion, reproduction, growth, and so forth) experienced by a free-ranging animal during the course of 24 hours. The doubly labeled water technique involves measuring the relative turnover of radioactively labeled hydrogen and oxygen injected into the animal as water. The hydrogen isotope measures the rate of water loss, and the oxygen isotope measures the sum of water and CO_2 loss (Kunz and Nagy, 1988; Nagy, 1987). The difference between the two turnover rates can be converted to field metabolic rate.

Using FMR data from many doubly labeled water experiments on a variety of vertebrate taxa, Nagy (1987) showed that FMR is strongly correlated with body mass (Fig. 21-4). The slope of the

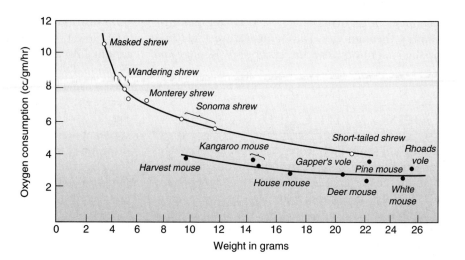

FIGURE 21-3   Oxygen consumption as a function of body weight in some small mammals. *(After Pearson, 1948, American Association for the Advancement of Science)*

relationship between FMR and body mass for 23 species of eutherians (0.81) was significantly higher than that measured for BMR (0.75), but the opposite was true for 13 metatherian species (slope = 0.58). These data showed that FMRs of medium-size mammals and birds (200 to 500 grams) are similar, suggesting that it costs the same amount of energy to live one day. Small eutherians (less than 200 grams) had lower FMRs than small metatherians, and large eutherians have relatively high FMRs. Thus, the FMR of a 67-kilogram mule deer (40,000 kilojoules per day [kJ/d]) was nearly four times higher than a 62-kilogram gray kangaroo (11,734 kJ/d; Nagy, 1987). Compared to a lizard (ectotherm) of similar size, a mammal expends approximately 17 times as much energy. The energetic difference is due to the increased cost of maintaining a high body temperature for the mammal. Nagy (1987) also showed that FMRs scaled differently with diet, season, and habitat. Desert-dwelling mammals typically have 30 percent lower FMRs than nondesert mammals. The rela-

tionship between body size and metabolic costs is further complicated by the ability to store fat and the type of insulation. Large size enhances fat storage, which may serve both as insulation and an effective metabolic fuel.

INSULATION

Individual mammals generally lack the ability to increase their body mass sufficiently for this to be an effective response to cold stress. A more useful strategy is to lower their thermal conductance. Effective insulation is, therefore, an important feature of cold-adapted mammals (Fig. 21-5). The insulative value of fur increases with its thickness. A half-centimeter-thick layer of shrew fur would have approximately the same insulative capacity as a half-centimeter layer of caribou fur. Shrew fur, however, is less than 0.3 centimeters thick; caribou fur is often 4 or 5 centimeters thick, and winter pelage can be as thick as 15 centimeters. Fur is so remarkably effective in some species that the thermal neutral zone may extend down to $-30°C$, as in the case of the arctic fox (*Alopex lagopus*). In many mammals active in the cold, the length of the woolly underfur and of the longer guard hairs varies seasonally; the summer pelage, which is acquired in spring, is short and has reduced insulating ability, but the winter coat, which replaces the summer pelage in autumn, is long and has great insulating ability. The hollow hair of some ungulates (such as the pronghorn) is remarkable insulation that allows winter activity under extreme conditions. In water, fur loses much of its insulative value because it is compressed by water (less thickness) and because water has higher thermal conductivity than air. The insulation provided by beaver fur drops from a value near 7 ($°Cm^2 \ sec \ J^{-1}$) to less than 1 when immersed in water. Seal fur has considerably less insulative value than an equivalent thickness of polar bear fur, yet seals inhabit some of the coldest waters on Earth. Seal pelage serves to trap a layer of still water rather than trapping a layer of air. This still water layer probably does not contribute much to insulation because skin temperatures of harp seals (*Phoca groenlandica*) are approximately the same as water temperatures. A thick layer of subcutaneous blubber, rather than fur, provides insulation in most aquatic mammals.

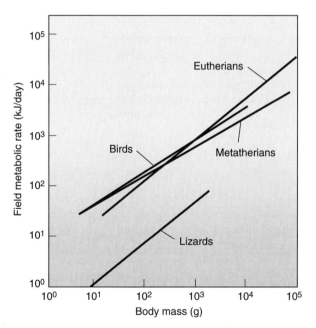

FIGURE 21-4 Scaling of field metabolic rates with body size in eutherian (23 species) and metatherian (13 species) mammals along with comparable data for birds (25 species) and lizards (25 species). Field metabolic rates of endotherms are approximately 17 times that of similar-size ectotherms. *(After Nagy, 1987)*

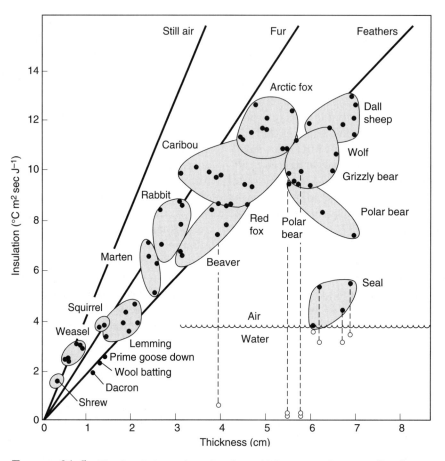

FIGURE 21-5 The insulative value of various thicknesses of mammalian fur. Dotted vertical lines illustrate the loss of insulation when fur is wet (open circles). Seal fur provides less insulation in air than the fur of other mammals, but retains most of its insulative value when wet. *(After Withers, 1992)*

BEHAVIORAL THERMOREGULATION

All animals that maintain a reasonably constant body temperature must balance heat gains and losses. Whereas lizards and many other ectotherms do this by practicing **behavioral thermoregulation,** mammals rely primarily on metabolic adjustments. Mammals use behavioral thermoregulation too, but their high resting metabolic rate and their activity under temperature extremes impose special thermoregulatory burdens. Nevertheless, behavior plays a critical role in reducing cold stress. Many small mammals curl up so that their body form is nearly spherical. This minimizes the ratio of surface area to volume and tends to protect such lightly insulated parts as the face and feet. Nest-building behavior is important for small mammals.

Nests, usually built in protected places, provide insulation that greatly augments that provided by the pelage. Group thermoregulation, involving several animals huddling together (West, 1977), reduces the exposed surface area of each animal and is apparently widespread in mammals. Social behavior may be an important energy-conserving strategy. Taiga voles (*Microtus xanthognathus*), which are active through severe interior Alaskan winters, enhance winter survival by living in groups of five to ten animals (Wolff and Lidicker, 1981). The communal nest of these voles is always occupied by one or several animals, and nest temperature is thus kept well above ambient temperature. Morton (1978) observed nest sharing in a small marsupial (*Sminthopsis crassicaudata*) and regarded it as an energy-saving behavior. The response of seeking shelter or a favorable place is also of importance. An animal foraging at ground level beneath a deep snowpack faces temperatures near 0°C, whereas ambient temperatures above the snow may be many degrees below zero.

INCREASING METABOLIC HEAT PRODUCTION

To maintain a constant body temperature when the ambient temperature is below body temperature, a mammal must balance heat lost to the environment with heat produced by metabolism. While behavioral thermoregulation is an important means of heat conservation in mammals, additional mechanisms for increasing heat production are often necessary. High metabolic rates (especially high peak metabolic capacities) are one solution typically found in cold-dwellers (Table 21-1). The high metabolic rates of voles and lemmings are generally regarded as an adaptation to activity through all seasons in temperate or boreal areas. The red fox and arctic fox, both of which forage throughout the winter, have metabolic rates that are considerably higher than expected for mammals of their respective sizes.

A second strategy to increase metabolic heat production is to use the elevated aerobic metabolic capacity of skeletal muscles or brown adipose tissue. Shivering, with attending heat production by skeletal muscles, is common in mammals inhabiting cold climates, and even Old World fruit bats (*Pteropus*) shiver conspicuously in cool temperatures. Production of heat by metabolizing fat is called **nonshivering thermogenesis.** Mammalian nonshivering thermogenesis typically involves the oxidation of a special type of fat called brown adipose tissue. Brown adipose tissue is found in limited deposits around the neck and between the shoulders in many cold-adapted eutherian mammals and some newborn mammals. Loudon et al. (1985) reported a similar type of fat tissue in a metatherian.

Brown adipose tissue rapidly produces heat when oxidized. Unlike normal fat metabolism, in which fat is first reduced to fatty acids that are transported in the bloodstream to other areas for subsequent oxidation, brown fat oxidation takes place within the fat cells themselves, which are densely packed with mitochondria (giving it the brown color) and richly vascularized. During the oxidation of brown fat, the adipose tissue generates significant amounts of heat that is then dispersed, via the vasculature, to the rest of the body. Brown adipose tissue can produce heat at ten times the rate possible for active skeletal muscle contraction (Withers, 1992). Not surprisingly, nonshivering thermogenesis via brown adipose tissue is especially important in infants (which often lack the insulation provided by a rich coat of fur) and during arousal from torpor or hibernation (discussed in the following section). In soricid shrews, nearly all body fat is in the form of brown adipose tissue, which can account for up to 20 percent of the animal's body weight (Hyvarien, 1994). The rapid heat production from these fat stores via nonshivering thermogenesis is partially responsible for the high degree of over-winter survivorship of shrews (Merritt, 1995).

REGIONAL HETEROTHERMY

Despite the adaptations mentioned above, at very low ambient temperatures the costs of endothermy may be unsupportably high. To offset these costs, cold-adapted mammals often practice **regional heterothermy,** the ability of some mammals to allow the temperatures of the skin or extremities to drop well below the core body temperature. Extremities, such as legs and ears, which are poorly insulated and dissipate heat rapidly, are allowed to become cool, thereby reducing heat loss by minimizing the temperature differential between these parts and the environment. In an Eskimo dog exposed to cold, the deep-body temperature was 38°C, the toe pads were 0°C, and the tops of the feet 8°C (Irving, 1966). This cooling of the extremities is due to **vasoconstriction** or to a **countercurrent heat exchange** system (Fig. 21-6). Arterial blood leaving the heart is at the core body temperature, while venous blood returning from the extremities is substantially colder. In a countercurrent heat exchange system, as warm arterial blood passes next to veins containing cool blood returning from the periphery, the warm arterial blood gives up heat to the venous blood. Consequently, the arterial blood is precooled before it reaches the skin and has little heat to lose to the environment. At the same time, the cold venous blood is warmed before it returns to the body core, minimizing a drop in core body temperature.

TABLE 21-1 Resting Metabolic Rates of Selected Mammals

Taxon	Common name	Body weight (g)	Metabolic rate (ml O_2/g/h)
Monotremata			
Ornithorhynchus anatinus	Platypus	1200	0.46
Tachyglossus aculeatus	Echidna	4200	0.22
Metatheria			
Sminthopsis crassicaudata	Dunnart	14	1.67
Sarcophilus laniarius	Tasmanian devil	6700	0.28
Trichosurus vulpecula	Brush-tail possum	1982	0.32
Insectivora			
Sorex cinereus	Masked shrew	3–5	16.8
Blarina brevicauda	Short-tailed shrew	14–18	5.3
Xenarthra			
Dasypus novemcinctus	Nine-banded armadillo	4000	0.20
Bradypus variegatus	Three-toed sloth	4500	0.18
Rodentia			
Heterocephalus glaber	Naked mole-rat	32	0.4
Peromyscus eremicus	Cactus mouse	22	2.2
Microtus pennsylvanicus	Meadow vole	32	3.2
Dicrostonyx groenlandicus	Collared lemming	76	3.9
Capromys pilorides	Cuban hutia	4300	0.23
Carnivora			
Vulpes vulpes	Red fox	4440	0.55
Vulpes zerda	Fennec fox	1106	0.36
Ursus americanus	Black bear	77,270	0.36
Proboscidea			
Elephas maximus	Indian elephant	2,730,000	0.15
Perissodactyla			
Equus caballus	Horse	260,000	0.25
Artiodactyla			
Sus scrofa	Pig	75,000	0.11
Rangifer tarandus	Caribou	105,000	0.36

Data from J. F. Eisenberg, 1981; Jarvis, 1978; and Noll-Banholzer, 1979a.

Regional heterothermy is a major cold adaptation of beavers (*Castor canadensis*). In fact, beavers employ two types of countercurrent heat exchangers, a rete system of small arteries and veins in the tail and a venae comitantes of arteries and veins in the hindlimbs (Cutright and McKean, 1979). In the beaver's hindlimbs, the venae comitantes consist of a central artery surrounded by a series of anastomosing veins. The rete system of the tail is even more specialized for heat exchange and consists of a series of interwoven arteries and veins. During the warm summer months, blood is shunted through a bypass to allow warm arterial blood to reach the skin and dissipate to the environment.

ADAPTIVE HYPOTHERMIA

Maintaining a relatively high and constant body temperature (endothermy) by endogenous heat production is metabolically expensive. Endothermy requires a relatively constant supply of high quality food for fuel. Unfortunately, energy resources (food and water) in the environment vary in time

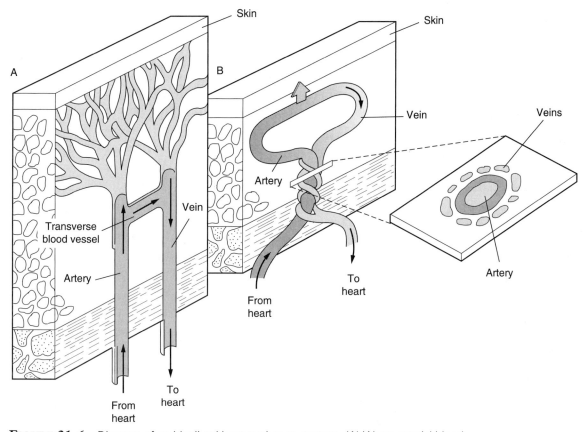

FIGURE 21-6   Diagram of an idealized heat exchange system. (A) Warm arterial blood moves into a network of capillaries near the skin surface, where heat can be lost through the skin. The cooled blood returns to the body core through the veins. In some mammals, this heat loss can be reduced by shunting the warm arterial blood through transverse blood vessels, thereby preventing heat from reaching the skin. (B) In mammals that live in very cold climates, a countercurrent heat exchange system may be used to conserve heat. Here, warm arterial blood is used to warm the cooler venous blood; the venous blood serves to precool the arterial blood before it reaches the skin. In this system, an artery surrounded by veins gives up some of its heat to the venous blood returning to the body core.

and location. The patchy distribution of food coupled with seasonal and daily fluctuations in ambient temperatures have selected for patterns of energy use that increase survival when food (or water) are scarce or when environmental temperatures are very high or very low. Adaptive hypothermia refers to a continuum of responses that allow energy to be saved by temporarily abandoning homeothermy (maintaining a constant T_b). At one end of the continuum is shallow hypothermia in which body temperature drops and is regulated within 10°C of **normothermia.** At the opposite end is torpor (or profound hypothermia). Torpor is defined by a suite of characteristics, including a T_b falling to within 1°C of ambient, reduced oxygen consumption, prolonged periods of apnea (suspended respiration), a markedly reduced heart rate, and the ability to arouse by mobilizing endogenous heat production (Gordon et al., 1977). Some mammals use adaptive hypothermia seasonally, some use it daily, and others use both patterns. Adaptive hypothermia occurs in monotremes, some metatherians, and some members of the orders Insectivora, Chiroptera, Primates, Carnivora, and Rodentia.

A number of energy-saving physiological changes occur during adaptive hypothermia or torpor. These include lowering of the heart rate, progressive vasoconstriction, suppression of shivering, reduction in breathing rate, and lowered oxygen consumption. These changes occur during entry into torpor and usually precede a decline in body temperature. Torpor in summer is called estivation, and winter torpor is hibernation.

Hibernation, a form of seasonal torpor lasting multiple days, is triggered by food unavailability and low environmental temperatures. Although described as a hibernator by many biologists, the black bear *(Ursus americanus)* actually undergoes shallow hypothermia. In many areas, this animal retires to a protected place into which it has carried insulative nesting material and remains there from October to April. Rogers (1981) found that bears hibernating in Minnesota maintained body temperatures above 31°C, some 7°C below their normal temperature. A wild, hibernating Alaskan black bear had a heart rate of 8 beats/minute when sleeping soundly in December, whereas the heart rate of active bears in summer is from 50 to 80 beats/minute.

Preparation for hibernation often (but not always) involves great increases in body weight resulting from fat storage. This gain ranges in sciurids from 80 percent of the fatfree weight in the golden-mantled ground squirrel *(Spermophilus lateralis)* to 30 percent of this weight in the yellow-pine chipmunk *(Tamias amoenus;* Jameson and Mead, 1964).

Although small mammals can exploit small food items (such as seeds) that are usually unavailable to larger animals and have access to a nearly limitless array of retreats, small size is a liability energetically. The high metabolic rates of small mammals must be sustained by high intakes of food, and seasonal changes in food availability present severe problems. Winters in the north and dry seasons in the deserts and in many tropical areas are times of potential food shortage for small mammals, and these are also periods when temperature or lack of moisture may limit activity. It is not surprising, therefore, that some small species have evolved means of surviving periods of food shortage and temperature stress and of using times of moderate temperatures and high food productivity to reproduce and to store food or fat.

Many small mammals (up to approximately 5 kilograms) periodically conserve energy by allowing the body temperature to drop to near that of the environment. This is not a primitive feature, a manifestation of some ancestral inability to sustain a steady temperature at all times, but is instead a highly adaptive ability. An extreme example of this ability is shown by a small pocket mouse *(Perognathus parvus;* MacMillen, 1983a). When hibernating at ambient temperatures down to 2°C, this mouse maintains a body temperature about 1°C above ambient temperature. At ambient temperatures between 2 and −5°C, it increases its metabolic rate just enough to maintain its body temperature at 2°C. Thus, despite a body and brain temperature near freezing, this mouse resets its T_b thermostat to maintain body temperature above lethal limits. Such adaptive hypothermia may well have been a factor important in furthering the success of the two largest mammalian orders, Rodentia and Chiroptera. Many small bats would not be able to forage only at night and fast throughout the day if they could not conserve energy in the day by means of daily torpor. Similarly, seasonally hostile areas would not be inhabited by some small rodents if these animals retained constant thermal homeostasis.

Torpor cycles can be seasonal (hibernation and estivation) or daily. Typically, torpor has three phases: a rapid entry phase, a prolonged period of torpor, and a relatively rapid arousal period (Fig. 21-7). Entry into torpor by some small mammals seems to be triggered by lack of food (Nestler et al., 1996). In others, entry is spontaneous. Several species of pocket mice and a kangaroo mouse *(Microdipodops pallidus)* enter torpor spontaneously (Brown and Bartholomew, 1969; French, 1977, 1993; Kenagy, 1973a; Meehan, 1976). Torpor in many small rodents (such as *Microdipodops,* some *Chaetodipus* and *Perognathus,* and some *Peromyscus*) is a circadian phenomenon: these animals are torpid by day and active and normothermic by night. In hibernators, periods of torpor become progressively longer (up to five days), with intervening arousals. During the shallow circadian torpor of some rodents, body temperature stays above roughly 15°C; during deep torpor in some rodents and some bats, body temperature drops to between 1 and 5°C.

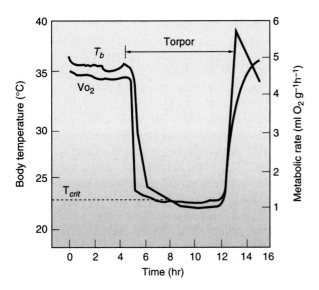

FIGURE 21-7 The body temperature and metabolic rate of a deer mouse *(Peromyscus maniculatus)* during daily torpor. *(After Nestler, 1990, Physiological Zoology, University of Chicago Press)*

It is clear that different animals respond to different stimuli for torpor, and many respond differently to the same stimulus. Many rodents enter torpor (seasonal or circadian) before experiencing energy deficits but in anticipation of surface conditions that will induce deficits if surface activity continues. These animals enter torpor either with substantial fat stores (larger rodents) or underground food hordes (smaller rodents), thus indicating continuing favorable energetic states. Some murid and heteromyid rodents become torpid rather quickly in response to low air temperatures and lack of food, but some, the golden hamster *(Mesocricetus auratus)*, for example, require two to three months of cold preparation before becoming torpid. Mammals with seasonal cycles, such as the arctic ground squirrel *(Spermophilus parryii)*, enter torpor in response to diminishing photoperiods and falling temperatures in the autumn. Laboratory studies by Dawe and Spurrier (1969), in which the injection of blood from hibernating thirteen-lined ground squirrels *(S. tridecemlineatus)* into active animals was followed by hibernation by the latter, suggest that some "trigger substance" in the blood may initiate hibernation in squirrels. The rate of entry into torpor is proportional to thermal conductance and therefore also to body mass. Consequently, small mammals enter torpor much more rapidly than large mammals (Table 21-2).

During torpor body temperature drops to a minimum critical level (T_{crit}) and is maintained at this level by endogenous heat production. Species differ in their lower critical body temperature. This temperature is fairly high in some heteromyids that undergo shallow torpor (12°C in *Chaetodipus hispidus*) but, as would be expected, is low in mammals using deep torpor. Critical body temperature is close to freezing in some bats of the genus *Myotis*, 2.8°C in the golden-mantled ground squirrel, and 4°C in the European hedgehog *(Erinaceus europaeus)*. The minimum critical

TABLE 21-2. The Time Required for Mammals of Various Body Masses for Entry Into and Arousal from Torpor at an Ambient Temperature of 15°C

Species	Body mass (g)	Entry time (min)	Arousal time (min)
Shrew *(Suncus)*	2	35	13
Honey possum *(Tarsipes)*	4	59	17
Echidna *(Tachyglossus)*	3500	1648	226
Marmot *(Marmota)*	4000	1766	237
Badger *(Taxidea)*	9000	2685	323
Bear *(Ursus)*	80,000	8307	741

Adapted from Withers, 1992.

temperature is generally above freezing, not to prevent tissue freezing, since torpid arctic ground squirrels can supercool to a T_{crit} of $-2.9°C$ (Barnes, 1989), but to reduce the costs of arousal from torpor.

Arousal from hibernation is energetically costly and is associated with the metabolism of energy-rich "brown fat" in some species and with shivering in others (Chaffee and Roberts, 1971). Etruscan shrews, the smallest mammals studied, use both mechanisms during arousal from daily torpor. During arousal, heat production is primarily from brown adipose tissue, augmented with shivering at body temperatures above 17°C (Fons et al., 1997). Arousal rates for Etruscan shrews averaged 0.83°C per minute, which is similar to rates reported for *Sorex cinereus* (Morrison et al., 1959) and among the highest rates found in mammals (Stone and Purvis, 1992). This high cost of arousal suggests that the most advantageous strategy for this hibernator would involve continuous deep torpor. However, periodic arousal during the hibernation period is the rule among many species studied. The broad pattern is one of progressively increasing periods of torpor through the early stages of hibernation and decreasing periods in the late stages. Among several rodents, the maximum periods of torpor were from 12 to 33 days, but a period of 80 days was recorded for the little brown bat *(Myotis lucifugus)*. The European hedgehog was found to be torpid 31 percent of the time at 10°C and 80 percent of the time at 4.5°C. The different patterns exhibited by insectivorans and rodents may reflect the fact that rodents can store energy (food or fat) to use during periodic arousals, and insectivores (shrews, bats, and so forth) cannot store food. (For a review of torpor in mammals, see Hudson, 1973.)

Periods of torpor are seemingly characteristic of the life cycles of some small mammals inhabiting hot regions. The Mohave ground squirrel *(Spermophilus mohavensis)* is a desert species that remains in its burrow from August to March, a period spanning both hot and cold seasons. Laboratory studies indicate that during this period the animals are intermittently torpid for periods of several hours to several days (Bartholomew and Hudson, 1960, 1961). This squirrel is able to elevate its temperature from 20 to 30°C in 20 to 35 minutes and, even during its active period in spring and summer, has an unusually variable body temperature from 31 to 41.5°C.

The tiny (4 grams) desert shrew *(Notiosorex crawfordi)* lives in the deserts of the southwestern United States and utilizes what Lindstedt (1980b) termed regulated hypothermia. Under laboratory temperatures between 20 and 25°C, the thermoregulatory pattern of this shrew involved two thresholds: the shrews maintained a body temperature near 38°C when active but, when food was restricted, allowed their diurnal temperature to drop to 28°C and regulated it at this level. In both euthermic and hypothermic states, the shrews regulated their temperatures with precision by adjusting metabolic heat production.

Studies of a variety of bats have both clarified and complicated the picture of temperature regulation in these animals. Among different species, contrasting reactions to temperature changes occur, and within the Chiroptera most mammalian styles of temperature regulation are represented. Seemingly, the larger megachiropterans are homeotherms. Those that have been studied are able to maintain body temperature within fairly narrow limits (35° to 40°C) over a range of ambient temperatures from approximately 0° to 40°C. Many pteropodid bats react to cold stress by shivering and by enveloping the body with the wings, which serve as blankets that provide considerable insulation for the body (Fig. 21-8; Bartholomew et al., 1964). Shallow diurnal torpor does occur in a few small Megachiroptera, including two tube-nosed bats *(Nyctimene)* from New Guinea (Bartholomew et al., 1970).

Compared with megachiropterans, microchiropteran bats are highly variable in their responses to temperature extremes. The Australian species *Macroderma gigas* (Megadermatidae), probably because of its large size (100 to 140 grams compared to less than 40 grams for most microchiropterans), is able to maintain a stable body temperature in the face of ambient temperatures as low as 0°C, and many of the reactions to temperature extremes in this bat are similar to those of megachiropterans (Leitner and Nelson, 1967). The Neotropical species *Desmodus rotundus,* a vampire bat (15 to 50 grams), was originally reported to be unable to regulate its body temperature in response to moderate changes in ambient temperature (Lyman and Wimsatt, 1966). However, more

FIGURE 21-8   An African epauletted bat
(Epomophorus wahlbergi, Pteropodidae) with its
wing membranes enshrouding its body. *(R. Bowker)*

recent studies of the three genera of vampire bats suggest that under natural conditions they are able to regulate their body temperature down to ambient temperatures approaching 0°C, and *Diphylla* and *Diaemus* may even become heterothermic under certain conditions (Hill and Smith, 1992).

Many tropical microchiropteran bats from the Old World and from the Neotropics are active at night and inactive during the day, and this activity cycle is reflected by their temperature cycle. In the Neotropical phyllostomid bats that have been studied, body temperatures are from 37° to 39°C during the night and 2° to 3°C lower during the day (Morrison and McNab, 1967). In general, these bats are able to maintain a high body temperature despite moderately low ambient temperatures. Broadly speaking, though, cold stress can usually be tolerated by tropical bats for only fairly short periods, after which the body temperature falls uncontrollably. Body temperatures below 20°C are often fatal.

Adaptive hypothermia, often involving (at different seasons) both daily torpor and hibernation, occurs in many vespertilionids of north temperate areas and seems to be the key to the survival of

some species in cool or cold regions. During the summer, some microchiropterans of temperate zones undergo daily torpor at low ambient temperatures. Tremendous metabolic savings are realized by microchiropteran bats that become hypothermic at low temperatures. Under experimental conditions, the average metabolic rate of six little brown bats *(Myotis)* kept at an ambient temperature of 35°C was 33 times that of these same bats when kept at 5°C (Henshaw, 1970). In addition, strikingly abrupt rises in metabolic rate occur during flight. The metabolic rate of the phyllostomid bat *Phyllostomus hastatus* was about 30 times greater in a flying individual than in those resting at a body temperature of 36.5°C (Thomas and Suthers, 1972). In the interest of saving energy, bats in temperate areas have the briefest possible periods of flight and, in bats with the ability, the longest possible daily periods of hypothermia.

Some bats that are homeothermic in summer abandon this pattern well before winter. Fat deposition is known to occur in the late summer or early fall in some species of vespertilionids that hibernate (Baker et al., 1968; Ewing et al., 1970; Krzanowski, 1961; Weber and Findley, 1970), and three species of *Myotis* became hypothermic during this time (O'Farrell and Studier, 1970). In *M. thysanodes,* the metabolic rate for homeothermic individuals at an ambient temperature of 20.5°C is 6.93 ml^3 O_2/g/h (cubic milliliters of oxygen per gram per hour); but drops to 0.59 ml^3 O_2/g/h in hypothermic bats. This decrease results in the saving of 2.81 kilocalories per day as a bat becomes hypothermic. Fat is deposited in preparation for hibernation at the rate of 0.17 grams per day in the period of maximum fat accumulation. This requires an extra 1.60 kilocalories per day, which is available primarily because of the late summer-autumn shift to daily hypothermia (Ewing et al., 1970; Krzanowski, 1961).

Winter hibernation in bats differs from short-term torpor largely in the length of dormancy and in the levels to which the metabolic rate and temperature drop. The duration of hibernation for bats differs widely among species and within a species, depending on the area. In the northeastern United States, *M. lucifugus* remains in hibernation for six or seven months, from September or October to April or May (Davis and Hitchcock, 1965). Periods of hibernation for bats in warmer areas are probably considerably shorter. At ambient temperatures near 5°C, bats in deep hibernation maintain a body temperature about 1°C above ambient temperature. These bats are responsive to certain stimuli and will begin arousal when handled or when subjected to unusual air movement. As a defense against freezing to death, bats spontaneously raise the metabolic rate at dangerously low ambient temperatures (below roughly 5°C) and either arouse fully or regulate body temperature and remain in hibernation.

Not all north temperate bat species enter torpor in response to declining fall temperatures. At least some species of *Nyctalus, Vespertilio, Lasiurus, Lasionycteris, Pipistrellus,* and *Tadarida* migrate from north to south as winter approaches. Although the cost of flight is high (4.6 kilocalories per 100 kilometers), many species of tree-roosting bats (for example, *Lasiurus* and *Nyctalus*) make seasonal migratory flights of over 1500 kilometers (McNab, 1982). A number of temperate zone species of *Myotis* make shorter (200 to 500 kilometers) seasonal migratory flights between their hibernation caves and summer roosts or maternity colonies. These species use both migration and torpor to save on energetic expenses.

THERMOREGULATORY PROBLEMS OF AQUATIC AND SEMIAQUATIC MAMMALS

Temperature regulation is a demanding problem for mammals that inhabit cold water. The rate of heat loss by an endotherm in water is some 10 to 100 times as great as the rate of loss in air of the same temperature (Kanwisher and Sundnes, 1966). Arctic and Antarctic waters are near 0°C year-round, and high-latitude lakes and rivers approach this temperature in winter. Consequently, a temperature differential of about 35°C between deep-body temperature and ambient temperature is common in mammals swimming in these waters. Despite the thermal inhospitability of this environment, cold waters are permanently inhabited by cetaceans, and pinnipeds spend much of their lives in such waters.

In addition, the muskrat *(Ondatra zibethicus),* beaver, some shrews *(Sorex* spp.), some otters *(Lontra canadensis, Enhydra lutris),* and the mink *(Mustela vison)* spend considerable time in cold water. Although these semiaquatic mammals lose heat to the water most rapidly from the foot pads, the nose, and other bare surfaces, most of the body is insulated by a layer of air entrapped by the fur. Nonetheless, heat is lost far more rapidly during immersion in water than when the animal is in air. Calder (1969) found that, in two species of shrews *(Sorex palustris* and *S. cinereus)* and two species of mice *(Zapus princeps* and *Peromyscus maniculatus),* thermal conductance in the water when the fur had entrapped air was 4.5 times that in air. When the fur was wet to the skin, the conductance rose to 9 times that in air. Calder also measured heat loss in the water shrew *(S. palustris),* the smallest homeothermic diver. The body temperature of water shrews with air entrapped in the fur dropped an average of 1.4°C in 30 seconds during dives beneath the surface of the water, whereas shrews with fur wet to the skin had a temperature drop of 4.5°C in the same time. The meticulous grooming and drying of the fur by a shrew after a dive is clearly highly adaptive and is more important than the added energetic expenses of diving with buoyant air entrapped in the fur.

Hinds et al. (1993) showed that for a given body mass there is no difference in cold-induced metabolism between metatherians and eutherians. Prolonged immersion in cold water, however, is especially difficult for all small mammals. In some species, heat loss from the extremities is reduced in water by countercurrent heat exchange. When a muskrat *(O. zibethicus)* swims, vasoconstriction and countercurrent systems keep its limbs at ambient temperature in cool and cold water, but vasodilation in the limbs allows for heat dissipation at a water temperature of 30°C or above (Fish, 1979). Because of the very high rate of thermal conductance in the limbs of these semiaquatic mammals, they would lose heat extremely rapidly when swimming if the limbs were kept near body-core temperature.

Some nearly permanent inhabitants of the sea, such as otariid seals, similarly use entrapped air as insulation, but many marine mammals (cetaceans, phocid seals, and walruses) lack insulative fur, and their bodies are in contact with water that may in extreme cases be 40°C below their deep-body temperature. How they maintain a constant body temperature under such demanding conditions is of considerable interest.

These marine mammals have a thick layer of subcutaneous blubber that forms an insulating envelope around the deep, vital parts of the body. A substantial amount of the weight of a marine mammal may be contributed by blubber. For example, in the small (75 kilograms) harbor porpoise *(Phocoena)*, 40 to 45 percent of the weight is blubber, and only 20 to 25 percent is muscle (Kanwisher and Sundnes, 1966; Koopman, 1998). Studies of seals by Irving and Hart (1957) have shown that the skin temperature varies directly with the water temperature down to 0°C. The cooled surface of the body and the thick blubber are an effective insulation, as indicated by the fact that the lower critical temperature of some seals is 0°C.

The young of seals and polar bears *(Ursus maritimus)* face especially severe thermoregulatory problems. These young must face extreme cold but are far smaller than the adults and thus lack the heat-conserving advantages of large size. The baby harp seal *(Phoca groenlandicus)* is born on drifting ice in the North Atlantic in winter and must survive air temperatures of −20°C or below. The pup weighs only 11 kilograms, whereas its mother may weigh as much as 140 kilograms. The pup has long fur, which offers better insulation than the short fur of the adult, and thermogenic adipose tissue yields energy during shivering and helps maintain the core body temperature (Blix et al., 1979). This adipose tissue is transformed to insulative blubber when the pup is several days old. Young harp seals use hypothermia (lowered body temperature) to conserve energy during extreme cold, wind, or rain. Their overall tolerance to cold is largely due to a high metabolic rate supported by rich (high fat, high energy) milk, shivering thermogenesis, and vasoconstriction in the skin (Blix et al., 1979).

Some of the most extreme thermal demands faced by endotherms are those met by cetaceans. Whales and porpoises live their entire lives in the water, and some species continuously occupy water at or near the freezing point. All cetaceans have insulating layers of blubber, but an extreme situation is faced by a small porpoise, which must maintain a deep-body temperature some 40°C higher than that of the sea, from which it is insulated by only 2 centimeters of blubber. An inflexible pattern of thermoregulation is inadequate even in inhabitants of the sea, which offers a relatively constant thermal environment. Some cetaceans migrate seasonally from cold waters to warm tropical seas. Because of the high thermal conductivity of water, skin temperatures generally equal water temperatures, and variations in water and skin temperatures of roughly 20 to 30°C may occur seasonally. The temperature of the body core, however, remains constant, and insulation requirements therefore may vary fivefold.

Gigantic differences in the ability of cetaceans to keep warm result from differences in body size and in thickness of blubber. The biggest whale is 10,000 times as heavy as the smallest porpoise, has roughly a 10 times greater mass-to-surface-area ratio that favors heat retention, and has a much thicker shell of blubber. Because of these differences, the whale has approximately a 100-fold advantage over the small porpoise in its ability to keep warm. The very factors working in favor of heat retention in the large cetaceans, however, are obviously disadvantageous under conditions of great activity or warm water. Because of the vast bulk of these animals, dissipation of heat is an acute problem. Sperm whales can dump heat directly by taking in water through the blowhole (Fig. 21-9). The water is then shunted down the right nasal passage to the nasofrontal sac. Along this route, the water passes by the highly vascular spermaceti organ. Excess heat from the blood in the capillaries of the spermaceti organ is dumped to the cold seawater flowing in the adjacent nasal passage, which is then expelled from the blowhole. The control of water flow is accomplished by the maxillonasalis muscle. Clearly, cetaceans must have considerable thermal versatility. How is this versatility achieved?

Although much remains to be learned, several points seem well established. First, metabolic rates of cetaceans differ markedly from species to species. The small porpoises have much higher basal metabolic rates than do large whales, far higher, in fact, than what would be predicted on the basis of weight. The harbor porpoise *(Phocoena phocoena)*, for example, metabolizes at about 1.6 times the predicted rate.

Second, blood flow through the well-developed vascular system in the flippers, dorsal fin, and flukes of cetaceans allows these structures to function effectively as heat dissipators under conditions of heat stress. The flow can apparently be shut

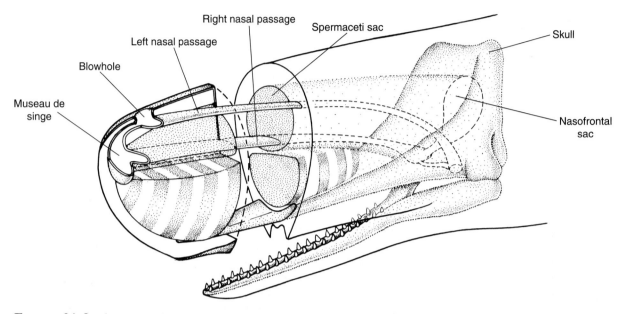

FIGURE 21-9   Sperm whales regulate their body temperature by directly dumping excess heat into the sea. Water can be taken through the blowhole into the nasal passages. As the seawater enters it passes through the spermaceti organ, which is highly vascularized with warm blood. Excess heat is dissipated from the blood to the seawater in the nasal passages and the warmed seawater is then ejected from the blowhole. The flow of water in the nasal passages is controlled by the large maxillonasalis muscle. *(After Clarke, 1979)*

down during cold stress, allowing for a minimum of heat loss from these surfaces.

Third, a remarkable series of vascular specializations allows for great variations in the thermal resistance offered by the blubber. A system of countercurrent heat exchangers in the vascular network supplying the blubber minimizes heat loss to the blubber and skin and hence to the environment (Fig. 21-6). In cetaceans, a second venous system in the blubber bypasses the countercurrent system during heat stress and allows considerable heat loss to the environment when heat dissipation is of prime importance. Similar countercurrent and by-pass systems occur in the flippers and fins. The extremities and much of the surface of the body can thus dissipate heat or can be maintained under an altered vascular supply that provides for maximal heat retention.

The great quantities of blubber on large whales (up to 20 centimeters thick) are seemingly not primarily useful as insulation. Because of their size, these animals could probably maintain a constant deep-body temperature with much less insulation.

These fat deposits may be useful primarily as food stores that can support an animal during periods of migration and fasting. It has been estimated that consumption of only half of a whale's blubber could fuel the basal metabolic rate for four to six months (Parry, 1949). Koopman (1998) showed that blubber in the harbor porpoise can be divided into two compartments, based on blubber thickness. The thoracic-abdominal blubber varies little in thickness around the girth of the animal and probably serves as insulation and energy storage. Blubber posterior to the anus, however, forms thick dorsal and ventral ridges but is very thin laterally. This unusual distribution suggests that post-anal blubber may serve primarily to maintain a favorable hydrodynamic shape of the peduncle.

COPING WITH HEAT

Some of the most severe problems in thermoregulation are those faced by mammals living in hot regions. In many low-latitude deserts, daytime surface and air temperatures in the summer rise well above the body temperature of most mammals. Under such conditions, heat from the environment is absorbed while the animals are producing considerable metabolic heat. In order to maintain thermal homeostasis, these animals must avoid the absorption of heat from the environment, dissipate such heat as it is absorbed, and lose endogenous heat. Unless the body temperature is elevated, these heat transfers must occur against a thermal gradient, from the relatively cool animal to the relatively hot environment. Such heat transfers invariably involve evaporative cooling, a luxury that most desert organisms cannot afford since they live in a region where water is in critically short supply. Nonetheless, even extremely hot and arid deserts are occupied by mammals, and some kinds, notably rodents, are quite common in such areas. A variety of physiological, anatomical, and behavioral adaptations have allowed mammals to inhabit these seemingly inhospitable regions.

BEHAVIORAL THERMOREGULATION

Most desert animals are never subjected to extremely high diurnal temperatures, nor are they able to survive them; their success is based on the ability to avoid extremely high temperatures rather than to cope with them. Perhaps the saving grace

of the desert is the great daily and seasonal fluctuation in temperature. Temperatures typically drop markedly at night, and winters are usually cool or cold. As a result, soil temperatures below the surface are never high, even in the summer, and nearly all desert rodents retreat to this refuge of coolness and relatively high humidity during the day. All but a very few desert rodents are strictly nocturnal, and all are more or less fossorial; these mammals are active above ground in the part of the circadian cycle when temperatures are lowest.

Consider, for example, the thermoregulatory behavior of the Namib Desert golden mole *(Eremitalpa granti),* which inhabits one of the hottest places on Earth. These golden moles spend the day buried in the loose dune sand and the nights actively foraging on the dune surface (Fielden et al., 1990). Daytime surface temperatures can exceed 50°C at midday during the summer months (November through February in the Namib Desert). Subsurface sand temperatures drop dramatically in the first 10 centimeters and are further reduced beneath clumps of perennial dune grass (Fig. 21-10). Like other chrysochlorids, *Eremitalpa* have lower than expected basal metabolic rates, maintain low body temperatures, and have higher than expected thermal conductances (Table 21-3; Fielden et al., 1990). Thus, heat exchange is enhanced while the moles are submerged in the

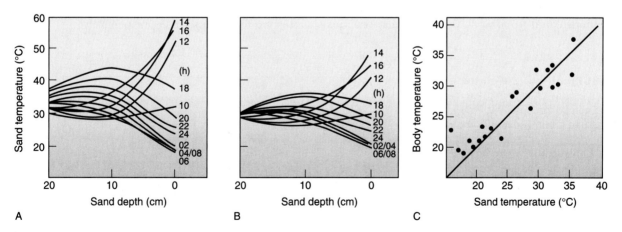

FIGURE 21-10   (A) The temperature of sand from the Namib Desert as a function of sand depth at various times of the day. (B) Sand temperature and depth beneath a clump of dune grasses at various times during the day. Both plots are for data collected in March. (C) The body temperature of Namib Desert golden moles *(Eremitalpa granti)* as they were removed from the sand and the temperature of the sand at their resting sites. *(From Fielden et al., 1990a)*

TABLE 21-3   Body Mass, Basal Metabolic Rate (BMR), Thermal Conductance (C), and Body Temperature (T_b) for Three Species of Golden Moles (Chrysochloridae)

Species	Mass (g)	BMR ($cm^3\ O_2g^{-1}\ h^{-1}$)	% Expected BMR	C ($cm^3\ O_2g^{-1}\ h^{-1}\ °C^{-1}$)	% Expected C	T_b (°C)
Eremitalpa granti[a]	26	0.52	22	0.30	158	34.7
Amblysomus hottentotus[b]	69	1.37	103	0.15	120	33.5
Chrysochloris asiatica[c]	36	1.20	62	0.25	151	35.0

[a]From Fielden et al., 1990.
[b]From Kuyper, 1979 and 1985.
[c]From Withers, 1978.

sand. These golden moles also choose submerged sites at depths that are within their thermal neutral zone and tend to favor the cooler sites underneath dune grasses. Fielden and his coworkers (1990) argue that the physiological characteristics of *Eremitalpa* are not adaptations to avoid thermal stress, which is alleviated by behavioral means, but evolved in response to a combination of factors. These factors include low oxygen levels and high CO_2 levels encountered while buried in loose sand, the high costs of burrowing, and the limited and often patchy distribution of prey.

Various other means of avoiding daytime heat are used by desert mammals. During the day, the white-throated woodrat (*Neotoma albigula*) remains in burrows insulated by piles of sticks and other debris. Frequently these houses are built in the shade of vegetation. Bighorn sheep (*Ovis canadensis*) and javelinas (*Pecari tajacu*) often take shelter in rock grottos or in the shade of steep rock outcrops, where for much of the day their body temperature is above air temperature and they can dissipate heat. Bovid horns are richly vascularized, and considerable heat can be lost across horn surfaces during hot weather (Bubenik and Bubenik, 1990). Horns also lose heat during cold months, and Picard et al. (1994) estimate that Barbary sheep (*Ammotragus lervia*) can suffer energy losses that exceed by 20 percent and 29 percent the energy output of a resting female and male, respectively.

FIGURE 21-11   Daily patterns of body temperature change in three mammals subjected to desert heat. Note that the antelope ground squirrel (*Ammospermophilus leucurus*) goes through a series of heating-cooling cycles during the day, while the camel slowly becomes hyperthermic as the day progresses. *(After Bartholomew, 1964)*

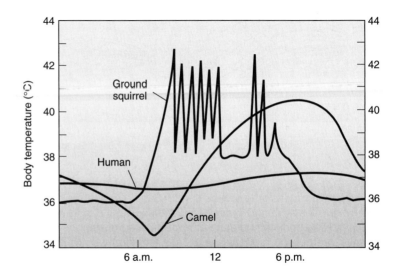

Several devices for lowering body temperature in bats were observed in animals under heat stress (Bartholomew et al., 1970). Vasodilation occurred in surfaces such as the scrotum, wing membranes, and ears. These naked surfaces are seemingly efficient heat dissipators. Other reactions to high temperatures were extension of the wings, fanning of the wings, and panting. Under intense heat stress, the animals salivated copiously and licked their bodies.

Seals and sea lions appear to bask in the sun, but, as might be expected, the very adaptations that enable a seal or sea lion to reduce heat dissipation in cold water make the animals unable to stand high temperatures. Whittow (1974) and his coworkers at the University of Hawaii studied the terrestrial thermal budget of California sea lions *(Zalophus californianus),* which inhabit some arid coasts of Mexico and the Galapagos Islands, where high temperatures occur regularly. At air temperatures of about 30°C, sea lions were unable to maintain a constant body temperature after the seawater had evaporated from their body surfaces. With continued exposure under experimental conditions, body temperature rose to slightly over 40°C. When the sea lions slept, their heat production dropped 24 percent, an obvious advantage for energy conservation and temperature regulation, but they were unable to dissipate sufficient heat in direct sunlight to avoid heat stress. Under stress, the animals fanned their flippers, which are known to sweat, thus increasing evaporative cooling. They also urinated and wet the underside of the body, thus further increasing evaporation. Under experimental conditions, however, these behaviors were inadequate and the animals were increasingly hyperthermic. Only when they were able to wet their bodies in the sea did body temperatures drop and stabilize.

Terrestrial heat production by a sea lion is dissipated approximately as follows: 2 percent is lost by respiratory evaporative cooling, 12 percent by evaporation from the skin, 52 percent by nonevaporative heat loss (conduction and convection) from the skin, and 15 percent by conduction from the parts of the body against the sand. Nineteen percent of the metabolic heat is stored, leading eventually to an elevation of body temperature. It is obvious why sea lions and other pinnipeds have difficulty staying out of the water for long periods

on a warm day, and increased activity on land at night is understandable. Although their physiological makeup limits the amount of time they can spend on land on a warm or hot day, by choosing the windy side of an island and by basking at sites where spray from breaking waves repeatedly wets them and increases evaporative cooling, sea lions can considerably extend their resting time on land.

EVAPORATIVE COOLING

Typical **panting,** which involves rapid and shallow respiration, is used entirely for heat dissipation and is an effective aid to temperature regulation. Laboratory studies of dogs, for example, indicated a tolerance of an ambient temperature of 43°C for at least seven hours (Robinson and Lee, 1941). Panting utilizes evaporative cooling of the mouth, tongue, and probably most important, the nasal mucosa (Schmidt-Nielsen et al., 1970). In the dog, and in many other mammals with a long snout and excellent olfaction, the turbinal bones of the nasal cavity are intricately rolled and provide a large surface area of nasal mucosa. This moist surface is ideal for evaporative dissipation of heat. The tongue is probably also important as a site of heat loss during panting. Blood flow to it increases sharply at the onset of panting and during heat stress increases six times over normal.

The resting respiratory rate of a dog is roughly 30 per minute, but this rate rises abruptly, with virtually no intermediate rate, to over 300 per minute during panting. The lateral nasal glands, which open some 2 centimeters inside the opening of each nostril, supply a major share of the water used in evaporative cooling during panting in dogs (Blatt et al., 1972). Under experimental conditions, the rate of secretion of one of these glands in a dog rose from no secretion at 10°C to 9.6 grams per hour at 50°C. Between 20 and 40 percent of the evaporative cooling during panting at high temperatures results from evaporation of the fluid from these glands. Because the glands are situated anterior to the turbinals, they tend to keep the nasal mucosa moist when air is drawn rapidly in through the nostrils during panting, enhancing evaporation.

Panting has several advantages over sweating. There is minimal loss of salt during panting, whereas salt loss during sweating (except probably in donkeys and camels) is always appreciable. In addition, adequate ventilation of evaporative surfaces always occurs during panting; in still air, however, sweating is seemingly not equally efficient. One potential disadvantage of panting is that the increased activity increases metabolism, thereby contributing more heat to be dissipated. Studies of respiratory frequency in dogs (Crawford, 1962) indicated that these animals pant at the resonant frequency of oscillation of the diaphragm (the natural frequency of vibration of this structure) and may therefore economize energy output. Considering water loss relative to total body surface area of a mammal, the amounts of water lost in sweating and panting are probably similar in many mammals. One apparent exception to this occurs in heteromyid rodents. Approximately 84 percent of the water lost by *Dipodomys merriami* is lost via the respiratory tract, compared to only 16 percent from **cutaneous evaporation** (Chew and Dammann, 1961; French, 1993). However, because desert heteromyids have lower metabolic rates (up to 25 percent lower) than many other rodent species and consequently pass less air over the respiratory surfaces per unit time, respiratory water loss is reduced. Both panting and sweating are obviously not effective means of cooling at high humidities. Because they spend the days in cool, humid burrows, the necessity of evaporative cooling is seldom experienced by *D. merriami*.

Thermoregulation at high temperatures in the rat kangaroo (*Potorous tridactylus*), a small and rather generalized macropodid, involves a specialized system of evaporative cooling (Hudson and Dawson, 1975). This metatherian weighs about 1 kilogram, and its metabolic rate and body temperature (36°C) are low relative to those of eutherian mammals of similar size. Thermal conductance from the well-furred body is low. (Studies by Dawson and Hulbert, 1970 and MacMillen and Nelson, 1969 have shown that a number of metatherians have body temperatures equivalent to those of eutherians but have metabolic rates that are about two thirds those of placentals of comparable size.) During exercise, at ambient temperatures below body temperature, heat is dissipated by the rat kangaroo primarily by panting, but at ambient temperatures approaching and exceeding body temperature, the bare tail, which contributes 9.4

percent of the total surface area, is a major route for heat loss. Vasodilation in the skin of the tail allows for increased nonevaporative heat loss, and at temperatures near and above body temperature, profuse sweating of the tail, but not of the body, produces rapid evaporative cooling. Constant side-to-side movement of the tail further facilitates evaporation. The maximum rate of sweating in the tail is extremely high, reaching 620 to 650 g/m^2/h, roughly double the highest measured rates in eutherians such as horses and cows.

HYPERTHERMIA

Some mammals have evolved physiological, anatomical, and behavioral strategies for tolerating long exposure to air temperatures higher than body temperatures. Large size itself is advantageous to mammals that must tolerate high temperatures. Because of the volume-to-surface-area ratio discussed earlier, the larger the animal the greater its ability to withstand exposure to high temperatures due to a relatively reduced surface area for heat gain. Stated differently, large animals have greater thermal inertia than do small ones. Of additional importance, just as insulation in the form of thick pelage slows the loss of body heat in low ambient temperatures, fur slows the penetration of heat to the body surface when temperatures are high.

Studies of temperature regulation in the camel (*Camelus dromedarius*) by Schmidt-Nielsen (1959) revealed a carefully regulated and highly adaptive circadian cycle of changes in body temperature. Fully hydrated camels in the Sahara Desert in winter, when cool temperatures (from roughly 0° to 20°C) prevailed, had fairly constant body temperatures that varied between 36° and 38°C. The fluctuations in body temperature were not random but followed the same pattern day after day, regardless of weather. In the summer, variations in body temperature were considerably greater; generally body temperature was between 34° and 35°C in the morning and reached a peak of approximately 40°C late in the day (Fig. 21-11). The camels seemed to be able to regulate their temperature but did so only above or below these extremes. When body temperatures reached 40.7°C, evaporative cooling in the form of sweating was used to dissipate heat and stabilize body temperatures.

Thus, during the day the camel accepted a heat load that sharply elevated its temperature. But during the relative coolness of the desert night, the heat stored during the day was passively dissipated and the body temperature dropped. Schmidt-Nielsen (1964) estimated that 5 liters of water would be required to dissipate the camel's load by evaporative cooling during a hot day. For an animal that does not have frequent access to water, such daily water loss would lead to fairly rapid dehydration. An additional advantage of high body temperature during the day results from narrowing the gap between environmental and body temperature; the smaller this temperature differential, the lower the rate of heat flow from the environment to the body.

Similar patterns of temperature fluctuation occur in the oryx *(Oryx)* and some other African antelope as a physiological adaptation to intense heat in desert or savanna areas (Fig. 21-12). The oryx frequently inhabits areas where no shade is available, and its ability to withstand a diurnal heat load is exceptional. Under experimental conditions, the oryx could withstand exposure to an ambient temperature of 45°C for 12 hours (Taylor, 1969a). During this period, the body temperature rose above 45°C and was sustained at this level for up to 8 hours with no injury to the animal. Rather than gaining heat from the environment, the oryx was actually losing heat. Such high body temperatures would kill most mammals fairly quickly, but circulatory specializations apparently allow the oryx to survive such extreme "overheating" by protecting delicate brain tissue from excessive heat.

Behavioral thermoregulation is a conspicuous part of the daily routine of most African rock-dwelling hyraxes. Field studies by Sale (1970) indicate the importance of behavior in adjusting to heat or coolness, and laboratory studies by Bartholomew and Rainy (1971) attest to the unusual system of body temperature regulation in the rock hyrax *(Heterohyrax brucei)*. The body temperature of normally active individuals varies from 35° to 37°C and is affected by ambient temperatures. The standard metabolic rate is some 20 percent below that expected on the basis of body mass, and the mean minimum heart rate (118 beats per minute) is 52 percent below the expected level.

Outside their nocturnal retreats, hyraxes adjust posture and location to exploit the environment to

FIGURE 21-12 Oryxes are adapted to life in arid zones. They have an exceptional ability to withstand overheating. *(Corel Corp.)*

maintain an appropriate body temperature. When the animals first emerge from their nocturnal retreats in deep rock crevices, they avoid extensive contact between their ventral surfaces and the cool rock, turn broadside to the first rays of the sun, and bask while maintaining a semispherical body form, which presumably prevents excessive heat dissipation. As the air and the rock begin to warm, the hyraxes sprawl on the rock, presenting a large surface area to the sun (Fig. 21-13A). Bartholomew and Rainy (1971) found that the lowest body temperatures of hyraxes were reached shortly before sunrise, despite their huddling together during the night. The basking utilizes solar rather than metabolic heat to raise body temperature. After basking, the hyraxes generally feed.

A   B

FIGURE 21-13   Hyraxes behaviorally thermoregulating. (A) Two *Procavia capensis*
(above) and two *Heterohyrax brucei* (below) basking with their bodies broadside to the
early morning sun. (B) Animals of both species huddled together on a cool day. *(H. Hoeck)*

As the temperature rises abruptly during the morning, the hyraxes often move first to the dappled shade beneath the sparse foliage of trees or bushes and then, when ambient temperatures reach 30°C or above, to deep shade. During these hot times, the hyraxes often lie full length on the rock. At peak ambient temperatures during the afternoon, they tend to remain on deeply shaded rock. Before dark, they may again be sprawled out, but this time on warm rock in the open. Solar radiation and shaded sites seemingly provide means of passively absorbing and dissipating heat, respectively.

Of particular importance is the fact that the rocks on which the hyraxes live provide an auxiliary means of adjusting body temperature. The rock outcrops form massive heat sinks with vastly more thermal inertia than air. The hyraxes then use the heat from the rock to compensate for body heat lost to the air. On cool, cloudy days, hyraxes huddle tightly together, thus reducing the surface area exposed to the air (Fig. 21-13B).

EXERTION AND HEAT STRESS

During heavy exercise, mammals produce heat much faster than it can be dissipated. Indeed, for moderate-size mammals (5 to 200 kilograms), the most intense thermal stress encountered is during heavy exertion. The rise in deep-body temperature during exertion in mammals is far more rapid than that in resting mammals in desert heat. Taylor (1974) calculated that the rate of excess heat production in the domestic dog during heavy exercise is ten times the highest possible heat gain the dog could face in the hottest desert.

During running, when muscular effort increases drastically, metabolic heat production may greatly exceed the ability of the animal to dissipate heat to the environment. Under these conditions, the ability to store heat by elevating body temperature can be highly advantageous. Cheetahs (*Acinonyx jubatus*), the fastest land mammals, inhabit many hot parts of Africa. During a sprint to capture a gazelle, a cheetah produces nearly 50 times the heat produced at rest (Gordon et al., 1977). This tremendous heat load cannot be dissipated by evaporative cooling alone, as it would be at rest, and must, therefore, be temporarily stored, causing body temperature to rise. Consequently, cheetahs can pursue prey only over short distances before they must stop to prevent overheating. Similarly, the gazelle fleeing from a cheetah stores heat. Gazelles running at low speeds store relatively little heat because they can dissipate much of the heat produced by muscular work by evaporative cooling (sweating) while they run. At speeds of at least 80 kilometers per hour, such as those encountered when a gazelle is pursued by a cheetah, over 70 percent of the metabolic heat is stored and body temperatures rise. Bovids can tolerate elevated body temperatures in part because they regulate their brain temperatures at a lower set point than their body temperature.

Because the mammalian brain begins to function abnormally at temperatures only 4° to 5°C above resting body temperature, shielding the brain from high temperatures during exercise is vital. In the oryx, in Thompson's gazelles *(Gazella thompsonii),* and probably in many other antelopes, the brain is provided with a specialized countercurrent cooling system of its own in the sinus cavernosus (Taylor, 1969a). The external carotid artery, on its way to the brain, divides into many branches in this sinus, and these branches are in close proximity to veins returning from the nasal passages (Fig. 21-14). This system is called a **carotid rete.** These veins carry relatively cool blood because evaporative cooling of the nasal mucosa cools the blood supplying these surfaces. Countercurrent heat exchange in the carotid rete assures that the blood supply of the brain is cooler than that of most of the rest of the body. During rapid running, the gazelle is able to keep its brain at a temperature 2.7°C below that of the blood in the

carotid artery (Fig. 21-15). A similar countercurrent system involving a carotid rete occurs in the domestic sheep, goat, cat, and dog and probably occurs widely. In free-ranging wildebeest *(Connochaetes gnou)* and springbok *(Antidorcas marsupialis),* selective brain cooling serves to help modulate thermoregulation but may not play a significant role in protecting the brain from thermal stress (Mitchell et al., 1997). Selective brain cooling occurred only during moderate increases in body temperature and was abandoned entirely when physical activity was highest. These results from free-ranging animals suggest that selective brain cooling does not protect the brain during strenuous activity.

Hyperthermia during running may be an important adaptation in a number of mammals. Taylor et al. (1971), found that the body temperature of a running African hunting dog *(Lycaon pictus)* is higher than that of the domestic dog (41.2°C versus 39.2°C) and the percentage of heat lost by respiratory evaporation produced by running is much lower (25.1 percent versus 49.7 percent). Taylor and his coworkers suggest that hyperthermia and the greatly reduced pulmonary water loss of the

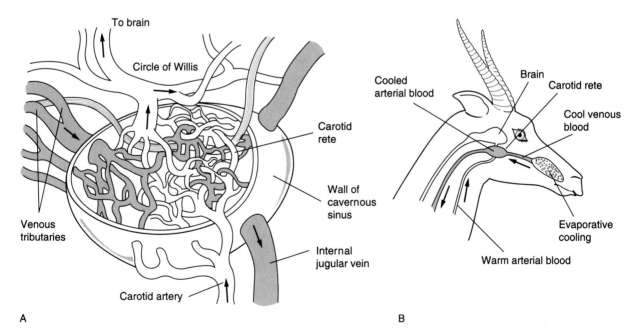

FIGURE 21-14 Cutaway diagram of the carotid rete of a gazelle (A) and a head of a gazelle (B) showing the location of the carotid rete used in cooling the blood to the brain.

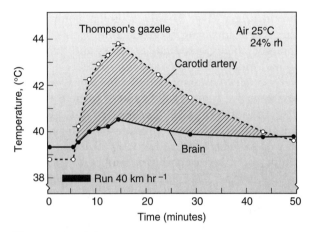

FIGURE 21-15 Temperature of the brain and carotid artery of a Thompson's gazelle *(Gazella thompsonii)* before, during, and after a run of 7 minutes at 40 kilometers/hour. The temperature of the carotid artery rose sharply after the gazelle began to run and exceeded the brain temperature until 40 minutes after the run. *(After Taylor and Lyman, 1972)*

(1979) found CO_2 levels in burrows to be ten times atmospheric levels. A dispersed or scarce food supply may pose added problems. In mammals, the demands of burrowing have favored small size, a compact, fusiform shape, specialized digging structures, and modified sensory organs. Less obvious, but of decisive importance, are physiological and behavioral adaptations.

A fascinating example of such adaptations is offered by the naked mole-rat *(Heterocephalus glaber,* Bathyergidae). This eusocial mammal (p. 473; see Fig 18-26) lives in large colonies, of usually 70 to 80 animals, in semideserts of East Africa. The workers average only 32 grams. These rodents have high thermal conductance (because they are naked), a basal metabolic rate less than 60 percent of the expected rate, low and labile body temperatures, and are essentially **poikilothermic.** The narrow thermoneutral zone (31° to 34°C) is within the range of dry-season burrow temperatures.

African hunting dog allow this animal to conserve water and maximize the distance it can chase prey.

FOSSORIAL MAMMALS: PHYSIOLOGICAL PROBLEMS

Several families of insectivorous mammals (Notoryctidae, Chrysochloridae, Talpidae) and a number of rodent families (Geomyidae, Spalacidae, Muridae, Octodontidae, Ctenomyidae, Bathyergidae) have fossorial members. Fossorial life in sealed burrows offers several advantages, including stable (and often moderate) temperatures, high humidities, and safety from predators. But there are major liabilities, such as the high energetic cost of burrowing. Vleck (1979) estimated that burrowing a given distance consumes 360 to 3400 times as much energy as moving the same distance on the surface. A North American mole *(Scapanus townsendii)* has been estimated to expend over 5000 joules of energy to tunnel one meter but only 9 joules to walk a comparable distance (Gorman and Stone, 1990). Also, microclimate factors can be more variable in a burrow than above ground: in burrows rapid changes in oxygen levels (6 to 21 percent) and carbon dioxide concentrations (0.5 to 4.8 percent) can occur after heavy rain or rapid digging (Withers, 1978). Schaefer and Sadleir

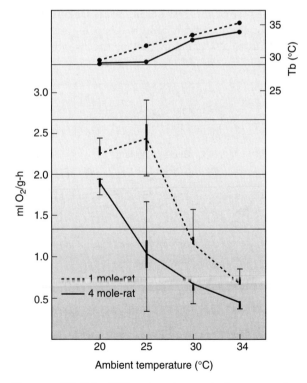

FIGURE 21-16 Differences in metabolic rates and body temperatures of naked mole-rats *(Heterocephalus glaber),* resting alone or huddling together in groups of four, at different ambient temperatures. *(After Jarvis, 1978, vol. 6,* Bulletin of the Carnegie Museum of Natural History)

Jarvis (1978) found that the metabolic rate of naked mole-rats increases modestly in response to temperatures below 30°C (Fig. 21-16). However, the decline in body temperatures and metabolic rates of individual mole-rats resting at ambient temperatures between 20° and 25°C indicates that these animals abandoned their attempts at physiological thermoregulation.

Jarvis (1978) views this animal's behavioral and physiological peculiarities as part of a strategy for surviving under high temperatures and limited food. The low metabolic rate and high thermal conductance may be prerequisites for energy-saving behavioral thermoregulation. When heat stressed from burrowing, the mole-rats can passively unload heat from their naked bodies by moving to a cool section of the burrow; when cold, they can warm passively by "basking" in a warmer section. Mole-rats reduce the cost of thermoregulation by huddling together when resting.

One type of "subterranean rodent syndrome" (low basal metabolic rate, high and extremely narrow thermoneutral zone, high thermal conductance, limited thermoregulatory ability) is well illustrated by some members of the bathyergid genus *Cryptomys* (Bennett et al., 1993; Bennett et

al., 1994b). These social mole-rats range from equatorial Africa to 35° south latitude, occupy semiarid deserts and mesic tropics, and differ markedly in size between species (60 to 272 grams); yet most share the above suite of adaptations (Table 21-4). Such characteristics also occur among a diverse group of other rodent families or subfamilies (Spalacinae, Rhizomyinae, Geomyidae, Ctenomyidae; McNab, 1966; Reig, 1970).

Viewed broadly, then, the trend is toward fossorial rodents of warmer areas displaying the "typical" subterranean rodent physiology described above for *Cryptomys,* but at least some of those from colder areas depart variously from this pattern. Yet even among warm-area species there are departures: some of the social bathyergid mole-rats have high resting metabolic rates (Bennett et al., 1993; Buffenstein and Yahau, 1991). The frequent deviations from the central pattern suggest that there is no single subterranean rodent syndrome.

Fossorial insectivores display diverse adaptations. The talpid moles that are fossorial occupy temperate areas and have high metabolic rates and low thermal conductance; they are homeotherms that neither hibernate nor estivate. But the southern African golden moles (Chrysochloridae) are at

TABLE 21-4 Thermoregulatory Characteristics of Mole-Rats of the Genus *Cryptomys*

Species or subspecies	Mean body mass (g)	Mean body temperature (°C)	RMR $(cm^3O_2g^{-1}h^{-1})$	TNZ[a] (°C)	Conductance $(cm^3O_2g^{-1}h^{-1}°C^{-1})$	Social status	Habitat
C. hottentotus darlingi	60	33	0.98	28–31.5	0.19	social	mesic
C. h. hottentotus	75	34	0.90	27–30	0.13	social	semiarid
C. h. amatus	77	33.8	0.63	28–32	0.12	social	mesic
C. bocagei	94	33.7	0.74	31.5–32.5	0.12	social	mesic
C. h. natalensis	102	33.8	0.80	30–31.5	0.13	social	mesic
C. damarensis	131	35	0.66	28–31	0.065	social	arid, semiarid
C. mechowi	272	33.7	0.60	29–30	0.09	?	mesic

From Bennett et al., 1994.
RMR = resting metabolic rate; TNZ = thermal neutral zone.
[a]Note the extremely narrow thermal neutral zones.

the opposite thermoregulatory pole. The Namib golden mole has a metabolic rate only 22 percent of the predicted rate, a high and narrow thermoneutral zone (31° to 35°C), and an extremely labile body temperature (19° to 38°C). The high thermal conductance favors diurnal torpor by allowing the body temperature to assume rapidly the temperature of the surrounding sand. Temperatures below 15°C are lethal. Low oxygen concentrations in the sand and a sparse and clumped food supply (mainly termites) have favored energy conservation in this golden mole (Fielden et al., 1990). In the temperate talpids, by contrast, a dependable, yearlong supply of soil invertebrates seems to pay the higher energetic costs of homeothermy.

Many fossorial mammals are convergent regarding respiratory and tissue adaptations (Nevo, 1979). Typically, the blood has a high affinity for oxygen. In the blind mole-rats (*Spalax;* Muridae), this is due to a high erythrocyte count and small corpuscle volume (Ar et al., 1977). In both the North African *Spalax* and the North American pocket gophers (Geomyidae), the myoglobin content of skeletal muscles is high (as in marine mammals), favoring rapid oxygen diffusion from capillaries to mitochondria (Lechner, 1978). Respiratory rates of *Spalax* and *Talpa* (an insectivore) are 40 percent below expected levels, and both undergo greatly increased rates at high carbon dioxide concentrations (Stahl, 1967). The resting heart rates of these animals are also low, but rise sharply under low oxygen levels. These respiratory and tissue adaptations are essential to mammals that regularly experience adverse concentrations of low oxygen and high carbon dioxide and are probably common to most fossorial mammals.

ENERGY COSTS OF LOCOMOTION

A 10-kilogram coyote and a 400-kilogram horse have similar top speeds, but at such speeds how do these animals compare with regard to the expenditure of energy? Some time ago, Hill (1950) made a series of predictions as to how energy use during running would change with the size of the runner. (Energy consumption by the muscles of a running animal is generally regarded as the result of the transformation of chemical energy to mechanical

energy.) Hill reasoned that, although large and small runners could often reach similar top speeds, the rates of work and energy use at these speeds would be higher in the small runners. His logic was that, whether an animal is small or large, each gram of muscle performs the same amount of work and consumes the same amount of energy during a stride, but the short legs of the small animal have to take many strides to cover the same distance covered in one stride by a large animal. When large and small mammals run at the same speed, then, the small ones should have the higher stride rates and should consume more energy per unit of body weight.

These proposed relationships have been studied experimentally by C. Taylor and his associates (Fedak et al., 1982; Heglund et al., 1982; Taylor et al., 1970; Taylor et al., 1982), who used mammals ranging in weight from 21 grams (a house mouse, *Mus musculus*) to 254 kilograms (African cattle, *Bos taurus*). Several important relationships were demonstrated by these studies:

- At the trot-gallop transition speed, the amount of energy used per stride per gram of muscle is nearly constant over a wide range of body sizes (Table 21-5). Mouse, baboon, and horse all expend nearly the same amount of energy per gram of muscle during a stride.
- The metabolic cost of muscle action in running animals increases linearly with speed. As shown in Fig. 21-17, the amount of energy a mammal expends increases as running speed increases.
- The mass-specific (per gram) use of energy by a running animal decreases as a function of weight (Fig. 21-17), varying as the -0.3 power of body mass. Thus, when a chipmunk and a horse are running at the same speed, each gram of chipmunk uses 15 times more energy than each gram of horse.

Several probable explanations of these relationships are available. Seemingly, muscular force is generated and dissipated more rapidly as an animal runs faster. With increasing speed, more muscle fibers that have a rapid contraction-relaxation cycle are brought into play. Each cycle uses a unit of energy, and the increase in the cost of rapid locomotion perhaps results from the increased use of rapid-cycling muscle fibers.

Muscular force must be generated and dissipated more rapidly in small mammals than in

TABLE 21-5 Speed, Stride Frequency, and Metabolic Energy Consumed at the Trot-Gallop Transition by Mammals of Three Size Classes

Body mass (kg)	Speed at trot-gallop transition (m/s)	Stride Frequency at trot-gallop transition (stride/s)	Energy used (J/kg/stride)
0.01	0.51	8.54	5.59
1.0	1.53	4.48	5.00
100	4.61	2.35	5.53

Values calculated by Taylor et al. (1982) from equations given by Heglund et al. (1974).

larger ones because, at comparable speeds, small mammals have higher stride rates. The muscles of small mammals contain "faster" fibers that have rapid contraction-relaxation cycles and use energy at a high rate. The decrease in the use of "fast" muscle fibers accompanying increased body size may partially account for the mass-specific decrease in the cost of running in large mammals.

Fedak et al. (1982) found that, during high-speed running by large mammals, the energy ex-

pended by the muscles was not sufficient to provide the work necessary for the total kinetic energy (energy of motion) developed. The authors concluded that the storage of energy by muscles and tendons and its release by elastic recoil provide a significant part of the total kinetic energy. In large mammals, elastic recoil is of considerable importance in locomotion; its contribution to locomotor efficiency in small mammals is unknown (Alexander, 1992).

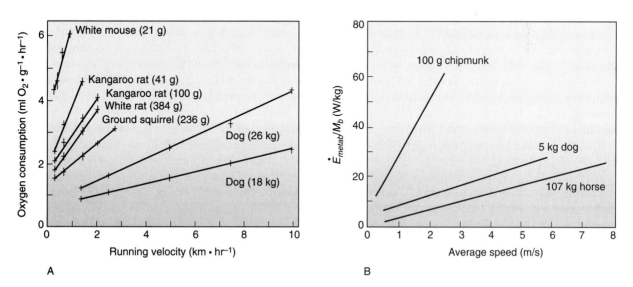

FIGURE 21-17 (A) Oxygen consumption by mammals of different sizes at various running speeds. (B) Mass-specific use of energy (use of energy per unit of weight) at various speeds by animals of different sizes. The notation E_{metab}/M_b (W/kg) = metabolic energy consumed in watts per kilogram of body weight. A watt is a unit of power equal to about 1/746 of an English horsepower. *(A from Taylor et al., 1970, p. 1105, by permission of the* American Journal of Physiology; *B from Fedak et al., 1982,* Journal of Experimental Biology, *Company of Biologists, Ltd.)*

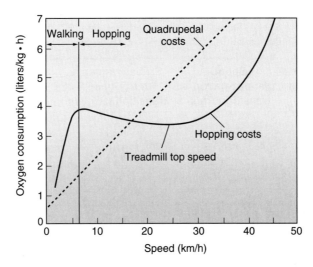

FIGURE 21-18   The relationship between metabolic rate and hopping speed in red kangaroos. Dotted line shows the relationship in a similarly-sized quadrupedal mammal. *(After Dawson, 1995)*

These relationships provide at least a partial explanation for the scaling of size in cursorial mammals. Artiodactyls, perissodactyls, and cursorial carnivorans generally weigh at least 10 kilograms and thus expend less energy per gram of body weight during running than smaller mammals. The mechanical problems associated with the support and propulsion of great weight may set upper limits on the size of runners, but some quite heavy mammals (such as 500-kilogram horses) are rapid and enduring runners. The swiftest cursors, however, generally weigh from about 50 to 125 kilograms (antelope, for example), and the cheetah, arguably the fastest of all runners, weighs between 50 and 65 kilograms.

Bipedal locomotion, involving leaping or bounding, has evolved independently many times in the class Mammalia. Questions as to the relative energy cost of bipedal versus quadrupedal locomotion have inspired much controversy. Some researchers have claimed that, over a considerable range of hopping speeds in bipeds (such as kangaroos), there is no increase in energy cost attending increasing speed (Fig. 21-18). This aerobic plateau is thought to represent an energetically more efficient mode of locomotion in bipeds, involving elastic storage in the hindlimb tendons. Research shows that kangaroos and wallabies can store and

recover up to 60 percent of the energy required to hop via elastic recoil of the large Achilles tendons in the hindlimbs (Alexander, 1982; Biewener and Baudinette, 1995).

A female kangaroo has the added burden of transporting a young joey in her pouch for several months. At the time the young permanently exits the pouch, the joey can weigh approximately 4.5 kilograms or about 20 percent of the mother's weight. The smaller Tammar wallaby (*Macropus eugenii*) carry pouch young equivalent to 15 percent of the mother's body weight. When 15 percent of each mother's body weight was artificially added to the pouch and the females were trained to run on a treadmill, they showed no increase in metabolic rate over speeds up to 4.5 meters per second (Baudinette and Biewener, 1998). Thus kangaroos and wallabies carry their young at no extra energetic cost because the added weight is recovered by the elastic storage of the tendons.

Other research indicates, however, that no such energy savings is associated with bipedal hopping in smaller bipeds (less than 3 kilograms; Thompson et al., 1980). MacMillen and Hinds (1992) compared metabolic patterns and running speeds of size-matched pairs of biped and quadrupedal heteromyid rodents running on a treadmill. Unlike their quadrupedal kin, bipedal heteromyids show a plateau in oxygen consumption at higher running speeds (over 3 to 4 km/h). However, this plateau is an anaerobic plateau accompanied by high blood lactate levels and does not contribute to energetic savings (MacMillen, 1983; MacMillen and Hinds, 1992). In addition, there is no energetic cost difference among running bipedal and quadrupedal heteromyids. Bipeds (*Dipodomys* and *Microdipodops*), however, do appear much more "willing" to run anaerobically at higher speeds, thereby creating an artificial "bipedal plateau." Although hopping is no less costly than running, the erratic turns made by hopping small mammals to avoid predators may add to the energy costs of hopping.

One of the most energetically expensive forms of locomotion is burrowing. Rodents that burrow through compacted soils (such as *Thomomys*, *Cryptomys*, or *Heterocephalus*) expend considerable energy to scrape away the soil and remove it from the burrow. The net cost of transport for the sand-swimming Namib mole (*Eremitalpa granti*), on the other hand, is much less than that for burrowing rodents (Seymour et al., 1998). Sand-swimming involves pushing through loose sand, which collapses behind the animal, leaving no tunnel. The energetic cost of sand-swimming is approximately 80 times higher than the cost of running on the surface (Fig. 21-19). Nevertheless, sand-swimming in the Namib mole is still an order of magnitude less expensive than burrowing through compacted soil is for *Thomomys bottae* (Vleck, 1979 and Seymour et al., 1998).

The metabolic costs of aquatic and aerial locomotion depend on very different principles. Aquatic mammals swim in an environment that is 800 times as dense as air. The fluid environment, therefore, provides sufficient buoyancy to offset much of the gravitational force. Because of the high density of water, however, drag in water is far higher than that in air (drag is the fluid force acting opposite to the direction of thrust or forward movement). Drag is reduced when the body form is **streamlined** for all organisms with moderate to high **Reynolds numbers** (for all mammals). The Reynolds number is a dimensionless ratio relating the force of inertia of the medium to its viscosity (Vogel, 1988). Velocity is squared in the calculation of drag and again in determining the Reynolds number, and, consequently, the drag on moderate

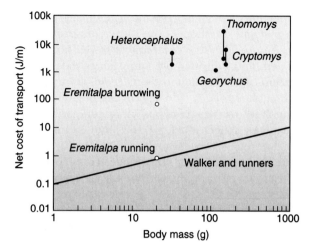

FIGURE 21-19 Net cost of transport of burrowing via sand-swimming and running on the dune surface in the Namib mole *(Eremitalpa)*. Similar costs for several species of mole-rat and one species of pocket gopher *(Thomomys)*, which burrow through more compacted soils, are shown for comparison. The solid line denotes the cost of transport for mammals in general. (*After Seymour et al., 1998*)

to fast swimmers increases very rapidly with each increment of speed. Because metabolic rate must be nearly doubled every time swimming speed increases by one body length/sec, there are metabolic limits to swimming speed. Drag can be reduced by streamlining. Seals, sea lions, dolphins, and cetaceans all have streamlined body shapes in water. The turbulence at the body surface that contributes to drag is further reduced because the bodies of aquatic mammals are relatively compliant (deformable at the surface). Streamlined, highly compliant bodies reduce drag to a minimum. The result is that large aquatic mammals develop more power relative to drag and swim faster than smaller mammals of the same body shape.

The energetic cost of swimming also depends on the mode of swimming employed and the position of the body in the water column. Mammals such as the platypus swim using a rowing motion of the forelimbs, while muskrats and beavers paddle with their hindlimbs underneath their bodies (Fish, 1993). Both rowing and paddling are relatively inefficient because thrust is generated only during half the stroke. Sea lions also use rowing motions of the foreflippers to generate thrust, resulting in maximum mechanical efficiencies of

80 percent (Feldkamp, 1987b). The surface area of the platypus forefoot is over 13 percent of its total surface area, a value comparable to sea lion foreflippers and considerably larger than the 4 to 6 percent of semiaquatic rodents (Feldkamp, 1987a; Fish et al., 1997). In addition to the greater thrust generated by the platypus' forefeet, swimming platypuses spend much of the time submerged, which reduces drag (Evans et al., 1994). Bow waves associated with surface swimming increase drag up to fivefold, resulting in considerably higher metabolic costs. Williams (1989), for example, reported a 41 percent savings in oxygen consumption and a 35 percent reduction in drag for a sea otter swimming submerged compared to one swimming at the surface.

Aerial locomotion involves some of the same physical principles described for aquatic locomotion, except that the major cost is due to gravity and not to the density of the medium. Velocity for gliding flight, such as that exhibited by flying squirrels (Rodentia), honey possums (Metatheria), and colugos (Dermoptera), is produced by gravity. Unlike flapping flight, there is relatively little metabolic cost to gliding because muscular effort is only needed to hold the gliding surface rigid and the animal loses altitude at rates between 1 and 2.5 m s^{-1} (Alexander, 1992). Flapping flight is discussed in Chapter 10.

WATER REGULATION

Roughly 35 percent of the earth's land surface is desert, where water is the primary limiting factor for plant and animal life. Desert areas are characterized by intense solar radiation by day and maximal heat loss by night, resulting in large daily variation in air temperature (commonly up to 30°C) in the summer, extremely low humidity through most of the year, and small amounts of precipitation, often at irregular intervals. On a summer day, the searing dry winds, radiation, and reflection of heat from the hot, pale soil add to the harshness of the desert environment. Few environments on earth are as hostile to life, and to the casual observer the desert gives the impression of overwhelming sterility. This impression is deceptive, however, for, in reality, the desert supports a great variety of life. Most mammals who live on the

desert remain hidden in shelters by day and forage in the relative cool of night.

Water is absolutely essential to life; to all mammals, life depends on the maintenance of an internal **water balance** within fairly narrow limits. (Water balance results when water intake, through drinking, eating, and production of **metabolic water,** equals water loss by evaporation from skin and lungs, defecation, and urination.) Mammals are approximately two-thirds water by mass. Most mammals are stressed when water loss reduces their body weight by as little as 10 or 15 percent, and death occurs in many mammals when such loss reduces the body weight by 20 percent. Loss of water occurs rapidly on the desert: water loss in a human on a hot summer day in the southwestern deserts of the United States has been recorded as 1.41 percent of body weight per hour; comparable figures for the donkey and dog are 1.24 and 2.62, respectively (Schmidt-Nielsen, 1964). Deprived of drinking water, a human or a dog can survive only a day or two of exposure in the summer. Nonetheless, some small desert rodents live without drinking water and must satisfy their water needs by utilizing water in their food and water derived from the metabolism of food. Similarly, some large mammals of desert areas must maintain water balance with only occasional access to drinking water. Although much remains to be learned about mammalian adaptations for water conservation in arid environments, excellent studies have provided a solid base of knowledge.

A number of solutions to the problem of maintaining water balance are used by desert mammals. These solutions depend on seasonal weather patterns, size of the animal, timing of activity cycles, diet, and a variety of behavioral, structural, and physiological features. The following discussions do not cover the subject of water conservation in mammals exhaustively but consider the adaptations that permit some mammals to maintain water balance in dry environments.

PERIODIC DRINKERS

In many arid or semiarid regions, scattered water holes or widely separated rivers offer water to mammals that can move long distances. The extensive grasslands of Africa form such an area, as did the North American Great Plains before the coming of settlers. Most large mammals in such areas probably drink every day or two in hot weather and seemingly are unable to survive for long periods without drinking. A few ungulates, such as the camel, however, occupy an intermediate position with regard to water needs. Although they can go for moderate periods without drinking, these mammals are not independent of drinking water, as are some desert rodents, and must drink water periodically.

Our present knowledge of the water metabolism of the camel is largely a result of the work of Schmidt-Nielsen et al. (1956, 1957). Their work, done in the northwestern Sahara on local domesticated camels, substantiated the popular idea that camels can tolerate long periods without drinking water, but more important, Schmidt-Nielsen and his group explained the adaptations allowing this tolerance. The ability of their experimental animals to tolerate dehydration was remarkable. One camel went without water for 17 days in the winter on a diet of dry food. During this period, it lost 16.2 percent of its body weight. In some areas, camels that foraged on native vegetation in the winter were never watered. Two camels kept without water for seven days in the heat of the summer lost slightly over 25 percent of their body weight. All of these animals drank tremendous amounts of water after their periods of dehydration, and none showed ill effects.

The camel economizes on water in several ways. Its body temperature drops sharply at night and then rises slowly during the heat of the day (Fig. 21-11). It is able to tolerate considerable hyperthermia, and typically the day is largely over before the animal's body temperature rises to levels at which evaporative cooling, in the form of sweating, must occur. Thus, relative to humans under similar conditions, very little moisture is expended each day in cooling the camel. Like the oryx, excess heat gained by day is lost passively at night by the camel. Further water saving results from the modest ability of the kidneys to concentrate urine and from the absorption of water from fecal material. Despite these important water-saving adaptations, however, the camel loses water steadily through evaporation from lungs and skin and in the urine and feces. Its ability to tolerate tremendous water loss (up to 27 percent of body weight) during these periods of dehydration is remarkable.

Apparently the proportions of water lost from various parts of the body differ in humans and camels. When a person in the desert has lost water equal to about 12 percent of body weight, the blood becomes viscous. As a result, the heart has difficulty moving the blood and the rate of blood circulation decreases. This leads to a marked reduction in the rate of dissipation of metabolic heat, to a sudden rise in body temperature, and death. However, in a camel that has lost 20 percent of its body weight due to water loss, water content of the plasma is nearly normal, but large amounts of water are lost from **interstitial fluid** and from intracellular water. In a camel deprived of water for eight days, interstitial fluid volume decreased 38 percent and intracellular water volume fell 24 percent. However, plasma volume decreased only 10 percent (Schmidt-Nielsen, 1964). Although the camel becomes strikingly dehydrated during periods without water, the blood apparently retains its fluidity and its ability to contribute to heat dissipation without straining the circulatory system. The donkey, which was also studied by Schmidt-Nielsen (1964), proved to be as capable as the camel of tolerating dehydration. The donkey lost water 2.5 times faster than the camel, however, and could not be independent of water for more than a few days.

DIETARY MOISTURE

A number of mammals that occupy deserts or semiarid areas are no better adapted to surviving without considerable moisture in their diet than are mammals of fairly moist areas. Even in some areas with fairly high precipitation, small mammals do not have regular access to drinking water and, as in the case of some desert rodents, satisfy their water requirements by eating moist food.

Succulent plants provide water for some desert rodents, such as the white-throated woodrat (*Neotoma albigula*), which occupies the hot deserts of the southwestern United States and northern and central Mexico. Paradoxically, this rodent needs large amounts of water, which it obtains largely from cactus. The desert woodrat (*N. lepida*) and the cactus mouse also utilize large quantities of cactus (*Opuntia*) as a source of both food and water (MacMillen, 1964a). These mammals have evolved the ability to cope metabolically with oxalic acid, a compound abundant in cactus and toxic to some mammals (Schmidt-Nielsen, 1964). The ability to obtain water from cacti and to deal with oxalic acid is not limited to the rodents mentioned above, all of which belong to the family Muridae, but also occurs in the rodent family Geomyidae, the pocket gophers. The northern pocket gopher *(Thomomys talpoides)*, inhabiting fairly dry short-grass prairies of Colorado, obtains water by eating prickly pear cactus (Vaughan, 1967).

Some desert rodents obtain water from succulent plants that contain high salt concentrations. These rodents have kidneys that are able to produce highly concentrated urine (urine that has little water relative to the contained solutes). The North African sand rat (*Psammomys obesus*, Muridae) is such an animal. The sand rat obtains water from the fleshy leaves of **halophytic** plants (plants that grow in salty soil), which grow along dry river beds in the desert (Mares et al., 1997; Schmidt-Nielsen, 1964). These leaves are 80 to 90 percent water but contain higher concentrations of salt than seawater and also have large amounts of oxalic acid. In order to utilize this water source, the sand rat must produce urine with extremely high concentrations of salt and must be able to metabolize large quantities of oxalic acid. The Australian hopping mouse, *Notomys cervinus* (Muridae) and a South American desert-dwelling rodent, *Eligmodontia typus,* have remarkably well-developed abilities to concentrate salts in their urine and probably use the succulent but highly saline leaves of halophytic plants as a water source (MacMillen and Lee, 1969; Mares, 1977).

Most deserts support a number of carnivorous and insectivorous mammals whose moisture requirements are seemingly met by the water in their food. The grasshopper mouse *(Onychomys)*, a small rodent widely distributed in the deserts and semi-arid sections of the western United States and Mexico, is almost exclusively insectivorous at some times of the year. This mouse has thrived in the laboratory on an entirely meat diet, with no drinking water (Schmidt-Nielsen, 1964). Schmidt-Nielsen (1964) found that the desert hedgehog (*Hemiechinus auritus,* an insectivoran) and the fennec (a fox), both inhabitants of North African deserts, could get adequate water from a predominantly carnivorous diet, as could the mulgara *(Dasycercus cristicauda),* an Australian dasyurid metatherian (Schmidt-Nielsen and Newsome, 1962). The fennec can maintain water balance for at least 100 days on a diet of mice and no drinking water. This small animal has an unusually low rate of evaporative water loss and equals water-independent desert rodents in its ability to concentrate urine (Noll-Banholzer, 1979b; Table 21-6), a capacity which enables it to excrete little water with the large amounts of urea produced by its high-protein diet.

Few large ungulates inhabit barren deserts where no drinking water or cover is available. One notable exception is the oryx, or gemsbok (Fig. 21-12), a large antelope that occurs in arid and semi-arid sections of Africa and even lives in the extremely dry Namib Desert. More remarkable than the amazing ability of the oryx to withstand intense desert heat (p. 385) is the animal's lack of dependence on drinking water. Careful studies by Taylor (1969a) showed that the water needs of the oryx are probably satisfied by its food, which consists of grasses and leaves of shrubs that by day may contain as little as 1 percent water. After nightfall, as the temperature drops and the humidity rises, these parched leaves absorb moisture from the air and probably contain approximately 30 percent water during much of the night (Fig. 21-20). By feeding at night, therefore, the oryx can manage a nightly intake of some 5 liters of water with its forage. This is a minimal amount of water for a 200-kilogram mammal living in shelterless desert that is sufficient for the oryx only because of a combination of mechanisms that favor water conservation.

Strategies for conserving water differ between quite similar ungulates, as indicated by studies of Grant's gazelle *(Gazella granti)* and Thompson's gazelle *(G. thompsonii)* by Taylor (1968b, 1972). These antelopes occur together in East Africa, but the range of Grant's gazelle extends into the harsh deserts of northern Kenya and Thompson's gazelle is restricted to less arid areas. Unexpectedly, under experimental conditions involving peak temperatures of about 40°C, Grant's gazelle had a rate of **pulmocutaneous evaporation** per kilogram of body weight about one-third higher than that of Thompson's gazelle. This apparent paradox was resolved when Taylor (1972) considered the performance of these antelope under extreme heat. At air temperatures above 42°C, Thompson's gazelle used panting to increase evaporative cooling; at an air temperature of 45°C, it maintained a body temperature of 42.5°C. Grant's gazelle, by

TABLE 21-6   Relative Urine-Concentrating Ability of Some Mammals as Indicated by Osmotic Concentration of Urine from Dehydrated Animals

Species	Common name	Urine osmolality (mOsm/liter)	Diet
Dipodomys merriami	Merriam's kangaroo rat	3165	Granivorous
Peromyscus crinitus	Canyon mouse	3047	Omnivorous
Perognathus longimembris	Little pocket mouse	1675	Granivorous
Onychomys torridus	Grasshopper mouse	2733	Insectivorous
Neotoma lepida	Desert woodrat	2436	Herbivorous
Lepus californicus	Black-tail jackrabbit	3600	Herbivorous
Vulpes zerda	Fennec	4022	Carnivorous
Canis lupus	Wolf	2608	Carnivorous
Felis catus	Cat	3118	Carnivorous
Madoqua sp.	Dik-dik antelope	4300	Herbivorous
Camelus dromedarius	One-humped camel	3100	Herbivorous
Oryx gazella	Oryx	2900	Herbivorous
Equus caballus	Donkey	1500	Herbivorous
Bos taurus	Zebu cattle	1400	Herbivorous

Data from MacMillen, 1972, Nagy et al., 1976, Noll-Banholzer, 1979a, 1979b, and Maloiy, 1973.

contrast, did not resort to evaporative cooling but allowed its body temperature to rise to 46°C and was thus able to dissipate heat to the air. As in the case of the oryx, the major source of water for desert-dwelling Grant's gazelles may be leaves that absorb water at night.

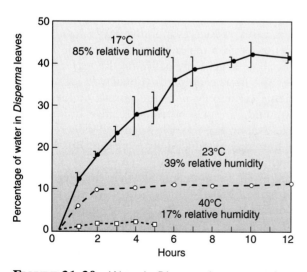

FIGURE 21-20   Water in *Disperma* leaves at various humidities and temperatures. The hygroscopic leaves may be an important water source for Grant's gazelles *(Gazella granti)* in some areas. *(After Taylor, 1968a)*

METABOLIC WATER

Many rodents that inhabit deserts, seasonally dry chaparral, or woodlands, must survive for extended periods without access to preformed water (water already in the form of H_2O, in contrast to water formed as a byproduct of chemical reactions such as the oxidation of starches). Among the rodent members of these habitats are species that primarily eat seeds. These rodents, some of which are saltatorial and can move rapidly over considerable distances in search of seeds, occupy even the most barren and inhospitable deserts of the world. They represent the families Heteromyidae, Dipodidae, and Muridae. These "water-independent" rodents share two life-history features: they are nocturnal, and they are semifossorial.

As a basis for further discussion, the routes of water intake and loss in water-independent rodents must be reviewed. Sources of water intake include succulent foods, metabolic water (water released as a byproduct of metabolism; seeds high in carbohydrates have high yields of metabolic water), and drinking water. Water is lost by lactation, defecation, urination, and pulmocutaneous evaporation. Seed-eating desert rodents do not have regular access to drinking water and eat little succulent food; their major water source is metabolic water. Water lost in lactation is important periodically to females. Urinary water loss is typically reduced in these rodents by the concentration of urine, and water is absorbed from fecal material. The primary channel for water loss is pulmocutaneous evaporation. Such loss may account for 90 percent of total water loss, and MacMillen and Grubbs (1976) and MacMillen (1972) demonstrated that there is no difference in the rate of such water loss between desert and nondesert rodents. Nagy and Peterson (1980) confirmed these results for free-living mammals. Because water-independent rodents do not sweat and may have reduced cutaneous water loss, pulmonary evaporation is of greatest importance.

Water intake in kangaroo rats and other seed-eating desert rodents can be accounted for fairly easily. Many seeds are high in carbohydrates, which yield large amounts of water when they are oxidized. For example, for every 100 grams of dry barley metabolized, 53.7 grams of water is produced. This may be augmented by preformed water in the food: seeds in the soil or on the surface at night absorb moisture, and seeds stored in nests in burrows may contain as much as 20 percent water (Morton and MacMillen, 1982).

Of central importance to desert rodents, then, is the balance between water lost via pulmonary evaporation (the principal route of water loss) and water gained from the metabolism of food (the major source of water). This balance as it relates to ambient temperature and body size has been studied by MacMillen and his associates (MacMillen, 1972; MacMillen and Christopher, 1975; MacMillen and Grubbs, 1976; MacMillen and Hinds, 1983a, b). Their research demonstrates that pulmocutaneous (evaporative) water loss is independent of ambient temperatures below thermal neutrality. Small desert mammals, breathing in dry air, always heat that air in the lungs to 35 to 38°C and saturate it with water vapor. When expired, this saturated air loses as much water at ambient temperatures of 30°C as it does at 5°C. These data suggest that nasal heat exchange mechanisms are not particularly effective in desert rodents. This relationship probably holds for many mammals. Despite the use of both heat exchange and hygroscopic moisture exchange, the camel resting in warm air (38°C) would not reduce evaporative water loss as well as the kangaroo rat resting in cool air (15°C) and using only nasal heat exchange. Since desert seed-eating rodents are nocturnal, they are active in the coolest part of the daily cycle when pulmonary water loss is lowest. Except in summer, temperatures in the desert drop dramatically at night. For much of the year, foraging desert rodents are faced with temperatures below their thermal neutral zone and must raise their metabolic rate accordingly. As metabolic rate rises, so does production of metabolic water. To desert rodents, therefore, the relationship between evaporative water loss, production of metabolic water, and ambient temperature is critically important.

Through careful analyses of data on simultaneous measurements of evaporative water loss and oxygen consumption (an indicator of the level of production of metabolic water), MacMillen and Hinds (1983a, b) showed that, at ambient temperatures below 16.6°C, many rodents produce more water metabolically than they lose by evaporation. At even lower temperatures, metabolic water production still further exceeds evaporative water loss (Fig. 21-21). There is, however, considerable interspecific variation in the ability to limit evaporative water loss. Merriam's kangaroo rat, for example, far surpasses average performance (MacMillen and Hinds, 1983a). MacMillen and coworkers conclude that nocturnality provides desert rodents with a favorable relationship between evaporative water loss and metabolic water production.

By being semifossorial, water-independent rodents vastly reduce evaporative water loss. Desert rodents spend a major share of their time in burrows; some species are active on the surface for as little as 1 hour each night (Kenagy, 1973a). These rodents commonly plug the mouth of their burrows, and, in such burrows, relative humidity is high. Evaporative water loss markedly decreases as the humidity of inhaled air increases. In addition, evaporative water loss is affected more by humidity changes in small mammals than in large ones (Christian, 1978). Temperatures in burrows are moderate, even when surface temperatures are high. For most of their lives, therefore, desert rodents live under conditions of high humidity and moderate temperatures and are well insulated from the rigors of the desert environment.

Although under laboratory conditions most species of kangaroo rats *(Dipodomys)* and all species

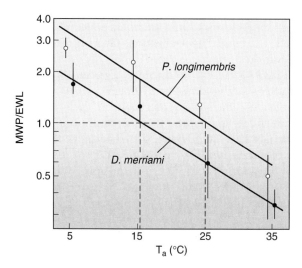

FIGURE 21-21 Relationship between metabolic water production and evaporative water loss (MWP/EWL) in *Dipodomys merriami* and *Perognathus longimembris* while on a diet of dry millet seeds. An extension of horizontal line at MWP/EWL = 1.0 indicates the temperature at which MWP = EWL (dotted lines). Note that this equality occurs at a much higher temperature in *P. longimembris* (about 25°C) than in *D. merriami* (about 17°C). *(From MacMillen and Hinds, 1983a, by permission of the American Institute of Biological Sciences)*

is always in a more favorable state of water balance than *D. merriami,* as indicated by the more dilute urine of *P. longimembris.*

MacMillen and Hinds (1983) regarded urine concentration, the traditional criterion of water regulatory efficiency in desert rodents, to be misleading and proposed that a more meaningful indicator for granivorous rodents is the ambient temperature at which MPW = EWL. Greater water regulatory efficiency of *P. longimembris* at higher temperatures is demonstrated by the fact that the ambient temperature at which MWP = EWL is 10°C higher in *P. longimembris* than in *D. merriami.* The demonstrated relationship between metabolism and evaporative water loss led MacMillen and Hinds to hypothesize that water regulatory efficiency in heteromyids should vary inversely with body weight. This hypothesis was confirmed by experiments on five genera and 13 species of heteromyids.

MacMillen and Hinds speculated that under selective pressures associated with increasing aridity, heteromyids became granivorous, improved water regulatory ability, minimized energy needs by reducing metabolic rate, and became smaller. At a body weight threshold of 35 to 40 grams, a divergence in locomotor style occurred: the larger kangaroo rats (more than about 40 grams) became bipedal hoppers and the smaller pocket mice (less than 40 grams) maintained quadrupedal locomotion. They proposed that the ecological importance of the mass-related differences in water regulatory efficiency favors survival in the small species that have limited locomotor ability. For example, bipedal hopping provides kangaroo rats with a rapid means of locomotion that allows erratic escape from predators in open situations. In contrast, the slower quadrupedal locomotion of pocket mice leaves these animals more vulnerable to predators on open ground and restricts the foraging area. Thus, kangaroo rats can forage widely and selectively and can choose seeds high in carbohydrates (seeds that have high yields of metabolic water), and their water regulatory ability became fixed at an intermediate level. Pocket mice, in contrast, must forage less widely and eat a variety of seeds, some of which, because of relatively low carbohydrate content, yield relatively less metabolic water and more urea to eliminate via the kidneys.

of pocket mice *(Chaetodipus* and *Perognathus)* remain healthy on a dry-seed diet, the urine of pocket mice has roughly half the osmotic concentration of that of kangaroo rats. Pocket mice, then, should be in a more favorable state of water balance than kangaroo rats. MacMillen and Hinds (1983) tested this hypothesis by measuring the primary route of water input (MWP; metabolic water production) and output (EWL; evaporative water loss) in *Dipodomys merriami* (36 grams) and *Perognathus longimembris* (8.0 grams) fed millet seeds and subjected to a wide temperature range. In these heteromyids, EWL is independent of ambient temperature at and below the rodent's temperatures of thermoneutrality. MWP is inversely related to ambient temperature: the lower the temperature the higher the metabolic rate, hence the greater the MWP. In both species, the ratio MWP:EWL (an expression of the state of water balance) is invariably inversely related to ambient temperature (Fig. 21-21). Because of its smaller size and associated higher metabolic rates, *P. longimembris*

URINE-CONCENTRATING ABILITY

An important factor in reducing water loss is the ability of the kidneys of water-independent rodents to concentrate urine. Kangaroo rat urine is roughly five times more concentrated than that of a human. Therefore, in excreting comparable amounts of urea, the kangaroo rat uses one fifth as much water as do humans. The concentration of dissolved compounds in the urine of kangaroo rats may be roughly twice that of seawater, and, in the laboratory, these animals have maintained water balance by drinking seawater. The urine of humans, on the other hand, has a concentration of dissolved compounds lower than that of seawater. When a human drinks seawater, the excretion of the dissolved salts requires the withdrawal of water from body tissues, resulting in severe dehydration.

Among desert rodents of the Old World, jerboas *(Jaculus jaculus)* and several gerbils of two genera *(Meriones* and *Gerbillus)* can live on dry food, and some surpass kangaroo rats in their ability to concentrate urine. Adaptations to a dry diet have clearly evolved independently in several rodent families (Heteromyidae, Dipodidae, Muridae). Striking convergent evolution in these families has led to saltato-rial adaptations in some members of each family, as well as to similar specializations favoring water conservation.

Two species of spiny mice *(Acomys,* Muridae) studied by Shkolnik and Borut (1969) in the desert of Israel are remarkable in their unusual pattern of adaptation to arid conditions. These animals have highly specialized kidneys that can concentrate urine to a greater degree than can the kangaroo rat kidney, but the spiny mice have an evaporative water loss two to three times as great as that in Merriam's kangaroo rat. Probably due to high water loss through the skin, spiny mice are unable to subsist on a diet of dry seeds (Table 21-7). Apparently, the high cutaneous water loss is important as a means of dissipating heat in a hot climate, and the great ability of the kidney to concentrate urine, coupled with a diet high in land snails (which have a high water content), compensates for the extravagant use of water in thermoregulation.

The kidneys of some bats are specialized to concentrate urine, but these animals are seemingly not independent of drinking water. Carpenter (1969) found that two desert-dwelling insectivo-

TABLE 21-7 Physiological Characteristics of Four Desert Rodents on a Natural Diet

Characteristic	Dipodomys merriami	Acomys cahirinus	Ammospermophilus leucurus	Psammomys obesus
Diet	Granivore	Omnivore	Omnivore	Herbivore
Body mass (g)	35	49	88	135
Dry matter intake (g/day)	3.57	3.14	6.55	12.11
Water influx (ml/day)	2.13	4.99	14.21	46.47
Metabolic water (ml/day)	1.91	1.29	2.42	3.45
(% water influx)	89.7	25.9	17.0	7.4
Preformed water (ml/day)	0.22	3.7	11.8	42.8
(% water influx)	10.3	74.1	83.0	92.2
Water efflux (ml/day)	2.13	4.99	14.21	46.47
Fecal water (ml/day)	0.07	0.44	1.30	7.57
(% water efflux)	3.2	8.8	9.2	16.3
Urinary water (ml/day)	0.49	2.11	6.73	18.9
(% water efflux)	23.0	42.3	47.4	41.6
Evaporative water loss (ml/day)	1.57	2.44	6.26	20.0
(% water efflux)	73.7	48.9	44.0	43.1

Data from Degen, 1996.

rous bats produced concentrated urine. Their need for water was increased by high evaporative water losses during flight and when they were not torpid. He estimated that these bats lost 3.1 percent of their body weight through evaporation per hour of flight. Carpenter concluded that they were not independent of drinking water but that their ability to fly long distances to drink water enabled them to maintain water balance in desert areas. Urine concentrations in the little brown bat reach peak levels during high evaporative cooling and just after feeding, but by drinking after feeding this species avoids water stress (Geluso and Studier, 1979). A marine fish- and crustacean-eating bat (*Myotis vivesi,* Vespertilionidae) that inhabits the arid islands and coasts of the Gulf of California has the ability to concentrate urine to the extent that it can utilize seawater as a water source (Carpenter, 1968). Because of high evaporative water losses, particularly during flight, the water gained from this bat's food probably is not sufficient to meet its water requirements, and, presumably, it must drink seawater.

Desert insectivores face slightly different problems. Insectivores inhabiting arid zones are not limited by their ability to obtain water but by their ability to conserve water by forming highly concentrated urine. Compared to granivores, insectivores obtain more preformed water from their food. The Namib Desert golden mole obtains sufficient water from its diet of termites and insect larvae, which contain 60 to 80 percent water (Redford and Dorea, 1984; Fielden et al., 1990a, b). Water loss is minimized by the golden mole's nocturnal foraging habits and by a low metabolic rate during diurnal torpor deep in the sand, where humidities are high. Although insects have a high moisture content, they also are high in protein. Wastes from protein catabolism must be excreted as urea in the urine, generally resulting in higher urinary water losses for insectivores than for granivores (Lindstedt, 1980).

The osmotic concentration of urine in desert insectivores far exceeds that of more **mesic-adapted** insectivores (Table 21-8). Namib golden moles, for example, lose only 11 percent of their total water budget via their urine, compared to over 30 percent in the similar-size short-tailed shrew (*Blarina brevicauda;* Deavers and Hudson, 1979; Fielden et al., 1990b).

TABLE 21-8 Average Osmotic Concentration of Urine from Insectivorous Mammals Exposed to Water Stress in the Laboratory

Species	Urine osmolarity (mosmol/kg)	Source
Arid Zone		
Pipistrellus hesperus	4340	Geluso, 1978
Hemiechinus auritus	4010	Yaakobi and Sholkni, 1974
Antrozous pallidus	3980	Geluso, 1975
Eremitalpa granti	3820	Fielden et al., 1990
Macrotis lagotis	3566	Hulbert and Dawson, 1974
Onychomys torridus	3180	Schmidt-Neilsen and Haines, 1964
Mesic Zone		
Erinaceus europaeus	3062	Yaakobi and Sholkni, 1974
Myotis volans	2910	Geluso, 1978
Planigale maculata	2317	Morton, 1980
Blarina brevicauda	1820	Deavers and Hudson, 1979

NASAL COUNTERCURRENT HEAT AND WATER EXCHANGE

The observation by Schmidt-Nielsen that kangaroo rats exhale air that is cooler than body temperature led to studies showing that the nasal passages of many mammals serve as heat exchange systems (Langman et al., 1979; Schmidt-Nielsen et al., 1970; Schmidt-Nielsen et al., 1980). These systems result in a significant reduction of pulmonary water loss.

The nasal passages of rodents function as heat exchangers with alternating flow in opposite directions in a single tube rather than steady flow in opposite directions in adjacent tubes (as in retia mirabilia). Inhaled air that is below body temperature cools the moist nasal mucosa, which is further cooled when water evaporates from it. Inhaled air then becomes warmed and saturated with water in the airways and lungs. During expiration, this humid, relatively warm air passes back through the narrow nasal passages and over the cool mucosa. The expired air is cooled, and thus moisture condenses on the mucosa. This moisture is subsequently absorbed back into the animal's system. This pattern, repeated with every respiratory cycle, results in expired air that is far below body temperature and substantially below the temperature of the inhaled (or ambient) air. Although the exhaled air is saturated with water vapor, it is far cooler than the air in the lungs and thus contains considerably less water. A kangaroo rat with a body temperature near 38°C, resting in air at 30°C and 25 percent relative humidity, exhaled air that was 27°C. In this case, 54 percent of the water used to humidify the inhaled air was recovered from the exhaled air by condensation on the mucosa. Ambient temperature influences the rate of water recovery: the cooler the inhaled air, the more the nasal passages are cooled. Consequently, the more the air being exhaled is cooled, the more water is recovered. At an ambient temperature of 15°C and 25 percent relative humidity, the kangaroo rat mentioned above would recover up to 88 percent of the water used to humidify the inhaled air. Moreover, in rodents in general, evaporative water loss is independent of ambient temperature below thermal neutrality, likely due to the influences of nasal mucosal cooling of expired air (MacMillen and Hinds, 1983; Hinds and MacMillen, 1985).

A dehydrated camel, however, not only cools exhaled air but also desaturates it. This is due to a **hygroscopic** (water-absorbent) layer of dried mucous and cellular debris that coats the nasal passages. This layer absorbs water rapidly during exhalation, and, as a result, the exhaled air is dried. K. Schmidt-Nielsen (1981) estimated that, if a camel in an air temperature of 28°C exhaled air that was at this same temperature but at only 75 percent relative humidity, it would lose by pulmonary evaporation only one third as much water as it would be forming as a byproduct of metabolism. These estimates are not intended to be full accounts of the camel's water intake and loss, but they illustrate the effectiveness of the coupling of a heat exchange system and hygroscopic moisture exchange.

LACTATION AND WATER BALANCE

The loss of water by a lactating female is substantial, and it is not surprising that some mammals recycle such water. Cat breeders have long been familiar with behaviors only recently discussed in the scientific literature. The lactating domestic cat *(Felis catus)* stimulates urination and defecation in her young by licking their genital areas. She then ingests the urine and feces. This keeps the nest sanitary and recycles to the mother much of the water lost in the milk. Similar behavior occurs in several Australian murid rodents, the dingo *(Canis familiaris),* and two species of kangaroos (Baverstock and Green, 1975). These species recovered about 30 percent of the water lost in milk. A laboratory study indicated that most of the water lost by female house mice *(Mus musculus)* during lactation, with the exception of that dissipated by evaporative water loss from the young, was recovered in a similar manner by the mother (Baverstock et al., 1979).

Recycling of water is far more important for desert dwellers than for mesic species. The importance of maternal ingestion of urine was studied by Oswald et al. (1993) in lactating females of mesic red-backed voles *(Clethrionomys gapperi),* white-footed mice *(Peromyscus leucopus)* that occupy a variety of habitats, and two species of desert-dwelling gerbils *(Gerbillus)*. For all these species, water requirements during lactation increased more than 100 percent, and all species recycled comparable absolute amounts of their young's urine. But although recycled water contributed only 3.2 percent to the total water budget for the mesic-adapted vole, this figure was 11.5 percent for the widespread white-footed mice and 39 percent for the xeric-adapted gerbils. In contrast to the other species, the gerbils' highly developed ability to minimize water loss in urine and feces results in a minimal water budget. Thus, recycled water is a major part of this budget during lactation. The milk of the desert-dwelling Merriam's kangaroo rat contains only 50 percent water, making it one of the most concentrated types of milk known and comparable to those of pinnipeds and cetaceans (Kooyman, 1963; Boness and Bowen, 1996). Behaviors that result in recycling water lost in milk are probably widespread among mammals that bear altricial young or have limited access to fresh water.

22 ECHOLOCATION

Animals that echolocate use echoes of sounds they produce to locate objects in their path. Because the mammals most familiar to us depend largely on vision for perceiving their environment, it is surprising to note that at least 18 percent of the known species of mammals probably use echolocation as their primary means, or at least as important secondary means of "viewing" their surroundings. Microchiropteran bats, two megachiropterans, some members of the order Insectivora, and probably all odontocete cetaceans echolocate. Future research may demonstrate that the use of echolocation among mammals is even more widespread.

An accurate picture of the bat's use of acoustical orientation was long in emerging. As early as 1793, Lazaro Spallanzani performed experiments that suggested that bats use acoustical rather than visual perception when avoiding obstacles and when feeding. A few years later (1798), Louis Jurine showed that bats could not avoid obstacles if their ear canals were plugged with wax, suggesting that hearing played an important role in bat orientation and prey capture. Not until the early 1940s, however, was the use of echolocation by bats conclusively demonstrated by the careful laboratory experiments of Griffin and Galambos (1940, 1941) and by the observations of Dijkgraaf (1943, 1946). Continued research, aided by electronic equipment, has contributed to our present detailed, if incomplete, knowledge of echolocation.

INTRODUCING ECHOLOCATION

THE REALITIES OF ECHOLOCATION SOUNDS

Mammals produce two types of echolocation calls. Odontocetes produce clicks in the nasal passage, and two species of megachiropteran bats make clicks with the tongue, whereas the signals of microchiropteran bats and shrews are made in the larynx (Novick, 1955). The tonal signals of bats are more intense than clicks, a factor of key importance because the more intense the signal the greater the echolocation range and thus the more time to react to prey or to obstacles. Echolocation calls can be described in terms of time (duration, repetition rate), frequency (pitch), and intensity (an expression of signal strength). The following discussions deal primarily with the signals of bats.

Echolocation calls are brief pulses of sound that vary in duration from about 100 milliseconds (a millisecond is 1/1000 of a second) to 0.25 milliseconds. Some bats (such as vespertilionids) typically give short signals; others (such as rhinolophids) emit longer signals (Fig. 22-1). An individual changes the duration and the rate (number per unit of time) of its calls during a search and capture sequence (Fig. 22-2).

The **frequency** (pitch) of echolocation calls varies widely among species, and seems to be controlled by varying tension on the vocal cords (Suthers and Fattu, 1982). The echolocation signals of most bats are **ultrasonic,** that is to say, above the range of human hearing, which reaches about 20 kilohertz (a kilohertz [kHz] equals 1000 cycles per second). Most bats emit echolocation signals within the range of about 20 to 100 kilohertz, but some use frequencies above 200 kilohertz. Although conventional wisdom holds that echolocation is based on ultrasonics, a number of bats use echolocation signals that are below 20 kilohertz and are therefore audible to many humans (Fenton, 1982; Rydell and Arlettaz, 1994; Fenton and Griffin, 1997). These bats represent three families and were observed at such widely scattered places as Arizona, British Columbia, Colombia, Europe, and Africa. Odontocete cetaceans also use echolocation signals that are audible to humans. Therefore, echolocation clearly does not depend on ultrasonics. Many echolocation calls are complex, consisting of a **fundamental frequency** (the lowest, or root tone of a chord) and several **harmonics** (frequencies that are integral multiples of the fundamental frequency).

Echolocation calls also vary in intensity of sound (Fig. 22-3). Intensity indicates signal strength, usually expressed in decibels (dB), whereas loudness refers to our perception of sound. The signal of a smoke detector and the echolocation call of a little brown bat have the same intensity (110 decibels measured at 10 centimeters from the source), but to us the smoke detector is loud and we hear no sound from the bat. "Whispering bats" and "loud bats" were recognized by Griffin (1958), and other bats have signals of intermediate intensity. Using broad-band microphones, Simmons et al. (1979) found that whereas the intense signals of some bats could be detected at 30 meters or more, those of others were only detectable at less than 0.5 meters.

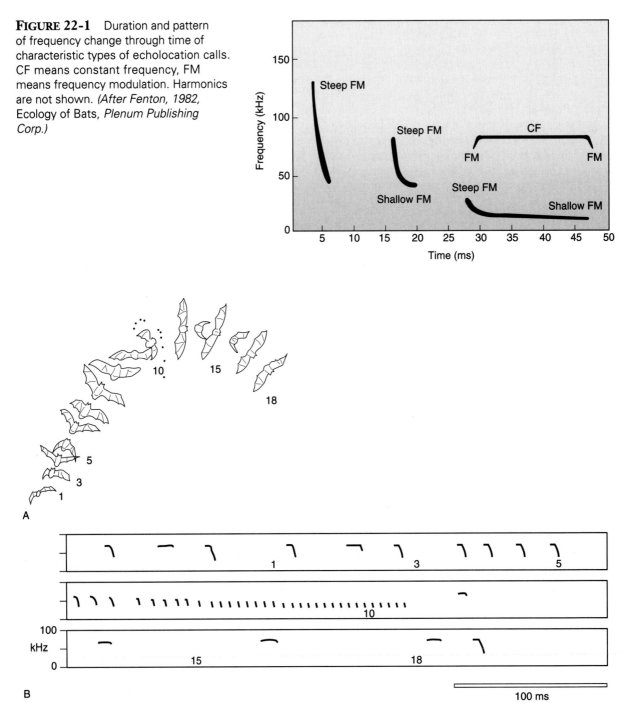

FIGURE 22-1 Duration and pattern of frequency change through time of characteristic types of echolocation calls. CF means constant frequency, FM means frequency modulation. Harmonics are not shown. *(After Fenton, 1982, Ecology of Bats, Plenum Publishing Corp.)*

FIGURE 22-2 (A) Drawings of eighteen stages in the flight path of *Noctilio albiventris* during an attack on a flying insect. The insect is denoted by a small dot. The bat attempts to capture the insect at stage 10. (B) The corresponding sequence of echolocation pulses (sonograms) recorded as frequency versus time during the attack. The numbers correspond to the stages illustrated in (A). Notice that the characteristics of the echolocation pulses changed as the bat switched from search phase to attack phase at stage 5. The pulse duration and pulse intervals are reduced during the attack phase. *(From Kalko et al., 1998 © Springer-Verlag, New York, Inc.)*

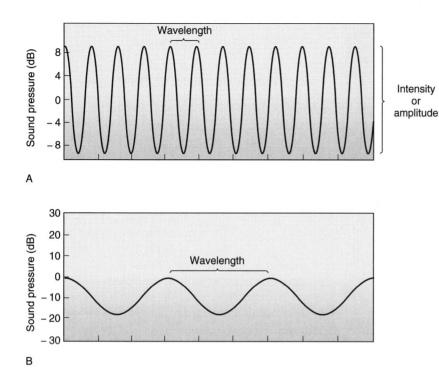

A

B

FIGURE 22-3 The details of a soundwave: (A) a high-frequency (short-wavelength) high-intensity sound, (B) a low-frequency (long-wavelength), low-intensity sound.

Among different species of bats, echolocation signals differ in **bandwidth** (breadth of frequencies produced) and in the information they provide. Narrow-band calls (those that span less than 10 kilohertz) are often designated as **constant frequency (CF) signals,** in contrast to broad-band, **frequency-modulated (FM)** calls (those that span more than 10 kilohertz). Narrow-band signals are useful for detecting a target, but do not provide details as to its position. Many bats use such calls (search-phase calls) when searching for prey. By increasing the bandwidth of their signals, bats increase the precision with which they can pinpoint a target. Some broad-band calls are shallow, covering a narrow range of frequencies relatively slowly, and others are steep, covering a wider range of frequencies more rapidly (Fig. 22-1). Although the shallow, broad-band (FM) calls are excellent for target detection and are used as search-phase signals by many bats, they are sensitive to distortion by **Doppler shift** and do not provide precise information on target location. Steep, broad-band (FM) signals, however, provide information for accurately localizing a target and are used when a bat attacks its prey (the attack phase). Typically a bat detects prey with narrow-band (CF) or shallow broad-band (shallow FM) calls, but switches to steep broad-band (steep FM) calls during the last instant of the attack (Fig. 22-2).

LIMITATIONS OF RANGE

Mention of the "last instant" introduces the realities of time and distance. Compared to vision, echolocation in air is an extremely short-range system. Flying bats must react rapidly to echoes from objects that are close and rapidly getting closer. Spheres 10 millimeters in diameter were first detected by echolocating *Eptesicus fuscus* (a bat with intense, long-range calls) at 5 meters (Kick, 1982). Using this figure and the observed foraging-flight speed for *E. fuscus* of 3.5 meters per second, Fenton (1990) estimated that these targets were first detected only 1.5 seconds in front of the bat. If the target is an edible flying insect, then, this bat has only 1.5 seconds to locate the insect precisely, to track its trajectory, and to maneuver for the capture. Novick (1970) found that the pursuit and capture of an insect takes a mustached bat (*Pteronotus parnellii*, Fig. 22-4) only 0.25 to 0.33 seconds. Table 22-1 shows the estimated maximum ranges of target detection for some bats. Most insectivorous bats are small, have low wing loadings, fly at slow or moderate speeds, and are highly

FIGURE 22-4 Close-up of the face of a mustached bat *(Pteronotus parnellii)*, showing the lips formed like a megaphone during echolocation. *(O. Henson)*

maneuverable, features that accord with the demands of short-range perception.

ECHOLOCATION IN WATER AND AIR

Because water and air differ drastically in density, they transmit sound differently. Sound travels over four times faster in water than in air (1541 meters per second versus 361 meters per second), the intensity of a given signal is greater in water than in air, and sound is **attenuated** (reduced in intensity) less rapidly in water. Sound waves in air spread from a source and rapidly lose intensity according to the inverse square law (Lawrence and Simmons, 1982; see Fig. 22-5). Relative to an echolocating mammal in air, then, one in water will receive information from echoes much faster, will use less energy to produce a signal of given intensity, and will transmit a signal farther. An additional complication for terrestrial echolocators is that temperature and humidity strongly influence the abil-

TABLE 22-1 Estimated Maximum Ranges of Target Detection for Some Bats

	Interpulse Interval (ms)	Range[a] (m)	Source
Rhinopoma hardwickei	100	17.0	Simmons et al., 1984
Taphozous mauritianus	110	18.7	Fenton et al., 1980
Cormura brevirostris	81	13.8	Barclay, 1983
Saccopteryx bilineata	53	9.0	Barclay, 1983
Nycteris grandis	20	2.4	Fenton et al., 1983
Nycteris thebaica	22	3.7	Fenton et al., 1983
Noctilio leporinus	100	17.0	Suthers & Fattu, 1973
Trachops cirrhosus	24	4.1	Barclay et al., 1981
Myotis adversus	90	15.3	Thompson & Fenton, 1982
Myotis daubentoni	15	2.6	Jones & Rayner, 1988
Lasionycteris noctivagans	167	28.4	Barclay, 1986
Eptesicus fuscus	100	17.0	Simmons et al., 1979
Chalinolobus variegatus	128	21.2	Obrist, 1989
Lasiurus cinereus	303	51.5	Barclay, 1986
Euderma maculatum	365	62.1	Leonard & Fenton, 1984

From Fenton, 1990.
[a]The estimates are based on intervals between signals, assuming that echolocating bats do not tolerate overlap between the echo from one signal and the next signal.

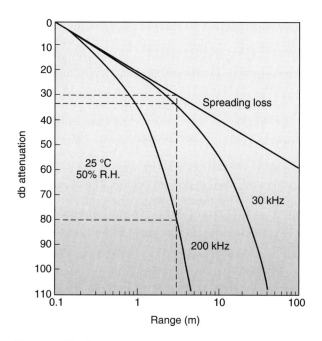

FIGURE 22-5 Rate of attenuation of echolocation calls in air. The dotted lines show the attenuation of calls at 30 kilohertz and 200 kilohertz. A target 3 meters in front of an echolocating bat will be weaker by 30 decibels (spreading loss line). Because bats listen for the echo, sounds must make a roundtrip of 6 meters and will be attenuated by nearly 60 decibels. In addition to spreading losses, the atmosphere also absorbs sound (curved lines) and thereby adds to the overall attenuation. Therefore, a bat 3 meters from a target, echolocating at 30 kilohertz will perceive an echo that is 64 decibels weaker than when it was emitted (60 decibels for roundtrip spreading loss and 4 decibels for atmospheric attenuation). A bat echolocating at 200 kilohertz will perceive a total attenuation of about 110 decibels (with atmospheric losses accounting for 50 decibels). *(After Lawrence and Simmons, 1982)*

ity of air to transmit sound (Harris, 1996). Because of the relatively poor sound transmission qualities of air, high-intensity tonal signals are important for bats. The less intense clicks of odontocetes are adequate because of the superior sound-transmission qualities of water.

For an aquatic mammal echolocation can provide long-range information. Goold and Jones (1995) estimated that the maximum echolocation range of the sperm whale *(Physeter catodon)* was at least 1500 meters. But for terrestrial mammals, echolocation yields only short-range information. Fenton (1990) estimated maximum range of target detection for 16 species of bats (Table 22-1): the

greatest range was 62 meters, the mean was only 19 meters, and the range for some bats was only about 2.5 meters.

HIGH-FREQUENCY SOUND

Because high frequencies are more rapidly attenuated in air than are low frequencies (Fig. 22-5), one might wonder why bats typically use high-frequency echolocation calls. Perhaps of prime importance is the relationship between prey size and the wavelength of echolocation pulses. The higher the frequency of a sound, the shorter its wavelength. Frequencies of roughly 30 kilohertz have a wavelength of approximately 11.5 millimeters, roughly the size of a small moth; this balance between prey size and wavelength is ideal, for objects approximately the size of a given wavelength reflect that wavelength particularly well. Low-frequency sounds have long wavelengths and tend to bend around small objects without producing an echo. Some species of bats can detect wires with as small a diameter as 0.08 millimeters (1/30 of a wavelength), but, in general, the wavelengths of the pulses emitted by bats are in the range that is most efficient for the detection of small-to-medium-size insects.

Frequency also influences directionality of hearing. All echolocating bats depend on accurately localizing the source of sound (echoes). In all bats studied by Obrist et al. (1993), hearing became increasingly directional with increasing frequency of echolocation calls.

SELF-DEAFENING

As a protection for the inner ear against outgoing echolocation sounds, bats have several "self-deafening" adaptations. In bats, as in all mammals, sound waves (vibrations in air) strike the tympanic membrane, where they are converted to mechanical vibrations that are amplified and transmitted by the ear ossicles to the oval window of the inner ear. Here vibrations are set up in the fluid in the inner ear, are transmitted to the basilar membrane, and are converted to nerve impulses that pass to the brain. Two muscles of the mammalian middle ear dampen the ability of the ossicles to transmit vibrations when an individual is subjected to unusually loud sounds or when it is vocalizing. These muscles — the tensor tympani, which changes the tension on the tympanic membrane, and the stapedius, which changes the angle at which the stapes contacts the oval window — are extremely well developed in bats, and their contractions reduce the bat's sensitivity to its own pulses (Fig. 22-6). Jen and Suga (1976), using electronic equipment, found that action potentials of the cricothyroid laryngeal muscles were followed 3 milliseconds later by action potentials of the middle ear muscles. This coordination of laryngeal and middle ear muscles ensures the contraction of the latter just prior to vocalization and attenuates (weakens in intensity) by some 25 percent the auditory self-stimulation (Suga and Shimozawa, 1974). In Brazilian free-tailed bats *(Tadarida brasiliensis)*, the stapedius muscle contracts approximately 10 milliseconds prior to pulse generation, pulling the stapes away from the oval window and partly preventing emitted sounds from reaching the inner ear. A few milliseconds later, the stapedius relaxes, re-engaging the stapes with the oval window to allow detection of the echo (Hill and Smith, 1992).

Additional neural attenuation of the direct reception of echolocation signals occurs in the brain. Nerve impulses arising from direct reception and passing from the cochlea to the inferior colliculus of the brain are attenuated by the neurons of the lateral lemniscus of the brain. This change, plus that effected by the middle ear muscles, attenuates the neural events by 40 percent. Suga and Shi-mozawa suggested that similar attenuating mechanisms occur in humans and keep the sounds of our own speech from becoming disturbingly loud. Direct reception of echolocation calls in bats is doubtless reduced also by the beaming of sounds by the lips and the complex noses and nose leaves.

An additional structural refinement is that the bones housing the middle and inner ear are insulated from the rest of the skull. This bony otic capsule does not contact other bones of the skull (Fig. 22-7) and is insulated from the skull by blood-filled sinuses or fatty tissue. During the emission of signals, the conduction of sound from the larynx and the respiratory passages through the bones of the skull is thus greatly reduced.

ORAL AND NASAL PULSE EMISSION

Echolocation signals of bats are emitted through either the mouth (oral emitters) or the nose (nasal emitters). Most microchiropteran bats are oral emitters; only members of the Nycteridae, Megadermatidae, Rhinolophidae, and Phyllostomidae are nasal emitters. A basic dichotomy in skull shape is associated with these two modes of sound emission (Freeman, 1984; Pedersen, 1998). The skulls of oral-emitting bats are similar to those of most terrestrial mammals, but in nasal-emitting bats, the rostral part of the skull is rotated ventrally, below the level of the braincase, resulting in the alignment of the nasal cavity, instead of the

FIGURE 22-6 Cutaway drawing of the middle and inner ear of a typical bat. Sound waves pass down the external auditory canal to the tympanic membrane. Vibrations in the tympanic membrane are amplified and transferred to the oval window of the cochlea by the three middle ear ossicles. Note the two middle ear muscles, tensor tympani and stapedius, which facilitate self-deafening in bats.

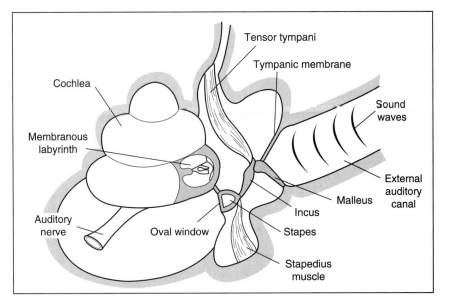

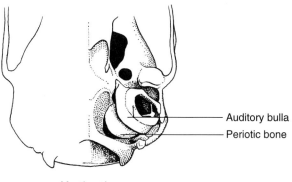

Myotis volans

FIGURE 22-7   Ventral view of the posterior part of the skull of a vespertilionid bat, showing how loosely the periotic bone and auditory bulla are attached to the skull.

Auditory bulla
Periotic bone

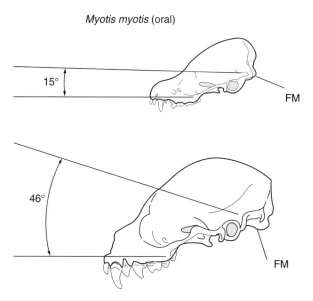

Myotis myotis (oral)

15°

FM

46°

FM

Artibeus jamaicensis (nasal)

FIGURE 22-8   Comparison of the cranial characteristics in (A) *Myotis myotis,* an oral emitter and (B) *Artibeus jamaicensis,* a nasal emitter. The angle from the plane of the semicircular canals in the ear to the palate is much greater in nasal emitters. The foramen magnum (FM) opens ventrally resulting in a head posture during flight which, coupled with the displacement of the rostrum below the level of the basicranium, aligns the nasal area, rather than the mouth, with the direction of flight in nasal emitters such as *A. jamaicensis.*

oral cavity, with the direction of flight (Fig. 22-8). Oral emitters keep their mouths open when echolocating and typically produce high-intensity pulses (Fig. 22-4). Nasal emitters, on the other hand, usually keep their mouths closed when echolocating; some (Rhinolophidae) give high-intensity pulses; others produce relatively low-intensity signals ("whispering bats"; Griffin, 1958). Viewed broadly, bats that emit high-intensity signals, whether nasal or oral emitters, usually catch flying insects, whereas those nasal emitters that have low-intensity calls feed on fruit, nectar, small animals taken from the ground or from vegetation, or combinations of these foods.

TWO CONTRASTING APPROACHES TO ECHOLOCATION

Most bats avoid self-deafening due to signal-echo overlap by separating the signal and its echo in time. They produce signals a small percentage of the overall time and do not broadcast signals and receive echoes at the same time. This is the **low-duty-cycle** approach to echolocation. Duty cycle refers to the percentage of time that a signal is produced: a low-duty-cycle bat emits signals less than 20 percent of the time, less than 200 milliseconds out of 1000 milliseconds. Because the time between a signal and its echo becomes progressively shorter as a bat approaches its target, a bat must drastically shorten its signals during an attack sequence to continue to avoid signal-echo overlap. During the final pursuit of an insect, a low-duty-cycle bat not only shortens the signals but increases the repetition rate (Fig. 22-2), which probably enables the bat to track the final-instant movements of the prey (Fenton and Bell, 1979). This drastic increase in signal rate produces the "feeding buzz."

A fundamentally different echolocation strategy enables some bats to detect the fluttering wings of insects and to track precisely the trajectories of prey by perceiving Doppler-shifted echoes. Echolocation in these bats — *Pteronotus parnellii* (Fig. 22-4), a mormoopid, and probably all species of the Rhinolophidae — is characterized by **high duty cycles** (greater than 60 percent). That is to say, these bats emit signals for more than 600 milliseconds out of 1000 milliseconds. These bats can produce pulses and receive echoes at the same time. The signals are dominated by long, constant-frequency components, and the bats separate signal from echo by using frequency (Doppler-shift compensation). These bats perceive Doppler shifts in the echoes from targets approaching or moving away (Simmons, 1974), and shifts generated by fluttering insect wings in echoes of the constant-frequency part of the echolocation signal (Schnitzler et al., 1983; Schnitzler, 1987). Some flutter-detecting bats studied by Schnitzler (1987) could discriminate between the wing-beat signatures of different insects (Fig. 22-9). They thus minimize self-deafening during signal emission by tuning the inner ear to

different frequencies than those of emitted signals. In *Rhinolophus ferrumequinum,* for example, the organ of Corti, which transduces vibrations into neural impulses to the brain, is most sensitive to frequencies just above and below those of the outgoing signal (Vater, 1987, 1998). This style of echolocation, seemingly present as early as the Eocene (Habersetzer and Storch, 1987), enables bats to detect fluttering insects against a complex but nonfluttering background (Bell and Fenton, 1984; Vater, 1987). The reliance on flutter-detection by high-duty-cycle bats is suggested by observations in Italy of foraging *R. ferrumequinum* that only attacked insects whose wings were beating (Griffin and Simmons, 1974) and by comparable observations of *P. parnellii* in the laboratory (Goldman and Henson, 1977).

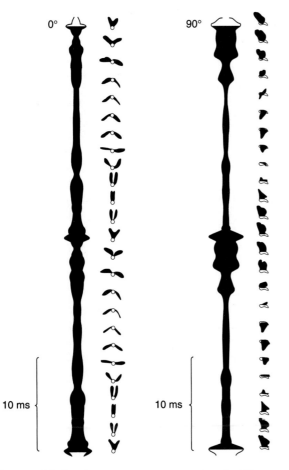

FIGURE 22-9 Changes in echo intensity of a CF pulse during the wing-beat cycle of a flying moth. The bat can use the amplitude changes to determine the wing-beat frequency of the prey, as well as its orientation with respect to the bat. Large, high-intensity glints are produced when the moth's wing surface area is greatest. *(After Schnitzler et al., 1983 © Springer-Verlag, New York, Inc.)*

THE ROLE OF ECHOLOCATION IN THE LIVES OF MAMMALS

Echolocation in land mammals probably evolved from communication calls and first became a means of detecting obstacles in dim or unpredictable lighting (Gould, 1970, 1971). This seemingly remains the sole function of echolocation in shrews. Tactile and olfactory cues and sounds made by prey are used by shrews for the recognition and tracking of prey. Odontocetes clearly use echolocation for general obstacle avoidance, and some may use echolocation for locating and tracking prey. The periodically increased click rates of deep-diving sperm whales are probably associated with the pursuit of prey (Goold and Jones, 1995; Watkins, 1977). But the importance of echolocation to odontocetes and its specific use during foraging is poorly known.

Regarding the importance of echolocation to bats, however, there is little uncertainty. Echolocation plays a central role in the lives of all microchiropteran bats that have been studied. The foraging strategies of bats that eat flying insects are built around the ability to echolocate. Echolocation in these bats has become highly specialized, allowing them not only to orient themselves and to locate obstacles but to detect, assess (as to size, wing-beat frequency and so on), and track flying insects. These bats can discriminate between types of insects (moth versus beetle, for example) and can track flying insects against a close and complex background of vegetation. Some bats that do not concentrate on flying insects but glean insects or small vertebrates from vegetation can perceive details of the texture of prey (Schmidt, 1988).

Some insectivorous bats may use other means than echolocation for locating prey. Sounds made by movements of their prey are used by some bats, and some may use flight sounds, calling sounds, or smells as cues (Fiedler, 1979; Kolb, 1976; Rentz, 1975; Thies et al., 1998). Little is known about the use of vision in prey detection by insectivorous bats. Bell (1982a) found that *Macrotus californicus* can use vision to find prey, and *Lavia frons* (Megadermatidae) seems to use vision to detect large, high-flying insects silhouetted against a twilight sky (Vaughan and Vaughan, 1986).

ECHOLOCATION PERFORMANCE

One aspect of echolocation performance in air is its strictly short-range usefulness. Another aspect is acuity, considered here as the smallest detectable differences in location, size and texture of a target. Simmons (1973) tested the range-detecting ability of one species each from the families Vespertilionidae, Phyllostomidae, Mormoopidae, and Rhinolophidae. All of the bats could discriminate range differences of from 1 to 2 centimeters, and this remarkable performance was apparently achieved by cross-correlation of the transmitted pulse with the returning echo. The essential variable from which the bats estimate distance is the time it takes for a pulse to reach a target and the echo to return. By comparing the relative **target-ranging** abilities of the bats in relation to the bandwidths of their echolocation pulses, Simmons determined that the FM components of the pulses are used for target ranging. Laboratory studies on a limited number of species have further demonstrated the ability to discriminate minor differences in target size, shape, and direction (Pollak and Casseday, 1989). In the laboratory, *E. fuscus* detected differences of less than 1 millimeter in the depth of small holes in a target (Simmons et al., 1974). The holes modified the spectrum of the echo from the target by absorbing sound energy at certain frequencies in the bandwidth of the bat's FM sweep; with changes in hole depth the absorption peaks shifted to different frequencies.

Schmidt (1988) did laboratory experiments on texture discrimination by echolocating *Megaderma lyra*, a bat that gleans prey from vegetation or the ground. This bat distinguished depth differences of only 0.2 millimeters, an ability seemingly based on the perception of frequency modifications in the spectra of echoes. This ability would be extremely useful in detecting differences in textures of mice or frogs, common prey of this bat. Apparently, bats associate features of the echo spectrum with target shape and texture when selecting prey.

Bats that use the high-duty-cycle approach to echolocation (*P. parnellii* and rhinolophids) precisely track the flight trajectories of insects. The longer the CF component of the pulse the greater the sensitivity to target velocities. Computed velocity resolutions listed by Simmons et al. (1975) indicate that the European *R. ferrumequinum,* with its very long narrow-band signal (up to 60 milliseconds), can perceive relative target velocities of less than 0.04 meters per second; *P. parnellii,* with its fairly long signals (up to 28 ms) can detect target velocities as low as 0.10 meters per second. These bats concentrate on the trajectory of prey while remaining aware of clutter (echoes returning from other than the target of interest). The ability to track prey with precision enables some high-duty-cycle bats to intercept rather than pursue prey, thus reducing energy expended on foraging.

FACES AND EARS AND ECHOLOCATION

The strange faces of bats, always a source of amazement to those unfamiliar with these animals, may have an important function in connection with echolocation. When mormoopids are echolocating, their lips are formed into megaphones that seemingly focus echolocation calls (Fig. 22-4). Möhres (1953) showed that the complex horseshoe-like structure surrounding the nostrils of rhinolophids serves as a diminutive megaphone to focus the short-wavelength pulses emitted by these bats into a beam; the 80-to-100 kilohertz pulses have wavelengths of only 3 or 4 millimeters. In addition, because the nostrils are situated almost exactly 0.5 wavelength apart, the pulses emitted through the nostrils undergo interference and reinforcement that tends to beam the pulses (see also Schnitzler and Grinnell, 1977; Pye, 1988). Prominent noseleaves occur in four chiropteran families: Rhinolophidae, Nycteridae, Megadermatidae, and Phyllostomidae (Fig. 22-10). In nasal emitters, noseleaves enhance the directionality of sound emissions (Hartley and Suthers, 1987). The facial patterns of many bats may well function similarly to direct pulses such that some species can scan their surroundings with a concentrated beam of sound, much as we probe the darkness with a flashlight beam. Arita (1990) surveyed noseleaf morphology of 46 phyllostomid species and hypothesized that noseleaf morphology is correlated with foraging behavior. Bogdanowicz et al. (1997) showed that noseleaf morphology correlates with diet (fruit versus nectar) but not with the type of echolocation calls.

The **tragus** (Fig. 22-10A), an often bladelike structure that is a prominent part of the external ear in most microchiropterans (but lacking in the Rhinolophidae), plays an important role in localizing sound (Lawrence and Simmons, 1982). Laboratory experiments using *E. fuscus* demonstrated that sounds reflected from the pinna to the tragus produce a secondary echo of sounds entering the ear canal. The time between this echo and the sound directly entering the ear canal encodes the vertical direction of a sound. Bats with the tragus intact could perceive changes of 3 degrees of arc in vertical angles separating targets; discrimination in bats with the tragus folded down was reduced to 12 to 14 degrees.

Obrist et al. (1993), who studied the external ears of 47 species of bats, found that, in general, the pinnae serve to focus and amplify sound and to increase the directionality of hearing. In many species the pinnae are "tuned" to the bats' echolocation calls. That is to say, for a given species, the frequencies of the echolocation calls with the most sound energy are those most strongly amplified by the pinnae. The structure of the pinnae in relation to echolocation calls is discussed in the section below on echolocation calls and foraging strategies.

Some high-duty-cycle bats, adapted to flutter detection and to tracking trajectories of prey, enhance their ability to scan their surroundings by coordinating body, head, and ear movements. As an example, when the African species *Hipposideros commersoni* hangs from an acacia branch and scans for insects, its pendant body revolves continuously back and forth through an arc of approximately 180 degrees; the head is in constant motion up and down, and the tips of the ears vibrate forward and backward. These movements seem to allow the

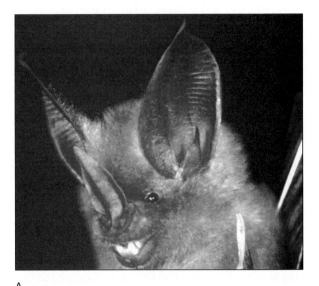

A

B

FIGURE 22-10   (A) The large, spear-shaped noseleaf of a *Mimon bennettii,* Phyllostomidae; (B) the distinctive noseleaf of a horseshoe bat, *Rhinolophus ferrumequinum,* Rhinolophidae. *(A by B. Fenton and B by M. Andera)*

bat to use its beamed signals to meticulously scan its surroundings. But why the ear movements? Simmons et al. (1975) proposed that, by moving the direction the ears aim, these bats scan the vertical plane. The rapid ear movements are out of phase: The tip of one ear moves forward while the other moves backward. These movements are in approximate synchrony with the signals emitted by the bat. These ear movements, alternately toward and away from the target, may heighten the Doppler shift of echoes from moving targets and improve the bat's discrimination of movement (Simmons et al., 1975; Valentine and Moss, 1998). One cannot help but marvel at the elaborate neural coordination of stereotyped body, head, and ear movements with signal emission rate.

EVOLUTION OF ECHOLOCATION

Gould (1970, 1971) hypothesized that the sonar pulses of bats were derived originally from vocalizations that established or maintained spacing or contact between individuals. The repetitive communication sounds used by infant bats and similar pulses, perhaps used originally during flight to maintain adequate spacing of foraging individuals, may have become important secondarily for detecting prey and avoiding obstacles. According to Gould (1971), "The prominence with which continuous, graded signals pervade the lives of such social and nocturnal mammals as bats suggests that echolocation is an inextricable and integral part of a communication system." This author suggests that some of the vocalizations used by early bats may have been inherited from their insectivorous ancestors, in which auditory communication was perhaps as important as it has been shown to be in some living insectivorans (Eisenberg and Gould, 1970; Gould, 1969).

More recently, Fenton et al. (1995) proposed that the ancestors of bats were small, nocturnal gliders that produced low-duty-cycle clicks (short pulses separated from the returning echoes by relatively long interpulse intervals). These clicks were used primarily for orientation as the animals glided from tree to tree. Later, the ancestors of bats replaced these simple orientation clicks with tonal signals of greater intensity. Tonal signals increased the distance at which flying insects or obstacles could be detected. Fenton and his colleagues hypothesized that only after echolocation has been optimized for the detection of flying prey did flapping flight evolve. The evolution of flight, coupled with sophisticated echolocation, led to the rapid diversification of bats in the Eocene. Further advances in echolocation design, such as high-duty-cycle pulses (pulses that last from 500 to 800

milliseconds each) and the ability to achieve self-deafening during pulse emission, allowed bats to specialize on a variety of food sources.

An alternative hypothesis has been offered by Simmons (Simmons, 1994; Simmons and Geisler, 1998). She suggests that the common ancestor of both Microchiroptera and Megachiroptera could already fly, but showed no specializations for echolocation. Megachiropteran bats diverged early on from this flying ancestor and independently evolved a less sophisticated form of echolocation in a few species. Studies of fossil bats, such as *Icaronycteris* and *Palaeochiropteryx,* reveal that early Microchiroptera already had enlarged hyoid bones for the attachment of large throat muscles and an enlarged cochlea in the inner ear typical of modern Microchiroptera (Simmons and Geisler, 1998). These features seem to indicate that flight evolved before echolocation in the Microchiroptera.

ECHOLOCATION CALLS AND FORAGING STRATEGIES

Adaptive radiation in foraging styles among insectivorous bats has been accompanied by the evolution of a variety of echolocation signal types. The following discussion considers how different echolocation strategies enable bats to exploit contrasting modes of foraging and the degree to which bats that forage similarly and occupy similar habitats share similar styles of echolocation and certain morphological features.

OPEN-HABITAT BATS

A number of bats within the Molossidae, Emballonuridae, and Vespertilionidae forage for insects in open situations such as meadows, above canopies of vegetation, or, in some cases, up to 600 meters above the ground (Fenton and Griffin, 1997). These bats are typically fast fliers and, when foraging, encounter no obstacles except insects and other bats. Most of these bats have narrow wings with pointed tips (high-aspect-ratio wings; Fig. 22-11). The energetic cost of flight in these bats is low (Norberg and Rayner, 1987), and all seem to remain in flight while foraging. Some are remarkably enduring fliers, remaining on the wing continuously for up to 6 or 7 hours (Table 22-2).

Most of these bats use long, high-intensity, low-frequency, narrow-band search-phase calls (a, i, and l in Fig. 22-12) that have low duty cycles (less than 20 percent). Low frequencies are relatively immune to atmospheric attenuation and are therefore ideal for maximizing the range of target detection. In general, these search-phase calls enhance long-range (for echolocation) detection of large prey. All six species of high-flying molossids studied in Zimbabwe had search-phase signals that conformed to this pattern (Fenton and Bell, 1981; Fenton and Griffin, 1997). The calls were long (10 to 20 milliseconds), of low frequency (mostly below 20 kilohertz) and had fairly narrow bandwidths (5 to 10 kilohertz). The calls of five of these species were within the range of human hearing. Among eight sympatric Neotropical emballonurids

studied by Kalko (1995), the search-phase calls with the lowest frequencies and longest durations were those of the two species that foraged above the forest canopy (Table 22-3). Because smaller insects are undetectable by low frequency signals, open-habitat bats probably capture mostly large insects. But some small vespertilionids and emballonurids that forage in the open use moderately high-frequency signals and can detect smaller insects. During the attack-phase of foraging, open-habitat bats (like other bats that capture flying insects) increase the rate and bandwidth of their echolocation signals.

Many open-habitat bats are flexible with regard to echolocation calls (Obrist, 1995). One such species is the northern bat *(Eptesicus nilssoni)* of Europe: in Sweden this bat used higher frequency

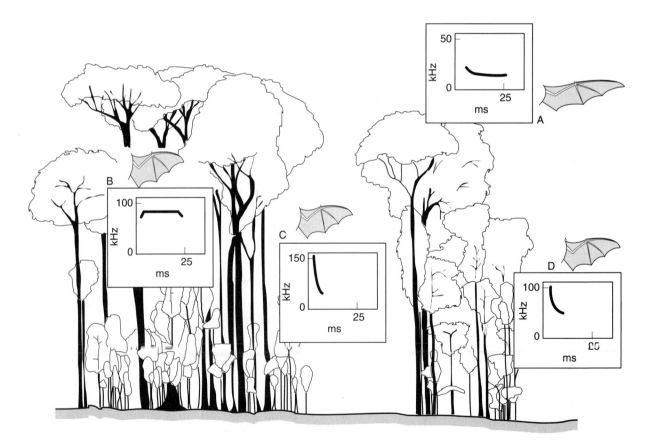

FIGURE 22-11 Habitat partitioning in a bat community. Bats foraging in open spaces above the canopy typically have high-aspect-ratio wings and produce low-frequency FM calls (A). Those species that forage in dense, cluttered spaces within the forest tend to have low-aspect-ratio wings and produce either long CF pulses or steep FM pulses of relatively high frequency (B and C). Bats that forage along the edges of vegetation often have long, rounded wings and emit steep FM calls of intermediate frequency (D).

TABLE 22-2   Aspects of Foraging Behavior of Some Animal-Eating Bats

	Number of bats	Longest flight (min)	One-way distance (km)	Foraging strategy	Source
Megaderma lyra	17	4	45	CF, SF	Audet et al., 1988
Nycteris grandis	4	2	27	CF, SF	Fenton et al., 1987
Rhinolophus hildebrandti	10	3	113	CF, SF	Fenton & Rautenbach, 1986
Myotis myotis	21	9	307	CF	Audet, 1990
Eptesicus fuscus	78	5	120	CF	Brigham & Fenton, 1986
Scotophilus borbonicus	9	5	62	CF	Fenton & Rautenbach, 1986
Nyctalus noctula	32	3	270	CF	Kronwitter, 1988
Lasiurus cinereus	23	20	436	CF	Barclay, 1989
L. borealis	4	0.6	206	CF	Hickey, 1988
Euderma maculatum	3	12	400	CF	Wai-Ping & Fenton, 1989
Mops midas	10	15	100	CF	Fenton & Rautenbach, 1986

After Fenton, 1990. *Can. J. Zool.*, 68: 411–422.
Data from field observations of known individuals carrying radio transmitters.
CF = denotes constant flight, SF = short flights from perches.

calls when foraging close to the ground than when foraging high above the ground (Rydell, 1993).

Maximizing the range of target detection is of prime importance to rapid-flying bats. This is done in part by increasing echo intensity. This is one function of the pinnae, which act like acoustic magnifying glasses: by focusing (amplifying) sound, the pinnae increase the sound pressure arriving at the tympanic membranes (**pinna gain**). In open-habitat bats, the pinnae are "tuned" (in shape and size) to enhance reception of the main frequency of the search-phase call and to improve the localization of targets (Obrist et al., 1993). In *Miniopterus schreibersi*, for example, the area of best

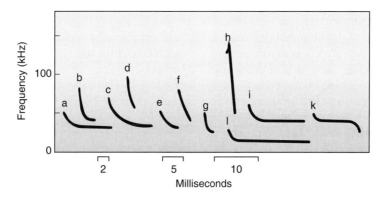

FIGURE 22-12   Diagnostic echolocation calls of members of a bat community near Portal, Arizona.   a, *Lasiurus cinereus;* b, *Myotis volans;* c, *Eptesicus fuscus;* d, *Pipistrellus hesperus;* e, *Myotis thysanodes;* f, *Myotis californicus;* g, *Antrozous pallidus;* h, *Myotis auriculus;* i, and k, *Tadarida brasiliensis* (i, an unmodified echolocation call; k, a honk); and l, *Nyctinomops macrotis. (From Fenton, 1984,* Quarterly Review of Biology, *University of Chicago Press)*

TABLE 22-3   Characteristics of Echolocation Search-Phase Signals
of Six Neotropical Emballonurids

Species	CF components (kHz)	Sound duration (ms)	Pulse interval (ms)
Diclidurus albus[a]	24.3 (20)[b]	11.3 (17)	183.3 (17)
Peropteryx sp.[a]	25.4 (28)	9.3 (17)	84.0 (17)
Cormura brevirostris	42.0 (60)	5.2 (54)	102.3 (48)
Saccopteryx bilineata	45.1 (60)	9.4 (41)	85.8 (41)
S. leptura	52.7 (41)	5.3 (31)	60.1 (25)
Rhynchonycteris naso	102.5 (70)	5.0 (90)	53.5 (82)

Data from Kalko, 1995.
[a]The first two species forage in the open, whereas the others forage near vegetation.
[b]Sample sizes are in parentheses.

hearing is at the lowest range of the echolocation call, where most of the signal energy is concentrated (Obrist et al., 1993). Large ears increase low-frequency sound pressure gain but are not similarly effective with high frequencies. Thus, the high-flying molossids that use low-frequency search-phase signals tend to have large ears. The hoary bat (*Lasiurus cinereus*), a vespertilionid, has such a signal but has rather small ears. Obrist et al. (1993) found that the pinnae of this bat are tuned to the second harmonic of its call and speculate that because this bat occurs seasonally at high latitudes and roosts in foliage small ears may be important in reducing heat loss.

The range of target detection in open-habitat bats with narrow-band search-phase signals may be further extended by the detection of acoustic "glints." These are produced by an echolocation signal that hits a flying insect at the instant in its wing-beat cycle when its wings are perpendicular to impinging sound waves (Fig. 22-9). The sound pressure levels of echoes from flying insects can thus be increased up to 20 decibels relative to echoes from nonflying insects (Kober and Schnitzler, 1990). Kalko (1995) estimated that some Neotropical emballonurids with high-duty-cycle, narrow-band signals could detect acoustic glints at the rate of 9 to 15 per second.

To summarize, open-habitat bats optimize long-range target perception by using intense, long, low-frequency, narrow-band calls and by having pinnae tuned to the main frequency of the search-phase call. Some species may enhance long-range echolocation by the perception of acoustic glints. These bats tend to have narrow wings and are rapid, enduring fliers that may forage in edge habitats but never forage in closed habitats.

EDGE-HABITAT BATS

Many, perhaps most microchiropteran bats forage in edge situations (Fig. 22-11). These bats forage in such places as the borders of woodlands, along streams, or along cliff faces or the edges of arroyos. In a typical perceptual setting faced by these bats, obstacles and clutter are on one side and largely uncluttered space is on the other. Insectivorous bats using edges often capture prey near vegetation, but usually do not fly within its clutter. Flight in these bats is typically slow, highly maneuverable, and energetically economical. The wings are usually long, with short, rounded wingtips, low wing loadings, and moderately high aspect ratios (Norberg and Rayner, 1987). Many of these bats forage while continuously in flight, but probably most flights are no longer than 120 minutes. At least some members of the Mormoopidae, Phyllostomidae, Rhinolophidae and Vespertilionidae are edge-habitat bats.

The search-phase signals of these bats tend to be short and intense, with a combination of broadband and narrow-band components. Many of these bats use shallow FM search-phase calls that allow for medium or short-range detection of targets and obstacles (b, c, d, e, f, g in Fig. 22-12). By increasing the bandwidth of their signals, bats can increase the precision of target localization (Simmons and Stein, 1980). Some bats increase bandwidth by adding harmonics; other bats do this by increasing the range of frequencies through which the signals sweep. The latter strategy is used by many vespertilionids. The echolocation call of *Myotis auriculus,* an extreme example, sweeps steeply through a range of roughly 100 kilohertz (h in Fig. 22-12). Some edge-habitat species, such as the mormoopid *P. parnellii* and some rhinolophids, use Doppler-shift flutter detection; their calls are dominated by constant frequency components. The vespertilionid *Antrozous pallidus* is remarkable because it often stops giving echolocation signals when hunting insects (Bell, 1982a).

Many species of bats, including many that forage along edges, adapt their echolocation behavior to suit different situations (Obrist, 1989). Such a bat is *Lasiurus borealis,* which uses narrow-band search-phase and broad-band attack-phase calls. In this bat, and perhaps many other vespertilionids, the pinnae enhance perception of the dominant frequency of the search-phase call (Obrist et al., 1993). Among rhinolophids, many of which forage along edges, matching of external ear characteristics with the dominant echolocation frequency is especially striking. The pinna gain in some species is as high as 30 decibels.

Many species that use echolocation calls adapted primarily to one habitat can forage in others. Phyllostomids such as *Artibeus* and *Glossophaga* are good examples: these bats have low-intensity echolocation calls and can forage amid clutter but also forage along edges.

CLOSED-HABITAT BATS

A number of bats, particularly those of tropical forests, forage amid or close to the clutter of branches and foliage. Thies et al. (1998) observed that for many such bats "the returning echoes of the food are often buried in a multitude of clutter — echoes from leaves, branches, and surfaces on which the food rests." Among these bats are at least some members of the families Emballonuridae, Nycteridae, Megadermatidae, Rhinolophidae, Phyllostomidae, and Vespertilionidae. Their diets range from invertebrates and small vertebrates to fruit and nectar. Flight is slow and highly maneu-

verable. The wings are broad with rounded wingtips and have low wing loadings and low aspect ratios (Fig. 22-11); this design is associated with energetically expensive flight (Norberg and Rayner, 1987). High energetic cost is the price these bats pay for the ability to hover and to make complex, split-second maneuvers. Although details of foraging behavior for most species are unknown, most seem not to remain in flight for long periods. Some species are sit-and-wait predators with brief foraging flights. The African megadermatid *Cardioderma cor,* during a night of foraging, made flights lasting an average of less than 5 seconds and was in flight for a total of less than 11 minutes (Vaughan, 1976). Some closed-habitat species, however, make flights of up to 113 minutes (for example, *Rhinolophus hildebrandti;* Table 22-2).

Most of these bats use short (less than 2 milliseconds), broad-band, steep FM calls of low intensity (less than 70 decibels). This signal design yields precise, close-range information of obstacles and targets (Simmons and Stein, 1980). Low-intensity calls minimize echoes from all but nearby clutter and may also partially avoid alerting prey (Fullard, 1987). Among closed habitat bats are many nasal emitters, including phyllostomids, rhinolophids, nycterids, and megadermatids. Rhinolophids that use closed habitats have narrow-band echolocation calls and use flutter detection.

Of particular interest is the dependence by some of these bats on olfaction or prey-generated sounds, rather than echolocation, for finding food. Thies et al. (1998) found that in flight cages the Neotropical frugivores *Carollia perspicillata* and *C. castanea* flew slowly (2 to 3 meters per second) and used echolocation continuously while foraging but located fruit by olfaction (see also Kalko and Condon, 1998). Realistically shaped artificial fruit that would have been indistinguishable by echolocation from real fruit was ignored. The high-frequency, multiharmonic calls of these bats provide detailed, close-range information. *Megaderma spasma* of India uses echolocation for perceiving its surroundings but can locate flying insects by using prey-generated sounds (Tyrrell, 1988). The Neotropical *Trachops cirrhosus* uses mating calls of male frogs for locating prey, but, during most attacks, the bat continues to give echolocation calls (Tuttle and Ryan, 1981). This bat attacked a tape recorder broadcasting frog calls, indicating that echolocation was not used for target detection. Two African species of *Nycteris* use this combination of echolocation and prey-generated sounds during foraging (Fenton et al., 1983). Both *Megaderma lyra* (Fiedler, 1979) and *Macrotus californicus* (Bell, 1982a) can locate nonflying prey by using prey-generated sounds; under light conditions equaling bright starlight, the latter species uses vision rather than echolocation when searching for nonflying prey (Bell, 1982b).

Many closed-habitat bats are gleaners, taking prey from the surfaces of vegetation, rocks, or the ground. These species tend to have large ears that amplify low-frequency sounds (below 15 kilohertz), such as those made by prey rustling through leaves. The pinnae of these bats are remarkable in their ability to localize low frequency sounds (Obrist et al., 1993). Among species that forage at short range and use broad-band echolocation calls with no dominant frequency, such as fruit- or nectar-eating phyllostomids, pinna gain and ability to localize sounds are only moderately developed.

DETECTION OF PREY BY FISHING BATS

The Neotropical fishing bat *Noctilio leporinus* locates prey by detecting ripples caused by small fish or by recognizing parts of the fish that break the water surface. During attacks on prey, the sequence is from short CF/FM to FM signals and only one harmonic is used. This fairly simple, single-harmonic pattern seems well adapted to this bat's unique foraging style. Of central importance is the detection of a disturbance on a fairly uniform background, a problem similar to that facing a bat flying high above obstacles.

VARIABILITY AND MULTIPLE USES OF ECHOLOCATION CALLS

Echolocation signals are used by bats for communication, as well as for orientation and locating prey. In addition, modifications in signal design allow some bats to exploit a variety of habitats. It is to be expected, then, that echolocation is flexible.

Echolocation calls display considerable geographic variation within a species. Thomas et al. (1987) found (in an extreme case) that *L. cinereus* in Arizona had calls with a minimum frequency 53.8 percent higher than the minimum frequency

of calls in Manitoba (26.0 kilohertz versus 16.9 kilohertz). Eight of 12 species showed geographic variation of at least 3 kilohertz in minimum echolocation frequency.

Individual variation in echolocation calls is known for some species. For example, a male *Euderma maculatum* foraging not far from a female of this species had calls of lower frequencies and longer intervals (437 milliseconds versus 352 milliseconds; Obrist, 1995). In British Columbia, *Myotis evotis* calls varied in frequency range from 30 to 86 kilohertz to 54 to 97 kilohertz (Thomas et al., 1987).

Much of the variability in call design is probably a response to conspecifics. When conspecifics flew by, all four species of vespertilionids studied by Obrist (1995) decreased call duration and increased the intervals between calls. Signals were "personalized" by three of these species by increasing frequency, and one species drastically increased the intensity of its calls when foraging near other lasiurines. One of these species *(Lasiurus borealis)* sharply altered its signal design in the presence of conspecifics (Fig. 22-13). Under certain circumstances, some bats add an extra pulse to the echolocation calls. On a collision course toward another bat, an *N. leporinus* lowered its call frequency and added a warning "honk" to the signal (Suthers, 1965). Fenton and Bell (1981) recorded a similar honk given by a *T. brasiliensis* when near other bats. *Rhinopoma hardwickei* consistently maintains CF signals when flying alone but changes frequencies when flying in a group (Habersetzer, 1981).

Echolocation calls communicate several kinds of information. Barclay (1982) reported that the gregarious *Myotis lucifugus* relied on signals of conspecifics to locate day roosts, mating sites, hibernation sites, and feeding areas. The solitary forager *E. maculatum*, in contrast, reacted aggressively to playbacks of calls of a conspecific individual. This bat either attacked the speaker or abruptly moved away (Leonard and Fenton, 1984). Apparently this species' low-frequency calls, with most of the sound energy at about 10 kilohertz, provide fairly long-range warnings that maintain spacing between individuals.

Echolocation calls also provide vocal signatures. This is the basis for mother-young recognition in a number of species (for example, *R. ferrumequinum*,

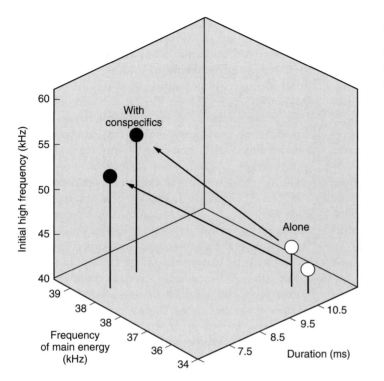

FIGURE 22-13 Echolocation calls of two *Lasiurus borealis* changed (arrows) when conspecifics were foraging nearby. Calls in the presence of conspecifics were relatively shorter in duration, had higher frequencies for the main energy component, and began at higher frequencies. *(After Obrist, 1995)*

Matsumura, 1981; *A. pallidus,* Brown, 1976). In a captive colony of *R. ferrumequinum,* individuals recognized each other by listening to echolocation calls and had clear roostmate preferences (Möhres, 1967). Vocal signatures and individual recognition may be important also during interactions among foraging bats (Obrist, 1995).

An additional kind of flexibility allows some bats to optimize echolocation performance in different habitats (Simmons et al., 1978). A number of species are known to alter their signal design (by adding steep FM components) when changing from foraging in open areas to those closer to obstacles (for example, *Eptesicus nilssoni,* Rydell, 1990; *Lasiurus borealis,* Obrist, 1995; *Nyctalus noctula,* Zbinden, 1989).

BATS VERSUS MOTHS

Studies by Roeder and his colleagues (Roeder and Treat, 1961; Dunning and Roeder, 1965; Roeder, 1965; Dunning, 1968; summarized by Fullard, 1987) have revealed a remarkable series of adaptations by certain moths in response to predation by bats. Nocturnal moths of the families Noctuidae,

Ctenuchidae, Geometridae, and Arctiidae have an ear on each side of the rear part of the thorax. Each ear is a small cavity within which is a transparent membrane. The ears are sensitive to a wide range of frequencies and allow the insects to detect the ultrasonic pulses of foraging bats. Upon detecting the approach of a bat, the moths alter their level flight and adopt various erratic flight patterns. Some members of these families of moths have carried the business of evading bats to an even greater extreme and have a noise-making organ on each side of the thorax. When the moths are disturbed, these organs produce trains of clicks with prominent ultrasonic components.

Under laboratory conditions, flying bats about to capture mealworms tossed into the air regularly turned away from their targets when confronted with recorded trains of moth-produced pulses. These pulses apparently protect moths from bats; probably these signals interfere with the echoes returning to the bat and deafen them temporarily. Alternatively, some arctiid moths, and perhaps some noctuids, are unpalatable to bats, and their identification by a bat having previous experience with them might be aided by the moth's ultrasonic clicks. Captive *Myotis lucifugus* avoided three species of arctiid moths when the moths were sounding off but bit into them when they were quiet (Dunning, 1968). Two of the moths were rejected when tasted, but the third, presumably a palatable form mimicking the sounds of the unpalatable species, was eaten. Additional evidence suggests, however, that some moths jam a bat's echolocation system by producing sounds that resemble the terminal buzz of a bat closing in on an insect (Fullard et al., 1979).

Investigations of the strategies bats use to hear and feed on moths have shown that predator-prey interactions between these animals are complex, with considerable fine-tuning of adaptations favoring prey detection, predator avoidance, and inconspicuousness on the part of both predator and prey (Fenton and Fullard, 1981). In moth-bat confrontations, the relative times or distances at which each animal detects the other is critical. Bats using loud signals in the 20- to 50-kilohertz range can be detected by some moths at distances up to 40 meters, whereas bats probably detect moths at no more than roughly 20 meters. Neither bats nor moths do consistently well, however, for some bats detect insects at ranges of only 1 meter, and experiments using an African slit-faced bat *(Nycteris macrotis)* showed that moths cannot detect this bat at distances greater than 0.2 meter. Nevertheless, the fact that ears are present in more than 95 percent of the moths of Ontario and in more than 85 percent of those in southern Africa suggests that sound detection is important to moths.

Many moth species have apparently become tuned to bat signals. In most areas, the echolocation frequencies used by bats are those to which the ears of moths are most highly sensitive. Most bats of North America have echolocation signals at frequencies between 15 and 60 kilohertz, and the ears of moths of this region are tuned to these frequencies. Since some bats obviously eat moths, however, we might well ask how such bats avoid the early warning systems of moths.

One major bat strategy is to use frequencies to which moths are relatively insensitive. Bats that use echolocation signals of low intensity and those that have extremely high-frequency signals can escape detection except at close range. Frequencies above and below the range used for detection by moths are called **allotonic frequencies.** In the Ethiopian, Oriental, and Australasian tropics, roughly one third of the species of insectivorous bats use allo-

tonic signals and thereby presumably partially avoid detection by moths. Fenton and his associates have found that, in many cases, bats with such signals tend to specialize on moths. Each of three moth-eating rhinolophids was found to use very high frequencies (100, 139, and 210 kilohertz). There are costs to using allotonic frequencies, however, because very high frequencies are severely attenuated, and low frequencies provide poor target discrimination. The benefits of remaining undetected by the prey probably outweigh the costs for many bats, including some North American vespertilionids. *Myotis evotis* and *M. keenii* both emit very high-frequency pulses that are not detected by their main prey, noctuid moths (*Catocala* spp.; Faure et al., 1990; 1993). On the opposite side of the auditory spectrum, the vespertilionid *Euderma maculatum* uses extremely low-frequency pulses (9 to 12 kilohertz) while foraging for moths (Woodsworth et al., 1981; Leonard and Fenton, 1984; Obrist, 1995). Fullard and Dawson (1997) tested the hypothesis that the use of allotonic frequencies by *E. maculatum* allows them to remain "acoustically inconspicuous" to sympatric eared moths. Their results indicated that the short-duration, low-frequency calls of *E. maculatum* are not easily detected by the moths. Fullard and Dawson estimated that eared moths would be unable to detect the presence of *E. maculatum* until the bats were within one meter of the moth, while *Eptesicus fuscus* would be detected at a distance of over 20 meters.

Another strategy that allows bats to remain inconspicuous to moths is that of partially abandoning echolocation. Some bats, such as some megadermatids, do not use echolocation for detecting prey but instead listen for sounds made by the prey while remaining silent themselves.

Bat-moth predator-prey interactions provide an interesting example of evolutionary gamesmanship. Echolocation is a major key to the great success of bats, allowing them to perceive and capture insects in darkness. Bat echolocation signals provide moths (and other insects) with a means of detecting foraging bats, however, and moths have developed antipredator behaviors that depend on this ability. A third countermove is seemingly in progress: the shift to allotonic echolocation among some moth-specialist, insectivorous bats may be a response to early detection by eared moths (Fenton and Fullard, 1981). Why, then, haven't eared moths responded by tuning their ears to these allotonic frequencies? Fullard (1982) and Fullard and Belwood (1988) suggested that eared moths have not adapted because their ears are tuned to the average "echolocation frequencies of all of the bats that present a significant risk." Common bats, such as *E. fuscus,* present a disproportionately greater risk to moths than the relatively rare *E. maculatum.*

ECHOLOCATION IN CETACEANS

Just as bats must cope with darkness, cetaceans frequently must perceive their underwater environment under conditions that render vision difficult, if not impossible. In some waters inhabited by cetaceans, suspended material such as soil particles or plankton limits visibility to a few meters or even a few centimeters. Water transmits light poorly, and, even under ideal conditions, visibility under water is limited. Also, some cetaceans forage at considerable depths, where there is not a trace of light. It is not surprising, then, that some cetaceans have developed echolocation.

Probably all odontocetes (toothed whales and dolphins) use echolocation for detecting obstacles and prey. Mysticetes (the baleen whales), however, are not known to echolocate. A tremendous variety of sounds, some having a fascinating, musical quality, are made by both mysticetes and odontocetes; the wailing, creaking, and squealing noises of cetaceans have become commonplace to sailors operating sonar equipment at sea (Table 22-4; Payne, 1970). Some of these underwater sounds may have a communication function (Clark and Clark, 1980). Biosonar emissions by odontocetes occur in two basic types: narrow-band continuous tones called whistles, used for intraspecific communication; and broad-band clicks used for echolocation. Acoustical energy moves through water very efficiently, making sonar ideal for the aquatic environment.

Since the account by Schevill and Lawrence (1949) of the underwater noises made by the white whale *(Delphinapterus),* considerable research has been done on the vocalizations of cetaceans. Much research has dealt with a common dolphin, *Tursiops truncatus* (Kellogg et al., 1953; Schevill and Lawrence, 1953; Wood, 1959; Norris et al., 1961; Lilly, 1962, 1963; Au et al., 1974). *Tursiops* is able to

TABLE 22-4 Characteristics of Sounds Produced by Representative Cetaceans

Suborder species	Common name	Sound type	Frequency range (kHz)	Frequency at max. energy (kHz)
Odontoceti				
Inia geoffrensis	River dolphin	click	25–200	95–105
Phocoena phocoena	Harbor porpoise	pulse	100–160	110–150
Delphinus delphis	Common dolphin	whistle	0.2–150	4–9
		click	0.2–150	30–60
Orcinus orca	Killer whale	scream	0.25–35	12
Stenella longirostris	Long-beaked spinner	click	1–160	60
		whistle	1–20	8–12
Tursiops truncatus	Bottlenose dolphin	click	0.2–150	60–80
		whistle	2–20	—
Physeter catodon	Sperm whale	coda	16–30	—
Mysticeti				
Balaenoptera musculus	Blue whale	moan	0.2–0.02	0.012–0.018
Balaenoptera physalus	Fin whale	call	0.16–0.75	0.02
Balaena mysticetus	Bowhead whale	call	0.1–0.58	0.14–0.34
Megaptera novaeangliae	Humpback whale	song	0.05–10	<4.0

From data summarized by Ketten, 1997.

detect obstacles and recognize food by means of echolocation; it uses short pulses resembling those of bats. *Tursiops* is capable of producing a great variety of sounds, but of primary importance for echolocation are the trains of clicks that it emits. The clicks are audible to humans but cover a wide spectrum of frequencies. The pulse rate rises as a porpoise approaches a target, and *T. truncatus* can distinguish between a piece of fish and a substitute water-filled capsule with a similar shape (Norris et al., 1961), or even between sheets of different thicknesses of the same metal (Evans and Powell, 1967). Pack and Herman (1995) presented bottlenose dolphins with a variety of complex shapes and showed that they were capable of immediately recognizing these shapes by echolocation.

Peak frequencies recorded in free-ranging dolphins by Au et al. (1974) were between 120 to 130 kilohertz, while those reported by Evans (1973) in an aquarium were close to 60 kilohertz. Au and his colleagues attributed the lower peak values in captive dolphins to possible interference reverberations from aquarium walls. Noisy environments may also affect the sonar signals used by cetaceans. The peak frequencies of click emission varied with location in the false killer whale *(Pseudorca)* from only 20 kilohertz to over 110 kilohertz in a noisy environment. These studies suggest that caution should be used in interpreting data gathered from animals held in tanks.

An intriguing alternative hypothesis concerning cetacean sonar has recently been advanced by Taylor et al. (1997). These authors suggest that some marine mammals may not use active sonar (hearing echos of sound pulses, such as bat echolocation or the ping of a submarine) or passive sonar (noisy objects are detected without the production of sound) at all. Active sonar advertises the position of the cetacean to the prey species. Indeed, there is evidence that dolphin echolocation clicks tend to alarm prey fish. Evidence that captive dolphins use something other than active or passive sonar comes from experiments in which eye cups were used to temporarily blind bottlenose dolphins *(Tursiops truncatus)* as they pursued live prey. Simultaneous recordings by the listening devices secured to the dolphin's head or by hydrophones suspended at various locations in the pool indicated that these dolphins did not use underwater clicks or other sounds to locate the prey, yet were

able to follow and capture 100 percent of prey fish in all trials (Taylor et al., 1997). These observations lead Taylor and her colleagues to propose that dolphins may be able to use ambient noise imaging to see their environments at close ranges.

Ambient noise imaging (ANI) is fundamentally different from both active and passive sonar systems and refers to the use of sound to see underwater. The physics of this type of underwater sound imaging are well beyond the scope of this discussion. However, Taylor and colleagues have calculated (using powerful computers and a model based on many years of research in dolphin hearing) what a dolphin might be able to detect using ambient noise imaging. These models predict that dolphins could use ambient noise imaging to "see" useful images for tens of meters underwater. If verified in living dolphins, it suggests that dolphins, and perhaps other marine mammals, have a whole new way of seeing with sound.

Where do the cetacean clicks and buzzes originate? The mechanism by which odontocetes produce their echolocation signals remains controversial. The two most likely sources of sound production are the larynx and the nasal air sac system. New imaging tools such as CAT, MRI, and RET scans, however, have enabled researchers to rule out the larynx as the source of echolocation signals in dolphins (Popper et al., 1997). Most of the recent evidence indicates that these signals are produced by vibrations of the nasal sac system near the nasal plugs (Fig. 22-14; Mackay and Liaw, 1981; Au, 1993; Popper et al., 1997). The nasal sac system consists of a series of muscular valves and compliant sacs associated with the blowhole. Muscles associated with these air sacs contract in synchrony with the echolocation clicks, while muscles surrounding the larynx do not (Ridgeway et al., 1980). Studies using x-ray computer assisted tomography (CAT scan) and magnetic resonance imaging (MRI) suggest that a pair of small, dorsal fatty projections (dorsal bursae) just posterior to the melon may be involved (Cranford et al., 1987). The paired dorsal bursae abut a liplike structure called the *museau de singe* (also called "monkey lips") that could control the passage of air through the system. Cranford (1988) hypothesized that the passage of pressurized air past the liplike structures or between the dorsal bursae produce sounds in much the same manner as the glottis in humans. This mechanism is identical to that proposed by Clarke (1979) for echolocating sperm whales. Exactly how cetaceans produce sonar signals is not definitely known, but the larynx has definitely been ruled out as a source of sound in dolphins (Au, 1993; Popper et al., 1997).

The role that the melon plays in focusing sounds is also poorly understood. It has been suggested, however, that sound waves produced by the nasal sac system are focused in a forward direction through the melon — a lens-shaped fatty structure that gives a domed profile to the forehead of many odontocetes (Fig. 22-14; Au, 1993). The lipid composition of the melon has been analyzed, and its acoustic properties suggest that it may serve as an "acoustic lens" to focus outgoing acoustic energy. Complex models of dolphin sound propagation developed by Aroyan (1990) show that the "melon is capable of some focusing, but by itself cannot explain the dolphin's directional beam."

The sperm whale (*Physeter catodon*) presents an interesting case of echolocation among cetaceans because it uses a unique mode of foraging that is probably made possible by echo ranging. (No proof that sperm whales echolocate is available, but this ability has been inferred on the basis of data from other odontocetes.) The click of a sperm whale is known to consist of a series of pulses (Backus and Schevill, 1966). The click lasts roughly 10 to 20 milliseconds and is composed of up to nine separate pulses. These vary in duration from 2 to 0.1 millisecond, and the interpulse intervals are 2 to 4 milliseconds. The clicks are repeated at rates from less than one click per second to 40 per second. One remarkable feature of sperm whale clicks is that a single pulse results in evenly spaced pulses of declining amplitude. Norris and Harvey (1972) hypothesized that this pulse pattern results from multiple reflection of the initial click from air sacs within the sperm whale's head. These authors believed that sonar clicks are produced in the front of the sperm whale's head by pneumatic action of the *museau de singe,* as described above for dolphins. According to this theory, air is forced forward down the right nasal passage, through the tightly opposed lips of the *museau de singe,* producing a high-intensity click. The multiple reflections of this original pulse result from reverberations reflected from the interconnected nasofrontal and distal air sacs along with the intervening oil-filled spermaceti organ. If true, the time delay between the multiple pulses should be directly proportional to the length of the spermaceti organ (which is also a function of total body length). Evidence from detailed studies of free-ranging sperm whales supports the idea

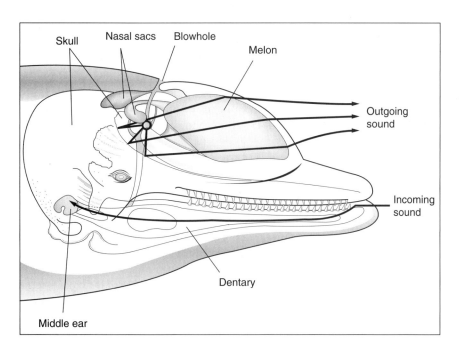

FIGURE 22-14 A dolphin generates clicks by forcing air through a set of "valves" located in the nasal sac region. These sounds are then reflected off the skull bones and focused by the melon on the dolphin's forehead. Returning echoes from the target are transmitted to the middle ear via the dentary.

that the sperm whale head acts as a sound reflector (Goold and Jones, 1995).

Although the exact mechanism of sound production is not certain, it is assumed that diving sperm whales use trains of clicks for echolocation of prey. Sperm whales feed largely on squid that they take at depths (down to at least 2000 meters; Watkins et al., 1993) at which prey is scarce and light is virtually absent. It appears, therefore, that the sperm whale is able to forage at great depths largely because it is able to use echo scanning to locate food under conditions that require efficient long-range echolocation. Goold and Jones (1995) estimated that sperm whales, by using click rates of 0.5 to 1 per second, should be capable of detecting prey 750 meters away. As they approach the potential prey, click rates must increase to provide information on target position and velocity. It has also been suggested that the rapid bursts of clicks refered to as "creaks" (up to 200 pulses per second) produced by sperm whales may be analogous to the terminal "buzz" of a flying bat in giving greater target resolution.

Some small cetaceans that inhabit turbid water have tiny eyes and presumably are dependent on echolocation. One of the most highly specialized of these is the blind river dolphin (*Platanista gangetica*), an inhabitant of the muddy and murky waters of the Ganges, Indus, and Brahmaputra river systems of India and Pakistan. This unusual dolphin habitually swims on its side with the ventralmost flipper either touching the bottom or moving within 2 or 3 centimeters of it. *Platanista* has greatly reduced eyes that are barely visible externally. The lens is absent, but the retina apparently retains the ability to detect light, although doubtless no image can be formed. The tiny eye opening is surrounded by a sphincter muscle, and another muscle opens the sphincter. Blind river dolphins in captivity continuously produced series of pulses at rates of from 20 to 50 per second, primarily in the frequency range between 15 and 60 kilohertz, and the animals have a remarkable ability to direct their pulses into a narrow beam (Herald et al., 1969).

Studies have demonstrated that the shape of the skull of *Platanista* affects the directional beaming of pulses (Evans et al., 1964). The pulses are reflected by the concave front of the skull and focused by the melon. The skull of *Platanista* is modified by large flanges from the maxillary bones.

These prominent flanges, rounded on the outside but with an intricate, radial pattern of latticework on the inside, probably serve as acoustical baffles that, with the melon, may concentrate the pulses into a beam (Herald et al., 1969). Observations of swimming dolphins and reception of these pulses with a hydrophone and amplifier system showed that they are indeed effectively beamed. When a dolphin's snout was directed as little as 10 degrees on either side of the receiver, the intensity of the pulses dropped markedly, and a far greater drop occurred when the angle was greater than 40 degrees. As the dolphin swims on its side, it moves its head constantly in a sweeping action close to the bottom. One is tempted to speculate that the dolphin is scanning the area ahead with a beam of pulses, using a system similar in some ways to that of the horseshoe bat, which also uses beamed pulses. Such scanning in the dolphin might serve effectively both in determining bottom contours and in finding food.

CETACEAN HEARING

Each ear of a cetacean functions as a separate hydrophone, allowing the animal to localize a sound source by discriminating (as we do) between the times the sound is received by each ear. The pressure that sound transmitted through water exerts on the bones of the entire skull causes vibrations to be transmitted by the skull. When the bone that houses the middle and inner ear is attached rigidly to the skull, as it is in most mammals, vibrations from water are transmitted through the bones of the skull and reach the ear from various directions. As a consequence, when a mammal with this type of skull is submerged, it is unable to localize accurately the source of a sound. Because sound localization is of great importance to cetaceans that use echolocation, these animals have evolved several structural features that insulate the bone surrounding the middle and inner ear (tympano-periotic bone) from the rest of the skull (Ketten, 1997).

First, the tympanic bullae (tympano-periotic bone; the fused auditory bulla and cochlea) are not fused to the skull in any cetacean, and, in the specialized porpoises and dolphins, the bullae are separated by an appreciable gap from adjacent bones of the skull. In addition, the bullae are insulated by an extensive system of air sinuses unique to cetaceans. These air sinuses surround the bullae and extend forward into the enlarged pterygoid fossae, and each sinus is connected by the eustachian tube to the cavity of the middle ear. The sinuses are filled with an oil-mucous emulsion, foamed with air, and are surrounded by fibrous connective tissue and venous networks. These sinuses can apparently retain air even when subjected to pressures of 100 atmospheres, pressures higher than those to which cetaceans are subjected during deep dives. Purves (1966) used a foam in which gelatin was substituted for mucous and oil and found that, at such pressures, the air became dispersed in the mixture as tiny bubbles. The foam in the air sinuses apparently forms a layer around the bullae that retains remarkably constant sound-reflecting and sound-insulating qualities through a wide range of pressures.

Norris (1964, 1968) regards the extremely thin back part of the lower jaw of delphinids as an acoustical window. He holds that sound passes into the skin and blubber overlying the dentary, through the thin part of this bone, which at its thinnest may be only 0.1 millimeter thick, to the intramandibular fat body, which leads directly to the wall of the auditory bulla, into which the sound presumably passes. Weight is given to the Norris hypothesis by experiments done by Bullock et al. (1968), who found that the jaw is the most acoustically sensitive area of the dolphin's head.

ECHOLOCATION BY INSECTIVORANS

The high-pitched sounds made by some insectivorans when they explore unfamiliar surroundings or objects are apparently used for echolocation (Swinhoe, 1870; Komarek, 1932; Reed, 1944; Crowcroft, 1957; Buchler, 1976). In a series of carefully controlled laboratory experiments, Gould et al. (1964) demonstrated that three species of *Sorex* could echolocate. These shrews searched around an elevated disk, found a lower platform, and jumped to it, all without the use of tactile, visual, or olfactory senses. While the shrews searched their environment, they emitted pulses with frequencies between 30 and 60 kilohertz; the pulse duration was from 5 to above 33 milliseconds. The shrews were unable to find the disk when their ears were plugged. The familiar short-tailed shrew of the eastern United States, *Blarina brevicauda,* produced similar pulses, and all shrews studied produced pulses with the larynx. Laboratory trials revealed that tenrecs (Tenrecidae) from Madagascar also echolocate (Gould, 1965). These primitive insectivorans produced pulses by clicking the tongue, and the pulses were of frequencies audible to humans (from 5 to 17 kilohertz).

WEB SITES

Mammalian Bioacoustic web site from Italy with downloadable bat sonograms

> http://cibra.unipv.it/mamm.html

Bat Ecology and Bioacoustics Lab at University of Bristol

> http://www.bio.bris.ac.uk/research/bats/batpage.htm

Bat Bioacoustic Information and explanations of bat call types

> http://www.biology.leeds.ac.uk/research/biomech/daw/bats/index.html

Auditory Neuroethology Lab at University of Maryland

> http://www.bsos.umd.edu/psyc/batlab/index.html

Southeast Australian Bat Call Library—with ANABAT 5.2 for downloading

> http://batcall.csu.edu.au/batcall/batcall1.html

The Cetacean Communication Project page with information on ambient noise imaging

> http://dsg.sbs.nus.edu.sg/comres.html

23 BEHAVIOR

The behavior of any animal is of great interest because natural selection favors behaviors that help the animal survive and reproduce. Consequently, the range of behaviors is wondrously diverse. In the case of the pronghorn, for example, great running speed became part of a unified functional system only because of a complex of behaviors that evolved in association with this ability. The formation of herds, sexual dimorphism, and systems of social behavior; the preference for open situations, the flashing of the white rump patch as a danger signal to other pronghorn, and the remarkable ability to detect enemies at a distance all allow the pronghorn to utilize its great speed effectively to escape

predators. How an animal uses its morphological and physiological equipment is of vital adaptive importance and forms the substance of behavior.

The behavior of mammals is of particular interest because of its flexibility and variability. Mammals learn much more rapidly than do other vertebrates and can modify behavior on the basis of past experience. This ability, superimposed on a rich array of innate (instinctive and unlearned) responses, or behaviors, makes for complex behavioral patterns that often differ widely among species. Remarkably well-developed sense organs, coupled with a brain capable of rapid evaluation of complex sensory information, have enlarged the perceptual sphere of mammals and have facilitated the evolution of communication and rich social behavior.

THE ETHOLOGICAL APPROACH

This chapter deals largely with ethology, the study of behavior in relation to structure and mode of life or, as put by Tinbergen (1963), "the biological study of behavior." One might suppose that behavior could be more readily observed and analyzed than could other aspects of biology and that detailed behavioral information on many species would have been assembled relatively early, but this is not the case. Indeed, little is known of the behavior of many animals that are well known morphologically. The study of mammalian ethology has reached the stage of synthesis (see, for example, Geist, 1974, and Jarman, 1974), and such excellent contributions as those of Ewer (1968, 1973), Wilson (1975), Eisenberg (1981), Eisenberg and Kleiman (1983), and Estes (1991) offer compilations and analyses of our knowledge of mammalian behavior.

Although mammals are remarkable in their ability to learn and to profit from experience, built-in patterns of behavior form an important part of their behavioral repertoire. Such innate behavior is individually variable within a species, particularly between sexes or age classes. These unlearned behaviors are genetically controlled and are shared with other members of the species. They are best regarded as simple sequences of movement elicited by specific stimuli. The term Erbkoordination, coined by Lorenz (1950), seems especially appropriate and refers to simple, but specific, hereditary movements or patterns of coordination. Such behavior in canids or hyaenids seems to be the lowering of the head, rump, and tail — assuming the "submissive posture" — in response to the menacing jaws, high head, cocked ears, and high tail of a dominant individual. Behavioral patterns have evolved along with structural and physiological features, and their adaptive importance is suggested by the prominent roles they play in the lives of many mammals.

Clearly, mammals are not completely unique behaviorally. They are set apart from other vertebrates by their superior ability to learn, remember, and innovate, but they resemble other vertebrates in their wide use of innate behaviors.

BEHAVIOR OF INDIVIDUALS

Certain behaviors serve to promote the survival of the individual, even though such behaviors often affect other members of the ecological community. Such behaviors include those resulting from circadian or circannian activity patterns, migration, foraging, and predator-avoidance behaviors. Many behaviors involve communication between two individuals that must each send and receive signals. These communication signals, be they visual displays, chemical pheromones, or vocalizations, also serve to increase individual survival. Territorial displays, courtship, and parental care fall into this category.

ACTIVITY RHYTHMS

A striking aspect of animal behavior is the rhythmic, or cyclic, pattern of activity. Some species are active at night (nocturnal) and some during the day (diurnal); others are active primarily at dawn and dusk (crepuscular). The activity periods tend to be at regular intervals. The time of emergence of a particular species of bat may differ by no more than 2 or 3 minutes night after night. Animals also exhibit other kinds of cyclic behavior. The timing of reproduction is cyclic, and in some mammals, such as some rodents and bats, daily or seasonal shifts occur between highly active and torpid states. Migratory movements are also cyclic. Daily activity rhythms, those based on a 24-hour cycle, are termed **circadian rhythms** and are better understood than are other types of rhythms.

Circadian rhythms differ markedly from one species to another. Most mammals are nocturnal, but even in two nocturnal species there are contrasts between the patterns of activity. In general, small mammals that are especially vulnerable to predation, such as rodents, tend to be nocturnal (most sciurid, arvicoline, and sigmodontine rodents are exceptions), whereas less vulnerable species, such as many ungulates, may be active during the day. The activity cycles of carnivorans seem to be geared to the circadian cycles of their prey or to the period when hunting is most rewarding.

Circadian rhythms are also influenced by interactions between species with similar environmental needs. In some cases, competition between species is reduced or eliminated because their activity cycles are out of phase. Two species of fishing bats *(Noctilio),* both of which feed over water, avoid interfering with one another partly by foraging at different times of the night and also by eating different prey (Hooper and Brown, 1968). Clearly, the circadian rhythm of an animal is part of its total adaptation to its particular mode of life and environment and has evolved just as have morphological characters.

The question of whether circadian cycles are **endogenous** (internally controlled) or **exogenous** (ultimately regulated by external stimuli) has occupied the attention of many biologists. Clearly, some strong endogenous control is present in many species. As an example, careful work on the flying squirrel *(Glaucomys volans)* by DeCoursey (1961) showed that, even under constant environmental conditions, including continuous darkness, flying squirrels maintained regular activity periods that deviated only ±2 minutes from the mean value for activity periods under natural conditions. When a laboratory animal whose circadian cycle is not in phase with the natural 24-hour light-dark cycle is again exposed to normal day and night conditions, its cycle rapidly shifts and becomes "synchronized," that is, it becomes adjusted and locked

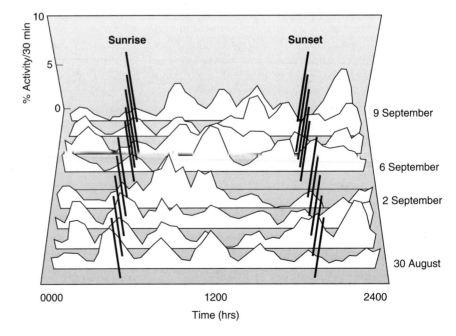

FIGURE 23-1 Summer activity patterns of a population of root voles *(Microtus oeconomus)* during an eight-day period in 1990. Vertical lines indicate the approximate hours of sunset and sunrise. *(After Halle, 1995)*

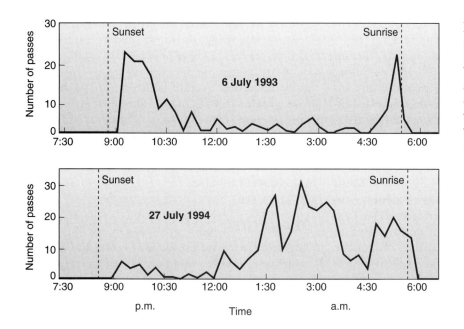

FIGURE 23-2   Activity patterns for bats recorded by echolocation monitoring. The peak activity, recorded as number of passes, changes seasonally due to variations in food supply and weather patterns. *(After Hayes, 1997)*

(entrained) to the 24-hour cycle (Bruce, 1960). Studies using large outdoor enclosures indicated that circadian activity of root voles *(Microtus oeconomus)* was not affected by weather conditions and suggested a lunar periodicity (Fig. 23-1; Halle, 1995). Circadian cycles and other animal behaviors are regulated by intricate and as yet poorly understood interactions between endogenous and exogenous factors (Fig. 23-2) including a molecular genetic clock that can be adjusted by retinal sensing of the light-dark cycle (Sassone-Corsi, 1998; Tosini and Menaker, 1996).

As might be expected if circadian cycles are adaptive, they often shift seasonally and depend on an animal's ability to track some environmental variable such as light or temperature. Attending the seasonal changes in environmental temperatures are changing metabolic demands put on small mammals, and some shifts in circadian rhythms may allow the animals to avoid activity during times of most intense temperature stress. Three types of seasonal changes in circadian rhythms were reported by Stebbins (1984) for small mammals in Canada: overall decreased winter activity, changes in percentage of nocturnal and diurnal activity, and changes in peak daily activity. Seasonal shifts in activity probably result in a considerable saving of energy.

Although less thoroughly studied than circadian rhythms, circannian rhythms play an equally prominent role in the lives of some mammals. Among vertebrates, such vital activities as breeding, migration, and hibernation are phased on an annual cycle, or a circannian rhythm. Pengelley (1967) defined a circannian rhythm as an endogenous cycle that has a length of approximately one year. Among mammals, such rhythms have been documented for the golden-mantled ground squirrel *(Spermophilus lateralis;* Pengelley and Fisher, 1963; Pengelley and Asmundson, 1970, 1971), the woodchuck *(Marmota monax;* Davis, 1967), and for two species of ground squirrels *(Spermophilus)* and four species of chipmunks *(Tamias;* Heller and Poulson, 1970).

Circannian rhythms are a major key to the survival of some temperate-zone and arctic mammals. In the words of Heller and Poulson (1970), these rhythms allow an organism "to anticipate, and thus prepare for, a future, annually occurring environmental condition such as cold weather, drought, food scarcity or optimal breeding time." The rhythm also ensures some flexibility of response to cyclic environments that may differ markedly from year to year. In addition, circannian rhythms enable "the organism to integrate a large number of environmental cues and, through phasing the rhythm, respond most favorably to conserve energy and to ensure reproductive success."

For hibernators in temperate regions, circannian rhythms make the animals sensitive to falling temperatures and declining body weight in the autumn so that the onset of hibernation may be hastened by unfavorable conditions or delayed by favorable temperatures and food supplies. In arctic areas, the extremely harsh environment and the sudden onset of winter, coupled with the brief time available for breeding and for putting on fat in preparation for hibernation, make flexibility non-adaptive. Here the adaptive premium shifts to a precise, inflexible, optimal schedule. Breeding at the optimal time each year, regardless of climatic conditions, probably ensures the greatest reproductive success, and precision in the onset of hibernation ensures maximum overwinter survival. Even in tropical areas, circannian rhythms may be highly adaptive. In Kenya, for example, the African false vampire bat *(Cardioderma cor)* and the giant leaf-nosed bat *(Hipposideros commersoni)* become pregnant well before the onset of the late March–April rainy season, seemingly in anticipation of the burst of insect abundance that accompanies the rains. This pattern of breeding is probably controlled by a circannian rhythm.

The factors controlling these remarkable rhythms are as yet poorly understood. For arctic hibernators, Heller and Poulson (1970) suggest that photoperiod is the most important environmental factor in phasing the underlying circannian rhythm. In temperate-zone hibernators, in contrast, although the timing of breeding seems inflexible, the animals are responsive to environmental conditions in the fall and may delay or hasten their entrance into hibernation. The situation is not entirely simple, however, for even temperate-zone hibernators that occupy the same area do not follow the same circannian rhythms. The activity of the golden-mantled ground squirrel *(S. lateralis)*, an inhabitant of mountains in the western United States, is controlled largely by an endogenous rhythm. This animal feeds into the fall and stores relatively nonperishable seeds in its burrow, and entrance into hibernation is relatively tightly scheduled, regardless of environmental cues (Pengelley and Asmundson, 1970, 1971). The Belding ground squirrel *(S. beldingi)*, which often lives almost side by side with the golden-mantled ground squirrel, feeds on green material that decomposes quickly if stored underground. This squirrel feeds as long as possible in the fall, putting on more and more fat, and stores no food. The golden-mantled ground squirrel can perhaps afford greater rigidity in the timing of its hibernation because of the cushion of stored food in the hibernaculum, but the Belding ground squirrel must depend entirely on food stored in the form of body fat and thus feeds as long as such activity is energetically feasible.

MIGRATION

Migratory behavior presumably evolved as a way to avoid undesirable conditions at a particular time of year. Migrations are energetically expensive and expose the participants to increased risks along the route. Clearly the costs are high, but the benefits of migration exceed the costs. Migration is beneficial in an area of declining resources, where individuals may benefit if they move to a resource-rich area. In other cases, evolution has favored migrations to sites that offer better protection from predators during the birthing season. Finally, migrations may bring large numbers of individuals together during the mating season from distant areas, ensuring the chances of mating.

Many mammals move away from breeding grounds to overwintering or feeding grounds on a regular, usually annual, cycle. Gray whales *(Eschrichtius robustus)* make annual migrations of over 10,000 kilometers, traveling from northern Pacific feeding areas to Mexican breeding areas and back. Northern elephant seals *(Mirounga angustirostris)* migrate biannually from the northern Pacific to California's Channel Islands and back (Fig. 23-3). These seals are sexually dimorphic with respect to their foraging areas and migratory routes and males tend to migrate greater distances (about 21,000 kilometers) than do females (Stewart and DeLong, 1995; Stewart, 1997).

The migratory journeys of terrestrial mammals are generally shorter than those of marine mammals, but some are impressive. The ungulate migrations of the Serengeti ecosystem are one spectacular example. The complex circuitous migration of the Serengeti wildebeest in 1960 was estimated to have covered at least 1700 kilometers (Talbot and Talbot, 1963). These migrations vary in length

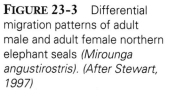

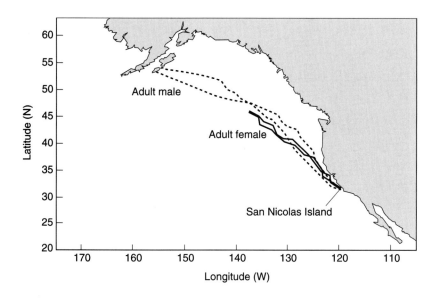

FIGURE 23-3 Differential migration patterns of adult male and adult female northern elephant seals *(Mirounga angustirostris). (After Stewart, 1997)*

and route from year to year and involve the movement of millions of animals of several species including: wildebeest *(Connochaetes),* zebra *(Equus),* and Thomson's gazelle *(Gazella thomsonii)* among others. Rainfall patterns are tightly linked with the patterns of ungulate migrations in the Serengeti. In Canada and Alaska, caribou *(Rangifer tarandus)* follow well-established routes between their calving and winter feeding grounds. Migration distances in bats vary from several hundred kilometers in *Myotis* to over 1500 kilometers in some populations of *Tadarida* (Hill and Smith, 1992). Some seasonal migrations involve movements from warmer summer foraging sites to winter **hibernacula** (hibernation sites). Some Old World fruit bats and many nectar feeding bats make short movements in response to the local abundance of their food resources.

FORAGING BEHAVIOR

Herbivores consume nonmobile prey and therefore are spared some of the problems that carnivores face. Herbivores are generally adapted to utilize efficiently only specific types of vegetation; they must often face seasonal food shortages, and must cope with chemical defenses and plant material that is difficult to digest. Specialized feeding behaviors, together with specializations of the dentition and digestive system, tend to maximize the return of energy from food relative to the energy expended in securing it.

Some of the most specialized foraging behaviors occur among rodents. Pocket gophers (Geomyidae), mole rats (Bathyergidae), and other fossorial rodents dig complex burrow systems, and part or all of their diet consists of underground parts of plants they encounter. Because of the tremendous energetic cost of burrowing, one would expect fossorial herbivores to have evolved behaviors favoring the most efficient system of finding food. The basic geometry of burrow spacing in one species of pocket gopher *(Thomomys bottae)* was found to be remarkably uniform both within one burrow system and from one system to another (Reichman et al., 1982). The basic building unit of the system, as well as the distance between forks and the lengths of the branches, was uniform within and among systems and is analogous to the nodes and internodes of plants. These units can be combined to increase overall burrow length, but the uniform spacing is maintained (Fig. 23-4). Such consistent burrow spacing suggests intense selection in pocket gophers for precise, uniform burrowing behavior.

The closely related kangaroo rats and pocket mice (Heteromyidae), by contrast, travel far from their burrows and gather seeds from the soil by using the long claws of the small forefeet. These rodents rapidly collect seeds in the cheek pouches but tend to gather seeds that are superior energetically to those randomly available in the soil. Typically, many loads are taken to the burrow in an evening. Later, in the safety of the burrow, the animals become even more selective: of the seeds

FIGURE 23-4 Aerial photographs and line drawings of pocket gopher burrows excavated at two sites near Cottonwood, Arizona. Burrow systems were marked with lime and photographed from the air. The diagram on the right designates burrows of adult males (M), adult females (F), nonreproductive males (m), and nonreproductive females (f). *(From Reichman et al., 1982, p. 688, by permission of the Ecological Society of America)*

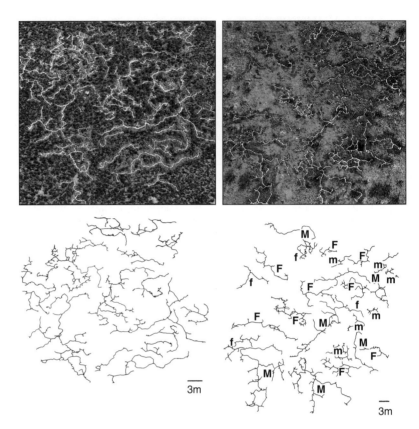

gathered, the rodents eat those richest in energy (Reichman, 1977).

Among mammals, food storage, or **caching,** is a widely used and seemingly highly adaptive behavior. (Caching, as used here, is the moving of food from one place to another for later consumption.) Caching by mammals takes many forms: a leopard drags its prey into a tree to protect the prey from hyenas or lions; the short-tailed shrew *(Blarina brevicauda)* buries a mouse immobilized by the shrew's toxic bite (Tomasi, 1978); and "haystacks" for winter consumption are stored beneath rocks by pikas. Caching is used by some insectivores (shrews, moles, and some hedgehogs), many carnivores (canids, ursids, some felids, and some mustelids), and by many rodents. Traveling back and forth from a foraging area to dependent young has probably favored the evolution of caching (Smith and Reichman, 1984). Among the advantages of this behavior are protection of food from competitors, protection from predators while eating, and a food

supply during lean times in an unpredictable environment.

The food most commonly cached by mammals is seeds, and seed-eating mammals cache food most frequently. Seeds provide a concentrated energy and nutrient reserve and can remain dormant in the soil for long periods. Studies of heteromyid rodents have revealed many remarkable behaviors associated with the caching and management of stored seeds.

Heteromyids have cached seeds for at least 10 million years (Voorhies, 1974), ample time for evolutionary fine-tuning of morphology and behavior associated with caching. In addition to developing anatomical features that further rapid gathering and transport of seeds (dexterous forefeet, external cheek pouches, and elongate hindlimbs), heteromyids have uncoupled food gathering and eating. Seeds are gathered in the cheek pouches, carried underground, then eaten — soon or even months later — in the safety of the burrow.

These rodents also have the remarkable ability to "manage" the seeds they collect (Reichman et al., 1985). The banner-tailed kangaroo rat *(Dipodomys spectabilis)* typically caches seeds beneath the north or northwest section of the mound at a depth near 30 centimeters or beneath 50 centimeters. Chambers in the soil are humid, and cached seeds rapidly become moldy. Over 100 species of fungi were found in the seed caches of several rodent species (Herrera et al., 1997). In the seed caches of *D. spectabilis,* seeds cached during rainy weather had the greatest abundance and diversity of fungi. In the laboratory, pocket mice *(Chaetodipus intermedius)* ingested slightly moldy seeds in preference to highly moldy or nonmoldy seeds, and wild *D. spectabilis* had similar preferences (Rebar and Reichman, 1983; Reichman and Rebar, 1985). In the laboratory, seeds were actually managed by the latter species for degree of fungal growth: sterile seeds were stored in the highest available humidities. Once seeds reached preferred levels of moldiness, they were moved to low humidities that inhibited further fungal growth (Reichman et al., 1986). Other experiments demonstrated that kangaroo rats are sensitive to extremely minor differences in the water content of seeds, that seeds with the highest water content are preferred, and that some fungal growth is tolerated to get more water (Frank, 1988a, b).

Frequent manipulation of cached seeds (in a way yet unknown) by *D. spectabilis* virtually eliminated germination of cached seeds (Reichman et al., 1985). Seeds germinated rapidly in unattended caches, while seeds rarely germinated in caches attended by resident animals. Chipmunks are known to stop germination and preserve the food value of cached beech seeds by biting off the tips of the embryos (Elliot, 1978).

The diet of a heteromyid rodent depending on a seed cache is limited and is not renewed daily; therefore this animal would be expected to optimize its diet for the long term rather than the short term. The rodent should not eat all of the preferred or most nutritious items first; it should instead diversify its diet in order to survive the longest possible time on the cached food items. Just such behavior was demonstrated in the laboratory by Reichman and Fay (1983). A caching pocket mouse *(C. intermedius)* began to diversify its diet by eating less preferred seeds in larger amounts after

several days of depending on its cache. Under the same conditions, a typically noncaching deer mouse *(Peromyscus maniculatus)* consumed the seeds in order of preference, as would be expected for an animal simply foraging for food.

The eastern woodrat *(Neotoma floridana),* which caches mostly vegetation other than seeds, discriminates between foods on the basis of perishability, tending to eat the more perishable foods and cache the less perishable items (Reichman, 1988; Post and Reichman, 1991). The short-tailed shrew makes similar decisions with different foods: seeds are cached before insects, and both of these are cached before mice, a sequence reflecting degrees of resistance to spoilage (Martin, 1984). Probably caching mammals, in general, can recognize items that will not decompose quickly when cached.

The caching behavior of the North American red squirrels *(T. hudsonicus)* is particularly notable. These squirrels depend for food on fir and pine seeds. Red squirrels use **scatterhoarding** and **larderhoarding.** Scatterhoards consist of a few food items stored at each of many locations throughout the territory, while larderhoards are large piles of food items located at a central site within the territory (Hurley and Lourie, 1997). Red squirrels cache cones in holes in large **middens** formed by the litter of cone fragments that accumulates beneath a squirrel's favorite feeding sites. The larderhoards are frequently 6 to 10 meters in diameter, contain from two to ten bushels of cones, and are in shady situations where the moisture retained in the midden aids in preserving the green cones (Finley, 1969). Small numbers of cones are commonly cached in logs or pools of water. The cones are harvested in late summer and fall and are cut, on occasion, at the rate of 29 per minute (Shaw, 1936). The squirrels are such effective harvesters that one pine in northern California lost 93 percent of its 926 cones to them (Schubert, 1953). Seeds from the cached cones are eaten during the winter. When snow is deep, access burrows through the snow into the midden are maintained.

Predators have evolved behaviors that facilitate the pursuit, capture, and killing of prey, which in most cases have defensive or predator-avoidance strategies. Some behavioral patterns are common to a wide array of carnivores. The neck bite is such a behavior. This killing technique was studied in the house cat by Leyhausen (1956), who presented the predators with normal and headless rats and with rats with the head fastened to the tail end. The cats aimed their bites at any constriction in the body; with normal prey this results in the neck bite. Grabbing prey across the back and shaking it violently is another pattern shared by many carnivores.

Most canids are solitary, cursorial hunters and capture prey by virtue of speed or, occasionally, endurance. Experience and learning are of great importance, and canids are highly adaptable. A coyote with access to a waterfowl marsh may learn to capture molting ducks, while its relative in the desert patrols the perimeters of sand dunes for kangaroo rats. Canids have an extraordinary sense of smell, and prey is often detected initially by wind-borne scent. Adaptability and diverse tastes, rather than a specialized style of hunting, are the keys to the success of solitary canids.

The cats employ more specialized hunting and killing techniques than canids. Cats are not long-distance runners but usually depend on short rushes directed against surprised prey. The sudden rushes of lions seldom cover more than 100 meters, and leopards and smaller felids frequently make only several bounds to reach their prey. The cheetah, an exceptional felid, may chase an antelope several hundred meters at speeds up to 95 kilometers per hour!

In order to use the typical feline hunting technique effectively, a cat must get close to its prey. The stalking of prey by felids involves a series of beautifully coordinated behaviors, described in detail by Leyhausen (1956). When prey are sighted, the cat crouches low to the ground and approaches, using the "slink-run" and taking advantage of every object offering concealment. At the last available cover, the cat stops and "ambushes." The brief rush to the prey ends in a spring; the forefeet clutch the animal, but the hind feet often remain planted and steady the cat for the possible struggle. The cat usually makes the kill not by belaboring the prey, as do many canids (Fig. 23-5A), but either by a powerful bite at the base of the skull or the neck, which crushes the back of the skull or some of the cervical vertebrae and the spinal cord, or, in the case of large prey, by strangulation. The shortening of the felid jaws is a specialization that contributes to the power of the bite.

Most cats are solitary, and cooperative effort in killing prey is rare. An exception is the African lion,

the only truly social felid, which often hunts in groups in which there is some cooperation between individuals, with adult females doing most of the killing. The lion typically stalks large prey, often as heavy as or heavier than itself (Fig. 23-5B). Although cooperative effort improves success, a prey animal is typically killed by a single lion. Lions attacking prey the size of a zebra or wildebeest attempt to bring the prey to the ground by clutching the rump, hind legs, or shoulders with the forepaws and throwing the prey off balance (Schaller, 1972). When the prey falls, the lion grabs with its jaws for the neck or nose and maintains a grip until the prey is suffocated. Schaller pointed out that, by centering the bite on the neck or nose, the lion immobilizes the horns, remains clear of thrashing hooves, and can easily keep the victim on the ground. This specialized killing behavior reduces the risk of serious injury from the powerful prey, although this is still a dangerous mode of foraging.

Unusual behavior patterns enable some carnivorans to break the exoskeletons of invertebrates and the shells of eggs. Some mongooses use the forefeet to throw objects against hard surfaces (Dücker, 1957; Ewer, 1968). Similarly, the spotted skunk (*Spilogale putorius*) breaks eggs by kicking them against rocks (Van Gelder, 1953). The sea otter (*Enhydra lutris*) smashes the sturdy shells of mollusks by using a tool: the otter floats on its back with a flat stone on its chest, grasps the mollusk with its forepaws, and pounds it against the stone (Fisher, 1939). An individual was observed to pound mussels (*Mytilus*) on a stone 2237 times during a feeding period lasting 86 minutes (Hall and Schaller, 1964). These otters are clearly selective in their choice of stones and may use the same one repeatedly.

DEFENSIVE BEHAVIOR

Threat behaviors are among the most familiar activities of mammals. A dog lifts its upper lip to expose the length of its upper canines, a cat opens its mouth and hisses, some rodents grind their teeth, and others perform footdrumming. These actions all signal a readiness to fight or to attack if the antagonist does not retreat or take other appropriate action. A threat is typical of a situation in which conflicting tendencies preclude either an immediate attack or a hasty retreat (Fig. 23-6).

A

B

FIGURE 23-5   (A) Group hunting effort of African hunting dogs *(Lycaon pictus)* used to subdue large prey usually involves harassing and weakening the prey until it collapses. (B) Typical attack behavior of a female lion *(Panthera leo)* involves grasping the prey by the neck or throat. *(From Schaller, 1972)*

Threats can be simple, as in animals that merely open the mouth wide to display the teeth, or complex, as in some horned artiodactyls in which both distinctive postures and movements are involved that usually seem to advertise the most important offensive weapons. Visual threats may be made more impressive or startling by such audible threats as explosive hisses or growls.

Some mammals have carried this type of defensive behavior a step further by discarding the pretense of defense. Such an **appeasement** posture or behavior is a complete surrender and contains no elements likely to trigger an opponent's aggression. Virginia opossums *(Didelphis virginiana)*, for example, feign death by becoming nearly catatonic as a defense of last resort. Complete vulnerability is

FIGURE 23-6 Antagonistic behavior of grizzly bears *(Ursus arctos)* typically results with one animal retreating. *(From Corel Corp.)*

emphasized, and the response on the part of the predator is to cease its hostile activity because the stimulus of a moving prey is absent. Wolves and many other mammals appease their dominant opponents by lying on the back with the vulnerable throat and underside unprotected. In several artiodactyls, lying down serves as appeasement (Burckhardt, 1958; Walther, 1966), and a subordinate black wildebeest *(Connochaetes gnou)* may roll on its side with its belly toward its superior and the side of its head on the ground (Ewer, 1968). In Grant's gazelle, a lowering of the head, the reverse of the high-headed threat posture, is adopted by a submissive animal (Walther, 1965). In some primates, the presenting of the rump as the female does prior to copulation is an appeasement gesture. The brightly colored skin on the rump of some Old World monkeys may serve in part to make the rump conspicuous and thus make "presenting" appeasement gestures more effective. A "grin" serves as an appeasement in some higher primates.

Appeasement behavior clearly helps both participants avoid further conflict. It allows an animal being defeated in a fight to avoid further injury and in many cases allows a subordinate animal to avoid a contest altogether. In highly social species, threat and appeasement behaviors foster the peaceful perpetuation of a dominance hierarchy and allow animals to be close to one another with a minimum of energy wasted on aggressive inter-

actions. Ritualized appeasement behavior may even be important in permitting a subordinate animal to seek social contact without risking attack (Schenkel, 1967).

SHELTER-BUILDING BEHAVIOR

Many mammals have evolved elaborate shelter-building behaviors that aid them in maintaining homeostasis. The nests, burrows, or houses of mammals provide insulation that augments the animal's own pelage and saves energy by reducing the rate of thermal conductance from the animal to the external environment or vice versa. The woodrat *(Neotoma)* collects a variety of materials with which it builds houses or improves the shelter provided by rock crevices or vegetation. Beavers construct the most elaborate shelters of any mammal other than humans, and lodges made of sticks and branches are often over 3 meters in diameter. Many terrestrial rodents construct nests beneath logs or rocks or in burrows. Arboreal rodents frequently build nests in the branches of trees or in hollows in trees. Some nest-building behaviors are perhaps common to many rodents, but the choice of nesting site seems to be species-specific. For example, red tree voles *(Arborimus longicaudus)* of the humid coastal belt of Oregon and California build their nests only in Douglas firs *(Pseudotsuga menziesii)*, the needles of which provide the primary food of the mice (Benson and Borell, 1931).

Fossorial rodents follow rather complex patterns of movement when digging. Probably many of the specific components of the total digging sequence arc innate behaviors. Pocket gophers (Geomyidae) use the forefeet to loosen the soil by powerful downward sweeps, and the hindlimbs kick the accumulated soil backward from beneath the animal. Pocket gophers periodically eject soil from a burrow entrance by pushing it with the chin and forelimbs. Kennerly (1971) has shown that the long, complex series of behavior patterns associated with mound building are basically innate but may be modified by learning. "Autoformulated releasers" probably play an important role in guiding mound building in rodents (and probably many other behavioral sequences in other mammals). Such a releaser is any alteration an animal makes to its perceptual environment that releases the animal's subsequent behavior. Thus, the mound of earth itself and changes in the mound resulting from the pocket gopher's activity release subsequent behaviors associated with mound building. The animal characteristically alternates direction in pushing soil from the burrow; it pushes a series of 5 to 20 loads to the right, a similar series to the left, and so on. The frequency distribution of directions of pushing soil indicates that efforts are mainly in three directions: either directly in front of the mouth of the burrow or at an angle of 90 degrees to either side. This results in the fan-shaped mound so typical of pocket gophers. That learning plays a part in burrowing and mound building is suggested by the fact that young animals are less successful in plugging the openings of burrows than are older animals.

The burrowing and mound-building behaviors of some African mole rats (Bathyergidae) differ markedly from those of pocket gophers. *Heliophobius,* for example, uses its incisors to excavate soil and pushes the dislodged soil in back of its body with its feet. The animal transports a load of soil to the surface by backing up against it with the rump and large hind feet. The forefeet push the animal backwards, and the head and upper incisors are braced against the roof of the burrow to gain purchase (Fig. 23-7). Unlike pocket gophers, which appear briefly at the surface each time they push a load of soil onto the mound, *Heliophobius* pushes a core of soil onto the mound without appearing on the surface (Jarvis and Sale, 1971).

Several kinds of Neotropical bats utilize leaf shelters, and among these the "leaf-tents" are of

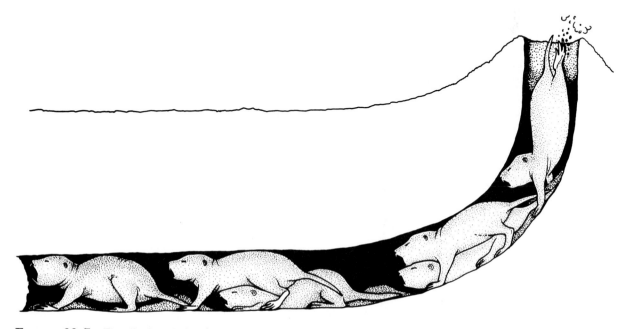

FIGURE 23-7   The digging chain of the naked mole rat *(Heterocephalus glaber)* of East Africa. *(After Jarvis and Sale, 1971)*

FIGURE 23-8   A group of fruit bats *(Uroderma bilobatum)* roosting on the underside of a tent made from a banana leaf. *(From Timm, 1987)*

special interest (Fig. 23-8). White bats *(Ectophylla alba),* for example, roost in groups of one to six individuals beneath the large leaves of several Neotropical plants (Timm, 1987). The bats cut the side veins extending from the midrib of the leaf in such a way that the end of the leaf bends down and forms a tent. One colony of bats uses several of these shelters alternately (Timm and Mortimer, 1976).

COMMUNICATION

Communication has often been broadly defined to include all interactions between animals that transmit information between them, but if all types of stimulus-reception sequences are regarded as communication, then essentially all behavior of one animal that can be perceived by another must be regarded as communication. For the purposes of discussion here, the definition of communication proposed by Otte (1974) will be used. Only a small segment of the multitude of stimuli received when an animal "views" its environment with its receptors is produced by other organisms and, through natural selection, has become modified to convey information. Communication signals, then, are "behavioral, physiological, or morphological characteristics fashioned or maintained by natural selection because they convey information to other organisms" (Otte, 1974). Each type of communication — visual, olfactory, auditory, and tactile — will be considered separately, but it should be stressed that usually a complex of several kinds of communication signals passes between animals (Table 23-1).

VISUAL SIGNALS. Visual signals involving displays were perhaps derived in vertebrates from move-

TABLE 23-1   Properties of Various Modes of Communication

Property	Visual	Auditory	Chemical	Tactile
Effective distance	moderate	long	long	very short
Ability to localize	high	moderate	variable	high
Ability to go beyond obstacles	poor	good[a]	good	—
Speed to transfer	fast	fast	slow	fast
Complexity of signal	high	high	low	moderate
Persistence of signal	variable	low	high	low

[a]True for moderate- to low-frequency sounds only.

ments showing intention (to flee, for example), from displacement activities (seemingly inappropriate actions that typically occur when two opposing "desires," such as to escape or to attack, are in conflict), and from such autonomic responses as the erection of hair (Hinde, 1970). The evolution of displays is in the direction of reduced ambiguity. They have tended to become simplified, exaggerated, and stereotyped (repeated without variation). Highly developed facial musculature, the ability of the body and ears to assume a variety of postures, the control many mammals have over the local erection of hair, and large and conspicuous secondary sexual characteristics, such as horns, allow mammals a remarkable breadth of visual communication.

Facial expressions are of great importance in communication, and natural selection has favored the development of distinctive facial markings that focus attention on the head. As described by Lorenz (1963), the facial expressions and ear postures of dogs signal degrees of aggressiveness or submissiveness. The posture of the head and the facial expression of many ungulates provide visual signals to other members of the herd or to territorial or sexual rivals. An elk *(Cervus elaphus)* ready to run from danger elevates its nose and opens its mouth (McCullough, 1969). Grant's gazelle *(Gazella granti)* holds its head high, elevates its nose, and pricks its ears forward when challenging another male (Estes, 1967). The head of both these animals is conspicuous: the elk's because the dark brown head and neck contrast strongly with the pale body, and the gazelle's because of bold black patterns. The intricate facial expressions of primates are frequently made more obvious by distinctive and species-specific patterns of pelage coloration and by brightly colored skin.

The stripe pattern of zebras *(Equus* spp.) is one of the most dramatic pelage patterns known. The high-contrast, black and white stripes probably serve several functions in zebra societies. A herd of fleeing zebra presents an abstract and possibly confusing picture to a pursuing predator. No two zebras have exactly the same pattern of stripes, leading to the suggestion that these patterns serve a function similar to fingerprints. Klingel (1967) provided the first evidence that individuals recognize one another, in part, based on the pattern of their stripes. Individual recognition may be partic-

ularly important in the mother-offspring bond (Estes, 1991).

The body is used for signaling in many species. This type of signaling is particularly well developed in ungulates that inhabit open areas and that gain an advantage from coordinated herd action. Grant's gazelle and Thompson's gazelle *(G. thompsonii)* of Africa, which have two warning displays (Estes, 1967), twitch the flank skin (conspicuously marked in *G. thompsonii)* just as they begin to run from a predator that has entered the minimum flight distance (the minimum distance at which an approaching enemy causes the animals to run). The most effective display is a stiff-legged bounding gait, called "stotting," used at times as the gazelles begin to run (Estes, 1991). The conspicuousness of this display is enhanced by the erection and flaring of the hair of the white rump patch. In some monkeys and apes, the presentation of the hindquarters as if inviting copulation is a social gesture symbolic of friendship and is accepted by a brief "token" mounting (Heinroth-Berger, 1959). Kangaroos threaten one another by standing bipedally at their maximum height, surely an impressive visual signal.

The effectiveness of visual signaling is heightened in many species of mammals by **weapon automimicry** in the form of striking (and to our eyes handsome) markings (Guthrie and Petocz, 1970). The ears are commonly used signaling devices in mammals. Artiodactyls use their ears in signaling, probably because of the proximity of the ears to the horns or antlers. The ears in many species of ungulates are marked or adorned with hair in such a way as to mimic the horns. This probably strengthens the visual signal given by the horns, as well as making the posture of the ears extremely obvious (Fig. 23-9). Facial markings may also play a role in automimicry by accentuating the horns. In the sable antelope, oryx, and Grant's gazelle, black markings create a design that extends the contours of the horns (Fig. 23-10).

OLFACTORY SIGNALS. Eisenberg and Kleiman (1972) defined olfactory communication as "the process whereby a chemical signal is generated by a presumptive sender and transmitted (generally through the air) to a presumptive receiver who by means of adequate receptors can identify, integrate, and respond (either behaviorally or

FIGURE 23-9 The drooping ear tips of the roan antelope *(Hippotragus equinus)* are an example of automimicry. *(R. Bowker)*

physiologically) to the signal." A chemical signal that elicits a response in a conspecific receiver is known as a **pheromone,** whereas an **allomone** conveys a message to a receiver of a different species. Olfactory communication is effective because specific chemicals can convey very specific messages, and a scent mark on an object will persist and yield a message long after it is deposited. Because scent is released into the air and disperses rapidly, however, a receiver must have a sense of smell acute enough to find the source by detecting concentration gradients. Also, olfactory signals broadcast in the air are as available to a predator as to

a conspecific. Pheromones of mammals have a variety of sources.

Urine and feces contain metabolic wastes that serve as chemical signals. Many kinds of mammals are highly specific in their choice of urination and defecation sites, and, in some species, a stereotyped routine is associated with urination and defecation. The dik-dik *(Madoqua kirkii),* a small, brush-dwelling, African bovid, deposits its feces in conspicuous piles at the borders of its territory, urinates on the piles, and makes scratch marks around them with its hoofs (Fig. 23-11). These obvious piles provide both olfactory and visual sig-

FIGURE 23-10 Facial markings of three antelopes act to extend the line of the horns onto the face.

Sable antelope
(Hippotragus niger)

Oryx
(Oryx beisa)

Grant's gazelle
(Gazella granti)

FIGURE 23-11 A male dik-dik *(Madoqua kirkii)* carefully smelling its dung pile (left) and then marking the pile by scratching the soil around it (right). *(M. Bowker)*

nals announcing territorial boundaries (Hendrichs and Hendrichs, 1971). All defecation and urination by the coyote can be regarded as scent marking. Marking is most common where intrusions into a home range occur (Wells and Bekoff, 1981; Gese and Ruff, 1997). Peters and Mech (1975) found that scent marking with urine by wolves was concentrated along the borders of pack territories. Wolves of one pack respect the territorial boundaries of another pack, and buffer zones between neighboring packs serve to reduce inter-pack conflict. Mech (1994) observed that wolf mortality was concentrated along these territorial boundaries. In aardwolves *(Proteles cristatus)*, scent marking serves to intimidate intruders and to synchronize mating (Sliwa and Richardson, 1998).

Urine and feces convey considerable information about an individual's physical condition (Endler, 1993). Males of most species of mammals can recognize when a female is in estrus by the smell of her urine, and usually copulation will not be attempted until this time. In coyotes *(Canis latrans)*, scent marking establishes reproductive synchrony between dominant (breeding) pack members, serves as an indicator of territorial boundaries, and may provide additional cues about spatial orientation to members of the group (Gese and Ruff, 1997).

Scent marking is used widely as a means of communication and is commonly an expression of dominance. Ralls (1971) indicated that scent marking is used by mammals "in any situation where they are both intolerant of and dominant to

other members of their species." Ewer (1973), in discussions of scent marking by carnivores, made a similar point, and the work on European rabbits *(Oryctolagus cuniculus)* by Mykytowycz (1968) and on Mongolian gerbils *(Meriones unguiculatus)* by Thiessen et al. (1971) raised several major points:

1. The maturation and use of scent glands are controlled by gonadal hormones produced at sexual maturity.
2. Most scent marking is done by dominant males.
3. Scent marking is associated with the possession of a territory.

Experimental verification of the relationship between social rank and scent marking has come from studies of house mice *(Mus musculus)*. Dominant males avidly marked the entire cage floor, whereas subordinate males voided urine in only a few places in the corners of a cage. Urination by the dominant male was regulated by interactions with another male: previously isolated dominant males immediately increased their urinary scent marking when caged with a subordinate male (Desjardins et al., 1973).

Also important as sources of pheromones are a variety of glands. Glands associated with the mouth, eyes, sex organs, anus, and skin are known to produce chemicals used in olfactory communication. Secretions from five locations on the body of the Australian honey glider *(Petaurus breviceps)* serve functions ranging from attracting newborn young, in the case of the pouch gland of the female,

FIGURE 23-12 Sources of scents used in intraspecific communication and pathways of social odors in the mule deer *(Odocoileus hemionus)*. The scents of the following are transmitted through the air: tarsal organ (1), metatarsal gland (2a), tail (4), and urine (5). When the animal lies down, the metatarsal gland marks the ground (2b). The hind leg is rubbed against the forehead (3a), and the forehead is rubbed against twigs (3b). Marked objects are sniffed and licked (3c). The interdigital glands (6) deposit scent on the ground. *(From Müller-Schwarze, 1971)*

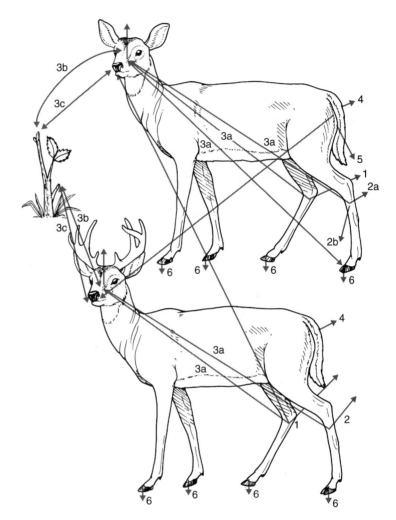

to contributing to a community odor within the social group in the cases of the frontal and sternal glands (Schultze-Westrum, 1965). Müller-Schwarze (1971) described a number of pathways of social odors in mule deer *(Odocoileus hemionus*; Fig. 23-12).

Reproductive behavior in some, and perhaps most, terrestrial mammals is strongly influenced by the sense of smell. The **vomeronasal organ** has been regarded by many as playing an important sexual role in the male. After smelling the genital area and urine of a female, some mammals, notably perissodactyls, artiodactyls, and some carnivorans, make a characteristic facial expression involving the upward curling of the upper lip and often the lifting of the head. This distinctive behavior, called **Flehmen** by K.M. Schneider (1930), has been

thought by some to be important in activating the vomeronasal organ and in perceiving sexual pheromones (Fig. 23-13). This hypothesis has gained support from laboratory studies on the hamster *(Mesocricetus auratus)* by Powers and Winans (1975), who found that destruction of the afferent nerves of the vomeronasal organ produced a disruption of copulatory behavior in one third of the altered animals. This procedure, coupled with destruction of the afferent nerves of the olfactory bulbs, completely eliminated copulatory behavior in all experimental animals. Powers and Winans suggested that input from both the vomeronasal organ and the olfactory bulbs is necessary for the arousal of sexual activity. Olfactory communication may also play an important role in the reproductive cycles of some primates. The

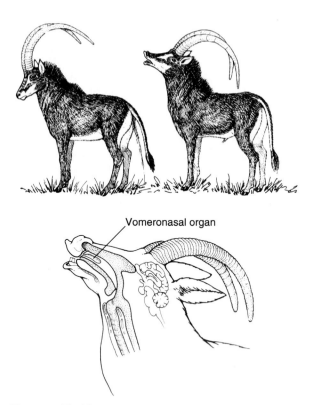

FIGURE 23-13   Male sable antelope *(Hippotragus niger)* exhibiting flehmen behavior to determine the estrus condition of a female. Cutaway drawing (lower) of the head region of a sable antelope showing the location of the vomeronasal organ used to detect airborne chemicals. *(After Estes, 1991)*

smell of vaginal secretions of rhesus monkeys *(Macaca mulatta)* in estrus is sexually stimulating to males and promotes copulation (Michael et al., 1971).

Ring-tailed lemurs *(Lemur catta)* make wide use of olfactory signals. Both sexes mark branches with secretions from the genitals, and, using the palms of the hands, males mark branches with other secretions. Scent glands occur on the chest and forearms of males. During aggressive confrontations, the tail is pulled between the forearms, anointed with scent, and then lifted high and waved to disperse the scent. Males indulge in "stink fights," which involve palmar marking, tail marking and waving, and often displacement of one animal by the other. The animals face each other when performing the scent marking, and the visual displays by the two animals, each using the conspicuously banded tail, seem to be mirror images of each other. The dominant animal moves forward while the other retreats. Vocalizations are also important during social interactions, and a variety of vocal signals are used. Kappeler (1998) showed that female scents probably function in mate attraction and maintaining female dominance hierarchies, while male scents are primarily used in male-male competition. Male scents were most often investigated by other males, and male rank and scent marking activity were positively correlated. One possible mechanism for intrasexual competition among males is illustrated by mouse lemurs *(Microcebus murinus):* scents from dominant males reduce testosterone levels and sexual activity in subordinant males (Perret, 1992). Finally, Kappeler (1998) showed that the recipient of the chemical signal may alter the message being sent. Fresh scent marks are rapidly countermarked by other males, possibly obscuring the original signal.

Self-anointing, an unusual signaling behavior shared by a number of species of hedgehogs (Erinaceidae), involves the smearing of saliva over the quills and results in a pungent smell easily detectable by humans. This behavior has been studied in the European hedgehog *(Erinaceus europaeus)* by Brockie (1976). When removed from their nests, nestlings self-anoint; this seems to aid the mother in recovering them. Self-anointing may also serve as a defensive behavior in hedgehogs. Brodie (1977) found that hedgehogs eat poisonous toads without becoming ill. Indeed, they recycle toxins from the toads parotid glands by mixing them with their saliva and licking the toxic mixture onto their spines. During laboratory tests, more than half of the Chinese rice-field rats *(Rattus rattus)* that encountered anal gland secretions of weasels *(Mustela sibirica)* anointed their bodies with the scent (Xu et al., 1995). The strong weasel scent may mask the rat's own odor and thereby avoid a weasel attack. Young raised in the laboratory with no contact with weasels displayed the same behavior, indicating that it is innate.

Some kinds of mammals are known to be able to discriminate between individuals of their species entirely by scent, and this ability is probably widespread. The mechanism for this recognition in the small Indian mongoose *Herpestes javanicus* has been studied by Gorman and coworkers (Gorman et al., 1974; Gorman, 1976). The anal pockets of this mongoose produce acids used in scent marking objects within the animal's home range. The glands contain a series of six volatile carboxylic acids derived from bacterial decomposition of sebaceous (oil gland) and apocrine (sweat gland) secretions in the pocket. The six acids occur in different relative amounts in different individuals, and animals recognize each other by scent on the basis of the unique carboxylic acid profiles. Bacterial production of these acids has been demonstrated in a number of mammals. Sebum (from sebaceous glands) and, to a limited extent, apocrine secretions are waterproofing agents deposited on the pelage of mammals. When metabolized by skin bacteria, these substances produce odor-yielding carboxylic acids. Gorman (1976) suggests that selection has favored the concentration of sebaceous and apocrine glands into discrete organs, where bacteria can produce the carboxylic scents used in olfactory signaling in relatively large amounts.

A reasonable summary statement is provided by Eisenberg and Kleiman (1972). They regarded scent "as a means of exchanging information, orienting the movement of individuals, and integrating social and reproductive behavior."

ACOUSTICAL SIGNALS. The sense of hearing in mammals is acute, and auditory communication is of great importance. Indeed, the sounds of some mammals that are rarely seen are commonly heard. Impressive choruses of howling coyotes may be heard nightly in some parts of the western United States, where the animals themselves are only occasionally seen. The importance of nearly constant auditory communication to a herd animal is difficult for one to imagine. Virtually continuous noises made by the members of a herd integrate the group by keeping individuals apprised of each other's location. In caribou *(R. tarandus),* the creaking and snapping of foot bones can be heard for considerable distances and enable scattered members of a herd to remain in auditory contact (Kelsall, 1970).

Vocal communication is widely used by mammals. In humans, of course, this type of communication reaches its most complicated development, but even in other primates some type of language can be recognized. Vervet monkeys *(Chlorocebus aethiops)* give different alarm calls to announce different predators, and to each call the monkeys have a different response. They look into the trees

in response to the leopard call, at the ground after the snake call, and at the sky after the eagle call (Seyfarth et al., 1980). Gunnison's prairie dog *(Cynomys gunnisoni)* gives alarm calls that are differentiated into local dialects (Slobodchikoff and Coast, 1980). The complexity of the call is related to the complexity of the habitat: More complex calls are given where shrubs, rocks, and tree stumps occur. Less complex calls are given by prairie dogs living in open grassland. The Japanese macaque *(Macaca fuscata)* and chimpanzee *(Pan troglodytes)* have complex repertoires of sound signals (Mizuhara, 1957 and Goodall, 1986). The more basic sounds used by the rhesus monkey may be linked by a series of intermediate sounds, and one basic sound may grade independently into other calls (Rowell, 1962). This yields a remarkably complex and rich vocal repertoire. The functional importance of some sounds can be recognized. The quiet grunt, for example, is used by many primates to maintain contact with each other (Marler, 1965), and vocal sounds are used by many mammals to announce their position or to maintain or re-establish contact with one another. This may be one function of howling choruses of canids and the calls of young in a variety of species.

The functions of the varied sounds made by cetaceans are as yet not well understood, but many are clearly used in communication. Some "vocal" sounds may keep members of a social group aware of one another's position or signal aggression. Some tail slapping and loud splashing by dolphins may provide long-distance communication between scattered members of a foraging group. The complex songs of humpback whales probably represent a form of social communication that we simply cannot decipher. Differences in the vocal repertoires among neighboring groups of cetaceans have been reported for killer whales *(Orcinus orca;* Ford, 1991) and sperm whales *(Physeter catodon;* Weilgart and Whitehead, 1997).

Many mammals have vocalizations that seem to serve primarily as territorial advertisements. Among primates, for example, male howler monkeys *(Alouatta* spp.) of the Neotropics, the woolly lemur *(Indri)* of Madagascar, and the gibbons *(Hylobates* spp.) of southeastern Asia (Marshall and Marshall, 1976) make loud, resonant territorial calls. Each species has a different song, the calls of the male and female differ, each individual has a slightly different set of calls, and a subadult will often join in with the female. Territorial songs by a family thus advertise the family's species and location, the sex and individual identity of each singer, and the presence of a subadult.

The vocal repertoire of some rodents consists basically of ultrasonic signals. Ultrasonics play an important role in the integration of the reproductive behavior of the laboratory rat (*Rattus norvegicus;* Barfield and Geyer, 1972): a 50-kilohertz call is associated with aggression and such aspects of sexual behavior as solicitation and mounting, and a 22-kilohertz signal is given by reproductively refractory or unreceptive individuals. Parent-young communication is also based on ultrasonics. Parents respond to the ultrasonic distress vocalizations of helpless young by bringing them back to the nest, and a decrease in the acoustical energy of the calls as the young grow older is associated with the development of homeothermy and the attending decrease in vulnerability to cold (Noirot, 1969). Some ultrasonic calls by rodents probably serve as territorial announcements (Sewell, 1968).

Some male rodents that vocalize after copulation may thereby increase their fitness. This may be the case with a Malaysian tree squirrel *(Callosciurus caniceps)*. Males of this species congregate within the home range of an estrus female, who mates with four to six males within a period of six hours. After copulation, each male makes barking sounds (like those given toward a terrestrial predator) for up to 35 minutes; during this barking the female and nearby males stop moving (Tamura, 1993). Because immobility after copulation seems to increase the probability of fertilization (Matthews and Adler, 1977), Tamura proposed that the vocalizing male may increase the chances of his sperm fertilizing the female by stopping the female's activity and delaying copulations by other males. Post-copulatory calls by male hamsters prolong the quiet, **lordosis** posture of the females (Cherry, 1989) and may increase the male's chances of **paternity.**

Vocalizations facilitate individual recognition in many kinds of mammals. Brown (1976) found that female pallid bats *(Antrozous pallidus)* and their young recognize each other on the basis of distinctive vocal signatures. To the attentive human ear, the "who-oop" call of the spotted hyena *(Crocuta crocuta)* differs from one individual to another

and probably facilitates individual recognition. There are pronounced differences (involving the spacing of pulses and the addition of snorts) among the threat calls of elephant seal *(M. angustirostris)* bulls. These differences give each bull a unique vocal signature that allows individual recognition among bulls competing for females on the breeding ground (Shipley et al., 1981). Complex underwater sounds made by dugongs *(Dugong dugon)* in Shark Bay, Western Australia, included not only signals associated with aggression and spacing behavior, but signals that probably facilitated individual recognition (Anderson and Barclay, 1995).

Matsumura (1981) analyzed in detail mother-infant acoustic communication in a horseshoe bat *(Rhinolophus ferrumequinum)*. Of the several kinds of acoustic signals used by this bat in mother-infant communication, some are primarily communicative, whereas others are virtually identical to echolocation sounds. The process of a mother's reuniting with its infant follows a consistent pattern. In the first stage, the constant frequency calls made by the flying mother overlap (in frequency) and are temporally out of phase with the broadband sounds of the infant. At a later stage, the mother shortens the duration of the "phrases" of her calls, which the infant's calls tend to follow. After many repetitions of mutual and alternate signaling, both mother and infant shift to single high-pitched, multi-harmonic syllables. Eventually, the sounds of the mother and infant overlap precisely and appear to be single sounds. When a mother and its infant are reunited and make body contact, vocalizations gradually cease. A mother and an infant other than her own are unable to synchronize their acoustic phrases and never make body contact. Each mother-infant pair seems to have a fixed and unique range of timing of their sounds during the synchronizing stage. This is perhaps the basis for precise mutual mother-infant recognition within colonies of many (often hundreds of) horseshoe bats. Of further importance, the repeated sequences of mutual signaling during an infant's preflight period may be crucial to the infant's development of vocalizations used for echolocation (Matsumura, 1981).

One of the most exciting discoveries in the realm of mammalian communication comes from Africa. Actually, it comes initially from the Washington

Park Zoo in Portland, Oregon, where Katherine Payne sensed that Asian elephants were making vocalizations that could not be heard but could be felt as a throbbing in the air. Recording equipment verified that these elephants were indeed making fairly intense, extremely low-frequency vocalizations (14 to 24 Hertz), mostly below the threshold of human hearing (**infrasound;** Payne et al., 1986). Payne found that similar infrasound was being made by African elephants in Amboseli National Park in Kenya.

Observations of African elephants by scientists and naturalists have revealed an almost science-fiction aspect to elephant communication. Low frequencies are attenuated far more slowly in air than are high frequencies: theoretically, elephant infrasounds could carry at least 6 miles. African elephants have a varied repertoire of infrasounds that announce locations of individuals or groups and that signal such conditions as alarm, aggressiveness, and reproductive readiness (Poole et al., 1988). Radio-tracking studies in Zimbabwe showed that families of different clans would suddenly alter their direction of travel when still several miles apart in an apparent attempt to avoid contact. Also in Zimbabwe, Garth Thompson observed a group of 80 elephants precipitously abandon their habitual home area the same day that many elephants were being shot in a culling operation 90 miles away in Hwange National Park. Several days later, Thompson found the 80 displaced elephants bunched together as far away from Hwange as they could get. Some message of danger and death had seemingly been relayed many miles from group to group of elephants. Long-range communication by infrasound adds yet another dimension to the complex social world of elephants.

Another remarkable discovery was made by William Barklow, who found that hippopotamuses (*Hippopotamus amphibius*) produce a wide variety of vocalizations above and beneath the water's surface. The physics of sound transmission in water and air are very different because water is more than 800 times denser than air. As a result, no sounds produced in air are heard underwater. Hearing sounds produced underwater is easy, but determining their direction is difficult. Mammals localize airborne sounds by turning the head until the sound waves reach both ears at the exact same time. Because of the greater density of water, waterborne sounds are transmitted to the middle ear as vibrations of the skull bones, reaching both ears simultaneously no matter what the position of the head (waterborne sounds appear to come from everywhere). Cetaceans solved this problem by suspending the middle and inner ears from ligaments, effectively isolating them from direct contact with the skull bones (p. 227). In dolphins, waterborne sounds are transmitted via the dentary to a tube of fat tissue on the medial surface of the dentary and via the fat tissue to the middle ear bones. Remarkably, Barklow found that several cranial features important for hearing underwater are shared by hippopotamuses and cetaceans: (1) both have a thin, dish-shaped area on the dentary bone; (2) in both the middle ear is suspended from ligaments; and (3) both share a fatty connection from the middle ear to the dentary (Schwartz, 1996). Barklow also found that approximately 80 percent of hippopotamuses vocalizations are given underwater (Barklow, 1997). These sounds include squeals, moans, and bursts of staccato clicks reminiscent of the sonar clicks of dolphins. The possibility that "click trains" are used in the murky river water as a form of echolocation was tested by Barklow and his colleagues, but no evidence of echolocation has yet been found.

Naked mole-rat (*Heterocephalus glaber*) colonies are one of the most complex and highly structured nonhuman vertebrate societies (Jarvis, 1981), and vocalizations are seemingly the most important mole-rat form of communication. Naked mole-rats, which have the richest vocal repertoire of any rodent (18 distinct vocalizations), give calls in a wide variety of behavioral contexts (Pepper et al., 1991). The function of one distinctive "chirp" is to bring a new food source to the attention of fellow colony members (Judd and Sherman, 1996). Having found the new food, a scout chirps as it carries a sample back to the nest, where it holds its head high and waves the food for all to smell. Colony members respond to this chirp-and-wave behavior by backtracking the signature scent of the scout to the newly found food. Naked mole-rats dig prodigious distances to find scattered patches of bulbs or tubers. A patch usually provides enough food to support the entire colony, at least for a time. By alerting fellow colony members to a new food patch, the colony is benefited and the inclusive fitness of the scout is increased, because the scout is typically closely related to all of its fellow colony members.

Acoustic communication by fossorial rodents is complicated by the fact that there are no air currents in the sealed burrows of many fossorial species and only a narrow range of frequencies can be transmitted efficiently (Credner et al., 1997; Brückmann and Burda, 1997). Under these conditions, low-frequency sounds are optimal, and many species of blind mole-rats are most sensitive to frequencies between 0.6 and 1 kilohertz. Some fossorial mammals that live in closed burrow systems, where visual and olfactory communication are useless, employ a combination of low-frequency vocalizations and **seismic signals** (vibrations transmitted through the soil). The solitary blind mole-rat, *Nannospalax ehrenbergi* (Muridae; Spalacinae), uses its head to rap out long-distance messages on the ceiling of its burrow (Heth et al., 1987). In addition, there is a vocal repertoire of some six calls, including a low-frequency (0.5 to 4.5 kilohertz) purr, made mostly by the male and occasionally by the female, that reduces agonistic behavior of the female and facilitates copulation (Heth et al., 1988). The solitary Cape mole-rat, *Georychus capensis* (Bathyergidae), produces seismic signals by drumming on the burrow floor with its hind feet (Narins et al., 1992). Although territorial drumming rates are similar in both sexes, in the breeding season the drumming rate of the male is nearly twice that of the female (26/second versus 15/second), and the communicating pair (in separate burrow systems) synchronize their "footrolls."

These seismic signals are propagated greater distances through the soil than are airborne signals, which are rapidly attenuated. Laboratory tests show that *Nannospalax* perceives and reacts to the seismic components of the head-thumping signals but ignores the auditory component transmitted through the air of the burrows (Rado et al., 1987). Banner-tailed kangaroo rats *(Dipodomys spectabilis)* also communicate by footdrumming (Randall, 1997; Randall and Matocq, 1997). Territorial ownership is typically communicated by airborne footdrumming sounds made by rats drumming on the surface of large seed cache mounds (Randall, 1993). Each kangaroo rat has a unique footdrumming signature that allows

these rodents to distinguish between neighbors and strangers (Randall, 1989, 1995). Kangaroo rats also often footdrum within their sealed burrows (when remaining in the burrows, the animals typically plug the entrance hole with dirt). Randall (1997) found that footdrumming by *D. spectabilis* produces seismic-borne energy nearly 40 decibels greater in peak intensity than the airborne sound that must first pass through the burrow walls. On windless nights, communication between distant territories is via airborne sounds produced by footdrumming on top of the mounds. On windy nights, however, communication among adjacent territories is via seismic signals made within burrows.

Footdrumming also serves as a warning and often deters snakes from pursuing kangaroo rats (Randall and Matocq, 1997). The pattern of footdrumming shifts dramatically when the kangaroo rat switches from territorial drumming to footdrumming in the presence of snakes (Fig. 23-14). Snakes can detect the seismic signals of footdrumming through low frequency vibration detectors in the skin (Hartline, 1971).

The low-frequency sensitivity of a number of fossorial or semifossorial mammals (*Georychus, Cryptomys, Nannospalax, Talpa, Dipodomys,* other heteromyids, and gerbils) suggests the widespread importance of seismic communications. Relative to what remains to be learned, we know little about acoustical communication in most mammals under natural conditions. This is a promising area for research.

TACTILE SIGNALS. The use of tactile communication by mammals is widespread. The sexual behavior of many mammals includes precopulatory activities by the male, such as laying the chin on the female's rump, nuzzling the genital area, or touching various other parts of her body. These behaviors presumably are sexually stimulating to the female or at least cause her to accept mounting by the male. Perhaps tactile stimuli are of greatest importance in connection with sexual activities in most mammals, but in primates they have assumed other roles.

Grooming of one individual by another is the most important form of tactile communication in primates (Sebeok, 1977), and this function far overshadows that of removing parasites. In

FIGURE 23-14
Footdrumming patterns of 15 male and female kangaroo rats *(Dipodomys spectabilis)*. Territorial drumming is indicated by the shaded ellipses, and footdrumming by the same individual in the presence of snakes is denoted by open ellipses connected by a line. *(From Randall and Matocq, 1997)*

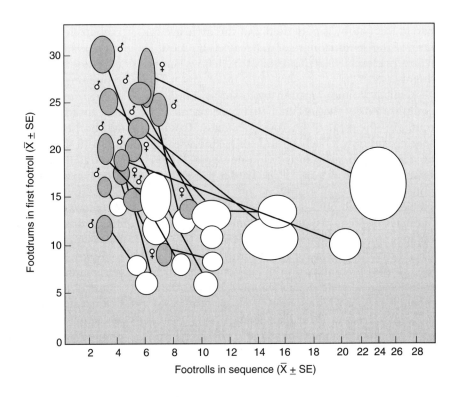

MATING SYSTEMS AND PARENTAL CARE

In general, females groom most often and for longer periods than males, and males groom each other much less often than they groom females. Grooming typically serves to reduce social tension, and, in many primates, is important in establishing and maintaining social contact between individuals. Grooming is regularly practiced after aggressive encounters, when it dissipates tensions and reestablishes "amicable" social contact. Often one animal will seek contact with another of higher social rank by grooming it, and mothers will distract young being weaned by grooming them.

Mutual grooming occurs also in collared peccaries *(Pecari tajacu)* and was described by Sowls (1974). Animals rub their heads against each other's flanks and rump, which bears a much-used scent gland, in a ritual that seems to serve as a greeting ceremony but also has a scent marking component. The African dik-dik uses its nose to touch various parts of another's body. Often the female rejoining her young or a male rejoining a female will perform this behavior. This seems to be a tactile reassurance and a reassertion of familiarity akin to mutual grooming (Bowker, 1977).

There is a fundamental conflict of interest between males and females with respect to their reproductive success. Males maximize their fitness by fathering as many offspring as possible. Females produce relatively few gametes compared to males and must nourish the developing fetus and the rapidly growing neonate. As a result, females have a greater investment in each offspring and should be selective in their choice of mates. Trivers (1972) proposed that sexual competition occurs because one sex invests much more than the other in the production of the young, a situation that may not be typical of all species. The conflict arises because females are a limiting resource for male reproductive success (Krebs and Davies, 1993). Males must compete among each other for the right to mate (**male-male competition**), while females seek to select the best mating partner from the many courting males (**female choice**).

It is not enough to produce the most offspring, however, as genes can only be passed on to subsequent generations if those offspring survive and re-produce themselves. Thus, parental care by one or both parents is essential in mammals. The pattern of parental care in mammals is often coupled to the type of mating system. Mating systems can be organized into several general categories:

1. **Monogamy** — A male and female form a tem-porary or permanent pair bond, and typically both parents provide care for the young.
2. **Polygyny** — One male mates with several fe-males, but each female mates with only one male per breeding season. In polygynous mat-ing systems, the female usually provides the parental care.
3. **Polyandry** — A female mates with several males during the breeding season, and typi-cally males provide the bulk of the parental care.
4. **Promiscuity** — Both sexes mate multiple times in a breeding season with different partners, and either sex may provide parental care.

In mammals, the females must not only carry and nourish the young during the gestation period but must also provide the young with milk for var-ious lengths of time after the birth. It is not sur-prising, then, that most mammals have evolved a polygynous mating system and females perform the bulk of the parental care. Monogamy occurs in only a few species of mammals. The following ex-amples illustrate the range of mating systems and parental care in mammals.

Clutton-Brock (1989) reviewed mammalian mat-ing systems and found that, in over 60 percent of mammals, the males defend territories that encom-pass one or more female home ranges. In cases where male territories are larger than female home ranges, males are often polygynous. In contrast, monogamy is the norm when female home ranges are larger than male territories.

Monogamy occurs in a variety of mammals, from some bats and rodents to some artiodactyls. Our knowledge of this mating system was reviewed by Kleiman (1977), who recognized two types. Type 1, termed **facultative monogamy,** occurs when the population densities of a species are so low that only one member of the opposite sex is available for mating. Type 2, termed **obligate monogamy,** occurs when the **carrying capacity** of a habitat is so low that only a single female can oc-cupy a home range and she cannot raise a litter without help from conspecifics. Monogamous mammals typically display little morphological or behavioral dimorphism, and the members of a pair interact infrequently except during the early stages of pair-bond formation and when rearing young. In obligate monogamy, sexual maturation in the young is delayed until after their association with the parents; older juveniles may help raise younger siblings, and the father often aids in the feeding, defense, and socializing of the young. Kleiman es-timated that fewer than 3 percent of mammalian species are monogamous, although this system is widespread among the mammalian orders. In some families of mammals — the marmosets (Cal-litrichidae) and dogs and foxes (Canidae), for ex-ample — monogamy is the major type of mating system. How long monogamous pair bonds last un-der natural conditions is difficult to determine, but the bonds clearly persist in some species through a number of reproductive seasons or as long as both members of the pair live.

Arvicoline rodents display several types of social and mating systems, from monogamy for the prairie vole (*Microtus ochrogaster;* Thomas and Bir-ney, 1979), to polygyny or promiscuity for the taiga vole (*M. xanthognathus;* Wolff, 1980) and meadow vole (*M. pennsylvanicus;* Madison, 1980). The abil-ity of some arvicolines to be socially flexible is probably associated with their tendency toward population fluctuations. The social biology of the California vole *(M. californicus)* is comparatively well known through the work of Lidicker (1973, 1976, 1979, 1980) and provides a model for what seems to be an intermediate system. A family group of California voles maintains and shares a runway system, to the exclusion of other individu-als and families. The family group has from 2 to 12 (mean = 6) individuals, including one male and one or more females and their young. Monogamy is the rule at low population densities, but at high densities polygyny is common. A pair bond proba-bly occurs, at least at low densities, and some pa-ternal care has been observed in this and some other species of *Microtus*. The inflexibility of the male-female pair system was demonstrated by Lidicker (1976, 1979), who introduced several pairs into a small enclosure. The fighting and mortality that ensued usually reduced the population to a single pair, which reproduced successfully. Over 25 years of research by Getz and his colleagues (Getz et al., 1993; Getz and Carter, 1996) revealed that the mating system of prairie voles is flexible and probably evolved in response to limited food supplies. Prairie voles can live in monogamous pairs or in large communal groups, depending on food supply and population density (Getz and McGuire, 1997).

Obligate monogamy occurs in some mammals when a male can defend a territory that includes only one female's home range. Male parental care is practiced in these cases because the male is unlikely to mate with any other females. Obligate monogamy is common in canids, and litter sizes tend to be large. Black-backed jackals (Canis mesomelas), for example, are monogamous, but some offspring may serve as helpers in rearing a new litter (Estes, 1991). Survival of the pups is directly associated with the presence of helpers in the Serengeti (Moehlman, 1983). Helpers regurgitate food for the pups and can spend as much time at the den guarding the litter as the parents. Bat-eared foxes (Otocyon megalotis) are probably monogamous but can form trios with a single male and two females (Van Lawick and Van Lawick-Goodall, 1970). Unlike other canids, bat-eared foxes are not territorial; home ranges often overlap those of neighboring family groups. Cubs are not provisioned at the den and mature earlier than many other canids. Lamprecht (1979) suggested that these differences in social organization result from their specialized insectivorous diet. At the other end of the extreme are canids such as African wild dogs (Lycaon pictus) and wolves (Canis lupus) that form large packs consisting of a breeding pair and several nonbreeding adults that assist in hunting and provisioning the young.

In species in which females tend to occur in small groups or have small home ranges, males can often monopolize several females, forming defended **harems** (referred to as **female defense polygyny**). Thanks to the careful work of Bradbury and Emmons (1974), the elaborate social organization of the small insectivorous bat, Saccopteryx bilineata, is well known. These bats occur widely in the Neotropics, where they live by day in colonies in the buttress cavities at the bases of large tropical trees. Each colony is organized into a number of harems, and each harem and its territory are maintained as a discrete social unit by a single male, with each unit containing from one to eight adult females. The harem territories are from about 0.10 to 0.36 square meters in area, and females are regularly spaced from 5 to 8 centimeters apart. This individual distance seems inviolate, except by young, which up to the age of about two months can approach their mother.

Territorial defense by males involves a remarkably intricate series of displays. Males bark, confront each other, patrol territorial boundaries, and open the scent gland in the wing and shake the wing as a visual and olfactory signal called "salting." Males salt both females in their own territory and another male's females across a territorial boundary, and will also salt another male across a boundary. Never did Bradbury and Emmons observe the antagonists sustain injury. The males are first to return to the colony sites at dawn. When females and their young begin to return, the males begin singing and performing various displays. The morning period of displaying and of sorting animals into their territories generally occupies several hours. A summary of behavior during a 60-minute period is shown in Fig. 23-15. Females vocalize, are highly aggressive toward one another, and rigidly maintain their individual distances. This tendency is probably a major factor limiting harem size.

Many bat species form harems. Small harems, consisting of a male and two to seven females, occur in the African vespertilionid Myotis bocagei. Harems of the southeast Asian club-footed bats (Tylonycteris pachypus) commonly include over a dozen females. The largest harems are found in tropical phyllostomid bats. Spear-nosed bat (Phyllostomus hastatus) harems can have over 100 females defended by a single male, but harems of 18 to 20 females are the norm (McCracken and Bradbury 1977, 1981; McCracken, 1987). The composition of females in these harems is stable from year to year, but the tenure of the male is variable. Harem males father most, but not all, of the offspring born to harem females (McCracken, 1987). Juvenile males disperse into groups of bachelor males, and yearling females disperse to form new harems within the same cave system. Juvenile dispersal serves to reduce the potential of inbreeding between father and daughters. In other social phyl-

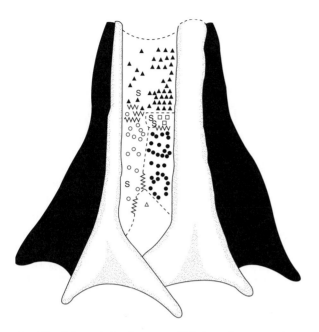

▲ – Hovers by male A - 1   ⋁⋁⋁ – Male boundary encounters
○ – Hovers by male A - 2   - - - – Male territorial boundaries
● – Hovers by male A - 3
□ – Hovers by male A - 4
△ – Hovers by male A - 5
s – Salts

FIGURE 23-15 The occurrence of various behaviors (hovers, saltings, and boundary displays) by the adult males of a colony of *Saccopteryx bilineata* during a 60-minute observation period. *(From Bradbury and Emmons, 1974)*

lostomid species, such as *C. perspicillata* and *Desmodus rotundus,* males do not defend harems. Harem formation depends on the degree to which males can control access to reproductive females, which is, in turn, determined by how long females maintain group tenure (Wilkinson, 1987; Fleming, 1988).

Many ungulates have breeding cycles that feature harems of females, each maintained by a dominant male. The area occupied by the females is the strongly defended territory of a single male, or in some cases, the male maintains a "breeding territory" that is defended even when females are not present. Breeding is done mostly by mature, vigorous, aggressive males. In an elk herd, only 12 percent of the bulls — the largest individuals — did

84 percent of the observed copulating (McCullough, 1969).

The rutting behavior of the elk (called red deer in Europe) is especially well known from the studies of Darling (1937), Graf (1955), McCullough (1969), Struhsaker (1967), and Clutton-Brock et al. (1982). McCullough recognized four main categories of bulls during the breeding season. Primary bulls are powerful, mature individuals that shed the velvet from their antlers early and are the first to establish harems (Fig. 23-16). Secondary bulls are large individuals that take over the harems by defeating the primary bulls as the latter become exhausted. Tertiary bulls assume control of the harems after the secondary bulls decline. Opportunist bulls are those whose only contact with cows is by chance. When a bull becomes exhausted through constantly herding cows together, driving rival bulls away, and copulating, all while unable to obtain adequate food and rest, it is beaten in a fight with a fresh bull, who takes over the harem from the deposed master.

In the pinnipeds that are polygynous — the otariids, some phocids, and the walrus — the males are extremely vocal, are much larger than the females, and maintain breeding territories. In the California sea lion (*Z. californianus*), large bulls establish territories adjacent to the water at sites favored as hauling-out places by females, which arrive at the rookery a few days before they give birth (Peterson and Bartholomew, 1967). Nonterritorial bulls usually form aggregations apart from the breeding rookery. Because the same females do not continuously occupy a male's territory and because males make no effective effort to herd females into territories, the term "harem" cannot be applied to the females in a territory. Roughly two weeks after parturition, females enter estrus and copulation occurs. Fighting between males occurs during the establishment of territories, and males on established territories signal their possession by incessant barking. Little actual fighting occurs after territories are established, but a boundary ceremony between males on adjoining territories periodically reaffirms boundaries. These ceremonies involve an initial charge toward one another, followed by open-mouthed head shaking as the animals confront each other at close quarters, and a final standoff in which the bulls stare obliquely at each other. The ceremony is so precisely ritualized

A

B

FIGURE 23-16 (A) A bull elk during the rut bugles to attract additional females to his harem. (B) Two young male elk fight to establish dominance. *(A from McCullough, 1969 and B from PhotoDisc, Inc.)*

that, should animals get uncomfortably close to one another, they adroitly avoid contact. Females are aggressive toward one another through much of the breeding season; again, however, injury is avoided by ritualized aggressive threats. Although males may be on territories in a rookery from May through August, each male maintains a territory for only one or two weeks; territories are thus occupied by a succession of males.

In contrast to the pinnipeds mentioned above, male elephant seals *(M. angustirostris)* establish a social hierarchy on the breeding ground, but are not territorial (Le Boeuf and Peterson, 1969). The highest ranking males stay close to breeding females, and breeding success is closely correlated

with social rank. On Año Nuevo Island off the coast of California, four of the highest ranking males, which constituted but 6 percent of the 71 bulls in the area, copulated with 88 percent of the 120 females. At another study area on the same island, the alpha bull (the bull at the top of the hierarchy) maintained its rank throughout the breeding season and was involved in 73 percent of the observed copulations.

While the advantages of harems to males is obvious, it has often been assumed that females benefit from this arrangement as well (Orians, 1969). In 20 years of careful observation and experimentation on yellow-bellied marmots *(Marmota flaviventris),* however, Armitage and his coworkers (Downhower and Armitage, 1971; Armitage, 1986, 1998) were able to show that reproductive success increases with harem size for males but not for females.

When females are not defensible and their daily movements are not predictable, males cannot rely on territories for access to mates, even in prime locations. Under these conditions males must resort to following the females or to forming seasonal harems. If females use predictable routes to water holes or foraging areas, males may form small territories along these well-worn paths and attempt to mate with the females as they pass through their small territory (resource defense polygyny). On open plains, where topi *(Damaliscus)* densities are high, males may form large or extremely small territories, depending on the habitat. When topi densities are high, males may form small territories of only 50 meters diameter (Jewell, 1972). Intense competition for the best locations leads to very small territories in kob *(Kobus kob),* lechwe *(K. leche),* and puku *(K. vardonii).* In extreme cases, an arena or **lek** is shared by 30 to 40 males, with only 15 to 30 meters between adjacent males (Leuthold, 1966; Floody and Arnold, 1975). Leks are arenas where males congregate (but where no resources are present) and females come to mate with the most aggressive or vigorous males. Leks form when males cannot defend the females or the resources they require. Male lechwe in leks gain higher mating rates than males that do not aggregate on the lek. When Nefdt and Thirgood (1997) experimentally reduced the number of females visiting the lek, males eventually abandoned their lek territories. In another experiment to test the hypothesis that female fallow deer *(Dama dama)* choose the "best" male and not the resources it defends, Clutton-Brock and coworkers (1989) forced the most successful males to move to another part of the arena. Despite their new locations, females still favored these males.

The amazingly specialized hammer-head bat *(Hypsignathus monstrosus)* forms leks. Males defend no resources and exhibit no parental care. Traditional lek sites may be used over and over again for over 60 years. Males select sites along streams, and the raucous vocalizations of the male are used not only for attracting females but for establishing dominance ranking and for maintaining a small territory in the lek. Males able to occupy the sites in the lek most favored by females do most of the breeding. Only 6 percent of the males were responsible for nearly 80 percent of the copulations in 1974 (Bradbury, 1977). Not surprisingly, males may take several years to establish themselves in these favored places.

The sex of the provider and the amount of parental care seem tightly coupled to the type of mating system. The paternity hypothesis, formulated to account for sexual differences in parental care, suggests that a parent's degree of care is correlated with the likelihood that this parent recognizes the offspring as its own (Trivers, 1972). Obviously, females can be absolutely certain of maternity, but males cannot be certain of their paternity.

Wilson (1975) regarded milk as the key to parental care in mammals. Mothers must invest considerably more energy in the care of postnatal young than in the growth and development of intrauterine young. Young are associated with their mother during much of their early lives, and, during this time, they are protected from a variety of dangers, including temperature extremes. Not only is avoidance of cold important when young are still poor thermoregulators but, in large ungulates, protecting young from heat stress may demand specialized maternal care. The mother-young group is the basic social unit in mammals. Even in species that are solitary as adults, the bond between young and mother is close. Because the care and nourishment of young demand from the females a tremendous amount of time and energy,

the females "are the limiting resource in sexual selection" (Wilson, 1975). Males, on the other hand, typically invest little time and energy in the young and are thus free to make behavioral adjustments, such as polygynous breeding or the holding of harems, that increase their fitness.

The amount of parental care given by mammals varies considerably. In the case of Malaysian tree shrews *(Tupaia tana),* mothers provide only minimal parental care (Emmons and Biun, 1991). Young are placed in "nursery" nests where the female comes every other day to nurse the young for a few minutes. Absentee parental care is rare in mammals and is taken to extremes in tree shrews: one female spent a total of less than 50 minutes with her young during the first month of their lives.

Male parental care is common in monogamous species. Care of offspring is also performed by individuals other than the biological parents (called **alloparental care**). In prairie voles, older littermates remain in the nest and groom and brood pups from subsequent litters (Wang and Novak, 1992). Alloparental care in mammals may even involve nursing unrelated offspring. Female bighorn sheep *(Ovis canadensis)* that have lost lambs due to predation allow unrelated lambs to nurse. In this case, the females may still be distantly related to the young they suckle because females tend to remain in their natal group (Hass, 1990). Alloparents tend to be related to the young they care for. The high risks associated with dispersal, establishing a territory of their own, and finding a mate probably serve to keep nonbreeders with parents (Emlen, 1982). The costs and benefits to the helpers is discussed in the section on altruism below (p. 469).

BEHAVIOR OF GROUPS

For many years, biologists assumed that some pervasive benefit derived from sociality, some automatic increase in survival and fitness resulting from a social life. In his consideration of the evolution of social behavior, Alexander (1974) stressed that, to the contrary, social living has important disadvantages. Competition for food, mates, and space is heightened, and the conspicuousness gained is disadvantageous to groups of prey and groups of predators alike. An additional liability for social animals is the rapid spread of disease or parasites.

Alexander (1974) also discussed three broad advantages of sociality. First, an individual's vulnerability to predation may be reduced by effective group defense or herd behavior. Defense of the group by dominant males is an important antipredator strategy of baboons, and a cohesive, running herd of ungulates presents a problem for predators. "The safety of the herd consists of the cohesive mass of animals running in an organized manner. The animals exposed are only those on the outside, and even these are protected by the number of flying hoofs and the ebbs and surges within the group. The vast array of movement has a disorienting effect on the observer's vision" (McCullough, 1969). Under a variety of situations, "hiding" within the herd is an effective means of escape; selection against the straggler or the individual who breaks from the herd is intense (referred to as the "selfish herd" by Hamilton, 1971). The conspicuous "misanthrope" is the animal most easily singled out and killed by predators. Second, the cooperative effort of a predatory group (such as hyenas or wolves) may be effective in bringing down large prey that could not be killed by a single predator. With baboons, scattered but rich sources of food can be found more often by many searchers than by a single animal. Finally, a paucity of safe nocturnal or diurnal retreats may have forced a partly social life on such animals as baboons and some bats.

After animal groups form, refinements in social behavior evolve. Alexander views these refinements as serving several functions. They may increase the advantages gained by group living. Such a behavior as the formation of a defensive ring of animals by musk oxen (Ovibos moschatus) tends to reduce the vulnerability of the herd to predation by wolves. Further advantages may also be gained by groups of predators. The precise positions and spacing maintained by individuals of some foraging groups of cetaceans may increase the ability of the group to perceive and capture prey.

Most important, in Alexander's view, is the fact that the evolution of social behavior affects reproductive competition among group members and the reproductive performance of the population at large. The social system of the baboon (Papio hamadryas) described by Kummer (1968), for example, is based on the one-male unit. An adult male maintains a group of one to several females, which are threatened or punished when they stray. This is a stable unit, and the male copulates with his females only, but by keeping his females with him constantly and by being aware as they come into estrus, he ensures his fitness. Young males are often unit followers, have opportunities to copulate with estrus females on occasion, and become familiar with the social behaviors that may later be used in gaining and maintaining their own unit. During the evolution of this system, the fitness of the socially integrated individuals was presumably greater than that of the individual who was solitary or did not learn the behavioral tactics associated with the social life.

Many ethologists agree that the evolution of social systems in animals is associated with increased **inclusive fitness,** but a social system in any group of animals is tested against the constraints imposed by a specific environment. Perhaps as strongly as any factor, the abundance and distribution of food limit evolutionary options. Indeed, the evolution of some social systems may have been influenced primarily by selective pressures imposed by the distribution of food in time and space. Using baboons as examples again, the small, one-male social unit serves well in dry areas, where productivity of the habitat is low and food has a patchy distribution but is nowhere abundant. Savanna-dwelling baboons, however, occupy a more productive area where food is scattered but a patch may provide abundant food; these animals forage in large social groups.

SOCIAL SYSTEMS

Patterns of social behavior in mammals are so diverse that broad summary statements are hard to frame. An additional problem is the incompleteness of our knowledge of the social behavior of many mammals. As might be expected, the large, more spectacular mammals and game species have been most thoroughly studied, but some of the most important groups of mammals remain poorly understood. Although rodents and bats together make up over half of the known species of mammals, we know relatively little of the social behavior in these groups. This section gives a necessarily cursory overview of the social systems of mammals. There are excellent treatments of this subject by Eisenberg (1966, 1981), Ewer (1968, 1973), Wilson (1975), and Eisenberg and Kleiman (1983).

Among metatherians, social behavior has evolved in two diprotodont families. In the Petauridae, several species are social to some extent, but sociality is perhaps best developed in the honey glider *(Petaurus breviceps),* in which cohesive family units are dominated by males. In the family Macropodidae, sociality is developed to varying degrees, and probably the most highly evolved marsupial social behavior occurs in the whiptail wallaby *(Macropus parryi).* The population studied by Kaufmann (1974) consisted of subunits called mobs. The members of a mob occupy a home range to the near exclusion of members of other mobs, but the area is not defended. The social organization of a mob is loose, but some structure is provided by a rather flexible dominance hierarchy between the males that is maintained by nonviolent, ritualized fighting (Dawson, 1995). (A dominance hierarchy is a fairly permanent social system based on dominance. Each individual recognizes its "position"; that is to say, it recognizes the animals that it can dominate and those that dominate it.) An estrus female is typically accompanied for

one to three days by her dominant-male consort, who has exclusive mating rights.

Although our knowledge of the sociobiology of bats is extremely fragmentary, there is no doubt that within this group a wide array of social systems occurs, and some species are known to have complex social behavior. A few species are completely solitary, except during copulation and when the mother-young bond is maintained briefly. A more common pattern, typical of vespertilionid bats and some species in other families, involves the separation of the sexes at the time of parturition. Females form nursery colonies exclusive of males; the sexes associate again when the young can fly and forage. Monogamous family groups are formed by a few bats in the families Emballonuridae, Nycteridae, Rhinolophidae, Megadermatidae, and Vespertilionidae (Fig. 23-17). Of great interest as examples of complicated social behavior are the vampire bats, which show reciprocal altruism (see p.471).

Widely divergent social systems occur in the Rodentia, from the solitary system of pocket gophers (Geomyidae) to the eusocial system of two species of mole rat (Bathyergidae; p. 473). Within the Bathyergidae alone, the complete range of systems — from solitary to eusocial — occurs. Social evolution in rodents has been influenced by their patterns of daily and seasonal activity, by the temporal

FIGURE 23-17   A colony of vespertilionid bats *(Myotis myotis)* roosting communally in a cave. *(M. Andera)*

and spatial distribution of food, by environmental constraints, and by predation. Simple social systems seem to prevail. In solitary rodents, such as pocket gophers, kangaroo rats (Heteromyidae), and New World porcupines (Erethizontidae), each individual is solitary except briefly when the male and female are together during mating.

The black-tail prairie dog (*Cynomys ludovicianus*) is highly social (Hoogland, 1995). Prairie dogs formerly occupied many parts of the western United States, where they occurred in large "towns," often including over 1000 animals and covering many acres. (Now, lamentably, prairie dog towns are rare in many parts of the west, owing partly to intensive, government-sponsored poisoning campaigns). The functional social units are **coteries,** which generally consist of an adult male, several adult females, and a group of young. No dominance hierarchy exists in the coterie. The paths, burrows, and food in the area held by a coterie are shared by its members, but hostility between coteries is the universal pattern. Members of the coterie become familiar with each other in part by grooming, playing, and "kissing" behaviors. During kissing, the mouth is open and the incisors are bared. This behavior is seemingly a ritualized method of distinguishing between friend and foe. Faced with the threatening expression of open mouth and bared teeth, a trespasser retreats, while a fellow coterie member meets its "friend" and kisses. A two-syllable territorial call is used to proclaim ownership of territory. A repetitive, high-pitched yelp is a warning of danger. During the spring, when females are pregnant or lactating, the coterie system partially dissolves and some yearlings and adults establish themselves beyond the territorial limits of their coterie. The personnel of coteries thus may change, but the territory itself is stable.

An individual gains several advantages from this social system. Many eyes are watchful for danger, and many voices are ready to sound a warning. The effect of the animals' foraging is to keep vegetation low over a wide area and to provide terrestrial carnivores with little concealment. Perhaps equally effective in providing for long-term occupancy of an area, the animals are kept spaced so that overuse of food plants is generally avoided.

Recent advances in genetic techniques, notably **DNA fingerprinting,** have allowed researchers to establish the degree of genetic relatedness of colony members. Travis et al. (1996) used such techniques to study mating systems in Gunnison's prairie dogs (*C. gunnisoni*) in Arizona. These animals are colonial and occupy stable territories, which are defended aggressively by resident males. Based on above-ground behavioral observations, the mating system is usually described as harem polygyny (Rayor, 1988; Travis and Slobodchikoff, 1993). DNA fingerprint analysis revealed that females within a territory tend to be related to one another, but males sharing territories are not. Furthermore, these analyses showed that females living on a territory often produce litters of mixed male parentage. Female Gunnison's prairie dogs mating with at least three males increase their chances of pregnancy and parturition (Hoogland, 1998). An important finding of the DNA fingerprint analysis was that 61 percent of all offspring were fathered by males from outside the female's territory. This level of extraterritorial paternity is higher than values reported for any other species (Travis et al., 1996). Resident males aggressively defend their territories from incursion by neighboring males, but seem incapable of preventing females from leaving the territory temporarily to seek copulations with other males. Territoriality may provide access to resources other than mates (Slobodchikoff and Schultz, 1988). Re-evaluation of the mating system of Gunnison's prairie dogs, in light of the DNA evidence, suggests that these rodents practice "overlap promiscuity," not harem polygyny as previously suggested (Boellstorff et al., 1994; Travis et al., 1996).

The evolution of sciurid social systems has been the subject of a recent review by Blumstein and Armitage (1998). They provide evidence that increasing social complexity among squirrels leads to social systems characterized by fewer breeding females, increased age at first reproduction, decreased litter size, and higher first-year survival rates of offspring (Table 23-2). The costs of fewer offspring are offset by the benefits of increased offspring survival. Increased survival in the first year in social sciurids (such as ground squirrels) results from retention of juveniles within the social unit (Armitage, 1981; Barash, 1989). Increasing social complexity should also foster some form of female reproductive suppression (as in the naked mole-

TABLE 23-2 Life History Traits and Social Complexity of Social Sciurids
Based on the Work of Blumstein and Armitage (1998)

Species	Social complexity	Minimum female mass (g)	% Females breeding	Time to first repro. (yrs)	Gestation time (days)	Lactation time (days)	Average litter size
Marmota olympus	1.46	1400	45	3	—	—	4.0
Marmota marmota	1.41	2811	48	3	37	45	2.4
Marmota caligata	1.35	3300	45	3	30	28	3.0
Marmota caudata	1.22	1400	14	3	30	—	4.2
Marmota monax	0.27	3314	95	1	34	46	4.0
Cynomys parvidens	1.23	703	80	1	—	—	3.9
Cynomys gunnisoni	1.03	600	66	1	29	39	4.6
Cynomys leucurus	0.84	575	88	1	30	35	5.6
Spermophilus columbianus	0.65	376	75	2	24	28	3.7
Spermophilus beldingi	0.40	211	95	1	27	25	4.2
Spermophilus richardsonii	0.39	269	95	1	22	29	6.8
Spermophilus beecheyi	0.26	486	90	1	28	38	7.5

From data cited in Blumstein and Armitage (1998).

rat, *H. glaber;* Faulkes et al., 1991). Blumstein and Armitage (1998) argued that reproductively suppressed females remain in the social unit because the costs of dispersal are high. Dispersal increases the risk of predation, and dispersers are subject to increased aggression from residents.

Even in the most primitive primates, the Lemuridae, sociality is well developed, but, relative to those of many other primates, lemurid social systems are rather simple. The mouse lemur *(Microcebus murinus),* studied by Petter (1962) and Martin (1973), occurs on Madagascar in dispersed "population nuclei." The proportion of females to males in these nuclei is four to one, and as many as 15 females may occupy the same nest. Surplus males not accompanying groups of females often occupy nests on the periphery of the area. Although mouse lemurs occupy nests together, perhaps because nest sites are at a premium, there is no organized social life, and animals forage alone. This primate, therefore, must be regarded as a basically solitary creature that is flexible enough to be able to occupy nests communally.

The ring-tailed lemur *(Lemur catta),* is far more advanced socially, being perhaps the most socially progressive of all lemuroid primates (Jolly, 1966,

1972). This lemur lives in troops that range in size from 10 to 20 or more animals. Adult males and females are equally represented in the troop, and their total numbers are usually equaled by the numbers of young. Troops occupy exclusive areas, and there is little intertroop contact. Social organization within a troop is based on dominance patterns. Females are dominant over males, a reversal of the usual primate system. A male dominance hierarchy is established, and dominant males seem to be able to remain for long periods with a troop but (surprisingly) do not always have first access to estrus females.

Although a variety of types of social organization occurs among higher primates, individuals must learn to be responsive to a complex social field. An individual remains aware simultaneously of the attitudes and displays of a number of members of the social group and of the social ranks of these animals. Manipulation of the social field becomes an important aspect of the behavior of many primates, and even the ranking of an individual may depend in part upon its effectiveness as a manipulator (Lee and Oliver, 1979). As a result of enduring social bonds between female baboons, two "friends" can put up an intimidating united front when one is threatened by a third individual. In some primate social systems, the rank of a female depends partly on her close association with a dominant male and on her ability to depend upon his help or protection during aggressive confrontations. Her status may abruptly decline if the male is deposed from his dominant position. In some baboon troops and in other primate societies in which group structure and cohesiveness are maintained by strong dominance patterns, the dominant male is the focal point of attention. An individual's behavior and the behavior of the entire social group are geared to the responses of this leader.

Baboons that live in savannas and pursue an almost entirely terrestrial life are often vulnerable to attack by predators. Food is frequently scattered, and a troop must forage over wide areas, thus increasing the chance of encounters with predators. A large and tightly organized social group has evolved in these baboons, perhaps largely in response to this pressure. These groups include from about a dozen to over 150 individuals. Each group occupies a largely exclusive home range; although home range boundaries are usually respected, they are seemingly not defended. When a group is mov-

ing (Fig. 23-18), males quickly respond to threats from any quarter, and their united action provides the primary defense of the troop.

Sexual dimorphism is pronounced in baboons and enhances the male's intimidating appearance, as well as his fighting ability. Male baboons are about twice as large as females, are more powerfully built, and have comparatively huge canines. The long fur over the crown, neck, and shoulders of the male accentuates the impression of size (Fig. 23-19).

The mating pattern of baboons seems to have a consistent relationship to dominance ranking. Whereas subadult, juvenile, and less dominant males copulate with females in the early stages of estrus, dominant males have exclusive rights to females during the period of maximal sexual swelling (the time when ovulation occurs). In some groups, only the highest ranking male copulated with females during the height of the swelling (Fig. 23-20), and, in a group observed by DeVore (1965) in Kenya, not one dominant male attempted copulation until the swelling was at its peak.

Habitat differences between savanna-dwelling baboons and desert-dwelling baboons are associ-

FIGURE 23-18 Positions of the members of a moving troop of baboons *(Papio hamadryas). (After Hall and DeVore, 1965)*

FIGURE 23-19   The threat "yawn" of a male baboon displays the large canines. *(From Corel Corp.)*

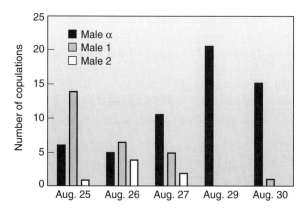

FIGURE 23-20   A comparison of the frequency of copulation by three male baboons with a female entering estrus. Note that the alpha (dominant) male copulated exclusively with the female on 29 August, at which time she was fully in estrus. *(After Hall and DeVore, 1965)*

ated with certain differences in social behaviors. Group size, for example, differs markedly. Savanna baboons depend on food patches that are typically rich enough to support large social groups: in Amboseli National Park of Kenya, groups contained an average of 51 individuals (Altmann and Altmann, 1970). Desert-dwelling baboons, on the other hand, have a relatively unpredictable and sparse food supply, and foraging groups are small, from several animals to perhaps 20. In addition, social organization differs between these baboons. Savanna baboons form only one type of social unit, the group, which includes many females, their young, and multiple mature males. The male-female pair bond is brief, lasting only a few hours or days, during the female's estrus period. Desert baboons, in sharp contrast, have four levels of social organization (Kummer, 1968, 1984). The smallest and most tightly knit unit is the family group, consisting of a single mature male, one to several adult females, their young, and often a young adult bachelor male "follower." Several such family groups band together to form a less tightly knit foraging unit called the clan, and a number of these clans form a fairly stable traveling unit called a band, which usually includes some 60 baboons. Many bands tolerate each other in order to sleep in safety on the same cliff at night; this loose aggregation is called a troop and contains several hundred baboons.

The social organization of a group of arboreal primates affects their foraging behavior. As an example, a consistent spacing is maintained by a group of Neotropical capuchin monkeys *(Cebus olivaceus)* moving through the trees. The front and center positions are occupied by the dominant male and female and by individuals tolerated by them, while peripheral positions toward the rear are occupied by individuals of lower social rank (Robinson, 1981). The monkeys near the center of the group can afford to be less alert for predators than those monkeys with peripheral positions and can therefore eat fruit more rapidly.

The great apes (Hominidae) have social systems that display no radical departures from basic primate patterns, but there are some unique features. Groups of the mountain gorilla *(Gorilla gorilla)* include from 2 to 30 animals. Social interplay between individuals is amiable, and assertions of dominance are low-key (Schaller, 1963, 1965a, b;

Fossey, 1972). Particularly notable is the age-graded male troop, with the nucleus of the group consisting of the dominant silver-backed male (ten years of age or older) and adult females and their young. Additional males, including less dominant silver-backs and black-backed males, attach themselves to the periphery of the group.

The chimpanzee *(P. troglodytes)* has been the subject of considerable field observation by a number of workers, including Izawa and Itani (1966), van Lawick-Goodall (1968, 1973), Izawa (1970), Nishida and Kawanaka (1972), Sugiyama (1973), Goodall (1983, 1986), and Stanford (1995). The basic social unit of chimpanzees is an often dispersed group of 30 to 80 animals that show considerable fidelity to a large home range. Particularly unusual is the looseness of the organization of the social group, with intricate patterns of establishment and dissolution of small parties. Highly evolved visual, tactile, and vocal communications are used. When a party of chimpanzees finds trees bearing fruit, their almost manic vocalizations and actions attract other parties to the bonanza. Male chimpanzees are capable of quick and coordinated hunting behavior (Teleki, 1973; Stanford, 1995; Stanford et al., 1994). Typically chimps hunt opportunistically, such as when they find a young baboon that has become separated from its troop. In this situation, some males will act to block its main path back to the troop while others block off other escape routes. Chimpanzee hunting may be more common than was previously thought. Stanford (1995) estimated that Gombe chimps consume approximately 1 ton of meat annually. Lower ranking males appear to barter meat for sex. Chimpanzees are clearly not always gentle and benign: on several occasions, males of one group have been observed systematically killing males of another group. Also, chimps occasionally kill and eat baby chimpanzees, but this behavior is considered aberrant (Goodall, 1986).

The closely related bonobo *(Pan paniscus)* has evolved a highly unusual complex of social behaviors (de Waal, 1995). Bonobo social systems are based on sexual contacts between individuals and include such behaviors as mouth-to-mouth "kissing," genito-genital rubbing between adult females, and pseudocopulation between adult males (de Waal, 1989, 1995). These behaviors appear to have evolved to minimize conflicts among group members.

Viewed broadly, primates differ from other mammals in the complexity of their social systems and in their highly developed vocal and visual communication. Much remains to be learned of primate sociobiology. This is especially true for the forest-dwelling species.

Our knowledge of the behavior of cetaceans is as yet extremely incomplete, but current evidence indicates that most species are social. Not only do some cetaceans travel and forage in social groups in which some consistent spatial organization is evident, but cooperative behavior is known. A recent review by Connor et al. (1998) compared the convergent evolution of social systems in odontocete whales with those of chimpanzees and elephants. Sperm whales *(Physeter catodon)* and the African elephant both have **matrilineal groups** comprised of up to a dozen females, males are largely solitary, and males do not generally have the opportunity to breed until relatively late in life. Chimpanzees *(P. troglodytes)* and bottle-nose dolphins *(Tursiops)* share a **"fission-fusion" social system** in which members of smaller groups freely move between groups.

Although most carnivorans are not social, highly organized social systems have evolved in some species. The lion pride and the spotted hyena clan are examples. The coyote *(C. latrans)* is socially flexible (Bekoff and Wells, 1980). When carrion provides large, defensible patches of food, coyotes form small, cohesive packs, but when depending on small, dispersed prey, such as rodents, coyotes are solitary. Some diurnal herpestids and procyonids are also social, with the packs of the banded mongoose *(M. mungo)*, which may include 40 animals, representing the extreme. The social structure of the white-nosed coati *(Nasua narica)* is unique within the Carnivora, consisting of groups of up to 30 females and their offspring, called "bands," and solitary males (Gompper et al., 1997). Coatis form female coalitions (groups of individuals acting as a unit to access resources or deter opponents) similar to those in primate societies. DNA analyses revealed that not all females within a band are related. Unrelated females receive more aggression and competition for food from subgroups of related females within the band, but stay with the band because the costs of being solitary are considerably higher. Gompper (1996) showed that solitary females were frequently displaced from fruit patches by larger, more aggressive males and suffered higher parasite loads and increased risk of predation. Coalitions of females were able to displace solitary males and gain greater access to the patchy food resources; simultaneously, they reduced their risk of predation by increased vigilance.

Among the most highly social canids (the wolf, *C. lupus;* the African hunting dog, *L. pictus;* the dhole of southern Asia, *Cuon alpinus*), a dominance hierarchy gives structure to the pack. By cooperative hunting, these cursorial animals can kill large prey.

Kruuk (1972), who studied spotted hyenas in the Ngorongoro Crater and the Serengeti Plains of Tanzania, observed a remarkable social system. The basic social unit of the spotted hyena is the clan, which may contain as many as 80 animals. The clan system is periodically disrupted in the Serengeti, where seasonal migrations of wildebeest and zebras result in drastic shifts in food supply. Each hyena clan defends a territory, the boundaries of which are maintained in part by systematic scent marking. Territorial disputes are often violent, and individuals are occasionally killed during border warfare. Females are larger than males and are dominant to them. A rather complex dominance hierarchy exists within a clan, and strong bonds develop between females. At a kill, dominance is often not asserted and competition is based largely on the ability to eat extremely rapidly (aided by profuse salivation) rather than on fighting prowess or size. Females tend their cubs at a central denning area, and here young receive early training (or practice) in the social rituals of the species.

In South Africa, spotted hyena clans have five social classes: resident females, cubs, and three classes of males (Henschel and Skinner, 1987). Some males stay with their natal clans (resident natal males). Alternatively, some males leave their natal clan and attempt to join neighboring clans, but the resulting social status of these males was generally low (peripheral immigrant males) unless they displayed persistent attention to the female clan members, in which case they occasionally gained breeding status within the clan (central immigrant males).

When two hyenas meet, they typically go through a meeting ritual, part of which involves mutual examination of the external genitalia. (The external genitalia of the female hyena mimic the male penis and scrotum: the female's clitoris is very large, resembles a penis, and is erectile, and two sacs filled with fibrous tissue form a false scrotum.) Kruuk concluded that the intraspecific mimicry evolved because of its importance during meeting ceremonies. This ritual probably enables individuals to be close to each other briefly and to "identify" each other while attention is attracted to the genitals and to a course of action other than fighting. This cooling-off period perhaps allows aggressive tendencies to subside. The leg-lifting action seems to be an appeasement gesture, hence its initiation typically by a subordinate animal. The adaptive value of such behavior to a spotted hyena is perhaps heightened by its flexible social life. Although often social, hyenas may be solitary for varying lengths of time, and peaceful meeting behavior and recognition of individuals is frequently important. In predators with such powerful offensive weapons as the teeth of hyenas, control of aggression is of critical importance, and the great array of scents, displays, and vocal signals serves to restrict the use of these weapons.

In the Serengeti, where prey biomass is low except for short periods of the year, hyena clans are large (median size of seven clans was 47 individuals). During part of the year, clans were forced to leave their territories and commute to migrating herds (Hofer and East, 1993a, b). Mean commuting distance was approximately 40 kilometers. During the commute, hyenas avoided contact with resident clans. When resident clans encountered commuters at kills, they responded aggressively, but they tended to ignore commuters in transit through their territories. The commuting system of Serengeti hyenas in response to a migratory food supply is unique among carnivores. Females with young cubs suffered disproportionately during the commuting season (46 to 62 percent of the year is spent commuting). Lactating females made briefer, but more frequent, commuting trips than did other members of the clan, but lactating females expended three times the travelling effort expended by nonlactating females (Hofer and East, 1993b).

Hyena clans in the Namib Desert are much smaller, consisting of three to five adults. Here the average home range was 570 square kilometers, compared to territories of approximately 50 square kilometers in the Serengeti (Tilson and Henschel, 1986; Hofer and East, 1993a). In the Namib Desert, home ranges did not overlap and hyenas did not exhibit territorial scent marking. The characteristics exhibited by Namib hyena clans are probably adaptations to the harsh environment and depauperate prey communities (Tilson and Henschel, 1986).

Long-term field studies on the lion *(Panthera leo)* in Serengeti National Park of Tanzania by Schaller (1972), Bertram (1973, 1975), Packer (1986), and by Heinsohn and Packer (1995), have provided a fascinating picture of the social life of this animal. The lion pride is a fairly stable social unit, usually with about 3 to 12 adult females. There is virtually no recruitment of outside females: all of the females are born and grow up within the pride, and all are closely related. A female reaching three years of age is either accepted as a member of the pride or is driven from it. Rejected females and males often become nomadic, follow migrating prey, and make up about 15 percent of the total lion population in the Serengeti. A pride is usually controlled by several adult males that defend the pride's largely exclusive and fairly stable territory. Although on occasion individuals may hunt alone, or part of a pride may separate from the rest, the members of a pride are familiar with each other, and social contacts are usually peaceful or seemingly affectionate. A member of the central sisterhood of the pride leads a stable, if at times violent, life, and her reproductive life is about 13 years.

A male, on the other hand, does not associate consistently with a single pride throughout life, and his reproductive life may be only two or three years. The life cycle of the male can be divided into several periods. Young males stay with their pride until they are about three years of age, when they are either forced from the pride or leave it voluntarily. Often several males leave the pride together; these males may be brothers or — because the females of a pride are all grandmothers, moth-

ers, sisters, or daughters — at least closely related. The young male outcasts become members of the nomadic population, and, unable to depend on the hunting prowess of the experienced females, often turn to scavenging. After roughly two years of nomadic life, these males are approaching the prime of life and are sexually mature and sufficiently formidable to take over a pride.

The pride they take over is almost never their natal pride. A pride lacking males may be taken over peacefully, or several males past the prime of life may be easily displaced, but violent fighting may accompany the displacement, and, because a group of males can successfully challenge males holding a pride whereas one or two males cannot, selection favors tight social ties between males. New males disrupt the life of the pride: pregnant females may abort, the cycle of females coming into estrus is interrupted, and the newcomer males may even kill cubs (infanticide). After a few months, however, females again begin coming into estrus, they are bred by the new males, and males often help take care of the cubs (Fig. 23-21). After only two or three years, however, when the males are aging, they are driven out by a new group of prime males. The reproductive life of displaced males is over, and, because of their declining physical condition, their life expectancy is not great.

FIGURE 23-21 A male lion mounts one of the females in his pride. Prides may have one adult male or a coalition of several males. *(Corel Corp.)*

All the females in a pride tend to synchronize the bearing of young, and a cub can nurse from lactating females other than its mother. This communal care of young can perhaps be explained by the close genetic relationships between females of a pride and by the increased survival of cubs with familiar companions (Bertram, 1975).

In general, however, lions are inefficient reproductively, and the mortality rate of cubs is high (about 80 percent). Bertram (1975) detailed the situation as follows: "Assuming that lions mate every 15 minutes for three days, that only one in five three-day mating periods results in cubs, that the mean size of litters is two and a half cubs and that the mortality among cubs is 80 percent, then a male must mate on the average some 3000 times for each of his offspring reared to the next generation." Because each copulation is relatively unimportant and because the males of a pride need each other to maintain control of the pride, pressure on the males to fight for the chance to copulate with an estrus female is reduced. The lion has few predators, and the size of the pride is perhaps controlled largely by periodic food shortages. The life span of lions is fairly long. Reproductive inefficiency, therefore, does not prejudice the survival of the pride. A critical factor may be the reduction of aggression in a pride to a level permitting the survival of at least some young. Bertram suggests that reproductive inefficiency and reduction of competition between males result in increased stability of the pride, fewer changes of the male guard, and hence greater chances for the survival of cubs.

In four genera of African herpestine carnivorans *(Suricata, Crossarchus, Mungos,* and *Helogale),* highly developed social systems have evolved, perhaps in response to predation. Studies by Rood (1978, 1980) have revealed the complex social organization of the dwarf mongoose *(Helogale parvula).* This small diurnal carnivore (about 320 grams) occurs in parts of eastern and southern Africa, usually in open woodland or scrubland with short grasses, where the dens are in termite mounds. The packs average about eight members, and include the dominant, breeding (alpha) pair, usually the oldest pack members, plus, typically, an additional male, two adult females, and several juveniles. All of the young born in the pack are those of the alpha pair; subordinates are behaviorally and endocrinologically suppressed (Creel et al., 1991, 1993). The mortality rate of young is well over 50 percent in their first year, mostly due to predation. Tight adherence to social rules and cooperation typify the social organization. The pack leader is usually the alpha female, and the pack sleeps together, forages together, and members share in watchfulness for, and defense against, predators. Group cohesion is maintained during foraging by repeated contact calls. An alarm call announces predators. Helpers (other than the alpha pair) do most of the grooming, baby-sitting, and carrying of the young between den sites, and helpers bring all of the young's food after they are weaned. Rood recorded eight cases of subordinate females lactating, perhaps after aborting their litters. In one case, a lactating subordinate saved a litter after the mother died. All of the lactating subordinates were related to the young of their respective packs. DNA fingerprinting demonstrated a high degree of inbreeding among pack members (Keane et al., 1996). Dwarf mongooses are unusual in that subordinates of both sexes often disperse, but this dispersal does not reduce the generally high degree of relatedness among pack members. Seemingly, dwarf mongoose packs do not suffer **inbreeding depression,** leading Keane et al. (1996) to hypothesize that many generations of mild "inbreeding have purged most of the deleterious recessive alleles from this population."

The social life of African elephants *(Loxodonta africana)* is known through the work of Laws and Parker (1968), Hendrichs and Hendrichs (1971), Douglas-Hamilton (1972, 1973), McKay (1973), Moss (1983), and Poole (1987, 1989). The elephant social system is structured at several levels. The first level is that of the family group, including an old matriarch and 10 to 20 females and their offspring. Because of the long life span of elephants, the family unit generally includes grandmothers, mothers, sons, daughters, grandsons, and granddaughters. Lifetime social bonds among females may last 50 years or more. The second level of the social system is the kinship group, consisting of several family groups that remain in the same vicinity and, on occasion, mingle peaceably. Under some conditions, as during migration, many kinship groups may band together to form clans, containing on occasion 100 or more animals. The clan probably has no social cohesion at any level above the kinship group.

Bulls leave the family units when they become sexually mature and assemble in all-male groups in which dominance is established by ritualized fighting and sparring. Dominant males are temporarily attached to family units with females in estrus. Males in **musth** (periods of heightened aggression brought about by elevated hormone levels) do most of the breeding and are given a wide berth by bulls not in musth.

Elephant societies are maintained by a complex series of tactile, olfactory, vocal, and visual behaviors. Deep rumbling sounds below human hearing thresholds are used to communicate over long distances. Greeting ceremonies involving inserting the trunk into the other's mouth often occur when animals are reunited. Especially remarkable is the importance of cooperative and apparently altruistic behavior within the family unit. A nursing elephant is allowed to suckle from any lactating female, young females approaching sexual maturity are solicitous of the well-being of small calves, and the safety of a calf seems to be the concern of the entire family unit. When threatened, the adult members of the family form a formidable defensive phalanx.

Ungulate social behavior is of particular interest for several reasons. Many species are large, occupy open situations, and can be observed easily, and have therefore been well studied. Because these open-country dwellers are probably the most highly social of all ungulates, we have a reasonably good understanding of ungulate social behavior. Further, a growing knowledge of the environments occupied by a variety of ungulates has provided a basis for a theoretical approach to the relationships between ecology and the evolution of social behavior, morphological features, and color patterns (Estes, 1974, 1991; Geist, 1974; Jarman, 1974).

Jarman (1974) and Estes (1991) gave order to our view of bovid behavior by relating the sociobiology of bovids to their ecology (Table 23-3). Jarman showed that, in contrast to forests, grasslands (open woodlands, savannas, and plains) support the higher diversity and biomass of bovids. In grasslands, the bulk of the food available to bovids is grass. Although a high percentage of each grass plant is edible, much less of a tree or shrub can be eaten. Grasslands therefore produce proportion-

TABLE 23-3   Behavioral and Ecological Comparison of a Forest Antelope and a Plains Antelope

	Forest Duiker	Plains Oryx
Habitat	lowland and montane forests and scrublands	arid zone plains and savannas
Body Size	small (4–64 kg)	large (150–200 kg)
Horns	short spikes in both sexes	long straight or curved in both sexes
Preorbital Glands	well-developed in both sexes	vestigial or absent
Feeding Habits	selective browsers	grazer
Water Requirements	water-dependent	water-independent
Coloration	concealing	conspicuous
Social System	solitary, monogamous sedentary, small home ranges	gregarious, polygynous nomadic, large territories
Breeding	perennial	seasonal
Defense	seek cover and hide	flee in open

From Estes (1991).

ally more food per growing season for bovids, but growing seasons are short. Although they produce less food for bovids, forests have moisture and plant productivity distributed more evenly throughout the year.

Associated with the differences in density and habitat between large and small bovids are sharp differences in social organization. The five categories of Jarman (1974), as outlined by Wilson (1975), are shown in Table 23-4 and indicate the relationships between habitat, feeding style, and social organization in African bovids.

Estes stressed a major structural and behavioral dichotomy within the family Bovidae (Estes, 1974). The ancestral bovids were probably small forest dwellers, perhaps resembling today's forest-dwelling duikers. The expansion of grasslands in the Miocene and Pliocene of Eurasia and Africa set the stage for the movement of bovids into open grassland or savanna habitats. This major evolutionary step brought some bovids under the influence of new suites of selective forces and led to their structural and behavioral divergence from the persistently forest-dwelling species. Viewed today, the dichotomy is between, on the one hand, the forest-dwelling browsers that are generally small, cryptically marked, and simple-horned and escape from predators by hiding and, on the other hand, the open-country grazers that are of medium or large size, are conspicuously marked, possess large and often complex horns, and use their speed in the open to escape predators. The behavioral dichotomy is also clearly delineated: the forest dwellers either are solitary or live in small family groups, and scent marking is the primary means of communication; the open-country grazers are typically highly gregarious and use primarily visual signals.

As noted and summarized in Table 23-4, the social organization of the classes of bovids recognized by Jarman and Wilson forms a progression from the small social units of the selective feeders of the bush and forest to the very large herds of the unselective feeders of the grassland. To flesh out the skeleton of this outline, comments on the social behavior of several species follow.

Blue duikers *(Cephalophus monticola)* are small forest-dwelling antelope that form monogamous pairs and defend small permanent territories aver-

aging 3 to 4 hectares in size. Territories are maintained in the forest by scent marking. Members of a pair stay close together, and pair bonds are reinforced by behaviors such as social licking and mutual pressing together of preorbital glands (Du-Bost, 1983). Duikers have several alarm calls and respond to danger by seeking cover.

Impala have a somewhat different social organization (Class C in Table 23-4; Jarman, 1970; Jarman and Jarman, 1974; Hart and Hart, 1987; 1988). Dominant male impalas, constituting about one third of the population of adult males, maintain territories in the breeding season in the most favorable habitat. These territories form a mosaic of adjoining areas, and each dominant male defends his area against other males of comparable social status. Females and bachelor males occupy home ranges that typically include a number of territories. A territorial male attempts to round up females that enter this area and keep them within it. In bachelor herds, the hierarchy is based partly on age distinctions, with older, larger horned animals dominating younger ones. Males at or near the top of the bachelor hierarchy challenge territorial males, and repeated encounters between a territorial male and his challenger might span several weeks. A male holding a prime territory much frequented by impala is kept busy herding females, checking for females in estrus, and keeping bachelor and competitive adult males at a distance. These males become exhausted and lose their territory more quickly than do males holding less preferable areas. In areas with seasonal precipitation, the territorial system is abandoned during the dry season.

The advanced social organization of the African buffalo (Class E in Table 23-4) was studied by Sinclair (1970, 1974, 1977) in Tanzania. This nonterritorial animal forms herds of from 50 to 2000 animals. The size of a given herd is rather constant, at a mean size of about 350 animals. For the first three years of its life, a young buffalo tends to remain near its mother, and bonds between mothers and daughters seem closer than those between mothers and sons. In the third year of life, males begin to leave their mother, and when they are four and five years old they form subgroups within the herd. Older adult males that remain with the herd establish a linear dominance hierarchy. The

TABLE 23-4 Behavioral/Ecological Classification of Some African Bovids

Social organization and feeding style	Body size (kg)	Antipredator behavior	Examples
Class A Solitary or in pairs or family groups; small, permanent home range; highly diversified diet, but selective feeders	1–20	Freeze, dash to cover and freeze, or lie down; do not outrun or counterattack predators	Dik-dik (*Madoqua*), duiker (*Cephalophus*)
Class B Several female-offspring units associate; group size 1–12; permanent home range; males solitary; diversified diet	15–100	Similar to Class A, but with some outrunning of predators for short distances	Reedbucks (*Redunca*), vaal rhebuck (*Pelea*), oribi (*Ourebia*), lesser kudu (*Tragelaphus imberbis*)
Class C Larger herds, of six to hundreds of members; males have breeding territories; selective browsers and grazers	20–200	Diverse; hiding used in heavy cover, running used in open situations; communication of alarm behavior important	Kob, waterbuck, lechwe (*Kobus*), gazelles (*Gazella*), impala (*Aepyceros*), greater kudu (*Tragelaphus*)
Class D During sedentary times, societies as in Class C species; gigantic herds developing during migration; feed on variety of grasses; selective as to plant parts eaten	100–250	Run from large predators or mount unified counterattack on smaller predators	Wildebeest (*Connochaetes*), hartebeest (*Alcelaphus*), topi (*Damaliscus*)
Class E Large, stable herds of females and young with males organized into dominance hierarchies; herd size up to 2000; no coalescing of herds during migration; unselective grazers or browsers	200–700	Run from predators or mount unified counterattack even on large predators; groups respond to distress calls of young	Buffalo (*Syncerus caffer*), oryx, gemsbok (*Oryx gazella*), probably eland (*Taurotragus*)

Based on Jarman, 1974, and Wilson, 1975.

repeated sparring typical of immature males may result in the formation of this hierarchy. The less dominant males are driven from the herd and form small bachelor groups that remain separate from the mixed herds. Old males, over about ten years of age, leave the herd and become extremely sedentary; they are either solitary or form small social units. The breeding is done largely by the dominant males of the herd, with the highest ranking ones having the greatest access to estrus females.

Especially remarkable is the way in which the herd functions as a tightly knit unit. An entire herd will rally to the defense of a young or adult animal in distress, and this formidable united front will discourage even the largest predators. A herd also moves and feeds as a closely massed unit. There is little attempt to maintain individual distance, and the bodies of herd members may, on occasion, be touching.

Geist (1974) pointed out that the widespread substitution by bovids of ritualized combat and aggressive displays for damaging physical contact has probably evolved under selection exerted by high densities and high diversities of predators. Bovids that attack and injure others invite damaging counterattack and are likely to be wounded, whereas those that use nondamaging, ritualized fighting are less likely to sustain injury. Because predators often concentrate on conspicuously wounded animals, selection by predators strongly favors the adherence by bovids to ritualized intraspecific contests. This great development of ritualized combat in African bovids, which must face many diverse predator populations, contrasts with the more damaging encounters between members of northern species, such as bighorn sheep *(O. canadensis),* that are under far less pressure from predators.

Degree of sexual dimorphism, which varies widely among ungulates, has been related to habitat and social behavior by some authors (Geist, 1974; Jarman and Jarman, 1979). In general, where high-quality habitat is patchy, there is intense competition among dominant males for territories within these patches and for the females that occupy them. The impala, which lives under these conditions, is extremely dimorphic: males are one and a half times as large as females and have flamboyant horns, whereas females are horn-

less. In woodland habitats, which can support only low densities of ungulates, the inhabitants are selective feeders and a male and a female may mate for life and share a territory. An example of such a type is the dik-dik, in which sexual dimorphism is slight. Dimorphism is also slight among species that occupy open grassland with a high carrying capacity. These ungulates occur in large herds, in which defense of females by males is not feasible, and males and females compete more or less on equal terms for food. African buffalo and wildebeest are examples.

KIN SELECTION

Natural selection favors individuals with life history strategies that maximize their genetic contribution to future generations. The particular life history strategy adopted will depend on the interaction between the individual and its environment (involving climate, food resources, predators, competitors, and nest sites). The individual's chance of surviving and reproducing will depend to a large extent on its behavior (Krebs and Davies, 1993). Because behavior has a genetic component, natural selection operates on genetically based behavioral variation in the population, just as it does on genetically based morphological variation. Like most morphological characteristics, the majority of behaviors are genetically determined to some degree. Often many genes influence the expression of a particular behavioral trait **(polygenic traits).** Those behaviors that increase the animal's foraging efficiency, predator avoidance, ability to find mates, and so on, will be favored by natural selection. Natural selection results in fitness differences tallied as the lifetime contribution to future generations of one individual's **genotype** relative to the contribution of other genotypes in the population.

Behaviors that appear to be **altruistic,** those that increase the fitness of the recipient while reducing the fitness of the donor, seem to contradict evolutionary theory. How could the frequencies of genes controlling altruistic behaviors have increased in populations via natural selection? One mechanism proposed by Hamilton (1964) is **kin selection.** Kin selection is a type of natural selection in which "the frequency of a gene in a population will be influenced not only by the effects that gene has on the survival and fertility of individuals carrying it, but also by its effects on the survival and fertility of relatives of that individual" (Maynard Smith, 1976). The term **inclusive fitness** refers to an individual's success in passing on genes to future generations, and includes the individual's genetic fitness via its own offspring plus any influences on the fitness of its other relatives. It is relatively easy to see how the fitness of an individual is increased by passing on copies of its genes via its own offspring who share, on average, 50 percent of their genes in common with each parent. Offspring, however, are only one type of kin. Siblings also share 50 percent of their genes in common, and cousins share 12.5 percent. The fitness gained by helping a sibling to survive and reproduce, therefore, is considerably greater than that gained by helping a cousin, but helping a cousin may still result in fitness benefits for the "altruist." Thus genes for altruistic behavior can increase in frequency in a population if

$$rB - C > 0$$

where r is the coefficient of relatedness between the donor and recipient, B is the fitness gained by the recipient, and C is the fitness cost to the donor of a behavior (Hamilton, 1964). If an individual helps a sister rear three more offspring (r = 0.25 for nieces) than she could have without the extra help, the fitness gained by the recipient ($B = 3$) is greater than the cost to the helper of raising one less young of her own ($C = 0.5$ for offspring). In this case, the gene for the altruistic behavior will increase in frequency in the population. As the geneticist Haldane (1932) said (anticipating Hamilton's theory), "I would give my life for three brothers or nine cousins!"

The importance of kin selection theory is that it allows the quantification of the costs and benefits of a particular behavior in terms of inclusive fitness and explains how apparently altruistic behaviors can evolve via natural selection. Apparently, altruistic behaviors in mammals include alarm calling to warn others of the approach of a predator, cooperative breeding **(alloparental care),** in which "helpers" assist in the rearing of another's young rather than reproducing themselves, coalition forming to cooperate in acquiring mates, and food sharing.

Consider the example of predator alarm calls given by Belding's ground squirrels *(S. beldingi).* These squirrels are social rodents that inhabit subalpine meadows. Females mate in the spring, establish territories around the nest burrow, give birth to three to six young, and males leave after mating and do not contribute to parental care. After weaning, juvenile males disperse, leaving the juvenile females with their mothers in the natal area. Females, therefore, are nearly always surrounded by close kin, while males seldom interact with close relatives. Sherman (1981a, b) found that closely related females cooperated to defend each other's young against territorial raids by unrelated conspecifics. (Undefended offspring were often killed by young males in search of an easy meal and by adult females attempting to take over nesting burrows.) Such cooperation in the defense of closely related young is what would be predicted by kin selection theory because an individual can increase its fitness by protecting close relatives.

Belding's ground squirrels also give alarm calls whenever predators are sighted. Individuals give two different types of alarm calls: one for aerial predators and one for terrestrial predators. In the case of aerial predators such as hawks and eagles, significantly more noncallers (28 percent) were captured than those that gave an alarm whistle (2 percent), regardless of their relationship to those nearby (Sherman, 1985). Here the caller clearly benefits from early detection of the predator, and this behavior is easily explained by individual selection. When terrestrial predators such as coyotes or snakes are considered, however, the outcome is very different. Callers suffered significantly higher predation than noncallers. A closer look at the genealogies of the ground squirrel population provides an explanation for the apparently altruistic behavior. When the frequency of alarm calls is broken down by sex and kinship (Fig. 23-22), it is clear that Belding's ground squirrels are playing favorites. Because daughters tend to stay in the vicinity of the natal territory, the offspring of the female giving the alarm call received the greatest benefit (individual selection), but females without young also benefited to a lesser extent. In the latter case, the evolution of terrestrial alarm calls can be explained by kin selection because adult females in the ground squirrel society are closely related.

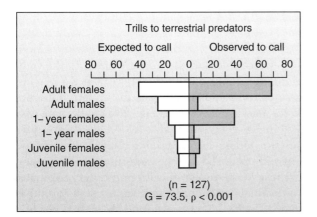

FIGURE 23-22 Observed and expected frequencies of alarm calls given in response to terrestrial predators by Belding's ground squirrels. Females gave more calls than would be expected by chance, and males gave many fewer calls than expected by chance alone. *(After Sherman, 1985)*

Many other examples of apparent altruism can be explained by kin selection. For example, cooperative breeding, in which nonbreeding helpers assist with the care and defense of the young, would also seem to reduce the fitness of the helper. Helpers perform a variety of tasks, including provisioning the young with food, defending the young against predators, or serving as extra baby-sitters, as in the case of the dwarf mongoose. Blackbacked jackals and African hunting dogs *(L. pictus)* deliver food to the den via regurgitation. In ring-tailed lemurs *(L. catta)* baby-sitters actually nurse infants (Pereira and Izand, 1989). Saddle-backed tamarins *(Saguinus fuscicollis)* give birth to twins, each of which can weigh up to 20 percent of the mother's weight. Several males other than the father help by carrying these juveniles (Terborgh and Goldizen, 1985). In most of these cases, the helpers are genetically related to the offspring they care for and, therefore, are increasing their own inclusive fitness by helping.

In addition to the benefits gained via inclusive fitness, helpers that forego breeding may receive benefits that are not obvious. African hunting dogs, for example, live in packs containing several adult females that do not breed. These nonreproductive females help rear the young of the dominant female and are often, but not always, closely related to her. Even for distantly related helpers, the short-term fitness cost of not breeding may be more than paid for by the long-term benefits of remaining in the pack. Solitary hunting is not a realistic option for these dogs because they rely on the pack to kill prey much larger than themselves.

Male lions often form long-lasting coalitions that cooperate in taking over prides. When coalition partners are closely related (for example, brothers), kin selection might account for the evolution of this cooperative behavior. However, by using DNA fingerprinting, Packer and coworkers (Packer and Pusey, 1982; Packer, 1986) demonstrated that approximately half of the Serengeti male coalitions included one or more unrelated males. Thus, certain types of altruistic or cooperative behavior do not appear to be directed at relatives and cannot be explained by kin selection alone. In lions, such unrelated males are accepted into a coalition because there is a direct benefit to all members of the coalition (individual selection). Larger coalitions have a better chance of taking over a pride and remain in control of the pride for longer periods than smaller coalitions. Because lion prides are loosely organized, any one of the male partners may be able to find and mate with a female in estrus, and thus all coalition males potentially benefit from cooperation. Coalition-forming males, therefore, probably produce more offspring over the long run than males that do not enter coalitions.

Cases in which the short-term cost of providing another with some resource is offset when the recipient returns the favor at a later time is called **reciprocal altruism.** As Trivers (1971) pointed out, reciprocity can evolve only if the donor and recipient can recognize one another, repayment is likely, and the benefit to the recipient is greater than the cost to the donor. Food sharing by vampire bats *(D. rotundus)* is typically considered a case of reciprocal altruism. Vampire bats share blood meals with their unrelated roostmates (Wilkinson, 1984, 1990). A bat that has failed to find a meal may beg for food from a roostmate. Begging is stereotypical, beginning with a short period of grooming followed by the recipient licking the donor's lips. The donor then regurgitates a few milliliters of blood, which is usually enough to sustain the recipient until the next night.

There are clear benefits to the recipient because vampire bats can starve to death if they do not find food in two consecutive nights. Wilkinson's (1984) experiments showed that food sharing occurred only between close relatives (explained by kin selection) or between unrelated bats that frequently shared the same roost. Furthermore, starved bats that received a meal were significantly more likely to reciprocate by donating a meal later on. The third condition for reciprocal altruism, that the benefit to the recipient be greater than the cost to the donor, was also demonstrated by Wilkinson's experiments. Weight loss after feeding in vampire bats declines exponentially (Fig. 23-23), with starvation occurring at 75 percent of their prefed weight. The shape of this weight loss curve results in relatively little cost to a donor who has recently fed. By donating a few milliliters of condensed blood, for example, a recently fed donor may lose 5 percent of its prefeeding weight, which is equivalent to approximately 6 hours worth of food for the donor. The recipient, on the other hand, not having fed for nearly 48 hours, is down to 80 percent of its normal prefeed-

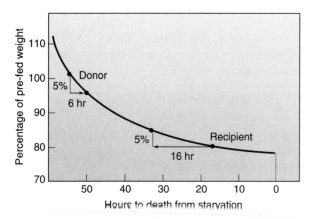

FIGURE 23-23   Model of the bloodsharing behavior of vampire bats *(Desmodus rotundus).* In this model, time to starvation is illustrated by the decreasing exponential curve. Donors, who give blood to a roostmate, lose 5 percent of their body weight and decrease their time to starvation by about 6 hours (right arrow). Recipients are already closer to starvation and may be at only 80 percent of their prefeeding weight. At this point on the curve, a gain of 5 percent from the donor pushes the time to starvation back by about 16 hours (left arrow). *(After Wilkinson, 1990)*

ing weight and is in danger of starvation. A blood meal equivalent to 5 percent of its body weight may result in an 16-hour reprieve from starvation. (A 5 percent loss to the donor shifts it to the right along the upper part of the curve, but a 5 percent gain to the recipient shifts it to the left along the lower part of the curve; Fig. 23-23). Perhaps this system has evolved because blood meals are difficult to obtain, because a bat that donates blood one night may be begging several nights later, and because of the long associations among roostmates (individuals may share the same roost for more than 12 years). Reciprocal food sharing, therefore, can evolve via natural selection acting at the level of the individual.

For kin selection or reciprocal altruism to operate, individuals must be able to recognize close relatives or roostmates. Familiarity may be one mechanism allowing kin recognition. Young golden-mantled ground squirrels *(Spermophilus lateralis)* prefer to play with siblings, even when the siblings were reared separately (Holmes, 1995). Siblings seemingly use visual or olfactory cues to recognize one another. Hamilton (1964) suggested that an animal might inherit the ability to distinguish kin via so-called "recognition alleles." This recognition mechanism would give kin a recognizable label and the ability to perceive others with that label. The alleles for this type of recognition may be associated with a family of genes called the **major histocompatibility complex** (MHC), a system used by the immune system for self-recognition. The immune system sends out specific types of cells to find and destroy foreign pathogens. It is the immune system's ability to distinguish self from nonself that enables it to selectively destroy these foreign molecules. MHC genes code for molecules embedded in each cell's membrane that act as molecular fingerprints to label the cell as "self." Mice *(Mus musculus)* that were inbred so that they differed by a single allele for one of the MHC genes were given a choice of mates. Males typically preferred females that differed in MHC alleles and therefore avoided inbreeding (Yamazaki et al., 1976, 1980; Penn and Potts, 1999). Mice in seminatural enclosures also prefer mates with different MHC genes (Potts et al., 1991). In rodents, genetic differences in MHC genes produce subtle differences in urinary odors that can be used to assess relatedness (Boyse et al., 1982). Thus, even a single genetic difference may act as a kin recognition label allowing animals to distinguish kin from nonkin.

EUSOCIALITY IN MAMMALS

Among the remarkably highly evolved social systems of insects are those in which only a single female colony member (the queen) breeds and the remainder of the colony is divided into social castes. (A caste is a group of morphologically distinctive individuals that perform specialized labor in a colony.) These eusocial insects typically have a long life span, overlapping generations, and cooperative care of eggs and young, and the reproductive female is the member of the colony least vulnerable to predation or accident. Insects with this system include some bees, some ants, social wasps, and termites. Only two vertebrate animals, the naked mole-rat and the Damara mole-rats *(Cryptomys damarensis),* are known to be eusocial (Jarvis, 1978, 1981; and Withers and Jarvis, 1980; Bennett and Jarvis, 1988; Sherman et al., 1991).

Naked mole-rats are small (25 to 50 grams), nearly hairless rodents that occupy the hot, dry parts of Kenya, Ethiopia, and Somalia. They live in large colonies of up to 40 individuals, with each colony occupying an extensive burrow system. The mole-rats eat enlarged roots and tubers of plants (geophytes) that are adapted to long dry seasons, and most burrowing is done when the usually hard soils are made friable by rain. These animals have a low metabolic rate and extremely limited thermoregulatory ability.

In many ways, the social system of the naked mole-rat parallels the eusocial systems of insects. The colony is composed of three castes. "Frequent workers" are small (25 to 30 grams), nonbreeding individuals that burrow cooperatively, forage, and build the communal nest. Members of this caste make many trips to the nest with food for the other castes. "Infrequent workers" are large (about 35 grams) relative to frequent workers and work at roughly half the rate of the latter. "Nonworkers" are the largest colony members (about 46 grams) and rarely work but do care for the young. They are brought food by the frequent workers, and when nonworkers sleep they are often joined by other mole-rats. Their huddling together reduces the energy expended by the colony. The single breeding female performs no colony tasks, breeds

with nonworker males, and produces from one to four very large litters (up to 24 young per litter) each year. Some individuals in each litter grow more rapidly than their siblings and become larger than the frequent workers; such individuals may replace infrequent workers or nonworkers that die. A laboratory colony studied by Jarvis had 16 frequent workers, nine infrequent workers, and eight nonworkers.

Suppression of reproduction in all but the single breeding female is probably under pheromonal control. Jarvis has shown that physical contact between the breeding female and other females is necessary for such suppression and suggests that the pheromone is carried in the urine and transmitted from the breeding female to others at the communal latrine. All males produce sperm, but the small frequent worker males have difficulty in copulating with the large breeding female. Females other than the single breeding female have ovaries that are seemingly quiescent; they contain many primordial and primary follicles but few more mature follicles. Jarvis has found that breeding females are extremely difficult to capture, suggesting that they are the individuals least vulnerable to predation.

The second eusocial mammal, the Damaraland mole-rat *(C. damarensis),* is also known through the fine studies of Jennifer Jarvis and her colleagues (Bennett and Jarvis, 1988; Jacobs et al., 1991; Jarvis and Bennett, 1991, 1993). This mole-rat is fully furred and averages 131 grams in weight. It inhabits the red Kalahari sands of some hot and arid deserts of southern Africa, where rain is highly unpredictable. Underground storage organs of various plants are the main food source for *C. damarensis.* Colonies of this mole-rat include up to 41 individuals, all of which share a common, extensive burrow system.

Evidence for eusociality in *C. damarensis* is conclusive. Each colony has but one reproductive female. She breeds three or four times a year, has a mean litter size of three, and remains the sole reproductive female throughout her time in the colony (up to ten years in a laboratory colony). She initiates courtship. Usually only one male per colony is reproductively active. The reproductive pair are typically the largest colony members and are dominant animals, with nonreproductive females often the lowest members in the hierarchy.

Reproduction is suppressed in nonreproductive females: ovarian follicles do not mature, seemingly due to blockage by the hypothalamus. Although spermatogenesis occurs in both reproductive and nonreproductive males, the latter have smaller testes and different behavior. DNA analysis of entire colonies show that each colony is composed of the reproductive pair and their offspring.

Nonreproductive members of both sexes are loosely differentiated into two classes, frequent and infrequent workers. The former class does most burrow-maintenance work. All colonies studied in the field had overlapping generations: at least one third of the nonreproductives remain in their natal colonies long enough to help care for four or more litters of siblings.

Eusociality has developed independently in the naked mole-rat and the Damara mole-rat, two evolutionarily divergent species (Faulkes et al., 1997), and in both cases ecological constraints probably played a critical role (Jarvis et al., 1994). Both species dig extensive burrows to find food (geophytes), and both occupy areas with low and erratic rainfall, where droughts of a year or more are common. The soil at the burrow depth of 25 centimeters is only workable when a rain of at least 25 millimeters either softens the hard soil, in the case of *H. glaber,* or makes the sand compact enough to retain a burrow, in the case of *C. damarensis.* At such times, the mole-rats must dig as fast as possible and locate enough food to last until the next heavy rain. During the brief moist period, a single animal could not dig fast enough to find sufficient food to last through a prolonged drought, whereas a group of animals could accomplish this. Jarvis et al. (1998) have shown that small colonies fail more frequently than do large ones, and that survival of newly formed colonies depends on the richness of the food supply and the size of the work force searching for food. Thus, in arid areas, ecological bottlenecks in the form of extended droughts and food scarcity have probably punctuated the evolutionary history of bathyergid mole-rats and may have favored sociality (Fig. 23-24). All five species of the three solitary bathyergid genera live in mesic areas, where the soil is friable for extended periods.

Although only the two bathyergid mole-rats described above are known to be eusocial, at least seven additional species of *Cryptomys* are colonial, and sociality is known in three other genera of sub-

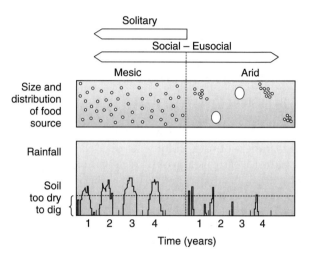

FIGURE 23-24 Factors thought to influence group size in mole-rats (Bathyergidae). Lack of sufficient rainfall in arid regions provides few opportunities to enlarge the tunnel system (dotted line) to new food patches. Together with the patchy distribution of food, these environmental factors may have led to the evolution of eusociality in mole-rats. *(From Jarvis, et al., 1994,* Trends in Ecology and Evolution, *with permission from Elsevier Science)*

terranean rodents. In some ways, the behavior of the social tuco-tuco (*Ctenomys sociabilis;* Ctenomyidae) of South America parallels that of the eusocial mole-rats (Lacey et al., 1997). Members of the colonies of *C. sociabilis* shared entire burrow systems, as many as 9 of the 13 colonies studied may have had multiple adults, colonies had from 1 to 18 juveniles, and all colony members shared a single nest. In contrast to the eusocial mole-rats, however, *C. sociabilis* colonies did not have reproductive division of labor, and probably not all colonies had an adult male. Among some 38 species of *Ctenomys,* only *C. sociabilis* is known to be social. Further research may show that this species lives under some of the same ecological constraints as those facing the eusocial bathyergid mole-rats.

Alexander and his colleagues (Alexander et al., 1991) summarized the combination of traits that probably favored the evolution of eusociality: extensive parental care, long-lasting and stable food resources that are easily defensible but have a patchy distribution, a subterranean lifestyle that reduces predation and allows young to remain in their natal groups, and high risks associated with dispersal and new-colony formation.

Web Sites

Animal Behavior Society with information and links

http://www.cisab.indiana.edu/ABS/

The Association for the Study of Animal Behaviour — a European site

http://www.hbuk.co.uk/ap/asab/

A site with links to articles, courseware, and other software reviews, and web pages concerned with teaching and learning animal behavior or ethology

http://www.liv.ac.uk/ctibiol/html/animal_behaviour_subject_area.html

24 ECOLOGY

One of the most remarkable attributes of humans is our ability to recognize relationships between disparate phenomena or events — to discover order, pattern, symmetry, predictability, and beauty in an apparently disordered world. The study of ecology demands such ability, for of central inter- est to ecologists is an understanding of the often complex relationships between living things. Kendeigh (1961) described ecology as "a study of animals and plants in their relations to each other and to their environment." Early ecological work featured descriptive field studies, and later investi-

gations added controlled field or laboratory experiments. These provide a base of knowledge on which the modern theoretical ecologists depend.

Our interest in ecology is not new. Early peoples understood many relationships between the major food animals and their environments. These peoples could predict where and when certain species could be found and used this knowledge to increase hunting success and hence survival. Many thousands of years later, we are belatedly understanding that our very survival may depend on an appreciation of basic ecological principles.

It is difficult to overemphasize the value of an ecological approach to the study of mammals. Knowledge of the biology of any mammalian species is clearly incomplete if the relationships of the animal to its environment are unknown. An understanding of mammalian ecology has been long in emerging, however, not because mammalogists lack interest in ecology but because a study of the ecology of even a single species involves detailed knowledge of many aspects of that species' biology, of its physical environment, and of the biology of species with which it is associated. The problem is more acute when one considers interactions among many species in a natural community.

Over an organism's lifetime, it will process materials taken in from its environment, thereby transforming energy as the organism grows and reproduces. In doing so, the organism alters the local environment and changes the availability of resources to other organisms. Even in death, an organism alters its environment by yielding energy to other organisms such as scavengers and decomposers. Ecological systems are often exceedingly complex, and ecologists typically organize this complexity into several levels: ecosystems, landscapes, communities, populations, and individuals. Ecosystems form the largest and most complex level because they include assemblages of organisms and their abiotic (nonliving) environments. The ecosystem approach necessarily includes hundreds or thousands of species (both plant and animal) over broad geographic areas. As a result, ecologists often resort to studying common "currencies," such as energy, when comparing ecosystems.

Ecological communities represent another level of organization, in which the focus is on interactions between populations of different species, such as predation, competition, mutualism, and trophic structure. With respect to mammals, it is here that many interesting and exciting interactions take place in the natural world. Thus, the competition for food among a lion pride, hyena clan, and a pack of hunting dogs during the dry season in the Serengeti are played out at the community level. Energy flow between the trophic levels of a community has a major influence on community organization.

Populations, on the other hand, consist of individuals of the same species, such as a pride of lions or a herd of caribou. Populations grow or decline over time, depending on the rate at which new members are added by birth or immigration and on the rate of mortality and emigration. Populations have limited geographic boundaries, determined by the available resources and barriers to dispersal. Populations are dynamic and potentially immortal, in that those members alive today are the descendants of those alive in past generations. Consequently, genetic structure, life history patterns, sex ratio, and age structure all play an important role in shaping populations of mammals.

The study of individuals and how they adapt to stressful and varied environments is traditionally the realm of physiology. Ecologists, however, have become interested in the interactions between individual organisms and their environment, a discipline called physiological ecology (see Chapter 21).

As the brief description above indicates, the scope of ecology is extremely broad. This chapter concentrates on ecological relationships and principles basic to an understanding of mammalian biology and is, therefore, necessarily incomplete and selective in its coverage.

GLOBAL CLIMATE PATTERNS

The environments of animals can be characterized in terms of physical and biotic factors. Physical factors include temperature, humidity, climatic patterns, precipitation, and soil types; biotic factors are those associated with interactions between organisms.

Solar radiation provides the energy, in the form of heat and light, on which living organisms depend. The intensity of solar radiation at the earth's surface is influenced largely by the directness with which the sun's rays strike the earth. The angle of these rays decreases, and the climate becomes progressively cooler, the farther north or south of the equator an area is situated.

Warm air holds more moisture than does cool air, and equatorial areas, especially areas near 25 degrees N and 25 degrees S latitude, receive relatively heavy precipitation. In addition, major global patterns of air circulation are set up as the warm air that rises from equatorial regions moves northward and southward. In a belt centered 30 degrees N and S of the equator, the equatorial air masses reach a stage of cooling at which they tend to sink. Cool air also carries less moisture. Some of the major deserts of the world, such as those in the southwestern United States and North Africa, are roughly 30 degrees north of the equator and are under the influence of this system of descending air (Fig. 24-1).

Over most of the world, even in tropical areas, rainfall is seasonal, and animals and plants must adapt to times of relative scarcity of food. Migrations of some tropical bats coincide with seasonal shifts in the abundance of insects, fruits, or flowers, and the dramatic migrations of wildebeest in East Africa are in response to seasonal changes in the availability of nutritious forage.

Superimposed on global climatic patterns are regional or local variations. Periodic increases in Pacific Ocean surface temperatures, referred to as El Niño Southern Oscillations, have pronounced effects on marine and terrestrial mammal populations. The 1991 to 1992 El Niño resulted in three times the normal annual rainfall in semiarid areas of northern Chile, superabundant food resources, and rapid increases in small mammal populations (Meserve et al., 1995). High mortality of some marine mammals along the Pacific coast has been associated with El Niño events (Trillmich and Ono, 1991). El Niño-caused increases in rodent densities may have been partly responsible for an outbreak of rodent-borne hemorrhagic fever viruses (Han-

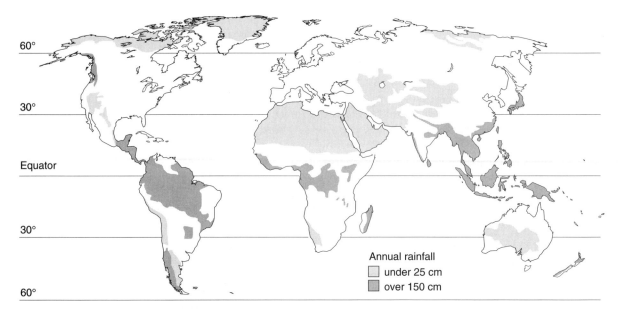

FIGURE 24-1   The important desert regions (those with less than 25 centimeters of precipitation annually) and the major wet regions (those with more than 150 centimeters of precipitation annually). *(After Espenshade, 1971)*

tavirus) in the southwestern United States (Childs et al., 1995).

Storms in the western United States typically sweep inland from the west. When storms move over north–south-oriented mountain ranges, the western slopes receive high precipitation, the eastern slopes receive lower precipitation, and the basins at the eastern bases of the mountains are often deserts. This "rain-shadow" effect strongly influences the distribution of both plants and animals (Fig. 24-2).

Even minor topographic features may affect precipitation and local distributions of plants and animals. When fog funnels through the passes in the Santa Ana Mountains of southern California, condensation dripping from the needles of the knob-cone pine *(Pinus attenuata)* can total a remarkable 10.2 centimeters of precipitation per month. This supplies sufficient moisture to allow the knob-cone pine to extend its growth into rainless periods (Vogl, 1973) and favors the local occurrence of the Coulter pine (*Pinus coulteri;* Pequegnat, 1951). Those pockets of relatively heavy precipitation may affect the distributions of small mammals.

Local topography may also strongly affect the amount of heat the surface of the earth receives and the distribution of plants and animals. The main axes of most mountain ranges lie north and south, the drainage systems are oriented more or less east and west, and the canyon walls face roughly north or south. In northern latitudes, because the sun's rays strike a south-facing slope more directly than a north-facing slope, south-facing slopes are considerably drier and warmer and have different biotas than do nearby north-facing slopes. In the precipitous, chaparral-covered mountains of southern California, for example, contrasting biotas occupy adjacent north- and south-facing slopes (Vaughan, 1954; Fig. 24-3).

Vertical temperature gradients are encountered as one ascends a mountain (or descends in an aquatic environment). The lowering of the temperature with increased elevation is of the magnitude of approximately 1° C for every 150 meters. This effect, coupled with increased precipitation at higher elevations, shorter growing seasons for plants, and drastic diurnal-nocturnal fluctuations in temperature, is associated with a distinct separation of climatic zones in high mountains throughout the world. In some areas of the western United States, an assemblage of "desert" mammals resembling those typical of arid lands as far south as central Mexico may occupy the arid or semiarid land at the foot of a mountain range, while the crests of the mountains a few miles away

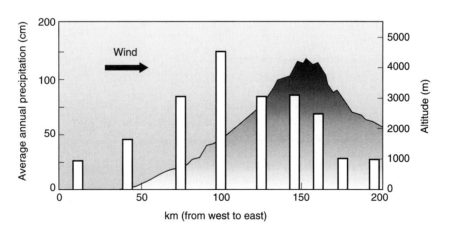

FIGURE 24-2   Effects of mountain ranges on local climate is illustrated by the rainfall patterns from the Pacific coast to the Sierra Nevada in California. Prevailing winds sweep moist air masses from the Pacific Ocean inland, where they are deflected upward by the Sierra Nevada. As the moist air rises, it cools and condenses, forming rain on the windward side of the mountain range. *(After E. Pianka, ©1988,* Evolutionary Ecology, *4/e reprinted by permission of Addison-Wesley Ed. Pub.)*

FIGURE 24-3
Assemblages of plants and mammals inhabiting (A) a south-facing slope and (B) a north-facing slope in lower San Antonio Canyon, San Gabriel Mountains, Los Angeles County, California. *(Data from Vaughan, 1954)*

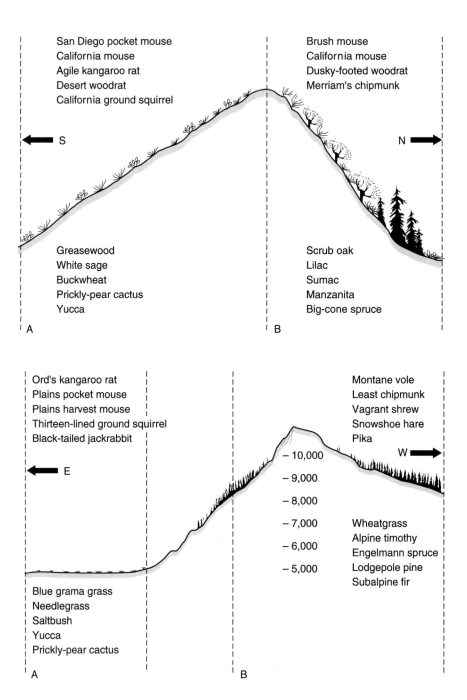

FIGURE 24-4
Assemblages of plants and mammals inhabiting (A) short-grass prairie and (B) subalpine habitats in northern Colorado (Larimer County).

may support boreal genera that occur as far north as northern Canada or Alaska (Fig. 24-4).

BIOMES

Climate obviously has a dramatic affect on the distribution of plants and animals across the earth's surface. Wet regions support dense vegetation, while very dry areas support little or no vegetation. The short growing season and low temperatures of alpine zones support unique communities of dwarf or slow-growing plants. Biological communities, although never exactly alike in plant and animal species, can be grouped into categories based on the dominant vegetation. These large-scale biological communities are called biomes. The world's major biomes include: tropical forests, savanna, deserts, polar regions, chaparral, temperate grass-lands, temperate deciduous forests, coniferous forests, and tundra (arctic and alpine) (Figs. 24-5 to 24-12). Biomes can, of course, be more finely subdivided. Temperate grasslands can be divided into short-grass or tall-grass prairies.

Biomes are also dynamic in both space and time. As the world's climate has changed over the millennia, the boundaries between adjacent biomes have shifted position (during ice ages, for example). The dynamic nature of the biomes has led to changes in the opportunities for speciation and reinvasion. The result is that groups of species living in the Sahara Desert of Africa are not identical to the species inhabiting the Gobi Desert of Asia. Both are desert biomes, but evolution has led to similar traits in independently evolved species (**convergent evolution**).

FIGURE 24-6   Savanna in the Serengeti Plains, Tanzania. This area supports large numbers of ungulates. Some of the typical kinds are zebra *(Equus burchellii)*, buffalo *(Syncerus caffer)*, wildebeest *(Connochaetes taurinus)*, and Thompson's gazelle *(Gazella thompsonii)*. Lions *(Panthera leo)*, hyenas *(Crocuta crocuta)*, and African hunting dogs *(Lycaon pictus)* prey on these ungulates. The African savanna supports a richer ungulate fauna than any other area in the world *(W. Robinette)*.

FIGURE 24-5   Tropical rain forest in the Chyulu Range, Tsavo West National Park, Kenya. In addition to a diverse group of smaller mammals, this forest supports African elephants *(Loxodonta africana)* and African buffalo *(Syncerus caffer)*.

FIGURE 24-7   Sandy dunes in the Namib Desert of Namibia, southern Africa. A gerbil *(Gerbillurus paeba)* and a golden mole *(Eremitalpa granti)* inhabit these nearly vegetationless dunes. *(T. Vaughan)*

Biomes typically grade into each other over fairly large areas, forming **ecotones**, and within each biome there is considerable local variation, resulting in a patchy appearance. Fire or floods may create openings in certain areas, **ecological succession** may result in one community being

FIGURE 24-8 Temperate shrubland (chaparral) near San Antonio Canyon, Los Angeles County, California. Typical mammals in this area are the western pipistrelle *(Pipistrellus hesperus)*, Merriam's chipmunk *(Tamias merriami)*, brush mouse *(Peromyscus boylii)*, California mouse *(P. californicus)*, dusky-foot woodrat *(Neotoma fuscipes)*, gray fox *(Urocyon cinereoargenteus)*, bobcat *(Lynx rufus)*, and mule deer *(Odocoileus hemionus)*. *(T. Vaughan)*

replaced by another, and human activity adds to the patchy appearance of most landscapes and biomes.

Succession occurs in places denuded or mostly denuded of life, such as newly formed river sandbars or land cleared by humans. Succession in a place that has never before supported life, such as a lava flow, is primary succession. Where the area previously supported a community, such as abandoned cropland or burned-over brushland, secondary succession occurs. Secondary succession is responsible for the mosaic of habitat patches that one commonly observes in agricultural areas or in many chaparral-dominated foothills of California.

The initial successional community on land typically consists of sun-tolerant, annual herbs with seeds adapted to wide dispersal. This pioneer community is usually short lived and is replaced by perennial herbs, which maintain themselves by spreading vegetatively. Succession might then proceed by shrubs replacing perennial herbs and shrubs eventually giving way to trees. In the course of tens or hundreds of years, a stable **climax community** is established. Plants in this community tolerate or are favored by their own effects on their environment (shading of the ground, leaf litter and added nitrate in the soil, in the case of forest trees). The climax is dynamic, but its species interactions, structure, and energy flow tend to perpet-

uate the community. The different communities in the succession are called **seral stages.** Succession in different areas obviously involves different seral stages and different climax communities. For example, the climax community in the Chihuahuan Desert consists of creosote bush *(Larrea)* and other xeric shrubs; on the high plains of eastern Colorado, short grasses form the climax; the climax community in the mountains of Colorado is typically coniferous forest.

Succession has pronounced effects on the distribution and densities of mammals. White-tail deer are favored by early successional stages. In Pennsylvania, in areas in the "brush" stage of succession, 24 species of plants yielded over 91 kilograms of deer food per acre. But in later successional stages, seven species of plants produced only 16 kilograms of food per acre (Gerstell, 1938). In Massachusetts, an upsurge in the production of deer food followed the abandonment of cultivated land; when a mature hardwood forest became established, however, production of deer food declined to nearly zero (Gould, 1937). As might be expected, white-tail deer are uncommon in mature northern forests, and the main factor allowing their spread and increased abundance in parts of northeastern United States was widespread logging (Hosley, 1956). Most small mammals are restricted to certain successional stages and are absent or scarce in others.

FIGURE 24-10   Subalpine coniferous forest near Rabbit Ears Pass, Routt County, Colorado. The following mammals are common in this community: shrews *(Sorex vagrans* and *S. cinereus)*, the red squirrel *(Tamiasciurus hudsonicus)*, least chipmunk *(Tamias minimus)*, pocket gopher *(Thomomys talpoides)*, montane vole *(Microtus montanus)*, red-backed vole *(Clethrionomys gapperi)*, beaver *(Castor canadensis)*, porcupine *(Erethizon dorsatum)*, red fox *(Vulpes vulpes)*, mule deer *(Odocoileus hemionus)*, and elk *(Cervus elaphus)*.

FIGURE 24-9   Temperate deciduous forest in the summer. Mammals typical of this type of community are the short-tailed shrew *(Blarina brevicauda)*, eastern chipmunk *(Tamias striatus)*, gray squirrel *(Sciurus carolinensis)*, flying squirrel *(Glaucomys volans)*, white-footed mouse *(Peromyscus leucopus)*, gray fox *(Urocyon cinereoargenteus)*, and white-tail deer *(Odocoileus virginianus)*. *(From PhotoDisc, Inc.)*

COMMUNITY ORGANIZATION

It has long been recognized that animals and plants with similar environmental requirements form identifiable communities. A perceptive description of a community was given by Mobius (1877), who considered an oyster bed as "a community of living beings, a collection of species and a massing of individuals, which find here everything necessary for their growth and continuance." A community is characterized not only by its unique plant and animal assemblage, however, but also by complex interactions between organisms and by the effects the physical environment has on the biota. The term "community" has been used to designate plant-animal assemblages of differing scale and importance (Odum, 1971), from the biota of a woodrat nest, for example, to that of the deciduous forests of the eastern United States.

The biota of a community is determined in part by such physical attributes as soil type, temperature, moisture, availability of light, nutrient supplies, and topography. At the interface where one community adjoins another, as where a grassy meadow meets a coniferous forest, an "edge effect" is produced. At this edge, a greater diversity and density of animals may occur than within either adjacent community, owing to the increased diversity of vegetation and types of shelter. Bowers et al. (1996) found that female meadow voles *(Microtus pennsylvanicus)* living at edges of fragmented habitat were larger, reproduced more often, and remained resident longer than females in continuous habitats, suggesting that edge situations are high-quality habitats.

FIGURE 24-11 A herd of pronghorn *(Antilocapra americana)* feed on a short-grass prairie. Other typical mammals inhabiting this community include the white-tail jackrabbit *(Lepus townsendii)*, the black-tail jackrabbit *(L. californicus)*, the thirteen-lined ground squirrel *(Spermophilus tridecemlineatus)*, prairie dog *(Cynomys ludovicianus)*, Ord's kangaroo rat *(Dipodomys ordii)*, northern grasshopper mouse *(Onychomys leucogaster)*, coyote *(Canis latrans)*, and badger *(Taxidea taxus)*. *(P. Stapp)*

FIGURE 24-12 A moose *(Alces alces)* forages in the taiga, a zone of scattered trees between the tundra and boreal forests. The taiga is inhabited by several species of shrew *(Sorex)*, collared lemming *(Discrostonyx groenlandicus)*, red-backed vole *(Clethrionomys gapperi)*, weasels and mink *(Mustela)*, gray wolf *(Canis lupus)*, arctic fox *(Alopex lagopus)*, lynx *(Lynx canadensis)*, and black and brown bears *(Ursus)*. *(T. Bowyer)*

Often there are no sharp dividing lines between adjacent communities. The zone of gradation between communities (ecotone) may be broad in such areas as in western Mexico, where one may travel for many kilometers through the transitional region between the Sonoran Desert community and the tropical thorn forest community. Despite such complexities, the major terrestrial communities are usually readily recognized (Brown et al., 1998).

Biological attributes, such as species abundance and diversity coupled with interactions among species, also influence community structure. In some cases, a single dominant species defines the community. The dominant species need not be the most numerous. The dominant species might be scarce, yet control access to the most important resources. Among the many species that make up a community, few are abundant. Most communities contain a few common species and many less abundant ones. Ecologists attempt to describe the biological attributes of a community by such measures as species richness (the number of species), evenness (relative abundance of individuals), and species diversity (a combination of richness and evenness). Diversity measures take into account both the number of species and how evenly distributed individuals in those species are across the whole community. Diversity indices can be used to compare the composition and complexity of different communities.

Diversity is determined not only by biotic factors but by a combination of climatic history, geography, and latitude.

The greatest species diversity occurs in tropical areas, with decreasing diversity in areas progressively closer to the poles (Fig. 24-13). As an example of this pattern, an area of some 40 square kilometers near Point Barrow, Alaska, supports only about 16 species of land mammals, whereas comparable figures for eastern Kansas and an area near Panama City, Panama, are 55 and 140, respectively (Hall and Kelson, 1959).

The broad trend toward equatorial species richness and decreasing species diversity with increasing latitude (the latitudinal diversity gradient) has long been recognized and well documented. In the tropics, temperatures are moderate, growing seasons are long, rainfall is fairly predictable and often generous, and biological productivity, habitat, and species diversity are high. By comparison, temperatures at high latitudes are extreme, precipitation is low, growing seasons are short, and there is little habitat or species diversity. Although high biological productivity in the tropics and low productivity in the far north is the rule, the positive relationship between productivity and species diversity does not hold through the latitudinal gradient. In fact, among many animal taxa, the greatest diversity occurs at intermediate levels of productivity (Rosenzweig, 1992). It seems, then, that the latitudinal diversity gradient is not directly related to productivity.

Rosenzweig relates the latitudinal diversity gradient to land area. He points out that the tropics are vastly more extensive than are other biomes, and that the broad tropical areas afford more chances for very extensive geographic ranges among plants and animals. In general, the larger an organism's geographic range the greater the chance for **allopatric speciation**: the tropics would thus be expected to spawn relatively profuse speciation (Terborgh, 1973). Further, Rosenzweig proposes that the fine subdivision of habitats by tropical species is the result of **coevolution** over considerable evolutionary time. Also, the extensiveness of tropical areas probably reduces the probability of extinction of species.

Kaufman (1995) offers a different hypothesis. She explains the latitudinal gradient in species richness in New World mammals in terms of latitudinal shifts in the relative importance of biotic versus abiotic factors. She uses the term "bauplan" for a suite of ecological and structural-functional features that characterize such higher taxa as genera, families, and orders. Members of the order Rodentia, for example, would share a common bauplan. Kaufman proposes that in tropical areas (where abiotic conditions are benign) nearly all bauplans can survive, but biotic interactions limit the **niche** space of each species and result in specialization and species packing. In areas near the pole, in contrast, abiotic conditions are severe and limit the number of bauplans and species that can exist (Fig. 24-13). Within the tropics (from the

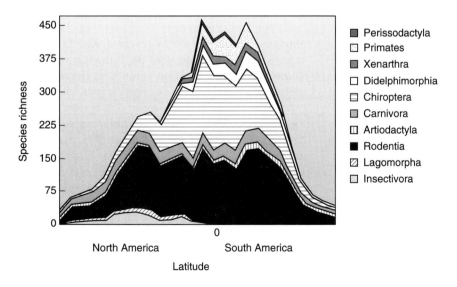

FIGURE 24-13 Latitudinal distribution of species richness of mammalian orders for North and South America. Zero latitude indicates the equator. *(From Kaufman, 1995)*

equator to roughly 26 degrees N and S latitude), the importance of biotic interactions remains high. Away from the tropics, abiotic constraints become increasingly important with each degree of latitude, with a reciprocal decrease in the importance of biotic interactions. Abiotic conditions thus set the higher latitude boundaries of species' ranges, whereas biotic interactions limit the distributions of species near the equator. No single hypothesis to explain the latitudinal diversity gradient has gained universal acceptance, but each provides new insights that broaden our biological perspective and put us closer to a resolution of the problem.

Large land masses are also more biologically diverse than are small land masses. Although this general concept was recognized by field biologists such as Darwin and Wallace, MacArthur and Wilson's (1963) theory of island biogeography formalized the relationship between island size and species diversity. This theory proposes that the number of species on an island is determined by a dynamic equilibrium between the rate of immigration of new species and the rate of species extinction. As used in this context, "islands" need not be surrounded by water. A patch of rain forest surrounded by farms or mountain tops may serve as an island if it is isolated from other similar habitats. The equilibrium theory of island biogeography has largely been replaced by the theory of **metapopulations.** Metapopulations consist of clusters of populations semi-isolated in suitable habitat patches, with some level of migration and gene flow among them. Some of the individual populations may go extinct, but the habitat patch can be recolonized by immigrants from other populations (Meffe and Carroll, 1994). This idea has gained wide acceptance among ecologists and conservation biologists (Hanksi and Gilpin, 1997).

The stability of a community also seems to depend in part on its complexity (Naeem et al., 1994; Tilman and Downing, 1994). The greater the species diversity among both plants and animals and the greater the number of energy pathways, the more resistant the community is to such changes as strong shifts in the densities of common species. MacArthur (1955) suggested that "stability increases as the number of links increases," and Goodman (1975) has discussed the positive relationship between species diversity and

community stability. These relationships between species diversity, community complexity and island biogeography have important implications for conservation biology (Wilson, 1992; Reid, 1998).

SPECIES INTERACTIONS

The more diverse a community, the more potential interactions there are among the members of that community. Considerations of interactions between one species of mammal and another and between plants and mammals are essential to an understanding of mammalian ecology. A biotic community is a tremendously complex functional unit within which animals live, feed, reproduce, and die. It has an evolutionary history and, through time, maintains a state of dynamic equilibrium. The role of an organism in a community depends on that organism's interactions with other members of the community and with the physical environment.

The fabric of a community depends on myriad interactions. Consider a typical oak forest community of eastern North America. Oak trees (*Quercus*) produce large numbers of acorns (mast) every few years. These seeds support many animals, including insects, birds, and several mammal species (Fig. 24-14). White-footed mice (*Peromyscus leucopus*), chipmunks (*Tamias striatus*), and white-tail deer (*Odocoileus virginianus*) consume large quantities of acorns during years of heavy acorn production, and rodent density appears to be tightly correlated with acorn production (Ostfeld et al., 1996). The deer and rodents are also the main hosts for parasitic deer ticks (*Ixodes scapularis*) and play an important role as reservoirs of Lyme disease. Tick populations also peak during summers following mast production when host densities are highest. Gypsy moth (*Lymantria dispar*) outbreaks can cause serious defoliation in oak forests, but evidence indicates that small mammal predators, including white-footed mice, play an important role in preventing gypsy moth outbreaks (Ostfeld et al., 1996; Ostfeld, 1997). Deer populations also increase in years after heavy mast production, resulting in overbrowsing of seedlings and saplings, which can eventually alter the species composition of the forest understory. Mast production by oak trees, therefore, has both direct and indirect effects on a number of plant and animals species

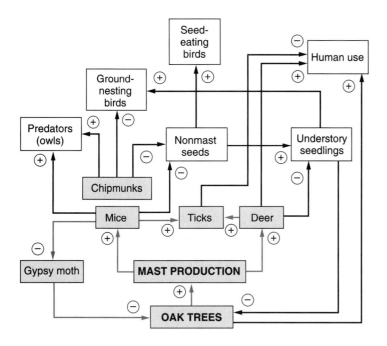

FIGURE 24-14 A model of the relationships between the members of an eastern deciduous forest community in North America. Biomass of a member of an interacting pair will increase if the arrow pointing toward the recipient is positive. For example, acorn (mast) production will increase the biomass of deer and mice, but mice will cause a decrease in biomass of gypsy moths. *(After Ostfeld et al., Bioscience, ©1996, American Institute of Biological Sciences)*

in the community and on a disease that affects humans.

Interactions in an ecosystem involving the effect one organism has on another are termed **coactions;** they include parasitism, competition, commensalism, and mutualism. In parasitism, one organism, such as a tapeworm, lives at the expense of another, but this coaction could also refer to predators and pathogens (disease producers such as viruses). Competition occurs when two or more species utilize a resource in short supply. Commensalism refers to two species that live together; one species gains some advantage, but the other neither gains nor loses. White-footed mice *(Peromyscus)* and shrews *(Notiosorex)* that take shelter in woodrat houses would be **commensals,** as would amphibians and reptiles that seek refuge in the burrows of pocket gophers. In mutualism, both species benefit: tick birds *(Buphagus)* eat ticks from the bodies of large ungulates, and the birds and the ungulates profit. This coaction often shifts to parasitism, however, when tick birds keep wounds of their hosts open by pecking and eating wound tissue and blood (Maclean, 1993).

Mutualisitic interactions are important aspects of the lives of some mammals. One of the most remarkable instances of mutualism concerns a bird, the African honey guide *(Indicator indicator)*, and

the African honey badger *(Mellivora capensis)*. The bird attracts the badger's attention by raucous chattering and then leads the way to a bee's nest. After the nest is torn apart by the mammal (African tribesmen sometimes perform this service), the honey guide eats bees and their larvae and wax, which it can digest, while the badger eats the honey. Of particular interest are the close and seemingly mutually beneficial associations that occur between two species of mammals. Herds of impala *(Aepyceros)* often stay with baboons *(Papio)* through much of the day. DeVore and Hall (1965) judged that the excellent eyesight of the baboons supplemented the acute senses of smell and hearing of the impalas and made the mixed group difficult for a predator to approach undetected.

LANDSCAPE ECOLOGY

Landscape ecology is a conceptual approach that recognizes the importance of spatial patterns of different biological communities in affecting everything from genes to entire ecosystems (Bowers, 1997). Landscape ecology deals with the spatial patterns of ecological systems that include patches of two or more community types. The basic unit of a landscape is the habitat patch (Hanski and Gilpin, 1997), which is defined by Kotliar and

Weins (1990) as "a surface area differing from its surroundings in nature or appearance." But, as pointed out by Bowers and Matter (1997), for a biologist, a patch must be defined on the basis of "the habitat and spatial requirements of individual species and may, depending on the scale, be used by foraging individuals or by whole populations." A **population** can be defined as a group of individuals occupying a local area that interact with one another, and a metapopulation, as a group of populations connected by dispersal (Hanski and Gilpin, 1997).

The importance of habitat heterogeneity as it affects spatial patterns of distribution of mammals was recognized early in the 20th century by such field biologists as Joseph Grinnell and Aldo Leopold. Important early studies on rodents and disease transmission by the Soviet ecologist Naumov (1936, 1948) demonstrated that rodent population densities were influenced by both habitat and topography, features regarded today as within the realm of landscape ecology. In a landmark publication, Anderson (1970) described the occurrence in granaries of genetically isolated populations of house mice *(Mus musculus)*. He speculated that spatial and genetic fragmentation of populations is typical of small mammals and that immigration and emigration are crucial factors influencing the genetics and dynamics of populations. Howard (1960) and Lidicker (1962, 1975) were the first researchers to focus attention on the nature and evolution of dispersal and their critical importance in population regulation.

Research within the realm of population ecology is wide ranging, from considerations of biotic processes involving populations to the reactions of individuals to features within their home ranges. Ultimately, habitat choices made by individual organisms govern broad-scale processes within a landscape. In the view of Bowers and Matter (1997), "An emphasis on individual behavior, as it is constrained and modified by landscape elements, would do much to advance our understanding of spatial processes and spatial patterns."

A fundamental question is how biological processes vary spatially (among habitat patches) and how they relate to large-scale, landscape-level dynamics involving metapopulations. Bowers and Matter (1997), who compiled data from published studies on the relationship between mammal densities and patch size, concluded that no uniform relationship between density and patch area occurs among all systems of patches. They found instead that these relationships were scale-dependent. Frequently movements by individuals selecting patches in small-scale landscapes (those with smaller, less isolated patches) resulted in higher densities in smaller patches (a negative density-area relationship), whereas the longer-term processes of colonization and extinction in large-scale landscapes (those with larger, more isolated patches) created higher densities in larger patches (positive density-area relationships). The authors concluded that patches are not biological entities with a predictable set and kind of processes but are convenient human constructs.

Spatial and temporal distributions of mammals are influenced by contrasting reactions of different species to the same habitat. For example, meadow voles at edges of small patches were larger, stayed in residence longer, and reproduced more often than voles in continuous habitats (Bowers et al., 1996). But prairie deer mice in the same study area avoided edge habitats where risk of predation was high (Bowers and Dooley, 1993).

The distribution and density of mammals over a heterogeneous landscape depends also on inter-specific differences in vagility and habitat requirements. A habitat that is a formidable barrier to one species may be crossed readily by another. In Belgium, for example, the forest-dwelling bank vole *(Clethrionomys glareolus)* seldom occurs in forest fragments surrounded by inhospitable (to the vole) open fields. The wood mouse *(Apodemus sylvaticus)*, in contrast, crosses open fields readily and occurs commonly in isolated forest fragments (Geuse et al., 1985). Eastern chipmunks move between forest fragments along fencerows and shelterbelts, but open fields are barriers (Henderson et al., 1985). An excellent example of the reaction of a small mammal to barriers and a patchy landscape is afforded by pikas *(Ochotona princeps)* living in abandoned mine tailings on the eastern slopes of the Sierra Nevada of California. Pikas are restricted to talus or rock jumbles adjacent to grasses or alpine tundra. Smith (1974) studied an 8-kilometer area with 24 islands of mine tailings ranging in size from perimeters of a few meters to more than 300 meters and separated by 20 to 300 meters. Two large islands supported permanent populations and provided dispersers that maintained the metapopulation. At a given time, only some of the other islands were occupied, and their pika populations were seemingly below carrying capacity. Pikas did not colonize islands separated from others by more than 300 meters. Extinction was negatively correlated with size of the island, and recolonization was negatively correlated with distance to the nearest island. A fluctuating balance was maintained between extinction and recolonization.

Our perspective of resource management has been broadened by landscape-scale studies of mammalian herbivores. A simulation model by Moen (1997) considered landscape-scale interactions among the population dynamics of moose, the energetics of individual moose, and the spatial distribution of plants and their responses to foraging by moose. The author concluded that different foraging decisions by moose resulted in different landscape patterns, which, in turn, determined whether moose gained enough energy to survive. Long-term survival of moose was threatened when the animals chose only the most nutritious patches of plants and avoided other patches because, in time, these latter patches grew out of the reach of the moose and were no longer available. Patches of the more nutritious plants were browsed so heavily that they grew slowly. The end result was that a landscape developed that was dominated by large patches of plants with insufficient available forage to support moose. Alternate foraging strategies, those that spread foraging more uniformly over the landscape, favored wide distribution of food that, in the long term, supplied enough food for moose survival. Where moose did not forage, succession continued to stages that might not support a moose population. Thus a fascinating conclusion emerges: where moose do not forage, ecosystem processes are affected at least as strongly as where moose do forage.

The spectacular ungulate migrations and grazing succession in the Serengeti of Africa are landscape-scale processes. Successive waves of grazers, each species affecting the availability of the preferred food of the next species, play a vital role in the maintenance and productivity of the Serengeti ecosystem (Maddock, 1979; McNaughton and Georgiadis, 1986). By grazing, Serengeti gazelles and hartebeets also accelerate nutrient cycling (sodium and nitrogen) in the soils and plants in a way that enhances their own carrying capacity (McNaughton et al., 1997). Excellent discussions of mammalian landscape ecology appear in a special feature of the *Journal of Mammalogy* (1997; 78:997–1052).

SPECIES COEXISTENCE

The relatively simple ecosystems of deserts have attracted many ecologists studying biotic communities. These studies have largely concentrated on North American deserts and have not only extended the breadth of our knowledge of desert ecology in general but have advanced our understanding of mammalian community structure and organization in particular.

In their fine treatment of community ecology of heteromyid rodents, Brown and Harney (1993) define community organization by "the number, identity, abundance, and morphological, physiological, and behavioral attributes of coexisting species." The species composition of communities is controlled by the ability of each species to satisfy its own needs and to adjust to the patterns of interactions among coexisting community members. As stated by Reichman (1991), "selective pressures in a mature, relatively stable community should consist of continuous and subtle evolutionary thrusts and parries between community members." Topographic diversity, with associated local shifts in climate, geology, and plant composition, result in great local variation in rodent species assemblages. Among 202 local patches of desert in the Southwestern United States, were 137 different combinations of 29 species of granivorous rodents (Brown, 1987; Brown and Kurzius, 1987). Despite this complexity, these assemblages seem to share a series of factors favoring coexistence.

Experimental field studies in which one species was removed and responses of the other species were monitored have shown that competition is an important factor in desert rodent communities. In southeastern Arizona, after exclusion of the large kangaroo rats, four of the five species of small granivorous rodents increased threefold (Munger and Brown, 1981; Brown and Munger, 1985). With kangaroo rat removal in southern New Mexico, the density of a common pocket mouse *(Chaetodipus penicillatus)* increased two and a half times (Freeman and Lemen, 1983). Experimental studies indicate that seed-eating rodents and ants compete for food and that their populations are food limited.

What factors, then, favor coexistence? Habitat diversity is an important factor. Heterogeneous habitats — those with diverse species and sizes of plants and with open spaces of varying sizes — allow for microhabitat segregation among rodents. Quadrupedal pocket mice and harvest mice *(Perognathus, Chaetodipus,* and *Reithrodontomys)* forage beneath or near shrubs, whereas bipedal kangaroo rats and kangaroo mice *(Dipodomys* and *Microdipodops)* prefer open spaces (Fig. 24-15). Precipitation, as it affects plant productivity (seed production), markedly influences coexistence. In the central Namib Desert of southwestern Africa, for example, the bushveld gerbil *(Tatera leucogaster)* survives in river valley **refugia** in dry years, but extends its range and becomes an important compo-

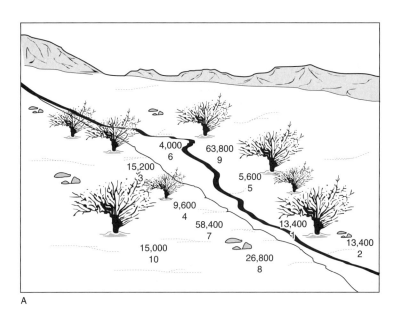

A

B

FIGURE 24-15   (A) Densities of seeds (no./m²) in several microhabitats in the Sonoran Desert of Arizona. Seeds were sampled to a depth of 20 millimeters. The seed numbers listed are from small areas and do not imply that such densities occurred over an entire square meter. (B) Photograph of the Sonoran Desert habitat. *(A is from Reichman, ©1984,* J. Biogeography, *Blackwell Science Ltd., and B is from T. Vaughan)*

nent of other communities in wet years (Griffin, 1990). Differences in habitat selection (Rosenzweig, 1973; Larsen, 1986), or partitioning of food resources on the basis of food distribution or foraging efficiency may also favor coexistence (Mares and Williams, 1977; Price, 1983). The kangaroo rat *D. merriami* learns the locations of seed patches far more quickly than does the pocket mouse *C. intermedius* (Rebar, 1995). This may enable the kangaroo rat to forage effectively in open spaces, where seeds occur in scattered clumps. Aggressive interactions involving the use of space are also part of the coexistence equation. Although relatively unstudied, predator-mediated habitat selection must be an additional factor (Kotler, 1984; Price et al., 1984). The use of dormancy by some species and different approaches to water conservation may also have an influence.

Coexistence among rodents is seemingly promoted by differences in body size (Brown, 1973; Bowers and Brown, 1982; Hopf and Brown, 1986). Body size in coexisting desert rodents is not randomly distributed but tends to be evenly arranged along a size gradient (Fig. 24-16). Such size differences occur also among small mustelid carnivores that live sympatrically and depend on small prey such as mice. Size divergence among sympatric mammals with similar ecological requirements has often been regarded as indirect evidence of interspecific competition (Brown and Wilson, 1956; Simberloff and Boecklen, 1981).

Dayan et al. (1990) examined community-wide **character displacement** in weasel communities in North America and in the mustelid-viverrid **guild** in Israel (a guild is a group of animals that share a common food resource). Because the canines do most of the damage when these animals bite and kill their prey, Dayan et al. (1990) reasoned that interspecific and intraspecific differences in the diameter of the upper canine would reflect competition for food. They found that, with respect to canine diameter, within each community there was pronounced sexual dimorphism, virtually no size overlap among sympatric morphospecies (each sex of each species was considered a separate morphospecies), and that size ratios between adjacent morphospecies were remarkably equal. Ratios of canine diameters among members of a Minnesota mustelid community illustrate this latter point: males of the smallest species *Mustela nivalis,* dif-

fered from females of this species by a ratio of 1.14:1, for female *M. erminea* and male *M. nivalis* the ratio was 1.08:1, and comparable ratios for other adjacent morphospecies were 1.29:1 for male and female *M. erminea*, 1.26:1 for female *M. frenata* and male *M. erminea,* and 1.34:1 for male and female *M. frenata.* Size ratios for mustelid communities at other sites (Alaska, Idaho, Oregon, Maine, Vermont), and for a mustelid-viverrid community in Israel, were remarkably similar. Viewed in light of the feeding behavior of mustelids, these results "seem consistent with a hypothesis of competitive character displacement" (Dayan et al., 1990).

Similar character displacement was present in two other guilds. Community-wide character displacement in upper canine diameter occurred among three species of small cats in Israel, and each species was strongly sexually dimorphic (Dayan et al., 1990). As with the mustelid species discussed above, size ratios between adjacent morphospecies were nearly equal. Among coexisting heteromyid rodents in North American deserts, community-wide character displacement occurred in upper incisor width; again size ratios between adjacent species were similar. The authors suggest that competition may have favored the use of different seed sizes by each rodent species, hence the differences in upper-incisor width (Dayan and Simberloff, 1994).

The consistent results of these studies support the view that character displacement in mammals is in response to competition among sympatric, ecologically similar species.

A TALE OF TWO DESERTS

The defining feature of deserts is aridity, to which all desert plants and animals must adapt. Thus, strategies for coping with aridity — such as periodic dormancy and the ability to concentrate urine — are shared by many desert mammals. Because deserts occur on every continent, however, their geological histories and amounts and seasonal distributions of rainfall vary widely, as do the geographic and taxonomic origins of their biotas. Accordingly, mammalian community organization differs markedly among different deserts. With this in mind, let us consider the Sonoran Desert of North America and the Namib Desert of the southwestern African coast.

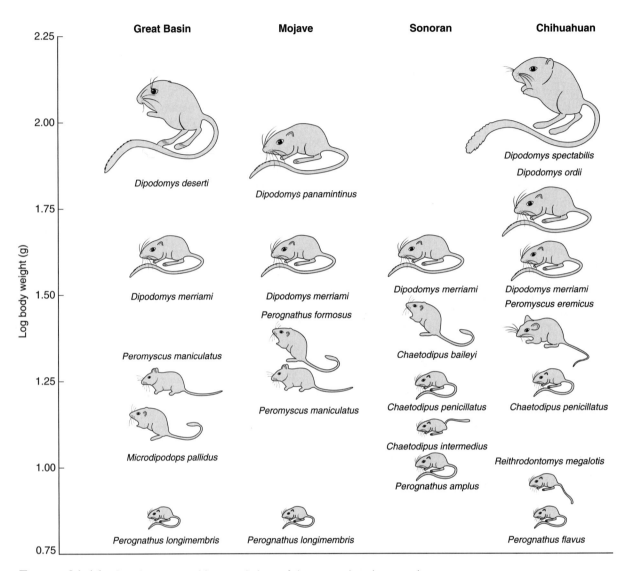

FIGURE 24-16 Species compositions and sizes of the most abundant granivorous rodents that coexist at selected sites in the four subdivisions of the North American Desert: Great Basin Desert, Fish Lake Valley, Nevada; Mojave Desert, near Johannesburg, California; Sonoran Desert, near Silverbell, Arizona; Chihuahuan Desert, near Portal, Arizona. *(From Brown and Harney, 1993)*

The Sonoran Desert is not a "typical" desert in part because it supports a relatively lush and diverse flora. Brief bursts of seed production occur in response to winter and/or summer rains, which average roughly 100 to 300 millimeters per year. Although rainfall is unpredictable, the standing crop of seeds in the soil — especially those of ephemeral forbs — provides a fairly stable re-

source on which most Sonoran Desert rodents depend (Brown, 1973). Windblown seeds collect in large numbers in depressions or on the lee sides of obstacles (Fig. 24-15). Seeds formed from 78 to 94 percent of the diets of four species of rodents (all heteromyids) that occupied a Sonoran Desert community in Arizona (Reichman, 1975). Insects and greenery were not major foods. They were eaten only seasonally, and their use fluctuated widely from year to year. For these heteromyids, collecting seeds in the cheek pouches and carrying seeds to scattered, shallow caches (scatter hoarding) or to chambers associated with the burrow system (larder hoarding) is the primary surface activity. Eating is done underground. Clearly these species belong to the granivorous guild. All of these four species are nocturnal, all are independent of free water, and the two smallest species remain underground, presumably in torpor, during the coldest months.

Compared to the Sonoran Desert, Africa's Namib Desert appears overwhelmingly barren (see Fig. 24-7). Except after the infrequent rains, the stark gypsum flats seem lifeless, with lichen crusts forming the dominant growth; only occasional patches of grass dot the great Namib "dune sea." The extremely low rainfall of this desert (about 10 to 125 millimeters per year) is a result of the South Atlantic anticyclone pressure system and the associated cold Benguela Current that sweeps up the coast from the south. Upwelling of cold, coastal waters give rise to persistent winds from the sea, maintaining an inversion layer (a layer of cooler air beneath a layer of warmer air) that reduces the turbulence and cloud formation necessary for rain. The Namib is an ancient desert, having fluctuated from semiarid to arid since about 85 million years ago, when the separation of Africa and South America formed the South Atlantic Ocean. But the Namib's rich biota has a remarkable source of water in addition to the scant rainfall. Fog forms when onshore winds drive warm, moist Atlantic air masses over the cold, coastal Benguela Current. Wind-driven fog sweeps as far inland as 100 kilometers for about 60 nights a year. Many plants and some animals can utilize condensed fog water (Seely, 1981). The frequent winds transport small fragments of plant and animal material over the dunes and flats. This detritus is the major food source for many Namib animals (Seely, 1979).

Seemingly in response to the diversity of arthropods that feed on detritus, and the lack of a stable standing crop of seeds in the soil, the gerbils inhabiting the Namib are omnivorous, with arthropods their most important food (Griffin, 1990). About 70 percent of the diets of two Namib gerbils studied by Perrin et al. (1992) was arthropods;

seeds formed up to 33 percent of the diets in part of the austral winter (May to August), and greenery was of varying importance. Gerbils lack external cheek pouches, and gerbils of the Namib rarely cache food, but remains of insects and plants in their burrows indicate that they sometimes eat food there (Downs and Perrin, 1989). These gerbils seem to lack the urine-concentrating abilities of those heteromyids known to be fully independent of free water. The gerbil's insect diet yields considerable free water and protein, and they can remove the byproducts of protein metabolism with minimal water loss. But when insects are scarce, these gerbils must eat succulent vegetation (Downs and Perrin, 1991). Seeds, which yield considerable metabolic water, are only sporadically available. That fog affects the water balance of some Namib rodents (including gerbils) is indicated by increased rates of water turnover in these rodents after wind-borne fogs (Withers et al., 1979). Precipitation (either fog, dew, or rain) occurs at least once each month of the year, a pattern critical to the survival of arthropods, the gerbils' main food. Perhaps this relatively frequent precipitation has also reduced the gerbils' need for physiological adaptations favoring extreme water conservation for long periods.

Viewed broadly, then, differences in rodent community organization between the Sonoran Desert and the Namib Desert seem to have resulted from different selective pressures associated with different climates and contrasting food bases. Whereas "in North American deserts seed abundance and distribution serve as the cornerstone of community organization among rodents" (Reichman, 1991), and rains are seasonal and unpredictable, "the extensive use of fog water and detritus largely characterize the biology of the Namib." (Seely, 1987).

The false assumption is often made that, in each major desert of the world, specialized seed-eating rodents have evolved. Clearly this does not hold for the Namib Desert or all other deserts. The Namib gerbils are convergent in some ways with heteromyids (large auditory bullae, long hindlimbs, and the ability to concentrate urine), but the lack of a dependable seed crop precludes granivory and favors omnivory. No one dietary mode is of dominant importance in all deserts.

EFFECTS OF MAMMALS ON THEIR ENVIRONMENT

If the importance of a group of animals is equated with its visible effect on the environment, mammals are clearly the most important terrestrial animals. The ability of humans to modify, or in many cases devastate, their environment is of ever-increasing importance as human populations continue to increase. Other mammals also strongly affect their environment, however, and in "natural" areas, where human interference has been minimal, the activities of wild mammals often drastically alter the character of the vegetation, the availability of water, the patterns of erosion, and the diversity and nature of the vertebrate and invertebrate fauna.

The great impact of mammals on their environment is largely a result of endothermy, with its attending high energy requirements. This impact results from a variety of activities, including feeding, patterns of migration or daily movement, the quest for water, and the construction of shelters or refuges. Some species of animals affect their communities so strongly that they can be viewed as **keystone species.** According to Paine (1969), such species "are the keystone of the community's structure, and the integrity of the community and its unaltered persistence through time, that is, stability, are determined by their activities and abundance."

The wildebeest *(Connochaetes taurinus)* is a keystone species. The annual migration of wildebeest from the Serengeti Plains in Tanzania, northward and westward toward the bush country of northern Tanzania and the Masai-Mara Game Reserve of Kenya, is a justly famous spectacle. The movement begins at the end of the rainy season, in May or June, and the animals return to the short grass in November. During a four-day period in May of 1974, grazing by approximately one-half million wildebeest reduced the green biomass of this *Themeda-Pennisetum* grassland by 84.9 percent and the height of the vegetation by 56 percent (McNaughton, 1976; Table 24-1). This apparent devastation markedly affected the subsequent growth of grasses and, indirectly, the dry-season distribution of the abundant Thompson's gazelle *(Gazella thompsonii).* This antelope, next to the wildebeest, is the most abundant ungulate in the Serengeti Plains. During the 28-day period following the main migration, areas grazed by wildebeest had a net productivity of green vegetation of 2.6 grams per square meter per day, whereas in experimental plots protected from grazing, green biomass declined at a rate of 4.9 grams per square meter per

TABLE 24-1 Effect on Grassland Vegetation (Largely *Themeda* and *Pennisetum*) of the Four-Day Passage of Wildebeest Herds in the Serengeti Plains

	Biomass (g/m²)	Height (cm)	Biomass concentration (mg/10cm²)
Fenced vegetation, wildebeest excluded			
Before passage	501.9	64	7.9
After passage	449.2	63	7.1
	N.S.	N.S.	N.S.
Vegetation subject to wildebeest grazing			
Before passage	457.2	66	6.9
After passage	69.0	29	2.4
	$p = .005$	$p = .005$	$p = .05$

After McNaughton, 1976.
p = level of significance; N.S. = not significant.

day. A dense mat of new and nutritious vegetation was produced in the grazed area, while in the ungrazed plots the bulk of the biomass was tall, nonnutritious grass stems.

One month after the exodus of the wildebeest, the area was occupied by Thompson's gazelles, which selectively grazed the areas of vigorous regrowth, that is, the areas previously grazed heavily by wildebeest. As an indication of the high selectivity of the gazelles, consumption of vegetation in these areas averaged 1.05 grams per square meter per day, whereas virtually no grazing by gazelles occurred in the plots where wildebeest had not grazed. McNaughton (1979, 1985) concluded that the wildebeest transformed a senescent grassland into a productive community. This grazing optimization hypothesis has generated much discussion (Belsky, 1987; Bergström, 1992). The impact of mammals on grasslands and savannas remains an active area of research.

Along the coastlines of the Aleutian Islands of Alaska, the sea otter *(Enhydra lutris)* seems to play a critical role in structuring nearshore marine communities (Estes and Palmisano, 1974, Estes and Duggins, 1995; Steinberg et al., 1995). The Rat Islands of the Aleutian archipelago support high populations of sea otters, and the nearshore community is characterized by abundant beds of macrophytes, consisting mostly of brown algae (kelp) and red algae. Filter feeders, such as barnacles and mussels, are scarce, as are motile herbivores, such as sea urchins and chitons. In the Near Islands, 400 kilometers to the northwest, sea otters are absent from Shemya Island and only recolonized Attu Island in the mid 1960s. Where sea otters were absent, macrophytes were scarce below the lower intertidal zone, and barnacles, mussels, sea urchins, and chitons were many times denser than along the Rat Islands. These differences seem related to the activities of otters. These animals prey heavily on sea urchins *(Strongylocentrotus* spp.), which graze on macrophytes. When high populations of otters drastically reduce the abundance of urchins, kelp beds flourish and the complexion of the nearshore community is conspicuously altered (Fig. 24-17).

Kelp forests in Australasia differ dramatically from those in the North Pacific. Australasian kelp communities lack sea otters, the primary predator of macroinvertebrate herbivores that graze on kelp.

As a result, Australasian kelps are under stronger selective pressure to evolve chemical defenses, and many species have evolved secondary metabolites called phlorotannins to deter herbivory. In the North Pacific kelp forests along the Alaskan coast, sea otter predation on the primary herbivores reduces herbivory on kelp, resulting in kelp communities with reduced levels of phlorotannins (Steinberg et al., 1995).

Productive kelp beds and a stable nearshore community on the Pacific coast of the United States may be maintained only in the presence of their keystone species, the sea otter. However, a

A

B

FIGURE 24-17   (A) A pair of sea otters *(Enhydra lutris)* forage on sea urchins along the coast of Alaska. (B) A pair of killer whales *(Orcinus orca)* from a larger pod forage for seals and possibly sea otters along the Alaskan shoreline. *(From PhotoDisc, Inc.)*

complex web of events has recently altered the structuring of nearshore kelp communities. The typical diet of killer whales along the Alaskan coast is Steller sea lions and harbor seals. However, seal and sea lion populations have declined in recent years as their food supply dwindled from overfishing by humans. In response to declines in seal populations, killer whales appear to have shifted their diet from seals to sea otters (Estes et al., 1998). Sea otter populations have declined precipitously in the 1990s. Estes and colleagues studied a sea otter population in a lagoon not accessible to killer whales and another population of otters in an open bay frequented by killer whales. Over a two-year period, the disappearance rate of sea otters from the open bay was five times that of otters in the inaccessible lagoon. As predicted, in coastal areas where sea otter populations have been decimated, sea urchin populations have increased and kelp beds have become depleted. These alterations in the nearshore community demonstrate the sea otter's role as a keystone species.

In some of the national parks and game preserves of Africa, there are high elephant populations, and because of the encroachment of people and agriculture, the elephants are no longer free to range widely when pressed by local or seasonal shortages of food or water. Studies of elephants in various parts of East Africa document the impact elephants can have on the landscape under these conditions (Laws, 1970). The diet of elephants seems ideally to consist of a mixture of grass and browse from trees and shrubs (Laws and Parker, 1968), and the preferred habitat is thus forest edge, the woodland, or bush-grass mosaic. During the dry seasons in the drier areas, such as Tsavo National Park in Kenya, sources of water are not generally distributed. Because elephants at these times need water daily, they concentrate within a radius of 20 to 30 kilometers of water. One hundred elephants can eat approximately 13,600 kilograms of vegetation a day. Thus vegetation is rapidly destroyed in the dry season in the vicinity of a water hole where hundreds of elephants congregate.

In Tsavo National Park, where in the 1960s 17,000 square miles were occupied by over 40,000 elephants, vast areas of bushland were transformed into grassland. Aerial photographic transects studied by Watson (1968) indicated that, in a period of

five years, elephants killed from 26 to 28 percent of the trees above 65 centimeters in crown diameter. But drought and poaching in the 1970s reduced the elephant population to about 10,000, and the trees have been regenerating. To the south, in Lake Manyara National Park of Tanzania, Douglas-Hamilton (1973) observed similar destruction: in one area of especially acute damage, elephants killed 8 percent of the umbrella trees *(Acacia tortilis)* in one year. This national park has been greatly enlarged since Douglas-Hamilton's observations, however, and pressure on the acacias has been reduced (Moss, 1982).

In Tsavo, as elsewhere, elephants have not affected the vegetation alone. Ungulates such as zebras *(Equus burchellii)*, Grant's gazelles *(Gazella granti)*, and oryx *(Oryx gazella)* have been favored by the shift toward grassland. Other effects include enlargement of wallows to form water holes in the wet season and local compaction of riverbeds resulting in the surfacing of previously subsurface water (Sheldrick, 1972).

Although not all large herbivores are keystone species, their feeding can strongly affect the growth forms and nutrient content of plants, factors that have a reciprocal effect on the herbivores (Charnov et al., 1976; Danell, 1983; Danell et al., 1994). Plant responses to herbivory are varied and complex (Whitham et al., 1991), but an example involving a mammal is pertinent. Studies of Swedish moose *(Alces alces)* and some of their food plants (such as birch, *Betula* spp., and willow, *Salix* spp.) have shown that these plants respond to winter browsing by lower total biomass, increased branching and shoot size, and increased concentrations in leaves of potassium, calcium, and nitrogen. In addition, the growth form of the tree is altered: height is reduced and biomass of shoots is increased within heights the moose can reach. It is advantageous, therefore, for the moose to browse the plant again after an interval that allows maximum regrowth. Increased fiber content and increased levels of secondary compounds in such regrowth can reduce digestibility, however, thereby deterring long-term use by the herbivore (Bryant et al., 1983).

Sixty-three years of aerial photographic and soil chemistry data reveal that beavers *(Castor canadensis)* affect the drainage systems of boreal ecosystems (Naiman et al., 1994). In northern Minnesota, beavers converted over 13 percent of the area studied from forest to meadows and ponds. Dams built to retain pond water also prevent the downstream transport of minerals leached from upland soils. Standing water saturated the soils, resulting in anaerobic conditions, which altered subsequent biogeochemical pathways to the extent that the standing stock of some elements increased by over 250 percent. As Naiman and his colleagues noted, the influence of beavers on boreal ecosystems has been "spatially extensive and long-lasting, affecting fundamental environmental characteristics of boreal forest drainage networks for decades to centuries."

Small mammals may also have a marked effect on vegetation. Twelve years after three species of kangaroo rats *(Dipodomys)* were removed from experimental plots in the Chihauhuan Desert, the species composition of desert plants was dramatically altered (Brown and Haske, 1990). What was once desert shrubland was converted to grassland in the absence of kangaroo rats. The change in plant species composition resulted from reduced seed predation and lack of soil disturbance, which allowed tall grasses to colonize the study plots. Change to a tall-grass community, in turn, allowed colonization by granivorous harvest mice *(Reithrodontomys)*. Removal of a single species of kangaroo rat did not cause large changes in plant composition. Brown and Haske (1990) used the term "keystone guild" to describe the three seed-eating species that affected community structure so strongly.

Fossorial mammals are widely distributed in North and South America, Eurasia, and Africa. Because they often occur in semiarid areas where vegetation is patchy, their mounds of earth are usually a conspicuous part of the landscape. The mounds built by fossorial rodents are of the greatest ecological importance. Mounds often cover 25 to 30 percent of the ground area each year. Mole-rat burrows underlie 11 percent of some South African fields (Reichman and Jarvis, 1989). The combined effects of burrows and mounds in a field occupied by mole-rats in South Africa was a reduction by 30 percent of standing plant biomass (Reichman and Jarvis, 1989). Mounds offer patches of bare, soft soil ready for plant colonization. Plant composition on newly vegetated mounds typically differs from that of surrounding areas. Although

low survivorship of plants occurs on mounds, the sizes of the plants and their seed crops are unusually large (Koide et al., 1987; Reichman, 1988). Anderson and MacMahon (1985) showed that northern pocket gopher mounds played a significant role in plant succession following the 1980 eruption of Mount St. Helens, Oregon. Burrowing by pocket gophers brought fresh soil to the surface of the ash (tephra), where seeds were able to take hold and germinate. Plant species diversity and abundance were greater on gopher mounds than at other nearby sites (Andersen and MacMahon, 1985).

PLANT-MAMMAL INTERACTIONS

Through many millions of years of plant-herbivore interactions, many plants have adapted to mammals by taking advantage of them as agents of seed dispersal or **pollination**. There is growing evidence that a diversity of mammals (including didelphimorphs, bats, primates, rodents, ungulates, and some carnivorans) improve the reproductive success of plants (Fleming and Sosa, 1994). This ability of mammals is due largely to their mobility, the effects of their digestive tracts on seeds, and to caching behavior. The Cretaceous ecological revolution that resulted in the domination of terrestrial floras by angiosperms (flowering plants) was perhaps facilitated by the diversification of birds and mammals and their effectiveness as seed dispersers (Regal, 1977).

For a particularly interesting example of mutually beneficial interactions between mammals and a plant, we return to Serengeti National Park. Here, and over wide areas of Africa, the umbrella tree *(Acacia tortilis)* is a conspicuous, picturesque, and important savanna tree. The green, leathery pods of the plant are eaten avidly by a variety of herbivores, ranging from the tiny dik-dik to the elephant, and elephants often feed heavily on the foliage. Lamprey et al. (1974) found that the germination rates of seeds that had been ingested and eliminated by herbivores were strikingly higher than the germination rates of uningested seeds. In some vertebrate-adapted seeds, the digestive process is known to erode the seed coat and hasten germination, but the high germination rate in ingested acacia seeds is thought by Lamprey and his coworkers not to depend primarily on this effect.

Beetles of the family Bruchidae lay their eggs on acacia seed pods, and the larvae feed and grow within the seeds. If the larvae damage the embryo or destroy a large amount of the cotyledon material, the seed will not germinate, but if the seed pods are eaten by herbivores soon after they fall to the ground, as is typically the case in the Serengeti, the digestive process kills the larval bruchids at an early stage of development, before they have killed the seeds. Some 500 seed samples were collected from the ground and stored for one year; over 95 percent of these seeds had bruchid damage, and the germination rate was only 3 percent. In seeds eaten by impalas *(Aepyceros melampus),* however, the damage rate was 26 percent and the germination rate was 28 percent; in seeds eaten by dik-diks, these figures were a 45 percent damage rate and 11 percent germination rate. The interactions seem clearly to be mutually advantageous. From the acacia, the herbivores get a seasonally important food, and from the mammals, the acacia gains an effective means of escape from a seed predator and wide dissemination.

In tropical areas, many frugivorous mammals disperse seeds. The Neotropical frugivorous bat *Carollia perspicillata* is an effective seed disperser (Fleming, 1988), and this function would be expected of many other frugivorous phyllostomid bats. In a Mexican rainforest, howler monkeys *(Alouatta palliata)* doubled the dispersal rates of seeds of some trees, and some seeds were dispersed 800 meters (Estrada, 1991). In Nepal, the one-horned rhinoceros *(Rhinoceros unicornis)* eats quantities of fruit of the successional tree *Trewia nudiflora,* and seeds germinate vigorously from the animal's manure. Because these fruits are too large and hard for bats, birds, or monkeys, the rhinoceros may be the main seed disperser of this tree (Dinerstein and Wemmer, 1988).

A remarkable case of plant/mammal coevolution involves a **geophyte** (a plant with underground storage organs) and mole-rats (Lovegrove and Jarvis, 1986). This plant *(Micranthus;* Iridaceae) lives in extremely nutrient-poor soils of the western Cape of South Africa, where there are over 1200 species of geophytes. Geophytes increase their nutrient stores (especially nitrogen and phosphorus) over many growing seasons (Dixon, 1981). Growing seasons are brief, and the plants are dormant the rest of the year. Geophytes are high in calories, are highly digestible (Bennett and Jarvis, 1995), and are a preferred food of mole-rats. Because of their great importance to the plant's survival, storage organs of many species are defended against herbivores by toxic or unpalatable compounds. The geophyte/mole-rat association probably began at least 20 million years ago (early Miocene), and mole-rats have developed tolerance to even the most toxic geophytes. But *Micranthus* has taken evolutionary gamesmanship in another direction: although its corms are invested by a thick, fibrous, and spiny tunic, the corm segments (up to 30) are highly palatable to mole-rats and are often stored in large numbers (3043 were found in one *Georychus capensis* cache). Mole-rats must peel the tunic from the entire corm, often dislodging segments that become lost. Because corms are so palatable, they are frequently eaten along the burrows rather than being carried to the cache, thus maximizing chances for dispersal of corm segments. *Micranthus* further favors its survival by producing a cluster of "cormlets" at a depth of 2 to 10 centimeters, well above the larger corms and above the foraging burrows of the mole-rats (at depths of about 95 to 200 centimeters). Also of interest, at sites where geophytes and mole-rats were studied, *Micranthus* was the most abundant geophyte, and here too is an exception to the nearly worldwide lack of sympatry among subterranean rodents: in the Cape three species of mole-rats commonly coexist.

Some mutually advantageous plant-animal associations are long-standing; in some cases, remarkable mutual adaptations have evolved. Especially noteworthy examples of such diffuse coevolution are offered by nectar-feeding bats and the plants on which they feed. In the Old World, a number of bats of the family Pteropodidae feed on nectar; in the Neotropics, nectar feeding occurs in the family Phyllostomidae. **Chiropterophily** (the dependence of a plant largely on bats for pollination) has been discussed by Alcorn et al. (1959), Baker (1961, 1973), Faegri and Van Der Pijl (1966), Fleming (1992), and Ostfeld (1992). Laboratory studies by Howell (1974b) have demonstrated several physiological adaptations of a bat *(Leptonycteris curasoae)* that feeds from flowers. Pollination of plants by mammals may originally have involved marsupials and lemurs, but nectar-feeding bats probably displaced these animals early in the Tertiary (Sussman and Raven, 1978). Only in areas where nectar-feeding bats are uncommon or absent, as in Madagascar and parts of Australia, does this archaic type of pollination by lemurs or marsupials persist.

Rarely does a single species of animal pollinate a plant; more commonly, pollination is by many agents, including perhaps a number of insects and birds. Such plants are termed **polyphilous.** But some are pollinated by relatively few agents, and these plants would be expected to evolve features making their flowers especially attractive to a small group of pollinators. This seems to have occurred in a number of bat-pollinated plants, but in no case is the bat the sole pollinating agent.

Most chiropterophilous plants occur in the tropics and subtropics and have the following features: the flowers occur on spreading, often leafless branches; they are whitish and offer visually conspicuous targets at night; they have a strong, often rank, smell; copious amounts of nectar are produced in early evening when the bats forage; the plants produce flowers over long periods, but only a few buds open each night (this favors **traplining,** a nightly patrolling of the same series of plants by the bats).

Two general types of flowers are chiropterophilous. The first type, an example of which is the large flower of the African baobab tree, accommodates large bats. The feeding bat clutches the ball of stamens of this flower while lapping nectar from the pillar-like stamen column. An African species of *Parkia* studied by Baker and Harris (1957) displays a division of labor between nectar-producing and reproductive stamens. Bats of the genera *Epomophorus* and *Nanonycteris* were observed lapping nectar from the circular depression in the top of the ball of the flowers, which was grasped by the bat's feet. Most of the flowers of a tree are in a staminate condition one night and a pistillate condition the next, thus ensuring cross-fertilization.

The second group of "bat flowers" includes many Neotropical types that have corolla tubes adapted to smaller bats. When a bat pushes its head into the tube, its face is liberally dusted with pollen.

Sanborn's long-nosed bat *(L. curasoae)* is a New World phyllostomid that feeds on nectar. Nectar is rich in carbohydrates but usually has no more than trace amounts of protein, a food essential to mammals. Howell (1974b) showed that, under natural conditions, the source of protein for these bats (and presumably for most nectar-feeding bats) is pollen. These bats ingest large quantities of pollen, probably mostly when grooming their fur, to which pollen adheres readily. Pollen of the plants visited by *Leptonycteris* (saguaro, *Carnegeia gigantea,* and agave, *Agave palmeri*) is much higher in protein than is the pollen of closely related plants that are pollinated mostly by other agents. Of additional interest, "bat pollens" contain at least 18 amino acids and are unusually high in the amino acid proline, which makes up over 80 percent of the protein collagen, the connective tissue that braces the wings and tail membranes of bats (Howell, 1974b). *Leptonycteris* has high concentrations of hydrochloric acid-secreting glands in its stomach and often ingests its urine. Both urea and hydrochloric acid are known to extract protein from pollen.

In the Sonoran Desert, where some bats are effective pollinators of certain columnar cacti, *L. curasoae* has seemingly affected the evolution of **gynodioecy** (separate female and hermaphrodite plants) and **trioecy** (separate males, females, and hermaphrodites) in the columnar cactus *Pachycereus pringlei.* The fitness of males and females of this cactus depends more on pollinators than does fitness in hermaphrodites, thus the distribution of these reproductive types is not random. Within flight distance of the bats' roosts, trioecy occurs; away from roosts, this cactus displays gynodioecy (Fig. 24-18; Fleming et al., 1998).

Herbivory is popular among mammals and has many advantages: the biomass of plants vastly exceeds that of animals, plants need not be pursued, and they are equally available throughout the 24-hour cycle. Because plants with high levels of secondary compounds must be detected and avoided, however, and because the energy yield of plant tissue per unit of weight is low relative to that of animal material, herbivores must invest a great deal of time in feeding. The pronghorn *(Antilocapra americana),* an herbivore that feeds primarily on forbs and small shrubs, utilizes a strategy common to many ungulates. Most of the animal's time is spent eating or processing food. It alternately feeds and beds down through the day and night, and ruminating (chewing the cud) occupies 60 to 80 percent of the bedding time (Kitchen, 1974).

Although many herbivores are selective in their feeding, a wide variety of food is generally utilized,

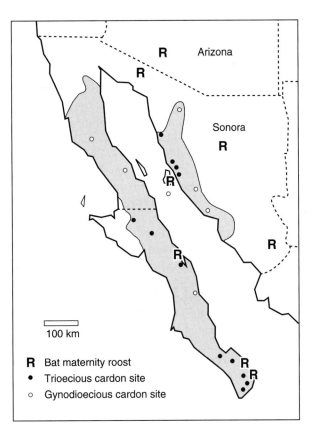

FIGURE 24-18 A map of the Baja Peninsula and northwestern Mexico showing the localities of the nectar-feeding bat *Leptonycteris curasoae* maternity roosts in relation to trioecious and gynodioecious cactus sites. Note that trioecious cactus tend to be located closer to maternity roosts. *(After Fleming et al., 1998)*

and seasonal variations in feeding habits are typical of temperate-zone species. A number of studies have shown that, given a wide variety of plants to choose from, most herbivorous mammals are selective foragers (see, for example, Yoakum, 1958; Ward and Keith, 1962; Zimmerman, 1965). Accordingly, a herbivore may show a great preference for one of the least abundant plants in its habitat.

Plant defensive chemicals are of two major types. Ephemeral plant parts (those available briefly), such as flowers, fruits, or new leaves, are typically protected by toxins that have probably evolved in response to pressure from dietary generalists. Toxins exhibit little individual variation from plant to plant within a species. Mature leaves, on the other hand, are a predictable and abundant food that is available over a relatively long period of time. Probably in response to dietary specialists, mature leaves have evolved secondary compounds with a high degree of individual variation (Rhoades, 1979). These chemicals reduce digestibility (Feeny, 1975; Cates and Rhoades, 1977). Tannins occur in the leaves of a wide array of plants and combine chemically or denature many mammalian digestive and nondigestive enzymes (Pridham, 1965). The volatile oils of conifers contain terpenoids. Some juniper terpenoids are known to have an antimicrobial action in the rumen of deer (Schwartz et al., 1980a). Because microbial fermentation in the rumen yields the largest share (50 to 70 percent) of the energy required by ruminants (Annison and Lewis, 1959), antimicrobial secondary compounds markedly inhibit ruminant digestion (Nagy et al., 1964) and may also inhibit digestion in the caecum of rodents (Vaughan, 1982).

Plant tissues with high levels of defensive chemicals are not popular foods among mammals. Conifer foliage, for example, is generally either an emergency food or one taken in small amounts with an array of other plants. In North America, only two species of mammals, the red tree vole (*Arborimus longicaudus;* Hamilton, 1962) and Stephens' woodrat (*Neotoma stephensi;* Vaughan, 1982) are conifer-leaf specialists. Plants other than conifers can also produce such high levels of defensive chemicals that herbivores are deterred. After intense browsing by snowshoe hares *(Lepus americanus),* four nonconiferous species of Alaskan trees produced adventitious shoots (shoots initiated from more or less mature tissue without connection with apical meristem) that had exceptionally high levels of terpenes and phenolic resins. Because these resins are repellent to snowshoe hares, the animals avoid the adventitious shoots (Bryant, 1981). These browsing-induced chemical defenses drastically reduce the forage available to the hares at the very time when high populations require large amounts of food. This may help trigger the population crash. These plants do not produce usable forage until three years after intense browsing, and only after this time can hare populations begin to recover. Bryant speculated that the chemical defenses of the food plants strongly influence the ten-year population cycle of the snowshoe hare.

Because defensive chemicals in plants are of broad occurrence, they have forced herbivores to evolve countermeasures. One such measure is microbial breakdown. Oxalates occur in many plants important in the diets of herbivores, and to nonadapted mammals oxalates can be lethal. Some mammals, including rabbits, rodents, pigs, horses, some ruminant artiodactyls, and humans, are known to degrade oxalates by microbial action in the large intestine or rumen. Oxalates in the diet favor intestinal or rumen bacteria that utilize these compounds for growth; higher populations of such bacteria result in increased rates of oxalate breakdown (Allison and Cook, 1981). Herbivores adapted in this way can tolerate levels of dietary oxalates that would kill nonadapted individuals. The ability of a number of wild rodents to eat large quantities of plants containing oxalates is probably due to adaptation of their intestinal or caecal bacteria.

A second countermeasure is selective foraging, the ability to discriminate among individual plants and eat only those with relatively low levels of defensive chemicals. Such finely tuned selective foraging depends on intraspecific variation in secondary compounds among plants. This type of variation has been well documented (Hanover, 1966, 1971; Schwartz et al., 1980), as has selective foraging by mammals. Glander (1977) found that, in Costa Rica, leaf-eating howler monkeys (*Alouatta palliata*) were forced by defensive chemicals to be selective of the leaves of trees. The leaves of only certain individuals of some species of trees were eaten, and the monkeys generally ate the petioles (leaf stalks), which have lower concentrations of defensive chemicals than do the leaves. The price of carelessness is high: three out of six dead howler monkeys that Glander examined had been eating the leaves of either of two trees with toxic leaves, as had a female that went into convulsions and fell from a tree.

A third way in which mammals avoid the effects of defensive chemicals is by having a diverse diet. By eating a great variety of plants, an herbivore can keep the levels of defensive chemicals ingested low enough to be tolerated.

The life histories of some folivorous (leaf-eating) mammals probably evolved under selective pressures associated with a low-energy diet and with chemicals in leaves that reduce digestibility. The low metabolic rate of Neotropical tree sloths and howler monkeys is perhaps an adaptation to their folivorous diet. The two North American conifer-leaf specialists offer interesting examples. Their life histories depart markedly from the norm for rodents. These departures may be the price paid for eating food that is abundant and predictable but protected by digestibility-reducing chemicals. Red tree voles, which feed on the leaves of Douglas fir, never reach high population densities, have small litters (usually two young), and their young grow slowly. Most other voles attain high densities, have large litters, and their young grow rapidly (Hamilton, 1962). Similarly, Stephens' woodrats, which eat juniper leaves, are never abundant, usually have but a single young per litter, and the young grow and reach sexual maturity slowly (Vaughan, 1985).

Halophytic (salt-loving) plants of many desert regions concentrate salts in their above-ground tissues. The halophyte *Atriplex*, the desert saltbush, common in African, Australian, and North and South American deserts, concentrates salt on the outside of its leaves. Several mammals have convergently evolved morphological, behavioral, and physiological mechanisms of dealing with the high salt loads of *Atriplex* (Mares et al., 1997). Chisel-toothed kangaroo rats (*Dipodomys microps*) of the Great Basin Desert use their specialized lower incisors to scrape off the hypersaline outer coating of *Atriplex* leaves (Kenagy 1972, 1973). The North African fat sand rat (*Psammomys obesus*) also eats *Atriplex* and other native halophytes (Degen, 1988). This rat is able to reduce salt intake by using its incisors to scrape away external salt from the leaves before eating the inner tissues and by renal adaptations that remove salt from the blood (Degen, 1988; Mares et al., 1997). In the deserts of Argentina, the red vizcacha rat (*Tympanoctomys barrerae*) also specializes on *Atriplex* (Mares et al., 1997). These rats use a combination of chisel-shaped lower incisors and a set of specialized hairs just posterior to the upper incisors to strip the salty layer off the leaves before consuming the inner parts.

Frugivorous mammals tend to process and digest fruits relatively rapidly, resulting in the excretion of the seeds with the feces. Seeds that have been subject to the gentle digestion inside the alimentary tract of a frugivore often germinate at higher rates than seeds that are not eaten. Although fruit is relatively easy to find and digest, it is usually protein deficient. Moermond and Denslow (1985) demonstrated that frugivorous mammals must eat approximately 2 grams of fruit for every gram of body weight in order to receive sufficient protein. Low levels of fats and vitamins occur in fruit pulp, and some amino acids may be lacking entirely (Rasweiler, 1977). Fruit specialists have adapted by simplifying the digestive tract to speed the passage of seeds through the system. Fleming (1988) estimated that the short-tailed fruit bat *(Carollia perspicillata)* needs about 4.7 kilocalories per night but requires approximately 14 milligrams of nitrogen (from protein mostly) to maintain a positive energy balance (Herbst, 1985). In order to satisfy these requirements, the bats would have to eat six fruits of *Chlorophora tinctoria* or 80 fruits of *Ficus ovalis*. These bats meet their energy and nitrogen needs by selecting a combination of fruits and reduce their overall foraging time by feeding on high-protein (nitrogen source) fruits early in the evening. They switch to fruits with higher overall energy content later in their foraging bouts (Fleming, 1988).

Obvious physical defenses of plants include thorns and seeds with hard or thick coats. In Africa, a great variety of plants, including the diverse and widespread acacias, bear imposing thorns, which probably evolved in response to browsing by a diversity of ungulates. The leaves of African acacias have a higher protein content than that of most trees and shrubs (Sauer at al., 1982), and, at least seasonally, are a preferred food of a number of African artiodactyls (Skinner and Smithers, 1990).

In summary, mammalian herbivores have been forced to adapt to the nearly ubiquitous defenses of plants. Adaptation of intestinal, caecal, or rumen microbes increases the rate of degradation of some defensive chemicals, and highly selective foraging enables some mammals to feed on those individual plants within a species that have relatively low levels of these chemicals. Some herbivorous mammals avoid ingesting large amounts of any defensive chemical by eating a broad array of plants.

ENERGY FLOW AND FOOD WEBS

Just as energy transfers from one part of an organism to another are vital to life, complex patterns of energy transfer within a biotic community main-

tain the system of interdependent and interacting species. The organisms involved in the transfer of energy within a community — from photosynthetic plants that utilize solar energy and inorganic materials to produce protoplasm, to animals that eat the plants, and thence to animals that eat animals — constitute the food chain. Typically, the transfer in a food chain goes from photosynthetic plants (primary producers) to herbivores (primary consumers) to primary carnivores (secondary consumers) that eat the herbivores, to secondary or perhaps tertiary carnivores in some extended chains.

The term **food web** has been used to describe the complex pathways of energy transfer that usually occur in nature, often involving predators and primary consumers that figure importantly in more than one food chain (Fig. 24-19). Animals that occupy comparable functional positions in the food chain — the position of primary consumers, for example — are at the same trophic level. Green plants occupy the first trophic level and are referred to as **autotrophs** (self-feeders). Using mammals as examples, the second trophic level is occupied by herbivorous rodents, rabbits, and ungulates. Small carnivores, such as weasels, occupy the third trophic level, and large carnivores are in the third or fourth level.

The food chain is often depicted as a pyramid in an attempt to stress the relationships of biomass (the total weight of organisms of a given type in the community), numbers of organisms, and available energy at the different trophic levels. Food chains typically rest on a broad food base of plant material, but the amount of energy available to animals in each successively higher trophic level becomes progressively reduced because energy is lost through respiration and through the death of organisms, when their energy goes to decomposer food chains. Also, energy is lost because of inefficient transfer from one level to another. Herbivores such as voles and lagomorphs, for example, assimilate only about 65 percent of the energy available from the fibrous plants they eat (Grodzinski and Wunder, 1975). Consider, for example, the pyramids of numbers, caloric content, and energy utilization shown in Figure 24-20 and Table 24-2.

The typical relationship of size and abundance of animals in a food chain involves small but numerous primary consumers, larger but much less

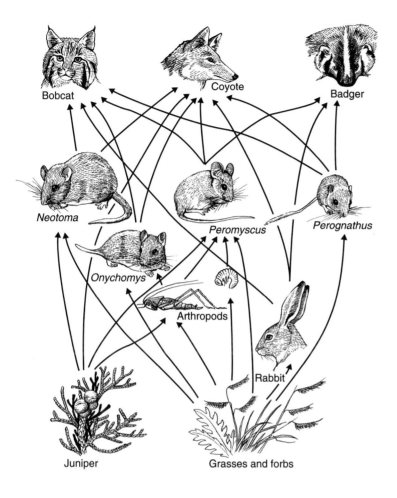

FIGURE 24-19 A simplified food web involving the mammals of a piñon-juniper community in Coconino County, Arizona. The arrows indicate the foods utilized by the mammals. The plants support arthropods, rabbits, and rodents. The rodents *(Neotoma stephensi, Onychomys leucogaster, Peromyscus truei, Perognathus flavus)* and the rabbits *(Lepus californicus* and *Sylvilagus audubonii)* are preyed upon by bobcats, coyotes, and badgers.

abundant secondary consumers, and still larger but relatively scarce tertiary consumers. Top predators occupy precarious positions because they depend on animals from trophic levels with low productivity.

Because of the great loss of energy accompanying food transfer between successive trophic levels, the total available energy is largest for consumers at the lower levels. Thus, predators feeding on primary consumers have more energy available to them than do predators feeding on secondary consumers. Considered in this light, the adaptive importance of filter feeding in some marine mammals becomes clear. Plankton feeders, such as the baleen whales and the enormously abundant Antarctic crab-eater seal *(Lobodon carcinophagus),* bypass some trophic levels by exploiting the tremendously larger sources of energy in plankton rather than feeding on larger fish at secondary or

tertiary levels. Only 10 to 20 percent of the energy entering a trophic level can be utilized by the next higher level; this factor limits the length of food chains.

Diagrams of food chains or food webs, although valuable for purposes of illustration, usually over-simplify what is really an extremely intricate mesh-work of interactions. A broadly adapted carnivore like a coyote, or an omnivore such as the opossum *(Didelphis),* may function in all trophic levels above that of the primary producer. For a coyote, the fruit of the prickly pear cactus or juniper berries may form one meal, while a jackrabbit or deer fawn may be the next. More frequently, seasonal differences in the position of an animal in the food chain may occur. Johnson (1961, 1964) found that the deer mouse *(P. maniculatus)* in Colorado and Idaho became strongly insectivorous in the summer and thus functioned during this season as

FIGURE 24-20 Pyramids of numbers, calories, and energy utilization for 1 acre of annual grassland near Berkeley, California. The pyramid showing calories is also approximately to scale for biomass. *(After Pearson, 1964)*

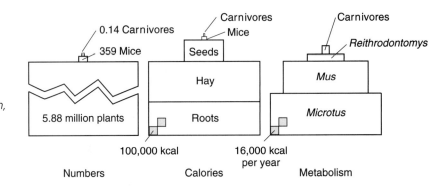

a secondary consumer, whereas it ate largely plant material during the cooler seasons, functioning as a primary consumer at those times.

There are important differences in small-mammal energetics in various ecosystems and in the proportions of energy utilized by rodents from different trophic levels. French et al. (1976) compared the small-mammal energetics in a series of grassland ecosystems: a tall-grass site in Oklahoma, a mid-grass site in South Dakota, a northern short-grass site in Colorado, a southern short-grass site in Texas, and a desert grassland site in New Mexico. The rodent communities of the ecosystems differed strikingly in composition and in biomass. Of particular interest is the fact that foods from different trophic levels were utilized to different extents by the rodents at different sites. Arvicoline herbivores specializing on green primary produc-

tion dominated (in terms of biomass) at the tall-grass site. Sciurid omnivores utilizing both primary production and primary consumers (insects) dominated in the northern short-grass prairie, and heteromyid seed eaters dominated the desert site. Broadly speaking, small mammals depended chiefly on herbage in the tall-grass and mid-grass site and on herbage, seeds, and invertebrates in the desert site.

Although the energy consumed by rodents was highest in the tall-grass prairie (where 172×10^3 kcal per hectare supported 935 grams live weight of small mammals per hectare), the amount of small mammal biomass relative to consumption by rodents was greatest in the northern short-grass prairie (where 32×10^3 kcal per hectare supported 277 grams of small mammals per hectare). This was probably due largely to the high assimi-

TABLE 24-2 Standing Crop of Plants, Prey, and Predators on 1 Acre of California Grassland and Rate of Use of Vegetation by Rodents and of Prey by Carnivores at Peak Population Levels

	Standing crop		Annual rate of use	
	Kg (dry wt)	Kcal	Kcal	% of Crop
Roots	2131	7,269,000		
Hay	2097	8,141,000		
Seeds	442	1,920,000		
Microtus	1.24	6,402	1,368,750	71
Mus	0.88	4,543	876,000	46
Reithrodontomys	0.084	434	81,650	4
Other prey	0.13	671	27,000	
Carnivores	0.126	650	11,700	97

After Pearson, 1964.

TABLE 24-3   Fraction of Available Energy Utilized by Small Mammals, According to Diet

Grassland site	Year	Herbage	Seed	Animal
Tall-grass	1970	0.063	—	0.88
(Oklahoma)	1971	0.010	—	0.38
	1972	0.047	0.080	All?
Mid-grass	1970	0.002	—	0.08
(So. Dakota)	1971	0.001	—	0.08
	1972	0.002	—	0.02
Northern short-grass	1970	—	0.002	All?
(Colorado)	1971	—	—	0.42
	1972	0.005	—	0.83
Southern short-grass	1970	—	—	0.96
(Texas)	1971	0.013	—	0.61
	1972	0.091	—	0.11
Desert	1970	0.032	—	All?
(New Mexico)	1971	0.190	—	All?
	1972	0.004	—	0.68

From French et al., 1976.

lation and digestion rates for granivores and omnivores. Usually less than 10 percent of the primary production at the various sites was utilized by small mammals, but, in striking contrast, invertebrates were heavily utilized, in extreme cases at close to 100 percent (Table 24-3). This information suggests that herbivorous voles underutilize their food resources and seed-eating heteromyids overexploit food reserves, an idea presented by Baker (1971).

French et al. (1976) pointed out that where mesic grasslands are receding northward, as they are in some parts of the mid-grass areas in the Great Plains of the United States, herbage-eating arvicolines are being replaced by specialized seed eaters. Sharp differences in the relative importance of herbivorous, omnivorous, granivorous, and insectivorous small mammals between different habitats can also be observed in areas other than grasslands (Fig. 24-21) and reflect differences in the composition of the vegetation.

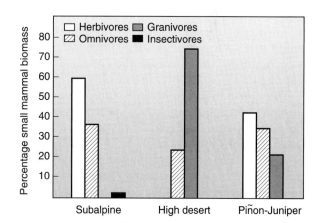

FIGURE 24-21   Relative biomass of herbivorous, omnivorous, granivorous, and insectivorous small mammals at a subalpine site (Rabbit Ears Pass, Routt County, Colorado), a high desert site, and a piñon-juniper site (both in Wupatki National Monument, Coconino County, Arizona. *(Subalpine data from Vaughan, 1974; biomass for other sites based on unpublished data from the summer of 1976, courtesy G. Bateman)*

POPULATIONS

A population is a group of individuals that occupy a localized area and interact with one another (Hanski and Gilpin, 1997). The boundaries of the area may be arbitrarily set by the ecologist or may be determined by the geographic ranges of species. Populations are characterized by such features as density, spacing patterns, age structure (ratio of one age class to another), rates of growth or decline, and genetic structure. These characteristics change over time as a result of changing selective pressures. Natural selection shapes the life history strategies of populations and plays a vital role in regulating population size.

DISTRIBUTION AND HABITAT SELECTION

Just as no two species of animals are structurally identical, no two are functionally identical or have exactly the same environmental requirements. The very morphological, physiological, and behavioral characters that determine the distinctness of a species also determine the distinctness of its habitat requirements. Each species requires a specific environment — a particular combination of physical and biotic factors — and each is functionally unique, pursuing a particular mode of life within its environment. The specific environmental setting a species occupies and the functional role it plays in this habitat constitute the animal's ecological niche.

The distribution of animals occurs at two levels, the geographic distribution of a species and the local distribution of individuals. A species' distribution represents the sum of the many local population distributions. The distribution of a particular species (or group of populations) is determined by appropriate climate, availability of suitable resources, barriers to dispersal, and the interspecific interactions with the many other organisms sharing the same area. Dispersal ability (and barriers to dispersal) also determine which species are likely to occur in a given area even when suitable habitat is present. Historical accidents may also play a role in shaping geographic distributions in some cases. Home ranges, territories, and microhabitats reflect the local distribution of individuals within an area of suitable habitat and are governed by access to important resources such as food, shelter, and availability of mates.

Although often depicted as continuous in field guides, the distribution of a species is not uniform throughout its geographic range. Rather, it is a mosaic of the distributions of the local populations. Nor are individuals in the population equally abundant in all regions. The natural world is heterogeneous, resulting in a patchwork distribution of habitats of varying quality.

Many factors contribute to the selection of appropriate habitats, and each species has physical and behavioral traits that are optimally suited to its habitat. Degrees of habitat tolerance differ among species, and local populations may adapt to local conditions. In the western Sierra Nevada of central California, four forest communities are delineated by altitude. Each species of shrew (*Sorex*) occupies a different major forest community (Williams, 1991). *Sorex ornatus* occupies the low elevation (500 to 1000 meters) digger pine/blue oak woodlands and ponderosa pine forests, and is replaced by *S. trowbridgii* in intermediate elevation (1500 to 2000 meters) mixed coniferous forests. As mixed coniferous forests give way to high elevation (>2000 meters) red fir and lodgepole pine forests, *S. trowbridgii* is replaced by *S. monticolus*. The water shrew (*S. palustris*) was associated with riparian habitats at elevations over 1500 meters. Although sympatry occurs, habitat segregation between species occurs on a local scale such that interspecific contacts are probably rare in shrews (Kirkland, 1991; Sheftel, 1994).

The environment at a terrestrial locality is not uniform but consists of a complex mosaic of microenvironments. As a general rule, few terrestrial mammals can withstand the most extreme temperatures (or other conditions) that occur in the habitats they occupy, but many are able to select microenvironments in which temperature extremes are moderated or eliminated. Rather than being a source of winter hardship for small mammals, snow is actually a blessing. It forms an insulating mantle that provides a microenvironment at the surface of the ground where activity, including breeding in some species, continues through the winter. To these small mammals, such as shrews (*Sorex* and *Blarina*), pocket gophers (*Thomomys*), voles (*Microtus, Clethrionomys, Phenacomys*), and lemmings (*Lemmus, Dicrostonyx*), the most stressful periods are in the fall, when intense cold descends but snow has not yet moderated temperatures at the surface of the ground (Formozov, 1946), and in the spring, when rapid melting of deep snow often results in local flooding (Jenkins, 1948; Ingles, 1949; Vaughan, 1969).

A group of beavers occupying a beaver lodge in the winter is not subjected to the extreme air temperatures outside the lodge (Fig. 24-22). Similarly, shrews forage beneath litter, under logs or rocks, or beneath dense foliage; not only is their food abundant in such places, but temperature and humidity are moderated by such cover. During the dead of winter, short-tailed shrews *(Blarina brevicauda)* maintain a relatively constant core body temperature of approximately 38°C, despite ambient temperatures above the snow as low as −21°C, by utilizing subsurface runways where temperatures hover around 1 to 2°C (Merritt and Adamerovich, 1991; Merritt and Bozinovic, 1994). These animals cannot tolerate the general climatic conditions of

the regions they occupy but are instead adapted to a limited set of conditions that occur in a chosen microenvironment.

Even seemingly unimportant characteristics can sometimes have a profound effect on the microhabitat of certain species. Abert's squirrels *(Sciurus aberti),* for example, have been shown to select nest trees based in part on the chemical characteristics of the trees' phloem (Snyder and Linhart, 1994). Trees chosen for nest sites tended to have lower levels of trace metals such as iron and copper and higher levels of sodium than non-nest trees. Lower concentrations of some monoterpenes may also be associated with choice of nest trees by some squirrels (Snyder, 1993).

TERRITORIALITY AND HOME RANGE

Because individuals of the same species are potentially competing for the same environmental resources at the same place and time, intraspecific competition should be intense. Indeed, this seems to be the case, with the result that individuals of the same species often occupy separate, or nearly separate, **home ranges.** Burt (1943) described the home range of a mammal as "that area traversed by the individual in its normal activities of food gathering, mating, and caring for the young." Home ranges may have irregular shapes and may partially overlap (Fig. 24-23). Within the home range of some animals is an area that is actively defended against other members of the species. This area, which usually does not include the peripheral parts of the home range, is called the **territory,** and species that apportion space in this fashion are termed territorial. A home range or a territory may be occupied by one individual, by a pair, by a family group, or by a social group consisting of a number of families.

To solitary animals or to members of a group, the occupancy of a home range has several important advantages. Each home range provides all the necessities of life for an individual or group, permitting self-sufficiency within as small an area as possible. The less extensively the animal must range, the less chance of encounters with predators. The home range quickly becomes familiar to the individual, who can then find food and shelter with the least possible expenditure of energy and can escape predators more effectively because es-

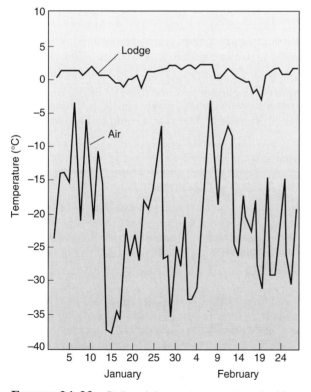

FIGURE 24-22 Daily minimum temperatures inside and outside a beaver lodge in Algonquin Park, Ontario, Canada. *(After Stephenson, 1969)*

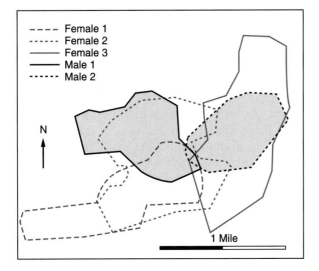

FIGURE 24-23 Distributions of red fox *(Vulpes vulpes)* home ranges at the University of Wisconsin Arboretum. Male home ranges are shaded and overlap with several females. *(After Ables, 1969)*

cape routes and retreats are familiar and no time or movement is lost in seeking shelter. Some mammals, such as rabbits and meadow voles *(Microtus),* maintain trails that serve as routes to food and as avenues of escape. To some male mammals that guard harems, the territory is an exclusive mating area.

An animal's reproductive success may be increased by its knowledge of areas adjoining its home range that are occupied by other animals of

the same species (in the case of solitary species) or by familiarity with animals sharing its home range (in the case of social species). During early life, young can develop under parental care largely free from interference by other individuals of their own species. Infanticide that involves killing of unrelated offspring may also play an important role in the evolution of female territoriality (Wolff, 1997). According to Wolff's hypothesis, females that have helpless, nonmobile young that are reared in a burrow should evolve territoriality to reduce the risks of infanticide from strangers and to ensure an exclusive food supply (Fig. 24-24).

Home range size varies tremendously (Table 24-4). Many mammals in the orders Insectivora, Chiroptera, Primates, Rodentia, Lagomorpha, Carnivora, Perissodactyla, and Artiodactyla are known to be territorial. The recognition of territorial boundaries in some species depends on scent marking and other means of territorial marking, and much remarkable behavior is associated with the maintenance of territories (some of this behavior is discussed in Chapter 23).

Some territorial species are distributed according to a pattern of home ranges that may persist through many generations. Hansen (1962) found such a pattern to be typical of northern pocket gophers *(Thomomys talpoides)* in Colorado. Each animal occupies an area of raised ground called a mima mound (Fig. 24-25), which is some 10 meters in diameter. The mima mounds are more pro-

FIGURE 24-24 Flow diagram of the relationship between female territoriality and the type and behavior of young. According to this model, females producing altricial, nonmobile young that are restricted to a burrow or den site should evolve territorial behavior. *(After Wolff, 1997)*

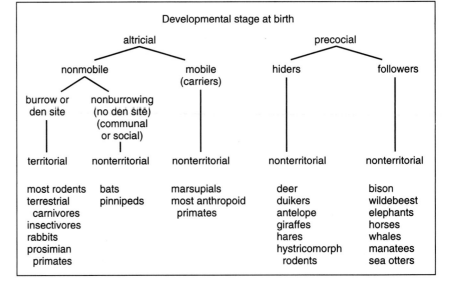

TABLE 24-4 Sizes of Home Ranges of Some Mammals

Species	Home range (acres)	Source
Common shrew (Sorex araneus)	0.7	Buckner, 1969
Varying hare (Lepus americanus)	14.5	O'Farrell, 1965
Mountain beaver (Aplodontia rufa)	0.3	P. Martin, 1971
Least chipmunk (Tamias minimus)	2.1–4.7 (summer only)	Martinson, 1968
Yellow-pine chipmunk (T. amoenus)	3.89 (males); 2.49 (females)	Broadbooks, 1970
White-footed mouse (Peromyscus)	0.08–10.66	Redman and Sealander, 1958; Blair, 1951
Red-backed vole (Clethrionomys gapperi)	0.25 (winter only)	Beer, 1961
Prairie vole (Microtus ochrogaster)	0.11 (males); 0.02 (females)	Harvey and Barbour, 1965
Timber wolf (Canis lupus)	23,040 (pack of 2) 345,600 (pack of 8)	Stenlund, 1955 Rowan, 1950
Red fox (Vulpes vulpes)	1280	Ables, 1969
Grizzly bear (Ursus arctos)	50,240 (1 mother + 3 yearlings)	Murie, 1944
Russian brown bear (U. arctos)	6400–8320	Bourliere, 1956
Raccoon (Procyon lotor)	13.3–83.4	Shirer and Fitch, 1970
Badger (Taxidea taxus)	2100	Sargent and Warner, 1972
Mountain lion (Puma concolor)	9600–19,200 (males) 3200–16,000 (females)	Hornocker, 1970b
Lynx (Lynx canadensis)	3840–5120	Saunders, 1963
Blacktail deer (Odocoileus hemionus)	90 (winter); 180 (summer)	Leopold, et al., 1951
Mule deer (O. hemionus)	502–2534	Swank, 1958
White-tail deer (O. virginianus)	126–282	Ruff, 1938
Pronghorn (Antilocapra americana)	160–480	Bromley, 1969

ductive of food than are the relatively narrow intermound areas, which usually have shallow soil. Except in the winter, the intermound areas are used little by pocket gophers, and the chances of survival are slim for an animal that is unable to establish itself in a mima mound. Likewise, woodrat houses may be used over periods of thousands of years (Wells and Jorgensen, 1964; Betancourt et al., 1990), as indicated by the presence in them of plants that no longer occur in the area but did thousands of years ago.

POPULATION GROWTH

Virtually all populations have the capacity for rapid rates of growth under ideal conditions (known as the **biotic potential**). Natural populations rarely reach their biotic potential even briefly because many factors restrict such rapid rates of increase.

Collectively, these factors are referred to as environmental resistance or **carrying capacity.** Food shortages, lack of suitable cover, disease, and predation all serve to limit an animal's biotic potential.

In 1908, for example, a small group of moose (Alces) colonized Isle Royale in Lake Superior from the adjacent mainland (Mech, 1966). The moose found ample food, no competitors, and no effective predators, because wolves did not become established on the island until the 1940s (Peterson, 1977). Under these nearly ideal conditions, the moose population soon increased to nearly 3000 by 1929. As moose numbers grew, browse became depleted and starvation ensued. A second census by Hickie (1936) in the mid 1930s revealed that the moose population had fallen to only a few hundred animals. As this example illustrates, each environment is capable of supporting only a limited number of animals at any given time, the environ-

FIGURE 24-25　Mima mounds in Mima Prairie, Thurston County, Washington. These mounds, some 10 meters in diameter, are probably formed by the burrowing activities of pocket gophers. *(V. Scheffer)*

ment's carrying capacity for that species (see Hengeveld, 1989 for other examples).

The number of individuals added to a population through reproduction depends on the **reproductive potential** of a species, which refers to the greatest number of individuals that a pair of animals or a population can produce in a given span of time. Reproductive potential is a function of age and sex ratios, age at which a female first bears young, litter size, and frequency of litters. Even species of the same genus occupying the same area

can have markedly different reproductive potentials, as indicated by litter size (Table 24-5). Reproductive performance is also responsive to environmental differences, such as temperature or rainfall, for sharp regional shifts in litter size occur in some species.

The ability of some species to vary reproductive performance in response to environmental conditions or population levels may be of considerable adaptive importance. The reproductive potential of mule deer is lower in poor habitats than in habitats with high-quality browse. Whereas well-nourished does may breed first at 17 months of age, those that occupy poor ranges may not breed first until 41 months of age (Taber and Dasmann, 1957). The litter size of carnivores is also affected by food supply. Stevenson-Hamilton (1947) reported that the litter size of the African lion dropped when food was scarce.

Survival rates of young also strongly affect population levels. Young are clearly a vulnerable part of a population and show the greatest fluctuations during population changes. A 74 percent decline in pocket gopher density in western Colorado in 1958 was associated with an extraordinary drop in the survival rate of young (Hansen and Ward, 1966), and in southern Colorado, Hansen (1962) found that high survival rates of young pocket gophers were characteristic of periods of high densities, whereas low survival rates of young were associated with a declining population. Low survival rates of young have also been found in declining vole populations.

Survivorship curves, such as those in Figure 24-26, graphically illustrate the generally high juvenile mortality. In this figure, Dall sheep *(Ovis)* have

TABLE 24-5　Differences in Reproductive Patterns of Three Species of *Peromyscus* That Are Sympatric in Some Areas

Characteristic	*P. maniculatus*	*P. truei*	*P. californicus*
Number of litters per season	4.00	3.40	3.25
Number of young per litter	5.00	3.43	1.91
Number of offspring per breeding female per season	20.00	11.66	6.21

From McCabe and Blanchard, 1950.

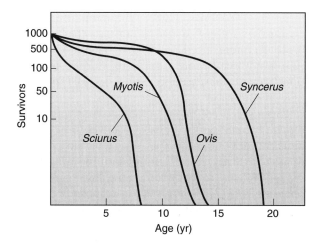

FIGURE 24-26 Survivorship curves for four mammals: gray squirrel (*Sciurus*; data from Barkalow et al., 1970), greater mouse-eared bat (*Myotis*; data from Gaisler, 1979), dall sheep (*Ovis*; data from Deevey, 1947), and the African buffalo (*Syncerus*; data from Sinclair, 1977). Juvenile mortality is high in all species but remains relatively high throughout the lifespan in *Sciurus*.

relatively low juvenile mortality relative to the gray squirrel (*Sciurus*). In the case of the African buffalo (*Syncerus*), the life expectancy at one year of age is considerably higher than at birth. Generally, life expectancy increases with age in large mammals because once they attain a certain percentage of their adult size they are more difficult for all but a few predators to kill. Bats, despite their relatively small size, also have relatively long lives (up to 30 or more years), and their survivorship curves resemble those of much larger species.

Life tables, which include survivorship data as well as **fecundity** data for each age class, illustrate how populations replace themselves over time. Life table data are usually based on females because paternity is often difficult to establish. Consider Fleming's (1988) life table data for female short-tailed fruit bats (*Carollia perspicillata*; Table 24-6). Mortality is expressed in a number of ways: survivorship, age-specific mortality, and life expectancy. A newborn *Carollia* has a 53.7 percent chance of surviving to its first birthday (l_x). Put another way, a cohort of newborns would be

TABLE 24-6 Life History Table for Female *Carollia perspicillata* From Costa Rica

Age (x yrs)	Probability of survival at year x l_x	Proportion dying during x th age interval d_x	Probability of an x yr old dying during age interval q_x	Number of females produced by each female m_x	$l_x m_x$	Life expectancy e_x
0	1.000	0.537	0.537	0.0		2.56
1	0.537	0.281	0.523	0.650	0.349	2.90
2	0.256	0.046	0.180	0.650	0.166	
3	0.210	0.047	0.224	0.650	0.137	$T = 3.14\text{yr}^{\text{a}}$
4	0.163	0.037	0.227	0.650	0.106	
5	0.126	0.026	0.206	0.650	0.082	
6	0.100	0.028	0.280	0.650	0.065	
7	0.072	0.026	0.361	0.650	0.047	
8	0.046	0.015	0.326	0.650	0.030	
9	0.031	0.013	0.419	0.650	0.020	
10	0.018			0.650	0.012	
				6.500	$R_o = \overline{1.013^{\text{b}}}$	

After Fleming, 1988.
[a] T = generation time in years.
[b] R_o = reproductive rate.

expected to suffer approximately 50 percent mortality in their first year of life (d_x). Age-specific mortality is designated by q. A five-year-old bat, for example, has a 20.6 percent chance of dying in its fifth year of life. Life expectancy (e_x) indicates the number of years a bat of a given age (x) is expected to live. A newborn, for example, is expected to live 2.56 years. Since this is a life table for females, the fecundity is also represented by m_x. From these data, it is possible to calculate the expected number of offspring a single female would produce in her lifetime (R_o). In the case of *Carollia,* a female will replace herself with a single daughter over her lifetime. For comparison, the net reproductive rate (R_o) of the African buffalo *(Syncerus caffer)* is 1.63 (Sinclair, 1977).

In mammals, the contribution made to a population by reproduction clearly depends on a variety of factors and is seldom constant within a species from year to year. As mentioned, variation in litter size, number of litters, and length of breeding season, age at which young animals breed, and survival rates of young are all important variables. In addition, the litter size and the percentage of females that become pregnant change with the age distribution of a species; the age composition of a population may therefore have a marked effect on its reproductive performance.

CONTROL OF POPULATION GROWTH

Under ideal conditions, with unlimited resources (no competition) and no predation, populations exhibit exponential growth leading to unimaginable numbers. In the natural world, however, conditions are never ideal for very long, as evidenced by the Isle Royale moose populations. The abundance of an animal species at a given time and a given locality depends on the carrying capacity of the habitat and the relationship between the rate at which the animals are added to the population (by reproduction or immigration) and the rate at which they are lost from the population (by death or emigration). Mammalian populations are in a state of dynamic equilibrium and tend to be stabilized within certain density limits by such interacting processes as competition, reproduction, predation, dispersal, and disease. These processes are density-dependent; that is, they change in relation to changes in population density. For example, the

numbers of rodents killed by predators usually increases as rodent density increases because of more frequent predator-rodent encounters. Disease can spread more rapidly in a dense population because increased crowding favors rapid disease transmission.

Factors that control population size irrespective of the population's density are referred to as density-independent factors. These factors, such as weather or fires, affect a population in ways and at rates unrelated to population size. For example, a series of large fires burned out of control in Yellowstone National Park in 1988, destroying 1.4 million acres of habitat. Likewise, the catastrophic eruption of the Mount Saint Helens volcano in Oregon in 1980 devastated an area of 600 square kilometers. Both of these natural disasters acted to reduce animal populations, regardless of density. Flooding and heavy storms are also known to result in significant mortality among mammals. A 70 percent decrease in the combined population of golden mice *(Ochrotomys nuttalli)* and cotton mice *(Peromyscus gossypinus)* was caused by a flood of three weeks' duration in eastern Texas (McCarley, 1959). In the Chihuahuan Desert of Arizona, a population of the keystone rodent species *Dipodomys spectabilis* abruptly declined to near extinction in the winter of 1983 to 1984. Heavy rains associated with tropical storm Octave seemingly damaged this kangaroo rat's seed stores and may have caused contamination of the seeds by toxic fungi (Valone et al., 1995).

FACTORS LIMITING POPULATIONS

Density-dependent factors depress the birth rate and/or increase mortality at high population densities. Competition for increasingly limited resources at high population densities and high levels of predation or parasitism are thought to be the most important density-dependent factors controlling population size.

Although predation is clearly a source of mortality among mammals, our understanding of the ability of predators to control or influence densities of prey species or to regulate population cycles remains incomplete. The degrees of impact that predators have on prey were summed up by Pearson (1971): "The effectiveness of predation varies from the relatively ineffective predation of rats on man, in which rats are able occasionally to kill infants or incapacitated adults, through the mink-muskrat system described by Errington (1967), in which mink take a significant proportion of homeless and stressed muskrats, to the almost total effectiveness of carnivore predation on *Microtus* until almost the last one has been killed." A predator-prey relationship must obviously have some stability; as indicated by Lack (1966), "only those predatory species which have not exterminated their prey survive today, hence we observe in nature only those systems which have proved sufficiently stable to persist, and many others were presumably terminated in the past by extinction."

Although the effectiveness of even a single species of predator seems to vary according to situation, some general predator-prey relationships that apply to mammals as well as to other animals can be recognized. The observed responses of a predator to changes in the density of a prey species indicate that predation is influenced by prey density and detectability. The number of a preferred prey taken by a carnivore increases as the density of the prey increases, because the greater the number of prey animals per unit area, the greater the opportunity for predators to encounter and capture them. This is a functional response on the part of the predator (Holling, 1959, 1961). Errington (1937) noted such a response in the predators of muskrats and suggested that the intensity of predation is a function of prey population levels. There may also be a numerical response, involving an increase in the predator density with a rise in the prey population. The numerical response may be the result of immigration of predators, as in the case of the pomarine jaeger responding to lemming abundance (Table 24-7), or it may be due to increased breeding success.

The relative abundance of predators and their prey have often been regarded as being in dynamic equilibrium. The degree to which a balance between prey population and predator population is reached and the extent to which the relationship is dynamic, vary widely in situations involving mammals. These factors depend in part on the ratio of predator density to prey density, the relative sizes of the predator and the prey, the ease with which prey can be captured, and the degree to which the prey populations are cyclic. Studies by Mech (1966) on Isle Royale in Lake Superior during the period from 1959 to 1961 indicated that the ratio of moose to wolves was roughly 30 to 1. Twenty wolves were supported by approximately 600 moose, the wolves' primary food. The weight differential between an adult moose and an adult wolf was 14 to 1 (450 kilograms to 33 kilograms),

TABLE 24-7 Densities of Breeding Pomarine Jaegers *(Stercorarius pomarinus)* and Nesting Success Near Point Barrow, Alaska

Year	Spring *Lemmus* density (no./acre)	# Pairs of jaegers	Census area (mi²)	Density (pairs/mi²)	Maximum density (pairs/mi²)	Breeding success (% of eggs)
1952	15–20	34	9	3.8	5–6	30–35
1953	70–80	128	7	18.3	25–26	20–25
1954	<1	0	—	—	—	—
1955	1–5	2	15±	0.13	—	0
1956	40–50	114	6	19.0	22–23	4
1957	<1	0	—	—	—	—
1958	<1	0	—	—	—	—
1959	1–5	3	15±	0.20	—	0
1960	70–80	118	5.75	20.5	25	55

After Maher, 1970.
Note the correlation between high densities of lemmings *(Lemmus sibiricus)* and high populations of nesting jaegers.

and the wolves had difficulty killing moose. A strongly contrasting situation was studied in California by Pearson (1971), who found that the ratio of voles to predators varied from 72 to 1 in 1962, during a period of low vole populations, to 5410 to 1 during a peak in vole numbers. In this case, the prey was a cyclic vole (*M. californicus*) with an adult weight of roughly 45 grams. The predators — feral cats, raccoons, gray foxes, and skunks — averaged about 2.25 kilograms in weight, yielding a rough estimate of prey-to-predator weight ratio of 0.02 to 1.

These examples are based on two very different patterns of predator-prey interaction. The wolf-moose interaction resulted in relatively stable predator and prey densities, whereas the situation involving the California vole was one of great instability. Because of the difficulty wolves have in bringing down moose, the pressure they exert on the moose population is highly selective in that primarily young or old animals are taken; adult moose in the prime of their reproductive life are not killed (Mech, 1966). The predators of the vole, on the other hand, show a high preference for this prey and find it easy to catch. Their kill includes voles of all ages, and the predator can kill almost every vole during times of vole scarcity (Pearson, 1966). As suggested by the above examples, predator-prey interactions are complex; so many variables are involved that few generalizations relating to such interactions can be universally applied.

Some predator-prey interactions involve removal of vulnerable individuals from the prey population. These are individuals that, because of inexperience, old age, injury, or sickness, are readily captured, or they are animals forced by intraspecific competition for space into marginal habitats in which their vulnerability to predators is increased. Careful field studies by Hornocker (1970a, b) on mountain lions in Idaho showed that, of 53 lion-killed elk and 46 lion-killed deer, 75 percent of the elk and 62 percent of the deer were young (less than 1.5 years) or old (more than 8.5 years). These percentages of young and old animals were considerably greater than would be expected if the mountain lions had killed randomly. Errington (1943, 1946, 1963) found that in Iowa, as muskrat populations rose above a "threshold of security," a number of animals were forced into marginal habitats by intraspecific competition for space. This "vulnerable surplus" was preyed upon heavily by mink and red foxes, which made a marked functional response to this available food source. This pattern of predation has been called compensatory predation.

But can predators control densities of mammals, or are they simply killing individuals that would quickly be removed from the population by other means? The answer clearly depends on the particular predator-prey interaction being considered (Erlinge, 1987; Meserve et al., 1993; Krebs, 1996; Korpimäki and Krebs, 1996). Murie (1944) concluded that in Alaska, wolf predation on Dall sheep lambs was the most important factor limiting numbers of sheep, and on Isle Royale the wolves seemingly kept moose densities below the level at which food supply would be the limiting factor (Mech, 1966). In Idaho, however, predation by mountain lions had little impact on the populations of deer and elk, which were controlled instead by the winter food supply (Hornocker, 1970b).

Heavy predation on populations of small mammals has been shown to affect population levels (Korpimäki and Krebs, 1996). Data reflecting the severity of predation on lemmings demonstrate how important this source of mortality may be locally. On the coastal plain near Point Barrow, Alaska, during times of high lemming densities, the combined impact of several predators deals a staggering blow to lemming populations, and is postulated to be an important factor causing the population crash (Pitelka et al., 1955; Pitelka, 1957a). The combined kill of lemmings by the major predators amounted to at least 49 per acre during a cycle of abundance (Table 24-8). Studies in California by Pearson (1963, 1964, 1966) and by Fitzgerald (1977) on *Microtus* have demonstrated that predators preying upon these cyclic rodents are unable to control a rising prey population but that "carnivore predation during a crash and especially during the early stages of the subsequent population low determines to a large extent the amplitude and timing of the microtine cycle of abundance" (Pearson, 1971).

Competition may also be a factor influencing population size. Competition occurs when two or more individuals occupying the same habitat at the same time are utilizing some environmental resource in short supply. Niche segregation allows

TABLE 24-8 Impact by Predators on High Population of Lemmings *(Lemmus sibiricus)* Near Point Barrow, Alaska

Predator	Age class	Density (ind./mi²)	Daily food consumption (g/ind.)	Season's lemming consumption per acre			
				Per ind.	25 May to 15 July	16 July to 31 Aug	Total
Pomarine jaeger	Adult	38	250	338	10	21	31
	Young	38	200	167	—	—	—
Snowy owl	Adult	2	250	350	1.3	1.6	3
	Young	7	150	160	—	—	—
Least weasel		64	50	100	5	5	10
Glaucous gull		20	250	125	0.7	—	1
Waste					4	—	4
Totals					21	28	49

After Maher, 1970.
Data for least weasel from Thompson, 1955; for snowy owl from Watson, 1958.

species to avoid competition or reduce it to limits that sympatric species can bear. Competition can be between members of the same species or between members of different species. Competition may also be direct or indirect (Miller, 1969). Individuals competing indirectly may never come in contact — a chipmunk may eat so many cutworms during the day that it becomes unprofitable for a white-footed mouse to search for cutworms at night. This sort of mutual use of the same resource is an important type of competition. Individuals competing directly, on the other hand, are in direct confrontation, as when a pair of lions take over a freshly killed wildebeest from a group of spotted hyenas. Direct competition often involves the defense of space. Typically, competition has one of the following outcomes: one species becomes extinct and is replaced by the other (a process repeated countless times during the history of life on earth); one species emigrates to another area; or one or both species change with regard to their use of the disputed resource.

Competition, because it has long been recognized as a selective force leading to ecological, structural, and physiological changes in species, has been the focal point of much research. The ecological consequences of competition between closely related species, between sexes of the same

species, and between distantly related taxa have been studied, in some cases experimentally.

Competition between closely related species may lead to several different types of segregation (Fig. 24-27). The niches of Merriam's kangaroo rat *(Dipodomys merriami)* and the rock pocket mouse *(Chaetodipus intermedius),* which live on common ground in the Sonoran Desert in Arizona, are separated in several ways (Brown and Harney, 1993). Merriam's kangaroo rat eats large numbers of insects but few creosote bush seeds, reduces its summer activity but is active all winter, and forages mostly in the open. The rock pocket mouse forages near shrub cover, eats many creosote bush seeds, and is inactive through much of the winter. In such demanding environments as subalpine meadows, where the growing season is short and the winter snow deep, dietary segregation may be strongly developed. Although the foraging areas of four species of subalpine rodents in Colorado overlap broadly, the diets of no two species are the same. The altitudinal separation of three species of chipmunks in the Sierra Nevada of California is the result of different physiological tolerances and levels of dominance (Chappell, 1978). The lodgepole chipmunk *(Tamias speciosus),* which is sensitive to heat, is dominant over the yellow pine chipmunk *(T. amoenus)* and the least chipmunk *(T. minimus)*

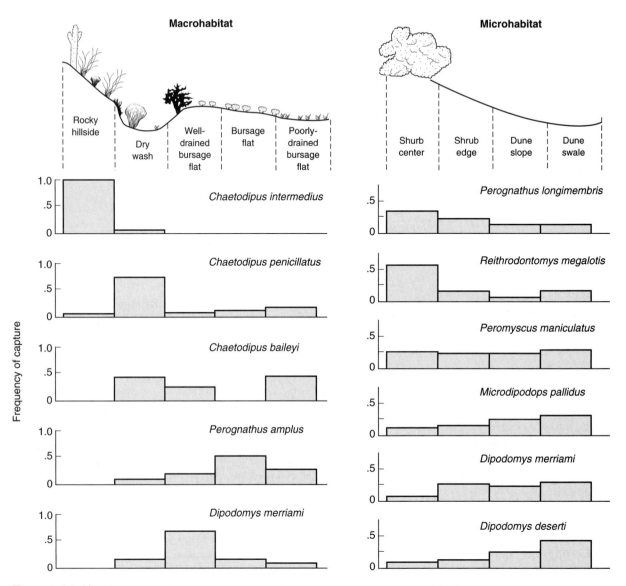

FIGURE 24-27   Examples of habitat and microhabitat segregation in a community of desert rodents. *Chaetodipus intermedius* are typically captured on the rocky hill sides; *Perognathus amplus* occupy the bursage flats habitats. Microhabitat utilization is illustrated on the right side of the figure. *(From Brown and Harney, 1993)*

and excludes these two species from its forest habitat. The yellow pine chipmunk tolerates dry, hot woodlands where the lodgepole chipmunk cannot live, and by being dominant over the least chipmunk forces this heat-adapted species to live in sagebrush habitats.

Competition between the sexes may be reduced in several ways. In heterogeneous habitats in Utah, female deer mice *(P. maniculatus)* are up to 15 per-

cent larger than males, are dominant over males, and occupy the moister microhabitats. Bowers and Smith (1979) hypothesized that females can occupy prime habitats because they are dominant over males, but other studies have indicated that this differential habitat use is primarily the result of reproductive constraints. Females choose habitats with safe nest sites and movements are restricted to foraging, whereas males improve their inclusive fitness by ranging widely and increasing their access to females (Morris, 1984; Scheibe, 1984; Scheibe and O'Farrell, 1995).

Competition between distantly related taxa is probably a major factor shaping community organization and may markedly influence population levels of competing organisms. Experimental studies suggest that rodents and ants compete for food and that their populations are food limited. The studies of Brown and his colleagues (Brown et al., 1979a, b; Davidson et al., 1980) have suggested the importance of competition between distantly related granivores (seed eaters) in structuring desert communities. Their study area in southern Arizona supported granivorous heteromyid rodents, which preferred large seeds, and harvester ants, which preferred small seeds. A series of experimental plots were manipulated by removing ants and leaving rodents, by removing rodents and leaving ants, or by removing both. By comparison with undisturbed control plots, numbers of ant colonies increased sharply (71 percent) in the plots with no rodents, and rodent biomass increased 29 percent in the antfree plots. Seed densities were roughly equal in the antfree, rodent-free, and undisturbed control plots, whereas they were four times higher in plots with neither ants nor rodents (Fig. 24-28). Plant density only doubled on the latter plots, indicating competition between seedlings. These results indicate that rodents and ants compete for seeds and affect seed and plant densities. They also show that plant densities are partly controlled by competition between plants and that the composition of the plant community may be influenced by the seed preferences of rodents and ants.

Further, rodents and ants interact indirectly. By concentrating on large-seed plants, rodents reduce the abundance of these plants and allow a reciprocal increase in the abundance of small-seed species, a shift that benefits ants. Clearly this system is dynamic. Even small year-to-year differences in rainfall, which are known to affect the germination and growth of desert annuals, would alter the food supply of the granivores and might favor either small-seed or large-seed plants. Such changes in food supply would be reflected by shifts in the populations of the granivores and would thus alter the intensity of seed harvesting and the crop of seeds that survive to germinate the following year.

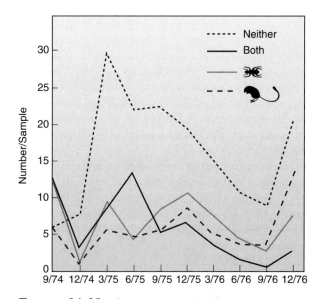

FIGURE 24-28 Comparisons of mean numbers of seeds in soil samples from plots where neither ants nor rodents were excluded, where both were excluded, where only ants were excluded, and where only rodents were excluded. Note that seed density increased markedly when granivores (seed eaters) were excluded. *(From Brown et al., ©1979, American Zoologist)*

Competition may also play a significant role in structuring microhabitats. The role of microhabitat in structuring desert rodent communities has been studied in experimental enclosures. Each of four species of heteromyid rodents occupied a different microhabitat, and each became more abundant when its particular microhabitat was increased. Each species shifted its microhabitat use when its competitors were manipulated experimentally. When its competitors were removed, a species used a greater breadth of habitat; with competitors present, it concentrated its activity in its preferred microhabitat (Price, 1978).

PARASITES AND DISEASE

Parasitism and disease are known to be significant causes of mortality among mammals (Elton, 1942) and may occasionally cause dramatic population crashes, as in the case of a die-off of prairie dogs *(Cynomys gunnisoni)* in Colorado caused by bubonic plague (Lechleitner et al., 1962). Talbot and Talbot (1963) estimated that 47 percent of the total mortality suffered by wildebeest was caused by diseases, of which rinderpest seemed most important. The blood parasite *Babesia* is a source of mortality among African lions. Disease in relation to population regulation, however, has been a difficult factor to assess (Chitty, 1954). Disease as the single cause of death may be relatively unimportant, but it may be important in contributing to the vulnerability of an animal to predation or to stressful environmental conditions.

Parasitism has been regarded periodically as an important cause of mortality, but careful observation indicates that otherwise healthy animals can often tolerate a moderately heavy parasite load. Heavy parasitism has been found to accompany a general health decline in rodents and rabbits during or following times of high density (Erickson, 1944; Batzli and Pitelka, 1971). Heavy infestations of botflies *(Cuterebra)* in white-footed mice are known to affect reproduction. Botfly larvae burrow beneath the skin and take up residence, breathing through an air hole to the outside. They mature within a month and exit via the air hole to pupate in the soil (Whitaker, 1968). These botflies often infest the inguinal and scrotal regions of white-footed mice, and up to 65 percent of the mouse population can be infested.

POPULATION CYCLES

Mammalian population cycles are among the most impressive biological phenomena (Fig. 24-29). Striking changes in density occur primarily in temperate, subarctic, and arctic areas. This difference is probably related to differences in species diversity in the two types of areas. High-latitude areas are characterized by biotic assemblages and food webs that are simpler than those of tropical areas. The typical boreal community has a limited biota and supports few species of vertebrates, but some species may, at least periodically, be remarkably abundant. The simplicity of the northern community is seemingly partly responsible for its instability, for, where so few kinds of organisms exist, any marked fluctuation in the density of one species seems to disrupt the entire community. In the complex tropical community, by contrast, the heterogeneity of the environment, the complexity of the food web, the diversity of carnivores, the intricate patterns of niche displacement and potential

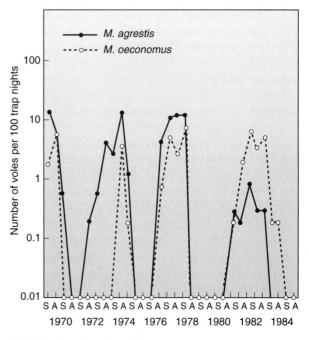

FIGURE 24-29 Cycles of population density in field voles *(Microtus agrestis)* and the root vole *(M. oeconomus)* from northern Finland. The abbreviations S and A represent spring and autumn, respectively. *(From Boonstra et al., 1998)*

competition, and the relatively small share of the energy resources available to any one species provide a buffer system against population outbreaks by any species.

CHARACTERISTICS OF POPULATION CYCLES

In areas where there are well-marked, multi-annual population cycles of voles, population peaks occur at intervals of two to five years. Some authors (Garsd and Howard, 1981) have objected to the term "cycle," but no one disputes that drastic fluctuations in vole populations do occur. The familiar term cycle is used here.

Not all species of voles are cyclic, and even within limited areas some populations may be more stable than others. In northern Alaska, population cycles are more pronounced in coastal areas, where only two species of vole occur, than in the foothills, where there are five species. Competitive interactions among the five species may reduce the probability of population fluctuations

(Pitelka, 1957a). Habitat heterogeneity has been regarded as favoring population stability (Lomnicki, 1980; Stenseth, 1980), but other factors may contribute. Whereas densities of California voles *(Microtus californicus)* were remarkably stable in a patchy environment in coastal northern California (Ostfeld and Klosterman, 1986; Krohne, 1982), three- to four-year population cycles occurred in a similarly heterogeneous habitat further inland (Cockburn and Lidicker, 1983). Stability at the coastal site was associated with reduced seasonality (more evenly distributed precipitation), an extended breeding season, and small litter sizes (Krohne, 1980). In northern Fennoscandia, the bank vole *(Clethrionomys glareolus)* is cyclic, but to the south, it is noncyclic (Bergstedt, 1965). As a final complication, there are even single populations of voles that have annual cycles at one time and multi-annual cycles at another (Getz et al., 1987).

On the coastal slopes of northern Alaska, oscillations in populations of lemmings *(Lemmus sibiricus)*, the chief herbivore, are characterized by a precipitous drop in density in the late summer and winter following a population peak, a period of one or two years of extremely low populations and localized distribution, and an upsurge in population in the winters following the low population, with peak numbers occurring early in the third or fourth summer (Pitelka, 1958). Short-term lemming cycles also occur in temperate and boreal parts of the Old World (Elton, 1942; Siivonen, 1954). At the peaks of cycles, voles reach amazing densities (Table 24-9). Longer term cycles, from 8 to 12 years, are known in populations of the varying hare *(Lepus americanus)* and the Canada lynx *(Lynx canadensis)*.

Population cycles in lemmings and voles can be divided into four phases: (1) increase, (2) peak, (3) decline, and (4) low density, with each phase being distinct to some extent. The increase phase is a time when densities rise markedly from one spring to the next. This increase may continue for several years and may be interrupted annually by short-term population declines (Hamilton, 1937), or, more typically, it may occur within one year, with extremely sharp increases over a period of three or four months (Fig. 24-30). Voles are flexible with regard to the duration of breeding, and during the increase phase, breeding begins early in the spring and often continues into the winter. At this phase, animals quickly reach sexual maturity. During an increase phase, some Norwegian lemmings *(Lemmus lemmus)* were found to be preg-

TABLE 24-9   Population Densities of Several Species of Arvicolines (*Microtus* and *Lemmus*)

Density (per acre)	Species	Region	Reference
1–20	M. pennsylvanicus	N. Minnesota	Beer et al., 1954
6–67	M. pennsylvanicus	New York	Townsend, 1935
3000	M. montanus	N.W. U.S.	Spencer, 1958a
200–4000	M. montanus	Oregon	Spencer, 1958b
25–81	M. californicus	N. California	Greenwald, 1957
425	M. californicus	N. California	Lidicker and Anderson, 1962
25–145	M. ochrogaster	Kansas	Martin, 1956
250–300	M. agrestis	England	Chitty and Chitty, 1962
1900	M. arvalis	France	Spitz, 1963
1004	M. guentheri	Israel	Bodenheimer, 1949
2400	M. spp.	U.S.S.R.	Hamilton, 1937
50–100	L. sibiricus	Alaska	Rausch, 1950
200–300	L. lemmus	Sweden	Curry-Lindahl, 1962

After Aumann, 1965.

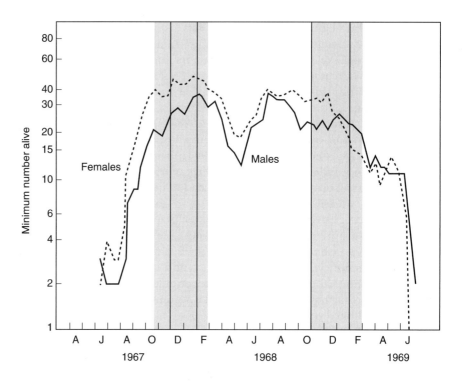

FIGURE 24-30 Changes in the densities of *Microtus pennsylvanicus* during a population cycle in southern Indiana. The shaded period is winter. *(After Gaines and Krebs, 1971)*

nant at 20 days of age (Koshkina and Kholansky, 1962). Koshkina (1965) found that the rate of sexual maturation was affected by population density, with early maturation being typical of an increasing population. Survival is generally high during the increase phase. There is more dispersal at this time than during the decline phase (Gaines et al., 1979). Dispersal during the increase phase (**presaturation dispersal**) occurs before the habitat is saturated (before it has reached its carrying capacity) and involves vigorous individuals with a good chance of survival (Lidicker, 1975).

The peak phase is a time of relatively little change in density. The population increase ceases, and the population may remain fairly stable for one year or may abruptly swing into a decline. During the peak phase, the breeding season in summer is typically brief and no winter breeding occurs. Animals attain sexual maturity late, and young born at peak times may not mature sufficiently to breed during their first summer. Growth rates are relatively high at this time, or survival is high, so larger sizes are achieved (Lidicker and Ostfeld, 1991). Mortality rates are relatively low during this phase, but dispersal is again high. Dispersal at this time, when the habitat is saturated with

voles, is termed **saturation dispersal.** This dispersal involves surplus animals (old, young, or social outcasts) that have little chance of survival (Lidicker, 1975).

The decline phase varies widely, from precipitous drops in density (population crashes) to uneven declines lasting one year or more. As in the peak phase, the decline phase is typified by brief summer breeding and no winter breeding, and animals reach sexual maturity late. Little dispersal occurs during the decline. Mortality rates are high within the relatively sedentary population.

The phase of low densities may last from one to three years, and annual shifts in abundance may occur (Krebs, 1966). At this time, the breeding season is short, animals reach sexual maturity late, mortality is high, and presaturation dispersal from refugia establishes new colonies.

POPULATION REGULATION IN SMALL MAMMALS

Although population cycles in mammals have been recognized and studied for many years, considerable controversy remains as to what factors control them. Many hypotheses have been advanced to explain the existence of mammalian population cycles. Stenseth and Ims (1993) reviewed the competing hypotheses and placed them into three categories: abiotic, biotic intrinsic, and biotic extrinsic. Abiotic factors, such as sunspots and weather conditions, are regarded by Korpimäki and Krebs (1996) as unlikely candidates for generating regular population cycles. Fuller (1967, 1969) and others, however, have considered weather to have important effects on vole populations. Lidicker (1988) described the specific effects of the length and intensity of the wet season in California's annual grasslands on vegetative productivity and demography of the California vole. Such demographic variables as litter size, growth rate, timing of reproduction, survival rate, and adult sex ratio are influenced by the responses of plants to patterns of precipitation. Vegetation in relation to vole cycles will be discussed later.

The self-regulation of populations by intrinsic biotic processes, such as genotypic, physiological, or behavioral changes occurring within a population, has been the focal point of much research. Stress is known to have pronounced effects on some mammals. Such pioneer works as those of Selye (1955) and Selye and Collip (1936), Green and his coworkers (1938, 1939), and Christian (1950) demonstrated increased adrenal activity and other physiological responses in some mammals living under conditions of high population densities. Christian proposed that these changes were important in controlling population cycles in mammals. Other studies, however, show that the physiological responses to crowding described by Christian are not consistently associated with high population levels under natural conditions (Clough, 1965; To and Tamarin, 1977). It seems, then, that although the stress syndrome occurs under some conditions, it is not important in controlling small-mammal population cycles (Lidicker, 1988; Krebs et al., 1992).

Behavioral and morphological changes during population cycles may also affect population densities. Paired experimental encounters between voles on neutral ground showed that the more aggressive animals were from peak populations (Krebs, 1970), and survival of large adults was greatest in such populations (Boonstra and Krebs, 1979). These populations live under conditions of small home ranges and large body size, and under

these conditions male voles are highly aggressive (Turner, 1971). Both male and female *Microtus longicaudus* in New Mexico were more aggressive at peak population levels than during population declines (Conley, 1971). The tendency of voles to disperse, on the other hand, is reduced at peak densities but is strong during the increase phase. Chitty (1958, 1960) hypothesized that selection for genetically determined behavioral features changed with changes in density. The occurrence of unusually large voles at times of peak densities (Chitty effect) has been widely accepted as part of the pattern of multi-annual cycles. These changes in size are regarded by some as being controlled by genetic polymorphism for body size and shifts in selection pressures during these cycles. Boonstra and Boag (1987), however, failed to find a genetic basis for this effect, and Lidicker and Ostfeld (1991) suggest that large individuals of the California vole are associated with high-quality food and high survival rates.

Social regulation of reproduction in voles by such responses as disruption of pregnancy may also influence demography and population cycles. The interruption of pregnancy after exposure to strange (unfamiliar) males has been reported for both Old World and New World voles, and Clulow et al. (1982) regarded this response as typical of voles in general. Heske (1987a,b), who studied voles under simulated natural conditions, found that when a strange male was introduced to an established male and pregnant female pair, or when the original male was replaced by a strange male, the female usually terminated pregnancy and conceived again. In the case of the introduction of the strange male, this intruder was nearly always defeated and banished by the established male. The female seemingly reacts to pheromones of the strange male. This estrus-induction syndrome is perhaps a ubiquitous feature of vole reproduction (Richmond and Stehn, 1976).

Regulation of populations by immunological dysfunction has been proposed repeatedly (Mihok et al., 1985; Folstad and Karter, 1992). The "immunocompetence selection hypothesis" (Lochmiller, 1996) derives from compilations of evidence from such fields as genetics, immunology, parasitology, and ecology. Lochmiller proposed that high population densities of herbivores deplete vegetation, leading to protein deficiency and malnutrition. In the decline phase of a cycle, such malnutrition is associated with high risk of infection by opportunistic parasites or disease. Through the various phases of the cycle selection affects heritable traits such as reproductive performance and immunity to pathogens. The sequence would thus involve selection favoring fecundity at the expense of lowered resistance to opportunistic parasites during the increase phase. Immune-response genes within the **major histocompatability complex** *(MHC)* are known to be associated with reproductive performance in a number of vertebrates. As an example, genes controlling body size and fertility are within the *MHC* of laboratory rats and mice. During the peak phase, the population might contain an unusually high frequency of animals with high reproductive ability but with low parasite and disease resistance. During the decline phase, then, normally nonpathogenic but opportunistic parasites could cause high mortality, mostly among the newborns. Neonate survival is widely regarded as the prime factor affecting reproductive success and subsequent population densities (Krebs and Myers, 1974; Loudan, 1985). Individuals stressed by infection may also be highly vulnerable to predation during the decline phase. Lochmiller's model focuses on the direct effects of the environment (mainly malnutrition) on survival rates as influenced by host immunity. A key assumption is that selection favoring rapid reproduction will be associated with high frequencies of genotypes confer-

ring reduced immunity to parasites and disease. This association has not been demonstrated for herbivores under natural conditions.

Studies of population dynamics of voles have led to intriguing speculation regarding rodent evolution. Evidence from studies of *Microtus* suggests that rates of evolution differ among populations with different demographic patterns. The frequencies of **allozyme** genotypes change with changes in vole density (Bowen, 1982; Gaines, 1985). In local populations of California voles, certain rare alleles were often lost during population crashes, indicating that the severity of selection declines during the optimal times that favor population growth. This permits rare genotypes to survive. Relative to stable populations, those given to pronounced cycles may have a greater probability of establishing "evolutionary innovation" in scattered founding populations that survive the crash (Lidicker, 1996). Some founding colonies are presumably established by presaturation dispersers with novel genotypes.

The quality and abundance of food has been studied in relation to population regulation. Lack (1954) regarded overexploitation of food as a major factor triggering changes in vole densities, and Pitelka (1958) suggested that reduced availability of food could cause poor reproductive success and population declines. Because voles are known to choose foods selectively, food might be expected to be limiting. But Schultz (1965, 1969) found that

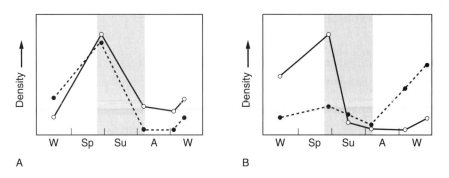

FIGURE 24-31   The effect of summer resources on densities of California voles. The summer (dry season) is indicated by the shaded region. (A) The delayed productivity effect. The dashed line represents a vole population exhibiting high winter densities and the solid line one with low winter density. (B) The damaged physiology effect. Here the dashed line represents a vole population that began the dry season with low densities and the solid line one with high densities. *(From Lidicker, 1988)*

neither the quantity nor the quality of tundra vegetation limited vole populations, and he observed greater vegetative productivity in places with voles than in experimental plots without them. Providing excess food at peak densities failed to slow the population crashes of voles and snowshoe hares (Krebs et al., 1986; Desy and Batzli, 1989). Furthermore, Batzli and Pitelka (1970, 1971) found that in one area the favorite food plant of California voles was ten times more abundant than in another area, yet the two populations of voles underwent similar declines at the same time.

In the annual grasslands of California, however, where rainfall and productivity of vegetation are strongly seasonal, plant productivity affects populations of the California vole. In these grasslands, breeding by this vole coincides with the growing season (winter), and floral composition, as it affects the availability of preferred food plants, influences mean litter size, survival rates, length of residency, and adult sex ratios (Brant, 1962; Lidicker, 1976; Cockburn and Lidicker, 1983; Heske, 1987). Vegetative productivity in the growing season interacts with vole density to determine survival during the dry season (summer). High densities of voles in the growing season reduce future plant

productivity (Ford and Pitelka, 1984), resulting in low dry-season vole survival and low densities the following wet season. Low winter densities, however, lead to good plant productivity, higher survival rates of voles the following summer, and higher vole densities the following winter (Fig. 24-31). Lidicker (1973) recognized the "damaged physiology effect": a population entering the dry season at high density has high mortality rates through this season, and survivors are physiologically damaged. Growth and reproduction are delayed for at least two months after the start of the following growing season. At this time, density continues to decline so that not only is the breeding season curtailed but it involves fewer breeders. Such plant-vole interactions clearly affect short-term population cycles and may influence multiannual cycles.

Some landscape-ecology research has dealt with "vegetation mosaics" as they influence population regulation. This term refers to the spatial arrangement of habitat patches of different quality. In heterogeneous environments, habitat patches often differ sharply in plant composition and cover (Fig. 24-32). For a species of vole, some patches may afford optimal habitat, while some may be

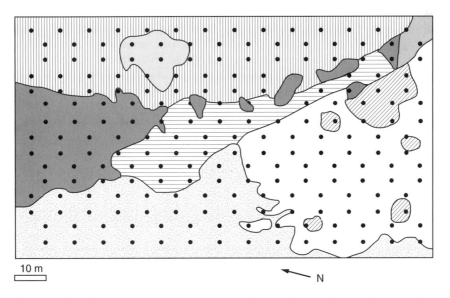

FIGURE 24-32 A vegetation mosaic near Bodega Bay, California. Patches of plants are indicated by dominant plant species: horizontal hatching, *Agrostis;* open, *Holcus;* stippled, *Bromus-Lolium;* vertical hatching, *Ammophila;* diagonal hatching, *Juncus;* medium gray, *Conium;* light gray, *Rhus;* dark gray, *Lupinus. (From Ostfeld and Klosterman, 1986)*

uninhabitable. High quality patches harbor "source" populations (where emigration exceeds immigration), whereas poor quality patches often function as "sinks," where immigration exceeds emigration. An example of this situation involves the California vole in an area with discrete vegetation patches. Relative to poor-quality (sparsely populated) patches, the high-quality (densely populated) patches were characterized by female-biased sex ratios, longer persistence of females, higher reproductive and emigration rates, and lower immigration rates (Ostfeld and Klosterman, 1986). Density of males was similar in all patches, but female density varied threefold. Laboratory experiments showed that vegetation from the densely populated patches supported rapid growth by voles, while voles lost weight on vegetation from sparsely populated patches. Patch quality seemingly depended primarily on food quality. Spacing behavior is thought to be a major factor regulating small mammal populations, and this study suggests that such behavior differs between the sexes.

Dispersal ranks with births and deaths as a major factor regulating vole populations. Presaturation dispersers leave their home areas before carrying capacity is reached. These animals seek to improve their inclusive fitness by avoiding inbreeding, improving access to high-quality food, and reducing pressure from predation and disease. Mobile voles, however, suffered significantly higher mortality from small mustelids than more sedentary individuals (Norrdahl and Korpimäki, 1998). The counterbalancing costs of dispersal include vulnerability to predation in marginal habitats, possible exclusion from established demes, and scarcity of prospective mates (Stenseth and Lidicker, 1992). Viewed broadly, dispersal tends to be a stabilizing influence on populations.

The complexity of herbivore cycles, their variability in time and space, and the weight of evidence pointing toward an array of regulating factors, has led many ecologists to adopt a multifactor perspective (Mihok et al., 1985; Stenseth, 1985; Hestbeck, 1986; Lidicker, 1988). The nature of the human mind is to seek simple, sweeping, universally applicable generalizations or single-factor answers. To many biologists, however, a multifactor approach is appealing because it accords with what we know of the real world, a world that is fascinating in large part because of its rich complexity.

A promising approach to understanding the regulation of herbivore cycles recognizes the importance of trophic (food chain) interactions and vegetation mosaics. Interactions among plants, herbivores, and predators, as affected by the distributions of habitat patches, are of central interest. The following examples illustrate research that takes this approach or that uses a multifactor perspective.

A landscape-ecology study of vole-predator dynamics in Finnish Lapland by Oksanen and Henttonen (1996) considered the varying importance of predation in regulating vole densities in a vegetation mosaic. These ecologists studied predator and prey populations from 1981 through 1987 in a landscape with such distinct communities as mesic spruce taiga, dry pine-spruce-birch woodlands, and timberline tundra. Densities of voles (three species of *Clethrionomys* and one of *Microtus*) were monitored in all seasons, and density estimates of small carnivores (mostly least weasels, *Mustela nivalis*, and stoats, *M. erminea*) were derived from data on their trackways in the snow. The two most widespread habitats had quite different patterns of vole density: in forb-rich, mesic taiga, densities fluctuated seasonally from 1981 to 1984, then crashed in 1985. In dry woodland, vole densities peaked in 1981 and declined to a crash in 1985 (Fig. 24-33). Stoat and weasel activity peaked in the winter of 1984 to 1985 and fell in late 1985. Both the stoat, a generalist predator weighing 125 to 300 grams, and the least weasel, a vole specialist weighing 45 to 120 grams, concentrated their hunting in vole-rich mesic taiga, but stoats ranged widely. In habitats adjacent to those where they concentrated their hunting, these mustelids preyed opportunistically on voles. This "spillover predation" may have markedly affected vole densities in marginal habitats.

Oksanen and Henttonen (1996) proposed first that for sustained vole cycles, periodic high densities of least weasels are vital. Larger, generalist predators, such as stoats, are capable vole predators primarily during snowfree times, whereas the least weasel (which is not much bigger than its prey) can hunt effectively beneath the surface of the snow. A population of least weasels that has reached high densities (for a carnivore) in response to high vole numbers can therefore put

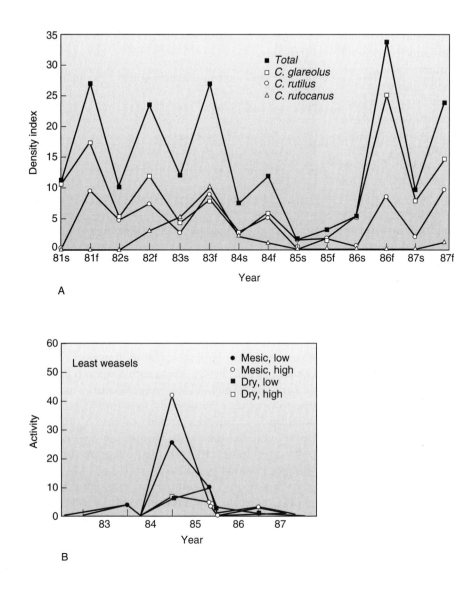

FIGURE 24-33 (A) Densities for three species of voles *(Clethrionomys glareolus, C. rutilus,* and *C. rufocanus)* in Finnish Lapland from 1981–1988. (B) Activity of least weasels in moist taiga and dry woodland in Finnish Lapland during 1982–1987. *(From Oksansen and Henttonen, ©1996, Munksgaard International Publishers, Ltd., Copenhagen, Denmark)*

heavy pressure on a declining vole population throughout the long boreal winter. The second proposal is that seasonal and irregular fluctuations in vole density result from predation by generalist predators, which are effective in snowfree times. The third proposal is that in barren tundra, where low plant productivity cannot support high vole densities, population crashes occur independent of predation. In this habitat, periodic depletion of food resources, rather than predation, regulates vole populations.

In a case involving snowshoe hares rather than voles, interactions among three trophic levels (plants, herbivores, predators) seem to regulate

herbivore densities. Keith (1981, 1990) postulated that snowshoe hare cycles were controlled by winter food shortages and predation. This hypothesis was tested during a ten-year study near Kluane Lake in the Yukon (Gilbert and Boutin, 1991; Hik, 1995; Krebs et al., 1995). In their "predation-risk hypothesis" these authors proposed that in a heterogeneous environment, food is never uniformly in short supply but that hares are forced by risk of predation to avoid the productive open areas in favor of the closed areas with abundant cover but poor food. As predators become more abundant and predation becomes more intense during the peak and decline phases of the cycle, the hares become progressively more restricted to safe, food-poor areas. The hares lose weight, and fecundity is low. During the decline phase of the cycle, snowshoe hares are chronically stressed — as indicated by high levels of free cortisol, reduced testosterone response, and overall poor body condition — by elevated predation risk, resulting in a marked deterioration in reproduction (Boonstra et al., 1998). Mortality from predation and the stress induced by predation risk cause the decline phase of the cycle and is regarded as the main driving force of this system. But, because the effects of the nutritional quality of the plants on the condition and reproductive performance of the hares are also vital factors, interactions among trophic levels, rather than predation alone, seem to regulate this cycle.

As a conceptual aid to understanding landscape influences on vole populations within patches, Lidicker (1985) used the ratio of optimal to marginal patch area *(ROMPA)*. This ratio indicates the proportion of the landscape that is optimal for a given species. The expectation was that this ratio would be useful in analyzing and predicting multiannual cycles. With little optimal habitat (low *ROMPA*) most of the landscape would serve as a sink, where survival would be brief, and densities in the source patch would be fairly stable. With much optimal habitat (high *ROMPA*), densities would be stable or undergo annual cycles. At intermediate values of *ROMPA,* however, multiannual cycles would be expected. In this case, the amount of suboptimal habitat would be large enough not to be filled with voles until after one or two breeding seasons. But when it was full of voles, densities would increase rapidly, dispersal would be frustrated, and the population would crash. Data on populations of grass-dwelling *M. arvalis* from eight sites in France support these predictions (Fig. 24-34). At sites where less than 5 percent of the landscape was grassland (low *ROMPA*), vole densities were low. At sites with more vole habitat (5 to 50 percent), multi-annual cycles were observed, and annual cycles occurred only at sites

FIGURE 24-34   Photo of a *Microtus arvalis.* *(M. Andera)*

with much vole habitat (greater than 50 percent) (Delattre et al., 1992). Areas with high percentages of forested land had low vole densities, perhaps due to predation by several forest-dwelling but opportunistic carnivores. Details of the *ROMPA* hypothesis aside, the ratio of optimal to marginal patch area is clearly a landscape feature affecting vole population regulation.

In his review of population ecology, Lidicker (1994) recognized several promising, interrelated themes that provide an overview of current thinking among ecologists. Spatial structuring of herbivore populations, as influenced by habitat heterogeneity, leads to genetic and social fragmentation of small mammal populations. Research on population cycles, therefore, increasingly deals with landscape ecology and vegetation mosaics, as does work in the vital discipline of conservation ecology (Soulé, 1986; Fiedler and Jain, 1992). The importance of varying degrees of isolation of **demes** relative to shifts in physiological, physical, and behavioral characteristics of herbivores is an area of great interest.

WEB SITES

A comprehensive list of ecology web sites

http://pbil.univ-lyon1.fr/Ecology/Ecology-WWW.html

The Ecological Society of America home page

http://esa.sdsc.edu/

United States Geologic Survey Biological Resources web site

http://www.nbs.gov/

United States Environmental Protection Agency web site

http://www.epa.gov/

The Association of Ecosystem Research Centers

http://culter.colorado.edu:1030/~aerc/

The Ecological Society of Australia home page

http://life.csu.edu.au/esa/

The Global Change Master Directory of information about climate change

http://gcmd.gsfc.nasa.gov/

Kennedy Space Center's ecology resource page including satellite images

http://atlas.ksc.nasa.gov/env.html

Organization for Tropical Studies web site

http://www.ots.ac.cr/

25 ZOOGEOGRAPHY

One of the most familiar kinds of biological information concerns **biogeography,** the study of the geographical distribution of living things. Children learn that lions and zebras live in Africa and not in North America, and that rain forests grow in the tropics and not in the arctic. This type of knowledge of the presence or absence of various kinds of animals in different parts of the world is the substance of **zoogeography,** the study of animal distribution and a subdiscipline of biogeography.

Considerations of zoogeography include two major approaches. The first is descriptive and static and seeks to delineate the distributions of living species. The second approach is ecological and his-

torical and attempts to explain the observed distributions. In either case, information is usually gained by field work and careful observation.

In the first approach, scientists have traditionally produced basic local or regional distribution maps and field guides. Regional guides and field guides are fundamentally important as a basis for education and research, as well as stimulating popular interest so crucial to conservation. High-quality field guides to mammals in many parts of the world have been available for decades (Hickman, 1981), and new or improved editions continue to appear regularly (for example, Corbet and Hill, 1992; Flannery, 1995; Gurung and Singh, 1996; Kingdon, 1997), but a few countries remain poorly known even in this regard, and field guides in their native language are unavailable to many local peoples.

Of course, species' distributions are not static, and the maps often found in guides are not intended to be immutable. As the **extirpation** of mammals from their potential or natural ranges continues, mappers frequently provide maps showing "historical" or "former range" and "present range" maps, while others study the dynamics of species' range boundaries. Studies of "range collapse," the dramatic shrinkage in geographic distribution that accompanies a species' endangerment and disappearance, have implications for the conservation and potential reintroduction of the species (Lomolino and Channell, 1995; Channell, 1998).

The second, historical and ecological, approach often involves syntheses based on diverse lines of evidence. **Historical** and **ecological zoogeography** have traditionally been separate. Historical biogeographers attempt to explain how things came to be via originations and extinctions over time. Those scientists studying what Udvardy (1969) called "dynamic zoogeography" ask the question: How, when, and from where did animals reach the areas they now occupy? Virtually every fauna consists of animals that reached the area at different times, from different regions, and by different means.

Our knowledge of the complex history of a fauna basically depends on the completeness of the worldwide fossil record and on our understanding of the geologic history and paleoecology of the major land masses. Ecological biogeographers investigate present relationships between organism and environment to explain geographic distribution. But the effects of ecology and history are not always distinguishable (Endler, 1982). Current biogeographic inquiry uses a synthetic perspective and eclectic approach (Anderson and Patterson, 1994; Riddle, 1995a; Brown et al., 1996). Increasingly, our knowledge improves as ecological and historical biogeographic data are integrated with new insights from fields such as molecular phylogeny. The result, sometimes called **phylogeography,** is typified by studies such as those of Riddle (1995b), Riddle et al. (1990), Hafner and Sullivan (1995), Hedges (1996), and Kay et al. (1997a).

An example of a phylogeographic study is that by Mustrangi and Patton (1997) of slender mouse opossums of the genus *Marmosops* in Brazil. These tiny, arboreal and terrestrial didelphids are found mainly in rain forests in northern South America. Most species inhabit the Amazon Basin northward to Central America, but a second, isolated group occupies the Atlantic rain forest in the coastal highlands of southeastern Brazil.

The two groups are separated by a large area of dry habitats uninhabitable by *Marmosops.* Fossils of related mouse opossums are known in the middle Miocene (Goin, 1997). Mustrangi and Patton investigated morphological and molecular variation in the eastern group, which was considered to contain just one species, *Marmosops incanus.* They discovered that, actually, two morphologically similar species, *M. incanus* and *M. paulensis,* are present in the Atlantic forest. The two species are very different from one another in the sequence of the mitochondrial cytochrome b gene, indicating a relatively low phylogenetic relationship where a close sister relationship would have been predicted. Closer investigation revealed that there were also subtle diagnostic morphological differences and differences in the habitat selected by each of the two species. One tended to live in relatively wetter, lower lying forests and slightly drier forests inland of the coastal mountains, while the other lived only at the south end of the Atlantic forest in higher montane forests. The authors noted similar geographic distributional patterns in a murid rodent and a bird. They also observed great genetic divergence between the Atlantic forest *Marmosops* species and the Amazonian *Marmosops* species, potentially dating to the Miocene, about the time of geological uplift of the coastal highlands.

Mustrangi and Patton concluded that, although further phylogenetic data are needed from the northern *Marmosops* group, the concordance of several factors (the common distribution of the opossums, rodent, and bird; the geological history of the coastal region; the known paleontological record; and the molecular sequence data) collectively hints at a probable Miocene divergence for the two Atlantic forest *Marmosops* species and indicates a common biogeographic history for the Atlantic forest biome. They predicted that other species of organisms will show a similar biogeographic pattern and that twice the known number of species exist in the genus *Marmosops*. The Atlantic rain forest is already known to have a large number of unique species of plants and animals, including new primates discovered in the last decade. The implication that there is an even greater diversity of unknown species to be discovered also has implications for conservation; the Atlantic rain forest faces imminent disappearance (Fonseca, 1985).

On a simple level, mammals occupy all continents, from far beyond the Arctic Circle in the north to the southernmost parts of the continents and large islands in the south. Antarctica presently has no land mammals, although it did in the distant past; today only seals and whales frequent its icebound margins. In the New World, the northernmost lands — the northern coasts of Greenland and of Ellesmere Island — are inhabited by the arctic hare *(Lepus arcticus)*, collared lemming *(Dicrostonyx torquatus)*, wolf *(Canis lupus)*, arctic fox *(Alopex lagopus)*, polar bear *(Ursus maritimus)*, short-tail weasel *(Mustela erminea)*, caribou *(Rangifer tarandus)*, and musk ox *(Ovibos moschatus)*. A similar group of mammals, lacking the musk ox, lives on the northern coast of the Taymyr Peninsula in the Soviet Union, which is the northernmost coast

of Asia (Berg, 1950). The southernmost part of Africa has a rich mammalian fauna. Tasmania, the southernmost part of the Australian region, supports two monotremes, many marsupials, several native rodents, and several bats. On Tierra del Fuego, at the southern tip of South America, occur a bat, several rodents, a fox, otters, and a llama. The chiropteran family Vespertilionidae occurs almost everywhere there is land except in arctic areas. The family Muridae is native to all continents except Antarctica. The families Leporidae, Sciuridae, Canidae, Mustelidae, and Felidae are native to all continents except Antarctica and Australia. (Leporids, canids, and felids were recently introduced to Australia.) All oceans, and all seas connected to the oceans, are inhabited by cetaceans; odontocetes also live in some large rivers and lakes.

One long-recognized phenomenon and popular topic of study in the geographical distribution of life on earth is the increase in the number of species of living things as one moves from the poles to the equator. The arctic regions are relatively poor in species and higher taxa as well as in morphological variety ("evolutionary novelty"; Jablonski, 1993), whereas the tropics are extraordinarily rich, both on land and in the seas. Given the harsher conditions for existence in arctic regions, it is perhaps not surprising that more species should exist in equatorial regions. Yet the earth has not always been so climatically polarized as it is at present. Glacial periods have taken place numerous times throughout earth history (notably in the late Precambrian, Ordovician-Silurian, Pennsylvanian-Permian, and Quaternary). As discussed later in this chapter, during much of the reign of mammals, global climates were uniformly warm and humid from poles to equator in the Cretaceous and early Tertiary, but a long, slow, general climatic cooling has taken place throughout the rest of the Cenozoic.

Explanations of the latitudinal diversity gradient are numerous and hotly debated (Brown and Lomolino, 1998). Researchers have suggested that the tropics (1) have higher rates of evolutionary origination (speciation) and act as a "diversity pump" (Darlington, 1957; Terborgh, 1973; Rohde, 1992); (2) have lower rates of extinction and thus tend to accumulate more species (Matthew, 1915; Stebbins, 1974); or (3) a combination of the two (Rosenzweig, 1992, 1995).

Students of mammalian geography, too, have frequently examined this high- to low-latitude increase in taxonomic and morphological diversity, especially in North America (Fig. 25-1; Simpson, 1964; Wilson, 1974; McCoy and Connor, 1980; Willig and Selcer, 1989; Pagel et al., 1991; Rosenzweig, 1992; Kerr and Packer, 1997; Shepherd, 1998). The phenomenon is less well examined for mammals on other continents, although Willig and Selcer (1989; for bats only), Willig and Sandlin (1989; for bats only), and Kaufman (1995) included both North and South America; Galliari and Goin (1993) mapped the phenomenon in Argentina. For marine mammals (baleen whales and

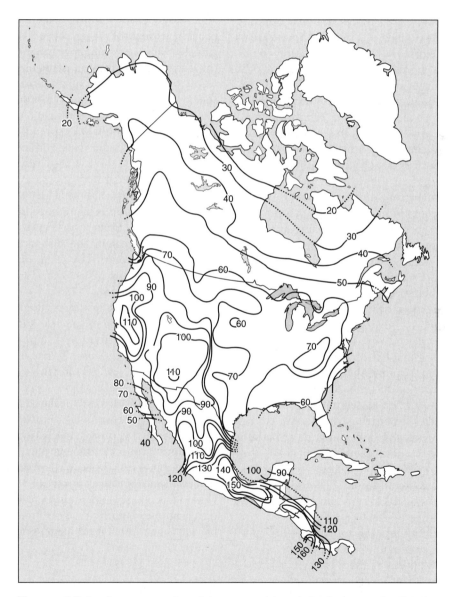

FIGURE 25-1 A representation of the geographic variation in the species density of terrestrial North American mammals. Contour lines enclose areas of equal species richness; numbers indicate the number of mammal species within the contour. Note that diversity increases toward the tropics. But at a given latitude, diversity is also greater in areas of greatest topographic relief, where habitats are the most diverse. *(After Simpson, 1964, from Brown and Lomolino, 1998)*

pinniped carnivorans), the pattern is reversed; greatest diversity occurs at high latitudes, as does the greatest abundance of their planktonic foods (Feldhamer et al., 1999).

DISPERSAL, VICARIANCE, AND FAUNAL INTERCHANGE

Dispersal occurs when an individual or a population moves from its place of origin to a new area. The ability to disperse is as basic as the ability to reproduce and is necessary to the survival of a species. A spacing of members of a population so that each individual can satisfy its environmental needs is critical to all organisms. Territoriality is one familiar means by which this spacing is ensured, and the young of territorial species usually establish home ranges largely separate from those of other individuals, including their parents. The pressures exerted by reproduction and the necessity for the spacing of individuals create a tendency of populations to occupy ever-increasing areas, to colonize unoccupied areas, and to repopulate areas where they were previously extirpated. The more widespread a species, the less likely it is to be forced into extinction by local mortality. As a result, natural selection has usually favored those species that have broad distributions. A high adaptive premium is placed on dispersal ability. Udvardy (1969) stated that "without evolved means of dispersal most animal populations would have succumbed, over a period of time, to the vicissitudes of the environment." A well-documented example of a single species colonizing new areas is the range expansion of the nine-banded armadillo from the Mexico–United States border along the Rio Grande, eastward to Florida and northward as far as Nebraska, between about 1850 and the present day (Humphrey, 1974; Taulman and Robbins, 1996).

The ability of a population to expand into new areas depends on its innate dispersal ability (which is greater, for example, in fliers than in burrowers), on the breadth of environmental conditions it can tolerate, and on the presence of barriers. Barriers may be ecological, for example, environmental conditions under which a species cannot survive, or physical, such as bodies of water, precipitous cliffs or mountains, or rough lava formations. If enough information were available, much

of the story of zoogeography could be told by considering the way in which animal dispersal patterns have been modified by the location, effectiveness, and longevity of barriers.

VICARIANCE

Barriers may affect the distribution and speciation of animals in a passive sense by **vicariance.** Vicariance biogeography is primarily historical in its approach. Instead of citing active dispersal by an organism, vicariance biogeographers seek to explain the observed distribution patterns of species as the result of the splitting of an area occupied by a species. If a barrier such as a mountain range, shallow sea, or river arises within the distribution of a species that is initially widespread, the species' distribution can become discontinuous or restricted. As a result, gene flow is also restricted or terminated and the separated populations may undergo divergence and speciation. Thus, the phylogeny of a taxonomic group is sometimes influenced by vicariance events. Some workers use **cladistic biogeography** to explain the historical biogeography of a region, primarily by examining the phylogenetic relationships of its taxa (for example, Page and Lydeard, 1994).

Either vicariance or dispersal might result in the **disjunct** distribution of species. A disjunction exists in the present distribution of tapirs; three modern species *Tapirus pinchaque, T. terrestris,* and *T. bairdii* occur in the New World tropics, whereas the closely related Malay tapir *T. indicus* of tropical southeast Asia is distantly isolated from the others. Distinguishing between vicariance and dispersal as the cause of the observed distribution of a given taxonomic group can be complex. Indeed, the difference between the two explanations becomes blurred (and perhaps semantic) when evidence is viewed from a geological time scale, and testability of hypotheses is difficult. Both mechanisms probably act upon animal (and plant) populations (Stace, 1989).

MIGRATION AND FAUNAL INTERCHANGE

Certain regions have apparently been major centers of origin of mammalian groups. Many orders and families first appear in the fossil record in Eurasia (Beard, 1998), and North America seems also to have been the place of origin for several groups. The present mammalian faunas of regions such as Africa and South America are partly **allochthonous** (some members originated outside the area where they now occur), derived from mammalian migrations from northern continents, but largely from **autochthonous** evolution (having originated in those areas in which they now occur). Many families are **endemic,** or unique to certain continents and islands. (A taxon is endemic to an area if it lives nowhere else.) Despite uncertainty as to the place of origin of many mammalian groups (where a group first appears in the fossil record is generally taken as its place of origin), movements of mammals from place to place are in some cases well documented by the fossil record.

Simpson (1940) recognized several avenues of faunal interchange. The **corridor** is a pathway that offers relatively little resistance to mammalian migration and along which considerable faunal interchange would be expected to occur. Such a continuous corridor now exists across Eurasia; interchange of animals between Europe and Asia is highly probable and has apparently occurred frequently. A **filter route** allows passage of certain animals, but stops others. Selective filtering has occurred at times along **Beringia,** the land bridge that has periodically connected Siberia and Alaska. When this bridge was present late in the Pleistocene, as an example, conditions were such that only animals adapted to cold climates and tundra habitats could migrate between these two continents. Mountain ranges, deserts, waterways, tropical areas, or abrupt changes in habitat may also form filter routes. Such physiographic features may simultaneously act as corridors, barriers, or filters to different species. The third and most restrictive route is the **sweepstakes route.** This is a pathway that will probably not be crossed by large numbers of any given type of animal but may be followed by an occasional individual. Such a pathway is that between Africa and Madagascar. Dispersal via a sweepstakes route must occur by swimming or flying or by such uncertain means as rafting from one land mass to another on floating vegetation or debris. The probability that an animal will

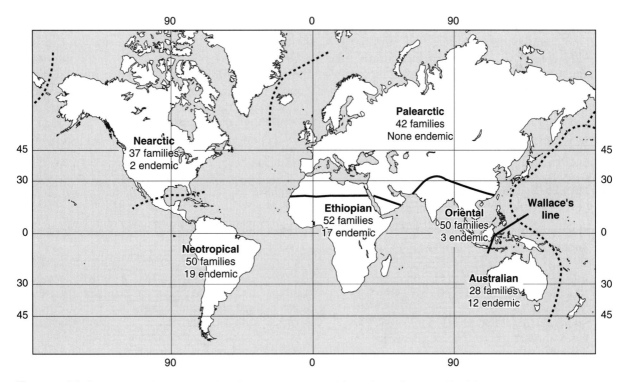

FIGURE 25-2   A map of the world showing the zoogeographic regions discussed in this chapter. The number of modern families of mammals in each region is given, as well as the number of families that are endemic to each.

follow a sweepstakes route is extremely low if the route is long, as, for example, from North America to Hawaii, but is increased if an animal is small and can cling to floating material, is aquatic, or can fly. (The only land mammals that reached Hawaii without the help of humans were bats.) Despite the unlikelihood of an animal's dispersal via a sweepstakes route, such dispersal was witnessed during recent (1998) hurricanes in the Caribbean that blew rafts of vegetation and debris carrying iguanas probably 300 kilometers between islands.

MAMMALS OF THE ZOOGEOGRAPHIC REGIONS

The zoogeographical realms shown in Fig. 25-2, which are the basis for the organization of this discussion, were proposed by Wallace (1876) and have been widely used in discussions of zoogeography. Biogeographic realms based on flowering plants (Good, 1974) tend to be concordant with these zoogeographic regions. Regional distribution patterns of mammals were recently summarized by Cole et al. (1994), with a view toward conservation priorities. Space does not allow us to discuss the important and complex historical, geologic, and biogeographic events in mammalian evolution that ultimately led to the regions discussed in the following sections. Instead, we have attempted to provide references as an entry into the literature pertinent to each region (for Antarctica, not included below, see Woodburne and Case, 1996).

PALEARCTIC REGION

The Palearctic Region includes much of the northern part of the Old World and is the largest terrestrial zoogeographic region. Included in this vast area are Europe, North Africa, Asia (except India, Pakistan, and southeastern Asia), and the Middle East. The climate is largely temperate, but contrasting conditions exist, from the intense heat of North Africa to the arctic cold of northern Siberia. Broad areas of coniferous forests, comparable in many ways to those of northern North America, are typical of much of the northern Palearctic Region, and deserts are widespread in the south. The Palearctic is at present separated from the Ethiopian Region by deserts, from the Oriental Region by the Himalayas, and from the Nearctic by the Bering Strait.

The Palearctic mammalian fauna includes 42 families and resembles most strongly the Oriental fauna, with which it shares 76 percent of its families (Table 25-1). Because of repeated faunal interchange across the land bridge that periodically spanned the Bering Strait, the Palearctic shares 46 percent of its mammalian families with the Nearctic. (Together, the Palearctic and Nearctic Regions or their shared mammal species are sometimes called "Holarctic.") Many genera, and a few species, of the families Soricidae, Vespertilionidae, Muridae, Canidae, Ursidae, Mustelidae, Felidae, and Cervidae occur in both regions. No family is endemic to the Palearctic Region. Historical zoogeography of Palearctic Region mammals is treated extensively in Beard and Dawson (1998).

TABLE 25-1   Comparison of the Mammals of the Faunal Regions

Region	Number of families	Number of endemic families	Percentage of families also found in					
			PA	NA	NT	ET	OR	AU
Palearctic (PA)	42	0	—	46	24	54	76	32
Nearctic (NA)	37	2	40	—	60	25	30	18
Neotropical (NT)	50	19	28	81	—	21	24	18
Ethiopian (ET)	52	17	67	35	22	—	66	32
Oriental (OR)	50	3	90	40	24	63	—	57
Australian (AU)	28	12	21	13	10	17	32	—

Based on data from Cole et al. (1994).

NEARCTIC REGION

The Nearctic Region includes nearly all of the New World north of the tropical sections of Mexico and contains habitats ranging from semitropical thorn forest to arctic tundra. The mammalian fauna includes 37 families, some of which are mostly tropical in distribution (for example, Emballonuridae, Phyllostomidae, and Tayassuidae), together with some primarily **boreal** families (typical of north-temperate forests: Dipodidae, Castoridae, and Ursidae). Only two Nearctic families (Aplodontidae and Antilocapridae) are endemic. The mammalian fauna of the Nearctic resembles most closely that of the Neotropical Region (Table 25-1). The transition between the Nearctic and Neotropical Regions was recently examined with respect to the distribution of bats by Ortega and Arita (1998). The Cenozoic history of mammals in North America is examined in detail in Woodburne (1987). Additional references for the Pleistocene include extensive coverage in Faunmap (Faunmap Working Group, 1994) and by Webb and Barnosky (1989).

NEOTROPICAL REGION

The Neotropical Region features great climatic and biotic diversity and includes all of the New World from tropical Mexico south. Much of the area is tropical or subtropical, and broad areas are covered with spectacular evergreen rain forest. Tropical savanna and grasslands occupy parts of the southern half of South America, and there are deserts in the south and along the western coast. The higher parts of the Andes support montane forests and alpine tundra. The South American part of the Neotropics was isolated from the rest of the world through most of the Cenozoic, but the Isthmus of Panama has provided a connection between South America and North America since the late Pliocene.

This region is second only to the Ethiopian Region in diversity of families of mammals. The Neotropical Region supports 50 families of mammals and has the largest number of endemic families (19). Especially characteristic of the Neotropical Region are marsupials, bats (including three endemic families), primates (two endemic families), xenarthrans (three endemic families), and hystricognath rodents (12 endemic or nearly en-

demic families). Two species of the genus *Lama* live in South America and are the only New World representatives of the family Camelidae. (Wild Old World camelids occur only in the Gobi Desert of Mongolia.) Tapirs are restricted to the Neotropical and Oriental Regions. The Neotropical mammalian fauna most strongly resembles that of the Nearctic, but it also shares one fourth of its families with the Oriental Region.

Historical biogeographic treatments of the Neotropics are many. Some of the most comprehensive and recent accounts include Simpson (1980), Pascual and Ortiz Jaureguizar (1990), Marshall and Sempere (1993), Woodburne and Case (1996), Kay et al. (1997b), and Hedges (1996). The Great American Interchange has been the focus of much research and is discussed in detail in Stehli and Webb (1985), Vrba (1992), Webb (1991), and Webb and Rancy (1996). A brief synopsis of the Great American Interchange is given later in this chapter.

ETHIOPIAN REGION

The Ethiopian Region includes Madagascar and Africa north to the Atlas Mountains, the Sahara, and the southern Arabian Peninsula. Deserts, tropical savannas, tropical forests, montane forests, and even alpine tundra are all represented, and the most extensive tropical savannas in the world occur in Africa.

The Ethiopian Region has the greatest number of mammalian families (52) of any faunal region and, next to the Neotropics, the greatest number of endemic families (17). The impressive array of ungulates that inhabits the savannas of Africa is unmatched elsewhere, and Africa is the last important stronghold of the families Equidae, Rhinocerotidae, Elephantidae, and Hippopotamidae. Although the only endemic artiodactylan family is Giraffidae, nearly all of the African genera of antelope (Bovidae) are endemic. The primitive lemuroid primates of Madagascar (5 families, all endemic to the island) and the diverse group of cercopithecid primates of Africa are especially typical of the region, and two of the five genera of great apes live only in Africa. Apart from South America, Africa is the only area with a fairly diverse hystricognath rodent fauna. Viverrid carnivores reach their greatest diversity in the Ethiopian Region, where most of the genera are endemic. The Ethiopian mammalian fauna most closely resembles those of the Palearctic and Oriental regions. Accounts of the historical biogeography of the Ethiopian Region include Maglio and Cooke (1978), Jolly et al. (1984); Kingdon (1989), and articles in Goldblatt (1993), especially Vrba (1993).

ORIENTAL REGION

Included in the Oriental Region are India, Indochina, southern China, the Malay Peninsula, the Philippine Islands, and the islands of Indonesia east to a line (**Wallace's Line**) between Borneo and Sulawesi and between Bali and Lombok. The area is dominated by tropical climates and once supported, before extensive land clearing and burning by humans, broad areas of tropical forests. Deserts occur in the Pakistan area. The Oriental Region is partly isolated from the Palearctic by deserts in the west and by the Himalayas to the north.

The mammalian fauna of the Oriental Region includes 50 families and resembles most strongly that of the Palearctic area, with which it shares 90 percent of its families of mammals. Many (63 percent) of the Oriental families of mammals also occur in the Ethiopian Region. The most distinctive elements of the Oriental mammalian fauna are all of tropical affinities. Five families of primates occur in this region. Three families of mammals — Tupaiidae (tree shrews), Cynocephalidae (flying lemurs), and Tarsiidae (tarsiers) — are endemic, and each occupies forested tropical areas. A few Oriental families occur elsewhere only in the Ethiopian Region (Manidae, Elephantidae, Rhinocerotidae, and Tragulidae). Some aspects of mammalian historical zoogeography in the Oriental Region are discussed by Corbet and Hill (1992) and Heaney (1986).

AUSTRALIAN REGION

The Australian Region includes Australia, Tasmania, New Guinea, Sulawesi, and many of the small islands of Indonesia (New Zealand and the Pacific area are not included). In the area are islands of various sizes and degrees of isolation. The island continent of Australia is connected with New

Guinea by a broad continental shelf under the Arafura Sea, but these land masses are presently separated by the Torres Strait, 160 kilometers wide. The northern part of the area, including New Guinea and parts of the eastern coast of Australia, are covered with tropical forest, but much of Australia is tropical savanna or desert. Some of the most arid deserts in the world occur in the interior of Australia.

The Australian Region is famous for its unusual mammalian fauna, and, to the popular imagination, Australia itself is an area supporting marsupials almost exclusively. Actually, Australia has 9 native terrestrial eutherian families (mostly bats) and 14 metatherian families. Over 50 percent of the extant native families of the Australian Region are marsupials, and 43 percent (the monotremes and marsupials) are endemic. The mammals of the Australian Region have their closest affinities with those of the Oriental Region. The historical biogeography of mammals of the Australian Region is vividly discussed by Archer et al. (1991), Woodburne and Case (1996), and Flannery (1995).

OCEANIC REGION

The oceans of the world compose the Oceanic Region. The oceans are much more homogeneous from basin to basin than are the continents because of their broad connections and the obvious potential for global dispersal of drifting and swimming organisms. In this region live the whales and porpoises, most of the seals, the sea lions and walruses, the sirenians, and the inhabitants of isolated oceanic islands (usually bats, introduced murid rodents, and other mammals associated with humans). New Zealand, included in the Oceanic Region, has two remaining native mammals; both are bats and one forms an endemic family (Mystacinidae; a second mystacinid species is recently extinct). Other large islands are included in the region with which their faunas have the most in common. Greenland, for instance, is included in the Nearctic Region, and Iceland is included in the Palearctic.

Marine mammals are usually excluded from lists of mammals that contribute to the continental zoogeographic regions. Nevertheless, one family each of sirenians (Trichechidae) and dolphins (Platanistidae) occur far upstream in rivers, as well as inhabiting brackish or salt water. Dugongs and manatees, the only living herbivorous marine mammals, primarily inhabit tropical and subtropical waters (although one species, extinct since about 1768, occurred in the Bering Sea) where they feed on sea grasses (angiosperms) and other plants that require much sunlight. Largely as a result of this diet, they usually remain in shallow waters close to coasts and islands where these plants are able to grow. Historical zoogeographic accounts of some marine mammals include Repenning et al. (1979), Muizon (1981, 1982), and Fordyce and Barnes (1994); see also Figure 13-9.

CONTINENTAL DRIFT, MAMMALIAN EVOLUTION, AND ZOOGEOGRAPHY

CONTINENTAL DRIFT

Ideas about continental drift were first developed early in the 20th century (Wegener, 1912, 1915, 1966), but an integrated theory of plate tectonics is relatively new. Until recent decades, teaching in geology and paleontology in North America was dominated by the view that the positions of the continents and the intervening oceans were fixed, that they have remained immutable back through the vast sweep of geologic time. Because they accepted these tenets, most North American paleontologists were forced to rely on often tenuous intercontinental land bridges or sweepstakes dispersal to account for intercontinental movements of terrestrial animals. Within the last 30 years, however, our geologic, paleontologic, and biogeographic perspective has been drastically transformed by convincing evidence in favor of the theory of continental drift.

The discovery by Colbert in 1969 of a Triassic fossil therapsid (*Lystrosaurus*) in Antarctica, put the capstone on the pyramid of evidence supporting continental drift. This nonaquatic dicynodont therapsid had previously been found in Triassic deposits on other southern continents; its distribution could be explained only by assuming that the continents had once been connected (Elliott et al., 1970). Wegener's wild theory was vindicated.

Today continental drift and its driving mechanism "plate tectonics" have been elaborated and supported by mountains of evidence and form a major unifying theory in geology that explains many other phenomena. The paleo-positions of the earth's continents and magnetic poles through much of geological time have been mapped in considerable detail (Smith et al., 1994).

During the early part of the 4.5 billion years of earth's history, the accreting masses that formed the early planet underwent a density separation. The densest materials sank to the core; the lightest materials where life evolved rose to the outer surface as the lithosphere (crust), hydrosphere (oceans), and atmosphere. The mostly granitic continents are the relatively lightest rocks forming the crust. Denser basalts form most of the ocean floor crust. In the last billion years or so, the crust became divided into a series of about 16 tectonic plates (Figure 25-3). Driven by convective undercurrents in the earth's mantle, these largely rigid plates move as molten magma wells up from the mantle, and new basaltic crust is added at spreading centers called midoceanic ridges (Figure 25-4). Riding atop the plates, the continents are passively carried along at rates of millimeters or centimeters per year. At the edge of some tectonic plates, opposite the spreading center, one plate may override another. The overriding plate is forced upward and its edge crushed, as evidenced by prominent geological faulting; in the process, ocean bottom sediments containing marine fossils and other rocks from low-lying areas may be pushed up to great elevations as mountains. The plate forced downward is "subducted" into the hot, dense mantle, where its materials (crustal basalts, ocean floor sediments, and water) are melted and recycled by rising up to the crust and erupting through it as volcanoes. Land masses have been rifted, or split apart, at spreading centers, and they have collided and united where adjacent plates converge. As crustal plates move, they buckle, bulge, and warp, causing epicontinental seas to advance onto low-lying areas and retreat from high areas. Entire ocean basins appear and disappear and alter oceanic circulation patterns, which in turn strongly affect climate. Likewise, major mountain chains of the world, formed in part by deformation of the earth's crust and volcanism, can alter atmospheric circulation and climate.

During the last half billion years, the early continents came together in the Paleozoic to form a temporary supercontinent called "Pangea" that

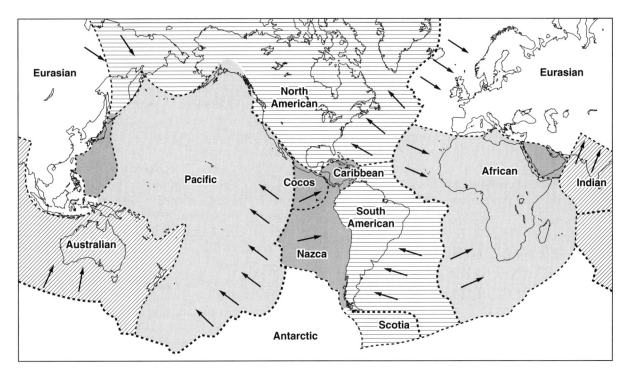

FIGURE 25-3 The present extent of the major tectonic plates and the direction of their movement. Plate movement is influenced by upwelling of molten rock from deep within the earth along the rift lines between some plates. *(Modified from Colbert, 1973)*

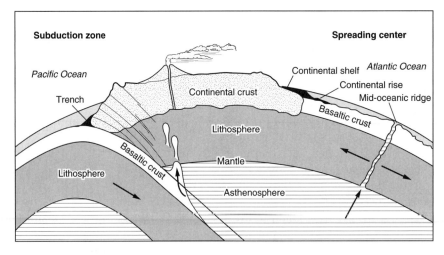

FIGURE 25-4 The dynamics of continental drift as demonstrated by a diagrammatic cross section through southern South America and the Nazca, South American, and African tectonic plates (not to scale). Rifting between two tectonic plates (African and South American) results from the welling up of molten rock from the depths, to form the Mid-Atlantic Ridge spreading center (right). The westward drift of the South American Plate, carrying the continental block, collides with the Nazca Plate beneath the Pacific Ocean. The Nazca Plate is forced down (subducted) into the mantle. The subduction zone is marked by the formation of a deep trench and the uplifting and volcanism of the Andes Mountains, with resultant frequent earthquakes. *(Modified from Colbert, 1973)*

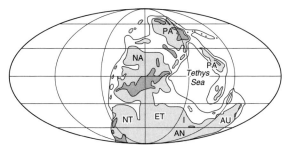

A. Late Carboniferous (306 million years ago)

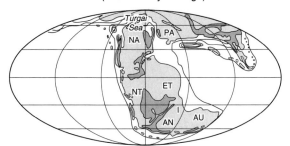

B. Early Jurassic (195 million years ago)

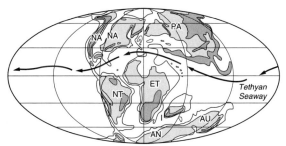

C. Early Late Cretaceous (94 million years ago)

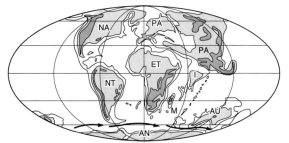

D. Middle Eocene (50 million years ago)

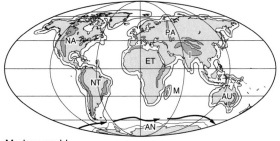

E. Modern world

Laurasia { PA = Palearctic
          NA = Nearctic

Gondwana { NT = Neotropical
           ET = Ethiopian
           I = Indian
           AN = Antarctic
           AU = Australian
           M = Madagascar

▨ Mountain ranges
▧ Lowlands
☐ Continental shelf

FIGURE 25-5 Paleogeographic maps of the earth's known land masses in selected geological time periods during the evolution of synapsids, including mammals. For ease of recognition, core areas of the continents are labeled according to the modern subcontinents or zoogeographic regions they ultimately become: in Laurasia, PA = Palearctic, NA = Nearctic; in Gondwana, NT = Neotropical, ET = Ethiopian, AU = Australian, AN = Antarctic, I = Indian, M = Madagascar. (A) The supercontinent Pangea, home to pelycosaurs, has formed from the aggregation of earlier Paleozoic continents. Note equatorial position of pre–North American highlands region, Tethys Sea in the East, and southern polar position of much of the land mass. (B) By the time morganucodontids and other early mammaliaforms exist, Tethys has widened and Pangea has drifted northward. (C) Circumequatorial ocean circulation through shallow seas may have helped to stabilize and equalize global climate. Dinosaurs exist from what is now northern Alaska to Antarctica, Gondwana breaks apart, angiosperms radiate, monotremes inhabit South America to Australia, and early metatherian and eutherian mammals begin their fundamental diversifications. Note broad epicontinental sea dividing western and eastern portions of North America. (D) Continued separation of former components of Gondwana closes Tethys Seaway and permits circum-Antarctic ocean circulation instead of circumequatorial. The switch is accompanied by the first Cenozoic appearance in Antarctica of ice, heralding a global general cooling trend lasting to present day. After this time, modern families of mammals rise to dominance — those in the southern hemisphere mostly in isolation. Mammals enter the oceanic realm for the first time (Sirenia and Cetacea). (E) The modern world, late in an ice age. Antarctica, former core of Gondwana and forested haven of diverse mammals, has become a deep freeze; amount of northern polar ice fluctuates. All land masses are dominated by a single, recently evolved species of primate. *(From Scotese after Brown and Lomolino, 1998)*

was surrounded by a single ocean, sometimes called "Panthalassa." As the plates continued to move, the continents began to separate again in the Mesozoic, at first into a mainly northern portion (sometimes called "Laurasia") and a southern portion (called "Gondwana") that were joined in the west but separated in the east by a sea called "Tethys" (Figure 25-5). Further fragmentation in the late Mesozoic resulted in modern continents that were essentially recognizable by the early Cenozoic.

To biologists, the dynamism of the earth is of tremendous importance. Just as continents have separated, collided, or drifted progressively farther apart, so too have terrestrial biotas been isolated or brought together, entire distribution patterns of marine biotas profoundly altered, and global ecological diversity shifted. To the evolutionary biologist, the movements of the earth's crust provide "the stage for all biological activity" (McKenna, 1972). It is obvious that, when one considers the biogeography of individual species or of entire bio-

tas, one must take into account plate tectonics and continental drift. The beauty of these concepts is that they often provide explanations for a diverse array of biogeographic patterns that have long appeared inexplicable.

As pointed out by Kurtén (1969), the fact that mammals evolved during a span of the earth's history when continents were moving apart may be a key to mammalian diversity. Kurtén believed that the greater diversity of mammals than of reptiles is a result of continental drift and that the main mammalian radiation occurred during the Cenozoic. Mammals evolved on several land masses under conditions of isolation or semi-isolation, whereas the less diverse reptiles developed before the continents had moved far apart and therefore developed under conditions allowing freer faunal interchange between evolving stocks. The idea of a post-Gondwanan diversification of mammals has been modified recently with better paleogeographic, paleontological, and molecular phyloge-

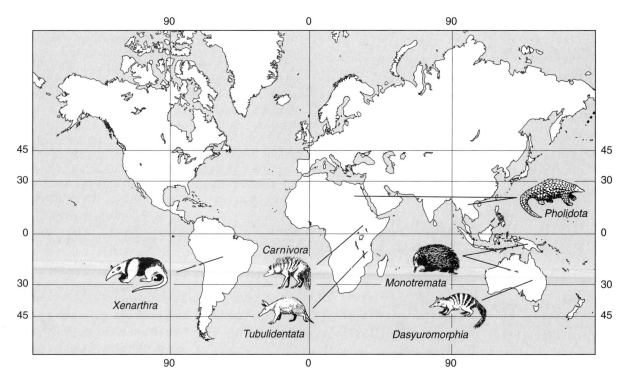

FIGURE 25-6 Members of at least six mammalian orders that occur in southern continents are adapted to eating ants and termites: Xenarthra (all members of the family Myrmecophagidae), Carnivora (the hyaenid *Proteles*), Tubulidentata (the aardvark *Orycteropus*), Monotremata (the echidnas *Tachyglossus* and *Zaglossus*), Dasyuromorphia (the numbat *Myrmecobius*), and Pholidota (the pangolin *Manis*).

netic evidence by Hedges et al. (1998). These authors, too, believe that the final fragmentation of Pangea was the mechanism responsible for the diversification of extant mammal orders. Based on molecular clock estimates of divergence timing, they push back the earliest divergences into the Cretaceous, 90 to 115 million years ago.

A striking feature of mammalian evolution has been duplication of functional and, to some extent, structural types in separate groups. Examples of such convergent evolution are abundant. Numerous unrelated rodents that live in deserts of different continents convergently evolved similar anatomical, physiological, and dietary lifestyles (Mares, 1993a, b). Perhaps more astounding is convergence among constituents of different higher taxa. Members of several orders specialize in eating ants and termites (Fig. 25-6). The orders Diprotodontia, Rodentia, Lagomorpha, Artiodactyla, and Perissodactyla all contain herbivorous, cursorial mammals that pursue basically similar modes of life. Small, terrestrial, insect-eating mammals have developed in at least seven orders (Didelphimorphia, Paucituberculata, Microbiotheria, Dasyuromorphia, Insectivora, Xenarthra, and Rodentia). The greatest duplication has occurred in southern land masses, which have been longer and more completely isolated than have the Nearctic and Palearctic areas. Mammalian diversity, then, may be as much a result of the progressive Mesozoic and Cenozoic separation of the continents as of the structural and functional adaptability of the mammals themselves.

CLIMATE AND MAMMALIAN DISTRIBUTION

The preceding chapters have mentioned the geologic periods of the Mesozoic and Cenozoic in relation to the evolutionary history of various mammalian families and orders. Cenozoic patterns of climatic change had a profound effect on the evolution, character, composition, and distribution of plant communities, and the evolutionary patterns and distributions of mammals, in turn, were influenced by these floral changes (Frakes et al., 1992). Although one cannot with assurance account for the myriad patterns of mammalian adaptation by recourse to climatic changes alone, there is no better point of beginning.

In the Cretaceous, climatic conditions were warm, humid, and stable worldwide. There was no pronounced latitudinal gradient in temperature, and the earth was probably icefree. Sea level was possibly the highest it has been in the last half billion years. Flora and fauna were remarkably uniform over much of the earth, with subtropical plants and animals existing at latitudes 70 degrees from the equator, although provincialism increased as Gondwana fragmented (Parrish, 1990; Taylor and Taylor, 1990; Krause et al., 1997; Hedges et al., 1996). Although angiosperms radiated rapidly during the late Cretaceous, they may have been restricted in their distribution to streamside areas, while ferns dominated other habitats, at least in the North American interior (Wing et al., 1993).

The uniformly subtropical climate of the Cretaceous continued through the Paleocene into the early part of the Eocene. In parts of North America, fossil root traces and soil profiles indicate that humid forests of the Cretaceous and Paleocene gave way to dry woodlands or wooded "savannas" or scrublands. Loss of the forest canopy of wind-pollinated conifers was followed by paleofloras dominated by wind-pollinated angiosperms (Wolfe, 1978). The early Eocene high-latitude forests and warm climates allowed forest-dwelling vertebrates to continue to dwell in arctic zones. For example, a fossil vertebrate fauna from Ellesmere and Axel Heiberg islands (the northernmost part of North America; paleolatitude about 75 degrees N) included catfish, bowfin, several kinds of turtles, crocodilians, and several kinds of mammals, including primitive primates (McKenna, 1980; Dawson et al., 1993).

However, by the middle Eocene (about 40 to 37 million years ago), a major rise in sea level, major changes in ocean circulation due in part to the breakup of Gondwana, and significant climatic deterioration began. Year-round ice might have begun to accumulate in Antarctica by this time. The northern continents also underwent a cooling and drying after the middle Eocene, resulting in changes in vegetation from warm subtropical forests to dry woodland by the latest Eocene. A diverse array of warmth-adapted organisms, both marine and terrestrial, underwent concomitant extinction in the late Eocene (Prothero, 1994a, b). The development of Antarctic ice and changes in ocean circulation during the middle Eocene to early Oligocene brought about increased seasonality and an overall cooling and drying to global paleo-climate. Grass pollen and grass fossils are absent from the Eocene record of North America during the early development of savannas, suggesting that Eocene-Oligocene drylands were dominated by low-growing woody scrub vegetation lacking grasses ("rangelands"), rather similar to modern, seasonally dry *Ephedra*-saltbush communities of the Great Basin or bluebush-saltbush communities of central Australia (Wolfe, 1994). In Europe, data from paleobotany and fossil vertebrates show no obvious Eocene drying trend, but such a trend is present by the Oligocene there. A drying trend is recorded in the middle Eocene fossil pollen record in Asia, where the Gobi Desert and parts of China were arid or subarid. No detailed record of Eocene paleoclimate has yet been developed in Africa, South America, or Antarctica, but Oligocene data indicate that the drying trend had occurred by then. For example, in South America, the beginning of Andean uplift changed vegetation from subtropical woodlands to scrubby, arid savanna woodlands; mammals that were adapted for grazing rather than browsing appeared earlier in South America than in other continents (Pascual and Ortiz Jaureguizar, 1990; MacFadden et al., 1996). In Australia, microphyllous and sclerophyllous plants greatly increased in importance in the Oligocene (Barker and Greenslade, 1982).

During at least the late Eocene, a land connection or near-connection still existed between South America and Antarctica, as evidenced by fossil mammals (microbiotheriid and polydolopid marsupials, a megatherioid sloth, a litoptern, and an astrapothere) on Seymour Island, Antarctic Peninsula, that are evolutionarily related to South American mammals (Bond et al., 1989; Carlini et al., 1990; Hooker, 1992; Vizcaino and Scillato-Yané, 1995; Goin and Carlini, 1995; Woodburne and Case, 1996; Flynn and Wyss, 1998). At this time, cool-temperate rain forests occupied the Antarctic Peninsula (Case, 1988).

Prior to the Oligocene, the Scotia Arc, an archipelago connecting the southern Andes and Antarctic Peninsula, diverted cool, high-latitude waters near Antarctica and allowed them to mix with warm, low-latitude waters. But in the Oligocene, tectonic movements of the Scotia Plate (see Figure 25-3) opened the Drake Passage between South America and Antarctica. This and the

widening separation of Australia and Antarctica allowed cold water circulation around Antarctica. The circum-Antarctic ocean circulation isolated Antarctica in the polar region and led to its more extensive glaciation and an attendant drop in global sea level (Zachos et al., 1992). However, a cool-temperate flora of low species diversity still grew on some parts of Antarctica throughout the Oligocene. Oligocene plant assemblages in Australia also reflect cool-temperate conditions, including plants indicative of greater seasonality; parts of the continent underwent increasing drying that gave a more open aspect to the forests. This sort of development of semiarid habitats probably continued or occurred in South America, too, based on the fossil record there of mammals adapted to grazing on coarse vegetation (Pascual and Ortiz Jaureguizar, 1990; Prothero and Berggren, 1992).

In the Northern Hemisphere, the Arctic region remained free of ice but the global cooling trend caused plant zones to shift such that the warm-, mesic-adapted ones narrowed and moved to lower latitudes and were replaced at high latitudes by cool-temperate forests. Increasing aridity in the rain shadow of the Rocky Mountains in central North America resulted in changes from the subtropical forests of the Eocene to dry woodland, wooded shrubland with gallery forest, and even open scrubland, or possibly grassland with gallery woodland, by the middle Oligocene. In Asia, Eocene forests were replaced with Oligocene woody savanna in Kazakhstan; in China, the deserts of the northwest were replaced by woody savanna.

Savanna woodlands continued to dominate North American floras in the Miocene, at least where fossil floras exist in the continental interior. The origins of such desert-adapted animals as kangaroo rats can be traced to the Miocene, when semideserts became widespread. After about 5 to 7 million years ago, grassland-steppe floras became widespread in North America. Grazing hoofed mammals reached their peak in North America and on most other continents in the Miocene (this occurred in the Oligocene in South America).

Global cooling continued through the Pliocene; glacial ice covered most or all of Antarctica by the middle Pliocene. At the same time, about 3.5 million years ago, extensive glaciation appeared also in South America and possibly northern Asia and Alaska, although plant fossils indicate that mixed boreal forests still grew in parts of Siberia, northern Alaska, and arctic Canada. For the first time in the Cenozoic, ice began to form in the arctic during the early to middle Pliocene. This north polar ice produced the Labrador Current, which forced the Gulf Stream southward, altering the climate of northern Europe. By 3 million years ago, glaciers covered parts of Greenland and Iceland. The tectonic closure of the Isthmus of Panama may have affected Atlantic Ocean circulation in such a way to produce sufficient atmospheric moisture to build the large volumes of ice in Greenland and Iceland. The uplift of the Tibetan Plateau, which affected atmospheric circulation in the Northern Hemisphere, may also have contributed to the global climate changes of the late Cenozoic. Vegetation in the arctic included taiga by about 5 million years ago and tundra by 2 million years ago. North American desert scrub communities developed only in the Pleistocene.

These climate changes can directly affect sea level (as do tectonic events), and sea level can affect the availability of low-lying areas and continental shelves as avenues for intercontinental dispersals. Using data from climatology, sea-level fluctuations, geochronology, and the mammalian fossil record, Woodburne and Swisher (1995) charted overland dispersals of mammals in the Cenozoic between North America and other continents as follows (numbered events correspond to those in Figure 25-7).

1. First immigration of rodents to North America (also several other extinct groups), presumably from Asia and/or Europe via Beringia or the North Atlantic.
2. Major dispersal between North America and western Europe via Greenland and Spitsbergen. Last great dispersal event between these continents via the North Atlantic corridor route. This dispersal route was subsequently lost as an ocean barrier was created by the tectonic widening of the North Atlantic Ocean. Generic similarity between North America and Europe was greater at this time (late Paleocene) than at any other time in the Cenozoic. Early perissodactyls *(Hyracotherium)*, artiodactyls *(Diacodexis)*, adapid primates, creodonts, rodents, and many other mammals of primitive aspect were involved.

3. Incursion of Asian mammals of modern aspect via trans-Beringian filter to corridor, especially perissodactyls (rhinocerotoids), but also early camels, rabbits, miacid carnivorans, others.
4. First North American appearance of certain early saber-tooth cats, canids, rabbits, tragulid-like deer, sciurids, tayassuids. Probably via Beringian filter route.
5. The Grand Coupure ("great cut"), a major mammalian faunal turnover event in Europe in which endemic and archaic taxa were abruptly replaced in the fossil record by a host of new forms. Sixty percent of indigenous European mammal genera went extinct, including certain primates, rodents, artiodactyls, perissodactyls, creodonts, condylarths, and others. With the retreat of the shallow Turgai Strait that previously separated Europe and Asia, most of the new arrivals entered Europe from Asia; fewer arrived from North America. The replacements consisted of rhinoceroses, titanotheres, hedgehogs, heterosoricids, murids, aplodontids, new sciurids, dipodids, castorids, mustelids, viverrids, lagomorphs, ursids, nimravids, felids, and many others. The Grand Coupure is associated with the major climatic cooling (and beginning of continental glaciers in Antarctica) during the late Eocene and early Oligocene. It is also associated with the tectonic uplift of the Alps as the African Plate contacted southwestern Eurasia, bringing with it Gondwanan mammals and other forms of life.
6. Major influx of Palearctic taxa into North America via Beringian corridor. Immigrants included the first North American soricine and other kinds of shrews, pika, bear-dogs (Amphicyonidae), bears, mustelids, procyonids, a rhinoceros, and others.
7. Another major immigration of Palearctic taxa via Beringian corridor. At this time, North America received its first true felid *(Pseudaelurus)*, first petauristine squirrels, an eomyid rodent, and new mustelids, including a lutrine. In an unrelated event, Xenarthrans (megalonychid sloths) dispersed to the Greater Antilles from South America, certainly by the middle Miocene, possibly earlier (MacPhee and Iturralde-Vinent, 1994).
8. A lesser but important immigration from Asia. Proboscideans crossed Beringia into North

FIGURE 25-7   Summary of North American land mammal ages and major intercontinental dispersals of mammals during the Cenozoic Era. See text for explanation of boldface numbered events 1 to 11. In the "Dispersal" column, the horizontal length of the bar indicates the relative number of taxa involved in the dispersal event. Numerous minor dispersal events are not shown. N.A. = North America; S.A. = South America. *(Modified from Woodburne and Swisher, 1995)*

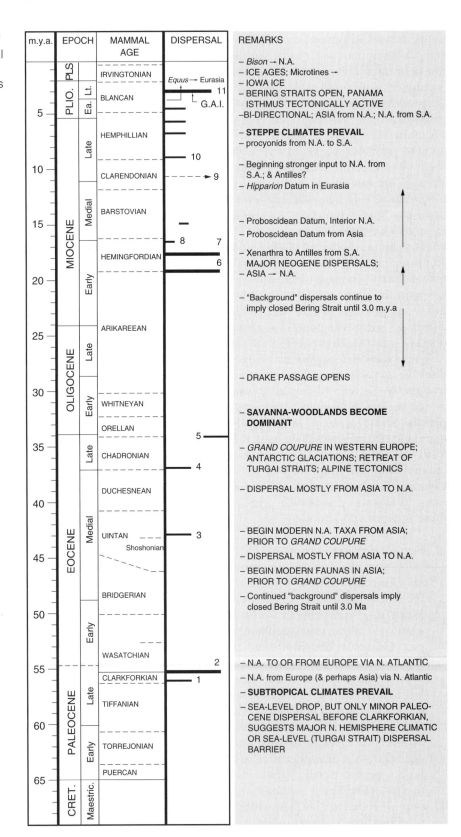

America but apparently they took another million years to reach the North American interior. At about this time, the first murid rodent *Copemys* also entered North America from Asia.

9. One of the few known Neogene dispersals from Nearctic to Palearctic, hipparionine horses left via Beringia and thereafter radiated widely in Eurasia, where horses had not previously existed.

10. Limited interchanges with Eurasia and South America over a few million years. North America's first arvicoline rodents arrived via Beringian filter. About 9 million years ago, before a land connection with South America was completely formed, two genera of large megalonychid sloths entered North America (possibly a sweepstakes dispersal), thus beginning a series of intermittent precursory exchanges between the two continents. (Note that sloths from South America reached the Antilles much earlier than North America.) Around 7.5 to 5.8 million years ago, procyonids strayed across the strait from North America to South America. About 6 million years ago or later, murid rodents entered South America. These dispersals may have taken place during brief low-stands of sea level (the Messinian low); in any case, they indicate a filter route. The isthmus between South America and Central America was not yet dry land but was "under construction." The area was (and is) tectonically active and could have provided an island-hopping route.

11. The Great American Interchange began by at least 2.7 million years ago. The Isthmus of Panama was established as a dry-land connection (filter route), as shown by hordes of immigrants moving from South America to North America and vice versa. North America regained metatherians (Didelphidae), which had become extinct there in the middle Miocene. North America also received its first hystricognath rodents (capybaras and porcupines), additional xenarthrans (glyptodonts, armadillos, more sloths), and toxodont notoungulates (although the huge toxodonts apparently did not move northward beyond Central America). Pliocene exchanges via Beringia continued with the bog lemming *Synaptomys* and spectacled bear *Tremarctos* arriving from

Asia, and the horse *Equus* exiting to Eurasia. Many other North American exchanges with Eurasia and South America continued intermittently through the Pleistocene.

Beginning about 2.5 million years ago, the late Pliocene and Pleistocene was a time of pronounced climatic shifts, when periods of lowered temperatures alternated with periods of relative warmth. Paleoclimatologists using oxygen isotope ratios and other data from ocean-bottom cores chart at least eleven major episodes of cool climates (four of which seem to have been accompanied in North America by major continental glacial advances), separated by warm intervals. Accompanying the periods of cooling, which were apparently worldwide, were a number of spectacular environmental changes. Precipitation increased everywhere, and with increased snowfall, continental glaciers developed and pushed southward.

At one time in the Pleistocene, over 25 percent of the land surface was covered with glaciers: Eurasia had 3.2 million square miles of ice; the Nearctic ice sheet covered 4.5 million square miles, and, during its greatest push southward, reached what is now Kansas. The weight of so much ice actually depressed the covered portions of the continents slightly. Glaciers on Mount Kenya in Africa extended about 1700 meters below the present vestigial snow fields (at 4500 meters), and New Guinea and Madagascar had montane glaciers. The distributions of floras were changed. Boreal vegetational zones were driven downward on mountainsides, and coniferous forests spread southward over areas that previously supported less boreal floras. Concurrently, tropical floras receded toward the equator, and deserts became far more restricted than they are today. In southwestern North America, juniper-piñon woodlands and sagebrush filled many intermountain basins that are occupied today by desert scrub (Betancourt et al., 1990).

The Pleistocene ended 10,000 years ago with the extinction in the Nearctic and Palearctic regions of such common Pleistocene mammals as mammoths, camels, woodland musk ox, shrub ox, ground sloths, horses, and the giant beaver (Kurtén and Anderson, 1980; Martin and Klein, 1984; Webb and Barnosky, 1989). Following the retreat of the last continental glaciers at about

10,000 years ago, the present interglacial period, called the Holocene, has been unusually stable compared to the wild fluctuations of at least the last 100,000 years of the Pleistocene.

Numerous Pleistocene episodes of faunal interchange between Siberia and North America occurred across the periodically emergent Beringia and had a profound effect on the North American mammalian fauna. At times, the Beringian climate was apparently cold temperate and permitted the passage of animals not adapted to arctic conditions; at other times it was arctic. We know what the habitat of the Bering land bridge was actually like only during at least the last glacial maximum

20,500 to 14,500 years ago. Pollen recovered in undersea cores from the bottom of the Bering and Chukchi Seas and in lake-bottom cores on land indicate that mesic tundra grew in the lowlands, and dry, sage-tundra grew in the uplands (Colinvaux, 1996). Mammals intolerant of tundra habitat and cold conditions were denied use of this route at that time. Arvicoline rodents have repeatedly used this route, but fewer other mammals have, indicating the importance of this bridge and its function as a filter (Bell, 1998).

During glacial advances, the ranges of boreal mammals extended well south of present limits. Remains of the musk ox *(Ovibos)*, arctic shrew

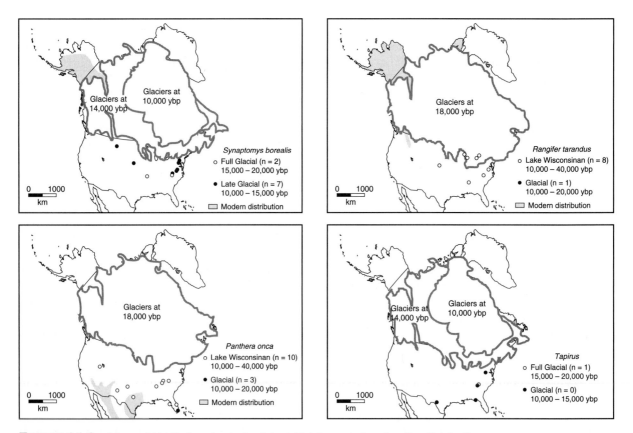

FIGURE 25-8   Maps of North America in the latest Pleistocene showing the distribution of continental glaciers and synchronous known-age fossil records of selected mammals. In the first two maps, the modern distributions of the mammal species are shown (where not covered in the Pleistocene by ice) by shading. These species and many others were restricted to Beringia and/or forced southward by the expanding ice sheets. Examples are *Synaptomys borealis,* the northern bog lemming and *Rangifer tarandus,* the caribou. Also shown are late Pleistocene records of the jaguar, *Panthera onca,* and tapirs, *Tapirus* spp. *(After Faunmap Working Group, 1994)*

(Sorex arcticus), collared lemming *(Dicrostonyx),* and many other species have been found well south of their present northern ranges (Fig. 25-8). Abundant evidence verifies the occurrence of northern assemblages of mammals during the Pleistocene as far south as Kansas and Oklahoma. There were reciprocal northward movements of subtropical or desert mammals during interglacial times, as indicated by the fossil occurrence of such animals as the hog-nosed skunk *(Conepatus)* and jaguar *(Panthera onca)* far north of their present ranges. A fossil record of the jaguar, for example, is from Tennessee, hundreds of kilometers north of the animal's present range. The lion *(Panthera leo atrox)* made its way from the Old World into North America and entered South America via the Panamanian Isthmus, temporarily achieving the title of the world's most widespread mammal species. The environmental fluctuations of the Pleistocene often brought together species that are found today in very different habitats. Like plants, mammal species seemed to respond individualistically to environmental change (Graham et al., 1996).

One of the most common and obvious patterns of mammalian distribution — the occurrence of isolated or semi-isolated populations of northern mammals on mountain ranges at fairly low latitudes — is the result of Pleistocene southward migrations of boreal faunas. During glacial advances, assemblages of boreal mammals were widespread in lowlands well south of their present ranges. Concurrent with the movements of these mammals northward during the retreat of cool climates were movements of boreal mammals into montane areas. Here, because of the effect of elevation on climate, cool refuges were available. Many of these montane populations have persisted in "boreal islands" far south of the northern stronghold of their closest relatives, and the diversity of mammals on some mountain ranges in the southwestern United States has resulted from a combination of Pleistocene dispersals, vicariant events, differentiations, and extinctions. The North American Southwest has been a favorite area for study among mammalian zoogeographers (Davis et al., 1988; Harris, 1990; Findley, 1996; Patterson, 1995; Patterson and Atmar, 1986; Sullivan, 1994).

Humans have obviously reduced the ranges of many mammals and probably contributed to the Quaternary extinction of others, at least in North

America (Martin, 1984). However, some species are extending their ranges northward today, perhaps in response to the present warm climatic cycle, but also in response to human alteration of the environment. The armadillo has extended its range from northern Mexico and southern Texas into much of the central and southern United States in recent years (Humphrey, 1974; Taulman and Robbins, 1996), and the cotton rat *(Sigmodon hispidus)* and opossum *(Didelphis virginiana)* are also moving northward.

SOUTH AMERICAN MAMMALS AND THE GREAT AMERICAN INTERCHANGE

The origins of the Neotropical mammalian fauna have long held the interest of distinguished scientists. Wallace (1876) was first to recognize the faunal interchanges that occurred between North and South America late in the Tertiary, and intensive paleontological field work in the late 19th century provided a more complete understanding of these events. In 1893, von Zettel wrote that "there was thus accomplished, toward the end of the Pliocene, one of the most remarkable migrations of faunas that geology has been able to record." The classic works of Simpson (1950, 1965a, 1965b, 1969, and especially 1980) did much to clarify our understanding of South American historical zoogeography, and recent field work and advances in our geologic knowledge have refined previously held views.

As discussed earlier in this chapter, North America experienced repeated immigrations from Europe and Asia in the Cenozoic (see Fig. 25-7). In strong contrast, South America was isolated from all other continents throughout most of the Cenozoic. An early connection of South America with North America was lost in the late Paleocene, and the last tenuous connection (possibly via an archipelago) with Antarctica was lost in the late Eocene. No reconnection with North America came about until the late Pliocene, 2.7 million years ago, when the Isthmus of Panama was established. For perhaps 35 million years the South American fauna evolved in isolation (with one influx of mammals — primates and hystricognath rodents — in the Oligocene).

The emergence of the Panamanian land bridge provided a gateway for an intermingling of North and South American faunas. This classic natural experiment has been called the Great American Interchange. The following discussion of this interchange is based largely on Marshall et al. (1982), Marshall (1988), Webb (1991), Vrba (1993), and Webb and Rancy (1996).

Before the Great American Interchange, most of the South American orders, families, and genera of mammals were autochthonous and endemic (Simpson, 1980). Metatherians arrived in the late Cretaceous to early Paleocene, before the eutherians (except for notoungulates), and began their fundamental radiation into major lineages (Woodburne and Case, 1996). The Tertiary marsupial radiation in South America — including marsupial saber-tooths, doglike borhyaenids, and others — has been discussed previously (Chapter 6), but the extremely impressive ungulate radiation also deserves mention. The condylarth stock were the only eutherians to reach South America early in the Tertiary, probably by the middle Paleocene. They radiated rapidly, and, by the end of the Paleocene, a diverse series of evolutionary lines were established. In isolation from North American ungulate stocks, the South American ungulates went their unique evolutionary ways. Although they clearly filled many of the same niches occupied by other lineages of ungulates in other parts of the world, many of the South American ungulates were anomalous-looking beasts unlike any ungulates elsewhere.

These South American ungulates spanned a considerable size range. There were rat-size little ones and tusk-bearing giants (order Astrapotheria) approaching the size of an elephant. Especially successful was the order Notoungulata, which included various herbivorous genera, one of the largest of which was *Toxodon,* a stubby-legged, rhinoceros-like beast some 3 meters in length (Fig. 25-9). Another group of notoungulates, the hegetotheres, included a number of small cursorial rabbit-like types. One advanced Miocene genus of the order Litopterna *(Thoatherium)* had one-toed feet that not only were much more specialized than those of the contemporary North American horses but were even more specialized than the feet of present-day horses (Figs. 25-10 and 25-11). Another litoptern had a camel-like body with heavy legs and with the snout lengthened into a proboscis (Fig. 25-9).

The distinctive South American ungulates reached their peak of diversity and numbers in the Oligocene and Miocene, but they declined in the Pliocene, and by the end of the Pleistocene only fossils remained. The decline of the South American ungulates was not due to the invasion of South America by Nearctic ungulates and carnivores. Major faunal shifts occurred before the emergence of the Panamanian land bridge. By this time, over half of the mammals occupying the adaptive zone of the large herbivores were not ungulates: of the

FIGURE 25-9 Life restorations of the South American notoungulate *Toxodon* (Toxodontidae) and the litoptern *Macrauchenia* (Macraucheniidae). *(From Alberdi et al., 1995)*

FIGURE 25-10 A life restoration of *Thoatherium* (Proterotheriidae), a highly cursorial South American ungulate from the Miocene.

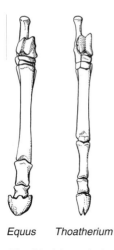

Equus   Thoatherium

FIGURE 25-11 The hind foot skeleton of a modern horse, *Equus,* and a Miocene South American litoptern, *Thoatherium*. Note that the vestiges of digits two and four are more strongly reduced in the litoptern than in the horse.

ten families of large Pliocene herbivores, four were xenarthrans and two were gigantic hystricognath rodents.

The tectonic evolution of Central America and the Caribbean region leading up to the Great American Interchange is complex (Pitman et al., 1993; Marshall and Sempere, 1993). After the Cretaceous–Paleocene dry-land connection between South America and North America was lost,

island arcs were probably intermittently available between South America–Antilles–Yucatan Peninsula (and possibly Florida Peninsula) from the Eocene to the present. The islands were pushed up by the eastward-moving Caribbean tectonic plate. By the early Miocene, most of Central America came into existence as the Chortis tectonic microplate and Middle American Arc combined and moved into place from the west (Pitman et al., 1993). By the late Miocene, a short island arc along Panama was emerging as the final link between South America and North America (Fig. 25-12). It was at this time that the forerunners of the interchange island-hopped between the two continents. By the late Pliocene, the Isthmus of Panama was established and hordes of mammals, other animals, and plants crossed (Fig. 25-13).

Not all mammalian taxa participated in the Great American Interchange, although our knowledge of the interchange suffers in part from a very poor fossil record in southern Mexico and Central America. As the fossil record of this region improves, ideas about the interchange will probably be modified. As far as is known, none of the endemic South American carnivorous metatherians crossed the land bridge. In fact, the only South American terrestrial predator to do so was *Titanis,* a 3-meter-tall, flightless phorusrhacid bird. This huge raptorial bird is known from fossils found along the Gulf of Mexico coast in what are now Texas and Florida. Ceboid monkeys do not seem to have entered Central America until some time after the main interchange had slowed in the middle Pleistocene.

The mammalian participants in the interchange can be divided into two groups on the basis of the means and timing of dispersal. The first group was made up of "waif immigrants" that dispersed along the island arc in the late Miocene. This group includes the Megalonychidae and Mylodontidae (extinct ground sloths), which dispersed from South America to North America, and the procyonids and murids, which moved into South America from North America. The second group includes mammals that dispersed across the Panamanian land bridge at various times after its emergence. A diversity of mammals belongs to this group, including North American taxa that immigrated to South America and South American taxa that entered North America.

FIGURE 25-12 Paleogeographic map of islands of the developing Central American Isthmus during (A) the late Miocene about 6 to 7 million years ago, when the first forerunners of the Great American Interchange crossed and (B) the middle Pliocene about 3 million years ago, immediately before the final emergence of the isthmus. Emergent land is represented by diagonal hatching; submerged continental shelf areas are indicated by stippling. To the northwest and southeast of this region, dry land was already continuous. The modern coastline (solid line) and international boundaries between Nicaragua, Costa Rica, Panama, and Colombia (dashed lines) are shown for reference. Arrows in B show the last remaining corridors for the dispersal of marine organisms between the Caribbean Sea and Pacific Ocean. *(Modified from Coates and Obando, 1996)*

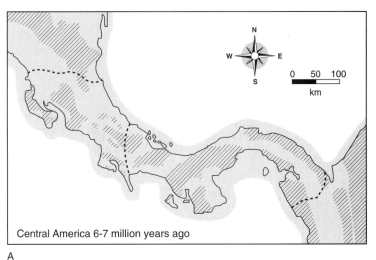

Central America 6-7 million years ago

A

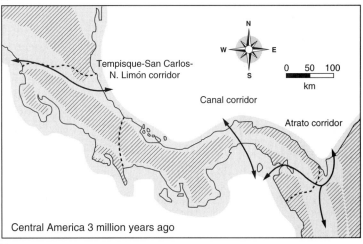

Tempisque-San Carlos- N. Limón corridor

Canal corridor

Atrato corridor

Central America 3 million years ago

B

This spectacular reciprocal interchange of land mammals (discounting bats and manatees) was roughly symmetrical at first but decidedly unbalanced later. On both sides, the number of families and genera increased and then declined due to extinctions. South American taxa generally diversified little in North America; after a million years, the impact of their intrusion had faded. In contrast, North American taxa radiated explosively in South America (Fig. 25-14) throughout the Pleistocene.

Reasons for the imbalance have been debated for decades. An early explanation was that the North American fauna, longer subjected to repeated immigrations, were better competitors and predators. More recently, speculative ecogeographic explanations for the imbalance have predominated. Late Cenozoic mountain-building ac-

tivity elevated the northern Andes to 4000 meters, affecting atmospheric circulation and climate in the isthmian region. Glacial phases, too, influenced the tide of dispersal. Webb and Rancy (1996) suggested a three-phase model for the Great American Interchange after the emergence of the isthmus. The first phase in the late Pliocene coincided with the onset of glacial conditions, during which the greater aridity would have caused forested habitats to shrink and allowed savanna habitats to expand into tropical latitudes (Fig. 25-15), establishing a corridor for savanna-adapted mammals along the slopes of the Andes and far into temperate regions on both continents. At such a time, movement was primarily from a larger North American source area southward into high latitudes in South America. Fossil faunas from the late Pliocene in Florida and

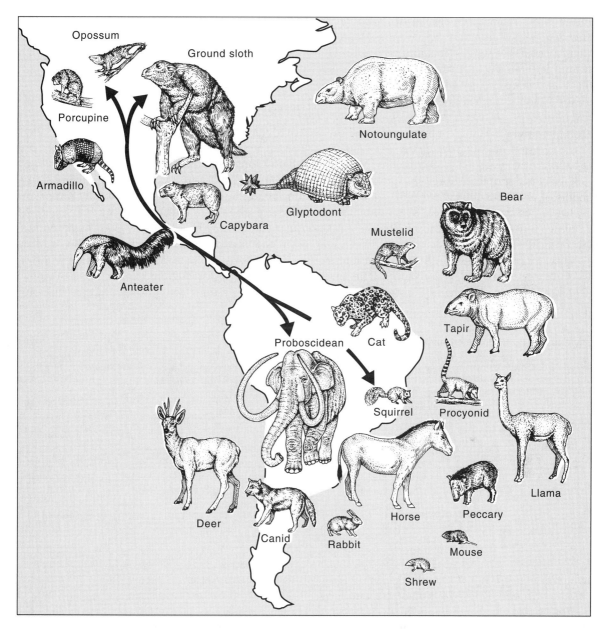

FIGURE 25-13   Representatives of the groups of mammals that were involved in the Great American Interchange. The interchange took place during the last 9 million years, beginning with a few early dispersals and reaching a climax after the Isthmus of Panama emerged as dry land in the Pliocene about 3 million years ago. Crossing the filter at different times into North America and Central America from South America were xenarthrans such as various ground sloths, glyptodonts, pampatheres (armadillo-like glyptodontoids), armadillos, and anteaters; rodents such as capybaras and porcupines; a notoungulate toxodont; and didelphid opossums. Reaching South America from North America at various times were carnivorans such as procyonids, mustelids, tremarctine bears, cats and saber-tooth cats; rodents such as sigmodontine murids and sciurids; proboscideans (gomphotheres); artiodactyls such as deer, camelids, and peccaries; perissodactyls such as tapirs and horses; rabbits; and shrews. *(Modified from Pough et al., 1998, and Marshall, 1988)*

Argentina show remarkable similarities with one another and are dominated by savanna-adapted mammalian taxa that suggest maximum continuity of savanna habitats between these regions. In the middle Pleistocene, a second phase occurred in which humid interglacial conditions caused closed-canopy rain forest to dominate in tropical America, thus providing a route for forest-adapted organisms to move primarily from the large reservoir of Amazonia northward. In a third phase in the late Pleistocene, tropical America suffered the extinction of about 56 genera of its mammals (54 of which were of large body size), presumably as a result of a combination of climate change and human overkill.

It is clear, then, that the South American mammalian fauna has a complex derivation. The unusually large number of endemic Neotropical taxa

is a reflection of the degree and duration of separation of South America from other continents and of the evolutionary success of Nearctic invaders there.

THE UNUSUAL MAMMALIAN FAUNA OF MADAGASCAR

Islands long isolated from continents frequently have an unusual mammalian fauna. Such a fauna may be dominated by a group equally important nowhere else, as in the case of the marsupials of Australia, or may be extremely poor in mammals, as in the case of New Zealand, where the only native mammals are bats. Madagascar is an interesting example of a **refugium** supporting a primitive mammalian fauna with little ordinal diversity,

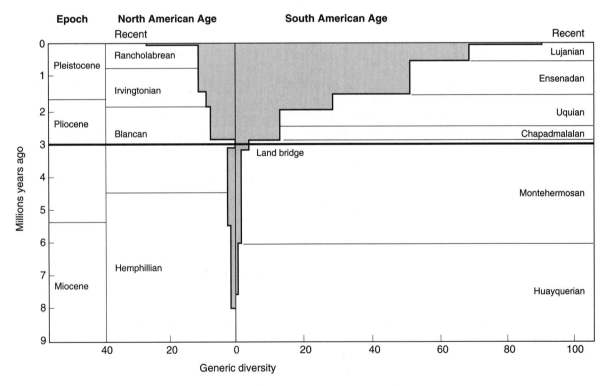

FIGURE 25-14   Graphic comparison of the number of genera of land mammals from North and South America participating in the Great American Interchange in the late Cenozoic. The initial invasion in each continent is approximately balanced up to and immediately after the appearance of the land bridge. However, the subsequent increase in North America of immigrants from South America faded by the middle Pleistocene; in South America the North American invaders continued to increase exponentially. *(After Marshall, 1988)*

FIGURE 25-15   Approximate distribution of general habitat types in South and southern North America during two phases of the late Cenozoic. In the aridity accompanying a glacial phase (A) dryland habitats such as savanna woodlands and other open landscapes predominated, creating a barrier to forest-adapted species and an avenue of dispersal for savanna-adapted mammals through the Central American Isthmus from temperate North America to temperate South America and vice versa. During such a phase, the slopes of the Andes Mountains and periphery of Amazonia might have held a mosaic of habitats including savannas. In an interglacial phase (B), Central America and the isthmus may have been clothed in rain forest, much like the present. This situation would have created a filter route across the isthmus, blocking savanna species and allowing passage into Central America of rainforest species from Amazonia that could cross or go around the northern Andes. *(After Webb, 1991)*

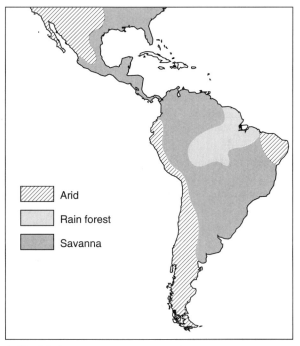

A. Glacial phase

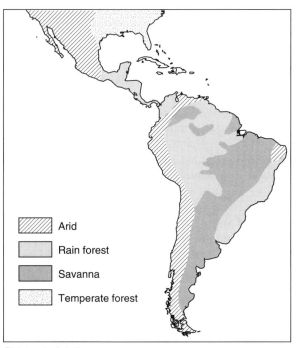

B. Interglacial phase

many endemics, and a seemingly incomplete exploitation of habitats.

Madagascar is a large island, 1600 kilometers in length, with a maximum width of 560 kilometers. It lies 420 kilometers east of the eastern coast of Africa. Madagascar separated from Gondwana as a fragment of the Indian subcontinent by 100 million years ago and has been isolated from other land masses throughout the Cenozoic. The island has supported six orders of mammals in Recent times: Insectivora, Chiroptera, Primates, Rodentia, Carnivora, and Artiodactyla. Most of the mammals are endemic. The most highly diversified groups are the lemuroid primates (with an astounding 32 species still surviving after humans arrived on the island; Mittermeier et al., 1994; Goodman and Patterson, 1997), which probably arrived in the Eocene, and the tenrecid insectivores, which perhaps dispersed to or from Africa as early as the Paleocene. Many of the Malagasy mammals occupy niches filled elsewhere by mammals of different taxa. There are viverrids that resemble cats and lemurs that are variously nocturnal or diurnal, terrestrial or arboreal, and the aye-aye *(Daubentonia)* that has woodpecker-like abilities. The tenrecs include hedgehog-like species, burrowing, molelike species, and web-footed, amphibious fish-eating species. The tiny, mouse-eared, termite-eating tenrec *Geogale* resembles the North American desert shrew *Notiosorex* and inhabits the arid southwestern part of Madagascar. The only artiodactyl present before the arrival of humans was the now extinct hippopotamus *(Hippopotamus lemelii),* and today the introduced river hog *(Potamochoerus)* is the only wild artiodactyl. The ungulate niche has largely gone unfilled, although a group of large Pleistocene lemurs, now extinct, may have been terrestrial herbivores. There is also a lack of frugivorous mammals compared to other tropical areas (Goodman et al., 1997).

THE ISLAND SYNDROME

Mammals isolated on islands typically face different selective pressures than do members of parental mainland stocks. On islands, competition is usually reduced or may be absent, predators are often absent or few kinds are present, and the flora may be depauperate. In some cases, as on some small desert islands off the eastern coast of Baja California, one or two species of rodents are the only mammalian inhabitants and just a fraction of the number of species of plants that occur on the mainland is present. Through time, island mammals tend to diverge from parental mainland stocks, but the pattern of divergence is not consistent for all species. Island mammals typically differ in size from mainland relatives, and Foster (1964) pointed out that, whereas some mammals become larger on islands, others become smaller. Island rodents and marsupials are generally larger than their mainland relatives; insectivores, lagomorphs, carnivores, and artiodactyls, however, are usually smaller. Examples are numerous. The *Peromyscus* inhabiting islands off the coast of British Columbia are unusually large, but the caribou that inhabited one of these islands (but is now extinct) was a dwarfed form. The gray fox *(Urocyon littoralis)* that lives on the Channel Islands off the coast of California is substantially smaller than the mainland *Urocyon cinereoargenteus,* and the mammoth that lived on these islands in the Pleistocene was dwarfed. However, evolutionary patterns on islands are not completely consistent. Counter to the usual trend, not all island insectivores are small. Unusually large insectivores (solenodonts; see p. 107) evolved on some of the islands of the

West Indies, and the largest insectivore of all time *(Deinogalerix koenigswaldi)* lived on the Mediterranean island of Gargano. This insectivore was larger than a fox and probably fed on rodents (Freudenthal, 1972).

The remarkable dwarfed Pleistocene mammals of the Mediterranean islands have been discussed by Sondaar (1977). In the Pleistocene, elephants *(Elephas)* and deer *(Cervus)* lived on many of these islands, and some islands supported hippopotami *(Hippopotamus).* These mammals must have reached the islands by sweepstakes routes, for the extent to which they diverged morphologically from the mainland stocks and the fact that generally not all types occurred on an island suggests that access to the islands was across water. All of the three types listed above are known to be strong swimmers, and a single pregnant female could have founded a population on an island. Of special interest are the similar evolutionary trends exhibited by large mammals on a number of islands between which passage of terrestrial mammals would have been impossible. Elephants on the islands became strongly dwarfed relative to the parental mainland stock of *Elephas namadicus. Elephas falconeri* of Sicily, an example of extreme dwarfism, was roughly 1 meter high, about one quarter the size of its mainland progenitor, and relative to mainland elephants, *E. falconeri* had short distal segments of the limbs, cheek teeth with fewer enamel ridges, and a much lower skull, with a reduction of the elaborate system of air sinuses. Short-leggedness was especially pronounced in the island deer, but the pig-size island hippopotami also became short-legged.

These patterns of parallel evolution probably resulted from similar selective pressures on the many isolated islands. No large predators were on the islands. Large size is an extremely effective adaptation to avoid predation; without large predators, great size was no longer of advantage. An unreliable food supply for herbivores may have favored smaller size, and, in the absence of predators, overpopulation might have triggered periodic heavy mortality. Beds of deer bones found on the island of Crete are interpreted by some paleontologists as evidence of mass mortality, and abnormalities of the bones suggest starvation as the cause of death. In the dwarf elephant *E. falconeri,* the reduction of the skull crest and the reduced number of enamel ridges on the molars were related to the general dwarfing (Maglio, 1973). The marked shortening of the limbs of the deer is thought by Sondaar (1977) to have been due to two factors: the absence of predators and the consequent lack of need for speed, and the need for sturdy and well-braced limbs with which to negotiate mountainous terrain.

Just as mammals on islands are divergent structurally, some have changed behaviorally. As an example, desert woodrats *(Neotoma lepida)* on Danzante Island in the Gulf of California have very large home ranges, and males are resource-defense polygynists, whereas this species on the nearby mainland does not have these behaviors (Vaughan and Schwartz, 1980). Mammals that live on islands and have no mammalian predators often show little fear of humans. Blake (1887) found gray foxes on Santa Cruz Island to be remarkably tame. They commonly approached to within 1 meter of a person and regularly visited Blake's camp for scraps of food. These foxes even approached sleeping persons and pulled at their blankets. In the face of human invasions and introductions of predators and competitors, island faunas around the world have suffered greatly.

WEB SITES

An atlas of the Ice Age earth. Paleovegetation maps for the continents since the last interglacial period

> http://www.esd.ornl.gov/projects/qen/nerc.html

Paleomap project; an atlas of plate tectonic reconstructions showing the paleo-positions of continents and coastlines during the last 1.1 billion years, with links to other sites

> http://www.scotese.com

FAUNMAP project; a database of maps, literature, radiometric ages, and other information about North American mammals during the late Pleistocene

> http://www.museum.state.il.us/research/faunmap/

26 MAMMALIAN CONSERVATION

As noted occasionally throughout this book, mammals face numerous threats to their continued existence including habitat degradation and destruction, overexploitation, loss of genetic diversity, endangerment, and extinction. The main problem confronting not only mammals but all the earth's **biodiversity** is clear and obvious everywhere: human overpopulation (Fig. 26-1) (Soulé, 1991; Ehrlich and Wilson 1991; Meffe et al., 1993; Forester and Machlis, 1996; Laurance and Bierregaard, 1997; Bazzaz et al., 1998). During the last 500 years or so, we humans have gone from a small global population to the dominant ecological force on the planet. For thousands of years after we evolved, we existed

in tenuous, small, isolated groups. Having "turned the corner" a century and a half ago, our population began an exponential increase. In a very short span of time (within the average lifespan of a person), we have converted or destroyed much of nature, except for a few relatively small parks and reserves isolated like islands in a sea of human-dominated ecosystems (Kingdon, 1989; Turner et al., 1990; Morowitz, 1991; Meyer, 1996; Vitousek et al., 1997). At present, only about 3 percent of the land surface worldwide is set aside in protected areas and nature reserves (Soulé, 1991); even less in the oceans.

The present human population is about 6 billion people. If growth continues at the present rate, the population could reach 10 billion to 17 billion in the next hundred years (Bongaarts, 1998). The demands of our growing population for simple basic resources such as shelter, water, fire, and food would consume the remaining large tracts of undeveloped nature. Accompanying this global encroachment and overconsumption is the expected mass extinction of one fourth to one half of all living species (Soulé, 1991; Wilson, 1992). Some biologists even contemplate our own extinction.

On a geological time scale, extinction is common. Most extinctions are individual or low-level "background" events, but large scale extinctions have periodically taken place. As noted by Jablonski (1991), "The most basic observation is simply that mass extinctions have happened: irreversible biotic upheavals have occurred repeatedly in the geological past. Marine and terrestrial biotas are not infinitely resilient, and certain environmental stresses can push them beyond their limits. . . . Survival of species or lineages during mass extinctions is not strictly random, but it is not necessarily closely tied to success during times of normal background extinction." Shall we be the cause of the next mass extinction? How can we stop or lessen this outcome? There are no easy answers. The problems and factors involved are immense and complex (Mares, 1986, 1992; Pimm, 1992; Wilson, 1992; Galliari and Goin, 1993; Goin and Goñi,

1993; Epstein, 1995; Heywood, 1996a, b; Patz et al., 1996; Costanza et al., 1997; Anonymous, 1998; Fullerton and Stavins, 1998; Laurance, 1998; Rapport et al., 1998), but our own survival may depend on our attempts to solve them (Koshland, 1991). A most encouraging trend in recent years is the increasing awareness of the problems involved in conserving biodiversity and the proliferation of conservationists and ideas to understand and solve them (Wilson, 1992; see, for example, various articles in Reaka-Kudla et al., 1997 and Laurance and Bierregaard, 1997). Many mammalogists are feverishly working to document basic biodiversity while simultaneously developing plans for its protection and management (for examples, Galliari and Goin, 1993; Peterson et al., 1993; Wemmer et al., 1993; Heaney, 1993; Heaney et al., 1997; Patton et al., 1997; Medellín, 1998). Another encouraging trend is a decrease in reproductive rate of humans around the world in recent years (Bongaarts, 1998).

Among the conservationists is Michael Soulé, who summarized several major factors of human interference, including the destruction of habitat; fragmentation of habitat; overexploitation; the spread of exotic (introduced and alien) species and diseases; air, soil, and water pollution; and climate change. He further attributed these proximate causes of the loss of biodiversity to still more fundamental factors of the human condition (Table 26-1). Because human nature is unlikely to change quickly, he proposed a number of tactics to ameliorate the attrition of biotic diversity: "The human condition is dynamic and unpredictable and will remain so for at least a century, if for no other reasons than the momentum of the population explosion and the unsatisfactory economic and social status for billions of people during the 21st century. The 'biotic condition,' therefore, will also be tenuous during this interval. Fortunately, conservationists have an increasing number of tools with which to deal with the crisis" (see Soulé, 1991). Limited space does not allow us to discuss them here, but the reader is encouraged to consult the

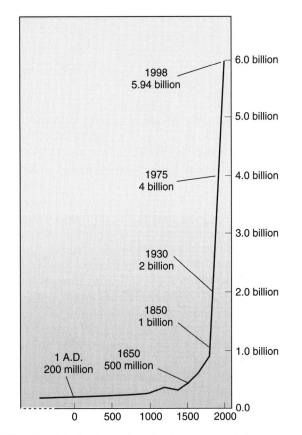

FIGURE 26-1 Growth of the human population during the last 2000 years. In 1999, about 6 billion people inhabit the earth, and every minute the population increases by about 162 people (net change of births and deaths). United Nations projections for the next 25 years predict an increase of another 2 billion, with a world population by the year 2100 ranging from a low estimate of 10.4 billion to a high estimate of 17.5 billion people *(United Nations, 1997, 1998).*

articles and books cited in this chapter, and the references therein, for a better understanding of the problems and potential solutions to preserving the earth's biota.

Many people point up the economic value of nature and natural resources as a rationale for their preservation (Geist, 1994; Daily, 1997; Reaka-Kudla et al., 1997), and there are different views of "economic value." There are those who believe that if forests and other wildlands are to be saved, they must "pay their own way," that is, provide basic sustenance for some humans (Robinson and Redford, 1991) or profits for others. For example,

the extraction of DNA from nature produces billions of dollars annually for biotechnology, agriculture, and public health (Myers, 1984). Sometimes these anthropocentric benefits can provide even more reason to exploit the resources, while little effort is reinvested in protecting them. Even selectively extracting resources may alter or erode the ecology and biodiversity of a habitat. Other people emphasize the ecological "services" of nature, from recreational opportunities to the regulation of carbon dioxide in the atmosphere (Repetto, 1992; Costanza et al., 1997; Daily, 1997), or the value of the genetic information content of species (Crozier, 1997).

Conservationists argue that cultural values are often in conflict with conservation policies and that "a new ethic or a revolutionary change in human consciousness is necessary" before significant progress is possible (Leopold, 1948; International Union for the Conservation of Nature, 1980; Devall and Sessions, 1985; Soulé, 1991; Morowitz, 1991; Goin and Goñi, 1991; Wilson, 1992; Shepard, 1996). One cultural value that has recently been questioned is the private ownership of land (Steinberg, 1995; Mitchell, 1998).

HUMAN IMPACT ON MAMMALS

We humans have long been interested in our fellow mammals and have long exploited them. As many as 4 million years ago, *Australopithecus* was killing and eating baboons and antelope, and the use of mammals for food remains characteristic of most cultures today. Many kinds of mammals have been domesticated, and some are taught to work for their owners. The trained Indian elephant lifts and drags teak logs in the remaining tropical forests of Sri Lanka, where the periodically saturated soil limits the usefulness of vehicles; dogs help some African hunters capture antelope and other game; trained rhesus monkeys pick coconuts from tall trees and drop them to their masters; and even the unruly camel has been trained. The raising of various kinds of mammals — from genetically engineered mice for medical research to cloned sheep and cattle — is an important enterprise today. The very distribution of early humans was probably influenced by their ability to kill their fellow mammals, for the skins and furs of mam-

TABLE 26-1 Fundamental Human Factors That Have Contributed To The Erosion of Biological Diversity During The Last Three Centuries

Factor	Example of impact on conservation
Population growth	Population pressures
Poverty	Hunger, deforestation, trade in rare and endangered species, failure of "grassroots" support
Misperception of environmental degradation	Desire for quick results and denial of long-term failures
Anthropocentrism	Lack of support for nonutilitarian causes
Cultural transitions	Unsustainable resource management during colonization and rapid social change
Economics	Failure of planning because of internationalization of markets and erratic pricing of commodities
Policy implementation	Civil disruption, wars, corruption, failure of law enforcement

From Soulé (1991) American Association for the Advancement of Science.

mals may have enabled primitive humans, probably endowed with hopelessly inadequate insulation, to penetrate cool or cold regions.

Wild mammals and most other organisms are under greater pressure than ever from humans in all parts of the world. In many less developed areas, the inhabitants hunt year-round; either they are primarily hunters who depend on mammals for much of their food or they hunt to supplement limited food supplies. In the developed countries, many people hunt for sport or trap for furs during regulated seasons. In the United States, where virtually all populations of game mammals are managed, the sale of hunting and trapping licenses contributes to wildlife management and conservation in general. Hunting remains popular in many parts of the United States, and the sale of firearms, ammunition, and other equipment associated with hunting is big business.

Our exploitation of the earth's mammals and other resources is not necessarily something to rebel against nor be ashamed of. Like all mammals, as obligate heterotrophic organisms and native inhabitants of the earth, we have no choice but to exploit responsibly the planet's resources for our own survival. The earth's natural resources are the ultimate origin for everything we need; there is no other source. Unfortunately, we are not presently using them conscientiously but are se-

verely overexploiting and abusing them when we barely fathom their ecological relationships. Even our efforts at conservation sometimes produce unexpected results because of other perturbations and unanticipated linkages between ecosystems (Chapin et al., 1998; for a mammalian example, see Estes et al., 1998).

In some ways our "overconsumption" of mammals and other wildlife is pure waste. On just four 0.5-kilometer stretches of a highway passing through rain forest in northeastern Queensland, Australia, during a 38-month period, Miriam Goosem (1997) recorded more than 4,000 road-killed vertebrates, including about 500 mammals. Roadkill data for roads in the United States do not seem to be available, but impromptu counts made during mammalogy class field trips are deplorably high in some areas of the country. When multiplied by the thousands of miles of roadways, the needless destruction is staggering. At the same time that the traffic on these roads kills some species, the roads themselves not only fragment the habitat but also present barriers to dispersal for other mammals. For example, Goosem (1997) also learned that musky rat-kangaroos completely avoided the vicinity of the highway; Lumholtz's tree kangaroos, green ringtail possums, and long-tailed pygmy possums avoided the highway or were reluctant to descend to the ground but were able to cross through the forest canopy as long as an "overpass" of contiguous branches was available.

Wild animals can be costly to agriculture and ranching. Pocket gophers, rabbits, meadow voles, ground squirrels, and even deer and elk may damage crops or rangeland, and efforts to combat these losses are frequently expensive. In addition, the United States federal government supports considerable research on mammals and, at the same time, finances the local control of virtually all carnivoran species for commercial ranching interests. According to the Predator Defense Institute (see web site at end of chapter), U.S. Department of Agriculture federal animal damage control programs cost U.S. taxpayers about $36 million per year. Many biologists deplore these losses and regard such programs as a misuse of federal monies.

The long-term exploitation of mammals by humans has had a devastating impact. In the last 500 years, about 82 species of mammals have become extinct (Cole et al., 1994; MacPhee and Flemming, 1999). According to a 1994 report, about 600 species (13 percent of existing mammal species, Cole et al., 1994) were considered at risk of extinction, but a new and more complete assessment of all mammal species by the International Union for Conservation of Nature and Natural Resources lists about 1,100 of 4,600 mammal species (nearly 25 percent!) that are threatened by extinction (Baillie and Groombridge, 1996; Baker, 1997). The conservation status of these species are rated according to various "categories of threat" by the IUCN as critically endangered, endangered, or vulnerable (see web site at end of this chapter; Baillie and Groombridge, 1996). Twenty-four of the 26 orders of mammals include threatened species. The greatest proportion of threatened species belong to the orders Primates (46 percent), Insectivora (36 percent), Artiodactyla (33 percent), Chiroptera (26 percent), Carnivora (26 percent), and Rodentia (17 percent). Many more species are potentially vulnerable; for many others, we know too little of their status to assess their conservation needs.

Some extinct mammals were disposed of remarkably summarily. Steller's sea cow of the Bering Sea was pushed to extinction only 27 years after its first discovery by whalers. Sea otters, which were hunted along the Pacific Coast of North America at least as early as 1786, were killed for their valuable fur; probably more than 200,000 were killed between 1786 and 1868 (Evermann, 1923). By 1900, these animals were rare over much of their range, and they were seemingly lucky to have survived until protected by legislation in the early 1900s. Not so lucky was the grizzly bear in California. In the 1890s, grizzlies still persisted in the San Gabriel Mountains near Los Angeles, but the last known southern California grizzly was killed in 1916, and the last verified occurrence in California was in 1922, in the foothills of the Sierra Nevada in central California (Grinnell et al., 1937). Only about 60 years were required to bring the grizzly in California from fair abundance to total extirpation. In Mexico, a population of grizzlies that survived in a small mountain range in central Chihuahua in 1957 was probably wiped out by 1963, the very year when funds were raised by the World Wide Fund for Nature to set aside a refuge for the animals. In Brazil, over 90 percent of the Atlantic rain forest has been destroyed by logging and other development where 17 of the 23 types of primates and many other mammals are endemic; no one knows how many unnamed species were lost in the process.

Mammals in the Eastern Hemisphere have fared no better. The quagga, a zebra that inhabited southern Africa, was extirpated in the wild about 1860, and another type of zebra was exterminated by roughly 1910. The Arabian oryx, well on its way to extinction, has been hunted in recent years with machine guns mounted on jeeps. The black rhinoceros of East Africa has been extirpated over broad areas, and its survival in the wild seems unlikely. The rate at which rhinoceros and elephant populations were once shot for sport in Africa is astonishing (Beard, 1977). Siberian tigers, which number only about 500 in the wild, are threatened by logging of old-growth forest in their prime habitat in Russia. The giant panda in China faces habitat destruction as well as a wide array of crucial but nonscientific problems from political upheavals and inadequate captive conditions to bureaucratic problems and poaching (Schaller, 1993). In Australia, introductions of foreign species by humans, whether intentional or inadvertent, have decimated populations of native, endemic species (Short and Smith, 1994; Smith and Quin, 1996). During the last 170 years European humans, European rabbits, and red foxes were brought to Australia; the spread of each of these alien species across the continent was similar to the spread of a contagion and was accompanied by dramatic contractions in the ranges of native mammals (Channell, 1998) and plants. Similar introductions of alien species (including humans) to many islands around the globe have threatened, wiped out, or replaced their unique and often highly endemic, but easily disrupted, ecosystems, even as mammalogists are conducting initial surveys and describing new species in these island ecosystems (Wiles, 1992; Atkinson and Cameron, 1993; Heaney, 1993; Heaney et al., 1997).

In a few cases, there is reason for hope that some species will be saved. Through persistent hunting by the hundreds of thousands, the blue whale, the largest animal of all times, was reduced to a total population of probably no more than several hundred individuals by the late 1960s. Since the hunting of blue whales was banned in 1968, their population has stabilized and may even be increasing (Baskin, 1993). Reintroductions of native species to their former ranges from captive breed-

ing populations offer hope for their future (Short et al., 1992; Lomolino and Channell, 1998); in Mongolia Przewalski's horse, extinct in the wild by about 1960, may be reintroduced into the Gobi Desert (Fig. 26-2). Other partial successes are also being achieved. Pesticides seemingly caused declines in bat populations in some areas of the United States, but since the use of DDT was banned the declines there may have been slowed. Bats have received a much-deserved increase in respect from the public, largely because of the efforts of organizations like Bat Conservation International.

It is obvious that the fate of wildlife depends on the persistence of appropriate habitat, which is being destroyed over broad areas at an ever-increasing rate by ever-expanding human populations (Fig. 26-3). In many developing countries, where the focal point of the inhabitants' existence is the day-to-day search for food and fuel, the pressure on land and wildlife is intense. This pressure becomes vastly more acute when revolution, struggles for independence, or strife between

FIGURE 26-2   Przewalski's horse (*Equus caballus przewalskii;* Perissodactyla) is on the brink of extinction. This animal is listed as endangered under the U.S. Endangered Species Act and as "extinct?" by the IUCN. *(Corel Corp.)*

FIGURE 26-3 Habitat destruction resulting from a clear-cut logging operation in the United States. *(PhotoDisc, Inc.)*

political factions is associated with reigns of lawlessness. Meanwhile, powerful corporations from the developed nations are a more insidious threat, driven by the addiction of our "advanced civilization" for incredible amounts of energy and materials from developing countries eager to become just like us.

The situation in Amazonia described by Laurance (1998) provides a tragic case in point: "Today, even the remotest areas of the Amazon are being influenced by human activities. Illegal gold-mining is widespread, with wildcat miners polluting streams with mercury (used to separate gold from sediments) and threatening indigenous Indians through intimidation and introductions of new diseases. A recent government census, for example, tallied more than 3000 illegal miners in the Yanomami Indian Reserve in northern Amazonia. There are also increasing numbers of major mineral, oil, and natural gas developments sanctioned by Amazonian governments. Much of the remote Peruvian Amazon — one of the world's most biologically important areas — has been opened up for oil and gas exploration, with multinational cor-

porations investing hundreds of millions of dollars in the region. Roads created for oil exploration and development in Ecuador have caused a sharp rise in forest colonization, land speculation, and commercial hunting.

"Hunting pressure is growing throughout the Amazon because of greater access to forests and markets and the common use of shotguns. Frequently exploited species include larger primates, deer, tapirs, peccaries, large rodents and top carnivores (such as jaguars and pumas). Intensive hunting can dramatically alter the structure of animal communities, extirpate species with low reproductive rates, and exacerbate effects of habitat fragmentation on exploited species. Hunting could potentially have diverse effects on rain forests — if top carnivores are eliminated, for example, populations of seed predators could increase rapidly and reduce the abundance of large-seeded tree species." (Reprinted from *Trends in Ecology and Evolution*, vol. 13, Laurance et al., "A crisis in the making." Copyright 1998, with permission from Elsevier Science.) Montane forest on the slopes surrounding the Amazon Basin are also being cleared at an alarming rate to grow coca for cocaine and opium poppies for heroin to supply the demand for these drugs in the United States and Europe (Goodman, 1993).

The need for the human race to set its own house in order is basic both to our own survival and to the perpetuation of the biological richness of the world. Clearly, our ability to solve social problems and the future of wildlife are tightly linked. People not under the pressures and stresses occasioned by high populations and limited resources, and at peace with one another, can work toward saving the biotas of the world — fearful people with empty stomachs make poor conservationists. On the success of the politicians, scientists, and teachers of the world in halting the rise in human populations and stopping strife between peoples hinges the survival of our biotic heritage. We cannot expect that many of the already threatened species will survive, but we can only hope that some small nuclei will escape the global destruction and will suffice to partially repopulate the few national parks and reserves. We have clearly reached the eleventh hour. If we do not learn from our mistakes, the wildlife of the world will pay a devastating price.

A NEW CONSERVATION ETHIC

Ecologists and conservationists have argued that each element of a biota plays an essential part in the ecosystem it occupies and that the loss of even a seemingly insignificant species might tip the delicate biotic balance. There is good evidence that diversity of biota in an ecosystem contributes to the stability and productivity of that ecosystem (Wilson, 1992; Tilman and Downing, 1994; Naeem et al., 1994; Reaka-Kudla et al., 1997). We do not begin to know the degree of pressure that most ecosystems can tolerate before collapse. Nor do we know the size of the human population that the earth can support (Cohen, 1996). Nevertheless, an open-ended, uncontrolled experiment is probably not the best way to find out.

Biologists have stressed practical problems: the pollution of water should be avoided not only because of the potential for serious public health problems but also because the ecosystem of a stream might be drastically altered and species of fish from which we derive pleasure or some monetary return might disappear. Range managers have emphasized economic problems: unwise grazing practices alter a grassland ecosystem to the point where its economic importance is reduced; in concrete terms, the weight that each head of cattle gains per day may be reduced to a point at which ranchers can no longer realize a profit. But can we, with due respect for honesty, justify conservation with only these kinds of arguments?

In his excellent discussion entitled *The Conservation of Non-resources*, Ehrenfeld (1976) points out that attempts to justify the conservation of many species on the basis of their economic importance are unjustifiable scientifically. The conservation doctrines lose force if a species is destroyed without the disruption of its ecosystem. If the ecosystem is destroyed following the loss of the species, however, not only is it too late to save the day, but the cause-and-effect relationship involved can never be proved and may not even be hypothesized.

We must look to a new conservation ethic. Few species can be proved essential to the survival of their ecosystems or to have great economic value; nonetheless, each species forms a part of a biological richness developed over millions of years, and each is worthy of perpetuation at least in part be-

cause of what Ehrenfeld terms its "natural art value." Two moving presentations of this view are quoted by Ehrenfeld. In his book, *Ulendo: Travels of a Naturalist in and out of Africa*, Carr (1974) states, "It would be cause for world fury if the Egyptians should quarry the pyramids, or the French should loose urchins to throw stones in the Louvre. It would be the same if the Americans dammed the Valley of the Colorado. A reverence for original landscape is one of the humanities. It was the first humanity. Reckoned in terms of human nerves and juices, there is no difference in the value of a work of art and a work of nature. There is this difference though. . . . Any art might somehow, some day be replaced — the full symphony of the savanna landscape never."

Regarding specific nonresource species, in this case, three small primates called lion tamarins, Coimbra-Filho et al. (1975) wrote: "In purely economic terms, it really doesn't matter if three Brazilian monkeys vanish into extinction. Although they can be (and previously were) used as laboratory animals in biomedical research, other far more abundant species from other parts of South America serve equally well or better in laboratories. Lion tamarins can be effectively exhibited in zoos, but it is doubtful that the majority of zoo-goers would miss them. No, it seems that the main reason for trying to save them and other animals like them is that the disappearance of any species represents a great esthetic loss for the entire world. It can perhaps be compared to the destruction of a great work of art by a famous painter or sculptor, except that, unlike a man-made work of art, the evolution of a single species is a process that takes many millions of years and can never again be duplicated."

Although effort expended on the perpetuation of some species can be justified on the basis of economic importance, we regard this natural art value of many species as their greatest importance. A more basic argument would insist that a species should be preserved because of a reverence for its vast evolutionary history and that each species has a right to play out its evolutionary role; we have no moral right to set ourselves up as the instrument of their destruction. Certainly, our present attitudes of intolerance and disregard for nature must change if our biotas are to survive, and an incipient rationale or conservation ethic may unite enough of humankind behind the conservation cause to turn the tide.

If the present decimation of the earth's wildlife is to be curtailed, at least the following steps must be taken:

1. Human population growth must be halted and the need for space and resources stabilized.
2. Large tracts of land undisturbed by humans must be maintained for wildlife.
3. Our exploitation of many species must be drastically reduced.
4. A broad understanding of the meaning of such biotic interactions as predation must underlie an interest in preserving balanced faunas.
5. Control of animals threatening crops and livestock must be local, with no attempt to exterminate a species over wide areas with little respect to the damage it is doing.
6. The use of biocides must be carefully controlled.
7. We must accept some types of economic losses and inconvenience caused by wildlife, and must feel that these are more than compensated for by our enjoyment of a rich and balanced biota.
8. We must realize that, in spite of our self-proclaimed superiority over nature, we are nevertheless a part of it and must treat it as an extension of ourselves.

We, as biologists, are among the most acutely aware of the problems facing the earth's biota. Although the human overpopulation problem has long been recognized, human population control remains controversial and unpopular in many cultures. Despite the recent global decline in human births, higher standards of living and better nutrition and health around the world have led to a decline in global mortality rates and greater longevity, ensuring continued population growth (Bongaarts, 1998). Biologists and others have dared to speak out to urge continued population control (Meffe et al., 1993; Hardin, 1993; Jayaraman, 1993a, b) and to plead for educating the world about the fundamental importance of the "balance of nature" and the inescapable fact that it includes us. As stated by Lovejoy (1997), "We must not only recognize that, but also behave as if, we live *within* ecosystems, rather than perceiving nature as something confined to a few protected areas isolated within a degraded, human-dominated landscape." What we do to the earth we do to ourselves. The most ethical and exemplary endeavor for present and future mammalogists is in conservation, not just of mammals but of all living things. We must take the lead, as a large group of ecologists recently publicly pledged to do (Bazzaz et al., 1998), in devoting part of our professional (and private) lives to deter environmental degradation and to teach the public about the importance of biodiversity. And on a personal level, we must each ask ourselves "What have I done today to lessen my own impact on the earth? Will my children's children be happy in the world I left them?"

Because the survival of the world's wildlife and indeed our very own survival are in our hands, one cannot help but fervently hope that the word *sapiens* (meaning wise) becomes a justly earned part of the name *Homo sapiens*.

WEB SITES

International Union for the Conservation of Nature and Natural Resources (IUCN), Species Survival Commission's red lists of threatened and extinct species

> http://www.iucn.org/themes/ssc/iucnredlists/redlist-index.htm

Searchable version of the IUCN Red List of Threatened Animals, including mammals, maintained by the World Conservation Monitoring Center

> http://www.wcmc.org.uk/species/animals/animal_redlist.html

Index of endangered species of vertebrates listed by the U.S. Fish and Wildlife Service

> http://www.fws.gov/r9endspp/vertdata.html

Home page of the U.S. Geological Survey's National Biological Service, with links to related web sites

> http://www.nbs.gov/

Caring for the Earth: A strategy for Sustainable Living, published in partnership by the International Union for the Conservation of Nature and Natural Resources, the United Nations Environment Programme, and the World Wide Fund for Nature, 1991

> http://coombs.anu.edu.au/~vern/caring/care-earth3.txt

The Nature Conservancy home page, with links to related sites

> http://www.heritage.tnc.org/

Predator Defense Institute home page

> http://arrs.envirolink.org/pdi/Default.htm

GLOSSARY

A

Abomasum The fourth and most posterior chamber of the ruminant stomach. It is homologous with the stomach of other mammals.

Acetabulum The lateral socket on the pelvis that accommodates the ball of the femur.

Acromion A continuation of the scapular spine that projects above the shoulder capsule.

Adaptive radiation The evolutionary diversification of numerous species from a common ancestor that colonized a new environment. A rapid cladogenetic event.

Adductors Muscles that act to move a skeletal element toward the midline of the body.

Age cohort Individuals in a population of a particular age group.

Alleles One of several alternative forms of a gene.

Allochthonous Having originated outside the area in which it now occurs.

Allomone A chemical released by one species that serves as a communication signal to another species.

Alloparental care The assistance in the care of offspring by individuals other than the parents.

Allopatric speciation A mode of speciation that occurs when the ancestral population becomes divided by a geographic barrier.

Allotonic frequencies Echolocation frequencies of bats that are above or below the range of frequencies used by moths to detect predators.

Allozyme A particular amino acid sequence of an enzyme produced by a given allele at a gene locus when there are different possible forms of the enzyme.

Altricial Neonates that are born relatively helpless and require extended periods of parental care.

Altruistic behavior A behavior that enhances the evolutionary fitness of an unrelated individual while simultaneously decreasing the fitness of the individual performing the behavior.

Amniote A vertebrate whose embryo is enclosed in a fluid-filled membrane called the "amnion."

Analogous Structures in two or more organisms that perform a similar function but the similarity is not due to descent from a common ancestor.

Ancestral Of, or pertaining to, or inherited from, a common ancestor.

Angle of attack The angle the leading edge of the wing or airfoil makes with the oncoming airstream.

Anticoagulant A substance that prevents or delays the clotting or coagulation of the blood.

Antigen Foreign substance that induces specific immunity. Any substance that may be specifically bound to an antibody.

Apnea The temporary cessation of breathing.

Apomorphy In cladistics, a derived character state, modified from the ancestral state.

Aponeurosis A broad, flat sheet of tendon that may serve as the origin or insertion of a muscle.

Appeasement A behavior that serves to reduce the potential for conflict between conspecifics.

Arboreal Living mainly within the crowns of trees.

Aspect ratio The ratio of the wing span to the average wing chord in an airfoil.

Attenuated To be reduced or diminished in force or intensity.

Autochthonous Having originated within the area or habitat in which it now occurs.

Autotroph An organism that is capable of assimilating energy from either sunlight or inorganic compounds (i.e., green plant).

B

Baculum A bone within the penis found in certain orders of mammals; os penis.

Baleen plates The keratinized straining plates that form from the integument of the upper jaw in mysticete whales. They are used to filter small invertebrates from the seawater.

Band width Of or pertaining to the breadth or range of frequencies produced.

Basal metabolic rate The minimum metabolic rate of an animal measured when the animal is resting, postabsorptive, and within its thermal neutral zone.

Behavioral thermoregulation The ability to change behaviors to maintain an appropriate thermal balance.

Beringia The geographic region now comprising eastern Siberia, Alaska, the Yukon Territory, and the continental shelf between them (sea floor of the Bering Strait) that was periodically dry land during vast intervals of the Cenozoic era, forming a land bridge between Asia and North America at times when sea level was lowered.

Bilophodont Cheek teeth having an occlusal pattern with paired transverse ridges or lophs.

Biodiversity The variety of living organisms considered at all levels in all habitats and ecosystems.

Biogeography The study of the geographical distributions of organisms, their habitats, and the historical and ecological factors that produced the distributions.

Biotic potential The exponential growth of a population growing in an ideal and unlimited environment.

Bipedal locomotion Walking, hopping, or running using only the two hindlimbs.

Blastocyst An early stage in the developing mammalian embryo consisting of an outer trophoblast layer and the inner cell mass.

Blastodisc A late stage in the development of the inner cell mass in which these cells differentiate into two layers of cells that resemble the flattened disk of cells sitting atop the yolk in avian embryos.

Boreal Refers to one of the major phytogeographical areas characterized by cold temperate regions of the Northern Hemisphere with a flora consisting of coniferous forests or taiga.

Boundary layer A region of fluid (or air) passing closest to and flowing over a body surface. This layer has different physical properties than the fluid stream farther away from the surface.

Brachiation Arboreal locomotion by an alternating series of arm swings and grasps in which the body is suspended below the branches.

Brachydont Pertaining to cheek teeth with low crowns. Typical of mammals with omnivorous diets.

Bruce effect An effect demonstrated in mice where the presence of a strange male or his odor causes a female to abort her pregnancy and become receptive.

Bulla A typically rounded series of bones that partly or completely cover the middle and inner ear region in mammals.

Bunodont Pertaining to low crowned teeth with rounded or blunt cusps, used for crushing.

C

Caching The storage of particular food items for future use when food is less abundant.

Caecum A blind diverticulum extending from the junction of the small and large intestines. Often contains symbiotic micro-organisms in herbivores that aid in digestion of cellulose.

Calcar A cartilaginous or bony process that projects medially from the ankle in many species of microchiropteran bats and helps support the uropatagium. Analogous to the uropatagial spur of megachiropteran bats.

Camber The front to back curvature of an airfoil.

Canine A tooth posterior to the incisors and anterior to the premolars that is usually elongated, single-rooted, and single-cusped, and which is rooted in the maxilla or dentary.

Caniniform Canine-shaped.

Cannon bone A bone in the hindlimb resulting from the fusion of metatarsals III and IV. Typically found in cursorial artiodactyls.

Carnassials The co-functioning pair of bladelike shearing teeth of carnivorans, including the last upper premolar and the first lower molar.

Carotid rete An interwoven network of blood vessels formed from the carotid artery.

Carpus (or carpals) The bones of the wrist.

Carrion Dead animal matter used as a food source by scavengers.

Carrying capacity The maximum population size that can be sustained by the available resources in the environment.

Castoreum The oily fluid secreted from the preputial glands of beavers.

Catarrhine A member of a group of primates that includes the Old World monkeys (Cercopithecidae), gibbons (Hylobatidae), apes, and humans (Hominidae).

Caudal Pertaining to the tail; toward the tail.

Cementum The relatively soft bony material covering parts of the crown of a tooth in some mammals.

Cervical Of or pertaining to the neck; of or pertaining to the cervix of the uterus.

Character displacement The divergence in the characteristics of two otherwise similar species in regions where the two species ranges meet or overlap, caused by the competition in the area of overlap.

Character state One of several alternative forms of a character. For example, incisor enamel (a character) with no pigmentation, slight orange pigmentation, or heavy reddish brown pigmentation (character states).

Character A feature of an organism that can be described, measured, or effectively communicated between scientists. For example, pigmentation of incisor enamel.

Cheek teeth The premolar and molar teeth in mammals.

Chiropatagium The portion of the wing membrane of a bat that extends between the digits.

Chiropterophily Phenomenon describing plants that rely wholly or in part on bats for pollination.

Chorioallantoic placenta A type of placenta, found in eutherians and to a lesser extent in peramelemorph metatherians, composed of an outer chorionic layer and an inner vascularized allantois. The highly vascularized villi aid in maternal-fetal nutrient exchange.

Chorionic villi Finger-like projections from the chorion that invade the maternal tissues and form the placenta.

Choriovitelline placenta A type of placenta, often called a "yolk sac placenta," found in metatherians (except bandicoots) in which there are no villi and there is only a weak connection to the uterus.

Circadian rhythm A biological rhythm or cycle with a periodicity of approximately 24 hours.

Clade A set of species derived from a single common ancestor.

Cladistics A method of reconstructing a phylogenetic hypothesis that is based on grouping taxa solely by their shared derived character states.

Cladistic biogeography An approach to biogeography that emphasizes only vicariance events in the splitting of taxa in separate regions by using cladistics as a means of analyzing their divisions through phylogenetic relationships.

Cladist A person who practices the cladistic approach to phylogenetic reconstruction.

Cladogram A branching diagram that illustrates hypothetical relationships between taxa and shows the evolution of lineages of organisms that have diverged from a common ancestor.

Climax community The end point in a successional sequence of ecosystems that has reached a steady state under a particular set of environmental conditions.

Cloaca A common chamber into which the gut, urinary tubes, and reproductive tubes empty their contents prior to leaving the body.

Cloning The process of producing many identical copies of a gene; also the generation of many genetically identical copies of an organism.

Coactions A series of interactions between species.

Cochlea The spiral portion of the bony labyrinth of the inner ear that contains the organ of Corti.

Coevolution The mutual influence on the evolution of two different species interacting with each other and reciprocally influencing each other's adaptations.

Commensals Organisms having a symbiotic relationship in which one organism benefits from the relationship and the other organisms is neither helped nor harmed by the relationship.

Constant Frequency Signal Those signals that typically span less than 10kHz.

Convergence The evolution of similar characteristics in unrelated animals as a result of adaptation to similar environmental conditions or for similar functions.

Convergent evolution See **Convergence.**

Coprophagy The eating of feces.

Copulation The sexual coupling of two individuals.

Coracoid A bone in the pectoral girdle of monotremes (and reptiles and birds) between the scapula and the sternum. In mammals, other than monotremes, a process anterior to the glenoid fossa on the scapula.

Corpus callosum A broad band of nerve fibers that interlinks the right and left cerebral hemispheres.

Corpus luteum The progesterone-secreting mass of follicle cells that develops in the ovary after the egg has been released at ovulation.

Corridor A broad and more or less continuous connection between adjacent land masses or habitat types that allows for dispersal of organisms between the adjacent areas.

Cortex The outer layer or portion of an organ, bone, or hair.

Coteries In the society of prairie dogs, the basic, small group of individuals that occupies communal burrows.

Cotyledonary placenta A type of chorioallantoic placenta in which the villi are grouped into tufts or balls separated by regions of smooth chorion.

Countercurrent heat exchange The exchange of thermal energy between two fluid streams that are traveling in opposite directions in adjacent conduits (or blood vessels).

Crepuscular Active mostly near dawn and dusk.

Crescentic Crescent shaped.

Crown groups Late-appearing or terminal groups resulting from a major evolutionary radiation that possess the group's synapomorphy.

Culling The organized removal by killing of a specific number of individuals in a population usually for management purposes.

Cursorial Adapted for running.

Cutaneous evaporation Water loss across the skin's surface.

Cuticular scale One of a series of keratinous plates covering the shaft of a hair.

D

Deciduous teeth Teeth that are replaced usually early in a mammal's life.

Deciduous placenta A type of placenta in which a portion of the uterine wall is lost at birth.

Decomposers Organisms (typically including fungi and bacteria) that ingest and digest nutrients from nonliving organic matter such as corpses, dead plant material, and animal wastes, and convert them into inorganic form.

Delayed development A type of embryonic development in which the growth rate of the embryo slows following implantation in the uterine lining.

Delayed fertilization A situation in which mating occurs and sperm are deposited in the uterine tract of the female, but ovulation and subsequent fertilization does not occur for several months. The sperm remain viable in the female's reproductive tract during this time.

Delayed implantation A delay in the embedding of the blastocyst into the uterine lining that may last several days or up to several months.

Deme A local population within which breeding occurs more or less at random.

Dental formula A shorthand expression of the characteristic number of each type of teeth on one side of the skull in mammals. For example, the dental formula for a soricid shrew is 3/1, 1/1, 3/1, 3/3 = 32, which is shorthand for three upper incisors over one lower incisor, one upper and one lower canine, three upper over one lower premolar, and three upper and lower molars on each side of the jaw, for a total of 32 teeth.

Dentary The single bone making up one half of the mandible, or lower jaw, of mammals.

Dentine A bonelike material that forms the body of a tooth. It is hard and made of hydroxyapatite crystals and collagen and differs from bone in that it lacks osteocytes and osteons.

Derived Refers to a character state that is a modified version of, and differs from, that in the ancestral stock.

Desertification A process by which land becomes increasingly more arid and desertlike over time. It is often caused by removal of vegetation from semiarid areas by humans.

Diaphysis The shaft of a long bone.

Diastema A space between the teeth, usually the incisors and cheek teeth. Typical of rodents and lagomorphs but also found in artiodactyls, perissodactyls, and other mammals.

Didactylous Refers to the condition in metatherians in which the digits are unfused. See **Syndactylous.**

Diestrus The last stage in the estrus cycle in which progesterone levels increase and then decline. The duration of this period can vary.

Diffuse placenta A type of chorioallantoic placenta in which the villi are spread over the entire surface of the chorion.

Digitigrade A foot posture in which the balls of the feet (metapodial pads) support the weight and the "heel" of the hand or foot is off the ground, as in cats and dogs.

Diphyodont A pattern of tooth replacement involving only two sets of teeth, typically a set of deciduous ("milk") teeth and a set of permanent teeth.

Diploid number The total number of chromosomes in the cell nucleus of a somatic cell.

Discoidal placenta A type of chorioallantoic placenta in which the villi are restricted to a disk-shaped region of the chorion.

Disjunct Distinctly separate or discontinuous ranges in which one or more populations are separated from other populations by sufficient distance to prevent gene flow between them.

Dispersal The one-way movement or spreading of organisms from the natal area to new areas.

Diurnal Active primarily during daylight hours.

DNA fingerprinting A technique that uses the unique pattern of DNA base pairs to identify a specific individual. Typically used to establish paternity.

Doppler shift The change in frequency of a sound produced by a moving object such that a pulse of sound hitting a target that is moving toward the source is compressed, resulting in a higher frequency than the original sound.

Drag Resistance to the motion of a body through water or air.

E

Echolocation The process of emitting sounds and using the information from the returning echoes to sense the surrounding environment.

Ecological succession The process by which species in a habitat are gradually replaced through a regular progression culminating in a stable climax community.

Ecological zoogeography The study of the distribution of animal species in relation to their life histories, dispersal abilities, and interspecific relationships.

Ecotone A habitat created by the juxtaposition of distinctly different habitats, such as a zone of transition between woodland and grassland.

Ectotherm An animal whose body temperature is determined primarily by passive heat exchange with its environment.

Egg tooth A temporary toothlike structure at the tip of the bill (in birds), used to crack and open the egg during hatching.

Eimer's organ A specialized touch receptor located on the snouts of moles and desmans.

Electroreceptor A sensory receptor that responds to changes in electric field intensity.

Embryonic diapause A period of arrested development of the blastocyst typical of macropods (kangaroos and wallabies).

Embryotroph Tissue debris resulting from the invasion of the blastocyst into the uterine lining that provides some nourishment for the embryo during the early stages of development.

Emigration Movement of individuals out of a population or away from a given region.

Enamel   Hard crystalline material in the teeth of vertebrates, similar in composition to bone, but without bone-forming osteocytes.

Endemic   Pertaining to a species that is native to a given geographic region and not found in any other regions.

Endogenous   Originating within the organism.

Endometrial gland   Secretory gland of the mucous-membrane lining of the uterus.

Endometrium   The mucous membrane lining the inner surface of the uterus into which the blastocyst implants during gestation.

Endotheliochorial placenta   An arrangement of the chorioallantoic placenta in which the chorion of the embryo is in direct contact with the maternal capillaries.

Endotherm   An animal whose body temperature is elevated substantially above the ambient temperature by internal, metabolic heat production.

Entoconid   A major cusp found in the lingual portion of the talonid of the lower molars.

Epiphysis   The head(s) of a bone, usually bearing an articular surface. A secondary center of ossification.

Epipubic bones   A pair of bones that extend anteriorly from the pubic bones of the pelvis in reptiles, monotremes, and most metatherians.

Epitheliochorial placenta   A type of chorioallantoic placenta in which the villi rest in pockets in the endometrium, but the fetal blood supply is separated from the maternal blood supply by six tissue layers.

Epizootic   Pertaining to disease that spreads rapidly through an animal population.

Estivation   A period of inactivity in response to hot, dry environments, characterized by lowered metabolic rates and lower heart rates.

Estrus   The period during which a female mammal will permit copulation. Also called "heat" in domesticated mammals.

Eupnea   Normal, quiet breathing.

Eusocial   Pertaining to a social system in which individuals live in the groups of overlapping generations and in which one or a few individuals produce all the offspring and the rest of the colony members serve as functionally sterile helpers in rearing young, procuring food, and defending the colony.

Evaporative cooling   Loss of heat through the evaporation of sweat or saliva from the skin, or of water vapor from the nasal mucosa or lungs.

Ever-growing teeth   Teeth that continue to grow throughout the life of the animal. Sometimes called "rootless" or "hypselodont."

Exogenous   Originating from outside the organism.

Extant   Currently in existence; not extinct.

Extension   Movement of an appendage so that the angle of the joint increases.

External auditory meatus   A passageway that leads from the base of the pinna or surface of the head to the tympanic membrane.

Extirpation   The extermination of a population or taxon from a given area.

F

Facultative monogamy   Monogamous depending on the circumstances or conditions.

Fecundity   The number of offspring that an individual produces during a given amount of time.

Female choice   The situation in which a female selects a mate based on his superior quality relative to those she does not select to mate with.

Female defense polygyny   A mating system in which males control access to females by directly interfering with other males.

Fenestrated   Perforated by one or many openings (referring to a bone surface). "Windowed."

Fertilization   Process by which the nucleus of the sperm fuses with the nucleus of the oocyte to produce a zygote.

Filter route   A narrow corridor of suitable habitat that acts as a selective filter blocking the passage of some but not all animals that attempt to disperse across it.

Fission-fusion social system   A social system found in some primate species in which group members leave temporarily and then may rejoin the group later.

Flehmen   A behavior done by male mammals in which the upper lip is retracted after smelling the urine (or pheromones) of a reproductively receptive female.

Flexion   A movement that reduces the angle between two articulating bones.

Fluke   The horizontal tail "fin" of a cetacean or sirenian.

Folivore   An animal whose primary diet consists of leaves.

Food web   A representation of the many paths of energy flow from one organism to another in a community by eating or being eaten.

Foramen   An opening or passage through a bone. Typically the opening transmits nerves or blood vessels.

Fossorial   Pertaining to an animal that digs burrows for shelter and forages underground.

Fovea centralis   The portion of the retina providing the sharpest vision, with the highest concentration of cones in those species with color vision.

Frequency   In acoustics, the number of wave lengths per second (expressed in Hertz or Kilohertz).

Frugivorous   Having a diet consisting primarily of fruit.

Frequency Modulated Signal   Acoustic signals that span more than 10 kilohertz.

Fundamental Frequency   The lowest or root tone of a chord.

Fusiform   A streamlined shape in which both ends are tapered.

G

Gametogenesis   The formation of the gametes (male and female reproductive cells).

Gape   The extent to which the mouth can be opened; the act of opening the jaws.

Gastric fermentation   The anaerobic digestion of food in the stomach chambers by microbes.

Genetic engineering   The artificial alteration of the genetic constitution of cells or individuals by direct modification, insertion, or deletion of individual genes. In some cases, new combinations of genes are made by joining DNA fragments from two or more different organisms.

Genotype   The genetic constitution of an individual or one of its cells, often referring to alleles of one or more genes.

Geochronology   The measurement of geological time as recorded in the earth's rock layers.

Geophyte   A plant that grows from and produces underground storage organs such as tubers, corms, or bulbs.

Gestation   The period of time from fertilization to birth in mammals that have embryos that develop within the body of the parent.

Graviportal   A mode of locomotion in large terrestrial mammals in which the limbs are straight and column-like in order to support the great mass of the animal.

Gregarious   Living in groups or herds.

Guano   Feces, especially that of bats or birds, that is harvestable for its nitrates and phosphates.

Guild   A group of populations that exploit a common set of resources in a similar manner.

Gynodioecy   A condition in some plant species in which there are separate females and hermaphrodites.

H

Hallux   The most medial (first) digit of the hind foot. "Big toe."

Halophytic   Refers to plants that can live in salty soils.

Harem   A group of females guarded by a male who prevents other males from mating with the females in his group.

Harmonics   Frequencies that are integral multiples of the fundamental frequency.

Hemochorial placenta   A type of chorioallantoic placenta in which the villi are in direct contact with the maternal blood supply.

Hemoendothelial placenta   A type of chorioallantoic placenta in which the fetal capillaries are surrounded by maternal blood. This condition shows the least separation between fetal and maternal blood supplies.

Hermaphrodite   An individual that has both male and female reproductive organs.

Heterodont   Pertaining to teeth that vary in structure in different parts of the jaws; for example, the teeth of a mammal are usually differentiated into incisors, canines, premolars, and molars.

Heterothermic   Referring to animals that normally have high body temperatures regulated by metabolic heat production, that can also allow their body temperatures to drop close to ambient temperatures.

Heterozygous   A genotype or individual that possesses different alleles at a particular gene locus.

Hibernaculum   The place in which an animal hibernates.

Hibernation   A period of inactivity normally induced by cold ambient temperatures characterized by lowered body temperatures and a depressed metabolic rate.

High duty cycle   Refers to acoustic signals that are emitted for more than 600 milliseconds out of 1000 milliseconds.

Historical zoogeography   The study of the distributions of animal species or higher taxa in terms of their past distributions and the physical history of the earth.

Holarctic   A zoogeographical region comprising the Palearctic and Nearctic regions (i.e., temperate and arctic Eurasia and North America).

Homeotherm   An animal that can maintain a constant body temperature by physiological means, regardless of the ambient temperature.

Home range   The area in which an individual or group spends the bulk of its time.

Homodont   Pertaining to teeth that do not vary in structure in different parts of the jaws, such as the numerous and identical conical teeth of many dolphins.

Homologous   Pertaining to structures or properties having a similar phylogenetic origin but not necessarily retaining a similar function, behavior, or identical structure.

Homoplasy   A similarity in a character in two different species that arises from evolutionary convergence or parallelism, not from common ancestry.

Hygroscopic   Able to extract moisture from the atmosphere.

Hyoid apparatus   A series of bones, derived from the gill arch supports, that support the muscles of the tongue.

Hyperphalangy   The presence of extra bones (phalanges) in the digits.

Hypoconid   The main cusp on the labial side of the talonid of the lower molars.

Hypoconulid   A prominent cusp on the posterior portion of the talonid of the lower molars.

Hypothermia   A condition of lower than normal body temperature in mammals.

Hypsodont   Pertaining to cheek teeth with high crowns.

Hystricognathous   A condition in certain rodents in which the angular process of the mandible is usually lateral to the plane of the alveolus of the lower incisor.

Hystricomorphous   A condition in certain rodents in which the infraorbital foramen is greatly enlarged and transmits a portion of the medial masseter muscle.

I

Ilium   The largest and most dorsal of the three bones of the pelvis. Its fusion with the sacrum solidly braces the hind limbs.

Immigration   The movement of individuals into a population, or into a given region.

Immune response   A response to a foreign substance or antigen by the body's immune system in which antibodies and immune system cells are mobilized to fight off the foreign invader.

Immunoglobin   One of a class of proteins comprising the antibodies.

Inbreeding   Mating among related individuals.

Inbreeding depression   The decrease in the average value of a characteristic, such as growth, health, fertility, and survival as a result of inbreeding.

Incisor   A tooth in mammals located anterior to the canines and rooted in the premaxillae (upper) or dentary (lower).

Inclusive fitness   The sum of an individual's own evolutionary fitness (as measured in descendant relatives) plus the indirect fitness of nondescendant relatives (i.e., cousins).

Induced ovulation   Release of an ovum triggered by the act of copulation.

Infanticide   The killing of offspring.

Infrasound   Sound of very low frequency (less than 20 Hz; below the range of human hearing), such as that produced by elephants for long distance communication.

Infundibulum   A funnel-shaped opening of the oviduct near the ovary that receives the oocytes at ovulation.

Ingroup   A group of organisms that is under cladistic analysis; a set of taxa that are presumed to be more closely related to one another than any is to an outgroup.

Interstitial fluid   The fluid in the tissues that fills the spaces between cells.

Intestinal fermentation   Anaerobic digestion of food by microbes in the caecum.

Interspecific   Between different species.

Intraspecific   Between individuals of the same species; within one species.

Ischial callosities   Pads of tough, horny skin attached to the flattened parts of the ischium.

Ischium   One of the three bones of the pelvis.

J

Joey   A young kangaroo that is still nursing but is no longer restricted to the pouch.

K

Karyotype   The characteristic chromosome complement of a cell, individual, or species.

Keystone species   A species that has a disproportionately large effect on other species in a community.

Kin   Related individuals.

Kin selection   Selection that affects the survival and reproductive success of genetically related individuals.

Kleptoparasites   An animal that steals food procured by other animals.

K-selection   Selection based on a population being maintained at or near the carrying capacity of the environment.

L

Lactation   A unique feature of mammals in which milk is formed and secreted by the female's mammary glands for nourishing the developing young after birth.

Lambdoidal crest   A bony crest or ridge between the parietal and occipital bones of the skull. Also called the "nuchal crest."

Land bridge   A dry land connection between land masses that forms a potential migration or dispersal route.

Larderhoarding   Storing large quantities of food at a central location within an animal's territory.

Lek   A communal courtship area that is regularly used by several males to attract and mate with females.

Lingual   Of, or pertaining to, the tongue. Sometimes used to refer to a structure that is on the tongue side of the toothrow.

Locus   The location of a gene on a chromosome; the location of a tooth in the toothrow.

Lophodont   Pertaining to cheek teeth in which there are a series of transverse ridges or lophs, on the occlusal (chewing) surface.

Lordosis   A behavior performed by females in which the lumbar curvature is exaggerated, lifting the genitalia and signaling a willingness to mate.

Low duty cycle   Refers to acoustic signals emitted for only a small percentage of the time.

Lower critical temperature   Ambient temperature below which the basal metabolic rate becomes insufficient to balance heat loss.

Lumbar   Pertaining to the lower back region and the vertebrae between the ribcage and the pelvis.

M

Major histocompatibility complex   A series of surface antigens that plays an important role in the coordination and activation of the immune response to foreign substances.

Male-male competition   Refers to competition between males for access to females during the breeding season.

Mammary gland   Milk-producing gland found in female mammals. The growth and activity of mammary glands are governed by several reproductive hormones.

Mandibular ramus   One of the major "branches" or portions of the dentary bone of the lower jaw. The horizontal ramus holds the teeth; the ascending ramus articulates with the skull.

Manus   The hand or forefoot.

Marsupium   An external pouch formed by folds of skin in the abdominal wall that encloses the mammary glands and serves as a protective incubation chamber for the young in many metatherians and some monotremes.

Mastication   The act of chewing.

Matriarchal   Pertaining to a society in which the bulk of activities and behaviors are centered around a dominant female.

Matrilineal group   Group in which rank in the dominance hierarchy is passed down from a mother to her offspring.

Medulla   The inner layer or core of an organ or hair.

Menstruation   A special type of estrus cycle found in some female primates in which the endometrial lining of the uterus regresses after ovulation and breaks down, resulting in the passage of blood and cellular debris through the vaginal opening.

Meroblastic   A type of embryonic cleavage in which the cleavage plane does not pass completely through the embryo before the next cleavage event begins, usually due to the large amount of yolk.

Mesaxonic   Pertaining to a type of foot in which the plane of symmetry passes through the third digit.

Mesic   Refers to habitats with moderate rainfall and humidity.

Metabolic water   Water produced by the aerobic breakdown of food.

Metacone   A main cusp on the posterior, labial side of the upper molars.

Metaconid   A main cusp on the posterior, lingual side of the trigonid in the lower cheek teeth.

Metaphysis   The region between the epiphysis and diaphysis where the epiphyseal plate is located in developing long bones.

Metapopulation   A network of semi-isolated populations with some level of regular gene flow and in which individual populations may go extinct but can be recolonized from one of the adjoining populations.

Metestrus   The third stage in the estrus cycle, in which the corpora lutea are fully formed and progesterone levels are high.

Microbial fermentation   The anaerobic breakdown of food by symbiotic protozoans and bacteria.

Midden   A refuse area in the nest of a wood rat or other mammal.

Molar   A cheek tooth located posterior to the premolars and having no deciduous precursor.

Molariform   Pertaining to teeth that have the shape and appearance of molars but which are not molars.

Molecular clock   A hypothetical means of measuring evolutionary time that is based in the assumption that the rate at which mutational changes accumulate is constant over time and therefore the changes are useful for dating the divergence of lineages. Also, a self-sustaining, autonomous, genetically regulated timekeeper within the hypothalamus of the brain that can be adjusted by daylight as sensed by the retina and perhaps by other environmental cues.

Molecular phylogeny   A hypothetical representation of the evolutionary history of a group of organisms based on characters defined at the molecular level.

Monestrus   A condition in which there is only one period of estrus per year.

Monodactyl   Having a single functional digit in the manus or pes.

Monogamous   Pertaining to a mating system in which one male and one female remain together to rear at least one litter.

Monophyletic   Refers to a group of organisms whose members are all descended from (and including) a common ancestor.

Morula   A cluster of cells (blastomeres) produced by the early mitotic division of the zygote.

Musk gland   A specialized gland, found in various bodily locations in different taxa, that secretes scent.

Musth   A period of heightened reproductive activity in male elephants.

Myometrium   The thick layer of smooth muscle in the wall of the uterus.

Myomorphous   A pattern of jaw musculature in rodents in which a slip of the medial masseter passes through an oval or V-shaped infraorbital foramen.

Myrmecophagous   Feeding primarily on social insects such as ants and termites. Sometimes called formicivorous.

Mystacial pad   The region on the snout from which most of the vibrissae originate.

N

Natal   Of or pertaining to birth.

Neonate   Newborn animal.

Neopallium   The expanded nonolfactory portion of the cerebral cortex in mammals.

Niche   The role of a species in an ecological community.

Nictitating membrane   A transparent membrane beneath the eyelid of some vertebrates that can cover and protect the eye.

Nocturnal   Active mainly during the nighttime.

Node   Refers to the place on a cladogram where two lineages diverge.

Nondeciduous placenta   A type of placenta that separates easily from the uterine wall, resulting in little or no damage to the uterus.

Nonshivering thermogenesis   A process in which fats are oxidized by enzymes to produce heat.

Normothermia   A state in which the body temperature is within the normal or preferred range of temperatures.

O

Obligate monogamy   A form of monogamy in which a male and female must pair in order to rear offspring.

Omasum   The third chamber of the stomach of a ruminant.

Omnivorous   A diet consisting of both animal and plant material.

Ontogenetic   Refers to the course of development from a zygote to an adult.

Oocyte   A developing ovum.

Organogenesis   The embryonic formation of organs beginning at the onset of gastrulation.

Os penis   See **Baculum.**

Ossicle   Any small bone, but usually refers to one of the three middle ear bones.

Osteoscute   A thin, bony scalelike plate.

Outgroup   A group used for taxonomic comparison that is related to but not part of the group under study.

Oviparous   A reproductive pattern involving the laying of eggs.

Ovoviviparous   A reproductive pattern in which the young hatch from eggs that are retained within the mother's uterus.

Ovulation   The process by which a mature ovum is released from the ovarian follicle.

P

Palmar   Of, or pertaining to, the ventral surface of the foot or hand.

Panting   A method of cooling by rapid, shallow breathing that increases the rate of evaporation of water from the respiratory surfaces.

Paracone   A main cusp anterior to the protocone and on the labial side of the upper molars.

Paraconid   A main cusp anterior to the protoconid and on the lingual side of the lower molars.

Parallelism   Evolutionary change in two or more lineages such that the corresponding features undergo equivalent alterations.

Paraphyletic   A taxonomic group in which some, but not all, members are descendants of a single common ancestor.

Parasagittal   A dorsoventrally oriented plane parallel to the long axis of the body and parallel to the midsagittal plane.

Paraxonic   A type of foot anatomy in which the plane of symmetry passes between a pair of similar-size digits.

Parturition   The process of giving birth.

Paternity   The state of being a father to an offspring.

Pedicel   A bony supporting structure for an antler.

Pelage   All of the hairs on an individual mammal.

Peristaltic   A travelling wave of contraction within a tubular structure such as the walls of the digestive tract.

Pes   The hind foot.

Pheromone   A volatile chemical signal used to communicate physiological or behavioral state to another member of the same species.

Photoperiod   A physiological response to daylength.

Phylogenetic relationship   See **Phylogeny.**

Phylogeny   The evolutionary history of a taxon or group of related taxa; also, a hypothesis graphically describing such a history.

Phylogeography   An approach to biogeography that integrates morphological, molecular genetic, historical geologic, paleontological, ecological, and phylogenetic data in interpreting the distribution of common units of taxa or biotas.

Phytoestrogen   A plant secondary compound that mimics the effects of estrogen hormones.

Pinna   The external ears that surround the external auditory canal and serve to funnel sound waves to the tympanic membrane.

Pinna gain   The ability of the pinna to focus and amplify sound pressures at the tympanic membrane.

Plagiopatagium   The membrane in a bat wing that extends between the body and hindlimbs to the arm and fifth digit.

Plantigrade   A style of locomotion in which the entire side of the foot (including the proximal ends of the metapodials) touches the ground.

Plesiomorphy   A character state that is ancestral for the taxa under study.

Poikilotherm   An animal whose body temperature tends to fluctuate more or less with the ambient temperature.

Polarity   The direction of evolutionary transformation in a character.

Pollex   The most medial digit of the manus (forefoot). "Thumb."

Pollination   The process by which pollen is placed on the stigma of the carpal of a flower, resulting in fertilization.

Polyandry   A mating system in which females acquire more than one male mate.

Polyestrus   The condition of having several estrus periods per year.

Polyphilous   The condition of a plant species having more than one pollinator.

Polygamous   A mating system in which both males and females mate with multiple members of the opposite sex.

Polygenic trait   Trait in which two or more genes have additive effects on a single phenotypic characteristic.

Polygynous   A mating system in which males mate with more than one female and females typically provide most of the parental care.

Population   Groups of individuals of the same species, inhabiting the same place and time.

Postpartum   Refers to the period after birth.

Precocial   Refers to young that are born relatively well developed and require minimal parental care.

Prehensile   Structures capable of grasping by wrapping around an object. Usually refers to tails that are capable of grasping and holding onto objects.

Premolar   Cheek tooth located anterior to the molars and posterior to the canines.

Preorbital vacuity   An opening in the skull bones anterior to the orbit. Found in bovids.

Prepuce   A fold of skin covering the glans penis.

Presaturation dispersal   Movement away from the natal area that occurs prior to the peak population density.

Proboscis   A long, more or less flexible, snout such as those found in tapirs and elephants.

Procumbent   Pertaining to teeth that project forward.

Proestrus   The initial stage in the estrus cycle when estrogen, progesterone, and luteinizing hormone are elevated.

Promiscuity   A mating system in which multiple matings by both sexes occur and there is no prolonged association between the mating pair.

Propatagium   The small membrane in the bat wing that extends between the arm and the occipitopollicalis muscle along the leading edge of the wing.

Protocone   The major cusp on the lingual side of upper cheek teeth.

Protoconid   A main cusp on the labial side of lower cheek teeth that is located at the apex of the trigonid.

Protrogomorphous   The ancestral condition in some rodents in which the masseter muscles arise solely from the zygomatic arch and do not penetrate the infraorbital foramen.

Protrusible   Capable of being thrust forward.

Puberty   The period during which an animal becomes sexually mature and capable of reproducing.

Pubis   One of the three bones of the pelvis.

Pulmocutaneous evaporation   Water loss that occurs across the pulmonary and skin surfaces.

R

Reciprocal altruism   A situation in which the short-term costs of providing some other individual with some resource or behavior is offset when the recipient returns the favor at a later time.

Refugium   Geographic area that provides temporary shelter or protection.

Regional heterothermy   The ability of some mammals to allow the temperatures of the skin or extremities to drop well below the core body temperature.

Regurgitation   A reverse movement of food from the stomach back to the buccal cavity, produced by reverse peristalsis.

Reproductive potential   See **Biotic potential.**

Rete mirabile   An extensive countercurrent arrangement of arterial and venous capillaries.

Reticulum   The second of four chambers in a ruminant stomach.

Reynolds numbers   A unitless number that predicts the tendency of a flowing liquid (or gas) to become turbulent, which is proportional to its velocity and density, but inversely proportional to its viscosity.

Rhinarium   An area of moist, hairless skin surrounding the nostrils.

Ricochetal   A style of saltatorial locomotion that involves quick changes in the direction of travel.

Riparian   Habitat along the banks of a waterway.

R-selected   Form of natural selection that favors rapid reproductive rates and growth rates and is typical of species found in unstable habitats.

Rumen   The first of four chambers in the ruminant stomach, actually formed from an enlargement of the lower part of the esophagus.

Ruminant   An artiodactyl that has a multichambered stomach and is able to rechew regurgitated plant material (cud) to improve its digestibility by symbiotic microbes.

S

Sacral   Of or pertaining to the vertebrae that are fused into a single sacrum to which the pelvic girdle attaches.

Sagittal crest   A vertical flange of bone on the dorsal midline of the braincase of a mammalian skull that increases the area of origin for the temporalis muscle.

Saltatorial   A style of locomotion involving repeated jumping or leaping.

Saturation dispersal   Movement out of the natal area that occurs when the population is at its peak density.

Scatterhoarding   The storage of a few food items at many scattered locations within an animal's territory.

Scent gland   Sweat or sebaceous gland modified for the production of odoriferous secretions.

Sciurognathous   A type of mandible in which the angular process is medial to or approximately in line with the plane of the incisor alveolus.

Sciuromorphous   A condition in which the masseter muscles originate on the zygomatic arch and from an expanded zygomatic plate up onto the rostrum, but no portion of the masseter passes through the infraorbital foramen.

Sclerotic cartilage   A ring of cartilage associated with the tough outer layer of the eye that forms the white area.

Scrotum   The pouch that encloses the testes in those species with descended testes.

Sebaceous gland   A skin gland that secretes the waxy or oily substance called "sebum."

Secondary palate   The bony partition comprised of parts of the premaxilla, maxilla, and palatine bones that separates the dorsal airway from the nostrils from the buccal cavity.

Sectorial   Pertaining to a tooth adapted for cutting.

Seismic signal   Communication signal produced by kangaroo rats that involves a series of low frequency vibrations that travel through the ground.

Selenodont   An occlusal pattern of a tooth in which longitudinally arranged crescent-shaped ridges or lophs are formed on the tooth surface. Typically occurs in artiodactyls.

Semifossorial   Refers to mammals that are partially, but not completely, adapted for life underground or for digging.

Seral stage   Successional stage on the way to formation of a climax community.

Serrate   Condition in which the edge of a structure is toothed or notched as in a saw blade.

Sex ratio   The ratio of males to females in a population.

Sexual dimorphism   The condition that exists when there are externally apparent differences between the males and females of a species.

Sister group   In a phylogeny, the monophyletic group most closely related to a monophyletic group under study. One of two clades that resulted from the splitting of a single lineage.

Spatial heterogeneity   Refers to a situation in which adjacent habitats over some area differ in the quality and quantity of the resources and species they contain.

Spatulate   Broad and rather flattened with a narrowed base.

Species   A named kind of organisms; the basic unit of biological classification. A group of potentially interbreeding natural populations that are capable of producing viable offspring, but not capable of reproducing with other such groups.

Spermaceti organ   An organ found in the head of certain odontocete whales that contains a waxy liquid.

Spermatogenesis   The series of cell divisions that leads to the production of mature sperm.

Spontaneous ovulation   Ovulation that occurs without a triggering event such as copulation.

Streamlined   See **Fusiform.**

Stridulating organ   An organ capable of producing sound by the rubbing together of modified hairs (in tenrecs) or parts of the exoskeleton (in insects).

Suckling   The process of taking nourishment from the breast or teat in mammals.

Subspecies   A relatively uniform and genetically distinct population of a species, often in a specific geographic region.

Suture   A contact line or tight joint between two bones such as the bones of the skull.

Sweat gland   Long tubular gland that extends from the dermis to the skin surface and secretes perspiration or scent.

Sweepstakes route   A chance, or accidental, crossing of a water barrier or other geographic barrier.

Symbiosis   An interaction between two species in which one benefits and the other either benefits, is harmed, or is unaffected.

Sympatric   Refers to two or more populations that occupy the same or overlapping geographic areas.

Symplesiomorphy   In phylogenetics, an ancestral character shared by two or more groups.

Synapomorphy   In phylogenetics, a derived, homologous character shared by two or more groups.

Syndactylous   A condition in which certain of the digits in metatherians (usually the second and third digits) are fused and share a common sheath of skin. See **Didactylous.**

Syndesmochorial placenta   A type of chorioallantoic placenta in which there is one less layer of tissue separating the maternal and fetal blood supply than is found in the epitheliochorial placenta.

Systematics   The study of patterns and processes of evolution that are used to construct phylogenies or classify organisms.

Systematist   One who practices systematics.

T

Talonid   The "heel" or back half of a tribosphenic lower molar that occludes with the protocone of an upper molar.

Tapetum lucidum   A reflective choroid layer in the eyes of nocturnal animals that aids in night vision.

Target ranging   The ability of bats to estimate the distance to a target by comparing the time it takes a pulse to reach the target with the time it takes the echo to return.

Tarsus (or tarsals)   The ankle bones.

Taxonomy   The practice of naming and classifying organisms.

Telolecithal   Refers to eggs that have large amounts of yolk at the vegetal pole relative to the amount of active cytoplasm at the animal pole.

Temporal fenestra   A hole or opening in the temporal region of the skull.

Territory   An area occupied and defended by an animal or group of animals.

Thegosis   A process in some mammals in which upper and lower teeth hone against one another, tending to sharpen or maintain an edge on the teeth as they wear away.

Thermal conductance   Heat loss from the skin surface to the environment.

Thermal inertia   A phenomenon by which animals of large body size are able to retain body heat longer than smaller animals.

Thermal neutral zone   A range of ambient temperatures within which an animal maintains its basal metabolic rate at a relatively constant and minimal level.

Thermogenesis   The generation of heat.

Thoracic   Pertaining to the region of the thorax.

Torpor   A type of adaptive hypothermia (dormancy) in which heart rate, body temperature, and respiration are reduced.

Tragus   The projection from the lower medial border of the pinna found in many microchiropteran bats.

Traplining   A foraging behavior in which the animal daily follows a particular path from one food resource to the next, stopping to feed at each.

Trenchant   Sharp; able to cut.

Tribosphenic   A type of molar in which the protocone of a three-cusped upper tooth fits into a basin in the talonid of a lower tooth (for crushing food), and in which crests between other cusps shear past one another (for cutting food).

Trichromatic vision   Color vision based on the presence of three main photopigments.

Trigonid   The front half of a tribosphenic lower molar, which includes a triangle formed by three cusps.

Trioecy   A reproductive pattern in plants in which individual plants can be either males, females, or hermaphrodites.

Trophoblast   The superficial layer of the blastocyst that will be involved with implantation, placenta formation, and hormone production.

Tuberculosectorial   A kind of tooth that has cusps (tubercles) for crushing and crests or ridges for shearing or sectioning food.

Turbinal bones   Convoluted or scroll-shaped bones within the nasal passage of the skull that provide increased surface area for moisturizing and warming inhaled air.

U

Ultrasonic   Sounds at frequencies beyond human hearing capability (i.e., above 20 kilohertz).

Ungulate   Mammal with hoofs.

Unguligrade   A type of locomotion in which only the hoofs (nails of the terminal phalanges) touch the ground when running.

Unicuspid   Teeth with a single cusp, such as a canine tooth.

Upper critical temperature   The temperature at which an animal must dissipate heat to maintain thermal homeostasis.

Uropatagial spur   A cartilaginous spur or rod that projects from the gastrocnemius tendon, found in megachiropteran bats. Analogous to the calcar of microchiropteran bats.

Uropatagium   A flight membrane in some bats that extends between the hindlimbs and the tail.

Uterine mucosa   The inner lining of the uterus.

V

Valvular Capable of being temporarily closed off, such as in the nostrils of sirenians.

Vasoconstriction The narrowing of a blood vessel.

Venter The belly surface.

Vermiform Having a long, thin, and flexible form.

Vibrissae The long, stiff hairs on the snout that serve as tactile receptors.

Vicariance In biogeography and evolution, the divergence of two populations from one original population that was split by the formation of a natural barrier.

Viviparity The ability to give birth to live young rather than laying eggs.

Vomeronasal organ A branch of the nasal epithelium that forms a pocket that may open to the mouth cavity. Important in sensing chemical behavioral cues.

W

Waif dispersal A chance movement to a new area. May be achieved, for example, by drifting on the ocean currents.

Wallace's line An imaginary line that separates the Oriental and Australian zoogeographical regions. The line passes between the Philippines and Maluku (Moluccas) in the north and between Sulawesi and Borneo and between Lombok and Bali in the south.

Warren A communal series of burrows used by a group of rabbits.

Water balance A suite of behavioral and physiological responses to prevent dehydration.

Weapon automimicry A feature in the appearance of an animal (such as facial stripes or ear shape) that mimics the possession of a "weapon" such as a horn.

Wing loading The body mass of a flyer divided by the area of its airfoils.

X

Xenarthrism A condition in which extra-articular surfaces are found on the posterior trunk vertebrae of xenarthran mammals (sloths, armadillos, and anteaters).

Z

Zona pellucida A thick, elastic covering of the egg cell.

Zonary placenta A type of chorioallantoic placenta in which the villi surround the fetus in a band.

Zoogeography The study of the geographical distribution of animals and their ecological communities.

Zygomatic arch An arch of bone on the side of a mammal skull that is formed by the jugal bone and a process of the squamosal bone.

Zygapophysis One of the articular surfaces between two adjacent vertebrae.

Zygote A fertilized egg.

BIBLIOGRAPHY

Note: This list of references is organized alphabetically by author's last name. Publications by the same author are listed in chronological order. Following the list of single-authored works by a particular author are the multiauthored publications listed in chronological order. In cases where several authors share the same last name, the references are listed alphabetically by author's first initial.

Ables, E. D. 1969. Home range studies of red foxes (*Vulpes vulpes*). J. Mamm., 50:108–120.

Adcock, E. W., F. Teasdale, C. S. August, S. Cox, G. Meschia, F. C. Battaglia, and M. A. Naughton. 1973. Human chorionic gonadotropin: Its possible role in maternal lymphocyte suppression. Science, 181:845–847.

Adkins, R. M., and R. L. Honeycutt. 1993. A molecular examination of archontan and chiropteran monophyly, 227–249, in *Primates and Their Relatives in Phylogenetic Perspective* (R. D. E. MacPhee, ed.). Plenum Press, New York.

Aiello, A. 1985. Sloth hair: unanswered questions, 213–218, in *The Evolution and Ecology of Armadillos, Sloths, and Vermilinguas* (G. G. Montgomery, ed.). Smithsonian Inst. Press, Washington, D.C.

Akersten, W. A. 1985. Canine function in *Smilodon* (Mammalia; Felidae; Machairodontinae). Contrib. Sci. Nat. Hist. Mus. Los Angeles Co., 356:1–22.

Alberdi, M. T., G. Leone, and E. P. Tonni (eds.). 1995. *Evolución Biológica y Climática de la Región Pampeana Durante los Últimos Cinco Millones de Años.* Un Ensayo de correlación con el Mediterráneo Occidental. Monografías Museo Nacional de Ciencias Naturales, Consejo Superior de Investigaciones Científicas, Madrid.

Albuja V. L., and B. D. Patterson. 1996. A new species of northern shrew-opossum (Paucituberculata: Caenolestidae) from the Cordillera del Cóndor, Ecuador. J. Mamm., 77:41–53.

Alcorn, S. M., S. E. McGregor, G. D. Butler Jr., and E. B. Kurtz Jr. 1959. Pollination requirements of the saguaro (*Carnegiea gigantea*). Cactus Succ. J. Amer., 31:39–41.

Aldridge, H. D. J. N., and I. L. Rautenbach. 1987. Morphology, echolocation and resource partitioning in insectivorous bats. J. Animal Ecol. 56:763–778.

Alexander, R. D. 1974. The evolution of social behavior. Ann. Rev. Ecol. Syst., 5:325–383.

Alexander, R. D., K. M. Noonan, and B. J. Crespi. 1991. The evolution of eusociality, 3–44, in *The Biology of the Naked Mole-rat* (P. W. Sherman, J. U. M. Jarvis, and R. D. Alexander, eds.). Princeton Univ. Press, Princeton, New Jersey.

Alexander, R. M. 1982. *Locomotion of Animals.* Blackie, Glasgow.

Alexander, R. M. 1992. *Exploring Biomechanics: Animals in Motion.* Sci. Amer. Library, New York.

Allard, M. W., M. M. Miyamoto, L. Jarecki, F. Kraus, and M. R. Tennant. 1992. DNA systematics and evolution of the artiodactyl family Bovidae. Proc. Nat. Acad. Sci., 89:3972–3976.

Allen, G. M. 1940. *The Mammals of China and Mongolia,* part 2. Amer. Mus. Nat. Hist., New York, pp. 621–1350.

Allin, E. F. 1975. Evolution of the mammalian middle ear. J. Morph., 147:403–437.

Allin, E. F., and J. A. Hopson. 1992. Evolution of the auditory system in Synapsida ("mammal-like reptiles" and primitive mammals) as seen in the fossil record, 587–614, in *The Evolutionary Biology of Hearing* (D. P. Webster, R. R. Fay, and A. N. Popper, eds.). Springer-Verlag, New York.

Allison, J. J., and H. M. Cook. 1981. Oxalate degradation by microbes of the large bowel of herbivores: the effect of dietary oxalate. Science, 212:675–676.

Altenbach, J. S. 1977. Functional morphology of two bats: *Leptonycteris* and *Eptesicus*. Spec. Publ. No. 5, Amer. Soc. Mamm.

Altenbach, J. S. 1979. Locomotor morphology of the vampire bat *Desmodus rotundus*. Spec. Publ., No. 6, Amer. Soc. Mamm., 1–137.

Altenbach, J. S., and J. W. Hermanson. 1987. Bat flight muscle function and the scapulo-humeral lock, 100–118, in *Recent Advances in the Study of Bats*. (M. B. Fenton, P. Racey, and J. M. V. Rayner, eds.). Cambridge Univ. Press, Cambridge.

Altman, P. L., and D. S. Dittmer. 1964. *Biology Data Book.* Fed. Amer. Soc. Exp. Biol., Washington, D.C.

Altmann, S. A., and J. Altmann. 1970. *Baboon Ecology.* Univ. Chicago Press, Chicago.

Ammerman, L. K., and D. M. Hillis. 1992. A molecular test of bat relationships: monophyly or diphyly? Syst. Biol., 41:222–232.

Andersen, D. C., and J. A. MacMahon. 1981. Population dynamics and bioenergetics of a fossorial herbivore, *Thomomys talpoides* (Rodentia: Geomyidae), in a spruce-fir sere. Ecol. Monogr., 51:179–202.

Andersen, D. C., and J. A. MacMahon. 1985. Plant succession following the Mount St. Helens volcanic eruption: facilitation by a burrowing rodent, *Thomomys talpoides*. Amer. Midl. Nat., 114:62–69.

Anderson, J. M. 1972. *Nature's Transplant: The Transplantation Immunology of Viviparity.* Appleton-Century-Crofts, New York.

Anderson, P. K. 1970. Ecological structure and gene flow in small mammals. Symp. Zool. Soc. London, 26:299–325.

Anderson, P. K. 1997. Shark Bay dugongs in summer. I. Lek mating. Behaviour, 134:433–462.

Anderson, P. K., and R. M. R. Barclay. 1995. Acoustic signals of solitary dugongs: physical characteristics and behavioral correlates. J. Mamm., 76:1226–1237.

Anderson, R. M. 1982. Fox rabies. 242–261, in *Population Dynamics of Infectious Diseases: Theory and Applications* (R. M. Anderson, ed.). Chapman and Hall, London.

Anderson, S., and J. K. Jones, Jr. 1984. *Orders and Families of Recent Mammals of the World*. John Wiley and Sons, New York.

Anderson, S., and B. D. Patterson. 1994. Biogeography, 215–233, in *Seventy-five Years of Mammalogy (1919–1994)* (E. C. Birney and J. R. Choate, eds.). Spec. Publ. No. 11, Amer. Soc. Mamm.

Andersson, A. 1969. Communication in the lesser bushbaby (*Galago senegalensis moholi*). Unpubl. M. S. thesis, University of Witwatersrand.

Andrews, P. 1988. A phylogenetic analysis of the Primates, 143–175, in *The Phylogeny and Classification of the Tetrapods*. Vol. 2 Mammals (M. J. Benton, ed.). Syst. Ass. Spec. Vol. 35B, Clarendon Press, Oxford.

Andrews, P., and L. Martin. 1987. Cladistic relationships of extant and fossil hominoids, 101–118, in *Primate Phylogeny* (F. E. Grine, J. G. Fleagle, and L. B. Martin, eds.). Academic Press, New York.

Annison, E. F., and D. Lewis. 1959. *Metabolism in the Rumen*. Methuen, London.

Anonymous. 1998. Urgent thinking required about development. Nature, 395:527.

Aplin, K. P., and M. Archer. 1987. Recent advances in marsupial systematics with a new syncretic classification, xv–lxxii, in *Possums and Opossums: Studies in Evolution* (M. Archer, ed.), vol. 1. Sydney, Australia, Surrey Beatty & Sons.

Ar, A., R. Arieli, and A. Shkolnik. 1977. Blood-gas properties and function in the fossorial mole-rat under normal and hypoxic-hypercapnic atmospheric conditions. Respir. Physiol., 30:201–218.

Archer, M. 1984. Origins and early radiations of marsupials, 585–625, in *Vertebrate Zoogeography and Evolution in Australasia* (M. Archer and G. Clayton, eds.). Hesperian Press, Victoria Park, Australia.

Archer, M., and L. Dawson. 1982. Revision of marsupial lions of the genus *Thylacoleo* Gervais (Thylacoleonidae, Marsupialia) and thylacoleonid evolution in the late Cainozoic, 477–494, in *Carnivorous Marsupials* (M. Archer, ed.). Royal Zool. Soc. New South Wales, Sydney, Australia.

Archer, M., and G. Clayton. 1984. *Vertebrate Zoogeography and Evolution in Australasia*. Hesperian Press, Victoria Park, Australia.

Archer, M., T. F. Flannery, A. Ritchie, and R. E. Molnar. 1985. First Mesozoic mammal from Australia—an early Cretaceous monotreme. Nature, 318:363–366.

Archer, M., S. Hand, and H. Godthelp. 1991. *Riversleigh: The Story of Animals in Ancient Rainforests of Inland Australia*. Reed Books, Balgowlah, New South Wales.

Archer, M., F. A. Jenkins, S. J. Hand, P. Murray, and H. Godthelp. 1992. Description of the skull and non-vestigial dentition of a Miocene platypus (*Obdurodon dicksoni* n. sp.) from Riversleigh, Australia, and the problem of monotreme origins, 15–27, in *Platypus and Echidnas* (M. L. Augee, ed.). Royal Zool. Soc., New South Wales, Sydney.

Archer, M., P. Murray, S. Hand, and H. Godthelp. 1993. Reconsideration of monotreme relationships based on the skull and dentition of the Miocene *Obdurodon dicksoni*, 75–94, in *Mammal Phylogeny, vol. 1. Mesozoic Differentiation, Multituberculates, Monotremes, Early Therians, and Marsupials* (F. S. Szalay, M. J. Novacek, and M. C. McKenna, eds.). Springer-Verlag, New York.

Archibald, J. D. 1996. Fossil evidence for a Late Cretaceous origin of "hoofed" mammals. Science, 272:1150–1153.

Arey, L. B. 1974. *Developmental Anatomy*. 7th ed. W. B. Saunders, Philadelphia.

Arieli, R., and A. Ar. 1981a. Heart rate responses of the mole rat (*Spalax ehrenbergi*) in hypercapnic, hypoxic, and cold conditions. Physiol. Zool., 54:14–21.

Arieli, R., and A. Ar. 1981b. Blood capillary density in heart and skeletal muscles of the fossorial mole rat. Physiol. Zool., 54:22–27.

Arita, H. T. 1990. Noseleaf morphology and ecological correlates in phyllostomid bats. J. Mamm., 71:36–47.

Arita, H. T., and M. B. Fenton. 1997. Flight and echolocation in the ecology and evolution of bats. Trends Ecol. Evol., 12:53–58.

Armitage, K. B. 1981. Sociality as a life-history tactic of ground squirrels. Oecologia, 48:36–49.

Armitage, K. B. 1986. Marmot polygyny revisited: determinants of male and female reproductive strategies, 303–331, in *Ecological Aspects of Social Evolution* (D. I. Rubenstein and R. W. Wrangham, eds.). Princeton Univ. Press, Princeton, New Jersey.

Armitage, K. B. 1998. Reproductive strategies of yellow-bellied marmots: energy conservation and differences between the sexes. J. Mamm., 79:385–393.

Armstrong, E. 1983. Relative brain size and metabolism in mammals. Science, 220:1302–1304.

Armstrong, R. B., C. D. Januzzo, and T. H. Kunz. 1977. Histochemical and biochemical properties of flight muscle fibers in the little brown bat, *Myotis lucifugus*. J. Comp. Physical. (B), 119:141–154.

Arnason, U., and A. Gullberg. 1994. Relationship of baleen whales established by cytochrome b gene sequence comparison. Nature, 367:726–728.

Aroyan, J. L. 1990a. Numerical simulation of dolphin echolocation beam formation. Unpubl. M. S. thesis, Univ. California at Santa Cruz.

Aroyan, J. L. 1990b. Supercomputer model of delphinid sonar beam formation. J. Acoust. Soc. Amer., 88:S4.

Asdell, S. A. 1964. *Patterns of Mammalian Reproduction.* Cornell Univ. Press, Ithaca, New York.

Atkinson, I. A. E., and E. K. Cameron. 1993. Human influence on the terrestrial biota and biotic communities of New Zealand. Trends Ecol. Evol., 8:447–451.

Au, W. W. L. 1993. *The Sonar of Dolphins.* Springer-Verlag, New York.

Au, W. W. L., R. W. Floyd, R. H. Penner, and A. E. Murchison. 1974. Measurement of echolocation signals of the Atlantic bottlenose dolphin, *Tursiops truncatus* Montagu, in open waters. J. Acoust. Soc. Amer., 56:1280–1290.

Audet, D. 1990. Foraging behavior and habitat use by a gleaning bat, *Myotis myotis.* J. Mamm., 71:420–427.

Audet, D., D. Krull, G. Marimuthu, S. Sumithran, and J. B. Singh. 1988. Foraging strategies and use of space by the Indian false vampire, *Megaderma lyra* (Megadermatidae). Bat. Res. News, 29:43.

Augee, M. L., and B. A. Gooden. 1992. Evidence for electroreception from field studies of the echidna, *Tachyglossus aculeatus*, 211–215, in *Platypus and Echidnas* (M. L. Augee, ed.). Royal Zool. Soc. New South Wales, Sydney.

Aumann, G. D. 1965. Microtine abundance and soil sodium levels. J. Mamm., 46:594–612.

Austad, S. N., and M. E. Sunquist. 1986. Sex-ratio manipulation in the common possum. Nature, 324:58–60.

Backus, R. H., and W. E. Schevill. 1966. *Physeter* clicks, 510–528, in *Whales, Dolphins and Porpoises* (K. S. Norris, ed.). University of California Press, Berkeley.

Bailey, W. J., J. L. Slightom, and M. Goodman. 1992. Rejection of the "flying primate" hypothesis by phylogenetic evidence from the ε-globin gene. Science, 256:86–89.

Baillie, J., and B. Groombridge. 1996. 1996 IUCN Red List of Threatened Animals. IUCN Species Survival Commission, Gland, Switzerland.

Baker, H. G. 1961. The adaptation of flowering plants to nocturnal and crepuscular pollinators. Quart. Rev. Biol., 36:64–73.

Baker, H. G. 1973. Evolutionary relationships between flowering plants and animals in American and African tropical forests, 145–159, in *Tropical Forest Ecosystems in Africa and South America: A Comprehensive Review* (B. J. Meggers, E. S. Ayensu, and W. D. Duckworth, eds.). Smithsonian Institution Press, Washington, D.C.

Baker, H. G., and B. J. Harris. 1957. The pollination of *Parkia* by bats and its attendant evolutionary problems. Evolution, 11:449–460.

Baker, J. R., and R. M. Ranson. 1933. Factors affecting the breeding of the field mouse (*Microtus agrestis*). Proc. Roy. Soc. London, 113B:486–495.

Baker, R. H. 1971. Nutritional strategies of myomorph rodents in North American grasslands. J. Mamm., 52:800–805.

Baker, R. H. 1997. American mammals need better care. J. Mamm., 78:977.

Baker, R. J., C. S. Hood, and R. L. Honeycutt. 1989. Phylogenetic relationships and classification of the higher categories of the New World bat family Phyllostomidae. Syst. Zool., 38:228–238.

Baker, R. J., M. J. Novacek, and N. B. Simmons. 1991. On the monophyly of bats. Syst. Zool., 40:216–231.

Baker, W. W., S. G. Marshall, and V. B. Baker. 1968. Autumn fat deposition in the evening bat (*Nycticeius humeralis*). J. Mamm., 49:314–317.

Balter, M. 1997. Morphologists learn to live with molecular upstarts. Science, 276:1032–1034.

Barash, D. P. 1989. *Marmots: Social Behavior and Ecology.* Stanford Univ. Press, Stanford, California.

Barclay, R. M. R. 1982. Interindividual use of echolocation calls: eavesdropping by bats. Behav. Ecol. Sociobiol., 10:271–275.

Barclay, R. M. R. 1983. Echolocation calls of emballonurid bats from Panama. J. Comp. Physiol., A, 151:515–520.

Barclay, R. M. R. 1986. The echolocation calls of hoary (*Lasiurus cinereus*) and silver-haired (*Lasionycteris noctivagans*) bats as adaptations for long- vs. short-range foraging strategies and the consequences of prey selection. Can. J. Zool., 64:2700–2705.

Barclay, R. M. R. 1989. The effect of reproductive condition on the foraging behavior of female hoary bats, *Lasiurus cinereus.* Behav. Ecol. Sociobiol., 24:31–37.

Barclay, R. M. R., M. B. Fenton, M. D. Tuttle, and M. J. Ryan. 1981. Echolocation calls produced by *Trachops cirrhosus* (Chiroptera: Phyllostomatidae) hunting for frogs. Can. J. Zool., 59:750–753.

Barfield, R. J., and L. A. Geyer. 1972. Sexual behavior: ultrasonic postejaculatory song of the male rat. Science, 176:1349–1350.

Barkalow, F. S. Jr., R. B. Hamilton, and R. F. Soots Jr. 1970. The vital statistics of an unexploited gray squirrel population. J. Wildl. Manag., 34:489–500.

Barker, W. R., and P. J. M. Greenslade. 1982. *Evolution of the Flora and Fauna of Arid Australia.* Peacock Publications, Adelaide.

Barklow, W. E. 1997. Some underwater sounds of the hippopotamus (*Hippopotamus amphibius*). Mar. Fresh. Behav. Physiol., 29:237–249.

Barnes, B. M. 1989. Freeze avoidance in a mammal: body temperatures below 0° C in an Arctic hibernator. Science, 244:1593–1595.

Barnes, L. G. 1976. Outline of eastern north Pacific fossil cetacean assemblages. Syst. Zool., 25:321–343.

Barnes, L. G., and A. E. Sanders. 1996. The transition from Archaeoceti to Mysticeti: late Oligocene toothed mysticetes from South Carolina, USA. J. Vert. Paleo., 16 (suppl. to 3):21A.

Barnosky, A. D. 1982. Locomotion in moles (Insectivora, Proscalopidae) from the middle Tertiary of North America. Science, 216:183–185.

Bartholomew, G. A. 1964. The roles of physiology and behavior in the maintenance of homeostasis in the desert environment. Symp. Soc. Exp. Biol., 18:7–29.

Bartholomew, G. A., and J. W. Hudson. 1960. Aestivation in the Mohave ground squirrel, *Citellus mohavensis*. Bull. Mus. Comp. Zool. Harvard, 124:193–208.

Bartholomew, G. A., and J. W. Hudson. 1961. Desert ground squirrels. Sci. Amer., 205(5):107–116.

Bartholomew, G. A., and N. E. Collias. 1962. The role of vocalization in the social behavior of the northern elephant seal. Anim. Behav., 10:7–14.

Bartholomew, G. A., W. R. Dawson, and R. C. Lasiewski. 1970. Thermoregulation and heterothermy in some of the smaller flying foxes (Megachiroptera) of New Guinea. Z. Vergleichende Physiologie, 70:196–209.

Bartholomew, G. A., and M. Rainy. 1971. Regulation of body temperature in the rock hyrax, *Heterohyrax brucei*. J. Mamm., 52:81–95.

Baskin, J. A. 1978. *Bensonomys, Calomys,* and the origin of the phyllotine group of neotropical cricetines (Rodentia: Cricetidae). J. Mamm., 59:125–135.

Baskin, Y. 1993. Blue whale populations may be increasing off California. Science, 260:287.

Bateman, J. A. 1959. Laboratory studies of the golden mole and the mole rat. African Wildlife, 13.

Bateman, G. C., and T. A. Vaughan. 1974. Nightly activities of mormoopid bats. J. Mamm., 55:45–65.

Bateson, G., and B. Gilbert. 1966. *Whaler's Cove Dolphin Community: An Interim Report.* Oceanic Inst. Makpuu Point, Waimanalo, Oahu, Hawaii.

Batzli, G. O., and F. A. Pitelka. 1970. Influence of meadow mouse populations on California grassland. Ecology, 51:1027–1039.

Batzli, G. O., and F. A. Pitelka. 1971. Conditions and diet of cycling populations of the California vole, *Microtus californicus*. J. Mamm., 58:141–163.

Bauchop, T. 1978. Digestion of leaves in vertebrate arboreal folivores, 193–204, in *The Ecology of Arboreal Folivores* (G. G. Montgomery, ed.). Smithsonian Inst. Press, Washington, D.C.

Baudinette, R. V., and A. A. Biewener. 1998. Young wallabies get a free ride. Nature, 395:653.

Baverstock, P. R., and B. Green. 1975. Water recycling in lactation. Science, 187:657–658.

Baverstock, P. R., B. Green, C. H. S. Watts, and L. Spencer. 1979. Water balance of small, lactating rodents: the total water balance picture of the mother-young unit. Comp. Biochem. Physiol., 63:247–252.

Bazzaz, F., G. Ceballos, M. Davis, R. Dirzo, P. R. Ehrlich, T. Eisner, et al. 1998. Ecological science and the human predicament. Science, 282:879.

Beard, K. C. 1998. East of Eden: Asia as an important center of taxonomic origin in mammalian evolution, 5–39, in *Dawn of the Age of Mammals in Asia* (K. C. Beard and M. R. Dawson, eds.). Bulletin of Carnegie Museum of Natural History 34, Pittsburgh.

Beard, K. C., B. Sigé, and L. Krishtalka. 1992. A primitive vespertilionoid bat from the early Eocene of central Wyoming. C. R. Acad. Sci. Paris, 314:735–741.

Beard, K. C., T. Qi, M. Dawson, B. Wang, and C. Li. 1994. A diverse new primate fauna from middle Eocene fissure-fillings in southeastern China. Nature, 368:604–609.

Beard, K. C., Y. Tong, M. R. Dawson, J. Wang, and X. Huang. 1996. Earliest complete dentition of an anthropoid primate from the late middle Miocene of Shanxi Province, China. Science, 272:82–85.

Beard, K. C., and M. R. Dawson. 1998. *Dawn of the Age of Mammals in Asia.* Bulletin of Carnegie Museum of Natural History 34, Pittsburgh.

Beard, P. H. 1977. *The End of the Game.* Doubleday and Company, Garden City, New York.

Bearlin, R. K. 1988. The morphology and systematics of Neogene Mysticeti from Australia and New Zealand. N. Z. J. Geol. Geophys., 31:257 (abstract).

Beatley, J. C. 1969. Dependence of desert rodents on winter annuals and precipitation. Ecology, 50:721–724.

Beatley, J. C. 1976. Rainfall and fluctuating plant populations in relation to distributions and numbers of desert rodents in southern Nevada. Oecologia, 24:21–42.

Beer, J. R. 1961. Seasonal reproduction in the meadow vole. J. Mamm., 42:483–489.

Beer, J. R., R. Lukens, and D. Olson. 1954. Small mammal populations on the islands of Basswood Lake, Minn. Ecology, 35:437–445.

Bekoff, M., and M. C. Wells. 1980. The social ecology of coyotes. Sci. Amer., 242(4):130–148.

Bel'kovich, V. M., and A. V. Yablokov. 1963. Marine animals "share experience" with designers. Nauka Zhizn', 30:61.

Bell, C. J. 1998. North American Quaternary land mammal ages and the biochronology of North American microtine rodents, 2–605 to 2–645, in *Dating and Earthquakes: Review of Quaternary Geochronology and its Application to Paleoseismology* (J. M. Sowers, J. S. Noller, and W. R. Lettis, eds.). U.S. Nuclear Regulatory Commission NUREG/CR 5562.

Bell, G. P. 1982a. Behavioral and ecological aspects of gleaning by the desert insectivorous bat, *Antrozous pallidus* (Chiroptera: Vespertilionidae). Behav. Ecol. Sociobiol., 10:217–223.

Bell, G. P. 1982b. Prey location and sensory ecology of two species of gleaning insectivorous bats, *Antrozous pallidus* (Vespertilionidae) and *Macrotus californicus* (Phyllostomatidae). Ph.D. thesis, Carleton Univ., Ottawa.

Bell, G. P., and M. B. Fenton. 1984. The use of Doppler-shifted echoes as a flutter detection and clutter rejection system: the echolocation and feeding behavior of *Hipposideros ruber* (Chiroptera: Hipposideridae). Behav. Ecol. Sociobiol., 15:109–114.

Bell, R. H. V. 1971. A grazing ecosystem in the Serengeti. Sci. Amer., 225:86–93.

Belsky, A. J. 1987. The effects of grazing: confounding of ecosystem, community, and organism scales. Amer. Nat., 129:777–783.

Bennett, A. F., and J. A. Ruben. 1979. Endothermy and activity in vertebrates. Science, 206:649–654.

Bennett, A. F., and J. A. Ruben. 1986. The metabolic and thermoregulatory status of therapsids, 207–218, in *The Ecology and Biology of Mammal-like Reptiles* (N. Hotton III, P. D. MacLean, J. J. Roth, and E. C. Roth, eds.). Smithsonian Inst. Press, Washington, D.C.

Bennett, N. C., and J. U. M. Jarvis. 1988. The social substructure and reproductive biology of colonies of the mole-rat, *Cryptomys damarrensis* (Rodentia: Bathyergidae). J. Mamm., 69:293–302.

Bennett, N. C., J. U. M. Jarvis, and F. P. D. Cotterill. 1993. Poikilothermic traits and thermoregulation in the Afrotropical social subterranean mole-rat (*Cryptomys hottentotus darlingi*) (Rodentia: Bathyergidae). J. Zool. London, 231:179–186.

Bennett, N. C., J. U. M. Jarvis, R. P. Millar, H. Sasano, and K. V. Ntshinga. 1994a. Reproductive repression in eusocial *Cryptomys damarensis* colonies: socially-induced infertility in females. J. Zool. London, 233:617–630.

Bennett, N. C., G. H. Aguilar, J. U. M. Jarvis, and C. G. Faulkes. 1994b. Thermoregulation in three species of Afrotropical subterranean mole-rats (Rodentia; Bathyergidae) from Zambia and Angola and scaling within the genus *Cryptomys*. Oecologia, 97:222–227.

Bennett, N. C., and J. U. M. Jarvis. 1995. Coefficients of digestibility and nutritional values of geophytes and tubers eaten by southern African mole-rats (Rodentia: Bathyergidae). J. Zool. London, 236:189–198.

Benson, S. B. 1933. Concealing coloration among some desert rodents of the southwestern United States. Univ. California Publ. Zool., 40:1.

Benson, S. B., and A. E. Borell. 1931. Notes on the life history of the red tree mouse, *Phenacomys longicaudus*. J. Mamm., 12:226–233.

Benton, M. J. 1988. *The Phylogeny and Classification of the Tetrapods*. Clarendon Press, Oxford.

Benton, M. J. 1990. *Vertebrate Palaeontology*. Unwin Hyman, Ltd. London.

Benton, M. J. 1997. *Vertebrate Palaeontology*. 2d ed. Chapman and Hall, London.

Berg, L. S., 1950. *Natural Regions of the U.S.S.R.* Macmillan, New York.

Berger, P. J., E. H. Sanders, P. D. Gardner, and N. C. Negus. 1977. Phenolic plant compounds functioning as reproductive inhibitors in *Microtus montanus*. Science, 195:575–577.

Berger, P. J., N. G. Negus, E. H. Sanders, and P. D. Gardner. 1981. Chemical triggering of reproduction in *Microtus montanus*. Science, 214:69–70.

Berger, P. J., N. Negus, and M. Day. 1997. Recognition of kin and avoidance of inbreeding in the montane vole *(Microtus montanus)*. J. Mamm., 78:1182–1186.

Berggren, W. A., D. V. Kent, C. C. Swisher III, and M.-P. Aubry. 1995. A revised Cenozoic geochronology and chronostratigraphy, 129–212, in *Geochronology Time Scales and Global Stratigraphic Correlation*. SEPM Spec. Publ. no. 54, SEPM (Society for Sedimentary Geology).

Bergstedt, B. 1965. Distribution, reproduction, growth and dynamics of the rodent species *Clethrionomys glareolus* (Schreber), *Apodemus flavicollis* (Melchior) and *Apodemus sylvaticus* (Linne) in southern Sweden. Oikos, 16:132–160.

Bergström, R. 1992. Browse characteristics and impact of browsing on trees and shrubs in African savannas. J. Veg. Sci., 3:315–324.

Bernard, H. J., and A. A. Hohn. 1989. Differences in feeding habits between pregnant and lactating spotted dolphins (*Stenella attenuata*). J. Mamm., 70:211–215.

Bernard, R. T. F., and G. S. Cumming. 1997. African bats: evolution of reproductive patterns and delays. Quart. Rev. Biol., 72:253–274.

Berta, A. 1994. What is a whale? Science, 263:180–181.

Bertram, B. C. R. 1973. Lion population regulation. E. Afr. Wildl. J., 11:215–225.

Bertram, B. C. R. 1975. The social system of lions. Sci. Amer., 232(5):54–65.

Bertram, G. C. L. 1940. The biology of the Weddell and crabeater seals, with a study of the comparative behavior of the Pinnipedia. Brit. Mus. (Nat. Hist.) Sci. Repts. Brit. Graham Land Exped. 1934–1937, 1:1–139.

Berzin, A. A. 1971. *The Sperm Whale*. Pacific Sci. Res. Inst. Fisheries Oceanogr. Trans. 1972, Israel.

Betancourt, J. L., T. R. Van Devender, and P. S. Martin. 1990. *Packrat Middens, the Last 40,000 Years of Biotic Change*. Univ. Arizona Press, Tucson, AZ.

Bieber, C. 1998. Population dynamics, sexual activity, and reproductive failure in the fat dormouse (*Myoxus glis*). J. Zool. London, 244:223–229.

Biewener, A. A., and R. V. Baudinette. 1995. In vivo muscle force and elastic energy storage during steady-speed hopping of tammar wallabies (*Macropus eugenii*). J. Exp. Biol., 198:1829–1841.

Blair, W. F. 1951. Evolutionary significance of geographic variation in population density. Texas J. Sci., 1:53–57.

Blake, E. W. Jr., 1887. The coast fox. West. Amer. Sci., 3:49.

Blatt, C. M., C. R. Taylor, and M. B. Habal. 1972. Thermal panting in dogs: the lateral nasal gland, a source of water for evaporative cooling. Science, 177:804–805.

Blix, A. S., H. J. Grav, and K. Ronald. 1979. Some aspects of temperature regulation in newborn harp seal pups. Amer. J. Physiol., 236:R188–197.

Bloedel, P. 1955. Hunting methods of fish-eating bats, particularly *Noctilio leporinus*. J. Mamm., 36:390–399.

Blumstein, D. T., and K. B. Armitage. 1998. Life history consequences of social complexity: a comparative study of ground-dwelling sciurids. Behav. Ecol., 9:8–19.

Bobek, B., K. Perzanowski, and J. Weiner. 1990. Energy expenditure for reproduction in male red deer. J. Mamm., 71:230–232.

Bodemer, C. W. 1968. *Modern Embryology*. Holt, Rinehart and Winston, New York.

Bodenheimer, F. S. 1949. *Problems of Vole Populations in the Middle East: Report on the Population Dynamics of the Levant Vole (Microtus guentheri D.)*. Azriel Print. Works, Jerusalem.

Boellstorff, D. E., D. H. Owings, M. C. T. Penedo, and M. J. Hersek. 1994. Reproductive behavior and multiple paternity of California ground squirrels. Anim. Behav., 47:1057–1064.

Bogdanowicz, W., R. D. Csada, and M. B. Fenton. 1997. Structure of noseleaf, echolocation, and foraging behavior in the Phyllostomidae (Chiroptera). J. Mamm., 78:942–953.

Bonaccorso, F. J., and B. K. McNab. 1997. Plasticity of energetics in blossom bats (Pteropodidae): impact on distribution. J. Mamm., 78:1073–1088.

Bonaparte, J. F., and G. W. Rougier. 1987. Mamíferos del Cretácico inferior de Patagonia. IV Congreso Latinamericano de Paleontología, Bolivia 1:343–359.

Bond, M., R. Pascual, M. Reguero, S. Santillana, and S. Marenssi. 1989. Los primeros "ungulados" extinguidos de la Antartida. Ameghiniana, 26:240.

Boness, D. J., W. D. Bowen, and O. T. Oftedal. 1994. Evidence of a maternal foraging cycle resembling that of otariid seals in a small phocid, the harbor seal. Behav. Ecol. Sociobiol., 34:95–104.

Boness, D. J., and W. D. Bowen. 1996. The evolution of maternal care in pinnipeds. BioSci., 46:645–654.

Bongaarts, J. 1998. Demographic consequences of declining fertility. Science, 282:419–420.

Bonner, G. N. 1991. Comparative hyoid morphology of nine chrysochlorid species (Mammalia: Chrysochloridae). Ann. Transvaal Mus., 35:295–311.

Bonner, G. N. 1995. Cytogenetic properties of nine species of golden moles (Insectivora: Chrysochloridae), J. Mamm., 76:957–971.

Bonner, W. N. 1990. *The Natural History of Seals*. Facts on File, New York.

Boonstra, R., and C. J. Krebs. 1979. Viability of large- and small-sized adults in fluctuating vole populations. Ecology, 60:567–573.

Boonstra, R., and P. T. Boag. 1987. A test of the Chitty hypothesis: inheritance in life-history traits in meadow vole *(Microtus pennsylvanicus)*. Evolution, 41:929–947.

Boonstra, R., D. Hik, G. R. Singleton, and A. Tinnikov. 1998. The impact of predator-induced stress on the snowshoe hare cycle. Ecol. Monogr., 79:371–394.

Bourliere, F. 1956. *The Natural History of Mammals*. Knopf, New York.

Bourliere, F. 1973. The comparative ecology of rain forest mammals in Africa and tropical America, 279–292, in *Tropical Forest Ecosystems in Africa and South America: A Comparative Review* (B. J. Meggers, E. S. Ayensu, and W. D. Duckworth, eds.). Smithsonian Inst. Press, Washington, D.C.

Bowen, B. S. 1982. Temporal dynamics of microgeographic structure of genetic variation in *Microtus californicus*. J. Mamm., 63:625–638.

Bowen, W. D., O. T. Oftedal, and D. J. Boness. 1992. Mass and energy transfer during lactation in a small phocid, the harbor seal (*Phoca vitulina*). Physiol. Zool., 65:844–866.

Bowers, M. A. 1997. Mammalian landscape ecology. J. Mamm., 78:997–998.

Bowers, M. A., and H. D. Smith. 1979. Differential habitat utilization by sexes of the deer mouse, *Peromyscus maniculatus*. Ecology, 60:869–875.

Bowers, M. A., and J. H. Brown. 1982. Body size and coexistence in desert rodents: chance or community structure. Ecology, 63:391–400.

Bowers, M. A., and J. L. Dooley, Jr. 1993. Predation hazard and seed removal by small mammals: microhabitat versus patch scale effects. Oecologia, 94:247–254.

Bowers, M., A. K. Gregario, C. J. Brame, S. F. Matter, and J. L. Dooley, Jr. 1996. Use of space and habitats by meadow voles at the home range, patch and landscape scales. Oecologia, 105:107–115.

Bowers, M. A., and S. F. Matter. 1997. Landscape ecology of mammals: relationships between density and patch size. J. Mamm., 78:999–1013.

Bowker, M. 1977. Behavior of Kirk's dik-dik, *Rhynchotragus kirki*. Unpublished Ph.D. dissertation, Northern Arizona University, Flagstaff.

Boyd, I. L. 1991. Environmental and physiological factors controlling the reproductive cycles of pinnipeds. Can. J. Zool., 69:1135–1148.

Boyse, E. A., G. K. Beauchamp, K. Yamazaki, J. Baid, and L. Thomas. 1982. A new aspect of the major histocompatibility complex and other genes in the mouse. Oncodevelopmental Biol. Med., 4:101–116.

Bradbury, J. W. 1977. Social organization and communication, 1–72, in *Biology of Bats* (W. Wimsatt, ed.). Volume 3. Academic Press, New York.

Bradbury, J. W., and L. H. Emmons. 1974. Social organization of some Trinidad bats. I: Emballonuridae. Z. Tierpsychol., 36:137–183.

Bradshaw, G. V. R. 1962. Reproductive cycle of the California leaf-nosed bat, *Macrotus californicus*. Science, 136:645–646.

Brain, C. 1992. Deaths in a desert baboon troop. Internat. J. Primatol., 13:593–599.

Brain, C., and R. Bohrmann. 1992. Tick infestation of baboons (*Papio ursinus*) in the Namib Desert. J. Wildl. Diseases, 28:188–191.

Bramble, D. M. 1989. Cranial specialization and locomotor habit in the Lagomorpha. Amer. Zool., 29:303–317.

Bramble, D. M., and F. A. Jenkins, Jr. 1998. Locomotor-respiratory integration: implications for mammalian and avian divergence. J. Vert. Paleo., 18(supplement to 3):28A.

Brant, D. H. 1962. Measures of the movements and population densities of small rodents. Univ. Calif. Publ. Zool., 62:105–184.

Brewer, R. 1979. *Principles of Ecology*. W. B. Saunders Comp., Philadelphia.

Brigham, R. M., and M. B. Fenton. 1986. The influence of roost closure on the roosting and foraging behavior of *Eptesicus fuscus* (Chiroptera: Vespertilionidae). Can. J. Zool., 64:1128–1133.

Broadbooks, H. E. 1970. Home ranges and territorial behavior of the yellow-pine chipmunk, *Eutamias amoenus.* J. Mamm., 51:310–326.

Brockie, R. 1976. Self-anointing by wild hedgehogs, *Erinaceus europaeus.* Anim. Behav., 24:68–71.

Brodie, E. D. 1977. Hedgehogs use toad venom in their own defense. Nature, 268:627–628.

Brody, S. 1945. *Bioenergetics and Growth.* Reinhold Publ., New York.

Bromley, P. T. 1969. Territoriality in pronghorn bucks on the National Bison Range, Moiese, Montana. J. Mamm., 50:81–89.

Bronson, F. H. 1971. Rodent pheromones. Biol. Reprod., 4:344–357.

Bronson, F. H. 1979. The reproductive ecology of the house mouse. Quart. Rev. Biol., 54:265–299.

Bronson, F. H. 1989. *Mammalian Reproductive Biology.* Univ. Chicago Press, Chicago.

Bronson, F. H., and J. A. Maruniak. 1975. Male-induced puberty in female mice: evidence for a synergistic action of social cues. Biol. Reprod., 13:94–98.

Brooke, A. P. 1994. Diet of the fishing bat, *Noctilio leporinus* (Chiroptera: Noctilionidae). J. Mamm., 75:212–218.

Brooks, D. R., and D. A. McLennan. 1991. *Phylogeny, Ecology, and Behavior: A Research Program in Comparative Biology.* Univ. Chicago Press, Chicago.

Brown, D. E., F. Reichenbacher, and S. E. Franson. 1998. *A Classification of North American Biotic Communities.* Univ. Utah Press, Salt Lake City.

Brown, J. H. 1973. Species diversity of seed-eating rodents in sand dune habitats. Ecology, 54:775–787.

Brown, J. H. 1987. Variation in desert rodent guilds: patterns, processes, and scales, 185–203, in *Organization of Communities: Past and Present* (J. H. R. Gee and P. S. Giller, eds.). Blackwell Scientific Publishers, Oxford, England.

Brown, J. H., and G. A. Bartholomew. 1969. Periodicity and energetics of torpor in the kangaroo mouse, *Microdipodops pallidus.* Ecology, 50:705–709.

Brown, J. H., D. W. Davidson, and O. J. Reichman. 1979a. An experimental study of competition between seed-eating desert rodents and ants. Amer. Zool., 19:1129–1143.

Brown, J. H., O. J. Reichman, and D. W. Davidson. 1979b. Granivory in desert ecosystems. Ann. Rev. Ecol. Syst., 10:201–227.

Brown, J. H., and J. C. Munger. 1985. Experimental manipulation of a desert rodent community: food addition and species removal. Ecology, 66:1545–1563.

Brown, J. H., and M. A. Kurzius. 1987. Composition of desert rodent faunas: combinations of coexisting species. Ann. Zool. Fennici, 24:227–237.

Brown, J. H., and E. J. Heske. 1990a. Mediation of a desert-grassland transition by a keystone rodent guild. Science, 250:1705–1707.

Brown, J. H., and E. J. Heske. 1990b. Temporal changes in a Chihuahuan rodent community. Oikos, 59:290–302.

Brown, J. H., and B. A. Harney. 1993. Population and community ecology of heteromyid rodents in temperate habitats, 618–651, in *Biology of the Heteromyidae* (H. H. Genoways and J. H. Brown, eds.). Spec. Publ. no. 10, Amer. Soc. Mammalogists.

Brown, J. H., G. C. Stevens, and D. M. Kaufman. 1996. The geographic range: size, shape, boundaries, and internal structure. Ann. Rev. Ecol. Syst., 27:597–623.

Brown, J. H., and M. V. Lomolino. 1998. *Biogeography.* 2d ed. Sinauer Associates, Sunderland, Massachusetts.

Brown, P. 1976. Vocal communication in the pallid bat, *Antrozous pallidus.* Z. Tierpsychol., 41:34–54.

Brown, W. L., and E. O. Wilson. 1956. Character displacement. Syst. Zool., 5:49–64.

Bruce, H. M. 1966. Smell as an exteroceptive factor, 83–87, in *Environmental Influences on Reproductive Processes* (Hansel, W. and R. H. Dutt, eds.). J. Applied Sci. (supplement) 25.

Bruce, V. G. 1960. Environmental entrainment of circadian rhythms. Cold Spr. Harb. Symp. Quant. Biol., 25:29–48.

Brückmann, G., and H. Burda. 1997. Hearing in blind subterranean Zambian mole-rats (*Cryptomys* sp.): collective behavioural audiogram in a highly social rodent. J. Comp. Physiol. A, 181:83–88.

Bryant, H. N. 1991. Phylogenetic relationships and systematics of the Nimravidae (Carnivora). J. Mamm., 72:56–78.

Bryant, J. P. 1981. Phytochemical deterrence of snowshoe hare browsing by adventitious shoots of four Alaskan trees. Science, 213:889–890.

Bryant, J. P., F. S. Chapin III, and D. R. Klein. 1983. Carbon/nutrient balance of boreal plants in relation to vertebrate herbivory. Oikos, 40:357–368.

Bubenik, G. A., and A. B. Bubenik. 1990. *Horns, Pronghorns, and Antlers: Evolution, Morphology, Physiology, and Social Significance.* Springer-Verlag, New York.

Buchler, E. R. 1976. The use of echolocation by the wandering shrew. Anim. Behav., 24:858–873.

Buckner, C. H. 1969. Some aspects of the population ecology of the common shrew. *Sorex araneus,* near Oxford, England. J. Mamm., 50:326–332.

Buffenstein, R., and S. Yahav. 1991. Is the naked mole-rat, *Heterocephalus glaber,* an endothermic yet poikilothermic mammal? J. Thermal Biol., 16:227–232.

Bullock, T. H., A. D. Grinnell, E. Ikezono, K. Kameda, Y. Katsuki, M. Nomoto, O. Sato, N. Suga, and K. Yanagisawa. 1968. Electrophysiological studies of central auditory mechanisms in cetaceans. Zeit. Vergleichende Phys., 59:117–316.

Burckhardt, D. 1958. Kindliches verhalten als ausdrucksvewegung im fortpflanzungszeremoniell einiger wiederkauer. Rev. Suisse Zool., 65:311–316.

Burns, J. J., J. J. Montague, and C. J. Cowles. 1993. *The Bowhead Whale*. Spec. Publ. No. 5, Soc. Marine Mamm.

Burns, V., H. Burda, and M. J. Ryan. 1989. Ear morphology of the frog-eating bat (*Trachops cirrhosus*, Family: Phyllostomidae): apparent specializations for low-frequency hearing. J. Morphol., 199:103–118.

Burrell, H. 1927. *The Platypus*. Angus and Robertson, Sydney.

Burt, W. H. 1943. Territoriality and home range concepts as applied to mammals. J. Mamm., 24:346–352.

Butler, P. M. 1969. Insectivores and bats from the Miocene of East Africa: new material, 1–37, in *Fossil Vertebrates of Africa* (L. Leakey, ed.), vol. 1. Academic Press, New York.

Butler, P. M. 1972. Some functional aspects of molar evolution. Evolution, 26:474–483.

Butler, P. M. 1990. Early trends in the evolution of tribosphenic molars. Biol. Rev., 65:529–552.

Calaby, J. H. 1971. The current status of Australian Macropodidae. Australian Zool., 16:17–29.

Calder, W. A. 1969. Temperature relations and underwater endurance of the smallest homeothermic diver, the water shrew. Comp. Biochem. Physiol., 30:1075–1082.

Cameron, R. D. 1994. Reproductive pauses by female caribou. J. Mamm., 75:10–13.

Camp, C. L., and N. Smith. 1942. Phylogeny and functions of the digital ligaments of the horse. Mem. Univ. Calif., 13:69–124.

Canivenc, R., and M. Bonnin. 1979. Delayed implantation is under environmental control in the badger (*Meles meles* L.). Nature, 278:849–850.

Carlton, A. 1936. The limb bones and vertebrae of the extinct lemurs of Madagascar. Proc. Zool. Soc. London, 110:281–307.

Carleton, M. D. 1984. Introduction to rodents, 255–265, in *Orders and Families of Recent Mammals of the World* (S. Anderson and J. K. Jones, eds.). Wiley & Sons, New York.

Carleton, M. D. 1994. Systematic studies of Madagascar's endemic rodents (Muroidea: Nesomyinae): revision of the genus *Eliurus*. Amer. Mus. Novitates, 3087:1–55.

Carleton, M. D., and G. G. Musser. 1984. Muroid rodents, 289–379, in *Orders and Families of Recent Mammals of the World* (S. Anderson and J. K. Jones, eds.). Wiley & Sons, New York.

Carleton, M. D., and D. F. Schmidt. 1990. Systematic studies of Madagascar's endemic rodents (Muroidea: Nesomyinae): an annotated gazetteer of collecting localities of known forms. Amer. Mus. Novitates, 2987:1–36.

Carleton, M. D., and S. M. Goodman. 1996. Systematic studies of Madagascar's endemic rodents (Muroidea: Nesomyinae): a new genus and species from the central highlands, 231–256, in *A Floral and Faunal Inventory of the Eastern Slopes of the Reserve Naturelle Integrale d'Andringitra, Madagascar: With Reference to Elevational Variation* (S. M. Goodman, ed.). Fieldiana: Zoology, 85:1–319.

Carleton, M. D., and S. M. Goodman. 1998. New taxa of nesomyine rodents (Muroidea: Muridae) from Madagascar's northern highlands, with taxonomic comments on previously described forms, 163–200, in *A Floral and Faunal Inventory of the Reserve Speciale d'Anjanaharibe-Sud, Madagascar: With Reference to Elevational Variation* (S. M. Goodman, ed.). Fieldiana: Zoology, 90:1–246.

Carlini, A. A., R. Pascual, M. A. Reguero, G. J. Scillato-Yané, E. P. Tonni, and S. F. Vizcaino. 1990. The first Paleogene land placental mammal from Antarctica: its paleoclimatic and paleobiogeographical bearings. 4th Internat. Cong. Syst. Evol. Biol., abstracts, p. 325. Univ. Maryland, College Park, Maryland.

Caroom, D., and F. H. Bronson. 1971. Responsiveness of female mice to preputial attractant: effects of sexual experience and ovarian hormones. Physiol. Behav., 7:659–662.

Carpenter, R. E. 1968. Salt and water metabolism in the marine fish-eating bat, *Pizonyx vivesi*. Comp. Biochem. Physiol., 24:951–964.

Carpenter, R. E. 1969. Structure and function of the kidney and the water balance of desert bats. Physiol. Zool., 42:288–302.

Carr, A. 1974. *Ulendo: Travels of a Naturalist in and out of Africa*. Knopf, New York.

Carroll, R. L. 1988. *Vertebrate Paleontology and Evolution*. W. H. Freeman & Co., New York.

Cartmill, M. 1974. Rethinking primate origins. Science, 184:436–443.

Cartmill, M. 1975. Strepsirhine basicranial structures and the affinities of the Cheirogaleidae, 313–356, in *Phylogeny of the Primates: A Multidisciplinary Approach* (W. P. Luckett and F. S. Szalay, eds.). Plenum Press, New York.

Case, J. A. 1988. Paleogene floras from Seymour Island, Antarctic Peninsula. Geol. Soc. Amer. Memoir, 169:523–530.

Case, J. A., M. O. Woodburne, and D. S. Chaney. 1988. A new genus of polydiploid marsupial from Antarctica. Geol. Soc. Amer. Memoir, 169:505–521.

Case, T. J. 1978. Endothermy and parental care in the terrestrial vertebrates. Amer. Nat., 112:861–874.

Catania, K. C. 1995. A comparison of the Eimer's organs of three North American moles: the hairy-tailed mole (*Parascalops breweri*), the star-nosed mole (*Condylura cristata*), and the eastern mole (*Scalopus aquaticus*). J. Comp. Neurol., 354:150–160.

Catania, K. C., and J. H. Kaas. 1996. The unusual nose and brain of the star-nosed mole. BioSci., 46:578–586.

Cates, R. G., and D. F. Rhoades. 1977. Patterns in the production of antiherbivore chemical defenses in plant communities. Biochem. Syst. Ecol., 5:185.

Cerchio, S., and P. Tucker. 1998. Influence of alignment on the mtDNA phylogeny of Cetacea: questionable support for a Mysticeti/Physeteroidea clade. Syst. Biol., 47:336–344.

Chaffee, R. R. J., and J. C. Roberts. 1971. Temperature acclimation in birds and mammals. Ann. Rev. Physiol., 33:155–202.

Chan, L.-K. 1995. Extrinsic lingual musculature of two pangolins (Pholidota: Manidae). J. Mamm., 76:472–480.

Channell, R. B. 1998. A geography of extinction: patterns in the contraction of geographic ranges. Unpubl. Ph.D. dissertation. Norman, University of Oklahoma.

Chapin, F. S., O. E. Sala, I. C. Burke, J. P. Grime, D. U. Hooper, W. K. Lauenroth, A. Lombard, H. A. Mooney, A. R. Mosier, S. Naeem, S. W. Pacala, J. Roy, W. L. Steffen, and D. Tilman. 1998. Ecosystem consequences of changing biodiversity. Bioscience, 48:45–52.

Chappell, M. A. 1978. Behavioral factors in altitudinal zonation of chipmunks (*Eutamias*). Ecology, 59:565–579.

Charles-Dominique, P. 1971. Eco-ethologie des prosimiens du Gabon. Biol. Gabonica, 7:121–228.

Charles-Dominique, P., and J. J. Petter. 1980. Ecology and social life of *Phaner furcifer,* 75–96, in *Nocturnal Malagasy Primates* (P. Charles-Dominique, H. M. Cooper, A. Hladik, C. M. Hladik, E. Pages, G. F. Pariente, A. Petter-Rousseaux, and A. Schilling, eds.). Academic Press, New York.

Charnov, E. L., G. H. Orians, and K. Hyatt. 1976. The ecological implications of resource depression. Amer. Nat., 110:247–259.

Cherry, J. A. 1989. Ultrasonic vocalizations by male hamsters: parameters of calling and effects of playback on female behavior. Anim. Behav., 38:138–153.

Chew, R. M., and A. E. Dammann. 1961. Evaporative water loss of small vertebrates, as measured with an infrared analyzer. Science, 133:384–385.

Childs, J. E., J. N. Mills, and G. E. Glass. 1995. Rodent-borne hemorrhagic fever viruses: a special risk for mammalogists? J. Mamm., 76:664–680.

Chitty, D. 1954. Tuberculosis among wild voles with a discussion of other pathological conditions among certain mammals and birds. Ecology, 35:227–237.

Chitty, D. 1958. Self-regulation of numbers through changes in viability. Cold Spr. Harb. Symp. Quant. Biol., 22:277–280.

Chitty, D. 1960. Population processes in the vole and their relevance to general theory. Can. J. Zool., 38:99–113.

Chitty, D., and H. Chitty. 1962. Population trends among the voles at Lake Vyrnwy, 1932–1960. J. Anim. Ecol., 35:313–331.

Chivers, D. L. 1974. *The Siamang in Malaya. A Field Study of a Primate in a Tropical Rain Forest.* Contrib. Primatol., vol. 4. S. Karger, Basel.

Choquenot, D., and D. M. J. S. Bowman. 1998. Marsupial megafauna, Aborigines and the overkill hypothesis: application of predator-prey models to the questions of

Pleistocene extinction in Australia. Global Ecol. Biogeogr. Letters, 7:167–180.

Chorn, J., and R. S. Hoffmann. 1978. *Ailuropoda melanoleuca.* Mamm. Species, 110:1–6.

Chow, M., and T. H. Rich. 1982. *Shuotherium dongi,* gen. et sp. nov., a therian with pseudotribosphenic molars from the Jurassic of Sichuan, China. Australian Mamm. 5:127–142.

Christian, D. P. 1978. Effects of humidity and body size on evaporative water loss in three desert rodents. Comp. Biochem. Physiol., 60:425–430.

Christian, D. P. 1979. Comparative demography of three Namib Desert rodents: responses to the provision of supplementary water. J. Mamm., 60:679–690.

Christian, J. J. 1950. The adreno-pituitary system and population cycles in mammals. J. Mamm., 31:247–259.

Churchfield, S. 1990. *The Natural History of Shrews.* Comstock Pub. Assoc., Cornell Univ. Press, Ithaca, New York.

Cifelli, R. L. 1993a. Theria of metatherian-eutherian grade and the origin of marsupials, 205–215, in *Mammal Phylogeny: Mesozoic Differentiation, Multituberculates, Monotremes, Early Therians, and Marsupials* (F. S. Szalay, M. J. Novacek, and M. C. McKenna, eds.). Springer-Verlag, New York.

Cifelli, R. L. 1993b. Early Cretaceous mammal from North America, and the evolution of marsupial dental characters. Proc. Natl. Acad. Sci. USA, 90:9413–9416.

Cifelli, R. L., T. B. Rowe, W. P. Luckett, J. Banta, R. Reyes, and R. I. Howe. 1996. Origins of marsupial pattern of tooth replacement: fossil evidence revealed by high-resolution X-ray CT. Nature, 379:715–718.

Cifelli, R. L., and C. de Muizon. 1997. Dentition and jaw of *Kokopellia juddi,* a primitive marsupial or near-marsupial from the medial Cretaceous of Utah. J. Mamm. Evol., 4:241–258.

Cifelli, R. L., J. I. Kirkland, A. Weil, A. L. Deino, and B. J. Kowallis. 1997. High-precision ^{40}Ar/^{39}Ar geochronology and the advent of North America's Late Cretaceous terrestrial fauna. Proc. Nat. Acad. Sci. USA, 94:11163–11167.

Cifelli, R. L., and C. de Muizon. 1998. Tooth eruption and replacement pattern in early marsupials. C. R. Acad. Sci. Paris, 326:215–220.

Cifelli, R. L., and S. K. Madsen. 1998. Triconodont mammals from the medial Cretaceous of Utah. J. Vert. Paleon., 18.

Ciochon, R. L., and A. B. Chiarelli. 1980. *Evolutionary Biology of the New World Monkeys and Continental Drift.* Plenum Press, New York.

Ciochon, R. L., and R. Corruccini. 1983. *New Interpretations of Ape and Human Ancestry.* Plenum Press, New York.

Clark, C. W., and J. M. Clark. 1980. Sound playback experiments with southern right whales (*Eubalaena australis*). Science, 207:663–665.

Clark, W. E. L. 1971. *The Antecedents of Man.* Quadrangle Books, New York.

Clarke, J. R., and J. P. Kennedy. 1967. Effect of light and temperature upon gonad activity in the vole *(Microtus agrestis)*. Gen. Comp. Endocrinol., 8:474–488.

Clarke, M. R. 1979. The head of the sperm whale. Sci. Amer., 240:128–141.

Clemens, W. A. 1968. Origin and early evolution of marsupials. Evolution, 22:1–18.

Clemens, W. A. 1970. Mesozoic mammalian evolution. Ann. Rev. Ecol. Syst., 1:357–390.

Clough, G. C. 1965. Lemmings and population problems. Amer. Sci., 53:199–212.

Clulow, F. V., E. A. Franchetto, and P. E. Langford. 1982. Pregnancy failure in the red-backed vole *Clethrionomys gapperi*. J. Mamm., 63:499–500.

Clutton-Brock, T. H. 1989. Mammalian mating systems. Proc. Royal Soc. London B., 236:339–372.

Clutton-Brock, T. H., F. E. Guinness, and S. D. Albon. 1982. *Red Deer: Behavior and Ecology of Two Sexes*. Univ. Chicago Press, Chicago.

Clutton-Brock, T. H., S. D. Albon, and F. E. Guinness. 1985. Parental investment and sex differences in juvenile mortality in birds and mammals. Nature, 313:131–133.

Clutton-Brock, T. H., M. Hiraiwa-Hasegawa, and A. Robertson. 1989. Mate choice on fallow deer leks. Nature, 340:463–465.

Coates, A. G., and J. A. Obando. 1996. The geologic evolution of the Central American Isthmus, 21–56, in *Evolution and Environment in Tropical America* (J. B. C. Jackson, A. F. Budd, and A. G. Coates, eds.). Univ. Chicago Press, Chicago.

Cockburn, A., and W. Z. Lidicker. 1983. Microhabitat heterogeneity and population ecology of an herbivorous rodent, *Microtus californicus*. Oecologia, 59:167–177.

Coe, M. J. 1967. Preliminary notes on the spring hare *Pedetes surdaster larvalis* in East Africa. E. Afr. Wildl. J., 5:174–177.

Cohen, J. E. 1996. *How Many People Can the Earth Support?* W. W. Norton & Co., New York.

Coimbra-Filho, A. F., A. Magananini, and R. A. Mittermeier. 1975. Vanishing gold: last chance for Brazil's lion tamarins. Animal Kingdom, Dec. 20.

Colbert, E. H. 1973. *Wandering Lands and Animals*. Dutton, New York.

Colbert, E. H. 1982. Personal communication.

Cole, F. R., D. M. Reeder, and D. E. Wilson. 1994. A synopsis of distribution patterns and the conservation of mammal species. J. Mamm., 75:266–276.

Cole, R. W. 1970. Pharyngeal and lingual adaptations in the beaver. J. Mamm., 51:424–425.

Colinvaux, P. A. 1996. Low-down on a land bridge. Nature 382:21–22.

Collins, L. R., and J. F. Eisenberg. 1972. Notes on the behavior and breeding of pacaranas *Dinomys branickii* in captivity. Internat. Zoo Yearbook, 12:108–114.

Condy, P. R. 1978. Annual cycle of the southern elephant seal *(Mirounga leonina)* at Marian Island. S. African J. Zool., 14:95–102.

Conley, W. H. 1971. Behavior, demography and competition in *Microtus longicaudus* and *M. mexicanus*. Ph.D. thesis, Texas Tech University, Lubbock.

Connor, R. C., J. Mann, P. L. Tyack, and H. Whitehead. 1998. Social evolution in toothed whales. Trends Ecol. Evol., 13:228–232.

Cooper, H. M., M. Herbin, and E. Nevo. 1993. Ocular regression conceals adaptive progression of the visual system in a blind subterranean mammal. Nature, 361:156–159.

Corbert, G. B. 1971. Family Macroscelididae. Part 1.5, 1–6, in *The Mammals of Africa: An Identification Manual* (J. Meester and H. W. Setzer, eds.). Smithsonian Inst. Press, Washington, D.C.

Corbert, G. B. 1988. The family Erinaceidae: a synthesis of its taxonomy, phylogeny, ecology and zoogeography. Mamm. Rev., 18:117–172.

Corbert, G. B., and J. Hanks. 1968. A revision of the elephant-shrews, Family Macroscelididae. Bull. British Mus. Nat. Hist. Zool., 16:1–111.

Corbet, C. B., and J. E. Hill. 1992. *The Mammals of the Indomalayan Region*. Oxford, Natural History Museum Publications and London, Oxford Univ. Press.

Costanza, R., R. d'Arge, R. De Groot, S. Farber, M. Grasso, B. Hannon, K. Limburg, S. Naeem, R. V. O'Neill, J. Paruelo, R. G. Raskin, P. Sutton, and M. van den Belt. 1997. The value of the world's ecosystem services and natural capital. Nature, 387:253–260.

Cott, H. B. 1966. *Adaptive Coloration in Animals*. Methuen, London.

Cranford, T. W. 1988. The anatomy of acoustic structures in the spinner dolphin forehead as shown by x-ray computed tomography and computer graphics, 67–77, in *Animal Sonar: Processes and Performances* (P. E. Nachtigall and P. W. B. Moore, eds.). Plenum Press, New York.

Cranford, T. W., M. Amundin, and D. E. Bain. 1987. A unified hypothesis for click production in odontocetes. Seventh Biennial Conf. Biol. Mar. Mamm., Miami Florida (abstract).

Crawford, E. C. Jr. 1962. Mechanical aspects of panting in dogs. J. Appl. Physiol., 17:249–251.

Credner, S., H. Burda, and F. Ludescher. 1997. Acoustic communication underground: vocalization characteristics in subterranean social mole-rats (*Cryptomys* sp., Bathyergidae) J. Comp. Physiol. A, 180:245–255.

Creel, S. R., N. M. Creel, D. E. Wildt, and S. L. Monfort. 1991. Behavioral and endocrine mechanisms of reproduction suppression in Serengeti dwarf mongooses. Anim. Behav., 43:231–245.

Creel, S. R., S. L. Monfort, and D. E. Wildt. 1993. Aggression, reproduction and androgens in wild dwarf mongooses: a test of the challenge hypothesis. Amer. Nat., 141:816–825.

Crelin, E. S. 1969. Interpubic ligament: elasticity in pregnant free-tailed bat. Science, 164:81–82.

Crompton, A. W. 1971. The origin of the tribosphenic molar, 165–180, in *Early Mammals* (D. M. Kermack and K. A. Kermack, eds.). J. Linn. Soc. (London) Zool., vol. 50.

Crompton, A. W. 1974. The dentitions and relationships of the southern African mammals *Erythrotherium parringtoni* and *Megazostrodon rudnerae.* Bull. Brit. Mus. Nat. Hist. (Geol.), 24:399.

Crompton, A. W. 1995. Masticatory function in nonmammalian cynodonts and early mammals, 55–75, in *Functional Morphology in Vertebrate Paleontology* (J. J. Thomason, ed.). Cambridge University Press, Cambridge.

Crompton, A. W., and F. A. Jenkins, Jr. 1968. Molar occlusion in Late Triassic mammals. Biol. Rev., 43:427.

Crompton, A. W., and K. Hiiemae. 1969. How mammalian molar teeth work. Discovery 5(1):23.

Crompton, A. W., and K. Hiiemae. 1970. Molar occlusion and mandibular movements during occlusion in the American opossum, *Didelphus marsupialis* L. J. Linn. Soc. (London) Zool., 49:21.

Crompton, A. W., and F. A. Jenkins, Jr. 1973. Mammals from reptiles: a review of mammalian origins. Ann. Rev. Earth Planetary Sci., Vol. I, Annual Reviews, Palo Alto.

Crompton, A. W., and F. A. Jenkins, Jr. 1979. Origin of mammals, 59–73, in *Mesozoic Mammals: The First Two-thirds of Mammalian History* (J. A. Lillegraven, Z. Kielan-Jaworowska, and W. A. Clemens, eds.). Univ. California Press, Berkeley.

Crompton, A. W., and Luo, Z. 1993. Relationships of the Liassic mammals *Sinoconodon, Morganucodon oehleri,* and *Dinnetherium,* 30–44, in *Mammal Phylogeny: Mesozoic Differentiation, Multituberculates, Monotremes, Early Therians, and Marsupials* (F. S. Szalay, M. J. Novacek, and M. C. McKenna, eds.). Springer-Verlag, New York.

Crompton, R. H., and P. M. Andau. 1987. Ranging, activity rhythms, and sociality in free-ranging *Tarsius bancanus:* a preliminary report. Inter. J. Primatol., 8:43–72.

Crowcroft, P. 1957. *The Life of the Shrew.* Max Reinhart, London.

Crozier, R. H. 1997. Preserving the information content of species: genetic diversity, phylogeny, and conservation worth. Ann. Rev. Ecol. Syst., 28:243–268.

Cullinane, D. M., D. Aleper, and J. E. A. Bertram. 1998a. The functional and biomechanical modifications of the spine of *Scutisorex somereni,* the hero shrew: skeletal scaling relationships. J. Zool., London, 244:477–452.

Cullinane, D. M., and D. Aleper. 1998b. The functional and biomechanical modifications of the spine of *Scutisorex somereni,* the hero shrew: spinal musculature. J. Zool., London, 244:453–458.

Curry-Lindahl, K. 1962. The irruption of the Norway lemmings in Sweden during 1960. J. Mamm., 43:171–184.

Cutright, W. J., and T. McKean. 1979. Countercurrent blood vessel arrangement in beaver (*Castor canadensis*). J. Morph., 161:169–176.

Czaplewski, N. J. 1987. Deciduous teeth of *Thyroptera tricolor.* Bat Research News, 28:23–25.

Czaplewski, N. J. 1997. Chiroptera, 410–431, in *Vertebrate Paleontology in the Neotropics: The Miocene Fauna of La Venta, Colombia* (R. F. Kay, R. H. Madden, R. L. Cifelli, and J. J. Flynn, eds.). Smithsonian Inst. Press, Washington, D.C.

Dagg, A. I. 1962. The role of the neck in the movements of giraffe. J. Mamm., 43:88–97.

Daily, G. C. 1997. *Nature's Services: Societal Dependence on Natural Ecosystems.* Island Press.

Dalquest, W. W., J. A. Baskin, and G. E. Schultz. 1996. Fossil mammals from a Late Miocene (Clarendonian) site in Beaver County, Oklahoma, 117–137, in *Contributions in Mammalogy. A Memorial Volume Honoring Dr. J. Knox Jones, Jr.* (H. H. Genoways and R. J. Baker, eds.). Mus. Texas Tech University, Lubbock.

Danell, K. 1983. Shoot growth of *Betula pendula* and *B. pubescens* in relation to moose browsing. Alces, 18:197–209.

Danell, K., R. Bergström, and L. Edenius. 1994. Effects of large mammal browsers on architecture, biomass, and nutrients of woody plants. J. Mamm., 75:833–844.

Daniel, J. C. Jr. 1970. Dormant embryos of mammals. BioSci., 20(7):411–415.

Daniel, M. J. 1979. The New Zealand short-tailed bat, *Mystacina tuberculata:* a review of present knowledge. New Zealand J. Zool., 6:357.

Daniel, M. J. 1990. Order Chiroptera, 114–137, in *The Handbook of New Zealand Mammals* (C. M. King, ed.). Oxford Univ. Press, Auckland.

Dapson, R. A. 1979. Phenologic influences on cohort-specific reproductive strategies in mice *(Peromyscus polionotus).* Ecology, 60:1125–1131.

Darling, F. F. 1937. *A Herd of Red Deer.* Oxford University Press, London.

Darlington, P. J. 1957. *Zoogeography: The Geographical Distribution of Animals.* Wiley, New York.

Dashzeveg, D. 1990. New trends in adaptive radiation of early Tertiary rodents (Rodentia, Mammalia). Acta Zoologica Cracoviensia, 33:37–44.

Dashzeveg, D., M. J. Novacek, M. A. Norell, J. M. Clark, L. M. Chiappe, A. Davidson, et al. 1995. Extraordinary preservation in a new vertebrate assemblage from the Late Cretaceous of Mongolia. Nature, 374:446–449.

Da Silva, M. N. 1998. Four new species of spiny rats of the genus *Proechimys* (Rodentia: Echimyidae) from the western Amazon of Brazil. Proc. Biol. Soc. Wash., 111:436–471.

Davidson, D. W., J. H. Brown, and R. S. Inouye. 1980. Competition and the structure of granivore communities. BioSci., 30:233–238.

Davis, D. E. 1967. The annual rhythm of fat deposition in woodchucks (*Marmota monax*). Physiol. Zool., 40:391–402.

Davis, R. B., C. F. Herreid Jr., and H. L. Short. 1962. Mexican free-tailed bats in Texas. Ecol. Monogr., 32:311–346.

Davis, R., C. Dunford, and M. V. Lomolino. 1988. Montane mammals of the American Southwest: the possible influence of post-Pleistocene colonization. J. Biogeography, 15:841–848.

Davis, W. H. 1970. Hibernation: ecology and physiological ecology, 265–300, *Biology of Bats* (W. A. Wimsatt, ed.). Academic Press, New York.

Davis, W. H., and W. Z. Lidicker, Jr. 1956. Winter range of the red bat. J. Mamm., 37:280–281.

Davis W. H. and H. B. Hitchcock. 1965. Biology and migration of the bat, *Myotis lucifugus*, in New England. J. Mamm., 46:296–313.

Dawe, A. R., and W. A. Spurrier. 1969. Hibernation induced in ground squirrels by blood transfusion. Science, 163:298–299.

Dawson, M. R. 1958. Later Tertiary Leporidae of North America. Univ. Kansas Paleo. Contrib. Vertebrata, Art. 6:1–750.

Dawson, M. R. 1967. Lagomorph history and stratigraphic record. In *Essays in Paleontology and Stratigraphy, Raymond C. Moore Commemorative Volume*. Spec. Publ. No. 2, Univ. Kansas, Dept. Geology.

Dawson, M. R. and L. Krishtalka. 1984. Fossil history of the families of Recent mammals, 11–57, in *Orders and Families of Recent Mammals of the World* (S. Anderson and J. K. Jones, eds.). Wiley & Sons, New York.

Dawson, M. R., C. K. Li, and T. Qi. 1984. Eocene ctenodactyloid rodents (Mammalia) of eastern and central Asia, 138–150, in *Papers in Vertebrate Paleontology Honoring Robert Warren Wilson* (R. M. Mengel, ed.). Spec. Publ. 9, Carnegie Mus. Nat. Hist.

Dawson, M. R., M. C. McKenna, K. C. Beard, and J. H. Hutchison. 1993. An early Eocene plagiomenid mammal from Ellesmere and Axel Heiberg Islands, Arctic Canada. Kaupia, Darmstädter Beiträge zur Naturgeschichte, 3:179–192.

Dawson, M. R., and K. C. Beard. 1996. New Late Paleocene rodents (Mammalia) from Big Multi Quarry, Washakie Basin, Wyoming. Palaeovertebrata, 25:301–321.

Dawson, T. J. 1995. *Kangaroos: Biology of the Largest Marsupials*. Cornell Univ. Press, Ithaca, New York.

Dawson, T. J., and A. J. Hulbert. 1970. Standard metabolism, body temperature, and surface areas of Australian marsupials. Amer. J. Physiol., 218:1233–1238.

Dawson, T. J., and C. R. Taylor. 1973. Energetic cost of locomotion in kangaroos. Nature, 246:313–314.

Dawson, W. W. 1980. The cetacean eye, 53–100, in *Cetacean Behavior: Mechanisms and Functions* (L. M. Herman, ed.). Wiley & Sons, New York.

Dawson, W., C. Adams, M. Barris, and C. Litzkow. 1979. Static and kinetic properties of the dolphin pupil. Amer. J. Physiol., 237:301–305.

Dayan, T., and D. Simberloff. 1994. Morphological relationships among coexisting heteromyids: an incisive dental character. Amer. Nat., 143:462–477.

Dayan, T., D. Simberloff, E. Tchernov, and Y. Yom-Tov. 1990. Feline canines: community-wide character displacement among the small cats of Israel. Amer. Nat., 136:39–60.

Deavers, D. R., and J. W. Hudson. 1979. Water metabolism and estimated field water budgets in rodents (*Clethrionomys gapperi* and *Peromyscus leucopus*) and an insectivore (*Blarina brevicauda*) inhabiting the same mesic environment. Physiol. Zool., 137–152.

Decker, D. M., and W. C. Wozencraft. 1991. Phylogenetic analysis of Recent procyonid genera. J. Mamm., 72:42–55.

DeCoursey, P. 1961. Effect of light on the circadian activity rhythm of the flying squirrel, *Glaucomys volans*. Z. Vergl. Physiol., 44:331–354.

Deevey, E. S. 1947. Life tables for natural populations of animals. Quart. Rev. Biol., 22:283–314.

Degen, A. A. 1988. Ash and electrolyte intakes of the fat sand rat, *Psammomys obesus*, consuming saltbush, *Atriplex halimus*, containing different water content. Physiol. Zool., 61:137–141.

Degen, A. A. 1996. *Ecophysiology of Small Desert Mammals*. Springer-Verlag, Berlin.

Dehnhardt, G., B. Mauck, and H. Bleckmann. 1998. Seal whiskers detect water movements. Nature, 394:235–236.

De Jong, W. W. 1998. Molecules remodel the mammalian tree. Trends Ecol. Evol., 13:270–275.

De Jong, W. W., J. A. M. Leunissen, and G. J. Wistow. 1993. Eye lens crystallins and the phylogeny of placental orders: evidence for a macroscelid-paenungulate clade? 5–12, in *Mammal Phylogeny; Placentals* (F. S. Szalay, M. J. Novacek, and M. C. McKenna, eds.). Springer-Verlag, New York.

Delany, M. J. 1971. The biology of small rodents in Mayanja Forest, Uganda. J. Zool. London, 165:85–129.

Delattre, P., P. Giraudoux, and J. Baudry. 1992. Land use patterns and types of common vole *(Microtus arvalis)* population kinetics. Agric. Ecosyst. Environ., 39:153–168.

Dellow, Q. W., and I. D. Hume. 1982. Studies on the nutrition of marcopodine marsupials: IV. digestion in the stomach and intestines of *Macropus giganteus, Thylogale thetis* and *Macropus eugenii*. Aust. J. Zool., 30:767–777.

Delson, E. 1985. *Ancestors: The Hard Evidence*. Alan R. Liss, New York.

de Meijere, J. C. 1894. Über die Haare der Säugetiere besonders über ihre Anordnung. Gegenaurs Morphol. Jahrb., 21:312–424.

D'Erchia, A. M., C. Gissi, G. Pesole, C. Saccone, and U. Arnason. 1996. The guinea-pig is not a rodent. Nature, 381:597–600.

Desjardins, C., J. A. Maruniak, and F. H. Bronson. 1973. Social rank in house mice: differentiation revealed by ultraviolet visualization of urinary marking patterns. Science, 182:939–941.

Desy, E. A., and G. O. Batzli. 1989. Effects of food availability and predation on prairie vole demography: a field experiment. Ecology, 70:411–421.

Devall, B., and G. Sessions. 1985. *Deep Ecology*. Gibbs M. Smith, Layton, Utah.

DeVore, I. 1965. *Primate Behavior*. Holt, Rinehart and Winston, New York.

DeVore, I., and K. R. L. Hall. 1965. Baboon ecology, 20–52, in *Primate Behavior* (I. DeVore, ed.). Holt, Rinehart and Winston, New York.

deWaal, F. 1982. *Chimpanzee Politics: Power and Sex Among Apes.* Harper and Row, New York.

de Waal, F. B. M. 1989. *Peacemaking Among Primates.* Harvard Univ. Press, Cambridge.

de Waal, F. B. M. 1995. Bonobo sex and society, Sci. Amer., March:82–88.

Diamond, J. 1993. *The Third Chimpanzee.* HarperPerennial Library, New York.

Dijkgraaf, S. 1943. Over een merkwaardige functie wan den gehoorzin bij vleermuizen. Verslagen Nederlandsche Akademie Wetenschappen Afd. Naturkunde, 52:622–627.

Dijkgraaf, S. 1946. Die Sinneswelt der Fledermäuse. Experientia, 2:438–448.

Dinerstein, E., and C. M. Wemmer. 1988. Fruits *Rhinoceros* eat: dispersal of *Trewia nudiflora* (Euphorbiaceae) in lowland Nepal. Ecology, 69:1768–1774.

Dinerstein, E., C. Wemmer, and H. Mishra. 1988. Adoption in greater one-horned rhinoceros *(Rhinoceros unicornis)* J. Mamm., 69:813–814.

Dixon, F. 1981. *The Natural History of the Gorilla.* Columbia Univ. Press, New York.

Dixon, K. W. 1981. Western Australian plants with underground fleshy storage organs. Ph.D. thesis, University of Western Australia.

Dobson, F. S., and G. R. Michener. 1995. Maternal traits and reproduction in Richardson's ground squirrels. Ecology, 76:851–862.

Dobzhansky, T. 1950. Mendelian populations and their evolution. Amer. Nat., 84:401–418.

Douglas-Hamilton, I. 1972. On the ecology and behavior of the African elephant: the elephants of Lake Manyara. Ph.D. thesis, Oxford University, Oxford.

Douglas-Hamilton, I. 1973. On the ecology and behavior of the Lake Manyara elephants. E. Afr. Wildl. J., 11:401–403.

Downhower, J. F., and K. B. Armitage. 1971. The yellow-bellied marmot and the evolution of polygyny. Amer. Nat., 105:355–370.

Downs, C. T., and M. R. Perrin. 1989. An investigation of the macro- and micro-environments of four *Gerbillurus* species. Cimbebasia, 11:41–54.

Downs, C. T., and M. R. Perrin. 1991. Urinary concentrating ability of four *Gerbillurus* species of southern Africa arid regions. J. Arid Environ., 20:71–81.

Dragoo, J. W., and R. L. Honeycutt. 1997. Systematics of mustelid-like carnivores. J. Mamm., 78:426–443.

Duangkhae, S. 1990. Ecology and behavior of Kitti's hog-nosed bat (*Craseonycteris thonglongyai*) in western Thailand. Nat. Hist. Bull. Siam Soc., 38:135–161.

Dubost, G. 1968. Aperçu sur le rythme annuel de reproduction des muridés du nord-est du Gabon. Biol. Gabonica, 4:227–239.

Dubost, G. 1983. La comportement de *Cephalophus monticola* Thunberg et *C. dorsalis* Grey, et la place des céphalophes au sein des ruminants, Part I. Mammalia, 47:141–177.

Dücker, G. 1957. Fard und Helligkeitssehen und Instinkte bei Viverriden und Feliden. Zool. Beitr. (Berl.), 3:25–99.

Ducrocq, S., J.-J. Jaeger, and B. Sigé. 1992. Late Eocene southern Asian record of a megabat and its inferences on the megabat phylogeny. Bat Research News 33:41–42.

Ducrocq, S., J.-J. Jaeger, and B. Sigé. 1993. Un megachiroptere dans l'Eocene superieur de Thailand; incidence dans la discussion phylogenique du groupe. Neues Jahrbuch für Geologie und Paläontologie Mh., 1993:561–575.

Dudley, R., and P. DeVries. 1990. Tropical rain forest structure and the geographical distribution of gliding vertebrates. Biotropica, 22:432–434.

Dunbar, R. I. M. 1984. *Reproductive Decisions: An Economic Analysis of Gelada Baboon Social Strategies.* Princeton Univ. Press, Princeton, New Jersey.

Dunning, D. C. 1968. Warning sounds of moths. Z. Tierpsychol., 25:129–138.

Dunning, D. C., and K. D. Roeder. 1965. Moth sounds and the insect-catching behavior of bats. Science, 147:173–174.

Durrell, G. M. 1954. *The Bafut Beagles.* Rupert Hart-Davies, London.

Eadie, W. R. 1952. Shrew predation and vole populations on a localized area. J. Mamm., 33:185–189.

Eckert, R., D. Randall, and G. Augustine. 1988. *Animal Physiology: Mechanisms and Adaptations.* W. H. Freeman, New York.

Ehrenfeld, D. W. 1976. The conservation of non-resources. Amer. Sci., 64:648.

Ehrlich, P. R., and E. O. Wilson. 1991. Biodiversity studies: science and policy. Science, 253:758–762.

Einarsen, A. S. 1948. *The Pronghorn Antelope.* Wildlife Management Institute, Washington, D.C.

Eisenberg, J. F. 1966. The social organization of mammals. Handb. Zool. (Berl.), 10:1–92.

Eisenberg, J. F. 1981. *The Mammalian Radiations: An Analysis of Trends in Evolution, Adaptation, and Behavior.* Univ. Chicago Press, Chicago.

Eisenberg, J. F. 1989. *Mammals of the Neotropics: The Northern Neotropics.* Vol. 1, Univ. Chicago Press, Chicago.

Eisenberg, J. F., and E. Gould. 1970. The tenrecs: a study in mammalian behavior and evolution. Smithsonian Contrib. Zool., 27:1–137.

Eisenberg, J. F., and D. G. Kleiman. 1972. Olfactory communication in mammals. Ann. Rev. Ecol. Syst., 3:1–32.

Eisenberg, J. F., N. A. Muckenhirn, and R. Rudran. 1972. The relation between ecology and social structure in primates. Science, 176:863–874.

Eisenberg, J. F., and D. E. Wilson. 1981. Relative brain size and demographic strategies in didelphid marsupials. Amer. Nat., 118:1–15.

Eisenberg, J. F., and D. G. Kleiman. 1983. *Advances in the Study of Mammalian Behavior.* Spec. Publ. No. 7, Amer. Soc. Mamm.

Ellerman, J. R. 1940. *The Families and Genera of Living Rodents, vol I, Rodents Other Than Muridae*. British Mus. Nat. Hist., London.

Elliott, D. H., E. H. Colbert, W. J. Breed, I. A. Jensen, and T. S. Powell. 1970. Triassic tetrapods from Antarctica: evidence for continental drift. Science 169:197–201.

Elliott, L. 1978. Social behavior and foraging ecology of the eastern chipmunk *(Tamias striatus)* in the Adirondack Mountains. Smithsonian Contrib. Zool., 265:1–107.

Elliott, V. A. 1976. Circadian rhythms and photoperiodic time measurement in mammals. Fed. Proc., 35:2339–2346.

Else, P. L., and A. J. Hulbert, 1981. Comparison of the "mammal machine" and the "reptile machine": Energy production. Amer. J. Physiol., 240:R3.

Elsner, R. 1969. Cardiovascular adjustments to diving, 117–145, in *The Biology of Marine Mammals* (H. T. Anderson, ed.). Academic Press, New York.

Elton, C. S. 1930. *Animal Ecology and Evolution*. Oxford Univ. Press, Oxford.

Elton, C. 1942. *Voles, Mice and Lemmings*. Clarendon Press, Oxford.

Emerson, S. B., and L. Radinsky. 1980. Functional analysis of sabertooth cranial morphology. Paleobiol., 6:295–312.

Emlen, S. T. 1982. The evolution of helping. I. An ecological constraints model. Amer. Nat., 119:29–39.

Emlong, D. 1966. A new archaic cetacean from the Oligocene of northwest Oregon. Univ. Oregon Bull. Mus. Nat. Hist., 142:1–51.

Emmons, L. H. 1981. Morphological, ecological, and behavioral adaptations for arboreal browsing in *Dactylomys dactylinus* (Rodentia, Echimyidae). J. Mamm., 62:183–189.

Emmons, L. H. 1990. *Neotropical Rainforest Mammals: A Field Guide*. Univ. Chicago Press, Chicago.

Emmons, L. H. 1991. Frugivory in treeshrews *(Tupaia)*. Amer. Nat., 138:642–649.

Emmons, L. H., and A. H. Gentry. 1983. Tropical forest structure and the distribution of gliding and prehensile-tailed vertebrates. Amer. Nat., 121:513–524.

Emmons, L. H., and A. Biun. 1991. Malaysian treeshrews. Nat. Geograph. Res. Explor., 7:70–81.

Endler, J. A. 1982. Problems in distinguishing historical from ecological factors in biogeography. Amer. Zool., 22:441–452.

Endler, J. 1993. Some general comments on the evolution and design of animal communication systems. Phil. Trans. Roy. Soc. London, B 340:215–225.

Engel, S. R., K. M. Hogan, J. F. Taylor, and S. K. Davis. 1998. Molecular systematics and paleobiogeography of the South American sigmodontine rodents. Mol. Biol. Evol., 15:35–49.

Englemann, G. F. 1985. The phylogeny of the Xenarthra, 51–64, in *The Evolution and Ecology of Armadillos, Sloths, and Vermilinguas* (G. G. Montgomery, ed.). Smithsonian Inst. Press, Washington, D.C.

Epstein, P. R. 1995. Emerging diseases and ecosystem instabilities: new threats to public health. Amer. J. Public Health, 1995:168–172.

Erickson, A. B. 1944. Helminth infections in relation to population fluctuations in snowshoe hares. J. Wildl. Manag., 8:134–153.

Erickson, C. J. 1991. Percussive foraging in the aye-aye, *Daubentonia madagascariensis*. Anim. Behav., 41:793–801.

Erlinge, S. 1987. Predation and noncyclicity in a microtine population in southern Sweden. Oikos, 50:347–352.

Errington, P. L. 1937. What is the meaning of predation? Smithsonian Rep. for 1936, p. 243.

Errington, P. L. 1943. An analysis of mink predation upon muskrats in north-central United States. Agric. Exp. Sta. Iowa State Coll. Res. Bull., 320:797–924.

Errington, P. L. 1946. Predation and vertebrate populations. Quart. Rev. Biol., 21:144–177, 221–245.

Errington, P. L. 1963. *Muskrat Populations*. Iowa State University Press, Ames.

Errington, P. L. 1967. *Of Predation and Life*. Iowa State University Press, Ames.

Espenshade, E. B. (ed.) 1971. *Goode's World Atlas*. 13th ed. Rand-McNally, Chicago.

Essop, M. F., E. H. Harley, and I. Baumgarten. 1997. A molecular phylogeny of some Bovidae based on restriction-site mapping of mitochondrial DNA. J. Mamm., 78:377–386.

Estes, J. A., and J. F. Palmisano. 1974. Sea otters: their role in structuring nearshore communities. Science, 185:1058–1060.

Estes, J. A., and D. O. Duggins. 1995. Sea otters and kelp forests in Alaska: generality and variation in a community ecological paradigm. Ecol. Monogr., 65:75–100.

Estes, J. A., M. T. Tinker, T. M. Williams, and D. F. Doak. 1998. Killer whale predation on sea otters linking oceanic and nearshore ecosystems. Science, 282:473–475.

Estes, R. D. 1967. The comparative behavior of Grant's and Thompson's gazelles. J. Mamm., 48:189–209.

Estes, R. D. 1974. Social organization of the African Bovidae, 166–205, in *The Behavior of Ungulates and its Relation to Management* (V. Geist and F. R. Walther, eds.). International Union for Conservation of Nature and Natural Resources, Morges, Switzerland.

Estes, R. D. 1991. *The Behavior Guide to African Mammals: Including Hoofed Mammals, Carnivores, Primates*. Univ. Calif. Press, Berkeley.

Estrada, A. 1991. Howler monkeys (*Alouatta palliata*), dung beetles (Scarabaeidae) and seed dispersal: ecological interactions in the tropical rainforest of Los Tuxtlas, Mexico. J. Trop. Ecol., 7:459–474.

Evans, B. K., D. R. Jones, J. Baldwin, and G. R. J. Gabbott. 1994. Diving ability of the platypus. Aust. J. Zool., 42:17–27.

Evans, W. E. 1973. Echolocation by marine delphinids and one species of fresh-water dolphin. J Acoust. Soc. Amer., 54:191–199.

Evans, W. E., W. W. Sutherland, and R. G. Beil. 1964. The directional characteristics of delphinid sounds,

353–372, in *Marine Bioacoustics* (W. N. Tavolga, ed.). vol. 1. Pergamon Press, Oxford.

Evans, W. E., and B. A. Powell. 1967. Discrimination of different metallic plates by an echolocating delphinid, 363–382, in *Animal Sonar Systems: Biology and Bionics* (R. G. Busnel, ed.). Laboratoire de Physiologie Acoustique, Jouy-en-Josas, France.

Evans, W. E., and J. Bastian. 1969. Marine mammal communication: social and ecological factors, 424–475, in *The Biology of Marine Mammals* (H. T. Anderson, ed.). Academic Press, London.

Evermann, B. W. 1923. The conservation of marine life of the Pacific. Sci. Mon., 16:521.

Ewer, R. F. 1968. *Ethology of Mammals.* Plenum Press, New York.

Ewer, R. F. 1973. *The Carnivores.* Cornell Univ. Press, Ithaca, New York.

Ewing, W. G., E. H. Studier, and M. J. O'Farrell. 1970. Autumn fat deposition and gross body composition in three species of *Myotis.* Comp. Biochem. Physiol., 36:119–129.

Faegri, K., and L. Van Der Pijl. 1966. *The Principles of Pollination Ecology.* Pergamon Press, New York.

Faulkes, C. G., D. H. Abbott, and J. U. M. Jarvis. 1991. Social suppression of reproduction in male naked mole-rats. J. Reprod. Fert., 91:593–604.

Faulkes, C. G., N. C. Bennett, M. W. Bruford, H. P. O'Brien, G. H. Aguilar, and J. U. M. Jarvis. 1997. Ecological constraints drive social evolution in the African mole-rats. Proc. R. Soc. Lond. B, 264:1619–1627.

Faunmap Working Group. 1994. Faunmap: A database documenting late Quaternary distributions of mammal species in the United States. Illinois State Museum Scientific Papers 25 (1 and 2): 690 pp. + diskette.

Faure, P. A., J. H. Fullard, and R. M. R. Barclay. 1990. The response of tympanate moths to the echolocation calls of a substrate-gleaning bat, *Myotis evotis.* J. Comp. Physiol. A, 166:843–849.

Faure, P. A., J. H. Fullard, and J. W. Dawson. 1993. The gleaning attacks of the northern long-eared bat, *Myotis septentrionalis,* are relatively inaudible to moths. J. Exp. Biol., 178:173–189.

Fay, F. H. 1981. Walrus—*Odobenus rosmarus,* 1–23, in *Handbook of Marine Mammals* (S. H. Ridgway and R. J. Harrison, eds.). Academic Press, London.

Fedak, M. A., N. C. Heglund, and C. R. Taylor. 1982. Energetics and mechanics of terrestrial locomotion. II: kinetic energy changes of the limbs and body as a function of speed and body size in birds and mammals. J. Exp. Biol., 97:23–40.

Feder, M. E. 1981. A cold look at paleophysiology. Paleobiology, 7:144–148.

Feeny, P. P. 1975. Biochemical coevolution between plants and their insect herbivores, 3–19, in *Coevolution of Plants and Animals* (L. E. Gilbert and P. H. Raven, ed.). University of Texas Press, Austin.

Feldhamer, G. A., L. A. Drickamer, S. H. Vessey, and J. F. Merritt. 1999. *Mammalogy: Adaptation, Diversity, and Ecology.* McGraw-Hill, Boston.

Feldkamp, S. D. 1987a. Swimming in the California sea lion: morphometrics, drag and energetics. J. Exp. Biol., 131:117–135.

Feldkamp, S. D. 1987b. Foreflipper propulsion in the California sea lion, *Zalophus californianus.* J. Zool., London, 212:43–57.

Fenton, M. B. 1982. Echolocation, insect hearing, and feeding ecology of insectivorous bats. 261–285, in *Ecology of Bats* (T. Kunz, ed.). Plenum Press, New York.

Fenton, M. B. 1984. Echolocation: Implications for ecology and evolution of bats. Quart. Rev. Biol., 59:33–53.

Fenton, M. B. 1990. The foraging behavior and ecology of animal-eating bats. Can. J. Zool., 68:411–422.

Fenton, M. B. 1994. Echolocation: its impact on the behaviour and ecology of bats. Écoscience 1:21–30.

Fenton, M. B. 1995. Natural history and biosonar signals, 37–86, in *Hearing in Bats* (A. N. Popper and R. R. Fry, eds.). Springer-Verlag Handbook of Auditory Research, vol. XI.

Fenton, M. B., and G. P. Bell. 1979. Echolocation and feeding behavior in four species of *Myotis* (Chiroptera). Can. J. Zool., 57:1271–1277.

Fenton, M. B., and G. P. Bell. 1981. Recognition of species of insectivorous bats by their echolocation calls. J. Mamm., 62:233–243.

Fenton, M. B., G. P. Bell, and D. W. Thomas. 1980. Echolocation and feeding behavior of *Taphozous mauritianus* (Chiroptera: Emballonuridae). Can. J. Zool., 58:1774–1777.

Fenton, M. B., and J. H. Fullard. 1981. Moth hearing and feeding strategies of bats. Amer. Sci., 69:266–275.

Fenton, M. B., C. L. Gaudet, and M. L. Leonard. 1983. Feeding behavior of the bats *Nycteris grandis* and *Nycteris thebaica* (Nycteridae) in captivity. J. Zool. London, 200:347–354.

Fenton, M. B., and I. L. Rautenbach. 1986. A comparison of the roosting and foraging behavior of three species of African insectivorous bats. Can. J. Zool., 64:2860–2867.

Fenton, M. B., D. H. M. Cumming, J. M. Hutton, and C. M. Swanepoel. 1987. Foraging and habitat use by *Nycteris grandis* (Chiroptera: Nycteridae) in Zimbabwe. J. Zool., Lond., 211:709–716.

Fenton, M. B., C. M. Swanepoel, R. M. Brigham, J. Cebek, and M. B. C. Hickey. 1990. Foraging behavior and prey selection by large slit-faced bats (*Nycteris grandis;* Chiroptera: Nycteridae). Biotropica, 22:2–8.

Fenton, M. B., I. L. Rautenbach, D. Chipese, M. B. Cumming, M. K. Musgrave, J. S. Taylor, and T. Volpers. 1993. Variation in foraging behaviour, habitat use, and diet of large slit-faced bats (*Nycteris grandis*). Z. Saugetierk., 58:65–74.

Fenton, M. B., D. Audet, M. K. Obrist, and J. Rydell. 1995. Signal strength, timing, and self-deafening: the evolution of echolocation in bats. Paleobiology, 21:229–242.

Fenton, M. B., and D. R. Griffin. 1997. High-altitude pursuit of insects by echolocating bats. J. Mamm., 78:247–250.

Ferkin, M. H. 1989. Adult-weaning recognition among captive meadow voles *(Microtus pennsylvanicus)*. Behaviour, 118:114–124.

Ferkin, M. H. 1990. Kin recognition and social behavior in microtine rodents, 11–24, in *Social Systems and Population Cycles in Voles* (R. Tamarin, R. S. Ostfeld, S. R. Pugh, and G. Bujalska, eds.). Birkhauser Verlag, Boston, Massachusetts.

Fiedler, J. 1979. Prey catching with and without echolocation in the Indian false vampire *(Megaderma lyra)*. Behav. Ecol. Sociobiol., 6:155–160.

Fiedler, P. L., and S. K. Jain (eds.). 1992. *Conservation Biology: The Theory and Practice of Nature Conservation, Preservation, and Management*. Chapman and Hall, New York.

Fielden, L. J., M. R. Perrin, and G. C. Hickman. 1990a. Feeding ecology of the Namib Desert golden mole, *Eremitalpa granti namibensis* (Chrysochloridae). J. Zool. London, 220:367–389.

Fielden, L. J., M. R. Perrin, and G. C. Hickman. 1990b. Water metabolism in the Namib Desert golden mole, *Eremitalpa granti namibensis* (Chrysochloridae). Comp. Biochem. Physiol., 96A:227–234.

Fielden, L. J., J. P. Waggoner, M. R. Perrin, and G. C. Hickman. 1990c. Thermoregulation in the Namib Desert golden mole, *Eremitalpa granti namibensis* (Chrysochloridae). J. Arid Environ., 18:221–237.

Findley, J. S. 1996. Mammalian biogeography in the American Southwest, 297–307, in *Contributions in Mammalogy: A Memorial Volume Honoring Dr. J. Knox Jones, Jr.* (H. H. Genoways and R. J. Baker, eds.). Museum of Texas Tech University, Lubbock.

Findley, J. S. and D. E. Wilson. 1974. Observations on the Neotropical disk-winged bat, *Thyroptera tricolor* Spix. J. Mamm., 55:562–571.

Fink, B. D. 1959. Observation of porpoise predation on a school of Pacific sardines. Calif. Fish and Game, 45:216–217.

Finley, R. B. Jr. 1969. Cone caches and middens of *Tamasciurus* in the Rocky Mountain region, in *Contributions in Mammalogy* (J. K. Jones Jr., ed.). University of Kansas, Mus. Nat. Hist. Misc. Publ. No. 51.

Fischer, M. 1989. Hyracoids, the sister group of perissodactyls, 37–56, in *The Evolution of Perissodactyls* (D. Prothero and R. M. Schoch, eds.). Oxford Univ. Press, New York.

Fischer, M., and P. Tassy. 1993. The interrelation between Proboscidea, Sirenia, Hyracoidea, and Mesaxonia: the morphological evidence, 217–234, in *Mammal Phylogeny, Placentals* (F. S. Szalay, M. J. Novacek, and M. C. McKenna, eds.). vol. 2, Springer-Verlag, New York.

Fish, F. E. 1979. Thermoregulation in the muskrat *(Ondatra zibethicus)*: the use of regional heterothermia. Comp. Biochem. Physiol., 64:391.

Fish, F. E. 1993. Influence of hydrodynamic design and propulsive mode on mammalian swimming energetics. Aust. J. Zool., 42:79–101.

Fish, F. E., R. V. Baudinette, P. B. Frappell, and M. P. Sarre. 1997. Energetics of swimming by the platypus *Ornithorhynchus anatinus:* metabolic effort associated with rowing. J. Exp. Biol., 200:2647–2652.

Fisher, E. M. 1939. Habits of the southern sea otter. J. Mamm., 20:21–36.

Fitch, H. S., R. Goodrum, and C. Newman, 1952. The armadillo in the southeastern United States. J. Mamm., 33:21–37.

Fitzgerald, B. M. 1977. Weasel predation on a cyclic population of the montane vole *(Microtus montanus)* in California. J. Anim. Ecol., 46:367–397.

FitzGibbon, C. 1995. Comparative ecology of two elephant-shrew species in Kenyan coastal forest. Mamm. Rev., 25:19–30.

FitzGibbon, C. 1997. The adaptive significance of monogamy in the golden-rumped elephant-shrew. J. Zool., London, 242:167–177.

Flannery, T. 1995. *Mammals of New Guinea*. Smithsonian Institution Press, Washington, D.C.

Fleagle, J. G. 1976. Locomotion and posture of the Malayan siamang. Folia Primatol., 26:245–269.

Fleagle, J. G. 1988. *Primate Adaptation and Evolution*. Academic Press, New York.

Fleagle, J. G., R. F. Kay, and M. R. L. Anthony. 1997. Fossil New World monkeys, 473–495, in *Vertebrate Paleontology in the Neotropics: The Miocene Fauna of La Venta, Colombia* (R. F. Kay, R. H. Madden, R. L. Cifelli, and J. J. Flynn, eds.). Smithsonian Inst. Press, Washington, D.C.

Fleming, T. H. 1970. Notes on the rodent faunas of two Panamanian forests. J. Mamm., 51:473–490.

Fleming, T. H. 1971. *Artibeus jamaicensis:* delayed embryonic development in a Neotropical bat. Science, 171:402–404.

Fleming, T. H. 1988. *The Short-tailed Fruit Bat: A Study in Plant-animal Interactions*. Univ. Chicago Press, Chicago.

Fleming, T. H. 1992. How do fruit- and nectar-feeding birds and mammals track their food resources? in *Effects of Resource Distribution on Animal-plant Interactions* (M. D. Hunter, T. Ohguchi, and P. W. Price, eds). Academic Press, New York.

Fleming, T. H. 1995. The use of stable isotopes to study the diets of plant-visiting bats. Symposia Zool. Soc. London, 67:99–110.

Fleming, T. H., E. T. Hooper, and D. E. Wilson. 1972. Three Central American bat communities: structure, reproductive cycles, and movement patterns. Ecology, 53:555–569.

Fleming, T. H., R. A. Nuñez, and L. Sternberg. 1993. Seasonal changes in the diets of migrant and non-migrant nectarivorous bats as revealed by carbon stable isotope analysis. Oecologia, 94:72–75.

Fleming, T. H., and V. J. Sosa. 1994. Effects of nectarivo-rous and frugivorous mammals on reproductive success of plants. J. Mamm., 75:845–851.

Fleming, T. H., S. Maurice, and J. L. Hamrick. 1998. Geographic variation in the breeding system and the evolutionary stability of trioecy in *Pachycereus pringlei* (Cactaceae). Evol. Ecol., 12:279–289.

Flexner, L. B., D. B. Crowie, L. M. Hellman, W. S. Wilde, and G. J. Vosburgh. 1948. The permeability of the human placenta to sodium in normal and abnormal pregnancies and the supply of sodium to the human fetus as determined with radioactive sodium. Amer. J. Obst. Gyn., 55:469–480.

Floody, O. R., and A. P. Arnold. 1975. Uganda kob *(Adenota kob thomasi)* territoriality and the spatial distribution of sexual and agonistic behaviors at a territorial ground. Z. Teirpsychol., 37:192–212.

Flynn, J., and H. Galiano. 1982. Phylogeny of early Tertiary Carnivora, with a description of a new species of *Protictis* from the Middle Eocene of Northwestern Wyoming. Amer. Mus. Novitates, 2725:1–64.

Flynn, J. J., and A. R. Wyss. 1998. Recent advances in South American mammalian paleontology. Trends Ecol. Evol., 13:449–453.

Flynn, J. J., N. N. Neff, and R. H. Tedford. 1988. Phylogeny of the Carnivora. 73–116, in *The Phylogeny and Classification of Tetrapods* (M. J. Benton, ed.). vol. 2. Clarendon Press, Oxford.

Fogden, M. 1974. A preliminary field study of the western tarsier, *Tarsius banacanus* Horsefield, 151–166, in *Prosimian Biology* (R. D. Martin, G. A. Doyle, and A. C. Walker, eds.). Duckworth, London.

Folk, G. E. Jr., A. Larson, and M. A. Folk. 1976. Physiology of hibernating bears, in *Bears—Their Biology and Management* (M. R. Pelton, J. W. Lentfer, and G. E. Folk, eds.). Publ. No. 40, Int. Union Conserv. Nat.

Folstad, I., and A. J. Karter. 1992. Parasites, bright males, and the immunocompetence handicap. Amer. Nat., 139:603–622.

Fons, R., S. Sender, T. Peters, and K. D. Jürgens. 1997. Rates of rewarming, heart and respiratory rates and their significance for oxygen transport during arousal from torpor in the smallest mammal, the Etruscan shrew *Suncus etruscus*. J. Exp. Biol., 200:1451–1458.

Fonseca, G. A. B. da. 1985. The vanishing Brazilian Atlantic forest. Biol. Conser., 34:17–34.

Ford, J. K. B. 1991. Vocal traditions among resident killer whales *(Orcinus orca)* in coastal waters of British Columbia. Can. J. Zool., 69:1454–1483.

Ford, R. G., and F. A. Pitelka. 1984. Resource limitation in populations of the California vole *Microtus californicus*. Ecology, 65:122–136.

Ford, S. M. 1986. Systematics of New World monkeys. 73–135, in *Comparative Primate Biology, Systematics, Evolution, and Anatomy* (D. R. Swindler and J. Erwin, eds.). vol. 1, Alan R. Liss, Inc., New York.

Fordyce, R. E., and L. G. Barnes. 1994. The evolutionary history of whales and dolphins. Ann. Rev. Earth Planetary Sci., 22:419–455.

Forester, D. J., and G. E. Machlis. 1996. Modelling human factors that affect the loss of biodiversity. Conserv. Biol., 10:1253–1263.

Formozov, A. N. 1946. The covering of snow as an integral factor of the environment and its importance in the ecology of mammals and birds. Material for Fauna and Flora of the USSR, New Series Zool. 5:1–141.

Formozov, A. N. 1966. Adaptive modifications of behavior in mammals of the Eurasian steppes. J. Mamm., 47:208–222.

Fossey, D. 1972. Living with mountain gorillas, in *The Marvels of Animal Behavior* (P. R. Marler, ed.). National Geographic Society, Washington, D.C.

Fossey, D. 1983. *Gorillas in the Mist*. Houghton Mifflin, Boston.

Foster, J. B., 1964. Evolution of mammals on islands. Nature 202:234.

Frakes, L. A., Francis, J. E., and Syktus, J. I. 1992. *Climate Modes of the Phanerozoic. The History of the Earth's Climate Over the Past 600 Million Years.* Cambridge University Press, Cambridge.

Francis, C. M., E. L. P. Anthony, J. A. Brunton, and T. H. Kunz. 1994. Lactation in male fruit bats. Nature, 367:691–692.

Frank, C. L. 1988a. The influence of moisture content on seed selection by kangaroo rats. J. Mamm., 69:353–357.

Frank, C. L. 1988b. The effects of moldiness level on seed selection by *Dipodomys spectabilis*. J. Mamm., 69:358–362.

Frank, S. A. 1990. Sex allocation theory for birds and mammals. Ann. Rev. Ecol. Syst., 21:13–55.

Freeman, P. W. 1981. A multivariate study of the family Molossidae (Mammalia: Chiroptera): morphology, ecology, evolution. Fieldiana Zoology, 7:1–173.

Freeman, P. W. 1984. Functional analysis of large animalivorous bats (Microchiroptera). Biol. J. Linn. Soc., 21:387–408.

Freeman, P. W. 1988. Frugivorous and animalivorous bats (Microchiroptera): dental and cranial adaptations. Biol. J. Linn. Soc., 33:249–272.

Freeman, P. W. 1995. Nectarivorous feeding mechanisms in bats. Biol. J. Linn. Soc., 56:439–463.

Freeman, P. W., and C. Lemen. 1983. Quantification of competition among coexisting heteromyids in the southwest. Southwest. Nat., 28:41–46.

French, A. R. 1977. Periodicity of recurrent hypothermia during hibernation in the pocket mouse, *Perognathus longimembris*. J. Comp. Physiol., 115:87.

French, A. R. 1993. Physiological ecology of the Heteromyidae: economics of energy and water utilization. 509–538, in *Biology of the Heteromyidae* (H. H. Genoways and J. H. Brown, eds.). Spec. Publ. No. 10, Amer. Soc. Mamm.

French, N. R., W. E. Grant, W. Grodzinski, and D. M. Swift. 1976. Small mammal energetics in grassland ecosystems. Ecol. Monogr., 46:201–220.

Freudenberger, D. O., I. R. Wallis, and I. D. Hume. 1989. Digestive adaptations of kangaroos, wallabies, and rat-kangaroos, 179–189, in *Kangaroos, Wallabies and Rat-kangaroos* (G. Grigg, P. Jarman, and I. Hume, eds.). Surrey Beatty and Sons, Chipping Norton, New South Wales, Australia.

Freudenthal, M., 1972. *Deinogalerix koenigswaldi* nov. gen., nov. spec.: A giant insectivore from the Neogene of Italy. Scripta Geologica, Leiden 14:1.

Friis, E. M., and W. L. Crepet. 1987. Time of appearance of floral features, 145–179, in *The Origins of Angiosperms and Their Biological Consequences* (E. M. Friis, W. G. Chalconer, and P. R. Crane, eds.). Cambridge Univ. Press, Cambridge.

Fullard, J. H. 1982. Echolocation assemblages and their effects on moth auditory systems. Can. J. Zool., 60:2572–2576.

Fullard, J. H. 1987. Sensory ecology and neuroethology of moths and bats: interactions in a global perspective, 244–272, in *Recent Advances in the Study of Bats* (M. B. Fenton, P. Racey, and J. M. V. Rayner, eds.). Cambridge Univ. Press, Cambridge, United Kingdom.

Fullard, J. H., M. B. Fenton, and J. A. Simmons. 1979. Jamming bat echolocation: the clicks of arctiid moths. Can. J. Zool., 57:647–649.

Fullard, J. H., and J. J. Belwood. 1988. The echolocation assemblage: acoustic ensembles in a neotropical habitat, 639–643, in *Animal Sonar* (P. E. Nachtigall and P. W. B. Moore, eds.). Plenum Press, New York.

Fullard, J. H., and J. W. Dawson. 1997. The echolocation calls of the spotted bat *Euderma maculatum* are relatively inaudible to moths. J. Exp. Biol., 200:129–137.

Fuller, T. K., and P. W. Kat. 1993. Hunting success of African wild dogs in southwestern Kenya. J. Mamm., 74:464–467.

Fuller, W. A. 1967. Ecologie hivernale des lemmings et fluctuations de leurs populations. Terre Vie, 114:97–115.

Fuller, W. A. 1969. Changes in numbers of three species of small rodents near Great Slave Lake, N.W.T., Canada, 1964–1967, and their significance for general population theory. Ann. Zool. Fennici, 6:113–144.

Fullerton, D., and R. Stavins. 1998. How economists see the environment. Nature, 395:433–434.

Gaines, M. S. 1985. Genetics, 845–883, in *Biology of the New World Microtus* (R. H. Tamarin, ed.). Spec. Publ. No. 8, Amer. Soc. Mamm.

Gaines, M. S., and C. J. Krebs. 1971. Genetic changes in fluctuating vole populations. Evolution, 25:702–723.

Gaines, M. S., A. V. Vivas, and C. L. Baker. 1979. An experimental analysis of dispersal in fluctuating vole populations: demographic parameters. Ecology, 60:814–828.

Gaisler, J. 1979. Ecology of bats, 281–342, in *Ecology of Small Mammals* (D. M. Stoddart, ed.). Chapman and Hall, London.

Galliari, C. A., and F. J. Goin. 1993. Conservación de la biodiversidad en la Argentina: el caso de los mamíferos, 367–399, in *Elementos de Política Ambiental. La Plata* (F. J. Goin and R. G. Goñi, eds.). H. Cámara de Diputados, Provincia de Buenos Aires.

Gallivan, G. J., and R. C. Best. 1986. The influence of feeding and fasting on the metabolic rate and ventilation of the Amazonian manatee (*Trichechus inunguis*). Physiol. Zool., 59:552–557.

Gambaryan, P. P. 1974. *How Mammals Run: Anatomical Adaptations.* Keter Publ., Jerusalem.

Gambaryan, P. P., and Z. Kielan-Jaworowska. 1997. Sprawling versus parasagittal stance in multituberculate mammals. Acta Palaeontologica Polonica 42:13–44.

Garber, P. A. 1984. Proposed nutritional importance of plant exudates in the diet of the Panamanian tamarin, *Saguinus oedipus geoffroyi.* Inter. J. Primatol., 5:1–15.

Gardner, A. L. 1993. Order Didelphimorphia, 15–23, in *Mammal Species of the World: A Taxonomic and Geographic Reference* (D. E. Wilson and D. M. Reeder, eds.). Smithsonian Inst. Press, Washington, D. C.

Garsd, A., and W. E. Howard. 1981. A 19-year study of microtine population fluctuations using time-series analysis. Ecology, 62:930–937.

Gasc, J. P., F. K. Jouffroy, S. Renous, and F. von Bloonitz. 1986. Morphofunctional study of the digging system of the Namib Desert golden mole (*Eremitalpa granti namibensis*): cinefluorographical and anatomical analysis. J. Zool., London (A), 208:9–35.

Gasparini, Z., X. Pereda-Suberbiola, and R. E. Molnar. 1996. New data on the ankylosaurian dinosaur from the Late Cretaceous of the Antarctic peninsula. Mem. Queensland Mus., 39:583–594.

Gatesy, J., D. Yelon, R. Desalle, and E. S. Vrba. 1992. Phylogeny of the Bovidae (Artiodactyla, Mammalia), based on mitochondrial ribosomal DNA sequences. Mol. Biol. Evol., 9:433–446.

Gaudin, T. J., and A. A. Biewener. 1992. The functional morphology of xenarthrous vertebrae in the armadillo *Dasypus novemcinctus* (Mammalia, Xenarthra). J. Morph. 214:63–81.

Gaudin, T. J., J. R. Wible, J. A. Hopson, and W. D. Turnbull. 1996. Reexamination of the morphological evidence for the cohort Epitheria (Mammalia, Eutheria). J. Mamm. Evol., 3:31–79.

Gauthier, J., A. G. Kluge, and T. Rowe. 1988. Amniote phylogeny and the importance of fossils. Cladistics 4(2):105–209.

Gawn, R. L. W. 1948. Aspects of the locomotion of whales. Nature, 161:44–46.

Geiser, F., and L. S. Broome. 1993. The effects of temperature on the pattern of torpor in a marsupial hibernator. J. Comp. Physiol. B, 163:133–137.

Geist, V. 1966. The evolution of horn-like organs. Behaviour, 27:175–214.

Geist, V. 1974. On the relationship of ecology and behavior in the evolution of ungulates: theoretical considerations, 235–246, in *The Behavior of Ungulates and its Relation to Management* (V. Geist and F. R. Walther, eds.). International Union for Conservation of Nature and Natural Resources, Morges, Switzerland.

Geist, V. 1994. Wildlife conservation as wealth. Nature, 368:491–492.

Geluso, K. N. 1975. Urine concentration cycles of insectivorous bats in the laboratory. J. Comp. Physiol., 99:309–319.

Geluso, K. N. 1978. Urine concentrating ability and renal structure of insectivorous bats. J. Mamm., 59:812–822.

Geluso, K. N., and E. H. Studier. 1979. Diurnal fluctuation in urine concentration in the little brown bat, *Myotis lucifugus,* in a natural roost. Comp. Biochem. Physiol., 62:471–473.

Genoud, M. 1985. Ecological energetics of two European shrews: *Crocidura russula* and *Sorex coronatus* (Soricidae: Mammalia). J. Zool., London (A), 207:63–85.

Genoways, H. H., and J. H. Brown. 1993. Biology of the Heteromyidae. Spec. Publ. No. 10, Amer. Soc. Mamm.

George, W., and B. J. Weir. 1972. The chromosomes of some octodontids with special reference to *Octodontomys* (Rodentia, Hystricomorpha). Chromosoma, 37:53–62.

Gerstell, R. 1938. The Pennsylvania deer problem in 1938. Penn. Game News, 9(5):12–13, 31; 9(6):10–11, 27, 32; 9(7):6–7, 29.

Gese, E. M., and R. L. Ruff. 1997. Scent-marking by coyotes, *Canis latrans:* the influence of social and ecological factors. Anim. Behav., 54:1155–1166.

Getz, L. L., J. E. Hofmann, B. J. Klatt, L. Verner, F. R. Cole, and R. L. Lindroth. 1987. Fourteen years of population fluctuations of *Microtus ochrogaster* and *M. pennsylvanicus* in east central Illinois. Can. J. Zool., 65:1317–1325.

Getz, L. L., C. M. Larson, and K. A. Lindstrom. 1992. *Blarina brevicauda* as a predator on nestling voles. J. Mamm., 73:591–596.

Getz, L. L., B. McGuire, J. Hofmann, T. Pizzuto, and B. Frase. 1993. Social organization of the prairie voles *(Microtus ochrogaster).* J. Mamm., 74:44–58.

Getz, L. L., and C. S. Carter. 1996. Prairie-vole partnerships. Amer. Sci., 84:56–62.

Getz, L. L., and B. McGuire. 1997. Communal nesting in prairie voles (*Microtus ochrogaster*)*:* formation, composition, and persistence of communal groups. Can. J. Zool., 75:525–534.

Geuse, P., V. Bauchau, and E. LeBoulenge. 1985. Distribution and population dynamics of bank voles and wood mice in a patchy woodland habitat in central Belgium. Acta Zool. Fennica, 173:65–68.

Gheerbrant, E., J. Sudre, and H. Cappetta. 1996. A Palaeocene proboscidean from Morocco. Nature, 383:68–70.

Ghiglieri, M. P. 1987. Sociobiology of the great apes and the hominid ancestor. J. Human Evol., 16:319–357.

Gibbons, A. 1998. Monogamous gibbons really swing. Science, 280:677–678.

Gilbert, B. S., and S. Boutin. 1991. Effect of moonlight on winter activity of snowshoe hares. Arctic and Alpine Res., 23:61–65.

Gillette, D. D. 1994. *Seismosaurus: The Earth Shaker.* Columbia Univ. Press, New York.

Gingerich, P. D. 1980. Eocene Adapidae: paleobiogeography and the origin of South American Platyrrhini, 123–138, in *Evolutionary Biology of the New World Monkeys and Continental Drift* (R. L. Ciochon and A. B. Chiarelli, eds.). Plenum Press, New York.

Gingerich, P. D., N. A. Wells, D. E. Russell, and S. M. Ibrahim Shah. 1983. Origin of whales in epicontinental remnant seas: new evidence from the early Eocene of Pakistan. Science, 220:403–406.

Ginsburg, L., and E. Heintz. 1968. La plus ancienne antilope, *Eotragus artenensis* du Burdigalien d'Artenay. Bull. Mus. Nat. Hist. Paris, 40:837–842.

Glander, K. E. 1977. Poison in a monkey's garden of Eden. Nat. Hist., 86:35–41.

Goin, F. J. 1997. New clues for understanding Neogene marsupial radiations, 187–206, in *Vertebrate Paleontology in the Neotropics: The Miocene Fauna of La Venta, Colombia* (R. F. Kay, R. H. Madden, R. L. Cifelli, and J. J. Flynn, eds.). Smithsonian Inst. Press, Washington, D.C.

Goin, F. J., and R. Pascual. 1987. News on the biology and taxonomy of the marsupials Thylacosmilidae (late Tertiary of Argentina). Anales de la Academia Nacional de Ciencias Ex. Fís. Nat., Buenos Aires, 39:219–246.

Goin, F. J., and R. Goñi. 1991. Naturaleza, naturalistas, tecnología e innovación, 211–220, in *Ciencia, Tecnología e Innovación: Perspectivas y Estrategias* (R. G. Goñi and F. J. Goin, eds.). La Plata, H. Cámara de senadores, Provincia de Buenos Aires.

Goin, F. J., and R. Goñi. 1993. Modernidad, progreso y medio ambiente. Aproximaciones a la crisis, 929–938, in *Elementos de Política Ambiental* (F. J. Goin and R. G. Goñi, eds.). La Plata, H. Cámara de Diputados, Provincia de Buenos Aires.

Goin, F. J., and A. A. Carlini. 1995. An early Tertiary microbiotheriid marsupial from Antarctica. J. Vert. Paleo., 15:205–207.

Goldblatt, P. 1993. *Biological Relationships Between Africa and South America.* New Haven, Yale Univ. Press.

Goldman, D. P., R. Giri, and S. J. O'Brien. 1989. Molecular genetic-distance estimates among the Ursidae as indicated by one- and two-dimensional protein electrophoresis. Evolution, 43:282–295.

Goldman, L. J., and O. W. Henson, Jr. 1977. Prey recognition and selection by the constant frequency bat, *Pteronotus parnellii.* Behav. Ecol. Sociobiol., 2:411–420.

Gompper, M. E. 1996. Sociality and asociality in white-nosed coatis (*Nasua narica*)*:* foraging costs and benefits. Behav. Ecol., 7:254–263.

Gompper, M. E., J. L. Gittleman, and R. K. Wayne. 1997. Genetic relatedness, coalitions and social behavior of white-nosed coatis, *Nasua narica*. Anim. Behav., 53:781–797.

Good, R. 1974. *The Geography of Flowering Plants*. 3rd ed. White Plains, Longman, New York.

Goodall, J. 1977. Infant killing and cannibalism in free-living chimpanzees. Folia Primat., 28:259–282.

Goodall, J. 1983. Population dynamics during a 15-year period in one community of free-living chimpanzees in the Gombe National Park, Tanzania. Primates, 21:545–549.

Goodall, J. 1986. *Chimpanzees of Gombe*. Harvard Univ. Press, Cambridge, Massachusetts.

Goodman, B. 1993. Drugs and people threaten diversity in Andean forests. Science, 261:293.

Goodman, D. 1975. The theory of diversity-stability relationships in ecology. Quart. Rev. Biol., 50:237–266.

Goodman, S. M., J. U. Ganzhorn and L. Wilmé. 1997. Observations at a *Ficus* tree in Malagasy humid forest. Biotropica 29:480–488.

Goodman, S. M., and B. D. Patterson (eds.). 1997. *Natural Change and Human Impact in Madagascar*. Smithsonian Inst. Press, Washington, D. C.

Goold, J. C., and S. E. Jones. 1995. Time and frequency domain characteristics of sperm whale clicks. J. Acoust. Soc. Amer., 98:1279–1291.

Goosem, M. 1997. Internal fragmentation: the effects of roads, highways, and powerline clearings on movements and mortality of rainforest vertebrates, 241–255, in *Tropical Forest Remnants: Ecology, Management, and Conservation of Fragmented Communities* (W. F. Laurance and R. O. Bierregaard, Jr., eds.). University of Chicago Press, Chicago.

Göpfert, M. C., and L. T. Wasserthal. 1995. Notes on echolocation calls, food and roosting behaviour of the Old World sucker-footed bat *Myzopoda aurita* (Chiroptera, Myzopodidae). Z. Säugetierk., 60:1–8.

Gordon, M. S., G. A. Bartholomew, A. D. Grinnell, C. B. Jorgensen, and F. N. White. 1977. *Animal Physiology: Principles and Adaptations*. 3d ed. Macmillan, New York.

Gorman, M. L. 1976. A mechanism for individual recognition by odor in *Herpestes auropunctatus* (Carnivora, Viverridae). Anim. Behav., 24:141–145.

Gorman, M. L., D. B. Nedwell, and R. M. Smith. 1974. An analysis of the anal scent pockets of *Herpestes auropunctatus* (Carnivora: Viverridae). J. Zool., London, 172:388–389.

Gorman, M. L., and R. D. Stone. 1990. *The Natural History of Moles*. Cornell Univ. Press, Ithaca, New York.

Goss, R. J. 1983. *Deer Antlers—Regeneration, Function and Evolution*. Academic Press, New York.

Gould, E. W. 1937. Occurance of low growing game foods during the old-field pine-mixed hardwood succession in the Harvard Forest. Master's thesis, Harvard Univ.

Gould, E. 1965. Evidence for echolocation in the Tenrecidae of Madagascar. Proc. Amer. Phil. Soc., 109:352–360.

Gould, E. 1969. Communication in three genera of shrews (Soricidae): *Suncus, Blarina*, and *Cryptotis*. Comm. Behav. Biol., Part A, 3:11–31.

Gould, E. 1970. Echolocation and communication of bats, 144–162, in *About Bats* (B. H. Slaughter and D. W. Walton, eds.). Southern Methodist University Press, Dallas.

Gould, E. 1971. Studies of maternal-infant communication and development of vocalization in the bats *Myotis* and *Eptesicus*. Comm. Behav. Biol., Part A, 5:263–313.

Gould, E. 1978. The behavior of the moonrat, *Echinosorex gymnurus* (Erinaceidae) and the pentail shrew, *Ptilocerus lowi* (Tupaiidae) with comments on the behavior of other Insectivora. Z. Tierpsychol., 48:1–27.

Gould, E., N. C. Negus, and A. Novick. 1964. Evidence for echolocation in shrews. J. Exp. Zool., 156:19–38.

Gould, E., and J. F. Eisenberg. 1966. Notes on the biology of the Tenrecidae. J. Mamm., 47:660–686.

Gradstein, F. M., F. P. Agterberg, J. G. Ogg, J. Hardenbol, P. Van Veen, J. Thierry, and Zehui Huang. 1995. A Triassic, Jurassic, and Cretaceous time scale, 95–126, in *Geochronology Time Scales and Global Stratigraphic Correlation*. SEPM Spec. Publ. No. 54, SEPM (Society for Sedimentary Geology).

Graf, W. 1955. The Roosevelt Elk. Port Angeles, Washington: Port Angeles Evening News.

Graham, R. W., E. L. Lundelius, Jr., M. A. Graham, E. K. Schroeder, R. S. Toomey III, E. Anderson, et al. 1996. Spatial response of mammals to late-Quaternary environmental fluctuations. Science, 272:1601–1606.

Grand, T. I., and R. Lorenz. 1968. Functional analysis of the hip joint in *Tarsius bancanus* (Horsefield, 1821) and *Tarsius syrichta* (Linnaeus, 1758). Folia Primatol., 9:161–181.

Grand, T., E. Gould, and R. Montali. 1998. Structure of the proboscis and rays of the star-nosed mole, *Condylura cristata*. J. Mamm., 79:492–501.

Gregory, J. E., A. Iggo, A. K. McIntyre, and U. Proske. 1989. Responses of electroreceptors in the snout of the echidna. J. Physiol., 414:521–538.

Gregory, W. K. 1910. The orders of mammals. Bull. Amer. Mus. Nat. His. New York, 27:1–524.

Green, B., J. Merchant, and K. Newgrain. 1997. Lactational energetics of a marsupial carnivore, the eastern quoll (*Dasyurus viverrinus*). Aust. J. Zool., 45:295–306.

Green, R. G., and C. L. Larson. 1938. A description of shock disease in the snowshoe hare. Amer. J. Hyg., 28:190–212.

Green, R. G., C. L. Larson, and J. F. Bell. 1939. Shock disease as the cause of the periodic decimation of the snowshoe hare. Amer. J. Hyg., 30B:83–102.

Greenwald, G. S. 1956. The reproductive cycle of the field mouse, *Microtus californicus*. J. Mamm., 37:213–222.

Greenwald, G. S. 1957. Reproduction in a coastal California population of the field mouse, *Microtus californicus*. Univ. California Publ. Zool., 54:421–446.

Griffin, D. R. 1958. *Listening in the Dark*. Yale Univ. Press, New Haven, Connecticut.

Griffin, D. R. 1970. Migrations and homing of bats, 233–264, in *Biology of Bats.* (W. A. Wimsatt, ed.). Academic Press, New York.

Griffin, D. R., and R. Galambos. 1940. Obstacle avoidance by flying bats. Anat. Rec., 78:95.

Griffin, D. R., and R. Galambos. 1941. The sensory basis of obstacle avoidance by flying bats. J. Exp. Zool., 86:481–506.

Griffin, D. R., and J. A. Simmons. 1974. Echolocation of insects by horseshoe bats. Nature, 250:731–732.

Griffin, M. 1990. A review of taxonomy and ecology of gerbilline rodents of the central Namib Desert, with keys to the species (Rodentia: Muridae), 83–98, in *Namib Ecology: 25 Years of Namib Research* (M. K. Seely, ed.). Transvaal Museum Monograph No. 7, Transvaal Museum, Pretoria.

Griffiths, M. 1978. *The Biology of the Monotremes.* Academic Press, New York.

Griffiths, M., R. T. Wells, and D. J. Barrie. 1991. Observations on the skulls of fossil and extant echidnas (Monotremata: Tachyglossidae). Aust. Mamm., 14:87–101.

Griffiths, T. A. 1997. Systematic relationship of the New Zealand short-tailed bat *Mystacina tuberculata* (Mystacinidae) to other bats, based on hyoid morphology. Program and Abstracts, 27th Annual North American Symposium on Bat Research, Tucson, Arizona.

Grigg, G. C., L. A. Beard, and M. L. Augee. 1989. Hibernation in a monotreme, the echidna *(Tachyglossus aculeatus).* Comp. Biochem. Physiol., 92A:609–612.

Grigg, G. C., M. L. Augee, and L. A. Beard. 1992. Thermal relations of free-living echidnas during activity and in hibernation in a cold climate. 160–173, in *Platypus and Echidnas* (M. L. Augee, ed.). Royal Zool. Soc. New South Wales, Sydney.

Grinnell, J. 1922. A geographical study of the kangaroo rats of California. Univ. Calif. Pub. Zool., 24:1–124.

Grinnell, J., and T. I. Storer. 1924. *Animal Life in the Yosemite.* Univ. California Press, Berkeley.

Grinnell, J., J. S. Dixon, and J. M. Linsdale. 1937. *Fur-bearing Mammals of California,* 2 vols. Univ. Calif. Press, Berkeley.

Grodzinski, W., and B. A. Wunder. 1975. Ecological energetics of small mammals, 173–204, in *Small Mammals: Their Productivity and Population Dynamics* (F. B. Golley, K. Petrusewicz, and L. Ryszkowski, eds.). Cambridge Univ. Press, Cambridge.

Grove, C. P. 1989. *A Theory of Human and Primate Evolution.* Oxford Univ. Press, New York.

Groves, C. P. 1993. Order Primates, 243–277, in *Mammal Species of the World: A Taxonomic and Geographic Reference* (D. E. Wilson and D. M. Reeder, eds.). Smithsonian Inst. Press, Washington, D.C.

Groves, C. P., and T. F. Flannery. 1990. Higher level systematics within the Peramelidae (Marsupialia), 37–42, in *Bandicoots and Bilbies* (J. H. Seebeck, P. R. Brown, R. L. Wallis, and C. M. Kemper, eds.). Surrey Beatty and Sons, Chipping Norton, New South Wales.

Groves, C., and P. Grubb. 1987. Classification of living cervids, 21–59, in *The Biology and Management of the Cervidae* (C. Wemmer, ed.). Nat. Zool. Soc. Symp. Vol., Smithsonian Inst. Press, Washington, D.C.

Grubb, P. 1993. Order Artiodactyla, 377–414, in *Mammal Species of the World* (D. E. Wilson and D. M. Reeder, eds.). Smithsonian Inst. Press, Washington, D.C.

Grzimek, B. 1990. *Grzimek's Encyclopedia of Mammals,* Vol. 1. McGraw-Hill Pub., New York.

Gurung, K. K., and R. Singh. 1996. *Field Guide to the Mammals of the Indian Subcontinent.* San Diego, Academic Press.

Gustafson, A. W., and D. A. Damassa. 1985. Annual variation in plasma sex steroid-binding protein and testosterone concentrations in the adult male little brown bat: relation to the asynchronous recrudescence of the testis and accessory reproductive organs. Biol. Reprod., 33:1126–1137.

Guthrie, M. J. 1933. The reproductive cycles of some cave bats. J. Mamm., 14:199–216.

Guthrie, R. D., and R. G. Petocz. 1970. Weapon automimicry among mammals. Amer. Nat., 104:585–588.

Guyton, A. C. 1976. *Textbook of Medical Physiology.* 5th ed. W. B. Saunders, Philadelphia.

Habersetzer, J. 1981. Adaptive echolocation sounds in the bat *Rhinopoma hardwickei:* a field study. J. Comp. Physiol. A, 144:559–566.

Habersetzer, J., and G. Storch. 1987. Klassifikation and functionelle Flügelmorphologie paläogener Fledermäuse (Mammalia, Chiroptera). Cour. Forschungsinst. Senckenb., 91:117–150.

Habersetzer, J., and G. Storch. 1989. Ecology and echolocation of the Eocene Messel bats, 213–233, in *European Bat Research 1987* (V. Hanak, T. Horacek, and J. Gaisler, eds.). Charles Univ. Press, Prague.

Habersetzer, J., G. Richter, and G. Storch. 1994. Paleoecology of early Middle Eocene bats from Messel, FRG: Aspects of flight, feeding, and echolocation. Hist. Biol., 8:235–260.

Hafner, D. J., and R. M. Sullivan. 1995. Historical and ecological biogeography of Nearctic pikas (Lagomorpha: Ochotonidae). J. Mamm., 76:302–321.

Haldane, J. B. S. 1932. *The Causes of Evolution.* Longman, London.

Hall, E. R., and K. R. Kelson. 1959. *The Mammals of North America.* Ronald Press, New York.

Hall, E. R., and W. W. Dalquest. 1963. *The Mammals of Veracruz.* Univ. Kansas Publ. Mus. Nat. Hist., 14:165–362.

Hall, K. R. L. 1968. Behavior and ecology of the wild patas monkey, in *Primates: Studies in Adaptation and Variability* (P. C. Jay, ed.). Holt, Reinhart, and Winston, New York.

Hall, K. R. L., and G. B. Schaller. 1964. Tool-using behavior of the California sea otter. J. Mamm., 45:287–298.

Hall, K. R. L., and I. DeVore. 1965. Baboon social behavior, 53–110, in *Primate Behavior* (I. DeVore, ed.). Holt, Rinehart and Winston, New York.

Halle, S. 1995. Effect of extrinsic factors on activity of root voles, *Microtus oeconomus*. J. Mamm., 76:88–99.

Hamilton, W. J., Jr., 1937. The biology of microtine cycles. J. Agric. Res., 54:779–790.

Hamilton, W. J. III. 1962. Reproductive adaptations of the red tree mouse. J. Mamm., 43:486–504.

Hamilton, W. D. 1964. The genetical evolution of social behavior. J. Theoret. Biol., 7:1–52.

Hamilton, W. D. 1971. Geometry for the selfish herd. J. Theoret. Biol., 31:295–311.

Hand, S. J., P. Murray, D. Megirian, M. Archer, and H. Godthelp. 1998. Mystacinid bats (Microchiroptera) from the Australian Tertiary. J. Paleo., 72:538–545.

Handley, C. O., Jr., D. E. Wilson, and A. L. Gardner. 1991. *Demography and Natural History of the Common Fruit Bat, Artibeus jamaicensis, on Barro Colorado Island, Panama.* Smithsonian Contrib. Zool., Washington, D.C.

Hanken, J., and B. K. Hall (eds.). 1993a. *The Skull: Development.* Volume 1. Chicago, Univ. Chicago Press.

Hanken, J., and B. K. Hall (eds.). 1993b. *The Skull: Patterns of Structural and Systematic Diversity.* Volume 2. Univ. Chicago Press, Chicago.

Hanken, J., and B. K. Hall (eds.). 1993c. *The Skull: Functional and Evolutionary Mechanisms.* Univ. Chicago Press, Chicago.

Hanover, J. W. 1966. Genetics of terpenes. I: gene control of monoterpene levels in *Pinus monticola.* Dougl. Heredity, 21:73–84.

Hanover, J. W. 1971. Genetics of terpenes. II: genetic variances in interrelationships of monoterpene concentrations in *Pinus monticola.* Heredity, 27:237–245.

Hansen, R. M. 1962. Movements and survival of *Thomomys talpoides* in a mima-mound habitat. Ecology, 43:151–154.

Hansen, R. M. 1978. Shasta ground sloth food habits, Rampart Cave, Arizona. Paleobiology, 4:302–319.

Hansen, R. M., and A. L. Ward. 1966. Some relations of pocket gophers to rangelands on Grand Mesa, Colorado. Colo. Agric. Exp. Sta., Tech. Bull., 88:1–20.

Hanski, I., and M. E. Gilpin. 1997. *Metapopulation Biology: Ecology, Genetics, and Evolution.* Academic Press, New York.

Happold, D. C. D. 1987. *The Mammals of Nigeria.* Clarendon Press, Oxford.

Hardin, G. 1993. *Living Within Limits: Ecology, Economics, and Population Taboos.* Oxford University Press, Oxford.

Harris, A. H. 1990. Fossil evidence bearing on southwestern mammalian biogeography. J. Mamm., 71:219–229.

Harris, C. M. 1996. Absorption of sound in air versus humidity and temperature. J. Acoustic Soc. Amer., 40:148–159.

Harrison, T. 1987. The phylogenetic relationships of the early catarrhine primates: a review of the current evidence, 41–80, in *Primate Phylogeny* (F. E. Grine, J. G. Fleagle, and L. B. Martin, eds.). Academic Press, New York.

Hart, J. A., and R. M. Timm. 1978. Observations on the aquatic genet in Zaire. Carnivore, 1:130–131.

Hart, L. A., and B. L. Hart. 1987. Species-specific patterns of urine investigation and flehman in Grant's gazelle (*Gazalla granti*), Thomson's gazelle (*Gazella thomsonii*), impala (*Aepyceros melampus*) and eland (*Taurotragus oryx*). J. Comp. Psychol., 101:229–304.

Hart, L. A., and B. L. Hart. 1988. Autogrooming and social grooming in impala. Ann. New York Acad. Sci., 525:399–402.

Hartenberger, J. L. 1985. The order Rodentia: major questions on their evolutionary origin, relationships and suprafamilial systematics, 1–33, in *Evolutionary Relationships Among Rodents: A Multidisciplinary Analysis* (W. P. Luckett and J. L. Hartenberger, eds.). Plenum Press, New York.

Hartley, D. J., and R. A. Suthers. 1987. The sound emission pattern and the acoustical role of the noseleaf in the echolocating bat, *Carollia perspicillata.* J. Acoust. Soc. Amer., 82:1892–1900.

Hartline, P. H. 1971. Physiological basis for detection of sound and vibration in snakes. J. Exp. Biol., 54:349–371.

Hartman, C. G. 1933. On the survival of spermatozoa in the female genital tract of the bat. Quart. Rev. Biol., 8:185–193.

Harvey, M. J., and R. W. Barbour. 1965. Home ranges of *Microtus ochrogaster* as determined by a modified minimum area method. J. Mamm., 46:398–402.

Hasler, J. F., A. E. Buhl, and E. M. Banks. 1976. The influence of photoperiod on growth and sexual function in male and female collared lemmings (*Dicrostonyx groenlandicus*). J. Repro. Fertil., 46:323–329.

Hass, C. C. 1990. Alternative maternal-care patterns in two herds of bighorn sheep. J. Mamm., 71:24–35.

Hatt, R. T. 1932. The vertebral column of ricochetal rodents. Bull. Amer. Mus. Nat. Hist., 63:599–738.

Hatt, R. T. 1934. The pangolins and aardvarks collected by the American Museum Congo expedition. Bull. Amer. Mus. Nat. Hist., 66:643–672.

Hatt, R. T. 1936. Hyraxes collected by the American Museum Congo expedition. Bull. Amer. Mus. Nat. Hist., 72:117–141.

Hausfater, G., and S. B. Hrdy. 1984. *Infanticide: Comparative and Evolutionary Perspectives.* Aldine Pub., Hawthorne, New York.

Hayes, J. P. 1997. Temporal variation in activity of bats and the design of echolocation-monitoring studies. J. Mamm., 78:514–524.

Hayman, D. L. 1977. Chromosome number—constancy and variation, 27–48, in *The Biology of Marsupials* (B. Stonehouse and D. Gilmore, eds.). Macmillan, London.

Heaney, L. R. 1985. Systematics of oriental pygmy squirrels of the genera *Exilisciurus* and *Nannosciurus* (Mammalia: Sciuridae). Misc. Publ. Mus. Zool. Univ. Michigan, 170:1–58.

Heaney, L. R. 1986. Biogeography of mammals in SE Asia: estimates of rates of colonization, extinction and speciation. Biol. J. Linn. Soc., 28:127–165.

Heaney, L. R. 1993. Biodiversity patterns and the conservation of mammals in the Philippines. Asia Life Sci., 2:261–274.

Heaney, L. R., D. S. Balete, and A. T. L. Dans. 1997. Terrestrial mammals, 141–168, in *Wildlife Conservation of the Philippines*. Philippine Red Data Book. Bookmark, Manila, Philippines.

Hedges, S. B. 1996. Historical biogeography of West Indian vertebrates. Ann. Rev. Ecol. System., 27:163–196.

Hedges, S. B., P. H. Parker, C. G. Sibley, and S. Kumar. 1996. Continental breakup and the diversification of birds and mammals. Nature, 381:226–229.

Hediger, H. 1950. Gefangenschaftsgeburt ein afrikanischen Springhasen. Zool. Gart. Leipzig, 17(5).

Heezen, B. C. 1957. Whales entangled in deep-sea cables. Deep Sea Res., 4:105–115.

Heglund, N. C., C. R. Taylor, and T. A. McMahon. 1974. Scaling stride frequency and gait to animal size: mice to horses. Science, 186:1112–1113.

Heglund, N. C., M. A. Fedak, C. R. Taylor, and G. A. Cavagna. 1982. Energetics and mechanics of terrestrial locomotion. IV: total mechanical energy changes as a function of speed and body size in birds and mammals. J. Exp. Biol., 97:57–66.

Heideman, P. D. 1989. Delayed development in Fischer's pygmy fruit bat, *Haplonycteris fischeri*, in the Philippines. J. Reprod. Fertil., 85:363–382.

Heideman, P. D., J. A. Cummings, and L. R. Heaney. 1993. Reproductive timing and early embryonic development in an Old World fruit bat, *Otopteropus cartilagonodus* (Megachiroptera). J. Mamm., 74:621–630.

Heideman, P. D., and K. S. Powell. 1998. Age-specific reproductive strategies and delayed embryonic development in an old world fruit bat, *Ptenochirus jagori*. J. Mamm., 79:295–311.

Heim de Balsac, H. 1954. Un genre inedit et inattendu de mammifera (Insectivore Tenrecidae) d'Afrique Occidentale. Compt. Rend. Acad. Sci. Paris, 239.

Heinrich, R. E., and K. D. Rose. 1995. Partial skeleton of the primitive carnivoran *Miacis petilus* from the early Eocene of Wyoming. J. Mamm., 76:148–162.

Heinroth-Berger, K. 1959. Beobachtungen an handaufgezogenen Mantelpavianen (*Papio hamadryas* L.), Z. Tierpsychol., 16:706–732.

Heinsohn, R., and C. Packer. 1995. Complex cooperative strategies in group-territorial African lions. Science, 269:1260–1262.

Heller, H. C., and T. L. Poulson. 1970. Circannian rhythms. II: endogenous and exogenous factors controlling reproduction and hibernation in chipmunks (*Eutamias*) and ground squirrels (*Spermophilus*). Comp. Biochem. Physiol., 33:357–383.

Heltne, P. G., J. F. Wojcik, and A. G. Pook. 1981. Goeldi's monkey, genus *Callimico*, 169–209, in *Ecology and Behavior of Neotropical Primates* (A. F. Coimbra-Filho and R. A. Mittermeier, eds.). Academia Brasiliera de Ciencias, Rio de Janeiro.

Henderson, M. T., G. Merriam, and J. Wegner. 1985. Patchy environments and species survival: chipmunks in an agricultural mosaic. Biol. Conser., 31:95–105.

Hendrichs, H., and U. Hendrichs. 1971. *Dikdik und Elephanten*. R. Piper, Munich.

Hengeveld, R. 1989. *Dynamics of Biological Invasions*. Chapman and Hall, New York.

Hennig, W. 1966. *Phylogenetic Systematics*. Univ. Illinois Press, Urbana.

Henschel, J. R., and J. D. Skinner. 1987. Social relationships and dispersal patterns in a clan of spotted hyaenas *Crocuta crocuta* in the Kruger National Park. S. Afr. Tydskr. Dierk., 22:18–24.

Henschel, J. R., and J. D. Skinner. 1990. The diet of spotted hyaenas *Crocuta crocuta* in Kruger National Park. Afr. J. Ecol., 28:69–82.

Henshaw, R. E. 1970. Thermoregulation in bats, 188–232, in *About Bats* (B. H. Slaughter and D. W. Walton, eds.). Southern Methodist University Press, Dallas.

Heppes, J. B. 1958. The white rhinoceros in Uganda. Afr. Wildlife, 12:273–280.

Herald, E. S., R. L. Brownell, Jr., F. L. Frye, E. J. Moris, W. E. Evans, and A. B. Scott. 1969. Blind river dolphin: first side-swimming cetacean. Science, 166:1408–1410.

Herbst, L. H. 1985. The role of nitrogen from fruit pulp in the nutrition of a frugivorous bat, *Carollia perspicillata*. Biotropica, 18:39–44.

Herman, L. H. 1980. Cognitive characteristics in dolphins, 363–429, in *Cetacean Behavior: Mechanisms and Functions* (L. H. Herman, ed.). Wiley & Sons, New York.

Herman, L. M., M. F. Peacock, M. P. Yunker, and C. J. Madsen. 1975. Bottlenosed dolphin: double-slit pupil yields equivalent aerial and underwater diurnal acuity. Science, 189:650–652.

Herman, L. M., and W. N. Tavolga. 1980. The communication systems of cetaceans, 363–429, in *Cetacean Behavior: Mechanisms and Functions* (L. M. Herman, ed.). Wiley & Sons, New York.

Hermanson, J. W., 1981. Functional morphology of the clavicle in the pallid bat, *Antrozous pallidus*. J. Mamm., 62:801–805.

Hermanson, J. W., and J. S. Altenbach. 1981. Functional anatomy of the primary downstroke muscles in the pallid bat, *Antrozous pallidus*. J. Mamm., 62:795–800.

Hermanson, J. W., W. A. La Framboise, and M. J. Daood. 1991. Uniform myosin isoforms in the flight muscles of little brown bats, *Myotis lucifugus*. J. Exp. Zool., 259:174–180.

Hermanson, J. W., M. A. Cobb, W. A. Schutt, F. Muradali, and J. M. Ryan. 1993. Histochemical and myosin composition of vampire bat (*Desmodus rotundus*) pectoralis muscle targets a unique locomotory niche. J. Morphol., 217:347–356.

Herrera, J., C. L. Kramer, and O. J. Reichman. 1997. Patterns of fungal communities that inhabit rodent food stores: effect of substrate and infection time. Mycologia, 89:846–857.

Hershkovitz, P. 1977. *Living New World monkeys (Platyrrhini), With an Introduction to the Primates.* vol. 1. Univ. Chicago Press, Chicago.

Hershkovitz, P. 1992. Ankle bones: the Chilean opossum *Dromiciops gliroides* Thomas, and marsupial phylogeny. Bonn. Zool. Beitr., 43:181–213.

Hertel, A. 1969. Hydrodynamics of swimming and wave-riding dolphins, 31–63, in *The Biology of Marine Mammals* (H. T. Anderson, ed.). Academic Press, New York.

Heske, E. J. 1987a. Spatial structuring and dispersal in a high density population of the California vole, *Microtus californicus.* Holarctic Ecol., 10:137–149.

Heske, E. J. 1987b. Pregnancy interruption by strange males in the California vole. J. Mamm., 68:406–410.

Hestbeck, J. B. 1986. Multiple regulation states in populations of the California vole, *Microtus californicus.* Ecol. Monogr., 56:161–181.

Heth, G., E. Frankenberg, A. Raz, and E. Nevo. 1987. Vibrational communication in subterranean mole rats (*Spalax ehrenbergi*). Behav. Ecol. Sociobiol., 20:31–33.

Heth, G., E. Frankenberg, and E. Nevo. 1988. "Courtship" calls of the subterranean mole rats (*Spalax ehrenbergi*): physical analysis. J. Mamm., 69:121–125.

Heyning, J. E., and J. G. Mead. 1997. Thermoregulation in the mouths of feeding gray whales. Science, 278:1138–1139.

Heywood, V. H. 1996a. *Global Biodiversity Assessment.* Cambridge University Press, New York.

Heywood, V. H. 1996b. *Global Biodiversity Assessment Summary for Policy Makers.* Cambridge University Press, New York.

Hickey, M. B. C. 1988. Foraging behavior and use of torpor by the hoary bat (*Lasiurus cinereus*). Bat Research News, 29:47.

Hickie, P. F. 1936. Isle Royale moose studies, Trans. N. Amer. Wildl. Conf., 1:396–399.

Hickman, G. C. 1981. National mammal guides: a review of references to Recent faunas. Mammal Rev., 11:53–85.

Higgins, L. V. 1993. The nonannual, nonseasonal breeding cycle of the Australian sea lion, *Neophoca cinerea.* J. Mamm., 74:270–274.

Hik, D. S. 1995. Does risk of predation influence population dynamics? Evidence from the cyclic decline of snowshoe hares. Wildlife Res., 22:115–129.

Hildebrand, M. 1959. Motions of the running cheetah and horse. J. Mamm., 40:481–495.

Hildebrand, M. 1960. How animals run. Sci. Amer., 202:148–156.

Hildebrand, M. 1965. Symmetrical gaits of horses. Science, 150:701–708.

Hildebrand, M. 1974. *Analysis of Vertebrate Structure.* Wiley, New York.

Hildebrand, M. 1985. Walking and running, 38–57, in *Functional Vertebrate Morphology* (M. Hildebrand, D. M. Bramble, K. F. Liem, and D. B. Wake, eds.). Belknap Press, Cambridge, Massachusetts.

Hildebrand, M. 1987. The mechanics of horse legs. Amer. Sci., 75:594–601.

Hill, A. V. 1950. The dimensions of animals and their muscular dynamics. Sci. Progr., 38:209.

Hill, C. J. 1941. The development of the Monotremata. Part V. Further observations on the histology and the secretory activities of the oviduct prior to and during gestation. Trans. Zool. Soc. London, 25:1–31.

Hill, J. E., and T. D. Carter. 1941. The mammals of Angola, Africa. Bull. Amer. Mus. Nat. Hist., 78:1–211.

Hill, J. E., and S. E. Smith, 1981. *Craseonycteris thonglongyai.* Mamm. Species, 160:1–4.

Hill, J. E., and M. J. Daniel. 1985. Systematics of the New Zealand short-tailed bat *Mystacina* Gray, 1843 (Chiroptera: Mystacinidae). Bull. British Mus. Nat. Hist. (Zool.), 48:279–300.

Hill, J. E., and J. D. Smith. 1992. *Bats: A Natural History.* Univ. Texas Press, Austin.

Hill, R. W. 1992. The altricial/precocial contrast in the thermal relations and energetics of small mammals, 122–159, in *Mammalian Energetics: Interdisciplinary Views of Metabolism and Reproduction* (T. E. Tomasi and T. H. Horton, eds.). Comstock Publ. Assoc. Cornell Univ. Press, Ithaca, NY.

Hind, A. T., and W. S. C. Gurney. 1997. The metabolic cost of swimming in marine homeotherms. J. Exp. Biol., 200:531–542.

Hinde, R. A. 1970. *Animal Behavior: A Synthesis of Ethology and Comparative Psychology.* 2d ed. McGraw-Hill, New York.

Hinds, D. S., and R. E. MacMillen. 1985. Scaling of energy metabolism and evaporative water loss in heteromyid rodents. Physiol. Zool., 58:282–298.

Hinds, D. S., R. V. Baudinette, R. E. MacMillen, and E. A. Halpern. 1993. Maximum metabolism and the aerobic factorial scope of endotherms. J. Exp. Biol., 182:41–56.

Hisaw, F. L. 1924. The absorption of the pubic symphysis of the pocket gopher, *Geomys bursarius* (Shaw). Amer. Nat., 58:93–96.

Hladik, C. M., P. Charles-Dominique, and J. J. Petter. 1980. Feeding strategies of five nocturnal prosimians in the dry forest of the west coast of Madagascar, 41–73, in *Nocturnal Malagasy Primates* (P. Charles-Dominique, H. M. Cooper, A. Hladik, C. M. Hladik, E. Pages, G. F. Pariente, A. Petter-Rousseaux, and A. Schilling, eds.). Academic Press, New York.

Hoeck, H. N. 1982. Population dynamics, dispersal, and genetic isolation in two species of hyrax (*Heterohyrax brucei* and *Procavia johnstoni*) on habitat islands in the Serengeti. Z. Tierpsychol., 59:177–210.

Hoeck, H. N., H. Klein, and P. Hoeck. 1982. Flexible social organization in hyrax. Z. Tierpsychol., 59:265–298.

Hoese, H. D. 1971. Dolphin feeding out of water in a salt marsh. J. Mamm., 52:222–223.

Hofer, H., and M. L. East. 1993a. The commuting system of Serengeti spotted hyaenas: how a predator copes with migratory prey. I. social organization. Anim. Behav., 46:547–557.

Hofer, H., and M. L. East. 1993b. The commuting system of Serengeti spotted hyaenas: how a predator copes with migratory prey. II. intrusion pressure and commuters space use. Anim. Behav., 46:559–574.

Holling, C. S. 1959. The components of predation as revealed by a study of small mammal predation of the European pine sawfly. Can. Entomol., 91:293–320.

Holling, C. S. 1961. Principles of insect predation. Ann. Rev. Entomol., 6:163–182.

Holmes, W. G. 1995. The ontogeny of littermate preferences in juvenile golden-mantled ground squirrels: effects of rearing and relatedness. Anim. Behav., 50:309–322.

Holmes, W. G., and P. W. Sherman. 1983. Kin recognition in animals. Amer. Sci., 71:46–55.

Honeycutt, R. L., and R. M. Adkins. 1993. Higher level systematics of eutherian mammals: an assessment of molecular characters and phylogenetic hypotheses. Ann. Rev. Ecol. Syst. 24:279–305.

Hoofer, S. R., and R. A. Van Den Bussche. 1998. Higher taxonomic relationships of Vespertilionidae based on 12S rRNA, tRNA^{VAL}, and 16S RNA sequence variation. Abstracts, 78th Annual Meeting, American Society of Mammalogists, Blacksburg, Virginia:106.

Hoagland, J. L. 1995. The black-tailed prairie dog. Univ. of Chicago Press, Chicago.

Hoogland, J. L. 1998. Why do female Gunnison's prairie dogs copulate with more than one male? Anim. Behav., 55:351–359.

Hooker, J. J. 1992. An additional record of a placental mammal (Order Astrapotheria) from the Eocene of West Antarctica. Antarctic Science, 4:107–108.

Hooper, E. T., and J. H. Brown. 1968. Foraging and breeding in two sympatric species of Neotropical bats, genus *Noctilio*. J. Mamm., 49:310–312.

Hopf, F. A., and J. H. Brown. 1986. The bulls-eye method for testing randomness in ecological communities. Ecology, 67:1139–1155.

Hopson, J. A. 1991. Systematics of the nonmammalian Synapsida and implications for patterns of evolution in synapsids, 635–693, in *Origins of the Higher Groups of Tetrapods. Controversy and Consensus.* (H.-P. Schultze and L. Trueb, eds.). Comstock Publishing Associates, Ithaca, New York.

Hopson, J. A. 1994. Synapsid evolution and the radiation of non-eutherian mammals, 190–219, in *Major Features in Vertebrate Evolution. Short Courses in Paleontology* (R. S. Spencer, ed.). No. 7. Paleontological Society, Univ. Tennessee, Knoxville.

Hopson, J. A., and H. R. Barghusen. 1986. An analysis of therapsid relationships, 83–106, in *The Ecology and Biology of Mammal-like Reptiles* (N. Hotton, P. D. MacLean, J. J. Roth, and E. C. Roth, eds.). Smithsonian Inst. Press, Washington, D.C.

Hopson, J. A., and G. W. Rougier. 1993. Braincase structure in the oldest known skull of a therian mammal: implications for mammalian systematics and cranial evolution. Amer. J. Sci., 293A:268–299.

Hornocker, M. G. 1970a. The American Lion. Nat. Hist., 79:40–49.

Hornocker, M. G. 1970b. An analysis of mountain lion predation upon mule deer and elk in the Idaho Primitive Area. Wildl. Monogr., No. 21:1–39.

Horst, R. 1969. Observations on the structure and function of the kidney of the vampire bat (Desmodus rotundus murinus), in *Physiological Systems in Semiarid Environments* (C. C. Hoff and M. L. Riedesel, eds.). Univ. New Mexico Press, Albuquerque.

Hosley, N. W. 1956. Management of the white-tailed deer in its environment, 187–260, in *The Deer of North America* (W. P. Taylor, ed.). Wildlife Management Institute, Telegraph Press, Harrisburg, Pennsylvania.

Howard, W. E. 1960. Innate and environmental dispersal of individual vertebrates. Amer. Midland Nat., 63:152–161.

Howell, A. B. 1932. The saltatorial rodent *Dipodomys:* functional and comparative anatomy of its muscular and osseous systems. Proc. Amer. Acad. Arts and Sci., 67:377–536.

Howell, A. B. 1944. *Speed in Animals.* Univ. Chicago Press, Chicago.

Howell, D. J. 1974. Bats and pollen: physiological aspects of the syndrome of chiropterophily. Comp. Biochem. Physiol., 48A:263–276.

Howell, D. J., and D. Burch. 1974. Food habits of some Costa Rican bats. Revista de Biologia Tropical, 21:281–294.

Hrdy, S. B. 1976. The care and exploitation of nonhuman primate infants by conspecifics other than the mother. Adv. Study Behav., 6:101–158.

Hrdy, S. B. 1977. *The Langurs of Abu: Female and Male Strategies of Reproduction.* Harvard Univ. Press, Cambridge, Mass.

Hu, Y., Y. Wang, Z. Luo, and C. Li. 1997. A new symmetrodont mammal from China and its implications for mammalian evolution. Nature, 390:137–142.

Hudson, J. W. 1973. Torpidity in mammals, 97–165, in *Comparative Physiology of Thermoregulation,* vol. III. Academic Press, New York.

Hudson, J. W., and T. J. Dawson. 1975. Role of sweating from the tail in the thermal balance of the rat-kangaroo *Potorous tridactylus.* Aust. J. Zool., 23:453–461.

Hugget, A. St.G., and W. F. Widdas. 1951. The relationship between mammalian foetal weight and conception age. J. Physiol., 114:306–317.

Hughes, R. L. 1974. Morphological studies on implantation in marsupials. J. Reprod. Fertil., 39:173–186.

Hughes, R. L. 1984. Structural adaptations of the eggs and the fetal membranes of monotremes and marsupials for respiration and metabolic exchange, 389–421, in *Respiration and Metabolism of Embryonic Vertebrates* (R. S. Seymour, ed.). Doordrecht, Junk.

Hughes, R. L., F. N. Carrick, and C. D. Shorey. 1975. Reproduction in the platypus, *Ornithorhynchus anatinus,* with particular reference to the evolution of viviparity. J. Reprod. Fertil., 43:374–375.

Hughes, R. L., L. S. Hall, M. Archer, and K. Aplin. 1990. Observations on placentation and development in *Echymipera kalubu*, 259–270, in *Bandicoots and Bilbies* (J. H. Seebeck, P. R. Brown, R. L. Wallis, and C. M. Kemper, eds.). Surrey Beatty & Sons, Sydney.

Hulbert, A. J., and T. J. Dawson. 1974. Standard metabolism and body temperature of perameloid marsupials from different environments. Comp. Biochem. Physiol., 47A:583–590.

Hult, R. 1982. Another function of echolocation for bottlenose dolphins *(Tursiops truncatus)*. Cetology, 47:1–7.

Humboldt, A. von, and A. Bonpland. 1852. *Personal Narrative of Travels to the Equinoctial Regions of America During the Years 1799–1804.* 3 vols. Henry G. Bohn, London.

Hume, I. D. 1982. *Digestive Physiology and Nutrition of Marsupials.* Cambridge Univ. Press, Cambridge, U.K.

Humphrey, S. R. 1974. Zoogeography of the nine-banded armadillo *(Dasypus novemcinctus)* in the United States. BioSci., 24:457–462.

Hunt, R. M., Jr. 1987. Evolution of the aeluroid Carnivora: significance of auditory structure in the nimravid cat *Dinictis.* Amer. Mus. Novitates, 2886:1–74.

Hunt, R. M., Jr., and R. H. Tedford. 1993. Phylogenetic relationships within the aeluroid Carnivora and implications of their temporal and geographic distribution, 53–73, in *Mammal Phylogeny, Placentals* (F. S. Szalay, M. J. Novacek, and M. C. McKenna, eds.), vol. 2. Springer-Verlag, New York.

Hunt, R. M., Jr., and L. G. Barnes. 1994. Basicranial evidence for ursid affinity of the oldest pinnipeds, 57–67, in *Contributions in Marine Mammal Paleontology Honoring Frank C. Whitmore, Jr.* (A. Berta and T. A. Deméré, eds.), vol. 29. Proc. San Diego Soc. Nat. Hist.

Hunter, J. P., and J. Jernvall. 1995. The hypocone as a key innovation in mammalian evolution. Proc. Natl. Acad. Sci. USA, 92:10718–10722.

Hurly, T. A., and S. A. Lourie. 1997. Scatterhoarding and larderhoarding by red squirrels: size, dispersion, and allocation of hoards. J. Mamm., 78:529–537.

Hurum, J. H. 1998. The inner ear of two Late Cretaceous multituberculate mammals, and its implications for multituberculate hearing. J. Mamm. Evol., 5:65–93.

Hyvärinen, H. 1994. Brown fat and the wintering of shrews, 139–148, in *Advances in the Biology of Shrews* (J. F. Merritt, G. L. Kirkland, Jr., and R. K. Rose, eds.). Spec. Pub. No. 18, Carnegie Mus. Nat. Hist.

Ingles, L. G. 1949. Ground water and snow as factors affecting the seasonal distribution of pocket gophers, *Thomomys monticola.* J. Mamm., 30:343–350.

International Union for the Conservation of Nature. 1980. *World Conservation Strategy: Living Resources Conserved for Sustainable Development.* Nairobi, Kenya.

Irvine, A. B. 1983. Manatee metabolism and its influence on distribution in Florida. Biol. Conserv., 25:315–334.

Irving, L. 1966. Adaptations to cold. Sci. Amer., 214:94–101.

Irving, L., and J. S. Hart. 1957. The metabolism and insulation of seals as bare-skinned mammals in cold water. Can. J. Zool., 35:497–511.

Izawa, K. 1970. Unit groups of chimpanzees and their nomadism in the savannah woodland. Primates, 11:1–45.

Izawa, K., and J. Itani. 1966. Chimpanzees in Kasakata Basin, Tanganyika: I. ecological study in the rainy season, 1963–64. Kyoto Univ. Afr. Studies, 1:73–156.

Jablonski, D. 1991. Extinctions: a paleontological perspective. Science, 253:754–757.

Jablonski, D. 1993. The tropics as a source of evolutionary novelty through geological time. Nature, 364:142–144.

Jachmann, H., P. S. M. Berry, and H. Imae. 1995. Tusklessness in African elephants: a future trend. African J. Ecol., 33:230–235.

Jacobs, D. S., N. C. Bennett, J. U. M. Jarvis, and T. M. Crowe. 1991. The colony structure and dominance hierarchy of the Damaraland mole-rat, *Cryptomys damarensis* (Rodentia: Bathyergidae), from Namibia. J. Zool., London, 224:553–576.

Jacobs, G. H. 1993. The distribution and nature of colour vision among the mammals. Biol. Rev., 68:413–471.

Jacobs, G. H., M. Neitz, J. F. Deegan, and J. Neitz. 1996. Trichromatic colour vision in New World monkeys. Nature, 382:156–158.

Jacobs, L. L., J. D. Congleton, M. Brunet, J. Dejax, L. J. Flynn, J. V. Hell, and G. Mouchelin. 1988. Mammal teeth from the Cretaceous of Africa. Nature, 336:158–160.

Jameson, E. W., Jr. 1953. Reproduction of deer mice (*Peromyscus maniculatus* and *P. boylei*) in the Sierra Nevada, California. J. Mamm., 34:44–58.

Jameson, E. W., Jr. 1988. *Vertebrate Reproduction.* John Wiley & Sons, New York.

Jameson, E. W., Jr., and R. A. Mead. 1964. Seasonal changes in body fat, water and basic weight in *Citellus lateralis, Eutamias speciosus* and *E. amoenus.* J. Mamm., 45:359–365.

Janis, C., and K. Scott. 1987. The origin of the higher ruminant families with special reference to the origin of the Cervoidea and relationships within the Cervoidea. Amer. Mus. Novitates, 2893:1–85.

Jarman, M. V. 1970. Attachment to home area in impala. E. Afr. Wildl. J., 8:198–200.

Jarman, P. J. 1989. On being thick-skinned: dermal shields in large mammalian herbivores. Biol. J. Linnean Soc., 36:169–191.

Jarman, R. J. 1974. The social organization of antelope in relation to their ecology. Behaviour, 48:215–267.

Jarman, R. J., and M. V. Jarman. 1974. Impala behavior and its relevance to management, 871–881, in *The Behavior of Ungulates and its Relation to Management* (V. Geist and F. R. Walther, eds.). International Union for Conservation of Nature and Natural Resources, Morges, Switzerland.

Jarman, R. J., and M. V. Jarman. 1979. The dynamics of ungulate social organization, in *Serengeti, Dynamics of an Ecosystem* (A. R. E. Sinclair and M. Norton-Griffiths, eds.). Univ. Chicago Press, Chicago.

Jarvis, J. U. M. 1978. Energetics of survival in *Heterocephalus glaber* (Rüppell), the naked mole-rat (Rodentia: Bathyergidae). Bull. Carnegie Mus. Nat. Hist., 6:81–87.

Jarvis, J. U. M. 1981. Eusociality in a mammal: cooperative breeding in the naked mole rat. Science, 212:571–573.

Jarvis, J. U. M., and J. B. Sale. 1971. Burrowing and burrow patterns of East African mole-rats *Tachyoryctes, Heliophobius and Heterocephalus*. J. Zool., London, 163:451–479.

Jarvis, J. U. M., and N. C. Bennett. 1991. Ecology and behavior of the family Bathyergidae, 66–96, in *The Biology of the Naked Mole-rat* (P. W. Sherman, J. U. M. Jarvis, and R. D. Alexander, eds.). Princeton Univ. Press, New Jersey.

Jarvis, J. U. M., and N. C. Bennett. 1993. Eusociality has evolved independently in two genera of bathyergid mole-rats but occurs in no other subterranean mammal. Behav. Ecol. Sociobiol., 33:253–260.

Jarvis, J. U. M., M. J. O'Riain, N. C. Bennett, and P. Sherman. 1994. Mammalian eusociality: a family affair. Trends Ecol. Evol., 9:47–51.

Jarvis, J. U. M., N. C. Bennett, and A. C. Spinks. 1998. Food availability and foraging by wild colonies of Damaraland mole-rats (*Cryptomys damarensis*): implications for sociality. Oecologia, 113:290–298.

Jayaraman, K. S. 1993a. Academies urge population control. Nature, 365:382.

Jayaraman, K. S. 1993b. Science academies call for global goal of zero population growth. Nature, 366:3.

Jen, P. H.-S., and N. Suga. 1976. Coordinated activities of middle-ear and laryngeal muscles in echolocating bats. Science, 191:950–952.

Jenkins, F. A., Jr. 1970a. Limb movement in a monotreme (*Tachyglossus aculeatus*): a cineradiographic analysis. Science, 168:1473–1475.

Jenkins, F. A. Jr. 1970b. Anatomy and function of expanded ribs in certain edentates and primates. J. Mamm., 51:288–301.

Jenkins, F. A. Jr. 1971. Limb posture and locomotion in the Virginia opossum (*Didelphis marsupialis*) and in other non-cursorial mammals. J. Zool., London, 165:303–315.

Jenkins, F. A. Jr. 1974. *Primate Locomotion*. Academic Press, New York.

Jenkins, F. A. Jr. 1990. Monotremes and the biology of Mesozoic mammals. Neth. J. Zool., 40:5–31.

Jenkins, F. A, Jr., and F. R. Parrington, 1976. The postcranial skeletons of the Triassic mammals *Eozostrodon, Megazostrodon*, and *Erythrotherium*. Phil. Trans. Royal Soc. London, B (Biol. Sci.), 273:387.

Jenkins, F. A., Jr., and D. W. Krause. 1983. Adaptations for climbing in North American multituberculates (Mammalia). Science, 220:712–714.

Jenkins, F. A., Jr., A. W. Crompton, and T. Downs, 1983. Mesozoic mammals from Arizona: new evidence on mammalian evolution. Science, 222:1233–1235.

Jenkins, F. A., Jr., S. M. Gatesy, N. H. Shubin, and W. W. Amaral. 1997. Haramiyids and Triassic mammalian evolution. Nature, 385:715–718.

Jenkins, H. O. 1948. A population study of the meadow mice *(Microtus)* in three Sierra Nevada meadows. Proc. Calif. Acad. Sci., ser. 4, 26:43–67.

Jenkins, P. D. 1992. Description of a new species of *Microgale* (Insectivora: Tenrecidae) from eastern Madagascar. Bull. British Mus. Nat. Hist. (Zoology), 58:53–59.

Jenkins, P. D., C. J. Raxworthy, and R. A. Nussbaum. 1997. A new species of *Microgale* (Insectivora, Tenrecidae), with comments on the status of four other taxa of shrew tenrecs. Bull. Nat. Hist. Mus. London, 63:1–12.

Jepsen, G. L., 1966. Early Eocene bat from Wyoming. Science, 154:1333–1338.

Jepsen, J. L. 1970. Bat origins and evolution, 1–64, in *Biology of Bats* (W. A. Wimsatt, ed.). Academic Press, New York.

Jewell, P. A. 1972. Social organization and movements of topi *(Damaliscus korrigum)* during the rut at Ishasha, Queen Elizabeth Park, Uganda. Zool. Afr., 7:233–255.

Johanson, D. C., and M. Edey. 1981. *Lucy: The Beginnings of Humankind*. Simon and Schuster, New York.

Johnson, D. R. 1961. The food habits of rodents on rangelands of southern Idaho. Ecology, 42:407–410.

Johnson, D. R. 1964. Effects of range treatment with 2,4-D on food habits of rodents. Ecology, 45:241–249.

Johnson-Murray, J. L. 1977. Myology of the gliding membranes of some petauristine rodents (Genera: *Glaucomys, Pteromys, Petinomys*, and *Petaurista*). J. Mamm., 58:374–384.

Johnson-Murray, J. L. 1987. The comparative myology of the gliding membranes of *Acrobates, Petauroides* and *Petaurus* contrasted with the cutaneous myology of *Hemibelideus* and *Pseudocheirus* (Marsupalia: Phalangeridae) and with selected gliding Rodentia (Sciuridae and Anomaluridae). Aust. J. Zool., 35:101–113.

Jolly, A. 1966. *Lemur Behavior: A Madagascar Field Study*. Univ. Chicago Press, Chicago.

Jolly, A. 1972. Troop continuity and troop spacing in *Propithecus verreauxi* and *Lemur catta* at Berenty (Madagascar). Folia Primat., 17:335–362.

Jolly, A., P. Oberlé, and R. Albignac. 1984. *Key Environments. Madagascar.* Pergamon Press, Oxford.

Jones, F. W. 1924. *The Mammals of South Australia. part II: The Bandicoots and Herbivorous Marsupials*. Government Printer, Adelaide.

Jones, G., and J. M. V. Rayner. 1988. Flight performance, foraging tactics and echolocation in free-living Daubenton's bats *Myotis daubentoni* (Chiroptera: Vespertilionidae). J. Zool. London, 215:113–132.

Jones, M. E., and D. M. Stoddart. 1998. Reconstruction of the predatory behaviour of the extinct marsupial thylacine (*Thylacinus cynocephalus*). J. Zool., London, 246:239–246.

Jones, R. B., and N. W. Nowell 1973a. Aversive effects of the urine of a male mouse upon the investigatory behavior of its defeated opponent. Anim. Behav., 21:707–710.

Jones, R. B., and N. W. Nowell. 1973b. The effect of urine on the investigatory behavior of male albino mice. Physiol. Behav., 11:35–38.

Judd, T. M., and P. W. Sherman. 1996. Naked mole-rats recruit colony mates to food sources. Anim. Behav., 52:957–969.

Julliot, C., S. Cajani, and A. Gautier-Hion. 1998. Anomalures (Rodentia, Anomaluridae) in central Gabon: species composition, population densities, and ecology. Mammalia, 62:9–21.

Kaczmarski, F. 1966. Bioenergetics of pregnancy and lactation in the bank vole. Acta Theriologica, 11:409–417.

Kalela, O. 1961. Seasonal change of habitat in the Norwegian lemming *Lemmus lemmus* L. Ann. Acad. Sci. Fennicae, Ser. A 4(55):1–72.

Kalko, E. K. V. 1995. Echolocation signal design, foraging habitats and guild structure in six Neotropical sheath-tailed bats (Emballonuridae). Symp. Zool. Soc. London, 67:259–273.

Kalko, E. K. V., E. A. Herre, and C. O. Handley, Jr. 1996. Relation of fig fruit characteristics to fruit-eating bats in the New and Old World tropics. J. Biogeogr., 23:565–576.

Kalko, E. K. V., and M. A. Condon. 1998. Echolocation, olfaction and fruit display: how bats find fruit of flagellichorous cucurbits. Funct. Ecol., 12:364–372.

Kalko, E. K., H.-U. Schnitzler, I. Kaipf, and A. D. Grinnell. 1998. Echolocation and foraging behavior of the lesser bulldog bat, *Noctilio albiventris*: preadaptations for piscivory? Behav. Ecol. Sociobiol., 42:305–319.

Kanwisher, J., and G. Sundnes. 1966. Thermal regulation in cetaceans, 397–409, in *Whales, Dolphins and Porpoises* (K. S. Norris, ed.). Univ. California Press, Berkeley.

Kappeler, P. M. 1998. To whom it may concern: the transmission and function of chemical signals in *Lemur catta*. Behav. Ecol. Sociobiol., 42:411–421.

Kappeler, P. M., and J. U. Ganzhorn. 1993. *Lemur Social Systems and Their Ecological Basis*. Plenum Press, New York.

Karasov, W. H., and J. M. Diamond. 1985. Digestive adaptations for fueling the cost of endothermy. Science, 220:202–204.

Kardong, K. V. 1998. *Vertebrates: Comparative Anatomy, Function, Evolution*. WCB/McGraw-Hill Comp., New York.

Kaufman, D. M. 1995. Diversity of New World mammals: universality of the latitudinal gradients of species and bauplans. J. Mamm., 76:322–334.

Kaufmann, J. H. 1974. Social ethology of the whiptail wallaby, *Macropus parryi*, in northeastern New South Wales. Anim. Behav., 22:281–369.

Kay, R. F. 1975. The functional adaptations of primate molar teeth. Am. J. Phys. Anthro., 43:195–216.

Kay, R. F. and B. A. Williams. 1994. The dental evidence for anthropoid origins. 361–446, in *Anthropoid Origins* (J. Fleagle and R. F. Kay, eds.). Plenum Press, New York.

Kay, R. F., R. H. Madden, C. Van Schaik, and D. Higdon. 1997a. Primate species richness is determined by plant productivity: implications for conservation. Proc. Nat. Acad. Sci. USA 94:13023–13027.

Kay, R. F., C. Ross, and B. A. Williams. 1997b. Anthropoid origins. Science, 275:797–804.

Kay, R. F., R. H. Madden, R. L. Cifelli, and J. J. Flynn. 1997c. *Vertebrate Paleontology in the Neotropics: The Miocene Fauna of La Venta, Colombia*. Smithsonian Inst. Press, Washington, D.C..

Keane, B., S. R. Creel, and P. M. Waser. 1996. No evidence of inbreeding avoidance or inbreeding depression in a social carnivore. Behav. Ecol., 7:480–489.

Keast, A. 1972. Australian mammals: zoogeography and evolution. 195–246, in *Evolution, Mammals, and Southern Continents* (A. Keast, F. C. Erk, and B. Glass, eds.). State Univ. New York Press, Albany.

Keith, L. B. 1981. Population dynamics of hares, 395–440, in *Proceedings of the World Lagomorph Conference* (K. Myers and C. D. Mac Innes, eds.). Univ. Guelph, Guelph, Ontario.

Keith, L. B. 1990. Dynamics of snowshoe hare populations. Current Mamm., 2:119–195.

Kellogg, W. N., R. Kohler, and H. N. Morris. 1953. Porpoise sounds as sonar signals. Science, 117:239–243.

Kelsall, J. P. 1970. Migration of the barren-ground caribou. Nat. Hist., 79:98–106.

Kemp, T. S. 1982. *Mammal-like Reptiles and the Origin of Mammals*. Academic Press, London.

Kenagy, G. J. 1972. Saltbush leaves: excision of hypersaline tissues by a kangaroo rat. Science, 178:1094–1096.

Kenagy, G. J. 1973a. Daily and seasonal patterns of activity and energetics in a heteromyid rodent community. Ecology, 54:1201–1219.

Kenagy, G. J. 1973b. Adaptations for leaf eating in the Great Basin kangaroo rat, *Dipodomys microps*. Oecologia, 12:383–412.

Kenagy, G. J. 1981. Effect of day length, temperature, and endogenous control on annual rhythms of reproduction and hibernation in chipmunks (*Eutamias* spp.). J. Comp. Physiol., 141:369–378.

Kenagy, G. J., and G. A. Bartholomew. 1981. Effects of day length, temperature, and green food on testicular development in a desert pocket mouse *Perognathus formosus*. Physiol. Zool., 54:62–73.

Kenagy, G. J., and D. Vleck. 1982. The data on daily rhythms of resting metabolic rate for 18 species of small mammals, 322–338, in *Vertebrate Circadian Systems* (Aschoff, Dann and Groos, eds.). Springer-Verlag, Berlin.

Kendeigh, S. C. 1961. *Animal Ecology*. Prentice-Hall, Englewood Cliffs, N.J.

Kennerly, T. R. 1971. Personal communication.

Kerr, J. T., and L. Parker. 1997. Habitat heterogeneity as a determinant of mammal species richness in high-energy regions. Nature, 385:252–254.

Ketten, D. R. 1997. Structure and function in whale ears. Bioacoustics, 8:103–135.

Khajuria, C. K., and G. V. R. Prasad. 1998. Taphonomy of a Late Cretaceous mammal-bearing microvertebrate assemblage from the Deccan inter-trappean beds of Naskal, peninsular India. Palaeogeog., Palaeoclimatol., Palaeoecol., 137:153–172.

Kick, S. 1982. Target-detection by the echolocating bat, *Eptesicus fuscus*. J. Comp. Physiol. A, 145:431–435.

Kielan-Jaworowska, Z. 1997. Characters of multituberculates neglected in phylogenetic analyses of early mammals. Lethaia, 29:249–266.

Kielan-Jaworowska, Z., T. M. Bown, and J. A. Lillegraven. 1979. Eutheria, 221–258, in *Mesozoic Mammals. The First Two-thirds of Mammalian History* (J. A. Lillegraven, Kielan-Jaworoska, Z., and W. A. Clemens, eds.). Univ. Calif. Press, Berkeley.

Kielan-Jaworowska, Z., A. W. Crompton, and F. A. Jenkins, Jr. 1987. The origin of egg-laying mammals. Nature, 326:871–873.

Kielan-Jaworowska, Z., and D. Dashzeveg. 1989. Eutherian mammals from the Early Cretaceous of Mongolia. Zoologica Scripta, 18:347–355.

Kielan-Jaworowska, Z., and P. P. Gambaryan. 1994. Postcranial anatomy and habits of Asian multituberculate mammals. Fossils and Strata, 36:1–92.

Kiltie, R. A. 1981. The function of interlocking canines in rain forest peccaries (Tayassuidae). J. Mamm., 62:459–469.

Kimbel, W. H., T. D. White, and D. C. Johanson. 1984. Cranial morphology of *Australopithecus afarensis*: a comparative study based on a composite reconstruction of the adult skull. Amer. J. Phys. Anthropol., 64:337–388.

King, J. E. 1983. *Seals of the World*. Cornell Univ. Press, Ithaca, New York.

Kingdon, J. 1971. *East African Mammals: An Atlas of Evolution*. Vol. 1. Academic Press, New York.

Kingdon, J. 1984a. *East African Mammals: An Atlas of Evolution*. Vol. 2A, *Insectivores and Bats*. Univ. Chicago Press, Chicago.

Kingdon, J. 1984b. *East African Mammals: An Atlas of Evolution*. Vol. 2B, *Hares and Rodents*. Univ. Chicago Press, Chicago.

Kingdon, J. 1989a. *East African Mammals: An Atlas of Evolution*. Vol. 3A, *Carnivores*. Univ. Chicago Press, Chicago.

Kingdon, J. 1989b. *East African Mammals: An Atlas of Evolution*. Vol. 3B, *Large Mammals*. Univ. Chicago Press, Chicago.

Kingdon, J. 1989c. *Island Africa*. Princeton Univ. Press, Princeton, New Jersey.

Kingdon, J. 1997. *The Kingdon Field Guide to African Mammals*. Academic Press, San Diego.

Kinnear, J. E., A. Cockson, P. Christensen, and A. R. Main. 1979. The nutritional biology of the ruminants and ruminant-like mammals: a new approach. Comp. Biochem. Physiol., 64:357–365.

Kirkland, G. L., Jr. 1991. Competition and coexistence in shrews (Insectivora: Soricidae), 15–22, in *The Biology of the Soricidae* (J. S. Findley and T. L. Yates, eds.). Spec. Pub. Mus. Southwestern Biology, 1:1–91.

Kirsch, J. A. W. 1977. The six-percent solution: second thoughts on the adaptedness of the Marsupialia. Amer. Sci., 65:276–288.

Kirsch, J. A. W., T. F. Flannery, M. S. Springer, and F. Lapointe. 1995. Phylogeny of the Pteropodidae (Mammalia: Chiroptera) based on DNA hybridisation, with evidence for bat monophyly. Aust. J. Zool., 43:395–428.

Kirsch, J. A. W., J. M. Hutcheon, D. G. P. Byrnes, and B. D. Lloyd. 1998. Affinities and historical zoogeography of the New Zealand short-tailed bat, *Mystacina tuberculata* Gray, 1843, inferred from DNA hybridization comparisons. J. Mamm. Evol., 5:33–64.

Kirschvink, J. L. 1990. Geomagnetic sensitivity in cetaceans: an update with live stranding records in the United States, 639–649, in *Sensory Abilities of Cetaceans* (J. Thomas and R. Kastelein, eds.). Plenum Press, New York.

Kirschvink, J. L., A. E. Dizon, and J. A. Westphal. 1986. Evidence from strandings for geomagnetic sensitivity in cetaceans. J. Exp. Biol., 120:1–24.

Kitchen, D. W. 1974. Social behavior and ecology of the pronghorn. Wildlife Monogr., 38:1–96.

Kleiman, D. G. 1977. Monogamy in mammals. Quart. Rev. Biol., 52:39–69.

Kleiman, D. G., and T. M. Davis. 1978. Ontogeny and maternal care. 387–402, in *Biology of Bats of the New World Family Phyllostomatidae* (R. J. Baker, J. K. Jones, Jr., D. C. Carter, eds.). Part III, Spec. Publ. Mus. Texas Tech Press, Lubbock, Texas.

Klingel, H. 1967. Soziale organisation und verhalten freilebender steppenzebras. Z. Tierpsychol., 24:580–624.

Klingener, D. 1964. The comparative myology of four dipodoid rodents (Genera *Zapus, Napeozapus, Sicista,* and *Jaculus*), Misc. Publ. Mus. Zool. Univ. Michigan, 124:1–100.

Klinowska, M. 1985a. Cetacean live stranding sites relate to geomagnetic topography. Aquatic Mamm., 11:27–32.

Klinowska, M. 1985b. Cetacean live stranding dates relate to geomagnetic disturbances. Aquatic Mamm., 11:109–119.

Klinowska, M. 1990. Geomagnetic orientation in cetaceans: behavioral evidence, 651–663, in *Sensory Abilities of Cetaceans* (J. Thomas and R. Kastelein, eds.). Plenum Press, New York.

Kober, R., and H. U. Schnitzler. 1990. Information in sonar echoes of fluttering insects available for echolocating bats. J. Acoustic. Soc. Amer., 87:882–896.

Kock, D., and G. Storch. 1996. *Thainycteris aureocollaris,* a remarkable new genus and species of vespertilionine bats from SE-Asia (Mammalia: Chiroptera: Vespertilionidae). Senckenbergiana Biologica, 76:1–6.

Koide, R. T., L. F. Huenneke, and H. A. Mooney. 1987. Gopher mound soil reduces growth and affects ion uptake of two annual grassland species. Oecologia, 72:284–290.

Kolb, A. 1976. Funktion und wirkungsweise der reichlaute der mausohrfledermaus, *Myotis myotis*. Z. Saugetierk., 41:226–236.

Komarek, E. V. 1932. Notes on mammals of Menominee Indian Reservation, Wisconsin. J. Mamm., 13:203–209.

Koopman, H. N. 1998. Topographic distribution of the blubber of harbor porpoises *(Phocoena phocoena)*. J. Mamm., 79:260–270.

Koopman, K. F. 1984. A synopsis of the families of bats. Part VII. Bat Research News, 25:25–27.

Koopman, K. F. 1993. Order Chiroptera, 137–241, in *Mammal Species of the World: A Taxonomic and Geographic Reference* (D. E. Wilson and D. M. Reeder, eds.). 2d ed. Smithsonian Inst. Press, Washington, D.C.

Koopman, K. F. 1994. *Chiroptera: Systematics. Handbook of Zoology*. Vol. VIII, Part 60, *Mammalia*. Walter de Gruyter. New York.

Kooyman, G. L. 1963. Milk analysis of the kangaroo rat, *Dipodomys merriami*. Science, 147:1467–1468.

Kooyman, G. L. 1968. An analysis of some behavioral and physiological characteristics related to diving in the Weddel seal, 227–261, in *Biology of the Antarctic Seals III*. Vol. 2. (W. L. Schmitt and G. A. Llano, eds.). Amer. Geophys. Union, Washington, D.C.

Kooyman, G. L. 1975. A comparison between day and night diving in the Weddell seal. J. Mamm., 56:563–574.

Kooyman, G. L. 1981. *Weddell Seal: Consummate Diver.* Cambridge Univ. Press, Cambridge.

Kooyman, G. L., and H. T. Andersen. 1969. Deep diving, in *The Biology of Marine Mammals* (H. T. Andersen, ed.). Academic Press, New York.

Kooyman, G. L., D. H. Kerem, W. B. Campbell, and J. J. Wright. 1971. Pulmonary function in freely diving Weddell seal, *Leptonychotes weddelli*. Resp. Physiol., 17:283–290.

Kooyman, G. L., R. L. Gentry, and D. L. Urquhart. 1976. Northern fur seal diving behavior: a new approach to its study. Science, 193:411–412.

Kooyman, G. L., E. A. Wahrenbrock, M. A. Castellini, R. W. Davis, and E. E. Sinnett. 1980. Aerobic and anaerobic metabolism during voluntary diving in Weddell seals; evidence of preferred pathways from blood chemistry and behavior. J. Comp. Physiol., 138:335–346.

Korpimäki, E., and C. J. Krebs. 1996. Predation and population cycles of small mammals, a reassessment of the predation hypothesis. BioSci., 46:754–764.

Korth, W. W. 1994. *The Tertiary Record of Rodents in North America*. Plenum Press, New York.

Koshkina, T. V. 1965. Population density and its importance in regulating the abundance of the red vole (Russian translated by W. A. Fuller). Bull. Moscow Soc. Nat. Biol., 70:5–19.

Koshkina, T. V., and A. S. Kholansky. 1962. Reproduction of the Norwegian lemming (*Lemmus lemmus* L.) on the Kola Peninsula (Russian translated by W. A. Fuller). Zool. Zh., 41:604–615.

Koshland, D. E., Jr. 1991. Editorial: Preserving Biodiversity. Science, 253:717.

Kotler, B. P. 1984. Predation risk and the structure of desert communities. Ecology, 65:689–701.

Kotliar, N. B., and J. A. Wiens. 1990. Multiple scales of patchiness and patch structure: a hierarchical framework for the study of heterogeneity. Oikos, 59:253–260.

Krassilov, V. 1973. Mesozoic plants and the problem of angiosperm ancestry. Lethaia, 6:163–178.

Krause, D. W. 1982. Jaw movement, dental function, and diet in the Paleocene multituberculate *Ptilodus*. Paleobiology, 8:265–281.

Krause, D. W., and J. F. Bonaparte. 1993. Superfamily Gondwanatherioidea: a previously unrecognized radiation of multituberculate mammals in South America. Proc. Nat. Acad. Sci. USA, 90:9379–9383.

Krause, D. W., G. V. R. Prasad, W. v. Koenigswald, A. Sahni, and F. E. Grine. 1997. Cosmopolitanism among Gondwanan Late Cretaceous mammals. Nature, 390:504–507.

Krause, W. J., and J. H. Cutts. 1984. Scanning electron microscope observations on the opossum yolk sac chorion immediately prior to uterine attachment. J. Anat., 138:189–191.

Krebs, C. J. 1966. Demographic changes in fluctuating populations of *Microtus californicus*. Ecol. Monogr., 36:239–273.

Krebs, C. J. 1970. *Microtus* population biology: behavioral changes associated with the population cycle in *M. ochrogaster* and *M. pennsylvanicus*. Ecology, 51:34–52.

Krebs, C. J. 1996. Population cycles revised. J. Mamm., 77:8–24.

Krebs, C. J., and J. H. Myers. 1974. Population cycles in small mammals, Adv. Ecol. Res., 8:267–399.

Krebs, C. J., B. S. Gilbert, S. Boutin, A. R. E. Sinclair, and J. N. M. Smith. 1986. Population biology of snowshoe hares. I. Demography of food-supplemented populations in the southern Yukon, 1976–1984. J. Anim. Ecol., 55:963 982.

Krebs, C. J., R. Boonstra, S. Boutin, M. Dale, S. Hannon, K. Martin, A. R. E. Sinclair, J. N. M. Smith, and R. Turkington. 1992. What drives the snowshoe hare cycle in Canada's Yukon? 886–896, in *Wildlife 2001: Populations* (D. M. McCullough and R. Barrett, eds.). Elsevier, London.

Krebs, C. J., S. Boutin, R. Boonstra, A. R. E. Sinclair, J. N. M. Smith, M. R. T. Dale, K. Martin, and R. Turkington. 1995. Impact of food and predation on the snowshoe hare cycle. Science, 269:1112–1115.

Krebs, J. R., and N. B. Davies. 1993. *An Introduction to Behavioral Ecology*. 3d. ed., Blackwell Sci. Pub., London.

Krebs, H. A. 1950. Body size and tissue metabolism. Biochem. Biophys. Acta, 4:249–269.

Krishtalka, L., and T. Setoguchi. 1977. Paleontology and geology of the Badwater Creek area, central Wyoming. Part 13. The Late Eocene Insectivora and Dermoptera. Annals Carnegie Mus. Nat. Hist., 46:71–99.

Krohne, D. T. 1980. Intraspecific litter size variation in *Microtus californicus*. II. Variation between populations. Evolution, 34:1174–1182.

Kronhe, D. T. 1982. The demography of low-litter size populations of *Microtus californicus*. Can. J. Zool., 60:368–374.

Kronwitter, F. 1988. Population structure, habitat use and activity patterns of the noctule bat, *Nyctalus noctula* Shreb., 1774 (Chiroptera: Vespertilionidae) revealed by radio-tracking. Myotis, 26:23–85.

Kruger, L. 1966. Specialized features of the cetacean brain, 232–254, in *Whales, Dolphins and Porpoises* (K. S. Norris, ed.). Univ. California Press, Berkeley.

Kruuk, H. 1972. *The Spotted Hyena. A Study of Predation and Social Behavior.* Univ. Chicago Press, Chicago.

Kruuk, H., and W. A. Sands. 1972. The aardwolf (*Proteles cristatus* Sparrman) 1783 as predator on termites. E. Afr. Wildl. J., 10:211–227.

Krzanowski, A. 1961. Weight dynamics of bats wintering in the cave at Pulway (Poland). Acta Theriol., 4:249–264.

Krzanowski, A. 1964. Three long flights by bats. J. Mamm., 45:152.

Kucera, T. E. 1991. Adaptive variation in sex ratios of offspring of nutritionally stressed mule deer. J. Mamm., 72:745–749.

Kulzer, E. 1965. Temperaturregulation bein fledermausen (Chiroptera) aus berschiedenen Klimazonen. Z. Vergl. Physiol., 50:1–34.

Kumar, S., and S. B. Hedges. 1998. A molecular timescale for vertebrate evolution. Nature, 392:917–920.

Kummer, H. 1968. *Social Organization of Hamadryas Baboons.* Univ. Chicago Press, Chicago.

Kummer, H. 1984. From laboratory to desert and back: a social system of hamadryas baboons. Anim. Behav., 32:965–971.

Kunz, T. H., and K. A. Nagy. 1988. Methods of energy budget analysis. 277–302, in *Ecological and Behavioral Methods for the Study of Bats* (T. H. Kunz, ed.). Smithsonian Inst. Press, Washington, D.C.

Kunz, T. H., and C. A. Diaz. 1995. Folivory in fruit-eating bats, with new evidence from *Artibeus jamaicensis* (Chiroptera: Phyllostomidae). Biotropica, 27:106–120.

Kurtén, B., 1969. Continental drift and evolution. Sci. Amer. 220(3):54.

Kurtén, B., and E. Anderson. 1980. *Pleistocene Mammals of North America.* New York, Columbia University Press.

Kurtén, L., and U. Schmidt, 1982. Thermoreception in the common vampire bat (*Desmodus rotundus*). J. Comp. Physiol., 146:223–228.

Kuyper, M. A. 1979. A biological study of the golden mole *Amblysomus hottentotus*. M. S. thesis. Univ. Natal, Pietermaritzburg, South Africa.

Kuyper, M. A. 1985. The ecology of the golden mole *Amblysomus hottentotus*. Mamm. Rev., 15:3–11.

Lacey, E. A., S. H. Braude, and J. Wieczorek. 1997. Burrow sharing by colonial tuco-tucos (*Ctenomys sociabilis*). J. Mamm., 78:556–562.

Lacher, T. E. 1981. The comparative social behavior of *Kerodon rupestris* and *Galea spixii* and the evolution of behavior in the Caviidae. Bull. Carnegie Mus. Nat. Hist., 1:1–71.

Lack, D. 1954b. Cyclic mortality. J. Wildl. Manag., 18:25–37.

Lack, D. 1966. *Population Studies of Birds.* Clarendon Press, Oxford.

Lacy, R. C. 1997. Importance of genetic variation to the viability of mammalian populations. J. Mamm., 78:320–335.

Lambertsen, R., N. Ylrich, and J. Straley. 1995. Frontomandibular stay of Balaenopteridae: a mechanism for momentum recapture during feeding. J. Mamm., 76:877–899.

Lamprecht, J. 1979. Field observations on the behavior and social system of the bat-eared fox (*Otocyon megalotis* Desmarest). Z. Tierpsychol., 52:171–200.

Lamprey, H. F., G. Halevy, and S. Makacha. 1974. Interactions between *Acacia*, bruchid seed beetles and large herbivores, E. Afr. Wildl. J., 12:81–85.

Lang, H., and J. P. Chapin, 1917. The American Museum Congo expedition collection of bats III: field notes. Bull. Amer. Mus. Nat. Hist., 37.

Lang, T. G. 1966. Hydrodynamic analysis of cetacean performance, 410–432, in *Whales, Dolphins and Porpoises* (K. S. Norris, ed.). Univ. California Press, Berkeley.

Langer, P. 1988. *The Mammalian Herbivore Stomach: Comparative Anatomy, Function and Evolution.* Gustav-Fischer, Stuttgart.

Langman, V. A., G. M. O. Maloiy, K. Schmidt-Nielsen, and R. C. Schroter. 1979. Nasal heat exchange in the giraffe and other large mammals. Resp. Physiol., 37:325–333.

Larsen, E. C. 1986. Competitive release in microhabitat use among coexisting desert rodents: a natural experiment. Oecologia, 69:231–237.

Laurance, W. F. 1998. A crisis in the making: responses of Amazonian forests to land use and climate change. Trends Ecol. Evol., 13:411–415.

Laurance, W. F., and R. O. Bierregaard, Jr. 1997. *Tropical Forest Remnants: Ecology, Management, and Conservation of Fragmented Communities.* Univ. of Chicago Press, Chicago.

Laurin, M. 1993. Anatomy and relationships of *Haptodus garnettensis*, a Pennsylvanian synapsid from Kansas. J. Vert. Paleo., 13:200–229.

Laurin, M., and R. R. Reisz. 1995. A reevaluation of early amniote phylogeny. Zool. J. Linn. Soc., 113:165–223.

Laurin, M., and R. R. Reisz. 1996. The osteology and relationships of *Tetraceratops insignis,* the oldest known therapsid. J. Vert. Paleo., 16:95–102.

Laursen, L., and M. Bekoff. 1978. *Loxodonta africana.* Mamm. Species, 92:1–8.

Lavocat, R. 1973. Les rongeurs du Miocene d'Afrique Orientale. I, Miocene inferieur. Trav. Mem. Inst. Ecole Practique des Hautes Etudes. Inst. de Montpellier Mem., 1:1–284.

Lavocat, R. 1976. Rongeurs caviomorphes de l'Oligocene de Bolivia. II, Rongeurs du bassin Deseadien de Salla-Luribay. Palaeovertebrata, 7:15–90.

Lavocet, R. 1980. The implications of rodent paleontology and biogeography to the geographical sources and origin of platyrrhine primates, 93–102, in *Evolutionary Biology of the New World Monkeys and Continental Drift* (R. L. Ciochon and A. B. Chiarelli, eds.). Plenum Press, New York.

Lawrence, B. D., and J. A. Simmons. 1982. Measurements of atmospheric attenuation at ultrasonic frequencies and the significance for echolocation by bats. J. Acoust. Soc. Amer., 71:585–590.

Laws, R. M. 1970. Elephants as agents of habitat and landscape change in East Africa. Oikos, 21:1–15.

Laws, R. M., and I. S. C. Parker. 1968. Recent studies on elephant populations in East Africa. Symp. Zool. Soc. London, 21:319–359.

Layne, J. N. 1958. Observations on freshwater dolphins in the upper Amazon. J. Mamm., 39:1–22.

Le Boeuf, B. J., and R. S. Peterson. 1969. Social status and mating activity in elephant seals. Science, 163:91–93.

Le Boeuf, B. J., and J. Reiter. 1988. Lifetime reproductive success in northern elephant seals, 344–362, in *Reproductive Success* (T. H. Clutton-Brock, ed.). Univ. Chicago Press, Chicago.

Le Boeuf, B. J., D. E. Crocker, S. B. Blackwell, P. A. Morris, and P. H. Thorson. 1993. Sex differences in diving and foraging behavior of northern elephant seals. Symp. Zool. Soc. London, 66:149–178.

Lechleitner, R. R., J. V. Tileston, and L. Kartman. 1962. Die-off of a Gunnison's prairie dog colony in central Colorado. I. Ecological observations and description of the epizootic. Zoonoses Res., 1:185–199.

Lechner, A. J. 1978. The scaling of maximal oxygen consumption and pulmonary dimensions in small mammals. Respir. Physiol., 34:29–44.

Lee, A. K., and R. W. Martin. 1988. *The Koala: A Natural History.* Univ. New South Wales Press, Sydney.

Lee, P. C., and J. I. Oliver. 1979. Competition, dominance, and the acquisition of rank in juvenile yellow baboons *(Papio cynocephalus).* Anim. Behav., 27:576–585.

Legendre, S. 1985. Molossidés (Mammalia, Chiroptera) cénozoiques de l'Ancien et du Nouveau Monde: statut systématique; intégration phylogénique de données. Neues Jahrbuch für Geologie und Paläontologie, Abhandlungen, 170:205–227.

Leighton, D. R. 1986. Gibbons: territoriality and monogamy, 135–145, in *Primate Societies* (B. B. Smuts, D. L. Cheney, R. M. Seyfarth, R. W. Wrangham, and T. T. Struhsaker, eds.). Univ. Chicago Press, Chicago.

Leitner, P., and J. E. Nelson. 1967. Body temperature, oxygen consumption and heart rate in the Australian false vampire bat, *Macroderma gigas.* Comp. Biochem. Physiol., 21:65–74.

Lenfant, C. 1969. Physiological properties of blood in marine mammals, 95–116, in *The Biology of Marine Mammals* (H. T. Anderson, ed.). Academic Press, New York.

Leonard, M. L., and M. B. Fenton. 1984. Echolocation calls of *Euderma maculatum* (Vespertilionidae): use in orientation and communication. J. Mamm., 65:122–126.

Leopold, A. 1949. *A Sand County Almanac and Sketches Here and There.* Oxford Univ. Press, London.

Leopold, A. S., T. Riney, R. McCain, and L. Tevis Jr. 1951. The jawbone deer herd. California Div. Fish and Game, Game Bull., 4:1–139.

Lettvin, J. Y., E. R. Gruberg, R. M. Rose, and G. Plotkin. 1982. Dolphins and the bends. Science, 216:650–651.

Leuthold, W. 1966. Variations in territorial behavior of Uganda kob, *Adenota kob thomasi* (Neumann 1896). Behaviour, 27:214–257.

Lever, C. 1985. *Naturalized Mammals of the World.* Longman, London.

Leyhausen, P. 1956. Verhaltensstudien an Katzen. Z. Tierpsychol., 2:1–120.

Li, C. K. 1977. Paleocene eurymyloids (Anagalidae, Mammalia) of Qianshan, Anhui. Vert. PalAsiat., 15:103–118.

Li, C. K., and S. Y. Ting. 1985. Possible phylogenetic relationships: eurymylid-rodent and mimotonid-lagomorph, 35–58, in *Evolutionary Relationships Among Rodents* (W. P. Luckett and J. L. Hartenberger, eds.). Plenum Press, New York.

Li, C. K., R. W. Wilson, M. R. Dawson, and L. Krishtalka. 1987. The origins of rodents and lagomorphs, 97–108, in *Current Mammalogy* (H. H. Genoways, ed.). Vol 1. Plenum Press, New York.

Li, W. H., H. Guoy, P. M. Sharp, C. O'Huigin, and Y. Y. Yang. 1990. Molecular phylogeny of Rodentia, Lagomorpha, Primates, Artiodactyla, and Carnivora and molecular clocks. Proc. Nat. Acad. Sci. USA, 87:6703–6707.

Lidicker, W. Z., Jr. 1962. Emigration as a possible mechanism permitting the regulation of population density below carrying capacity. Amer. Nat., 96:23–29.

Lidicker, W. Z., Jr. 1968. A phylogeny of New Guinea rodent genera based on phallic morphology. J. Mamm., 49:609–643.

Lidicker, W. Z., Jr. 1973. Regulation of numbers in an island population of the California vole: a problem in community dynamics. Ecol. Monogr., 43:271–302.

Lidicker, W. Z., Jr. 1975. The role of dispersal in the demography of small mammals, in *Small Mammals: Their Productivity and Population Dynamics* (F. B. Golley, K. Petrusewicz, and L. Ryzkowski, eds.). International Biol. Prog., vol. 5. Cambridge Univ. Press, Cambridge.

Lidicker, W. Z., Jr. 1976. Experimental manipulation of the timing of reproduction in the California vole. Res. Pop. Ecol., 18:14–27.

Lidicker, W. Z., Jr. 1979. Analysis of two freely-growing enclosed populations of the California vole. J. Mamm., 60:447–466.

Lidicker, W. Z., Jr. 1980. The social biology of the California vole. The Biologist, 62:46–55.

Lidicker, W. Z., Jr. 1985. Dispersal, 420–454, in *Biology of the New World Microtus* (R. H. Tamarin, ed.). Spec. Publ., Amer. Soc. Mamm., 8:1–893.

Lidicker, W. Z., Jr. 1988. Solving the enigma of microtine "cycles." J. Mamm., 69:225–235.

Lidicker, W. Z., Jr. 1994. Population ecology, 323–347, in *Seventy-five Years of Mammalogy 1919–1994* (E. C. Birney and J. R. Choate, eds.). Spec. Publ., Amer. Soc. Mammal., 11:1–433.

Lidicker, W. Z., Jr. 1996. Rodent evolution in evolutionary and ecological time: are there any connections? 203–210, in *Biodiversity and Adaptation* (A. Zaime, ed.). Proc. Fifth International Conference of Rodents and Spatium Actes Editions, Rabat, Morocco.

Lidicker, W. Z., Jr., and P. K. Anderson. 1962. Colonization of an island by *Microtus californicus,* analyzed on the basis of runway transects. J. Anim. Ecol., 31:503–517.

Lidicker, W. Z., Jr., and R. S. Ostfeld. 1991. Extra-large body size in California voles: causes and fitness consequences. Oikos, 61:108–121.

Lillegraven, J. A. 1974. Biogeographical considerations of the marsupial-placental dichotomy. Ann. Rev. Ecol. Syst., 5:263–283.

Lillegraven, J. A. 1975. Biological considerations of the marsupial-placental dichotomy. Evolution, 29:707–722.

Lillegraven, J. A. 1979b. Reproduction in Mesozoic mammals, 259–276, in *Mesozoic Mammals: The First Two-thirds of Mammalian History* (J. A. Lillegraven, Z. Kielan-Jaworowska, and W. A. Clemens, eds.). Univ. California Press, Berkeley.

Lillegraven, J. A., Z. Kielan-Jaworowska, and W. A. Clemens, (eds.) 1979a. *Mesozoic Mammals: The First Two-thirds of Mammalian History.* Univ. California Press, Berkeley.

Lillegraven, J. A., M. J. Kraus, and T. M. Bown. 1979b. Paleogeography of the world of the Mesozoic, 277–308, in *Mesozoic Mammals: The First Two-thirds of Mammalian History.* (J. A. Lillegraven, Z. Kielan-Jaworowska, and W. A. Clemens, eds.). Univ. California Press, Berkeley.

Lillegraven, J. A., M. C. McKenna, and L. Krishtalka. 1981. Evolutionary relationships of Middle Eocene and younger species of *Centetodon* (Mammalia, Insectivora, Geolabididae) with a description of the dentition of *Ankylodon* (Adapisoricidae). Univ. Wyoming Publ., 45:1–115.

Lillegraven, J. A., S. D. Thompson, B. K. McNab, and J. L. Patton. 1987. The origin of eutherian mammals. Biol. J. Linnean Soc., 32:281–336.

Lillegraven, J. A., and G. Krusat. 1991. Cranio-mandibular anatomy of *Haldanodon exspectatus* (Docodonta: Mammalia) from the Late Jurassic of Portugal and its implications to the evolution of mammalian characters. Contrib. Geol. Univ. Wyoming, 28:39–138.

Lillegraven, J. A., and G. Hahn. 1993. Evolutionary analysis of the middle and inner ear of Late Jurassic multituberculates. J. Mamm. Evol., 1:47–74.

Lilly, J. C. 1962. Vocal behavior of the bottle-nosed dolphin. Proc. Amer. Phil. Soc., 106:520–529.

Lilly, J. C. 1963. Distress call of the bottle-nosed dolphin: stimuli and evoked behavioral responses. Science, 139:116–118.

Lindberg, D. G. 1980. *The Macaques: Studies in Ecology, Behavior and Evolution.* Van Nostrand Reinhold, New York.

Lindstedt, S. L. 1980a. Energetics and water economy of the smallest desert mammal. Physiol. Zool., 53:82–97.

Lindstedt, S. L. 1980b. Regulated hypothermia in the desert shrew. J. Comp. Physiol., 137:173–176.

Lindstedt, S. L., J. F. Hokanson, D. J. Wells, S. D. Swain, H. Hoppeler, and V. Navarro. 1991. Running energetics in the pronghorn antelope. Nature, 353:748–750.

Linzey, D. W., and A. V. Linzey, 1967. Maturational and seasonal molts in the golden mouse, *Ochrotomys nuttalli.* J. Mamm., 48:236–241.

Lochmiller, R. L. 1996. Immunocompetence and animal population regulation. Oikos, 76:594–602.

Lomnicki, A. 1980. Regulation of population density due to individual differences and patchy environment. Oikos, 35:185–193.

Lomolino, M. V., and R. Channell. 1995. Splendid isolation: patterns of geographic range collapse in endangered mammals. J. Mamm., 76:335–347.

Lomolino, M. V., and R. Channell. 1998. Range collapse, reintroductions and biogeographic guidelines for conservation: a cautionary note. Conserv. Biol., 12:481–484.

Lorenz, K. 1950. The comparative method of studying innate behavior patterns. Symp. Soc. Exp. Biol., 4:229–269.

Lorenz, K. 1963. *Das Sogenannte Böse,* G. Borotha-Schoeler, Vienna. (English version, 1966. *On Aggression.* Methuen, London.)

Lorini, M., and V. G. Persson. 1990. Nova especie de *Leontopithecus* Lesson 1840, do sul do Brasil (Primates, Callitrichidae). Boletím do Museu Nacional, Rio de Janeiro, 338:1–14.

Loudan, A. S. I. 1985. Lactation and neonatal survival of mammals. Symp. Zool. Soc. Lond., 54:183–207.

Loudon, A., N. Rothwell, and M. Stock. 1985. Brown fat, thermogenesis and physiological birth in a marsupial. Comp. Biochem. Physiol., 81A:815–819.

Lovegrove, B. G., and J. U. M. Jarvis. 1986. Coevolution between mole-rats (Bathyergidae) and a geophyte, *Micranthus* (Iridaceae). Cimbebasia, 8:79–85.

Lovejoy, T. E. 1997. Foreword, ix–x, in *Tropical Forest Remnants: Ecology, Management, and Conservation of Fragmented Communities.* (W. F. Laurance and R. O. Bierregaard, Jr., eds.). Univ. Chicago Press, Chicago.

Loy, A., E. Dupré, and E. Capanna. 1994. Territorial behavior in *Talpa romana,* a fossorial insectivore from south-central Italy. J. Mamm., 75:529–535.

Lucas, S. G., and Luo, Z. 1993. *Adelobasileus* from the Upper Triassic of West Texas: the oldest mammal. J. Vert. Paleo., 13:309–334.

Luckett, W. P. 1980. Monophyletic or diphyletic origins of Anthropoidea and Hystricognathi: evidence from fetal membranes, 347–368, in *Evolutionary Biology of the New World Monkeys and Continental Drift* (R. L. Ciochon and A. B. Chiarelli, eds.). Plenum Press, New York.

Luckett, W. P. 1985. Superordinal and intraordinal affinities of rodents: developmental evidence from dentition and placentation, 227–276, in *Evolutionary Relationships Among Rodents* (W. P. Luckett and J. L. Hartenberger, eds.). Plenum Press, New York.

Luckett, W. P. 1993. An ontogenetic assessment of dental homologies in therian mammals, 182–203, in *Mammal Phylogeny: Mesozoic Differentiation, Multituberculates, Monotremes, Early Therians, and Marsupials* (F. S. Szalay, M. J. Novacek, and M. C. McKenna, eds.). Springer-Verlag, New York.

Luckett, W. P. 1994. Suprafamilial relationships within Marsupialia: resolution and discordance from multidisciplinary data. J. Mamm. Evol., 2:225–283

Luckett, W. P., and F. S. Szalay. 1975. *Phylogeny of the Primates, A Multidisciplinary Approach.* Plenum Press, New York.

Luckett, W. P., and J. L. Hartenberger. 1985. *Evolutionary Relationships Among Rodents: A Multidisciplinary Analysis,* Plenum Press, New York.

Luckett, W. P., and J. L. Hartenberger. 1993. Monophyly or polyphyly of the order Rodentia: possible conflict between morphological and molecular interpretations. J. Mamm. Evol., 1.127–147.

Lund, R. D., and J. S. Lund, 1965. The visual system of the mole, *Talpa europaea.* Exp. Neurol., 13:302–316.

Lundrigan, B. 1996. Morphology of horns and fighting behavior in the family Bovidae. J. Mamm., 77:462–475.

Luo, Z., and A. W. Crompton. 1994. Transformation of the quadrate (incus) through the transition from non-mammalian cynodonts to mammals. J. Vert. Paleo., 14:341–374.

Lyman, C. P., and W. A. Wimsatt. 1966. Temperature regulation in the vampire bat, *Desmodus rotundus.* Physiol. Zool., 39:101–109.

Lynch, G. R., H. W. Heath, and C. M. Johnston. 1981. Effects of geographical origin on the photographic control of reproduction in the white-footed mouse, *Peromyscus leucopus.* Biol. Repro., 25:475–480.

MacArthur, R. H. 1955. Fluctuations of animal populations and a measure of community stability. Ecology, 36:533–536.

MacArthur, R. H., and E. O. Wilson. 1963. An equilibrium theory of insular zoogeography. Evolution, 17:373–387.

MacDonald, D. 1984. *The Encyclopedia of Mammals.* Facts on File Publ., New York.

MacFadden, B. J. 1992. *Fossil Horses: Systematics, Paleobiology and Evolution of the Family Equidae.* New York: Cambridge Univ. Press.

MacFadden, B. J., T. E. Cerling, and J. F. Prado. 1996. Cenozoic terrestrial ecosystem evolution in Argentina: evidence from carbon isotopes of fossil mammal teeth. Palaios, 11:319–327.

MacKay, M. R. 1970. Lepidoptera in Cretaceous amber. Science, 167:379–380.

Mackay, R. S., and H. M. Liaw. 1981. Dolphin vocalization mechanisms. Science, 212:676–678.

Maclean, G. L. 1993. *Robert's Birds of Southern Africa.* New Holland Publ., London.

MacMillen, R. E. 1964. Population ecology, water relations, and social behavior of a southern California semidesert rodent fauna. Univ. Calif. Publ. Zool., 71:1–66.

MacMillen, R. E. 1972. Water economy of nocturnal desert rodents, in *Comparative Physiology of Desert Animals* (G. M. O. Maloiy, ed.). Symp. Zool. Soc. London. Vol. 31, Academic Press, New York.

MacMillen, R. E. 1983a. Water regulation in *Peromyscus.* J. Mamm., 64:38–47.

MacMillen, R. E. 1983b. The adaptive physiology of heteromyid rodents. Great Basin Nat., 7:65–76.

MacMillen, R. E., and A. K. Lee. 1969. Water metabolism of Australian hopping mice. Comp. Biochem. Physiol., 28:493–514.

MacMillen, R. E., and J. E. Nelson. 1969. Bioenergetics and body size in dasyurid marsupials. Amer. J. Physiol., 217:1246–1251.

MacMillen, R. E., and E. A. Christopher. 1975. The water relations of two populations of noncaptive desert rodents, 117–137, in *Environmental Physiology of Desert Organisms* (N. F. Hadley, ed.). Dowden, Hutchinson and Ross, Stroudsburg, PA.

MacMillen, R. E., and D. E. Grubbs. 1976. The effects of temperature on water metabolism in rodents, in *Progress in Animal Biometeorology* (D. H. Johnson, ed.). Vol. 1. Swetz and Zeitlinger, Lisse, The Netherlands.

MacMillen, R. E., and D. S. Hinds. 1983a. Water regulatory efficiency in heteromyid rodents: a model and its application. Ecology, 64:152–164.

MacMillen, R. E., and D. S. Hinds. 1983b. Adaptive significance of water regulatory efficiency in heteromyid rodents. BioSci., 33:333–334.

MacMillen, R. E., and D. S. Hinds. 1992. Standard, cold-induced, and exercise-induced metabolism of rodents. 16–33, in *Mammalian Energetics: Interdisciplinary Views of Metabolism and Reproduction* (T. E. Tomasi and T. H. Horton, eds.). Cornell Univ. Press, Ithaca, NY.

MacPhee, R. D. E. 1987. The shrew tenrecs of Madagascar: systematic revision and Holocene distribution of *Microgale* (Tenrecidae, Insectivora). Amer. Mus. Novitates, 2889:1–45.

MacPhee, R. D. E. 1993. *Primates and Their Relatives in Phylogenetic Perspective.* Advances in Primatology Series, Plenum Press, New York.

MacPhee, R. D. E., and M. J. Novacek. 1993. Definition and relationships of Lipotyphla, 13–31, in *Mammal Phylogeny: Placentals* (F. S. Szalay, M. J. Novacek, and M. C. McKenna, eds.). Springer-Verlag, New York.

MacPhee, R. D. E., and M. A. Iturralde-Vinent. 1994. First Tertiary land mammal from Greater Antilles: an Early Miocene sloth (Xenarthra, Megalonychidae) from Cuba. Amer. Mus. Novitates, 3094:1–13.

MacPhee, R. D. E., and C. Flemming. 1999. Requiem aeternum: The last five hundred years of mammalian species extinctions, in *Extinctions in Near Time: Contexts, Causes, and Consequences.* (R. D. E. MacPhee, ed.). Plenum Press, New York.

Maddison, W. P. and D. R. Maddison. 1992. *MacClade: Analysis of Phylogeny and Character Evolution.* Sinauer Assoc., Inc. Sunderland, Massachusetts.

Maddock, L. 1979. The "migration" and grazing succession, 104–129, in *Serengeti, Dynamics of an Ecosystem* (A. R. E. Sinclair and M. Norton-Griffiths, eds.). Univ. Chicago Press, Chicago.

Madison, D. M. 1980. A review of the social biology of *Microtus pennsylvanicus.* The Biologist, 62:20–33.

Madsen, O., P. M. T. Deen, G. Pesole, C. Saccone, and W. W. de Jong. 1997. Molecular evolution of mammalian aquaporin-2: further evidence that elephantshrew and aardvark join the paenungulate clade. Mol. Biol. Evol., 14:363–371.

Maglio, V. J., 1973. Origin and evolution of the Elephantidae. Trans. Amer. Phil. Soc., 63:1.

Maglio, V. J., and H. B. S. Cooke. 1978. *Evolution of African Mammals.* Cambridge, Massachusetts, Harvard Univ. Press.

Mahboubi, M., R. Ameur, and J. Y. Crochet. 1984. Earliest known proboscidean from Early Eocene of north-west Africa. Nature 308:543–544.

Maher, W. J. 1970. The pomarine jaeger as a brown lemming predator in northern Alaska. Wilson Bull., 82:130–157.

Maier, W., and F. Schrenk. 1987. The hystricomorphy of the Bathyergidae, as determined from ontogenetic evidence. Z. Säugetierk., 52:156–164.

Maley, L. E., and C. R. Marshall. 1998. The coming of age of molecular systematics. Science, 279:505–506.

Mallory, F. F., and R. J. Brooks. 1978. Infanticide and other reproductive strategies in the collared lemming (*Dicrostonyx groenlandicus*). Nature, 273:144–146.

Maloiy, G. M. O. 1973. The water metabolism of a small East African antelope: the dik-dik. Proc. Roy. Soc. London (B), 184:167–178.

Marenssi, S. A., M. A. Reguero, S. N. Santillana, and S. F. Vizcaino. 1994. Eocene land mammals from Seymour Island, Antarctica: paleobiogeographical implications. Antarct. Sci., 6:3–15.

Mares, M. A. 1977. Water economy and salt balance in a South American desert rodent *Eligmodontia typus.* Comp. Biochem. Physiol., 56A:325–332.

Mares, M. A. 1980. Convergent evolution among desert rodents: a global perspective. Bull. Carnegie Mus. Nat. Hist., 16:1–51.

Mares, M. A. 1986. Conservation in South America: problems, consequences, and solutions. Science, 233:734–739.

Mares, M. A. 1992. Conservation realities. Science, 257:111–112.

Mares, M. A. 1993a. Desert rodents, seed consumption, and convergence. BioScience, 43:373–379.

Mares, M. A. 1993b. Heteromyids and their ecological counterparts: a pandesert view of rodent ecology and evolution, 652–714, in *Biology of the Heteromyidae* (H. H. Genoways and J. H. Brown, eds.). Special Publication No. 10, American Society of Mammalogists.

Mares, M. A., and D. F. Williams. 1977. Experimental support for food particle size resource allocation in heteromyid rodents. Ecology, 58:1186–1190.

Mares, M. A., R. A. Ojeda, C. E. Borghi, S. M. Giannoni, G. B. Diaz, and J. K. Braun. 1997. How desert rodents overcome halophytic plant defenses. BioScience, 47:699–704.

Marler, P. R. 1965. Communication in monkeys and apes, 544–585, in *Primate Behavior* (I. DeVore, ed.). Holt, Rinehart and Winston, New York.

Marshall, L. G. 1972. Evolution of the peramelid tarsus. Proc. Royal Soc. Victoria, 85:51–60.

Marshall, L. G. 1977. A new species of *Lycopsis* (Borhyaenidae: Marsupialia) from the La Venta fauna (Miocene) of Colombia, South America. J. Paleont., 51:633–642.

Marshall, L. G. 1988. Land mammals and the great American interchange. Amer. Sci., 76:380–388.

Marshall, L. G., S. D. Webb, J. J. Sepkoski, Jr., and D. M. Raup. 1982. Mammalian evolution and the great American interchange. Science 215:1351.

Marshall, L. G., J. A. Case, and M. O. Woodburne. 1990. Phylogenetic relationships of the families of marsupials, 433–505, in *Current Mammalogy* (H. H. Genoways, ed.), vol. 2, Plenum Press, New York.

Marshall, L. G., and T. Sempere. 1993. Evolution of the Neotropical Cenozoic land mammal fauna in its geochronologic, stratigraphic, and tectonic context, 329–392, in *Biological Relationships Between Africa and*

South America (P. Goldblatt, ed.). Yale Univ. Press, New Haven.

Marshall, P. T., and G. M. Hughes. 1980. *Physiology of Mammals and Other Vertebrates.* 2d ed. Cambridge Univ. Press, Cambridge.

Martin, E. P. 1956. A population study of the prairie vole (*Microtus ochrogaster*) in northeastern Kansas. Univ. Kansas Publ. Mus. Nat. Hist., 8:361–416.

Martin, I. G. 1984. Factors affecting food hoarding in the short-tailed shrew *Blarina brevicauda.* Mammalia, 48:65-71.

Martin, L. D. 1980. Functional morphology and the evolution of the cats. Trans. Nebraska Acad. Sci., 7:141–154.

Martin, L. D. 1989. Fossil history of the terrestrial Carnivora, 536–568, in *Carnivore Behavior, Ecology, and Evolution* (J. L. Gittleman, ed.). Cornell Univ. Press, Ithaca, New York.

Martin, P. 1971. Movements and activities of the mountain beaver (*Aplodontia rufa*). J. Mamm., 52:717–723.

Martin, P. S. 1984. Prehistoric overkill, in *Quaternary Extinctions: A Prehistoric Revolution.* (P. S. Martin and R. G. Klein, eds.). Univ. Arizona Press, Tucson.

Martin, P. S., and R. G. Klein. 1984. *Quaternary Extinctions: A Prehistoric Revolution.* Univ. of Arizona Press, Tucson.

Martin, R. D. 1968. Reproduction and ontogeny in tree shrews (*Tupaia belangeri*), with reference to their general behaviour and taxonomic realtionships. Z. Tierpsychol., 25:409–495, 25:505–532.

Martin, R. D. 1973. A review of the behavior and ecology of the lesser mouse lemur (*Microcebus murinus,* J. F. Miller 1777), in *Comparative Ecology and Behavior of Primates* (R. P. Michael and J. H. Crook, eds.). Academic Press, New York.

Martin, R. D. 1986. Primates: a definition, 1–31, in *Major Topics in Primate and Human Evolution* (B. Wood, L. Martin, and P. Andrews, eds.). Cambridge Univ. Press, Cambridge, U.K.

Martin, R. D. 1990. *Primate Origins and Evolution.* Chapman & Hall, London.

Martin, R. D., and S. K. Bearder. 1979. Radio bushbaby. Nat. Hist., 88:77–81.

Martin, T. 1994. African origin of caviomorph rodents is indicated by incisor enamel microstructure. Paleobiology, 20(1):5–13.

Martinson, D. L. 1968. Temporal patterns in the home ranges of chipmunks. J. Mamm., 49:83–91.

Maschenko, E. N., and A. V. Lopatin. 1998. First record of an Early Cretaceous triconodont mammal in Siberia. Bulletin de L'Institut Royal des Sciences Naturelles de Belgique, Sciences de la Terre 68:233–236.

Matschie, P., 1899. *Beitrage zur Kenntnis von Hypsignathus monstrosus* Allen. Sitz. Ber. Ges. Naturf. Freunde, Berlin.

Matsumura, S. 1981. Mother-infant communication in a horseshoe bat *(Rhinolophus ferrumequinum nippon):* vocal communication in three-week old infants. J. Mamm., 62:20–28.

Matthee, C. A., and T. J. Robinson. 1997a. Mitochondrial DNA phylogeography and comparative cytogenetics of the springhare, *Pedetes capensis* (Mammalia: Rodentia). J. Mamm. Evol., 4:53–73.

Matthee, C. A., and T. J. Robinson. 1997b. Molecular phylogeny of the springhare, *Pedetes capensis,* based on mitochondrial DNA sequences. Mol. Biol. Evol., 14:20–29.

Matthew, W. D. 1910. The phylogeny of the Felidae. Bull. Amer. Mus. Nat. Hist., 28:289–316.

Matthew, W. D. 1915. Climate and evolution. New York Acad. Sci. Ann., 24:171–318.

Matthews, M. and N. T. Adler. 1977. Facultative and inhibitory influences on reproductive behavior and sperm transport in rats. J. Comp. Physiol. Psychol., 91:727–741.

Mattingly, D. K. and P. A. McClure. 1985. Energy allocation during lactation in cotton rats (*Sigmodon hispidus*) on a restricted diet. Ecology, 66:928–937.

Mayer, J. J., and P. N. Brandt. 1982. Identity, distribution, and natural history of the peccaries, Tayassuidae, 433–455, in *Mammalian Biology in South America* (M. A. Mares and H. H. Genoways, eds.). Special Publ. Series, Pymatuning Lab. Ecol. Vol. 6, Univ. Pittsburgh Press, Pittsburgh.

Maynard Smith, J. 1976. Evolution and the theory of games. Amer. Sci., 64:41–45.

Mayr, E., 1942. *Systematics and the Origin of Species.* Columbia Univ. Press, New York.

Mayr, E., 1963. *Animal Species and Evolution.* Harvard Univ. Press, Cambridge, Mass.

McBride, A. F. 1956. Evidence for echolocation in cetaceans. Deep-sea Res., 3:153.

McCabe, T. T., and B. D. Blanchard. 1950. *Three Species of Peromyscus.* Rood Associates, Santa Barbara, Calif.

McCarley, H. 1959. The effect of flooding on a marked population of *Peromyscus.* J. Mamm., 40:57–63.

McClure, P. A. 1981. Sex-biased litter reduction in food-restricted woodrats (*Neotoma floridana*). Science, 211:1058–1060.

McCoy, E. D., and E. F. Connor. 1980. Latitudinal gradients in the species diversity of North American mammals. Evolution, 34:193–203.

McCracken, G. F. 1987. Genetic structure of bat social groups, 282–298, in *Recent Advances in the Study of Bats* (M. B. Fenton, P. Racey, and J. M. V. Rayner, eds.). Cambridge Univ. Press, Cambridge.

McCracken, G. F., and J. W. Bradbury. 1977. Paternity and genetic heterogeneity in the polygynous bat, *Phyllostomus hastatus.* Science, 198:303–306.

McCracken, G. F., and J. W. Bradbury. 1981. Social organization and kinship in the polygynous bat, *Phyllostomus hastatus.* Behav. Ecol. Sociobiol., 8:11–34.

McCullough, D. R. 1969. The tule elk, its history, behavior, and ecology. Univ. Calif. Publ. Zool., 88:1–209.

McFarland, M. J. 1986. Ecological determinants of fission-fusion sociality in *Ateles* and *Pan.* 181–190, in *Primate*

Ecology and Conservation (J. G. Else and P. C. Lee, eds.). Cambridge Univ. Press, Cambridge.

McKay, G. M. 1973. The ecology and behavior of the Asiatic elephant in southeastern Ceylon. Smithsonian Contrib. Zool., 125:1–113.

McKean, T. A., and B. Walker. 1974. Comparison of selected cardiopulmonary parameters between the pronghorn and the goat. Respir. Physiol., 21:365–370.

McKenna, M. C., 1972. Possible biological consequences of plate tectonics. BioScience, 22:519.

McKenna, M. C. 1975. Toward a phylogenetic classification of the Mammalia, 21–46, in *Phylogeny of the Primates: A Multidisciplinary Approach* (W. P. Luckett and F. S. Szalay, eds.). Plenum Press, New York.

McKenna, M. C. 1980. Eocene paleolatitude, climate, and mammals of Ellesmere Island. Palaeogeogr. Palaeoclimat. Palaeoecol., 30:349–362.

McKenna, M. C., and S. K. Bell. 1997. *Classification of Mammals Above the Species Level.* Columbia Univ. Press, New York.

McLaren, A. 1970. The fate of the zona pellucida in mice. J. Embryol. Exp. Morph., 23:1–19.

McLean, D. C. 1944. The prong-horned antelope in California. Bureau Game Conserv., Calif. Div. Fish Game, San Francisco, 30:221–241.

McLeod, S. A., F. C. Whitmore, Jr., and L. G. Barnes. 1993. Evolutionary relationships and classification, in *The Bowhead Whale* (J. J. Burns, J. J. Montague, and C. J. Cowles, eds.). Spec. Pub. No. 2, Soc. Marine Mamm.

McNab, B. K. 1966. The metabolism of fossorial rodents: a study of convergence. Ecology, 47:712–733.

McNab, B. K. 1978. The evolution of endothermy in the phylogeny of mammals. Amer. Nat., 112:1–21.

McNab, B. K. 1979. The influence of body size on the energetics and distribution of fossorial and burrowing mammals. Ecology, 60:1010–1021.

McNab, B. K. 1980. Energetics and the limits to a temperature distribution in armadillos. J. Mamm., 61:606–627.

McNab, B. K. 1982. Evolutionary alternatives in the physiological ecology of bats, 151–200, in *Ecology of Bats* (T. H. Kunz, ed.). Plenum, New York.

McNab, B. K. 1984. Physiological convergence amongst ant-eating and termite-eating mammals. J. Zool. London, 203:485–510.

McNab, B. K., and F. J. Bonaccorso. 1995. The energetics of pteropodid bats, 111–122, in *Ecology, Evolution, and Behaviour of Bats*. Symp. Zool. Soc. London, 67:1–421.

McNaughton, S. J. 1976. Serengeti migratory wildebeest: facilitation of energy flow by grazing. Science, 191:92–94.

McNaughton, S. J. 1979. Grazing as an optimization process: grass-ungulate relationships in the Serengeti. Amer. Nat., 113:691–703.

McNaughton, S. J. 1985. Ecology of a grazing ecosystem: the Serengeti. Ecol. Monogr., 55:259–294.

McNaughton, S. J., and N. J. Georgiadis. 1986. Ecology of African grazing and browsing mammals. Ann. Rev. Ecol. Syst., 17:39–65.

McNaughton, S. J., F. F. Banyikwa, and M. M. McNaughton. 1997. Promotion of the cycling of diet-enhancing nutrients by African grazers. Science, 278:1798–1800.

McNeill Alexander, R. 1968. *Animal Mechanics.* Univ. Washington Press, Seattle.

Mead, R. A. 1968a. Reproduction in eastern forms of the spotted skunk (genus *Spilogale*). J. Zool. London, 156:119–136.

Mead, R. A. 1968b. Reproduction in western forms of the spotted skunk (genus *Spilogale*). J. Mamm., 49:373–389.

Mead, R. A. 1989. The physiology and evolution of delayed implantation in carnivores. 437–464, in *Carnivores Behavior, Ecology, and Evolution* (J. L. Gittleman, ed.). Cornell Univ. Press, Ithaca, NY.

Mech, L. D. 1966. *The Wolves of Isle Royale.* U.S. Nat. Park Serv., Fauna Ser. 7.

Mech, L. D. 1994. Buffer zones of territories of gray wolves as regions of intraspecific strife. J. Mamm., 75:199–202.

Medellín, R. A. 1998. True international collaboration: now or never. Conserv. Biol., 12:939–940.

Meehan, T. E. 1976. The occurrence, energetic significance and initiation of spontaneous torpor in the Great Basin pocket mouse (*Perognathus parvus*). Ph.D. dissertation, University of California, Irvine.

Meffe, G. K., A. H. Ehrlich, and D. Ehrenfeld. 1993. Human population control: the missing agenda. Conserv. Biol., 7:1–3.

Meffe, G. K., and C. R. Carroll. 1994. *Principles of Conservation Biology.* Sinauer Assoc., Sunderland, Massachusetts.

Meier, B., and Y. Rumpler. 1987. Preliminary survey of *Hapalemur simus* and a new species of *Hapalemur* in eastern Betsileo, Madagascar. Primate Conserv., 8:10–43.

Meier, B., R. Albignac, A. Peyrieras, Y. Rumpler, and P. Wright. 1987. A new species of *Hapalemur* (Primates) from south east Madagascar. Folia Primatol., 48:211–215.

Melton, D. A. 1976. The biology of aardvarks (Tubulidentata–Orycteropodidae). Mammal Rev., 6:75–88.

Meng, J., A. Wyss, M. R. Dawson, and R. Zhai. 1994. Primitive fossil rodent from Inner Mongolia and its implications for mammalian phylogeny. Nature, 370:134–136.

Meng, J., and A. R. Wyss. 1995. Monotreme affinities and low-frequency hearing suggested by multituberculate ear. Nature, 377:141–144.

Menkhorst, P. W. 1995. *Mammals of Victoria: Distribution, Ecology, and Conservation.* Oxford Univ. Press, Oxford.

Merritt, J. F. 1995. Seasonal thermogenesis and changes in body mass of masked shrews, *Sorex cinereus*. J. Mamm., 76:1020–1035.

Merritt, J. F., and A. Adamerovich. 1991. Winter thermoregulatory mechanisms of *Blarina brevicauda* as revealed by radiotelemetry, 47–64, in *The Biology of the So-*

ricidae (J. S. Findley and T. L. Yates, eds.). Spec. Pub. Mus. Southwestern Biology, Albuquerque.

Merritt, J. F., and F. Bozinovic. 1994. Thermal biology of free-ranging shrews as revealed by computer-facilitated radiotelemetry: energetic implications, 163–169, in *Advances in the Biology of Shrews* (J. F. Merritt, G. L. Kirkland, Jr., and R. K. Rose, eds.). Spec. Publ. No. 18, Carnegie Mus. Nat. Hist., Pittsburgh.

Merritt, J. F., G. L. Kirkland, Jr., and R. K. Rose. 1994. *Advances in the Biology of Shrews*. Spec. Publ. No. 18, Carnegie Mus. Nat. Hist., Pittsburgh.

Meserve, P. L., Gutierrez, J. R., and F. M. Jaksic. 1993. Effects of vertebrate predation on a caviomorph rodent, the degu *(Octodon degus)*, in a semiarid thorn scrub community of Chile. Oecologia, 94:153–158.

Meserve, P. L., J. A. Yunger, J. R. Gutiérrez, L. C. Contreras, W. B. Milstead, B. K. Lang, et al. 1995. Heterogeneous responses of small mammals to an El Niño southern oscillation event in northcentral semiarid Chile and the importance of ecological scale. J. Mamm., 76:580–595.

Messenger, S. L., and J. A. McGuire. 1998. Morphology, molecules, and the phylogenetics of cetaceans. Syst. Biol., 47:90–124.

Messer, M., A. S. Weiss, D. C. Shaw, and M. Westerman. 1998. Evolution of the monotremes: phylogenetic relationship to marsupials and eutherians, and estimation of divergence dates based on α-lactalbumin amino acid sequences. J. Mamm. Evol., 5:95–105.

Meyer, W. B. 1996. *Human Impact on the Earth*. Cambridge Univ. Press, Cambridge.

Michael, R. P., E. B. Keverne, and R. W. Bonsall. 1971. Pheromones: Isolation of male sex attractants from a female primate. Science, 172:964–966.

Michener, G. R. 1998. Sexual differences in reproductive effort of Richardson's ground squirrels. J. Mamm., 79:1–19.

Michener, G. R., and D. R. Michener. 1977. Population structure and dispersal in Richardson's ground squirrels. Ecology, 58:359–368.

Michener, G. R., and L. Locklear. 1990a. Differential costs of reproduction for male and female Richardson's ground squirrels. Ecology, 71:855–868.

Michener, G. R., and L. Locklear. 1990b. Over-winter weight loss by Richardson's ground squirrels in relation to sexual differences in mating effort. J. Mamm., 71:489–499.

Migula, P. 1969. Bioenergetics of pregnancy and lactation in the European common vole. Acta Theriol., 14:167–179.

Mihok, S. B., N. Turner, and S. L. Iverson. 1985. The characterization of vole population dynamics. Ecol. Monogr., 55:399–420.

Milinkovitch, M. C. 1992. DNA–DNA hybridizations support ungulate ancestry of Cetacea. J. Evol. Biol., 5:149–160.

Milinkovitch, M. C., G. Orti, and A. Meyer. 1993. Revised phylogeny of whales suggested by mitochondrial ribosomal DNA sequences. Nature, 361:346–348.

Milinkovitch, M. C., A. Myer, and J. R. Powell. 1994. Revised phylogeny of all major groups of cetaceans based on DNA sequences from three mitochondrial genes. Mol. Biol. Evol., 11:939–948.

Milinkovitch, M. C., and J. G. M. Thewissen. 1997. Even-toed fingerprints on whale ancestry. Nature, 388:622–624.

Miller, E. H., and D. J. Boness. 1983. Summer behaviour of the Atlantic walrus (*Odobenus rosmarus rosmarus*) on Coats Island, NWT. Z. Säugetierk., 48:298–313.

Miller, G. J. 1969. Man and *Smilodon*: a preliminary report on their possible coexistence at Rancho La Brea. Los Angeles Co. Mus. Contrib. Sci., vol. 163.

Miller, J. S. 1975. Tactics of energy partitioning in breeding *Peromyscus*. Can. J. Zool., 53:967–976.

Miller, R. S. 1969. Competition and species diversity. Brookhaven Symp. Biol., 22:63–70.

Mills, M. G. L. 1989. The comparative behavioral ecology of hyenas: the importance of diet and food dispersion, 125–142, in *Carnivores Behavior, Ecology, and Evolution* (J. L. Gittleman, ed.). Cornell Univ. Press, Ithaca, New York.

Mindell, D. P., C. W. Dick, and R. J. Baker. 1991. Phylogenetic relationships among megabats, microbats, and primates. Proc. Natl. Acad. Sci. USA, 88:10322–10326.

Mitchell, D., S. K. Maloney, H. P. Laburn, M. H. Knight, G. Kuhnen, and C. Jessen. 1997. Activity, blood temperature and brain temperature of free-ranging springbok. J. Comp. Physiol. (B), 167:335–343.

Mitchell, E. D. 1975. Review of biology and fisheries for smaller cetaceans. J. Fish. Res. Board Can., 32:875–1240.

Mitchell, J. H. 1998. *Trespassing: An inquiry into the Private Ownership of Land*. Addison-Wesley, New York.

Mittermeier, R. A., and A. F. Coimbra-Filho. 1981. *Ecology and Behavior of Neotropical Primates*. Academia Brasiliera de Ciencias, Rio de Janeiro.

Mittermeier, R. A., M. Schwarz, and J. M. Ayres. 1992. A new species of marmoset, genus *Callithrix* Erxleben, 1777 (Callitrichidae, Primates) from the Rio Maués region, state of Amazonas, central Brazilian Amazonia. Goeldiana Zoologia 14:1–17.

Mittermeier, R. A., I. Tattersall, W. R. Konstant, D. M. Meyers, and R. B. Mast. 1994. *Lemurs of Madagascar*. Conservation International, Washington, D. C.

Miyamoto, M. M., and M. Goodman. 1986. Biomolecular systematics of eutherian mammals: phylogenetic patterns and classification. Syst. Zool., 35:230–240.

Miyamoto, M. M., J. L. Slightom, and M. Goodman. 1987. Phylogenetic relations of humans and African apes from DNA sequences in the psi eta-globin region. Science, 238:369–373.

Miyamoto, M. M., and J. Cracraft. 1991. *Phylogenetic Analysis of DNA Sequences*. Oxford Univ. Press, New York.

Mizuhara, H. 1957. *The Japanese Monkey: its social structure.* San-ichi-syobo, Kyota (in Japanese).

Mobius, K. 1877. Die Auster und die Austernwirtschaft. Berlin. (Transl., 1880, The oyster and oyster culture.) Rept. U.S. Fish. Comm., 1880:683–751.

Moehlman, P. D. 1983. Socioecology of silverbacked and golden jackals (*Canis mesomelas* and *Canis aureus*), 423–453, in *Advances in the Study of Mammalian Behavior* (J. F. Eisenberg and D. G. Kleiman, eds.). Spec. Publ. No. 7, Amer. Soc. Mamm.

Moehlman, P. D. 1989. Intraspecific variation in canid social systems, 143–163, in *Carnivores Behavior, Ecology, and Evolution* (J. L. Gittleman, ed.). Cornell Univ. Press, Ithaca, New York.

Moen, R. 1997. A spatially-explicit model of moose foraging and energetics. Ecology, 78:505–521.

Moermond, T. C. and J. S. Denslow. 1985. Neotropical avian frugivores: patterns of behavior, morphology, and nutrition with consequences for fruit selection, 865–897, in *Neotropical Ornithology* (P. A. Buckley, M. S. Foster, E. S. Morton, R. S. Ridgely, and N. G. Smith, eds.). A.O.U. Monographs.

Mohr, E. 1941. Schwanzverlust und Schwanzregeneration bei Nagetieren. Zool. Anzeiger, 135:49–65.

Möhres, F. P. 1953. Uber die Ultraschallorientierung der Hufeisennasen (Chiroptera—Rhinolophidae). Z. Vergl. Physiol., 34:547–588.

Möhres, F. P. 1967. Communicative characters of sonar signals in bats. 939–945, in *Cours d'Ete' O.T.A.N. sur les Systèmes Sonars Animaux: Biologie et Bionique* (N.A.T.O. Advanced Study Institute). Vol. 2, Laboratoire de Physiologie Acoustique, Paris.

Mones, A. 1981. Sinopsis sistematica preliminar de la familia Dinomyidae (Mammalia: Rodentia, Caviomorpha). Proc. II Congresso Latino-Americano de Paleontologica, Rio Grande do Sul, Brasil, 605–619.

Montgelard, C., F. M. Catzeflis, and E. Douzery. 1997. Phylogenetic relationships of artiodactyls and cetaceans as deduced from the comparison of cytochrome b and 12S rRNA mitochondrial sequences. Mol. Biol. Evol., 14:550–559.

Moors, P. J. 1974. The foeto-maternal relationship and its significance in marsupial reproduction: a unifying hypothesis. Aust. Mamm., 1:263–266.

Mooser, O., and W. W. Dalquest. 1975. Pleistocene mammals from Aguascalientes, Central Mexico. J. Mamm., 56:781–820.

Morell, V. 1996. New mammals discovered by Biology's new explorers. Science, 273:1491.

Morgan, G. S., and C. A. Woods. 1986. Extinction and the zoogeography of West Indian land mammals. Biol. J. Linn. Soc., 28:167–203.

Morowitz, H. J. 1991. Balancing species preservation and economic considerations. Science, 253:752–754.

Morris, D. W. 1984. Sexual differences in habitat use by small mammals: evolutionary strategy or reproductive constraint. Oecologia, 65:51–57.

Morris, R. J. 1986. The acoustic faculty of dolphins, 369–399, in *Research on Dolphins* (M. M. Bryden and R. Harrison, eds.). Clarendon Press, Oxford.

Morrison, P., F. A. Ryser, and A. R. Dawe. 1959. Studies on the physiology of the masked shrew *Sorex cinereus*. Physiol. Zool., 32:256–271.

Morrison, P., and B. K. McNab. 1967. Temperature regulation in some Brazilian phyllostomid bats. Comp. Biochem. Physiol., 21:207–221.

Morton, S. R. 1978. Torpor and nest sharing in free-living *Sminthopsis crassicaudata* (Marsupialia) and *Mus musculus* (Rodentia). J. Mamm., 59:569–575.

Morton, S. R. 1980. Field and laboratory studies of water metabolism in *Sminthopsis crassicaudata* (Marsupialia, Dasyuridae). Aust. J. Zool., 28:213–227.

Morton, S. R., and R. E. MacMillen. 1982. Seeds as sources of preformed water for desert-dwelling granivores. J. Arid Environ., 5:61–67.

Moses, R. A., G. J. Hickling, and J. S. Millar. 1995. Variation in sex ratios of offspring in wild bushy-tailed woodrats. J. Mamm., 76:1047–1055.

Moss, C. J. 1982. *Portraits in the Wild.* 2d. ed. Univ. Chicago Press, Chicago.

Moss, C. J. 1983. Estrous behavior and female choice in the African elephant. Behaviour, 86:167–196.

Muizon, C. de. 1981. El gran viaje de las focas. Mundo Científico 1(6):663–665.

Muizon, C. de. 1982. Phocid phylogeny and dispersal. Annals South African Mus., 89:175–213.

Muizon, C. de. 1993. Walrus-like feeding adaptation in a new cetacean from the Pliocene of Peru. Nature, 365:745–748.

Muizon, C. de. 1994. A new carnivorous marsupial from the Palaeocene of Bolivia and the problem of marsupial monophyly. Nature, 370:208–211.

Muizon, C. de. 1998. *Mayulestes ferox*, a borhyaenoid (Metatheria, Mammalia) from the early Palaeocene of Bolivia: phylogenetic and palaeobiologic implications. Geodiversitas, 20:19–142.

Muizon, C. de., and I. M. Brito. 1993. Le bassin calcaire de São José de Itaboraí (Rio de Janeiro, Brésil): ses relations fauniques avec le site de Tiupampa (Cochabamba, Bolivie). Ann. de Paleontol., 79:233–269.

Muizon, C. de, and H. G. McDonald. 1995. An aquatic sloth from the Pliocene of Peru. Nature, 375:224–227.

Muizon, C. de, and B. Lange-Badre. 1997. Carnivorous dental adaptations in tribosphenic mammals and phylogenetic reconstruction. Lethaia, 30:353–366.

Muizon, C. de, R. L. Cifelli, and R. Céspedes Paz. 1997. The origin of the dog-like borhyaenoid marsupials of South America. Nature, 389:486–489.

Müller, F. 1969. Verhaltnis von körperentwicklung und cerebralisation in ontogenese und phylogenese der sänger: versuch einer libersicht des problems. Verh. naturf. Ges., Basel, vol. 80.

Müller-Schwarze, D. 1971. Pheromones in the black-tailed deer *(Odocoileus hemionus columbianus)*. Anim. Behav., 19:141–152.

Munger, J. C., and J. H. Brown. 1981. Competition in desert rodents: an experiment with semipermeable exclosures. Science, 211:510–512.

Murie, A. 1944. *The Wolves of Mount McKinley*. U.S. Dept. Inter. Nat. Park Service., Fauna Ser 5.

Musser, G. G., and M. D. Carleton. 1993. Family Muridae, 501–755, in *Mammal Species of the World: A Taxonomic and Geographic Reference* (D. E. Wilson and D. M. Reeder, eds.). 2d ed. Smithsonian Inst. Press, Washington, D.C.

Mustrangi, M. A., and J. L. Patton. 1997. Phylogeography and systematics of the slender mouse opossum *Marmosops* (Marsupialia, Didelphidae). Univ. Calif. Publ. Zool., 130:1–86.

Myers, N. 1984. *The Primary Source: Tropical Forests and Our Future*. Norton, New York.

Mykytowycz, R. 1968. Territorial marking by rabbits. Sci. Amer., 218:116–126.

Myrcha, A., L. Ryskowski, and W. Walkowa. 1969. Bioenergetics of pregnancy and lactation in white mice. Acta Theriol., 15:161–166.

Nadler, R. D. 1975. Sexual cyclicity in captive lowland gorillas. Science, 189:813–814.

Naeem, S., L. J. Thompson, S. P. Lawler, J. H. Lawton, and R. M. Woodfin. 1994. Declining biodiversity can alter the performance of ecosystems. Nature, 368:734–737.

Nagy, J. G., H. W. Steinhoff, and G. M. Ward. 1964. Effects of essential oils of sagebrush on deer rumen microbial function. J. Wildl. Manag., 28:785–790.

Nagy, K. A. 1987. Field metabolic rate and food requirement scaling in mammals and birds. Ecol. Monogr., 57:111–128.

Nagy, K. A., V. H. Shoemaker, and W. R. Costa. 1976. Water, electrolyte, and nitrogen budgets of jackrabbits (*Lepus californicus*) in the Mojave Desert. Physiol. Zool., 49:351–363.

Nagy, K. A., and C. C. Peterson. 1980. Scaling of water flux rates in animals. Univ. Calif. Publ. Zool., 120:1–172.

Naiman, R. J., G. Pinay, C. A. Johnston, and J. Pastor. 1994. Beaver influences on the long-term biogeochemical characteristics of boreal forest drainage networks. Ecology, 75:905–921.

Naples, V. L. 1982. Cranial osteology and function in the tree sloths, *Bradypus* and *Choloepus*. Amer. Mus. Novitates, 2739:1–41.

Naples, V. L. 1990. Morphological changes in the facial region and a model of dental growth and wear pattern development in *Nothrotheriops shastensis*. J. Vert. Paleo., 10:372–389.

Narins, P. M., O. J. Reichman, J. U. M. Jarvis, and E. R. Lewis. 1992. Seismic signal transmission between burrows of the Cape mole-rat, *Georychus capensis*. J. Comp. Physiol. A., 170:13–21.

National Research Council, Food and Nutrition Board. 1968 National Academy of Science Publications, 1968.

Naumov, N. P. 1936. On some peculiarities of ecological distribution of mouse-like rodents in southern Ukraine. Zool. Zhurnal, 15:675–696.

Naumov, N. P. 1948. *Sketches of the Comparative Ecology of Mouse-like Rodents*. Isd-vo-Akademii Nauk, SSSR, Moscow.

Nefdt, R. J. C., and S. J. Thirgood. 1997. Lekking, resource defense, and harassment in two subspecies of lechwe antelope. Behav. Ecol., 8:1–9.

Neff, N. A. 1983. The basicranial anatomy of the Nimravidae (Mammalia: Carnivora): character analyses and phylogenetic inferences. Ph.D. dissertation, City University of New York, New York.

Nessov, L. A., and A. A. Gureev. 1981. [A jaw of a most ancient shrew from the Upper Cretaceous of the Kizylkum desert]. Doklady Akademii Nauk SSSR, 257:1002–1004 (in Russian).

Nessov, L., and Z. Kielan-Jaworowska. 1991. Evolution of the Cretaceous Asian therian mammals. Fifth Symposium on Mesozoic Terrestrial Ecosystems and Biota, Extended Abstracts. Contributions from the Paleontological Museum, Univ. Oslo 364:51–52.

Nestler, J. R. 1990. Relationships between respiratory quotient and metabolic rate during entry to and arousal from daily torpor in deer mice (*Peromyscus maniculatus*). Physiol. Zool., 63:504–515.

Nestler, J. R., G. P. Dieter, and B. G. Klokeid. 1996. Changes in total body fat during daily torpor in deer mice (*Peromyscus maniculatus*). J. Mamm., 77:147–154.

Nevo, E. 1979. Adaptive convergence and divergence of subterranean mammals. Ann. Rev. Ecol. Syst., 10:269–308.

New, D. A. T., M. Mizell, and D. L. Cockroft. 1977. Growth of opossum embryos *in vitro* during organogenesis. J. Exp. Morph., 41:111–123.

Nicol, S., N. A. Andersen, and U. Mesch. 1992. Metabolic rate and ventilatory pattern in the echidna during hibernation and arousal, 150–159, in *Platypus and Echidnas* (M. L. Augee, ed.). Royal Zool. Soc. New South Wales, Sydney.

Nikolai, J. C., and D. M. Bramble. 1983. Morphological structure and function in desert heteromyid rodents. Great Basin Nat., 7:44–64.

Nishida, T., and K. Kawanaka. 1972. Interunit-group relationships among wild chimpanzees of the Mahali Mountains. Kyoto Univ. Afr. Stud., 7:131–169.

Noirot, E. 1969. Sound analysis of ultrasonic distress calls of mouse pups as a function of their age. Anim. Behav., 17:340–349.

Noll-Banholzer, U. 1979a. Body temperature, oxygen consumption, evaporative water loss and heart rate in the fennec. Comp. Biochem. Physiol., 62:585–592.

Noll-Banholzer, U. 1979b. Water balance and kidney structure of the fennec. Comp. Biochem. Physiol., 62:593–597.

Norberg, U. M. 1969. An arrangement giving a stiff leading edge to the hand wing in bats. J. Mamm., 50:766–770.

Norberg, U. M. 1972. Bat wing structures important for aerodynamics and rigidity. Z. Morph. Tiere, 73:45–62.

Norberg, U. M. 1981. Allometry of bat wings and legs and comparison with bird wings. Philos. Trans. Royal Soc. London (B) 292:359–398.

Norberg, U. M. 1994. Wing design, flight performance, and habitat use in bats, 205–239, in *Ecological Morphology: Integrative Organismal Biology* (P. C. Wainwright and S. M. Reilly, eds.). Univ. Chicago Press, Chicago.

Norberg, U. M., and J. M. V. Rayner. 1987. Ecological morphology and flight in bats (Mammalia: Chiroptera): wing adaptations, flight performance, foraging strategy, and echolocation. Phil. Trans. R. Soc. B, 316:335–427.

Norrdahl, K., and E. Korpimäki. 1998. Does mobility or sex of voles affect risk of predation by mammalian predators? Ecology, 79:226–232.

Norris, K. S. 1964. Some problems of echolocation in cetaceans, 316–336, in *Marine Bioacoustics* (W. N. Tavolga, ed.). Pergamon Press, New York.

Norris, K. S. 1968. The echolocation of marine mammals, 391–423, in *The Biology of Marine Mammals* (H. T. Andersen, ed.). Academic Press, New York.

Norris, K. S., A. Prescott, D. V. Asa-Doran, and P. Perkins. 1961. An experimental demonstration of echolocation behavior in the porpoise, *Tursiops truncatus* (Montagu). Biol. Bull., 120:163–176.

Norris, K. S., and G. W. Harvey. 1972. A theory for the function of the spermaceti organ of the sperm whale (*Physeter catodon*), in *Animal Orientation and Navigation* (S. R. Galler, K. Schmidt-Koenig, G. J. Jacobs, and R. E. Belleville, eds.). NASA Spec. Pub. 262.

Norris, K. S., and G. W. Harvey. 1974. Sound transmission in the porpoise head. J. Acoust. Soc. Amer., 56:659–664.

Norris, K. S., and T. P. Dohl. 1980. The structure and functions of cetacean schools, 211–261, in *Cetacean Behavior: Mechanisms and Functions* (L. M. Herman, ed.). Wiley & Sons, New York.

Norris, K. S., and B. Møhl. 1983. Can odontocetes debilitate prey with sound? Amer. Nat., 122:85–104.

Novacek, M. J. 1985. Cranial evidence for rodent affinities, 59–81, in *Evolutionary Relationships of Rodents* (W. P. Luckett and J. L. Hartenberger, eds.). Plenum Press, New York.

Novacek, M. J. 1986. The skull of leptictid insectivorans and the higher-level classification of eutherian mammals. Bull. Amer. Mus. Nat. Hist., 183:1–111.

Novacek, M. J. 1989. Higher mammal phylogeny: the morphological-molecular synthesis, 421–435, in *The Hierarchy of Life* (B. Fernholm, K. Bremer, and H. Jörn-vall, eds.). Elsevier Science Publishers B. V. (Biomedical Division).

Novacek, M. J. 1992a. Fossils, topologies, missing data, and the higher level phylogeny of eutherian mammals. Syst. Biol., 41:58–73.

Novacek, M. J. 1992b. Mammalian phylogeny: shaking the tree. Nature, 356:121–125.

Novacek, M. J. 1993a. Reflections on higher mammalian phylogenetics. J. Mamm. Evol., 1:3–30.

Novacek, M. J. 1993b. Genes tell a new whale tale. Nature, 361:298–299.

Novacek, M. J., M. C. McKenna, N. A. Neff, and R. L. Cifelli. 1983. Evidence from earliest known erinaceomorph basicranium that insectivorans and primates are not closely related. Nature, 306:683–684.

Novacek, M. J., T. M. Bown, and D. Schankler. 1985. On the classification of the early Tertiary Erinaceomorpha (Insectivora, Mammalia). Amer. Mus. Novitates, 2813:1–22.

Novacek, M. J., and A. R. Wyss. 1986. Higher-level relationships of the Recent eutherian orders: morphological evidence. Cladistics, 2:257–287.

Novacek, M. J., A. R. Wyss, and M. C. McKenna. 1988. The major groups of eutherian mammals, 31–71, in *The Phylogeny and Classification of the Tetrapods* (M. J. Benton, ed.). Vol. 2, Oxford Univ. Press, Oxford.

Novacek, M. J., G. W. Rougier, J. R. Wible, M. C. McKenna, D. Dashzeveg, and I. Horovitz. 1997. Epipubic bones in eutherian mammals from the Late Cretaceous of Mongolia. Nature, 389:483–486.

Novick, A. 1955. Laryngeal muscles of the bat and production of ultrasonic sounds. Amer. J. Physiol., 1 83:648.

Novick, A. 1958a. Orientation in paleotropical bats. II: Megachiroptera. J. Exp. Zool., 137:443–462.

Novick, A. 1958b. Orientation in paleotropical bats. I: Microchiroptera. J. Exp. Zool., 138:81–153.

Novick, A. 1970. Echolocation in bats. Nat. Hist., 79(3):32–41.

Nowak, R. M., and J. L. Paradiso. 1991. *Walker's Mammals of the World*, 5th ed. Vol. 1:1–642; Vol. 2:643–1629. Johns Hopkins Univ. Press, Baltimore.

Oates, J. F. 1984. The niche of the potto, *Perodicticus potto*. Inter. J. Primatol., 5:51–61.

O'Brien, S. J., W. G. Nash, D. E. Wildt, M. E. Bush, and R. E. Benveniste. 1985. A molecular solution to the riddle of the giant panda's phylogeny. Nature, 317:140–144.

Obrist, M. K. 1989. Individuelle Variabilität der Echoortung, Vergleichende Freilanduntersuchungen an vier vespertilioniden Fledermausarten Kanadas. Dissertation, Ludwig-Maximilians-Universität, München.

Obrist, M. K. 1995. Flexible bat echolocation: the influence of individual habitat and conspecifics on sonar signal design. Behav. Biol. Sociobiol., 36:207–219.

Obrist, M. K., M. B. Fenton, J. L. Egers, and P. A. Schlegel. 1993. What ears do for bats: a comparative study of pinna sound pressure transformation in Chiroptera. J. Exp. Biol., 180:119–152.

Odum, E. P. 1971. *Fundamentals of Ecology.* 3d ed. W.B. Saunders, Philadelphia.

O'Farrell, M. J., and E. H. Studier. 1970. Fall metabolism in relation to ambient temperatures in three species of *Myotis.* Comp. Biochem. Physiol., 35:697–703.

O'Farrell, T. P. 1965. Home range and ecology of snow-shoe hares in interior Alaska. J. Mamm., 46:406–418.

Oftedal, O. T., S. J. Iverson, and D. J. Boness. 1987. Milk and energy intakes of suckling California sea lion pups (*Zalophus californianus*) in relation to sex, growth and predicted maintenance requirements. Physiol. Zool., 60:560–575.

Oftedal, O. T., W. D. Bowen, and D. J. Boness. 1993. Energy transfer by lactating hooded seals and nutrient deposition in their pups during the four days from birth to weaning. Physiol. Zool., 66:412–436.

O'Gara, B. W. 1978. *Antilocapra americana.* Mamm. Species, 90:1–7.

O'Gara, B. W., R. F. Moy, and G. D. Bear. 1971. The annual testicular cycle and horn casting in the pronghorn (*Antilocapra americana*). J. Mamm., 52:537–544.

Oksanen, T., and H. Henttonen. 1996. Dynamics of voles and small mustelids in the taiga landscape of northern Fennoscandia in relation to habitat quality. Ecography, 19:432–443.

Onuma, M., T. Kusakabe, S. Kusakabe. 1998. Phylogenetic position of Insectivora in Eutheria inferred from mitochondrial cytochrome c oxidase subunit II gene. Zool. Sci., 15:139–145.

Orians, G. H. 1969. On the evolution of mating systems of birds and mammals. Amer. Nat., 103:589–603.

Ortega, J., and H. T. Arita. 1998. Neotropical–Nearctic limits in Middle America as determined by distributions of bats. J. Mamm., 79:772–783.

O'Shea, T. J. 1994. Manatees. Sci. Amer., 271:66–72.

O'Shea, T. J., G. B. Rathbun, R. K. Bonde, C. D. Buergelt, and D. K. Odell. 1991. An epizootic of Florida manatees associated with a dinoflagellate bloom. Marine Mamm. Sci., 7:165–179.

O'Shea, T. J., B. B. Ackerman, and H. F. Percival. 1995. *Population Biology of the Florida Manatee (Trichechus manatus latirostris).* National Biological Service, Information and Technical Report 1.

Ostfeld, R. S. 1992. Small-mammal herbivores in a patchy environment: individual strategies and population responses, 43–74, in *Effects of Resource Distribution on Animal-plant Interactions* (M. D. Hunter, T. Ohgushi, and P. W. Price, eds). Academic Press, New York.

Ostfeld, R. S. 1997. The ecology of lyme-disease risk. Amer. Sci., 85:338–346.

Ostfeld, R. S., and L. L. Klosterman. 1986. Demographic substructure in a California vole population inhabiting a patchy environment. J. Mamm., 67:693–704.

Ostfeld, R. S., C. G. Jones, and J. O. Wolff. 1996. Of mice and mast, ecological connections in eastern deciduous forests. BioScience, 46:323–330.

Oswald, C., and P. A. McClure. 1990. Energetics of concurrent pregnancy and lactation in cotton rats and woodrats. J. Mamm., 71:500–509.

Oswald, C., P. Fonken, D. Atkinson, and M. Palladino. 1993. Lactational water balance and recycling in white-footed mice, red-back voles, and gerbils. J. Mamm., 74:963–970.

Otte, D. 1974. Effects and functions in the evolution of signaling systems. Ann. Rev. Ecol. Syst., 5:385–417.

Oxnard, E. 1981. The uniqueness of *Daubentonia.* Amer. J. Phys. Anthropol., 54:1–21.

Pack, A. A., and L. M. Herman. 1995. Sensory integration in the bottlenosed dolphin: immediate recognition of complex shapes across the senses of echolocation and vision. J. Acoust. Soc. Amer., 98:722–733.

Packer, C. 1986. The ecology of felid sociality, 429–451, in *Ecological Aspects of Social Evolution* (D. J. Rubenstein and R. W. Wrangham, eds.). Princeton Univ. Press, Princeton, New Jersey.

Packer, C., and A. E. Pusey. 1982. Cooperation and competition within coalitions of male lions: kin selection or game theory? Nature, 296:740–742.

Padykula, H. A., and J. M. Taylor. 1982. Marsupial placentation and its evolutionary significance. J. Reprod. Fert., Suppl., 31:95–104.

Page, R. D. M., and C. Lydeard. 1994. Towards a cladistic biogeography of the Caribbean. Cladistics, 10:21–41.

Pagel, M. D., R. M. May, and A. R. Collie. 1991. Ecological aspects of the geographical distribution and diversity of mammalian species. Amer. Nat., 7:791–815.

Paige, K. N. 1995. Bats and barometric pressure: conserving limited energy and tracking insects from the roost. Funct. Ecol., 9:463–467.

Paine, R. T. 1969. A note on trophic complexity and community stability. Amer. Nat., 103:91–93.

Parker, P. 1977. An ecological comparison of marsupial and placental patterns of reproduction, 273–286, in *The Biology of Marsupials* (B. Stonehouse and D. Gilmore, eds.). Macmillan, London.

Parrish, J. T. 1990. Gondwanan paleogeography and paleoclimatology, 15–26, in *Antarctic Paleobiology: Its Role in the Reconstruction of Gondwana.* (T. N. Taylor and E. L. Taylor, eds.). Springer-Verlag, New York.

Parry, D. A. 1949. The structure of whale blubber and its thermal properties. Quart. J. Microbiol. Sci., 90:13–26.

Pascual, R., and Ortiz Jaureguizar, E. 1990. Evolving climates and mammal faunas in Cenozoic South America, 23–60, in *The Platyrrhine Fossil Record* (J. G. Fleagle and A. L. Rosenberger, eds.). Academic Press, London.

Pascual, R., M. Archer, E. Ortiz Jaureguizar, J. L. Prado, H. Godthelp, and S. J. Hand. 1992a. First discovery of monotremes in South America. Nature, 356:67–74.

Pascual, R., M. Archer, E. Ortiz Jaureguizar, J. L. Prado, H. Godthelp, and S. J. Hand. 1992b. The first non-Australian monotreme: an early Paleocene South American platypus (Monotremata, Ornithorhynchidae), 2–15, in *Platypus and Echidnas* (M. L. Augee, ed.). Royal Zool. Soc. New South Wales, Sydney.

Pascual, R., F. J. Goin, and A. A. Carlini. 1994. New data on the Groeberiidae: unique late Eocene–early Oligocene South American marsupials. J. Vert. Paleo., 14:247–259.

Patterson, B. 1965. The fossil elephant shrews (Family Macroscelididae). Bull. Mus. Comp. Zool. (Harvard Univ.), 133:295–335.

Patterson, B. 1975. The fossil aardvarks (Mammalia: Tubulidentata). Bull. Mus. Comp. Zool. (Harvard Univ.), 147:185–237.

Patterson, B., and R. Pascual. 1972. The fossil mammal fauna of South America, 247–309, in *Evolution, Mammals, and Southern Continents* (A. Keast, F. C. Erk, and B. Glass, eds.). SUNY Press, Albany, NY.

Patterson, B. D. 1994. Accumulating knowledge on the dimensions of biodiversity: systematic perspectives on Neotropical mammals. Biodiversity Letters, 2:79–86.

Patterson, B. D. 1995. Local extinctions and the biogeographic dynamics of boreal mammals in the Southwest, 151–176, in *Storm Over a Mountain Island: Conservation Biology and the Mount Graham Affair* (C. A. Istock and R. S. Hoffmann, eds.). University of Arizona Press, Tucson.

Patterson, B. D., and W. Atmar. 1986. Nested subsets and the structure of insular mammalian faunas and archipelagos. Biol. J. Linnean Soc., 28:65–82.

Patterson, B., W. Segall, and W. D. Turnbull. 1989. The ear region in xenarthrans (=Edentata: Mammalia). Part I. Cingulates. Fieldiana: Geology, 18:1–46.

Patterson, B., W. Segall, W. D. Turnbull, and T. J. Gaudin. 1992. The ear region in xenarthrans (=Edentata: Mammalia). Part II. Pilosa (sloths, anteaters), palaeanodonts, and a miscellany. Fieldiana: Geology, 24:1–79.

Patton, J. L., M. N. F. da Silva, M. C. Lara, and M. A. Mustrangi. 1997. Diversity, differentiation, and the historical biogeography of nonvolant small mammals of the Neotropical forests, 455–465, in *Tropical Forest Remnants: Ecology, Management, and Conservation of Fragmented Communities* (W. F. Laurance and R. O. Bierregaard, Jr., eds.). Univ. of Chicago Press, Chicago.

Patz, J. A., P. R. Epstein, T. A. Burke, and J. M. Balbus. 1996. Global climate change and emerging infectious diseases. J. Amer. Med. Assoc., 275:217–223.

Payne, K. B., W. R. Langbauer, Jr., and E. M. Thomas. 1986. Infrasonic calls of the Asian elephant (*Elephas maximus*). Behav. Ecol. Sociobiol., 18:297–301.

Payne, R. S. 1970. *Songs of the Humpback Whale*. An LP Record by CRM Records, Del Mar, California.

Payne, R. S., and S. McVay. 1971. Songs of the humpback whales. Science, 173:585–597.

Pearson, O. P. 1942. On the cause and nature of a poisonous action produced by the bite of a shrew (*Blarina brevicauda*). J. Mamm., 23:159–166.

Pearson, O. P. 1948. Metabolism of small mammals with remarks on the lower limit of mammalian size. Science, 108:44.

Pearson, O. P. 1959. Biology of the subterranean rodents, *Ctenomys*, in Peru. Memorias del Museo de Historia Natural "Javier Prado," 9:1–56.

Pearson, O. P. 1963. History of two local outbreaks of feral house mice. Ecology, 44:540–549.

Pearson, O. P. 1964. Carnivore-mouse predation: an example of its intensity and bioenergetics. J. Mamm., 45:177–188.

Pearson, O. P. 1966. The prey of carnivores during one cycle of mouse abundance. J. Anim. Ecol., 35:217–233.

Pearson, O. P. 1971. Additional measurements of the impact of carnivores on California voles (*Microtus californicus*). J. Mamm., 52:41–49.

Pedersen, S. C. 1998. Morphometric analysis of the chiropteran skull with regard to mode of echolocation. J. Mamm., 79:91–103.

Pengelley, E. T. 1967. The relation of external conditions to the onset and termination of hibernation and estivation, in *Mammalian Hibernation*, Vol. III (K. C. Fisher et al., eds.). Oliver and Boyd, London.

Pengelley, E. T., and K. C. Fisher. 1963. The effect of temperature and photoperiod on the yearly hibernating behavior of captive golden-mantled ground squirrels (*Citellus lateralis tescorum*). Can. J. Zool., 41:1103–1120.

Pengelley, E. T., and S. J. Asmundson. 1970. The effect of light on the free-running circannual rhythm of the golden-mantled ground squirrel, *Citellus lateralis*. Comp. Biochem. Physiol., 30:177–183.

Pengelley, E. T., and S. J. Asmundson. 1971. Annual biological clocks. Sci. Amer., 224:72–79.

Penn, D. J. and W. K. Potts. 1999. The evolution of mating preferences and major histocompatibility complex genes. Amer. Nat., 153:145–164.

Penny, D., and M. Hasegawa. 1997. The platypus put in its place. Nature, 387:549–550.

Pennycuik, P. R. 1969. Reproductive performance and body weights of mice maintained for 12 generations at 34°C. Aust. J. Biol. Sci., 22:667–675.

Pepper, J. W., S. H. Braude, E. A. Lacey, and P. W. Sherman. 1991. Vocalizations of the naked mole-rat, 243–274, in *The Biology of the Naked Mole-rat* (P. W. Sherman, J. U. M. Jarvis, and R. D. Alexander, eds.). Princeton Univ. Press, Princeton, New Jersey.

Pequegnat, W. E. 1951. The biota of the Santa Ana Mountains. J. Entomol. Zool., 42:1–84.

Pereira, M. E., and M. K. Izard. 1989. Lactation and care for unrelated infants in forest-living ringtailed lemurs. Amer. J. Primatol., 18:101–108.

Perret, M. 1992. Environmental and social determinants of sexual function in the male lesser mouse lemur (*Microcebus murinus*). Folia Primatol., 59:1–25.

Perrin, M. R., H. Boyer, and D. C. Boyer. 1992. Diets of the hairy-footed gerbils *Gerbillurus paeba* and *G. tytonis* from the dunes of the Namib Desert. Israel J. Zool., 38:373–383.

Peter, W. P., and A. Feiler. 1994. Hörner von einer unbekannten bovidenart aus Vietnam (Mammalia: Ruminantia). Faunistische Abhandlungen, Staatliches Mus. für Tierkunde Dresden, 19:247–253.

Peters, R. P., and L. D. Mech. 1975. Scent-marking in wolves. Sci. Amer., 63:628–637.

Peterson, A. T., O. A. Flores-Villela, L. S. León-Paniagua, J. E. Llorente-Bousquets, M. A. Luis-Martinez, A. G. Navarro-Sigüenza, et al. 1993. Conservation priorities in Mexico: moving up in the world. Biodiversity Letters, 1:33–38.

Peterson, R. O., 1977. *Wolf Ecology and Prey Relationships on Isle Royale.* Nat. Park Service Monogr. Ser. 11.

Peterson, R. S., and G. A. Bartholomew. 1967. *The Natural History and Behavior of the California Sea Lion.* Spec. Publ. No. 1, Amer. Soc. Mamm.

Petter, J. J. 1962. Ecological and behavioral studies of Madagascar lemurs in the field. Ann. N.Y. Acad. Sci., 102:267–281.

Pettigrew, J. D. 1986. Flying primates? megabats have the advanced pathway from eye to midbrain. Science, 231:1304–1306.

Pettigrew, J. D. 1991a. Wings or brain? convergent evolution in the origins of bats. Syst. Zool., 40:199–216.

Pettigrew, J. D. 1991b. A fruitful, wrong hypothesis? response to Baker, Novacek, and Simmons. Syst. Zool., 40:231–239.

Pettigrew, J. D., B. G. M. Jamieson, S. K. Robson, L. S. Hall, K. I. McAnally, and H. M. Cooper. 1989. Phylogenetic relations between microbats, megabats, and primates (Mammalia: Chiroptera and Primates). Phil. Trans. Royal Soc. London, B 325:489–559.

Pettigrew, J. D., and J. A. Kirsch. 1995. Flying primates revisited: DNA hybridization with fractionated, GC-enriched DNA. South African J. Sci., 91:477–482.

Philippe, H. 1997. Rodent monophyly: pitfalls of molecular phylogenies. J. Mol. Evol., 45:712–715.

Pianka, E. R. 1976. Natural selection of optimal reproductive strategies. Amer. Zool., 16:775–787.

Pianka, E. R. 1988. *Evolutionary Ecology.* 4th ed., Harper & Row, New York.

Picard, K., D. W. Thomas, M. Festa-Bianchet, and C. Lanthier. 1994. Bovid horns: an important site of heat loss during winter? J. Mamm., 75:710–713.

Pierson, E. D. 1986. Molecular systematics of the Microchiroptera: Higher taxonomic relationships and biogeography. Unpubl. Ph.D. dissertation, University of California, Berkeley.

Pierson, E. D., V. M. Sarich, J. M. Lowenstein, M. J. Daniel, and W. E. Rainey. 1986. A molecular link between the bats of New Zealand and South America. Nature, 323:60–63.

Pimm, S. L. 1992. *The Balance of Nature?* Univ. Chicago Press, Chicago.

Pine, R. H. 1994. New mammals not so seldom. Nature, 368:593.

Pine, R. H., J. E. Rice, J. E. Bucher, D. H. Tank, and A. M. Greenhall. 1985. Labile pigments and fluorescent pelage in didelphid marsupials. Mammalia, 49:249–256.

Pitelka, F. A. 1957a. *Some Characteristics of Microtine Cycles in the Arctic.* Eighteenth Ann. Biol. Coll., Oregon State College.

Pitelka, F. A. 1958. Some aspects of population structure in the short-term cycle of the brown lemming in northern Alaska. Cold Spr. Harbor Symp. Quant. Biol., 22:237–251.

Pitelka, F. A., P. Q. Tomich, and G. W. Treichel. 1955. Ecological relations of jaegers and owls as lemming predators near Barrow, Alaska. Ecol. Monogr., 25:85–117.

Pitman, W. C., III, S. Cande, J. LaBrecque, and J. Pindell. 1993. Fragmentation of Gondwana: the separation of Africa from South America, 15–34, in *Biological Relationships Between Africa and South America* (P. Goldblatt, ed.). Yale University Press, New Haven.

Pivorunas, A. 1979. The feeding mechanisms of baleen whales. Amer. Sci., 67:432–440.

Poinar, H. N., Hofreiter, W. G. Spaulding, P. S. Martin, B. A. Stankiewicz, H. Bland, R. P. Evershed, G. Possnert, and S. Paabo. 1998. Molecular coproscopy: Dung and diet of the extinct ground sloth *Nothrotheriops shastensis*. Science, 281:402–406.

Pollak, G. D., and J. H. Casseday. 1989. *The Neural Basis of Echolocation in Bats.* Springer-Verlag, Berlin.

Pond, C. M. 1977. The significance of lactation in the evolution of mammals. Evolution, 31:177–199.

Pook, A. G., and G. Pook. 1981. A field study of the socio-ecology of the Goeldi's monkey (*Callimico goeldii*) in northern Brazil. Folia Primatol., 35:288–312.

Poole, J. H. 1987. Elephants in musth, lust. Nat. Hist., 11:46–55.

Poole, J. H. 1989. Mate guarding, reproductive success and female choice in African elephants. Anim. Behav., 37:842–849.

Poole, J. H., and C. J. Moss. 1981. Musth in the African elephant (*Loxodonta africana*). Nature, 292:830–831.

Poole, J. H., K. Payne, W. R. Langbauer, Jr., and C. J. Moss. 1988. The social contexts of some very low frequency calls of African elephants. Behav. Ecol. Sociobiol., 22:385–392.

Popper, A. N., H. L. Hawkins, and R. C. Gisiner. 1997. Questions in cetacean bioacoustics: some suggestions for future research. Bioacoustics, 8:163–182.

Porter, C. A., I. Sampaio, H. Schneider, and M. P. C. Schneider. 1994. Evidence on primate phylogeny from ε-globin gene sequence and flanking regions. J. Mol. Evol., 40:30–55.

Post, D., and O. J. Reichman. 1991. Effects of food perishability, distance, and competitors on caching behavior by eastern woodrats. J. Mamm., 72:513–517.

Pough, F. H., C. M. Janis, and J. B. Heiser. 1998. *Vertebrate Life*. MacMillan Pub., New York.

Pournelle, G. H. 1968. Classification, biology, and description of the venom apparatus of insectivores of the genera *Solenodon, Neomys,* and *Blarina,* 31–42, in *Venomous Animals and Their Venoms* (W. Bucherl, E. A. Buckley, and V. Deulofeu, eds.). Academic Press, New York.

Powers, J. B., and S. S. Winans. 1975. Vomeronasal organ: critical role in mediating sexual behavior of the male hamster. Science, 187:961–963.

Prakash, I. 1959. Foods of the Indian false vampire. J. Mamm., 40:545–547.

Prasad, G. V. R., and B. K. Manhas. 1997. A new symmetrodont mammal from the Lower Jurassic Kota Formation, Pranhita-Godavari Valley, India. Geobios, 30:563–572.

Price, M. V. 1978. The role of microhabitat in structuring desert rodent communities. Ecology, 59:910–921.

Price, M. V. 1983. Laboratory studies of seed size and species selection by heteromyid rodents. Oecologia, 60:259–263.

Price, M. V., N. W. Waser, and T. A. Bass. 1984. Effects of moonlight on microhabitat use by desert rodents. J. Mamm., 65:353–356.

Pridham, J. 1965. *Enzyme Chemistry of Phenolic Compounds*. MacMillan, New York.

Prothero, D. R. 1994a. The Late Eocene–Oligocene extinctions. Ann. Rev. Earth Planet. Sci., 22:145–165.

Prothero, D. R. 1994b. *The Eocene–Oligocene Transition: Paradise Lost*. Columbia Univ. Press, New York.

Prothero, D. R., and W. A. Berggren. 1992. *Eocene–Oligocene Climatic and Biotic Evolution*. Princeton, Princeton Univ. Press, New Jersey.

Proctor-Grey, E. 1984. Dietary ecology of the coppery brushtail possum, green ringtail possum and Lumholtz's tree-kangaroo in North Queensland, 129–135, in *Possums and Gliders* (A. P. Smith and I. D. Hume, eds.). Aust. Mamm. Soc. and Surrey Beatty and Sons, Sydney.

Pryor, K., and K. S. Norris. 1991. *Dolphin Societies: Discoveries and Puzzles*. Univ. California Press, Berkeley.

Pucek, M. 1968. Chemistry and pharmacology of insectivore venoms, 43–50, in *Venomous Animals and Their Venoms* (W. Bucherl, E. A. Buckley, and V. Deulofeu, eds.). Academic Press, New York.

Purves, P. E. 1966. Anatomy and physiology of the outer and middle-ear in cetaceans, 320–380, in *Whales, Dolphins and Porpoises* (K. S. Norris, ed.). Univ. California Press, Berkeley.

Pye, J. D. 1988. Noseleaves and bat pulses, 791–796, in *Animal Sonar Systems: Processes and Performances* (P. A. Nachtgall and P. W. B. Moore, eds.). Plenum Press, New York.

Quilliam, T. A. 1966. The mole's sensory apparatus. J. Zool., London, 149:76–78.

Quinn, T. H., and J. J. Baumel. 1993. Chiropteran tendon locking mechanism. J. Morph., 216:197–208.

Quintana, R. D., S. Monge, and A. I. Malvárez. 1998. Feeding patterns of capybara *Hydrochaeris hydrochaeris* (Rodentia, Hydrochaeridae) and cattle in the non-insular area of the lower delta of the Paraná River, Argentina. Mammalia, 62:37–52.

Racey, P. A. 1973. Environmental factors affecting the length of gestation in heterothermic bats. J. Reprod. Fert., Suppl., 19:175–189.

Racey, P. A. 1982. The ecology of reproduction, 57–104, in *Ecology of Bats* (T. H. Kunz, ed.). Plenum Press, New York.

Radinsky, L. B. 1966. The adaptive radiation of the phenacodontid condylarths and the origin of the Perissodactyla. Evolution, 20:408–417.

Radinsky, L. B. 1984. Ontogeny and phylogeny in horse skull evolution. Evolution, 38:1–15.

Rado, R., N. Levi, H. Hauser, J. Witcher, N. Adler, N. Intrator, Z. Wollberg, and J. Terkel. 1987. Seismic signalling as a means of communication in a subterranean mammal. Anim. Behav., 35:1249–1251.

Rageot, R. 1978. Observationes sobre el monito del monte. Chile Min. Agric. Corp. Nac. For. Dept. Tecn. IX. 1–16.

Rahm, U. 1969. Notes sur la cri du *Dendrohyrax dorsalis* (Hyracoidea). Mammalia, 33:68–79.

Rahm, U. 1970. Note sur la reproduction des sciuridés et muridés dans forêt équatoriale au Congo. Rev. Suisse Zool., 77:635–646.

Ralls, K. 1971. Mammalian scent marking. Science, 171:443–449.

Ralls, K., J. D. Ballou, and A. Templeton. 1988. Estimates of lethal equivalents and the cost of inbreeding in mammals. Conserv. Biol., 2:185–193.

Randall, J. A. 1989. Individual footdrumming signatures in bannertailed kangaroo rats, *Dipodomys spectabilis*. Anim. Behav., 38:620–630.

Randall, J. A. 1993. Behavioral adaptations of desert rodents (Heteromyidae). Anim. Behav., 45:263–287.

Randall, J. A. 1995. Modification of footdrumming signatures by kangaroo rats: changing territories and gaining new neighbors. Anim. Behav., 49:1227–1237.

Randall, J. A. 1997. Species-specific footdrumming in kangaroo rats: *Dipodomys ingens, D. deserti, D. spectabilis*. Anim. Behav., 54:1167–1175.

Randall, J. A., and E. R. Lewis. 1997. Seismic communication between the burrows of kangaroo rats, *Dipodomys spectabilis*. J. Comp. Physiol. A, 181:525–531.

Randall, J. A., and M. D. Matocq. 1997. Why do kangaroo rats *(Dipodomys spectabilis)* footdrum at snakes? Behav. Ecol., 8:404–413.

Randolph, P. A., T. C. Randolph, K. Mattingly, and M. M. Foster. 1977. Energy costs of reproduction in the cotton rat, *Sigmodon hispidus*. Ecology, 58:31–45.

Rapport, D. J., R. Costanza, and A. J. McMichael. 1998. Assessing ecosystem health. Trends Ecol. Evol., 13:397–402.

Rasmussen, L. E. L., A. J. Hall-Martin, and D. L. Hess. 1996. Chemical profiles of male African elephants, *Loxodonta africana:* physiological and ecological implications. J. Mamm., 77:422–439.

Rasweiler, J. J. 1977. The care and management of bats as laboratory animals, 519–617, in *Biology of Bats* (W. A. Wimsatt, ed.). Academic Press, New York.

Rathbun, G. 1978. Evolution of the rump region in the golden-rumped elephant-shrew. Bull. Carnegie Mus. Nat. Hist., 6:11–19.

Rathbun, G. 1979. The social structure and ecology of the elephant-shrews. Z. Tierpsychol., 20:1–79.

Rathbun, G. 1984. Sirenians, 537–547, in *Orders and Families of Recent Mammals of the World* (S. Anderson and J. K. Jones, eds.). Wiley & Sons, New York.

Rathbun, G. 1992. The fairly true elephant-shrew. Nat. Hist., 9:55–61.

Rausch, R. 1950. Observations on a cyclic decline of lemmings *(Lemmus)* on the Arctic coast of Alaska during the spring of 1949. Arctic, 3:166–177.

Raven, P. H., and E. O. Wilson. 1992. A fifty-year plan for biodiversity surveys. Science, 258:1099–1100.

Rayor, L. S. 1988. Social organization and space-use in Gunnison's prairie dog. Behav. Ecol. Sociobiol., 22:69–78.

Reaka-Kudla, M. L., D. E. Wilson, and E. O. Wilson. 1997. *Biodiversity II. Understanding and Protecting Our Biological Resources.* Joseph Henry Press, Washington, D.C.

Rebar, C. E. 1995. Ability of *Dipodomys merriami* and *Chaetodipus intermedius* to locate resource distributions. J. Mamm., 76:437–447.

Rebar, C., and O. J. Reichman. 1983. Ingestion of moldy seeds by heteromyid rodents. J. Mamm., 64:713–715.

Redford, K. H., and J. G. Dorea. 1984. The nutritional value of invertebrates with emphasis on ants and termites as food for mammals. J. Zool. London, 203:385–395.

Redman, J. P., and J. A. Sealander. 1958. Home ranges of deer mice in southern Arkansas, J. Mamm., 39:390–395.

Reed, C. A. 1944. Behavior of a shrew mole in captivity. J Mamm., 25:196–198.

Reeve, N. 1994. *Hedgehogs.* T & AD Poyser, Ltd., London.

Regal, P. J. 1977. Ecology and evolution of flowering plant dominance. Science, 196:622–629.

Reichman, O. J. 1975. Relation of desert rodent diets to available resources. J. Mamm., 56:731–751.

Reichman, O. J. 1977. Optimization of diet through food preferences by heteromyid rodents. Ecology, 58:454–457.

Reichman, O. J. 1984. Spatial and temporal variation in seed distributions in Sonoran Desert soils. J. Biogeogr., 11:1–11.

Reichman, O. J. 1988a. Caching behavior by eastern woodrats, *Neotoma floridana,* in relation to food perishability. Anim. Behav., 36:1525–1532.

Reichman, O. J. 1988b. Comparisons of the effects of crowding and pocket gopher disturbance on mortality, growth and seed production of *Berteroa incana.* Amer. Midland Nat., 120:57–69.

Reichman, O. J. 1991. Desert mammal communities, 311–347, in *The Ecology of Desert Communities* (G. Polis, ed.). Univ. Arizona Press: Tucson, Arizona.

Reichman, O. J., and K. M. Van De Graaff. 1975. Influence of green vegetation on desert rodent reproduction. J. Mamm., 53:503–506.

Reichman, O. J., T. G. Whitham, and G. A. Ruffner. 1982. Adaptive geometry of burrow spacing in two pocket gopher populations. Ecology, 63:687–695.

Reichman, O. J., and P. Fay. 1983. Comparisons of the diets of a caching and a noncaching rodent. Amer. Nat., 122:576–581.

Reichman, O. J., and C. Rebar. 1985. Seed preferences by desert rodents based on levels of moldiness. Anim. Behav., 33:726–729.

Reichman, O. J., D. T. Wicklow, and C. Rebar. 1985. Ecological and mycological characteristics of caches in the mounds of *Dipodomys spectabilis.* J. Mamm., 66:643–651.

Reichman, O. J., A. Fattaey, and K. Fattaey. 1986. Management of sterile and moldy seeds by a desert rodent. Anim. Behav., 34:221–225.

Reichman, O. J., and J. U. M. Jarvis. 1989. The influence of three sympatric species of fossorial mole-rats (Bathyergidae) on vegetation. J. Mamm., 70:763–771.

Reid, R. T. 1970. The future role of ruminants in animal production, in *Physiology of Digestion and Metabolism in the Ruminant* (A. T. Phillipson, ed.). Oriel Press, Newcastle-upon-Tyne, England.

Reid, W. V. 1998. Biodiversity hotspots. Trends Ecol. Evol., 13:275–280.

Reig, O. A. 1970. Ecological notes on the fossorial octodontid rodent *Spalacopus cyanus* (Molina). J. Mamm., 51:592–601.

Reig, O. A. 1980. A new fossil genus of South American cricetid rodents allied to *Wiedomys,* with an assessment of the Sigmodontinae. J. Zool. London, 192:257–281.

Reiss, K. Z. 1997. Myology of the feeding apparatus of myrmecophagid anteaters (Xenarthra: Myrmecophagidae). J. Mamm. Evol., 4:87–117.

Reisz, R. R. 1986. *Pelycosauria. Handbuch der Paloherpetologie* 17A:1–102. Gustav Fischer, Stuttgart.

Reiter, R. J. 1983. The role of light and age in determining melatonin production in the pineal gland, 227–241, in *The Pineal Gland and its Endocrine Role* (J. Axelrod, F. Fraschini, and G. P. Velo, eds.). Plenum Press, NY.

Renfree, M. B. 1993. Ontogeny, genetic control, and phylogeny of female reproduction in monotreme and therian mammals, 4–20, in *Mammal Phylogeny.* Vol. 1, *Mesozoic Differentiation, Multituberculates, Monotremes,*

Early Therians, and Marsupials (F. S. Szalay, M. J. Novacek, and M. C. McKenna, eds.). Springer-Verlag, New York.

Renfree, M. B., D. W. Lincoln, O. F. X. Almeida, and R. V. Short. 1981. Abolition of seasonal embryonic diapause in a wallaby by pineal denervation. Nature, 293:138–139.

Rentz, D. C. 1975. Two new katydids of the genus *Melanonotus* from Costa Rica with comments on their life history strategies (Tettigoniidae: Pseudohyllinae). Entomol. News, 86:129–140.

Repenning, C. A. 1976. Adaptive evolution of sea lions and walruses. Syst. Zool., 25:375–390.

Repenning, C. A., C. E. Ray, and D. Grigorescuy. 1979. Pinniped biogeography, in *Historical Biogeography, Plate Tectonics, and the Changing Environment* (J. Gray and A. J. Boucot, eds.). Oregon State Univ. Press, Corvallis.

Repetto, R. 1992. Accounting for environmental assets. Sci. Amer., June:94–100.

Retief, J. D., C. Krajewski, M. Westerman, R. H. Winkfein, and G. H. Dixon. 1995. Molecular phylogeny and evolution of marsupial protamine P1 genes. Proc. Royal Soc. London (B), 259:7–14.

Reyes, A., G. Pesole, and C. Saccone. 1998. Complete mitochondrial DNA sequence of the fat dormouse, *Glis glis:* further evidence of rodent paraphyly. Mol. Biol. Evol., 15:499–505.

Rhoades, D. F. 1979. Evolution of plant chemical defenses against herbivores, 3–54, in *Herbivores: Their Interactions with Plant Secondary Metabolites* (G. A. Rosenthal and D. H. Janzen, eds.). Academic Press, New York.

Rice, D. W. 1984. Cetaceans, 447–490, in *Orders and Families of Recent Mammals of the World* (S. Anderson and J. K. Jones, eds.). Wiley & Sons, New York.

Rich, T. H., P. Vickers-Rich, A. Constantine, T. F. Flannery, L. Kool, and N. van Klaveren. 1997. A tribosphenic mammal from the Mesozoic of Australia. Science, 278:1438–1442.

Richard, P. B. 1982. La sensibilité tactile de contact chez le desman (*Galemys pyrenaicus*). Biol. Behav., 7:325–336.

Richmond, M. E., and R. A. Stehn. 1976. Olfaction and reproductive behavior in microtine rodents. 197–217, in *Mammalian Olfaction, Reproductive Processes, and Behavior* (R. L. Doty, ed.). Academic Press, New York.

Riddle, B. R. 1995a. Special feature: mammalian biogeography. J. Mamm., 76:281–282.

Riddle, B. R. 1995b. Molecular biogeography in the pocket mice (*Perognathus* and *Chaetodipus*) and grasshopper mice (*Onychomys*): the late Cenozoic development of a North American aridlands rodent guild. J. Mamm., 76:283–301.

Riddle, B. R., R. L. Honeycutt, and P. L. Lee. 1990. Mitochondrial DNA phylogeography on northern grasshopper mice (*Onychomys leucogaster*)—the influence of Quaternary climatic oscillations on population dispersion and divergence. Molecular Ecol., 2:183–193.

Ridgway, S. H., and R. Howard. 1979. Dolphin lung col-

lapse and intramuscular circulation during free diving: evidence from nitrogen washout. Science, 206:1182–1183.

Ridgway, S. H., D. A. Carder, R. F. Green, A. S. Gaunt, S. L. L. Gaunt, and W. E. Evans. 1980. Electromyographic and pressure events in the nasolaryngeal system of dolphins during sound production, 239–249, in *Animal Sonar Systems* (R. G. Busnel and J. F. Fish, eds.). Plenum Press, New York.

Ridgway, S. H., and R. Howard. 1982. Dolphins and the bends. Science, 216:651.

Rinker, G. C. 1954. The comparative myology of the mammalian genera *Sigmodon, Oryzomys, Neotoma*, and *Peromyscus* (Cridetinae), with remarks on their intergeneric relationships. Misc. Publ. Mus. Zool. Univ. Michigan, 83:1–124.

Rissman, E. F., and R. E. Johnson. 1985. Female reproductive development is not activated by male California voles exposed to family cues. Biol. Repro., 32:352–360.

Robichaud, W. G. 1998. Physical and behavioral description of a captive saola, *Pseudoryx nghetinhensis*. J. Mamm., 79:394–405.

Robinette, W. L., J. S. Gashwiler, D. A. Jones, and H. S. Crane. 1955. Fertility of mule deer in Utah. J. Wildl. Manag., 19:115–136.

Robinson, J. G. 1981. Spatial structure in foraging groups of wedge-capped capuchin monkeys (*Cebus nigrivittatus*). Anim. Behav., 29:1036–1056.

Robinson, J. G., and K. H. Redford. 1991. *Neotropical Wildlife Use and Conservation*. Univ. Chicago Press, Chicago.

Robinson, K., and D. H. K. Lee. 1941. Reactions of the dog to hot atmospheres. Proc. Royal Soc. Queensland, 53:159–170.

Rodger, J. C., T. P. Fletcher, and C. H. Tyndale-Biscoe. 1985. Active anti-parental immunization does not affect the success of marsupial pregnancy. J. Reprod. Immuno., 8:249–256.

Roe, L. J., J. G. M. Thewissen, J. Quade, J. R. O'Neil, S. Bajpai, A. Sahni, and T. Hussain. 1998. Isotopic approaches to understanding the terrestrial-to-marine transition of the earliest cetaceans, 399–422, in *The Emergence of Whales* (J. G. M. Thewissen, ed.). Plenum Press, New York.

Roeder, K. D. 1965. Moths and ultrasound. Sci. Amer., 212(4):94–102.

Roeder, K. D., and A. E. Treat. 1961. The detection and evasion of bats by moths. Amer. Sci., 49(2):135–148.

Rogers, E. 1986. *Looking at Vertebrates: A Practical Guide to Vertebrate Adaptations*. Longman Group, Ltd., Essex, England.

Rogers, L. 1981. A bear in its lair. Nat. Hist., 90:64–70.

Rohde, K. 1992. Latitudinal gradients in species diversity: the search for the primary cause. Oikos 65: 514–527.

Romer, A. S. 1966. *Vertebrate Paleontology*. Univ. Chicago Press, Chicago.

Romer, A. S. and T. S. Parsons. 1977. *The Vertebrate Body.* Saunders, Philadelphia.

Rood, J. P. 1970a. Notes on the behavior of the pygmy armadillo. J. Mamm., 51:179.

Rood, J. P. 1970b. Ecology and social behavior of the desert cavy (*Microcavia australis*). Amer. Midl. Nat., 83:415–454.

Rood, J. P. 1972. Ecological and behavioral comparisons of three genera of Argentine cavies. Anim. Behav. Monogr., 5:1–83.

Rood, J. P. 1978. Dwarf mongoose helpers at the den. Z. Tierpsychol., 48:277–287.

Rood, J. P. 1980. Mating relationships and breeding suppression in the dwarf mongoose. Anim. Behav., 28:143–150.

Rose, K. D. 1982. Skeleton of *Diacodexis*, oldest known artiodactyl. Science, 216:621–623.

Rose, K. D. 1996. On the origin of the order Artiodactyla. Pro. Nat. Acad. Sci. USA, 93:1705–1709.

Rose, K. D., and R. J. Emry. 1993. Relationships of Xenarthra, Pholidota, and fossil "edentates": the morphological evidence, 81–102, in *Mammal Phylogeny: Placentals* (F. S. Szalay, M. J. Novacek, and M. C. McKenna, eds.). Springer-Verlag, New York.

Rose, K. D., and E. L. Simons, 1977. Dental function in the Plagiomenidae: origin and relationships of the mammalian order Dermoptera. Univ. Michigan Contr. Mus. Paleo., 24:221–236.

Rose, K. D., A. Walker, and L. L. Jacobs. 1981. Function of the mandibular tooth comb in living and extinct mammals. Nature, 289:583–585.

Rosenberg, H. I., and K. C. Richardson. 1995. Cephalic morphology of the honey possum, *Tarsipes rostratus* (Marsupialia: Tarsipedidae); an obligate nectarivore. J. Morph., 223:303–323.

Rosenberger, A. L. 1984. Aspects of the systematics and evolution of the marmosets, 159–180, in *Primatologica no Brazil* (M. T. deMello, ed.). Anais do 1. Congresso Brasileiro de Primatologia, Sociedad de Primatologia.

Rosenzweig, M. L. 1973. Habitat selection experiments with a pair of coexisting heteromyid species. Ecology, 54:111–117.

Rosenzweig, M. L. 1992. Species diversity gradients: we know more and less than we thought. J. Mamm., 73:715–730.

Rosenzweig, M. L. 1995. *Species Diversity in Space and Time.* Cambridge Univ. Press, New York.

Rosevear, D. R. 1969. *The Rodents of West Africa.* British Mus. Nat. Hist., London.

Ross, G. J. B., P. B. Best, and B. G. Donnelly. 1975. New records of the pygmy right whale (*Caperea marginata*) from South Africa, with comments on distribution, migration, appearance, and behavior. J. Fish. Res. Board Can., 32:1005–1017.

Rougier, G. W., J. R. Wible, and J. A. Hopson. 1996a. Basicranial anatomy of *Priacodon fruitaensis* (Triconodontidae, Mammalia) from the Late Jurassic of Colorado, and a reappraisal of mammaliaform relationships. Amer. Mus. Novitates, 3183:1–38.

Rougier, G. W., J. R. Wible, and M. J. Novacek. 1996b. Middle-ear ossicles of the multituberculate *Kryptobaatar* from the Mongolian Late Cretaceous: implications for mammaliamorph relationships and the evolution of the auditory apparatus. Amer. Mus. Novitates, 3187:1–43.

Rowan, W. 1950. Winter habits and numbers of timber wolves, J. Mamm., 31:167–169.

Rowe, M. J., and R. C. Bohringer. 1992. Functional organization of the cerebral cortex in monotremes, 177–193, in *Platypus and Echidnas* (M. L. Augee, ed.). Royal Zool. Soc. New South Wales, Sydney.

Rowe, T. 1988. Definition, diagnosis and origin of Mammalia. J. Vert. Paleo., 8(3):241–264.

Rowe, T. 1993. Phylogenetic systematics and the early history of mammals, 129–145, in *Mammal Phylogeny: Mesozoic Differentiation, Multituberculates, Monotremes, Early Therians, and Marsupials* (F. S. Szalay, M. J. Novacek, and M. C. McKenna, eds.). Springer-Verlag, New York.

Rowe, T. 1996a. Brain heterochrony and origin of the mammalian middle ear, 71–95, in *New Perspectives on the History of Life* (M. T. Ghiselin and G. Pinna, eds.). Memoirs of the California Academy of Sciences Number 20, California Academy of Sciences, San Francisco.

Rowe, T. 1996b. Coevolution of the mammalian middle ear and neocortex. Science, 273:651–654.

Rowe, T., and J. Gauthier. 1992. Ancestry, paleontology, and definition of the name Mammalia. Syst. Biol., 41:372–378.

Rowell, T. E. 1962. Agonistic noises of the rhesus monkey (*Macaca mulatta*). Symp. Zool. Soc. London, 8:91–96.

Rudran, R. 1973. Adult male replacement in one-male troops of purple-faced langurs (*Presbytis senex senex*) and its effect on population structure. Folia Primat., 19:166–192.

Rudran, R. 1979. Demography and social mobility in a red howler monkey (*A. seniculus*) population, in *Vertebrate Ecology in the Northern Neotropics* (J. F. Eisenberg, ed.). Smithsonian Inst. Press, Washington, D.C.

Ruff, F. J. 1938. Trapping deer on the Pisgah National Game Preserve, North Carolina. J. Wildl. Manag., 2.151–101.

Rumpler, Y., S. Corvella, and D. Montagnon. 1994. Systematic relationships among Cheirogaleidae (Primates, Strepsirhini) determined from analysis of highly repeated DNA. Folia Primatol., 63:149–155.

Ryan, J. M. 1986. Comparative morphology and evolution of cheek pouches in rodents. J. Morph., 190:27–41.

Ryan, J. M. 1989. Comparative myology and phylogenetic systematics of the Heteromyidae (Mammalia, Rodentia). Misc. Publ. Mus. Zool. Univ. Michigan, 176:1–103.

Rydell, J. 1990. Behavioral variation in echolocation pulses of the northern bat (*Eptesicus nilssoni*). Ethology, 85:103–113.

Rydell, J. 1993. Variation in the sonar of an aerial-hawking bat (*Eptesicus nilssoni*). Ethology, 93:275–284.

Rydell, J., and R. Arlettaz. 1994. Low-frequency echolocation enables the bat *Tadarida teniotis* to feed on tympanate insects. Proc. Royal Soc. London, 257B:175–178.

Rylands, A. B. 1993. *Marmosets and Tamarins: Systematics, Behavior, and Ecology*. Oxford Univ. Press, Oxford.

Sadleir, R. M. F. S., and C. H. Tyndale-Biscoe. 1977. Photoperiod and the termination of embryonic diapause in the marsupial *Macropus eugenii*. Biol. Reprod., 16:605–608.

Sahley, C. T., and L. E. Baraybar. 1994. The natural history and population status of the nectar-feeding bat, *Platalina genovensium* in southwestern Peru. Abstract of paper presented at the 24th annual North American Symposium on Bat Research, Ixtapa, Mexico. Bat Research News, 35:113.

Sale, J. B. 1970. The behavior of the resting rock hyrax in relation to its environment. Zool. Afr., 5:87–99.

Sampson, S. D., L. M. Witmer, C. A. Forster, D. W. Krause, P. M. O'Connor, P. Dodson, and F. Ravoavy. 1998. Predatory dinosaur remains from Madagascar: implications for the Cretaceous biogeography of Gondwana. Science, 280:1048–1051.

Sanchez-Villagra, M. R., and R. F. Kay. 1997. A skull of *Proargyrolagus*, the oldest argyrolagid (late Oligocene Salla Beds, Bolivia), with brief comments concerning its paleobiology. J. Vert. Paleo., 17:717–724.

Sanders, E. H., P. D. Gardner, P. J. Berger, and N. C. Negus. 1981. 6-Methoxybenzoxazalinone: a plant derivative that stimulates reproduction in *Microtus montanus*. Science, 214:67–69.

Santini-Palka, M. E. 1994. Feeding behavior and activity patterns of two Malagasy bamboo lemurs, *Hapalemur simus* and *Hapalemur griseus* in captivity. Folia Primatol., 63:44–49.

Sassone-Corsi, P. 1998. Molecular clocks: mastering time by gene regulation. Nature, 392:871–874.

Sargent, A. B., and D. W. Warner. 1972. Movements and denning habits of a badger. J. Mamm., 53:207–210.

Sarich, V. M. 1985. Rodent macromolecular systematics, 141–170, in *Evolutionary Relationships Among Rodents* (W. P. Luckett and J. L. Hartenberger, eds.). Plenum Press, New York.

Sarich, V. M., and J. E. Cronin. 1980. South American mammal molecular systematics, evolutionary clocks, and continental drift, 399–421, in *Evolutionary Biology of the New World Monkeys and Continental Drift* (R. L. Ciochon and A. B. Chiarelli, eds.). Plenum Press, New York.

Sauer, J. J. C., J. D. Skinner, and R. Neitz. 1982. Seasonal utilization of leaves by giraffes (*Giraffa camelopardalis*) and the relationship of the seasonal utilization to the chemical composition of leaves. S. African J. Zool., 17:210–219.

Saunders, J. K., Jr. 1963. Movements and activities of the lynx in Newfoundland. J. Wildl. Manag., 27:390–400.

Savage, D. E. 1951. A Miocene phyllostomatid bat from Colombia, South America. Univ. Calif. Pub. Geol. Sci., 28:357–365.

Schaefer, V. H., and R. M. F. S. Sadleir. 1979. Concentrations of carbon dioxide and oxygen in mole tunnels. Acta Theriol., 24:267–276.

Schaller, G. B. 1963. *The Mountain Gorilla: Ecology and Behavior*. Univ. Chicago Press, Chicago.

Schaller, G. B. 1965a. The behavior of the mountain gorilla, 324–367, in *Primate Behavior: Field Studies of Monkeys and Apes* (I. De Vore, ed.). Holt, Rinehart and Winston, New York.

Schaller, G. B. 1965b. *The Year of the Gorilla*. Ballantine Books, New York.

Schaller, G. B. 1972. *The Serengeti Lion: A Study of Predator-prey Relations*. Univ. Chicago Press, Chicago.

Schaller, G. B. 1993. *The Last Panda*. Univ. Chicago Press, Chicago.

Schaller, G., Hu Jinchu, Pan Wenshi, and Zhu Jing. 1985. *The Giant Pandas of Wolong*. Chicago Univ. Press, Chicago.

Schaller, G. B., and A. Rabinowitz. 1995. The saola or spindlehorn bovid *Pseudoryx nghetinhensis* in Laos. Oryx, 29:107–114.

Schaller, G. B., and E. S. Vrba. 1996. Description of the giant muntjac (*Megamuntiacus vuquangensis*) in Laos. J. Mamm., 77:675–683.

Scheffer, V. B. 1958. *Seals, Sea Lions, and Walruses*. Stanford Univ. Press, Stanford, California.

Scheibe, J. S. 1984. The effects of weather, sex and season on the nocturnal activity of *Peromyscus truei* (Rodentia). Southwestern Nat., 29:1–5.

Scheibe, J. S., and M. J. O'Farrell. 1995. Habitat dynamics in *Peromyscus truei*: eclectic females, density dependence, or reproductive constraints? J. Mamm., 76:368–375.

Scheich, H., G. Langner, C. Tidemann, R. B. Coles, and A. Guppy. 1986. Electroreception and electrolocation in platypus. Nature, 319:401–402.

Schenkel, R. 1967. Submission: its features and functions in the wolf and dog. Amer. Zool., 7:319–329.

Schevill, W. E., and B. Lawrence. 1949. Underwater listening to the white porpoise, *Delphinapterus leucas*. Science, 109:143–144.

Schevill, W. E., and B. Lawrence. 1953. Auditory response of a bottle nosed porpoise, *Tursiops truncatus*, to frequencies above 100 kc. J. Exp. Zool., 124:147–165.

Schevill, W. E., W. A. Wadkins, and C. Ray. 1963. Underwater sounds of pinnipeds. Science, 141:50–53.

Schevill, W. E., W. A. Wadkins, and C. Ray. 1966. Analysis of underwater *Odobenus* calls with remarks on the development and function of the pharyngeal pouches. Zoologica, 51:103–106.

Schmid, J., and P. M. Kappeler. 1994. Sympatric mouse lemurs (*Microcebus* spp.) in western Madagascar. Folia Primatol., 63:162–170.

Schmidt, S. 1988. Evidence for a spectral basis of texture perception in bat sonar. Nature, 331:617–619.

Schmidt-Nielsen, K. 1959. The physiology of the camel. Sci. Amer., 201:140–151.

Schmidt-Nielsen, K. 1964. *Desert Animals: Physiological Problems of Heat and Water.* Oxford Univ. Press, New York.

Schmidt-Nielsen, K. 1981. Countercurrent systems in animals. Sci. Amer., 244:118–128.

Schmidt-Nielsen, K., B. Schmidt-Nielsen, T. A. Houpt, and S. A. Jarnum. 1956. The question of water storage in the stomach of the camel. Mammalia, 20:1–15.

Schmidt-Nielsen, K., B. Schmidt-Nielsen, T. A. Houpt, and S. A. Jarnum. 1957. Body temperature of the camel and its relation to water economy. Amer. J. Physiol., 188:103–112.

Schmidt-Nielsen, K., and A. E. Newsome. 1962. Water balance in the mulgara *(Dasycercus cristicauda),* a carnivorous desert marsupial. Aust. J. Biol. Sci., 15:683–689.

Schmidt-Nielsen, K., and H. B. Haines. 1964. Water balance in a carnivorous desert rodent, the grasshopper mouse. Physiol. Zool., 37:259–263.

Schmidt-Nielsen, K., F. R. Hainsworth, and D. E. Murrish. 1970. Counter-current heat exchange in the respiratory passages: effects on water and heat balance. Resp. Physiol., 9:263–276.

Schmidt-Nielsen, K., R. C. Schroter, and A. Shkolnik. 1980. Desaturation of exhaled air in camels. Proc. Royal Soc. London, B211:305–319.

Schneider, K. M. 1930. Das Flehmen. Zool. Gart. (Leipzig), 3:183–198; 4:349–364; 5:200–226, 287–297.

Schneider, R., H. Jurg Kugn, and G. Kelemen, 1967. *Die Larynx der Hypsignathus monstrosus Allen 1861.* Ein Unifum in der Morphologie des Kehlkopfes. Z. Wiss. Zool.

Schnitzler, H.-U. 1987. Echos of fluttering insects: information for echolocating bats, 226–243, in *Recent Advances in the Study of Bats* (M. B. Fenton, P. Racey, and J. M. V. Rayner, eds.). Cambridge Univ. Press, Cambridge, U.K.

Schnitzler, H.-U., and A. D. Grinnell. 1977. Directional sensitivity of echolocation in the horseshoe bat, *Rhinolophus ferrumquinum.* I. Directionality of sound emission. J. Comp. Physiol., 116:51–61.

Schnitzler, H. U., D. Menne, R. Kober, and K. Heblich. 1983. The acoustical image of fluttering insects in echolocating bats, 235–250, in *Neuroethology and Behavioral Physiology: Roots and Growing Pains* (F. Huber and H. Markl, eds.). Springer, Berlin.

Schröpfer, R., B. Klemer-Fringes, and E. Naumer. 1985. Locomotion patterns and habitat utilisation of the two jerboas *Jaculus jaculus* and *Jaculus orientalis* (Rodentia, Dipodidae). Mammalia, 49:445–454.

Schubert, G. H. 1953. Ponderosa pine cone cutting by squirrels. J. Forestry, 51:202.

Schulte, B. A. 1998. Scent marking and responses to male castor fluid by beavers. J. Mamm., 79:191–203.

Schultz, A. M. 1965. The tundra as a homeostatic system. Presented at A.A.A.S. Meeting, Dec. 1965 (abstract).

Schultz, A. M. 1969. A study of an ecosystem: the Arctic tundra. In *The Ecosystem Concept in Natural Resource Management* (G. M. Van Dyne, ed.). Academic Press, New York.

Schultze-Westrum, T. 1965. Nochweis differenzierter duftstoffe beim gleitbeulter *Petaurus breviceps papuanus* Thomas (Marsupialia, Phalangeridae). Naturwiss., 51(9):226–227.

Schutt, W. A., Jr., and J. S. Altenbach. 1997. A sixth digit in *Diphylla ecaudata,* the hairy legged vampire bat (Chiroptera, Phyllostomidae). Mammalia, 61:280–285.

Schutt, W. A., Jr., and N. B. Simmons. 1998. Morphology and homology of the chiropteran calcar, with comments on the phylogenetic relationships of *Archaeopteropus.* J. Mamm. Evol., 5:1–32.

Schwartz, C. C., J. G. Nagy, and W. L. Regelin. 1980a. Juniper oil yield, terpenoid concentration, and antimicrobial effects on deer. J. Wildl. Manag., 44:107–113.

Schwartz, C. C., W. L. Regelin, and J. G. Nagy. 1980b. Deer preference for juniper forage and volatile oil treated food. J. Wildl. Manag., 44:114–120.

Schwartz, D. M. 1996. Snatching scientific secrets from the hippo's gaping jaws. Smithsonian, March:91–105.

Schwartz, J. H. 1988. *Aspects of the Biology of the Orangutan.* Oxford Univ. Press, Oxford.

Schwartz, J., and I. Tattersall. 1987. Tarsiers, adapids and the integrity of Strepsirhini. J. Human Evol., 16:23–40.

Scillato-Yané, G. J. 1976. Sobre un Dasypodidae (Mammalia, Xenarthra) de edad Riochiquense (Paleoceno Superior) de Itaboraí, Brasil. An. Acad. Brasil. Ciênc., 48:527–530.

Scott, K. M., and C. M. Janis. 1993. Relationships of the Ruminantia (Artiodactyla) and an analysis of the characters used in ruminant taxonomy, 282–302, in *Mammal Phylogeny, Placentals* (F. S. Szalay, M. J. Novacek, and M. C. McKenna, eds.). Vol. 2. Springer-Verlag, New York.

Sebeok, T. A. 1977. *How Animals Communicate.* Indiana University Press, Bloomington, London.

Seely, M. K. 1979. Ecology of a living desert: twenty years of research in the Namib. S. Africa. J. Sci., 75:298–303.

Seely, M. K. 1981. Desert plants use fog water. Sci. Prog., 14:4.

Seely, M. K. 1987. *The Namib.* Shell Namibia Ltd.

Seligsohn, D., and F. S. Szalay. 1974. Dental occlusion and the masticatory apparatus in *Lemur* and *Varecia:* their bearing on the systematics of living and fossil primates, 543–562, in *Prosimian Biology* (R. D. Martin, G. A. Doyle, and A. C. Walker, eds.). Duckworth, London.

Selye, H. 1955. Stress and disease. Science, 122:625–631.

Selye, H., and J. B. Collip. 1936. Fundamental factors in the interpretation of stimuli influencing the endocrine glands. Endocrinology, 20:667–672.

Sewell, G. D. 1968. Ultrasound in rodents. Nature, 217:682–683.

Seyfarth, R. M., D. L. Cheney, and P. Marler. 1980. Vervet monkey alarm calls: semantic communications in a free-ranging primate. Anim. Behav., 28:1070–1094.

Seymour, R. S., P. C. Withers, and W. W. Weathers. 1998. Energetics of burrowing, running, and free-living in the Namib Desert golden mole (*Eremitalpa namibensis*). J. Zool., London, 244:107–117.

Sharman, G. B. 1970. Reproductive physiology of marsupials. Science, 167:1221–1228.

Shaw, W. T. 1936. Moisture and its relation to the cone-storing habit of the western pine squirrel. J. Mamm., 17:337–349.

Sheftel, B. I. 1994. Spatial distribution of nine species of shrews in the central Siberian taiga, 45–56, in *Advances in the Biology of Shrews* (J. F. Merritt, G. L. Kirkland, Jr., and R. K. Rose, eds.). Spec. Pub. No. 18, Carnegie Mus. Nat. Hist., Pittsburgh.

Sheldrick, D. 1972. Death of the Tsavo elephant. Sat. Rev., Sept. 30, p. 29.

Shepard, P. 1996. *Traces of an Omnivore*. Island Press, Washington, D.C.

Shepherd, U. L. 1998. A comparison of species diversity and morphological diversity across the North American latitudinal gradient. J. Biogeogr., 25:19–29.

Sherman, P. W. 1981a. Reproductive competition and infanticide in Belding's ground squirrels and other animals, 311–331, in *Natural Selection and Social Behaviour: Recent Research and New Theory* (R. D. Alexander and Q. W. Tinkle, eds.). Chiron Press, New York.

Sherman, P. W. 1981b. Kinship, demography and Belding's ground squirrel nepotism. Behav. Ecol. Sociobiol., 8:251–259.

Sherman, P. W. 1985. Alarm calls of Belding's ground squirrels to aerial predators: nepotism or self-preservation? Behav. Ecol. Sociobiol., 17:313–323.

Sherman, P. W., J. U. M. Jarvis, and R. D. Alexander. 1991. *The Biology of the Naked Mole-rat*. Princeton Univ. Press, New Jersey.

Shimamura, M., H. Yasue, K. Phshima, H. Abe, H. Kato, T. Kishiro, et al. 1997. Molecular evidence from retroposons that whales form a clade within even-toed ungulates. Nature, 388:666–670.

Shipley, K., M. Hines, and J. S. Buchwald. 1981. Individual differences in threat calls of northern elephant seal bulls. Anim. Behav., 29:12–19.

Shirer, H. W., and H. S. Fitch. 1970. Comparison from radiotracking of movements and denning habits of the raccoon, striped skunk, and opossum in northeastern Kansas. J. Mamm., 51:491–503.

Shkolnik, A., and A. Borut. 1969. Temperature and water relations in two species of spiny mice (*Acomys*). J. Mamm., 50:245–255.

Short, J., and A. Smith. 1994. Mammal decline and recovery in Australia. J. Mamm., 75:288–297.

Short, J., S. D. Bradshaw, J. Giles, R. I. T. Prince, and G. R. Wilson. 1992. Reintroduction of macropods (Marsupialia: Macropodoidea) in Australia—a review. Biol. Conser., 62:189–204.

Shoshani, J. 1986. Mammalian phylogeny: comparison of morphological and molecular results. Mol. Biol. Evol., 3:222–242.

Sibley, C. G., and J. E. Ahlquist. 1984. The phylogeny of the hominoid primates as indicated by DNA–DNA hybridization. J. Mol. Evol., 20:2–15.

Sibley, C. G., and J. E. Ahlquist. 1987. DNA hybridization evidence of hominoid phylogeny: results from an expanded data set. J. Mol. Evol., 26:99–121.

Sigé, B. 1971. Anatomie du membre anterieur chez un chiroptere molosside (*Tadarida* sp.) du Stampien de Cereste (Alpes-de-Haute-Provence). Palaeovertebrata, 4:1.

Sigé, B. 1974. Données nouvelles sur le genre *Stehlinia* (Vespertilionoidea, Chiroptera) du Paléogène d'Europe. Palaeovertebrata, 6:253–272.

Sigé, B., J.-Y. Crochet, and A. Insole. 1977. Les plus vieilles taupes. Géobios, Mém. Spéc., 1:141–157.

Sigé, B., H. Thomas, S. Sen, E. Gheerbrant, J. Roger, and Z. Al-Sulaimani. 1994. Les chiroptères de Taqah (Oligocene inférieur, Sultanat d'Oman). Premier inventaire systématique. Münchner Geowiss. Abhandlungen, 26:35–48.

Sigogneau-Russell, D. 1989. Discovery of the first Symmetrodonta (Mammalia) from the African continent. C. R. Acad. Sci. Paris, 309:921–926.

Sigogneau-Russell, D. 1991. New therian mammals from the Early Cretaceous of Morocco. C. R. Acad. Sci. Paris, 313:279–285.

Sigogneau-Russell, D. 1995. Two possibly aquatic triconodont mammals from the Early Cretaceous of Morocco. Acta Palaeontologica Polonica, 40:149–162.

Sigogneau-Russell, D., D. Dashzeveg, and D. E. Russell, 1992. Further data on *Prokennalestes* (Mammalia, Eutheria *inc. sed.*) from the Early Cretaceous of Mongolia. Zool. Scripta, 21:205–209.

Sigogneau-Russell, D., and P. Ensom. 1997. *Thereuodon* (Theria, Symmetrodonta) from the Lower Cretaceous of North Africa and Europe, and a brief review of symmetrodonts. Cretaceous Res., 19:1–26.

Siivonen, L. 1954. Features of short-term fluctuations. J. Wildl. Manag., 18:38–45.

Sikes, R. S. 1996. Tactics of maternal investment of northern grasshopper mice in response to postnatal restriction of food. J. Mamm., 77:1092–1101.

Silverman, H. G., and M. H. Dunbar. 1980. Aggressive tusk use by the narwhal (*Monodon monoceros*). Nature, 284:57–58.

Simberloff, D. S., and W. Boecklen. 1981. Santa Rosalia reconsidered: size ratios and competition. Evolution, 35:1206–1228.

Simmons, J. A. 1973. The resolution of target range by echolocating bats. J. Acoust. Soc. Amer., 54:157–173.

Simmons, J. A. 1974. Response of the doppler echolocation system in the bat, *Rhinolophus ferrumequinum.*
J. Acoust. Soc. Amer., 56:672–682.

Simmons, J. A., W. A. Lavender, B. A. Lavender, C. A.
Doroshow, S. W. Kiefer, R. Livingston, A. C. Scallet,
and D. E. Crowley. 1974. Target structure and echo
spectral discrimination by echolocating bats. Science,
186:1130–1132.

Simmons, J. A., D. J. Howell, and N. Suga. 1975. Information content of bat sonar echoes. Amer. Sci.,
63:204–215.

Simmons, J. A., W. A. Lavender, B. A. Lavender, J. E.
Childs, K. Huleback, M. R. Rigden, et al. 1978. Echolocation by free-tailed bats (*Tadarida*). J. Comp. Physiol.,
125:291–299.

Simmons, J. A., M. B. Fenton, and M. J. O'Farrell. 1979.
Echolocation and pursuit of prey by bats. Science,
203:16–21.

Simmons, J. A., and R. A. Stein. 1980. Acoustic imaging in
bat sonar: echolocation signals and the evolution of
echolocation. J. Comp. Physiol., 135:61–84.

Simmons, J. A., S. A. Kick, and B. D. Lawrence. 1984.
Echolocation and hearing in the mouse-tailed bat,
Rhinopoma hardwickei: acoustic evolution of echoloca-
tion in bats. J. Comp. Physiol. A, 154:347–356.

Simmons, N. B. 1993a. The importance of methods: Archontan phylogeny and cladistic analysis of morphological data, 1–61, in *Primates and Their Relatives in Phylogenetic Perspective* (R. D. E. MacPhee, ed.). Advances
in Primatology Series, Plenum Press, New York.

Simmons, N. B. 1993b. Morphology, function, and phylogenetic significance of pubic nipples in bats (Mammalia: Chiroptera). Amer. Mus. Novitates,
3077:1–37.

Simmons, N. B. 1994. The case for chiropteran monophyly. Amer. Mus. Novitates, 3103:1–54.

Simmons, N. B. 1995. Bat relationships and the origin of
flight. Symp. Zool. Soc. London, 67:27–43.

Simmons, N. B. 1998. A reappraisal of interfamilial relationships of bats, 3–26, in *Bat Biology and Conservation Biology* (T. H. Kunz and P. A. Racey, eds.).
Smithsonian Inst. Press, Washington, D.C.

Simmons, N. B., M. J. Novacek, and R. J. Baker. 1991. Approaches, methods, and the future of the chiropteran
monophyly controversy: a reply to J. D. Pettigrew. Syst.
Zool., 40:239–243.

Simmons, N. B., and T. H. Quinn. 1994. Evolution of the
digital locking mechanism in bats and dermopterans:
a phylogenetic perspective. J. Mamm. Evol., 2:231–254.

Simmons, N. B., and J. H. Geisler. 1998. Phylogenetic relationships of *Icaronycteris, Archaeonycteris, Hassianycteris,*
and *Palaeochiropteryx* to extant bat lineages, with comments on the evolution of echolocation and foraging
strategies in Microchiroptera. Bull. Amer. Mus. Nat.
Hist., 235:1–182.

Simons, R. S. 1996. Lung morphology of cursorial and
non-cursorial mammals: lagomorphs as a case study

for a pneumatic stabilization hypothesis. J. Morph.,
230:299–316.

Simpson, G. G. 1937. Skull structure of the Multituberculata. Bull. Amer. Mus. Nat. Hist., 73:727–763.

Simpson, G. G. 1940. Mammals and land bridges. J. Washington Acad. Sci. 30:137.

Simpson, G. G. 1945. The principles of classification and a
classification of mammals. Bull. Amer. Mus. Nat. Hist.,
85:1–350.

Simpson, G. G. 1950. History of the fauna of Latin America. Amer. Sci., 38:361.

Simpson, G. G. 1951. *Horses.* Oxford Univ. Press, New York.

Simpson, G. G. 1964. Species density of North American
Recent mammals. Syst. Zool., 13:57–73.

Simpson, G. G. 1965a. *Attending Marvels: A Patagonian Journal.* Time, New York.

Simpson, G. G. 1965b. *The Geography of Evolution.* Chilton
Books, Philadelphia.

Simpson, G. G. 1969. South American mammals, in *Biogeography and Ecology in South America.* (E. J. Fihkan et
al., eds.). Mono. Biol., 19. W. Junk, The Hague.

Simpson, G. G. 1970. The Argyrolagidae, extinct South
American marsupials. Bull. Mus. Comp. Zool.,
139:1–86.

Simpson, G. G. 1980. *Splendid Isolation: The Curious History
of South American Mammals.* Yale Univ. Press, New
Haven.

Simpson, C. D. 1984. Artiodactyls, 563–587, in *Orders and
Families of Recent Mammals of the World* (S. Anderson
and J. K. Jones, eds.). John Wiley & Sons, New York.

Sinclair, A. R. E. 1970. Studies of the ecology of the East
African buffalo. Ph.D. thesis, Oxford Univ., Oxford.

Sinclair, A. R. E. 1974. The social organization of the East
African buffalo, 676–689, in *The Behavior of Ungulates
and its Relation to Management* (V. Geist and F. R.
Walther, eds.). International Union for Conservation
of Nature and Natural Resources, Morges, Switzerland.

Sinclair, A. R. E. 1977. *The African Buffalo, A Study in Resource Limitation of Populations.* Univ. Chicago Press,
Chicago.

Skinner, J. D., S. Davis, and G. Ilani. 1980. Bone collecting
by striped hyaenas, *Hyaena hyaena,* in Israel. Paleontol.
Afr., 23:99–104.

Skinner, J. D., and R. H. N. Smithers. 1990. *The Mammals
of the Southern African Subregion.* University of Pretoria,
Pretoria, Republic of South Africa.

Skoczen, S. 1958. Tunnel digging by the mole (*T. europaea*
Linne). Acta Theriol., 2:235–249.

Slijper, E. J. 1962. *Whales.* Basic Books, New York.

Slijper, E. J. 1979. *Whales.* Cornell Univ. Press, Ithaca, New
York.

Sliwa, A. and P. R. K. Richardson. 1998. Responses of aardwolves, *Proteles cristatus,* Sparrman 1783, to translo-
cated scent marks. Anim. Behav., 56:137–146.

Slobodchikoff, C. N., and R. Coast. 1980. Dialects in the
alarm calls of prairie dogs. Behav. Ecol. Sociobiol.,
7:49–53.

Slobodchikoff, C. N., and W. S. Schulz. 1988. Cooperation, aggression, and the evolution of social behavior, 13–32, in *The Ecology of Social Behavior* (C. N. Slobodchikoff, ed.). Academic Press, New York.

Smith, A. B. 1994. *Systematics and the Fossil Record: Documenting Evolutionary Patterns*. Blackwell Scientific Publications, Boston.

Smith, A. G., D. G. Smith, and B. M. Funnell. 1994. *Atlas of Mesozoic and Cenozoic Coastlines*. Cambridge Univ. Press, New York.

Smith, A. P., and D. G. Quin. 1996. Patterns and causes of extinction and decline in Australian conilurine rodents. Biol. Conserv., 77:243–267.

Smith, A. T. 1974. The distribution and dispersal of pikas: consequences of insular population structure. Ecology, 55:1112–1119.

Smith, B. L., R. L. Robbins, and S. H. Anderson. 1996. Adaptive sex ratios: another example? J. Mamm., 77:818–825.

Smith, C. C., and O. J. Reichman. 1984. The evolution of food caching by birds and mammals. Ann. Rev. Ecol. Syst., 15:329–351.

Smith, G. W., and D. R. Johnson. 1985. Demography of a Townsend ground squirrel population in southwestern Idaho. Ecology, 66:171–178.

Smith, H. M. 1960. *Evolution of Chordate Structure*. Holt, Rinehart, and Winston, New York.

Smith, R. B. 1971. Seasonal activities and ecology of terrestrial vertebrates in a Neotropical monsoon environment. M.S. thesis, Northern Arizona University, Flagstaff.

Smith, R. H. 1975. Nitrogen metabolism in the rumen and the comparative and nutritive value of nitrogen compounds entering the duodenum, in *Digestion and Metabolism in the Ruminant* (I. W. McDonald and A. C. I. Warner, eds.). Univ. New England, Armidale, New South Wales, Australia.

Smithers, R. H. N. 1971. The mammals of Botswana. Museum Memoirs, National Museums and Monuments of Rhodesia, 4:1–340.

Smythe, N. 1970. On the existence of "pursuit invitation" signals in mammals. Amer. Midl. Nat., 104:491–494.

Smythe, N. 1978. The natural history of the Central American agouti (*Dasyprocta punctata*). Smithsonian Contrib. Zool., 257:1–52.

Snyder, M. A. 1993. Interactions between Abert's squirrel and ponderosa pine: the relationship between selective herbivory and host plant fitness. Amer. Nat., 141:866–879.

Snyder, M. A., and Y. B. Linhart. 1994. Nest-site selection by Abert's squirrel: chemical characteristics of nest trees. J. Mamm., 75:136–141.

Solounias, N. 1997. Remarkable new findings regarding the evolution of the giraffe neck. J. Vert. Paleo., 17(supplement to 3):78A.

Sondaar, P. Y. 1977. *Insularity and Its Effects on Mammal Evolution*. NATO Advanced Study Institute, Plenum Press, New York.

Sorenson, M. W., and C. H. Conaway. 1968. The social and reproductive behavior of *Tupaia montana* in captivity. J. Mamm., 49:502–512.

Soulé, M. E. 1986. *Conservation Biology: The Science of Scarcity and Diversity*. Sinauer Associates, Sunderland, Mass.

Soulé, M. E. 1991. Conservation: Tactics for a constant crisis. Science, 253:744–750.

Southern, H. N. 1964. *Handbook of British Mammals*. Blackwell, Oxford.

Sowls, L. K. 1974. Social behavior of the collared peccary *Dicotyles tajacu*, in *The Behavior of Ungulates and its Relation to Management* (V. Geist and F. R. Walther, eds.). International Union for Conservation of Nature and Natural Resources, Morges, Switzerland.

Speakman, J. R. 1993. The evolution of echolocation for predation. Symp. Zool. Soc. London, 65:39–63.

Spencer, A. W., and H. W. Steinhoff. 1968. An explanation of geographic variation in litter size. J. Mamm., 49:281–286.

Spencer, D. A. 1958a. Preliminary investigations on the northwestern *Microtus* irruption. U.S. Fish and Wildl. Serv., Denver Wildl. Res. Lab. Spec. Report.

Spencer, D. A. 1958b. Biological and control aspects, in *The Oregon Meadow Mouse Irruption of 1957–1958*. Federal Cooperative Extension service, Oregon State College, Corvallis.

Spitz, F. 1963. Étude des densities de population de *Microtus arvalis*. Pall. A Saint-Michal-en-L'Hern (Vendu). Mammalia, 27:497–531.

Spotorno, A. E., J. C. Marin, M. Yévenes, L. I. Walker, R. Fernández-Donoso, J. Pincheira, et al. 1997. Chromosome divergences among American marsupials and the Australian affinities of the American *Dromiciops*. J. Mamm. Evol., 4:259–269.

Sprankel, H. 1965. Untersuchungen an *Tarsius*. I: morphologie des schwanzes nebst ethologischen bemerkungen. Folia Primat., 3:153–188.

Springer, M. S. 1997. Molecular clocks and the timing of the placental and marsupial radiations in relation to the Cretaceous–Tertiary boundary. J. Mamm. Evol., 4:285–301.

Springer, M. S., and J. A. W. Kirsch. 1993. A molecular perspective on the phylogeny of placental mammals based on mitochondrial 12S rDNA sequences, with special reference to the problem of the Paenungulata. J. Mamm. Evol., 1:149–166.

Springer, M. S., M. Westerman, and J. A. W. Kirsch. 1994. Relationships among orders and families of marsupials based on 12S ribosomal DNA sequences and the timing of marsupial radiation. J. Mamm. Evol., 2:85–115.

Springer, M. S., J. A. W. Kirsch, and J. A. Case. 1997. The chronicle of marsupial evolution, in *Molecular Evolution and Adaptive Radiation* (T. J. Givnish and K. J. Sytsma, eds.). Cambridge Univ. Press, Cambridge.

Stace, C. A. 1989. Dispersal versus vicariance: no contest. J. Biogeogr., 16:200–201.

Stahl, W. R. 1967. Scaling of respiratory variables in mammals. J. Appl. Physiol., 22:453–460.

Stanford, C. B. 1995. Chimpanzee hunting behavior and human evolution. Amer. Sci., 83:256–261.

Stanford, C. B., J. Wallis, H. Matama, and J. Goodall. 1994. Patterns of predation by chimpanzees on red colobus monkeys in Gombe National Park, Tanzania, 1982–1991. Am. J. Phys. Anthro., 94:213–228.

Stanhope, M. J., J. Czelusniak, J.-S. Si, J. Nickerson, and M. Goodman. 1992. A molecular perspective on mammalian evolution from the gene encoding interphotoreceptor retinoid binding protein, with convincing evidence for bat monophyly. Molec. Phylo. Evol., 1:148–160.

Stanhope, M. J., W. J. Bailey, J. Czelusniak, M. Goodman, J. S. Si, J. Nickerson, J. G. Sgouros, G. A. M. Singer, and T. K. Kleinschmidt. 1993. A molecular view of primate supraordinal relationships from the analysis of both nucleotide and amino acid sequences, 251–292, in *Primates and Their Relatives in Phylogenetic Perspective* (R. D. E. MacPhee, ed.). Plenum Press, New York.

Stebbins, G. L. 1974. *Flowering Plants: Evolution Above the Species Level.* Harvard Univ. Press, Cambridge, Massachusetts.

Stebbins, L. L. 1984. Overwintering activity of *Peromyscus maniculatus, Clethrionomys gapperi, C. rutilus, Eutamias amoenus,* and *Microtus pennsylvanicus.* 301–314, in *Winter Ecology of Small Mammals* (J. F. Merritt, ed.). Spec. Pub. No. 10, Carnegie Mus. Nat. Hist. Pittsburgh, PA.

Stehli, F. G., and S. D. Webb. 1985. *The Great American Biotic Interchange.* Plenum Press, New York.

Stehn, R. A., and F. J. Jannett, Jr. 1981. Male-induced abortion in various microtine rodents. J. Mamm., 62:369–372.

Stein, B. R. 1990. Limb myology and phylogenetic relationships in the superfamily Dipodoidea (birch mice, jumping mice, and jerboas). Z. Zool. Syst. Evolut.-forsch., 28:299–314.

Steinberg, P. D., A. A. Estes, and F. C. Winter. 1995. Evolutionary consequences of food chain length in kelp forest communities. Proc. Nat. Acad. Sci. USA, 92:8145–8148.

Steinberg, T. 1995. *Slide Mountain, or the Folly of Owning Nature.* Univ. California Press, Berkeley.

Stenlund, M. H. 1955. *A Field Study of the Timber Wolf (Canis lupus) on the Superior National Forest, Minnesota.* Minn. Dept. Cons. Tech. Bull. 4.

Stenseth, N. C. 1980. Spatial heterogeneity and population stability: some evolutionary consequences. Oikos, 35:165–184.

Stenseth, N. C. 1985. Mathematical models of microtine cycles: models and the real world. Acta Zool. Fennica, 173:7–12.

Stenseth, N. C., and W. Z. Lidicker, Jr. 1992. The study of dispersal: a conceptual guide, 5–20, in *Animal Dispersal, Small Mammals as a Model* (N. C. Stenseth and W. Z. Lidicker, Jr., eds.). Chapman and Hall, London.

Stenseth, N. C., and R. A. Ims. 1993. *The Biology of Lemmings.* The Linnean Society of London, London, U.K.

Stephenson, A. B. 1969. Temperatures within a beaver lodge in winter. J. Mamm., 50:134–136.

Sterling, I. 1969. Ecology of the Weddell seal in McMurdo Sound, Antarctica. Ecology, 50:573–586.

Stevenson-Hamilton, J. 1947. *Wildlife in South Africa.* Cassell, London.

Stewart, B. S. 1997. Ontogeny of differential migration and sexual segregation in northern elephant seals. J. Mamm., 78:1101–1116.

Stewart, B. S., and R. L. DeLong. 1995. Double migrations of the northern elephant seal, *Mirounga angustirostris.* J. Mamm., 76:196–205.

Stock, C. 1949. Rancho La Brea: a record of Pleistocene life in California. Los Angeles County Mus., Sci. Ser., no. 13, p. 1.

Stone, G. N., and A. Purvis. 1992. Warm-up rates during arousal from torpor in heterothermic mammals: physiological correlates and a comparison with heterothermic insects. J. Comp. Physiol. B, 162:284–295.

Storer, T. I., and R. L. Usinger, 1965. *General Zoology.* McGraw-Hill, New York.

Storrs, E., H. P. Burchfield, and R. J. W. Rees. 1989. Reproduction delay in the common long-nosed armadillo, *Dasypus novemcinctus* L, in *Advances in Neotropical Mammalogy* (K. H. Redford and J. F. Eisenberg, eds.). Sandhill Crane Press, Gainesville, Florida.

Strahan, R. 1995. *Mammals of Australia.* Smithsonian Inst. Press, Washington, D.C.

Strickler, T. L. 1980. Downstroke muscle histochemistry in two bats. 61–68 in *Proceedings of the fifth International Bat Research Conference* (D. E. Wilson and A. L. Gardner, eds.). Texas Tech Press, Lubbock.

Struhsaker, T. T. 1967. Behavior of elk (*Cervus canadensis*) during the rut. Z. Tierpsychol., 24(1):80–114.

Struhsaker, T. T. 1977. Infanticide and social organization in the redtail monkey (*Cercopithecus ascanius schmidti*) in the Kibale Forest, Uganda, Z. Tierpsychol., 4:75.

Struhsaker, T. T., and L. Leland. 1979. Socioecology of five sympatric monkey species in the Kibale Forest, Uganda, 159–228, in *Advances in the Study of Behavior.* Vol. 9. Academic Press, New York.

Strum, S. C. 1981. Processes and products of change: baboon predatory behavior at Gilgil, Kenya, 255–302, in *Omnivorous Primates: Gathering and Hunting in Human Evolution* (R. S. O. Harding and G. Teleki, eds.). Columbia Univ. Press, New York.

Studier, E. H., and D. J. Howell. 1969. Heart rate of female big brown bats in flight. J. Mamm., 50:842–845.

Sudman, P. D., L. J. Barkley, and M. S. Hafner. 1994. Familial affinity of *Tomopeas ravus* (Chiroptera) based on protein electrophoretic and cytochrome B sequence data. J. Mamm., 75:365–377.

Sudre, J. 1979. Nouveaux mammiferes Eocene du Sahara occidental. Palaeovertebrata, 9:83–115.

Suga, N., and T. Shimozawa. 1974. Site of neural attenuation of responses to self-vocalized sounds in echolocating bats. Science, 183:1211–1213.

Sugiyama, Y. 1973. Social organization of wild chimpanzees, in *Behavioral Regulators of Behavior in Primates* (C. R. Carpenter, ed.). Bucknell Univ. Press, Lewisburg, Pa.

Sullivan, J., and D. L. Swofford. 1997. Are guinea pigs rodents? the importance of adequate models in molecular phylogenetics. J. Mamm. Evol., 4:77–86.

Sullivan, R. M. 1994. Micro-evolutionary differentiation and biogeographic structure among coniferous forest populations of the Mexican woodrat (*Neotoma mexicana*) in the American Southwest: a test of the vicariance hypothesis. J. Biogeogr., 21:369–389.

Sussman, R. W., and P. H. Raven. 1978. Pollination by lemurs and marsupials: an archaic coevolutionary system. Science, 200:731–736.

Sussman, R. W., and W. G. Kinzey. 1984. The ecological role of the Callitrichidae: a review. Amer. J. Phys. Anthro., 64:419–449.

Suthers, R. A. 1965. Acoustic orientation by fish-catching bats. J. Exp. Zool., 158:319–348.

Suthers, R. A. 1967. Comparative echolocation by fishing bats, J. Mamm., 48:79–87.

Suthers, R. A., and J. M. Fattu. 1982. Selective laryngeal neurotomy and the control of phonation by the echolocating bat, *Eptesicus*. J. Comp. Physiol., 145:529–537.

Swank, W. G. 1958. *The Mule Deer in Arizona Chaparral.* Arizona Game and Fish Dept., Wildl. Bull. No. 3.

Swartz, S. M., M. B. Bennett, and D. R. Carrier. 1992. Wing bone stresses in free flying bats and the evolution of skeletal design for flight. Nature, 359:726–729.

Sweeney, R. C. H. 1956. Some notes on the feeding habits of the ground pangolin *Smutsria temmincki* (Smuts). Ann. Mag. Nat. Hist., London, 12:893–896.

Swinhoe, R. 1870. On the mammals of Hainan. Proc. Zool. Soc. London, 1870:224–239.

Swofford, D. L. 1998. *PAUP 4.0 Phylogenetic Analysis Using Parsimony (and other methods).* Computer software and manual, Sinauer Associates.

Szalay, F. S. 1977. Phylogenetic relationships and a classification of the eutherian Mammalia, 315–374 in *Major Patterns in Vertebrate Evolution* (M. K. Hecht, P. C. Goody, and B. M. Hecht, eds.). Plenum Press, New York.

Szalay, F. S. 1982. A new appraisal of marsupial phylogeny and classification, 612–640, in *Carnivorous Marsupials* (M. Archer, ed.). Sydney, Royal Zool. Soc. New South Wales.

Szalay, F. S. 1993. Metatherian taxon phylogeny: evidence and interpretation from the cranioskeletal system, 216–242, in *Mammal Phylogeny: Mesozoic Differentiation,* *Multituberculates, Monotremes, Early Therians, and Marsupials* (F. S. Szalay, M. J. Novacek, and M. C. McKenna, eds.). New York, Springer-Verlag.

Szalay, F. S. 1994. *Evolutionary History of the Marsupials and an Analysis of Osteological Characters.* New York, Cambridge Univ. Press.

Szalay, F. S., and E. Delson. 1979. *Evolutionary History of the Primates.* Academic Press, New York.

Szalay, F. S., M. J. Novacek, and M. C. McKenna. (eds.). 1993a. *Mammal Phylogeny. Mesozoic Differentiation, Multituberculates, Monotremes, Early Therians, and Marsupials.* Springer-Verlag, New York.

Szalay, F. S., M. J. Novacek, and M. C. McKenna (eds.). 1993b. *Mammal Phylogeny. Placentals.* Springer-Verlag, New York.

Szalay, F. S., and B. A. Trofimov. 1996. The Mongolian Late Cretaceous *Asiatherium,* and the early phylogeny and paleobiogeography of Metatheria. J. Vert. Paleo. 16:474–509.

Taber, A. B., C. P. Doncaster, N. N. Neris, and F. H. Colman. 1993. Ranging behavior and population dynamics of the Chacoan peccary, *Catagonus wagneri.* J. Mamm., 74:443–454.

Taber, R. D., and R. F. Dasmann. 1957. The dynamics of three natural populations of deer *Odocoileus hemionus columbianus.* Ecology, 38:233–246.

Talamantes, F. 1975. Comparative study of the occurrence of placental prolactin among mammals. Gen. Comp. Endocrin., 27:115–121.

Talbot, L. M., and M. H. Talbot. 1963. *The Wildebeest in Western Masailand.* Wildl. Monogr., No. 12. The Wildlife Society, Washington, D.C.

Tamura, N. 1993. Role of sound communication in mating of Malaysian *Callosciurus* (Sciuridae). J. Mamm., 74:468–476.

Tassy, P., and J. Shoshani. 1988. The Tethytheria: elephants and their relatives, 283–316, in *The Phylogeny and Classification of the Tetrapods.* Vol. 2. (M. J. Benton, ed.). Oxford Univ. Press, Oxford.

Tate, G. H. H. 1933. A systematic revision of the marsupial genus *Marmosa.* Bull. Amer. Mus. Nat. Hist., vol. 66.

Tate, G. H. H., and R. Archbold. 1937. Results of the Archbold expeditions. 16: some marsupials of New Guinea and Celebes. Bull. Amer. Mus. Nat. Hist., 73:331–476.

Tattersall, I. 1982. *The Primates of Madagascar.* Columbia Univ. Press, New York.

Tattersall, I., and J. H. Schwartz. 1974. Craniodental morphology and the systematics of the Malagasy lemurs (Primates, Prosimii). Anthropol. Papers Amer. Mus. Nat. Hist., 52:141–192.

Taulman, J. F., and L. W. Robbins. 1996. Recent range expansion and distributional limits of the nine-banded armadillo (*Dasypus novemcinctus*) in the United States. J. Biogeogr., 23:635–648.

Taylor, C. R. 1968a. Hygroscopic food: a source of water for desert antelopes? Nature, 219:181–182.

Taylor, C. R. 1968b. The minimum water requirements of some East African bovids. Symp. Zool. Soc. London, 21:195.

Taylor, C. R. 1969. The eland and the oryx. Sci. Amer., 220:89–95.

Taylor, C. R. 1972. The desert gazelle: a paradox resolved, in *Comparative Physiology of Desert Animals* (G. M. O. Maloiy, ed.). Symp. Zool. Soc. London, 31. Academic Press, New York.

Taylor, C. R. 1974. Exercise and thermoregulation, in *Environmental Physiology* (D. Robertshaw, ed.). Internat. Rev. Science, Environ. Physiol., Butterworth, London.

Taylor, C. R., K. Schmidt-Nielsen, and J. L. Raab. 1970. Scaling of energetic cost of running to body size in mammals. Amer. J. Physiol., 219:1104–1107.

Taylor, C. R., K. Schmidt-Nielsen, R. Dmi'el, and M. Fedak. 1971. Effect of hyperthermia on heat balance during running in the African hunting dog. Amer. J. Physiol., 220:823–827.

Taylor, C. R., and C. P. Lyman. 1972. Heat storage in running antelopes: Independence of brain and body temperatures. Amer. J. Physiol., 222:114–117.

Taylor, C. R., N. C. Heglund, and G. M. O. Maloiy. 1982. Energetics and mechanics of terrestrial locomotion. I: metabolic energy consumption as a function of speed and body size in birds and mammals. J. Exp. Biol., 97:1–21.

Taylor, E., J. R. Potter, M. Chitre. 1997. Ambient noise imaging potential of marine mammals. Proceedings of the Underwater Bio-Sonar and Bioacoustics Symposium, Loughborough, U.K.

Taylor, J. M., and H. A. Padykula. 1978. Marsupial trophoblast and mammalian evolution. Nature, 271:588.

Taylor, T. N., and E. L. Taylor. 1990. *Antarctic Paleobiology. Its Role in the Reconstruction of Gondwana.* Springer-Verlag, New York.

Tedford, R. H. 1967. The fossil Macropodidae from Lake Menindee, New South Wales. Univ. Calif. Publ. Geol. Sci., 64:1–165.

Tedford, R. H. 1976. Relationships of pinnipeds to other carnivores (Mammalia). Syst. Zool., 25:363–374.

Teerink, B. J. 1991. *Hair of West-European Mammals: Atlas and Identification Key.* Cambridge, Cambridge Univ. Press.

Tejedor, M. F. 1998. The evolutionary history of platyrrhines: old controversies and new interpretations. Neotropical Primates, 6:77–82.

Teleki, G. 1973. *The Predatory Behavior of Wild Chimpanzees.* Bucknell Univ. Press, Lewisburg, Pennsylvania.

Terborgh, J. 1973. On the notion of favorableness in plant ecology. Amer. Nat., 107:481–501.

Terborgh, J. 1983. *Five New World Primates.* Princeton Univ. Press, Princeton, New Jersey.

Terborgh, J., and A. W. Goldizen. 1985. On the mating system of the cooperatively breeding saddle-backed tamarin (*Saguinus fuscicollis*). Behav. Ecol. Sociobiol., 16:293–299.

Terman, C. R. 1992. Reproductive inhibition in female white-footed mice from Virginia. J. Mamm., 73:443–448.

Tershy, B. R. 1992. Body size, diet, habitat use, and social behavior of *Balaenoptera* whales in the Gulf of California. J. Mamm., 73:477–486.

Thewissen, J. G. M. 1985. Cephalic evidence for the affinities of Tubulidentata. Mammalia, 49:257–284.

Thewissen, J. G. M. 1994. Phylogenetic aspects of cetacean origins: a morphological perspective. J. Mamm. Evol., 2:157–184.

Thewissen, J. G. M. 1997. Cetacea, 80–82, in *Yearbook of Science and Technology.* McGraw-Hill, New York.

Thewissen, J. G. M. 1998. Cetacean origins: evolutionary turmoil during the invasion of the oceans, 451–464, in *The Emergence of Whales* (J. G. M. Thewissen, ed.). Plenum Press, New York.

Thewissen, J. G. M., and S. K. Babcock. 1991. Distinctive cranial and cervical innervation of wing muscles: new evidence for bat monophyly. Science, 251:934–936.

Thewissen, J. G. M., and S. T. Hussain. 1993. Origin of underwater hearing in whales. Nature, 361:444–445.

Thewissen, J. G. M., S. T. Hussain, and M. Arif. 1994. Fossil evidence for the origin of aquatic locomotion in archaeocete whales. Science, 263:210–212.

Thewissen, J. G. M., and S. A. Etnier. 1995. Adhesive devices on the thumb of vespertilionoid bats (Chiroptera). J. Mamm., 76:925–936.

Thewissen, J. G. M., L. J. Roe, J. R. O'Neil, S. T. Hussain, A. Sahni, and S. Bajpal. 1996. Evolution of cetacean osmoregulation. Nature, 381:379–380.

Thies, W., E. K. V. Kalko, and H. U. Schnitzler. 1998. The roles of echolocation and olfaction in two Neotropical fruit-eating bats, *Carollia perspicillata* and *C. castanea*, feeding on *Piper*. Behav. Ecol. Sociobiol., 42:397–409.

Thiessen, D. D., K. Owen, and G. Lindzey. 1971. Mechanisms of territorial marking in the male and female Mongolian gerbils (*Meriones unguiculatus*). J. Comp. Physiol. Psychol., 77:38–47.

Thomas, D. W., G. P. Bell, and M. Brock Fenton. 1987. Variation in echolocation call frequencies recorded from North American vespertilionid bats: a cautionary note. J. Mamm., 68:842–847.

Thomas, J. A., and E. C. Birney. 1979. Parental care and mating system of the prairie vole, *Microtus ochrogaster*. Behav. Ecol. Sociobiol., 5:171–186.

Thomas, S. P., and R. A. Suthers. 1972. The physiology and energetics of bat flight. J. Exp. Biol., 57:317–335.

Thompson, D. Q. 1955. The role of feed and cover in population fluctuations of the brown lemming at Point Barrow, Alaska. Trans. North Amer. Wildl. Conf., 20:166–176.

Thompson, S. D., R. E. MacMillen, E. M. Burke, and C. R. Taylor. 1980. The energetic cost of bipedal hopping in small mammals. Nature, 287:223–224.

Thorington, R. W., Jr., and S. Anderson. 1984. Primates, 187–217, in *Orders and Families of Recent Mammals of the World* (S. Anderson and J. K. Jones, eds.). John Wiley & Sons, New York.

Thorington, R. W., Jr., K. Darrow, and C. G. Anderson. 1998. Wing tip anatomy and aerodynamics in flying squirrels. J. Mamm., 79:245–250.

Tilman, D., and J. A. Downing. 1994. Biodiversity and stability in grasslands. Nature, 367:363–365.

Tilson, R. L., and J. R. Henschel. 1986. Spatial arrangement of spotted hyaena groups in a desert environment, Namibia. Afr. J. Ecol., 24:173–180.

Timm, R. M. 1987. Tent construction by bats of the genera *Artibeus* and *Uroderma*. 187–212, in *Studies in Neotropical Mammalogy: Essays in Honor of Philip Hershkovitz* (B. D. Patterson and R. M. Timm, eds.). Fieldiana Zool. No. 39, Field Mus. Nat. Hist., Chicago.

Timm, R. M., and J. Mortimer. 1976. Selection of roost sites by Honduran white bats, *Ectophylla alba* (Chiroptera: Phyllostomatidae). Ecology, 57:385–389.

Timm, R. M., and S. E. Lewis. 1991. Tent construction and use by *Uroderma bilobatum* in coconut palms (*Cocos nucifera*) in Costa Rica. Bull. Amer. Mus. Nat. Hist., 206:251–260.

Tinbergen, N. 1963. On aims and methods of ethology. Z. Tierpsychol., 20:410–433.

To, L. P., and R. H. Tamarin. 1977. The relation of population density and adrenal gland weight in cycling and non-cycling voles (*Microtus*). Ecology, 58:928–934.

Todd, N. B., and S. R. Pressman. 1968. The karyotype of the lesser panda (*Ailurus fulgens*) and general remarks on the phylogeny and affinities of the panda. Carnivore Genetics Newsletter, 5:105–108.

Tomasi, T. E. 1978. Function of venom in the short-tailed shrew, *Blarina brevicauda*. J. Mamm., 59:852–854.

Tosini, G., and M. Menaker. 1996. Circadian rhythms in cultured mammalian retina. Science, 272:419–421.

Townsend, M. T. 1935. Studies on some small mammals of central New York. Roosevelt Wildl. Ann., 4:1–120.

Tracy, C. R., J. S. Turner, and R. B. Huey. 1986. A biophysical analysis of possible thermoregulatory adaptations in sailed pelycosaurs, 195–206, in *The Ecology and Biology of Mammal-like Reptiles* (N. Hotton III, P. D. MacLean, J. J. Roth, and E. C. Roth, eds.). Smithsonian Inst. Press, Washington, D.C.

Travis, S. E., and C. N. Slobodchikoff. 1993. Effects of food resource distribution on the social system of Gunnison's prairie dog (*Cynomys gunnisoni*). Can. J. Zool., 71:1186–1192.

Travis, S. E., C. N. Slobodchikoff, and P. Keim. 1996. Social assemblages and mating relationships in prairie dogs: a DNA fingerprint analysis. Behav. Ecol., 7:95–100.

Trillmich, F., and K. A. Ono. 1991. *Pinnipeds and El Niño: Responses to Environmental Stress*. Springer-Verlag, Berlin.

Trivers, R. L. 1971. The evolution of reciprocal altruism. Quart. Rev. Biol., 46:35–57.

Trivers, R. L. 1972. Parental investment and sexual selection, 136–179, in *Sexual Selection and the Descent of Man, 1871–1971* (B. Campbell, ed.). Aldine, Chicago.

Trivers, R. L., and D. E. Willard. 1973. Natural selection of parental ability to vary the sex ratio of offspring. Science, 179:90–92.

Troughton, E. 1947. *Furred Animals of Australia*. Charles Scribner's Sons, New York.

Turner, A. 1997. *The Big Cats and Their Fossil Relatives: An Illustrated Guide to Their Evolution and Natural History*. Columbia Univ. Press, New York.

Turner, B. L., II, W. C. Clark, R. W. Kates, J. F. Richards, J. T. Mathews, and W. B. Meyer. 1990. *The Earth as Transformed by Human Action. Global and regional changes in the biosphere over the past 300 years*. Cambridge Univ. Press, Cambridge.

Turner, B. N. 1971. The annual cycle of aggression in male *Microtus pennsylvanicus,* and its relation to population parameters. M.S. thesis, University of North Dakota.

Turner, J. S., and C. R. Tracy. 1986. Body size, homeothermy, and the control of heat exchange in mammal-like reptiles, 185–194, in *The Ecology and Biology of Mammal-like Reptiles* (N. Hotten III, P. D. MacLean, J. J. Roth, and E. C. Roth, eds.). Smithsonian Inst. Press, Washington, D.C.

Tuttle, M. D., and M. J. Ryan. 1981. Bat predation and the evolution of frog vocalizations in the Neotropics. Science, 214:677–678.

Tyndale-Biscoe, C. H. 1984. Mammals-Marsupials, 386–454, in *Marshall's Physiology of Reproduction* (G. E. Lamming, ed.). Churchill Livingstone, Edinburgh.

Tyndale-Biscoe, C. H., and M. Renfree. 1987. *Reproductive Physiology of Marsupials*. Cambridge Univ. Press, Cambridge, U.K.

Tyrrell, K. 1988. The use of prey-generated sounds in flycather-style foraging by *Megaderma spasma*. Bat Research News, 29:51.

Uchida, T. A., and T. Mori. 1987. Prolonged storage of spermatozoa in hibernating bats, 351–365, in *Recent Advances in the Study of Bats* (M. B. Fenton, P. Racey, and J. M. V. Rayner, eds.). Cambridge Univ. Press, Cambridge, U.K.

Udvardy, M. D. F. 1969. *Dynamic Zoogeography*. Van Nostrand Reinhold, New York.

United Nations. 1997. *World Population Prospects*, The 1996 Revision. United Nation, New York.

United Nations. 1998. *World Population Projections to 2150*. United Nations, New York.

Vachrameev, V. A., and M. A. Akhmet'yev. 1972. The development of floras on the boundary of the Late Cretaceous and the Paleocene (on data from the study of leaf remains), in *The Development and Replacement of the*

Organic World on the Boundary of the Mesozoic and Cenozoic (V. N. Shimansiy and A. N. Solov'yev, eds.). (Conference April, 1972.) Abstr. pap. methodol. mater., Moscow: Akad. Nauk SSSR, Moscow Soc. Natur. (In Russian.)

Valentine, D. E., and C. F. Moss. 1998. Sensorimotor integration in bat sonar, 220–230, in *Bat Biology and Conservation* (T. H. Kunz and P. A. Racey, eds.). Smithsonian Inst. Press, Washington, D.C.

Valone, T. J., J. H. Brown, and C. L. Jacobi. 1995. Catastrophic decline of a desert rodent, *Dipodomys spectabilis:* insights from a long-term study. J. Mamm., 76:428–436.

Van De Graaff, K., and R. P. Balda. 1973. Importance of green vegetation for reproduction in the kangaroo rat, *Dipodomys merriami merriami.* J. Mamm., 54:509–512.

van der Klaauw, C. 1931. The auditory bulla in some fossil mammals. Bull. Amer. Mus. Nat. Hist., 62:1–341.

Van Deusen, H. M., and J. K. Jones, Jr. 1967. Marsupials, 61–86, in *Recent Mammals of the World* (S. Anderson and J. K. Jones, Jr., eds.). Ronald Press, New York.

Van Gelder, R. G. 1953. The egg-opening technique of a spotted skunk. J. Mamm., 34:255–256.

Van Lawick, H., and J. Van Lawick-Goodall. 1970. *Innocent Killers.* Collins, London.

Van Lawick-Goodall, J. 1968. The behavior of free-living chimpanzees in the Gombe Stream Reserve. Anim. Behav. Monogr., 1:1–311.

Van Lawick-Goodall, J. 1973. The behavior of chimpanzees in their natural habitat. Am. J. Psychiatry, 130:1–12.

van Noordwijk, M. A. 1985. The socio-ecology of Sumatran long-tailed macaques (*Macaca fascicularis*), II: the behavior of individuals. Unpublished Ph.D. dissertation. Utrecht: Rijksuniversiteit.

Van Roosmalen, M. G. M., T. Van Roosmalen, R. A. Mittermeier, and G. A. B. da Fonseca. 1998. A new and distinctive species of marmoset (Callitrichidae, Primates) from the lower Rio Aripuanã, state of Amazonas, central Brazilian Amazonia. Goeldiana Zoologia 22:1–27.

Van Valen, L. 1967. New Paleocene insectivores and insectivore classification. Bull. Amer. Mus. Nat. Hist., 135:217–284.

Van Valen, L., and R. E. Sloan, 1966. The extinction of the multituberculates. Syst. Zool., 15:261–278.

Vater, M. 1987. Narrow-band frequency analysis in bats, 200–225 in *Recent Advances in the Study of Bats* (M. B. Fenton, P. Racey, and J. M. V. Rayner, eds.). Cambridge Univ. Press, Cambridge, U.K.

Vater, M. 1998. Adaptation of the auditory periphery of bats for echolocation, 231–245, in *Bat Biology and Conservation* (T. H. Kunz and P. A. Racey, eds.). Smithsonian Inst. Press, Washington, D.C.

Vaughan, T. A. 1954. Mammals of the San Gabriel Mountains of California. Univ. Kansas Publ., Mus. Nat. Hist., 7:513–582.

Vaughan, T. A. 1959. Functional morphology of three bats: *Eumops, Myotis, Macrotus.* Univ. Kansas Publ., Mus. Nat. Hist., 12:1–153.

Vaughan, T. A. 1967. Food habits of the northern pocket gopher on shortgrass prairie. Amer. Midl. Nat., 77:176–189.

Vaughan, T. A. 1969. Reproduction and population densities in a montane small mammal fauna, 51–74, in *Contributions in Mammalogy* (J. K. Jones Jr., ed.). Misc. Publ., Mus. Nat. Hist., University of Kansas, No. 51.

Vaughan, T. A. 1970a. Adaptations for flight in bats, 127–143, in *About Bats* (B. H. Slaughter and D. W. Walton, eds.). Southern Methodist Univ. Press, Dallas.

Vaughan, T. A. 1970b. The skeletal system. The muscular system. Flight patterns and aerodynamics, 97–138, in *Biology of Bats* (W. A. Wimsatt, ed.). Academic Press, New York.

Vaughan, T. A. 1974. Resource allocation in some sympatric, subalpine rodents. J. Mamm., 55:764–795.

Vaughan, T. A. 1976. Nocturnal behavior of the African false vampire bat (*Cardioderma cor*). J. Mamm., 57:227–248.

Vaughan, T. A. 1977. Foraging behavior of the giant leaf-nosed bat (*Hipposideros commersoni*). E. Afr. Wildl. J., 15:237–249.

Vaughan, T. A. 1982. Stephens' woodrat, a dietary specialist. J. Mamm., 63:53–62.

Vaughan, T. A., and G. C. Bateman, 1970. Functional morphology of the forelimbs of mormoopid bats. J. Mamm., 51:217–235.

Vaughan, T. A., and T. J. O'Shea. 1976. Roosting ecology of the pallid bat, *Antrozous pallidus.* J. Mamm., 57:19–42.

Vaughan, T. A., and M. M. Bateman. 1980. The molossid wing: some adaptations for rapid flight, 69–78, in *Proceedings of the Fifth International Bat Research Conference* (D. Wilson and A. Gardner, eds.). Texas Tech Press, Lubbock.

Vaughan, T. A., and S. T. Schwartz. 1980. Behavioral ecology of an insular woodrat. J. Mamm. 61:205.

Vaughan, T. A., and W. P. Weil. 1980. The importance of arthropods in the diet of *Zapus princeps* in a subalpine habitat. J. Mamm., 61:122 124.

Vaughan, T. A., and N. J. Czaplewski. 1985. Reproduction in Stephens' woodrat: the wages of folivory. J. Mamm., 66:429–443.

Vaughan, T. A., and R. P. Vaughan. 1986. Seasonality and the behavior of the African yellow-winged bat. J. Mamm., 67:91–102.

Vehrencamp, S. L., F. G. Stiles, and J. W. Bradbury. 1977. Observations on the foraging behavior and avian prey of the Neotropical carnivorous bat, *Vampyrum spectrum.* J. Mamm., 58:469–478.

Villa-Ramirez, B., 1966. *Los Murciélagos de Mexico.* Inst. Biol., UNAM.

Villa-Ramirez, B., and E. L. Cockrum, 1962. Migration in the guano bat, *Tadarida brasiliensis mexicana*. J. Mamm., 43:43–64.

Vitousek, P. M., H. A. Mooney, J. Lubchenco, and J. M. Melillo. 1997. Human domination of earth's ecosystems. Science, 277:494–499.

Vizcaino, S. F., and G. J. Scillato-Yané. 1995. An Eocene tardigrade (Mammalia, Xenarthra) from Seymour Island, West Antarctica. Antarctic Sci., 7:407–408.

Vleck, D. 1979. The energy cost of burrowing by the pocket gopher *Thomomys bottae*. Physiol. Zool., 52:122–136.

Vogel, P. 1983. Contribution a l'écologie et a la zoogéographie de *Micropotamogale lamottei* (Mammalia, Tenrecidae). Rev. Ecol. (Terre Vie), 38:37–49.

Vogel, S. 1988. *Life's Devices, The Physical World of Animals and Plants*. Princeton University Press, Princeton, New Jersey.

Vogl, R. J. 1973. Ecology of the knobcone pine in the Santa Ana Mountains, California. Ecol. Monogr., 43:125–143.

Vorhies, M. R. 1974. Fossil pocket mice burrows in Nebraska. Amer. Midland Nat., 91:492–498.

Vrba, E. S. 1992. Mammals as a key to evolutionary theory. J. Mamm., 73:1–28.

Vrba, E. S. 1993. Mammal evolution in the African Neogene and a new look at the Great American Interchange, 393–432, in *Biological Relationships between Africa and South America* (P. Goldblatt, ed.). Yale Univ. Press, New Haven.

Wahlert, J. H. 1985. Skull morphology and relationships of geomyoid rodents. Amer. Mus. Novitates, 2819:1–20.

Wai-Ping, V., and M. B. Fenton. 1989. Ecology of spotted bats (*Euderma maculatum*): roosting and foraging behavior. J. Mamm., 70:617–622.

Walker, A. 1969. The locomotion of the lorises, with special reference to the potto. E. Afr. Wildl. J., 8:1–5.

Walker, E. P. 1968. *Mammals of the World*. 2d ed., vol. 2. Johns Hopkins Univ. Press, Baltimore.

Walker, M. M., M. E. Bitterman, and J. L. Kirschvink. 1986. Experimental and correlational studies of responses to magnetic field stimuli by different species, 194–205, in *Biophysical Effects of Steady Magnetic Fields* (G. Maret, N. Boccara, and J. Kiepenheuer, eds.). Springer-Verlag, New York.

Walker, M. M., J. L. Kirschvink, G. Ahmed, and A. E. Dizon. 1992. Evidence that fin whales respond to the geomagnetic field during migration. J. Exp. Biol., 171:67–78.

Wall, C. E., and D. W. Krause. 1992. A biomechanical analysis of the masticatory apparatus of *Ptilodus* (Multituberculata). J. Vert. Paleo., 12:172–187.

Wallace, A. R. 1876. *The Geographical Distribution of Animals*. 2 vols. Harper, New York. Reprinted by Hafner, New York.

Walther, F. 1965. Verhaltensstudien an der Grantgazell (*Gazella granti* Brooke, 1872) im Ngorongoro-Krater. Z. Tierpsychol., 22:167–208.

Walther, F. 1966. Zum liegeverhalten des weissschwanzgnus (*Connochaetes gnou* Zimmerman, 1780). Z. Säugetierk., 31:1–16.

Wang, X., and M. Novak. 1992. Influence of the social environment on parental behavior and pup development of meadow voles (*Microtus pennsylvanicus*) and prairie voles *(M. ochrogaster)*. J. Comp. Psychol., 106:163–171.

Ward, A. L., and J. O. Keith. 1962. Feeding habits of pocket gophers on mountain grasslands. Ecology, 43:744–749.

Ward, S. C., and D. R. Pilbeam. 1983. Maxillofacial morphology of Miocene hominoids from Africa and Indo-Pakistan, 211–238, in *New Interpretations of Ape and Human Ancestry* (R. Ciochon and R. S. Corruccini, eds.). Plenum Press, New York.

Watkins, W. A. 1977. Acoustic behaviors of sperm whales. Oceanus, 20:50–58.

Watkins, W. A., M. A. Daher, K. M. Fristrup, T. J. Howald, and G. N. Disciara. 1993. Sperm whales tagged with transponders and tracked underwater by sonar. Marine Mamm. Sci., 9:55–67.

Watson, A. 1958. The behavior, breeding and food-ecology of the snowy owl *Nyctea scandiaca*. Ibis, 99:419–462.

Watson, R. M. 1968. Report on aerial photographic studies of vegetation carried out in the Tsavo area of Kenya. Typescript. (Cited by Laws, 1970).

Wayne, R. K., R. E. Benveniste, D. N. Janczewski, and S. J. O'Brian. 1989. Molecular and biochemical evolution of the Carnivora, 465–494, in *Carnivore Behavior, Ecology, and Evolution* (J. L. Gittleman, ed.). Cornell Univ. Press, Ithaca, New York.

Webb, S. D. 1985. The interrelationships of tree sloths and ground sloths. 105–112, in *The Evolution and Ecology of Armadillos, Sloths, and Vermilinguas* (G. G. Montgomery, ed.). Smithsonian Inst. Press, Washington, D.C.

Webb, S. D. 1991. Ecogeography and the Great American Interchange. Paleobiology, 17:266–280.

Webb, S. D., and B. E. Taylor. 1980. The phylogeny of hornless ruminants and a description of the cranium of *Archaeomeryx*. Bull. Amer. Mus. Nat. Hist., 167:117–158.

Webb, S. D., and A. D. Barnosky. 1989. Faunal dynamics of Pleistocene mammals. Ann. Rev. Earth Planet. Sci., 17:413–438.

Webb, S. D., and A. Rancy. 1996. Late Cenozoic evolution of the Neotropical mammal fauna, 335–358, in *Evolution and Environment in Tropical America* (J. B. C. Jackson, A. F. Budd, and A. G. Coates, eds.). Univ. Chicago Press, Chicago.

Weber, N. S., and J. S. Findley. 1970. Warm-season changes in fat content of *Eptesicus fuscus*. J. Mamm., 51:160–162.

Webster, F. A., and D. R. Griffin, 1962. The role of flight membranes in insect capture by bats. Anim. Behav., 10:332–340.

Wegener, A. 1912. Die entstehung der Kontinente. Geologische Rundschau, 3:276–292.

Wegener, A. 1915. *Die Entstehung der Kontinente und Ozeane.* Braunschweig, Vieweg.

Wegener, A. 1966. *The Origin of Continents and Oceans.* (Translation of 1929 edition by J. Biram.) Dover Publications, New York.

Weilgart, L., and H. Whitehead. 1997. Group-specific dialects and geographical variation in coda repertoire in South Pacific sperm whales. Behav. Ecol. Sociobiol., 40:277–285.

Weir, B. J. 1974. The tuco-tuco and plains viscacha, 113–130, in *The Biology of the Hystricomorph Rodents* (I. W. Rowlands and B. J. Weir, eds.). Academic Press, New York.

Wells, M. C., and M. Bekoff. 1981. An observational study of scent marking in coyotes, *Canis latrans.* Anim. Behav., 29:332–350.

Wells, P. V., and C. D. Jorgensen. 1964. Pleistocene wood rat middens and climatic change in the Mojave Desert: a record of juniper woodlands. Science, 143:1171–1174.

Wemmer, C., R. Rudran, F. Dallmeier, and D. E. Wilson. 1993. Training developing-country nationals is the critical ingredient to conserving global biodiversity. BioScience, 43:762–767.

West, S. D. 1977. Midwinter aggregation in the northern red-backed vole (*Clethrionomys rutilus*). Can. J. Zool., 55:1404–1409.

Westerman, M., and D. Edwards. 1992. DNA hybridization and the phylogeny of monotremes, 28–34, in *Platypus and Echidnas* (M. L. Augee, ed.). Royal Zool. Soc. New South Wales, Sydney.

Wetzel, R. M. 1977. The Chacoan peccary *Catagonus wagneri* (Rusconi). Bull. Carnegie Mus. Nat. Hist., 3:1–36.

Wetzel, R. M., R. E. Dubos, R. L. Martin, and P. Myers. 1975. *Catagonus,* an "extinct" peccary, alive in Paraguay. Science, 189:379–381.

Wever, E. G., J. G. McCormick, J. Palin, and S. H. Ridgway. 1972. Cochlear structure in the dolphin, *Lagenorhynchus obliquidens.* Proc. Nat. Acad. Sci. USA, 69:657–661.

Wharton, C. H. 1950. Notes on the life history of the flying lemur, *Cynocephalus volans.* J. Mamm., 31:269–273.

Whitaker, J. O., Jr. 1963. Food, habitat and parasites of the woodland jumping mouse in central New York. J. Mamm., 44:316–321.

Whitaker, J. O., Jr. 1968. Parasites, 254–311, in *The Biology of Peromyscus* (J. A. King, ed.). Spec. Pub. No. 2., Amer. Soc. Mammalogists.

White, J. L. 1997. Locomotor adaptations in Miocene xenarthrans, 246–264, in *Vertebrate Paleontology in the*

Neotropics. The Miocene Fauna of La Venta, Colombia (R. F. Kay, R. H. Madden, R. L. Cifelli, and J. J. Flynn, eds.). Smithsonian Inst. Press, Washington, D.C.

Whitham, T. G., J. Maschinski, K. C. Larson, and K. N. Paige. 1991. Plant responses to herbivory: the continuum from negative to positive and underlying physiological mechanisms, 227–256, in *Plant-animal Interactions: Evolutionary Ecology in Tropical and Temperate Regions* (P. W. Price, T. M. Lewinsohn, G. W. Fernandes, and W. W. Benson, eds.). John Wiley and Sons, New York.

Whittow, G. C. 1974. Sun, sand, and sea lions. Nat. Hist., 83:56–63.

Wible, J. R. 1991. Origin of Mammalia: the craniodental evidence re-examined. J. Vert. Paleo., 11(1):1–28.

Wible, J. R. 1993. Cranial circulation and relationships of the colugo *Cynocephalus* (Dermoptera, Mammalia). Amer. Mus. Novitates, 3072:1–27.

Wible, J. R., and H. H. Covert. 1987. Primates: cladistic diagnosis and relationships. J. Human Evol., 16:1–22.

Wible, J. R., and M. J. Novacek. 1988. Cranial evidence for the monophyletic origin of bats. Amer. Mus. Novitates, 2911:1–19.

Wible, J. R., D. Miao, and J. A. Hopson. 1990. The septomaxilla of fossil and Recent synapsids and the problem of the septomaxilla of monotremes and armadillos. Zool. J. Linnean Soc., 98:203–228.

Wible, J. R., and J. A. Hopson. 1993. Basicranial evidence for early mammal phylogeny, 45–62, in *Mammal Phylogeny: Mesozoic Differentiation, Multituberculates, Monotremes, Early Therians, and Marsupials* (F. S. Szalay, M. J. Novacek, and M. C. McKenna, eds.). Springer-Verlag, New York.

Wible, J. R., and J. A. Hopson. 1995. Homologies of the prootic canal in mammals and non-mammalian cynodonts. J. Vert. Paleo., 15:331–356.

Wible, J. R., G. W. Rougier, M. J. Novacek, M. C. McKenna, and D. Dashzeveg. 1995. A mammalian petrosal from the Early Cretaceous of Mongolia: implications for the evolution of the ear region and mammaliamorph interrelationships. Amer. Mus. Novitates, 3149:1–19.

Wickler, W. Von, and D. Uhrig, 1969. Verhalten und okologische Nische der Gelbflugelfledermaus, *Lavia frons* (Geoffroy) (Chiroptera, Megadermatidae). Z. Tierpsychol., 26:726–736.

Wiles, G. J. 1992. Recent trends in the fruit bat trade on Guam, 53–60, in *Pacific Flying Foxes: Proceedings of an International Conservation Conference* (D. E. Wilson and G. L. Graham, eds.). U.S. Fish and Wildlife Service, Biological Report 90(23).

Wiley, E. O. 1981. *Phylogenetics: The Theory and Practice of Phylogenetic Systematics.* John Wiley and Sons, New York.

Wilkinson, G. S. 1984. Reciprocal food sharing in the vampire bat. Nature, 308:181–184.

Wilkinson, G. S. 1987. Altruism and co-operation in bats, 299–323, in *Recent Advances in the Study of Bats* (M. B.

Fenton, P. Racey, and J. M. V. Rayner, eds.). Cambridge Univ. Press, Cambridge.

Wilkinson, G. S. 1990. Food sharing in vampire bats. Sci. Amer., 262(2):76–82.

Williams, D. F. 1991. Habitat of shrews (genus *Sorex*) in forest communities of the western Sierra Nevada, California, 1–14, in *The Biology of the Soricidae* (J. S. Findley and T. L. Yates, eds.). Spec. Pub. No. 1, Mus. Southwestern Biology, Albuquerque, New Mexico.

Williams, E. M. 1998. Synopsis of the earliest cetaceans Pakicetidae, Ambulocetidae, Remingtonocetidae, and Protocetidae, 1–28, in *The Emergence of Whales* (J. G. M. Thewissen, ed.). Plenum Press, New York.

Williams, G. C. 1967. Natural selection, the costs of reproduction, and a refinement of Lack's principle. Amer. Nat., 100:687–690.

Williams, T. C., L. C. Ireland, and J. M. Williams. 1973. High altitude flights of the free-tailed bat, *Tadarida brasiliensis,* observed with radar. J. Mamm., 54:807–821.

Williams, T. M. 1989. Swimming by sea otters: adaptations for low energetic cost locomotion. J. Comp. Physiol. A, 164:815–824.

Willig, M. R., and E. A. Sandlin. 1989. Gradients of species density and species turnover in New World bats: a comparison of quadrat and band methodologies, 81–96, in *Latin American Mammals: Their Conservation, Ecology, and Evolution* (M. A. Mares and D. J. Schmidly, eds.). Univ. Oklahoma Press, Norman.

Willig, M. R., and K. W. Selcer. 1989. Bat species gradients in the New World: a statistical assessment. J. Biogeogr, 16:189–195.

Wilson, D. E. 1979. Reproductive patterns, in *Biology of Bats of the New World Family Phyllostomatidae.* Part III (R. J. Baker, J. K. Jones, Jr., and D. C. Carter, eds.). Spec. Publ. Mus., No. 16. Texas Tech Press, Lubbock, Texas.

Wilson, E. O. 1975. *Sociobiology: The New Synthesis.* Univ. Chicago Press, Chicago.

Wilson, E. O. 1992. *The Diversity of Life.* Harvard Univ. Press, Cambridge, Massachusetts.

Wilson, J. A., and A. C. Runkel. 1991. *Prolapsus,* a large sciuravid rodent and new eomyids from the Late Eocene of Trans-Pecos Texas. Pearce-Sellards Series No. 48:1–30.

Wilson, J. W. 1974. Analytical zoogeography of North American mammals. Evolution, 28:124–140.

Wilson, R. W. 1949. Early Tertiary rodents of North America. Carnegie Inst. Washington Publ., 584:67–164.

Wilson, R. W. 1960. *Early Miocene Rodents and Insectivores from Northeastern Colorado.* Univ. Kansas Paleontol. Cont. Vertebrata, Art. 7.

Wilson, R. W. 1972. Evolution and extinction in early Tertiary rodents. Proc. Int. Geol. Congr., 24:217–224.

Wilson, D. E., and D. M. Reeder. 1993. *Mammal Species of the World: A Taxonomic and Geographic Reference.* Smithsonian Inst. Press, Washington, D.C.

Wimsatt, W. A. 1944. Further studies on the survival of spermatozoa in the female reproductive tract of the bat. Anat. Rec., 88:193–204.

Wimsatt, W. A. 1945. Notes on breeding behavior, pregnancy, and parturition in some vespertilionid bats of eastern United States. J. Mamm., 26:23–33.

Wimsatt, W. A. 1969a. Transient behavior, nocturnal activity patterns, and feeding efficiency of vampire bats (*Desmodus rotundus*) under natural conditions. J. Mamm., 50:233–244.

Wimsatt, W. A. 1969b. Some interrelations of reproduction and hibernation in mammals. Symp. Soc. Exp. Biol., 23:511–549.

Wimsatt, W. A., and B. Villa-Ramirez, 1970. Locomotor adaptations in the disc-winged bat. Amer. J. Anat., 129:89–119.

Wing, S. L., L. J. Hickey, and C. C. Swisher. 1993. Implications of an exceptional fossil flora for Late Cretaceous vegetation. Nature, 363:342–344.

Winge, H. 1941. *The Interrelationships of the Mammalian Genera.* Vol. 1. C.A. Reitzels Forlag, Copenhagen. (Danish translated by E. Deichmann and G. M. Allen.)

Wislocki, G. B. 1942. Studies on the growth of deer antlers. I: On the structure and histogenesis of the antlers of the Virginia deer (*Odocoileus virginianus borealis*). Amer. J. Anat., 71:371–415.

Wislocki, G. B., J. C. Aub, and C. M. Waldo. 1947. The effects of gonadectomy and the administration of testosterone proprionate on the growth of antlers in male and female deer. Endocrinology, 40:202–224.

Withers, P. C. 1978. Bioenergetics of a 'primitive' mammal, the Cape golden mole. S. Afr. J. Sci., 74:347–348.

Withers, P. C. 1992. *Comparative Animal Physiology,* Saunders College Publ., Philadelphia.

Withers, P. C., G. N. Louw, and J. Henschel. 1979. Energetics and water relations of Namib Desert rodents. S. Afr. J. Zool., 15:131–137.

Withers, P. C., and J. U. M. Jarvis. 1980. The effect of huddling on the thermoregulation and oxygen consumption for the naked mole-rat. Comp. Biochem. Physiol., 66:215–219.

Wolfe, J. A. 1978. A paleobotanical interpretation of Tertiary climates in the Northern Hemisphere. Amer. Sci., 66:694–703.

Wolfe, J. A. 1994. Tertiary climatic changes at middle latitudes of western North America. Palaeogeogr., Palaeoclimat., Palaeoecol., 108:195–205.

Wolff, J. O. 1980. Social organization of the taiga vole (*Microtus xanthognathus*). The Biologist, 62:34–45.

Wolff, J. O. 1997. Population regulation in mammals: an evolutionary perspective. J. Anim. Ecol., 66:1–13.

Wolff, J. O., and W. Z. Lidicker. 1981. Communal winter nesting and food sharing in taiga voles. Behav. Ecol. Sociobiol., 9:237–240.

Wolff, J. O., J. L. Lundy, and R. Baccus. 1988. Dispersal, inbreeding avoidance and reproductive success in white-footed mice. Anim. Behav., 36:456–465.

Wood, A. E. 1935. Evolution and relationships of the heteromyid rodents with new forms from the Tertiary of western North America. Ann. Carnegie Mus., 24:73–262.

Wood, A. E. 1957. What, if anything, is a rabbit? Evolution, 11:417–425.

Wood, A. E. 1958. Are there rodent suborders? Syst. Zool., 7:169–173.

Wood, A. E. 1965. Grades and clades among rodents. Evolution, 19:115–130.

Wood, A. E. 1980. The origin of caviomorph rodents from a source in Middle America: A clue to the area of origin of the playrrhine primates, 79–91, in *Evolutionary Biology of the New World Monkeys and Continental Drift* (R. I. Ciochon and A. B. Chiarelli, eds.). Plenum Press, New York.

Wood, F. G., Jr. 1959. Underwater sound production and concurrent behavior of captive porpoises, *Tursiops truncatus* and *Stenella plagiodon*. Bull. Mar. Sci. Gulf Caribbean, 3:120–133.

Woodall, P. F., and C. FitzGibbon. 1995. Ultrastructure of spermatozoa of the yellow-rumped elephant shrew *Rhynchocyon chrysopygus* (Mammalia: Macroscelidea) and the phylogeny of elephant shrews. Acta Zool. (Stockholm), 76:19–23.

Woodburne, M. O. 1987. *Cenozoic Mammals of North America: Geochronology and Biostratigraphy.* Univ. of California Press, Berkeley.

Woodburne, M. O., and C. C. Swisher, III. 1995. Land mammal high-resolution geochronology, intercontinental overland dispersals, sea level, climate, and vicariance, in *Geochronology Time Scales and Global Stratigraphic Correlation*, SEPM Special Publication No. 54, Society for Sedimentary Geology.

Woodburne, M. O., and J. A. Case. 1996. Dispersal, vicariance, and the Late Cretaceous to early Tertiary land mammal biogeography from South America to Australia. J. Mamm. Evol., 3:121–161.

Woodburne, M. E., and R. H. Tedford, 1975. The first Tertiary monotreme from Australia. Amer. Mus. Novitates, 2588:1–11.

Woodburne, M. O., and W. J. Zinsmeister. 1982. Fossil land mammal from Antarctica. Science, 218:284–286.

Woods, C. A. 1982. The history and classification of South American hystricognath rodents—reflections on the far away and long ago, 377–392, in *Mammalian Biology in South America* (M. A. Mares and H. H. Genoways, eds.). Spec. Publ. Pymatuning Lab. Ecol., Univ. Pittsburgh Press, Pittsburgh.

Woods, C. A. 1984. Hystricognath rodents, 389–446, in *Orders and Families of Recent Mammals of the World* (S. Anderson and J. K. Jones, eds.). John Wiley & Sons, New York.

Woods, C. A. 1993. Suborder Hystricognathi, 771–806, in *Mammal Species of the World* (D. E. Wilson and D. M. Reeder, eds.). Smithsonian Inst. Press, Washington, D.C.

Woods, C. A., J. A. Ottenwalder, and W. L. R. Oliver. 1985. Lost mammals of the Greater Antilles: summarized findings of a ten week field survey in the Dominican Republic, Haiti, and Puerto Rico. Dodo (Jersey Wildlife Preservation Trust), 22:23–42.

Woodsworth, G. C., G. P. Bell, and M. B. Fenton. 1981. Observations on the echolocation, feeding behavior and habitat use of *Euderma maculatum* in southcentral British Columbia. Can. J. Zool., 59:1099–1102.

Worthy, T. H., M. J. Daniel, and J. E. Hill. 1996. An analysis of skeletal size variation in *Mystacina robusta* Dwyer, 1962 (Chiroptera: Mystacinidae). New Zealand J. Zool., 23:99–110.

Wozencraft, W. C. 1984. A phylogenetic reappraisal of the Viverridae and its relationship to other Carnivora. Ph.D. dissertation. Univ. Kansas, Lawrence.

Wozencraft, W. C. 1989. The phylogeny of the Recent Carnivora, 495–535, in *Carnivores; Behavior, Ecology, and Evolution* (J. L. Gittleman, ed.). Cornell Univ. Press, Ithaca, New York.

Wozencraft, W. C. 1993. Order Carnivora, 279–348, in *Mammal Species of the World.* 2d ed. (D. E. Wilson and D. M. Reeder, eds.). Smithsonian Inst. Press, Washington, D.C.

Wrangham, R. W. 1987. The significance of African apes for reconstructing human social evolution, in *The Evolution of Human Behavior: Primate Models* (W. G. Kinzey, ed.). SUNY Press, Albany, N.Y.

Wright, S. 1977. *Evolution and the Genetics of Populations: Experimental Results and Evolutionary Deductions.* Univ. Chicago Press, Chicago.

Wynn, R. M. 1971. Immunological implications of comparative placental ultrastructure, in *The Biology of the Blastocyst* (R. J. Blandau, ed.). Univ. Chicago Press, Chicago.

Wyss, A. R. 1987. The walrus auditory region and the monophyly of pinnipeds. Amer. Mus. Novitates, 2871:1–31.

Wyss, A. R., and J. J. Flynn. 1993. A phylogenetic analysis and definition of the Carnivora, 32–53, in *Mammal Phylogeny, Placentals.* Vol. 2 (F. S. Szalay, M. J. Novacek, and M. C. McKenna, eds.). Springer-Verlag, New York.

Wyss, A. R., J. J. Flynn, M. A. Norell, C. C. Swisher, R. Charrier, M. J. Novacek, and M. C. McKenna. 1993. South America's earliest rodent and recognition of a new interval of mammalian evolution. Nature, 365:434–437.

Xu, Z., D. M. Stoddart, H. Ding, and J. Zhang. 1995. Self-anointing behavior in the rice-field rat, *Rattus rattoides.* J. Mamm., 76:1238–1241.

Yaakobi, D., and A. Shkolnik. 1974. Structure and concentrating capacity in kidneys of hedgehogs. Amer. J. Physiol., 226:948–952.

Yablokov, A. V. 1994. Validity of whaling data. Nature, 367:108.

Yacoe, M. E., J. W. Cummings, P. Myers, and G. K. Creighton. 1982. Muscle enzyme profile, diet, and flight in South American bats. Amer. J. Physiol. 242:189–194.

Yalden, D. W. 1966. The anatomy of mole locomotion. J. Zool. London, 149:5–64.

Yamazaki, K., A. E. Boyse, V. Mike, H. T. Thaler, B. J. Mathieson, J. Abbott, J. Boyse, Z. A. Zayas, and L. Thomas. 1976. Control of mating preferences in mice by genes in the major histocompatibility complex. J. Exp. Med., 144:1324–1335.

Yamazaki, K., M. Yamaguchi, E. A. Boyse, and L. Thomas. 1980. The major histocompatibility complex as a source of odors imparting individuality among mice, 267–273, in *Chemical Signals* (D. Müchwartze and R. M. Silverstein, eds.). Plenum Press, New York.

Yates, T. L. 1984. Insectivores, elephant shrews, tree shrews, and dermopterans, 117–144, in *Orders and Families of Recent Mammals of the World* (S. Anderson and J. K. Jones, eds.). John Wiley & Sons, New York.

Yates, T. L., and I. F. Greenbaum. 1982. Biochemical systematics of North American moles (Insectivora: Talpidae). J. Mamm., 63:368–374.

Yoakum, J. 1958. Seasonal food habits of the Oregon pronghorn antelope (*Antilocapra americana oregona* Bailey). Inter. Antelope Conf. Trans., 9:47–59.

York, A. E., and V. B. Scheffer. 1997. Timing of implantation in the northern fur seal, *Callorhinus ursinus*. J. Mamm., 78:675–683.

Young, R. A., 1976. Fat, energy, and mammalian survival. Amer. Zool., 16:699–710.

Zachos, J. C., Breza, J. R., and Wise, S. W. 1992. Early Oligocene ice-sheet expansion on Antarctica. Geology, 20:569–573.

Zbinden, K. 1989. Field observations on the flexibility of the acoustic behavior of the European bat *Nyctalus noctula* (Schreber, 1774). Revue Suisse Zool., 96:335–343.

Zeller, U., J. R. Wible, and M. Eisner. 1993. New ontogenetic evidence on the septomaxilla of *Tamandua* and *Choloepus* (Mammalia, Xenarthra), with a reevaluation of the homology of the mammalian septomaxilla. J. Mamm. Evol., 1:31–46.

Zhang, Ya-ping, and Shi Li-ming. 1991. The riddle of the giant panda. Nature, 352:572.

Zimmerman, E. G. 1965. A comparison of habitat and food of two species of *Microtus*. J. Mamm., 46:605–612.

Zimmermann, E., S. Cepok, N. Rakotoarison, V. Zietemann, and U. Radespiel. 1998. Sympatric mouse lemurs in north-west Madagascar: a new rufous mouse lemur species (*Microcebus ravelobensis*). Folia Primatol., 69:106–114.

Zortéa, M., and S. Lucena Mendes. 1993. Folivory in the big fruit-eating bat, *Artibeus lituratus* (Chiroptera: Phyllostomidae) in eastern Brazil. J. Tropical Ecol., 9:117–120.

INDEX

Classification	Common Name(s)
Hominidae	Apes, human
Hylobatidae	Gibbons
Callitrichidae	Marmosets
Cebidae	New World monkeys
Order Carnivora (271 species)	
Family Felidae	Cats
Viverridae	Civets
Herpestidae	Mongooses
Hyaenidae	Hyenas, aardwolf
Canidae	Wolves, foxes, jackals
Ursidae	Bears, giant panda
Otariidae	Eared seals, fur seals, sea lions
Phocidae	Earless seals
Odobenidae	Walrus
Mustelidae	Weasels, skunks, badgers, otters
Procyonidae	Raccoons, ringtail cats, coatis
Order Cetacea (78 species)	
Family Balaenopteridae	Rorquals
Eschrichtiidae	Gray whale
Balaenidae	Right whales
Neobalaenidae	Pygmy right whale
Physeteridae	Sperm whales
Ziphiidae	Beaked whales
Platanistidae	River dolphins
Delphinidae	Ocean dolphins
Monodontidae	Narwhal, beluga
Phocoenidae	Porpoises
Order Sirenia (5 species)	
Family Dugongidae	Dugongs, sea cows
Trichechidae	Manatees
Order Proboscidea (2 species)	
Family Elephantidae	Elephants
Order Perissodactyla (18 species)	
Family Equidae	Horses, asses, zebras
Tapiridae	Tapirs
Rhinocerotidae	Rhinoceroses
Order Hyracoidea (6 species)	
Family Procaviidae	Hyraxes
Order Tubulidentata (1 species)	
Family Orycteropodidae	Aardvark
Order Artiodactyla (220 species)	
Family Suidae	Swine
Tayassuidae	Peccaries, javelinas
Hippopotamidae	Hippopotami